Nanostructured Biomaterials for Overcoming Biological Barriers

RSC Drug Discovery Series

Editor-in-Chief
Professor David Thurston, *London School of Pharmacy, UK*

Series Editors:
Dr David Fox, *Pfizer Global Research and Development, Sandwich, UK*
Professor Salvatore Guccione, *University of Catania, Italy*
Professor Ana Martinez, *Instituto de Quimica Medica-CSIC, Spain*
Professor David Rotella, *Montclair State University, USA*

Advisor to the Board:
Professor Robin Ganellin, *University College London, UK*

Titles in the Series:
1: Metabolism, Pharmacokinetics and Toxicity of Functional Groups
2: Emerging Drugs and Targets for Alzheimer's Disease; Volume 1
3: Emerging Drugs and Targets for Alzheimer's Disease; Volume 2
4: Accounts in Drug Discovery
5: New Frontiers in Chemical Biology
6: Animal Models for Neurodegenerative Disease
7: Neurodegeneration
8: G Protein-Coupled Receptors
9: Pharmaceutical Process Development
10: Extracellular and Intracellular Signaling
11: New Synthetic Technologies in Medicinal Chemistry
12: New Horizons in Predictive Toxicology
13: Drug Design Strategies: Quantitative Approaches
14: Neglected Diseases and Drug Discovery
15: Biomedical Imaging
16: Pharmaceutical Salts and Cocrystals
17: Polyamine Drug Discovery
18: Proteinases as Drug Targets
19: Kinase Drug Discovery
20: Drug Design Strategies: Computational Techniques and Applications
21: Designing Multi-Target Drugs
22: Nanostructured Biomaterials for Overcoming Biological Barriers

How to obtain future titles on publication:
A standing order plan is available for this series. A standing order will bring delivery of each new volume immediately on publication.

For further information please contact:
Book Sales Department, Royal Society of Chemistry, Thomas Graham House, Science Park, Milton Road, Cambridge, CB4 0WF, UK
Telephone: +44 (0)1223 420066, Fax: +44 (0)1223 420247, Email: books@rsc.org
Visit our website at http://www.rsc.org/Shop/Books/

Nanostructured Biomaterials for Overcoming Biological Barriers

Edited by

Maria Jose Alonso and Noemi S. Csaba

Nanobiofar Group, Department of Pharmaceutical Technology, School of Pharmacy, Santiago de Compostela, La Coruna, Spain
E-mail: mariaj.alonso@usc.es; noemi.csaba@usc.es

RSCPublishing

RSC Drug Discovery Series No. 22

ISBN: 978-1-84973-363-2
ISSN: 2041-3203

A catalogue record for this book is available from the British Library

Published by The Royal Society of Chemistry,
Thomas Graham House, Science Park, Milton Road,
Cambridge CB4 0WF, UK

Registered Charity Number 207890

For further information see our web site at www.rsc.org

Printed in the United Kingdom by Henry Ling Limited, at the Dorset Press, Dorchester, DT1 1HD

Preface

The book *Nanostructured Biomaterials for Overcoming Biological Barriers* provides a terrific summary of how nanostructures can be helpful in the biomedical area. It starts with a historical perspective and then goes over how nanostructures can be used to overcome different barriers in the body. Such barriers include the intestinal barrier, the nasal barrier, the ocular barrier, the pulmonary barrier, the skin barrier and finally the blood brain barrier. In each case the book is very well organized with chapters in each section going over physiological considerations and mechanistic issues followed by various formulation strategies for delivering specific types of molecules. For example, in the section for nanostructures for overcoming the intestinal barrier, oral vaccines are considered and the same is true for the section on nasal barriers. In the sections for the ocular, pulmonary, skin and blood-brain barriers; drugs, including large molecular weight drugs, are primarily discussed.

The book continues by going over how nanostructures can overcome biological barriers related to parenteral drug delivery. Here the idea of using lipid nanocapsules for parenteral drug delivery is discussed as well as how one can overcome biological barriers with parenteral nanomedicines. Finally parenteral drug delivery using polymers is evaluated.

The next chapters go over biological barriers to tissue engineering and specifically discuss physiological and mechanistic issues as well as drug delivery related to this field. Next, regulatory issues such as nanotoxicology are reviewed. This is an important area that scientists want to understand more about for the practical use of nanomedicines. Finally, a clinically relevant case study by Bioalliance Pharma is presented and the book is tied together with thoughtful closing remarks.

RSC Drug Discovery Series No. 22
Nanostructured Biomaterials for Overcoming Biological Barriers
Edited by Maria Jose Alonso and Noemi S. Csaba
© The Royal Society of Chemistry 2012
Published by the Royal Society of Chemistry, www.rsc.org

Overall, the book provides a very good understanding of the biological barriers and discusses the most innovative current approaches to overcome these barriers through the use of nanotechnologies and biomaterials.

Robert S. Langer
Department of Chemical Engineering,
Massachusetts Institute of Technology,
Cambridge, USA

Contents

Section 1 Historical View

RSC Drug Discovery Series No. 22
Nanostructured Biomaterials for Overcoming Biological Barriers
Edited by Maria Jose Alonso and Noemi S. Csaba
© The Royal Society of Chemistry 2012
Published by the Royal Society of Chemistry, www.rsc.org

Section 2   Nanostructures Overcoming the Intestinal Barrier

Section 3   Nanocarriers Overcoming the Nasal Barriers

Section 4   Nanostructures Overcoming the Ocular Barrier

Section 5 Nanostructures for Overcoming the Pulmonary Barriers

Section 6 Nanostructures Overcoming the Skin Barrier

Section 7   Nanostructures Overcoming the Blood-Brain Barrier

Chapter 7.2   Drug Delivery Strategies: BBB–Shuttles   364
R. Prades, M. Teixidó and E. Giralt

**Chapter 7.3   Drug Delivery Strategies: Nanostructures for Improved
Brain Delivery   392**
*Maria de la Fuente, Maria V. Lozano, Ijeoma F. Uchegbu
and Andreas G. Schätzlein*

Section 8   Overcoming Biological Barriers with Parenteral Nanomedicines

Section 9   Drug Delivery in Tissue Engineering

Section 10   Nanomedicine and Nanotoxicology

Section 11   A Clinically Relevant Case Study

Section 12   Concluding Remarks

Section 1
Historical View

CHAPTER 1

Historical View of the Design and Development of Nanocarriers for Overcoming Biological Barriers

MARÍA JOSÉ ALONSO*[a] AND PATRICK COUVREUR[b]

[a] Department of Pharmacy and Pharmaceutical Technology, Faculty of Pharmacy, Campus Sur, University of Santiago de Compostela, 15706 Santiago de Compostela, Spain; [b] UMR CNRS 8612, Université Paris-Sud, UFR de Pharmacie 5, rue Jean-Baptiste Clément, 92296 Chatenay-malabry, France
*E-mail: mariaj.alonso@usc.es

1.1 Introduction

Despite the fact that the first nanocarriers described in the literature were discovered by serendipity, the search for ways to overcome biological barriers has been the major driving force for the significant development of nanocarriers over the last decades.[1,2] The barriers that drugs need to confront have been recognized, for a long time, in a generic form, however, the knowledge about their specific composition and biological behavior has been limited until quite recently. Fortunately, the important advances made in the last decades in the field of cellular and molecular biology have led to relevant information on the nature and mechanistic behavior of these barriers.

Figure 1.1 illustrates the input of knowledge required for the design and development of new nanomedicine products. As indicated, besides the necessity of an extensive knowledge on the biological barriers, important

RSC Drug Discovery Series No. 22
Nanostructured Biomaterials for Overcoming Biological Barriers
Edited by Maria Jose Alonso and Noemi S. Csaba
© The Royal Society of Chemistry 2012
Published by the Royal Society of Chemistry, www.rsc.org

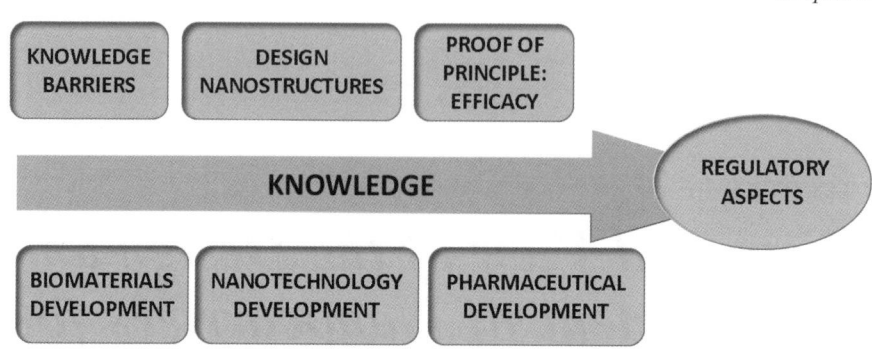

Figure 1.1   Diagram showing the knowledge required for the design and development
of a new nanomedicine product.

doses of imagination and creativity have been required for the successful
design of nanostructures and the adequate selection of biomaterials. In fact, a
critical limitation in the overall evolution of nanocarriers has been the
identification of materials that can be acceptable for the human body. In the
majority of cases, these materials are natural compounds that are well known
with regard to their biological behavior. Examples of these materials include
proteins, lipids and polysaccharides, which are present in our body. Other
materials, in particular polymers, have been taken from previous applications,
such as the preparation of prosthesis and other medical devices.

Another basic element for the development of nanocarriers has been the
availability of adequate methodologies to produce them with the sufficient
yield and efficiency. In this sense it is important to keep in mind, not only the
necessity for the technology to be scalable, and as simple and mild as possible,
but also the requisite for the nanocarriers to have an appropriate drug loading
and delivery rate. An additional hurdle is often related to the difficulties for
preserving the stability of the active compounds associated to the nanostructure. A final, and no less complex, issue is the one related to the
pharmaceutical presentation of nanocarriers and the stability during long-term
storage. In fact, preserving the colloidal stability of nano-matter is a very
difficult task due to its natural thermodynamic tendency to aggregate in order
to reduce the specific surface area. This situation often leads to the necessity of
adding stabilizers and/or converting them into a powder form, either by
lyophilization or spray-drying.

As nanocarriers are intended to solve critical problems of drugs, the
evolution of the drug discovery field and the specific pharmaceutical profile of
the new drugs has a crucial impact on the design of nanocarriers and, thus, in
the selection of biomaterials and nanotechnologies. In this sense, the
application of biotechnology within the pharmaceutical arena has led to an
increasing number of macromolecular drugs and antigens. Some of these
drugs, referred to as biopharmaceuticals, are currently being substituted by
low molecular weight peptides and RNA fragments, which are obtained by

chemical synthesis. Irrespective of their origin, these macromolecules are very vulnerable in the biological environment as they may suffer extensive degradation before reaching their target and have great difficulties in crossing epithelial barriers.

Finally, we should be aware that having acceptable knowledge of the biological barriers applying rational criteria to the design of nanocarriers is not sufficient in itself to reach the clinical development stage. As a matter of fact, besides the necessary proof-of-efficacy, a mandatory requirement from the regulatory point of view, is the assessment of the safety and mechanism of action. This is not a trivial task, as the analysis of the interaction of nanostructured materials with biological structures requires specific methodologies and validation procedures and standards.

Although relevant discoveries in drug delivery happened by chance or as a result of the enthusiasm of visionaries, the significant development of this field, especially over the last three decades, has been a consequence of important multidisciplinary efforts. Only by a multidisciplinary approach (Figure 1.2) involving knowledge from experts in chemistry, physics, engineering, biology, pharmacy and medicine could one understand the recent past and the even more promising future of nanocarriers and drug delivery field. These complementary efforts required for the advancement of nanotechnologies applied to medicine are well illustrated in the place that nanotechnology takes in the map of science.[3]

The intention of the various chapters and sections of this book is to present an overview of the advances in the knowledge of the biological barriers

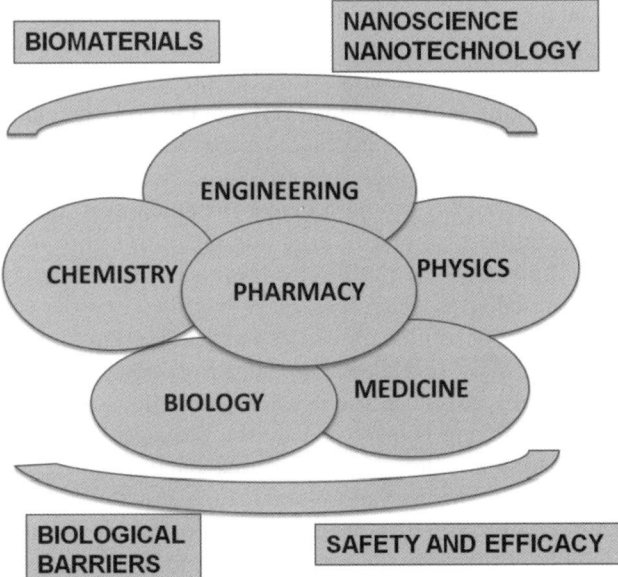

Figure 1.2   Multidisciplinary approach required for the development of nanomedicines.

associated to different modalities of administration and the corresponding development of nanoscience and nanotechnology-based alternatives available for confronting such barriers. In this initial chapter, our goal is to briefly summarize the critical information and to present how knowledge has been used throughout history for the specific design of nanocarriers. The chapter will end with the presentation of the current status of the nanocarriers from the pharmaceutical perspective and the prospects of future developments and cases of success. This initial chapter will be followed by a number of chapters (Sections 2 to 9) intended to describe in detail the limitations of specific barriers associated with different modalities of administration and examples of nanotechnology approaches to confront them. These chapters are described in section 1.3 of this chapter. Because of the intended pharmaceutical and practical projection of this book, it was found critical to include a chapter dealing with toxicological issues (Chapter 10) as well as case study of clinical pharmaceutical development (Chapter 11).

1.2 The Barriers Being Confronted Using Nanocarriers

The most important limitations of current therapies result from their limited success in a significant part of the population. This is largely determined by biological differences among individuals, including distinct disease markers, but also by the low specificity of many drug molecules for their specific targets and the emergence of resistance. The need to target drugs to their site of action was first presented by Paul Ehrlich in the early years of the 20th century.[4] He understood the necessity to devise ways to shuttle drugs in order to have them concentrated in the right place. The work carried-out throughout nearly a century has proven the great challenges involved in making the targeting concept a reality. This could be understood by the fact that humans have not evolved towards making the drug transit through the body easy but, mainly, to prevent the entrance of foreign entities, *i.e.* drugs, and to destroy them in case they manage to enter the body. The first critical barriers for a drug to reach the internal body compartments are the skin and the mucosal surfaces. Due to the highly restrictive nature of the skin, drug administration through this barrier has been mainly limited to the purpose of local action, although a number of nanotechnology-based approaches have been described in order to disrupt and facilitate the transport of solutes across this barrier (Section 6). In contrast, the intestinal mucosa, highly specialized in the absorption of nutrients, is also permissive to the transport of certain drugs as far as they can diffuse passively or are susceptible of being transported by the biological transporters of the epithelium. Unfortunately, despite this permissive nature, there are increasing numbers of drugs and antigens that are unable to cross this barrier, and many are additionally compromised in their stability in the harsh gastrointestinal environment. This is the case for polar compounds and macromolecules in general, such as peptides, proteins and antigens. As presented in Section 2, there are currently a number of nanocarriers which hold great promise, as they have displayed a capacity to overcome the barriers associated with oral administration. Alternative mucosal modalities of administration have also been explored with some positive results, in particular for the nasal and pulmonary routes. The definition and development of nanocarriers for overcoming these destructive barriers are of key interest for the growing biotech industry. The most relevant efforts towards this goal are described in Sections 3 and 5. There are other hard to access sites on the body, *e.g.* the eye or the brain, which have attracted the attention of nanotechnologists as well. The effective delivery of drugs either to the surface or to the inner eye or across the blood brain barrier (BBB) has been the goal of a number of nanostructures and large devices with nanostructural patterns (see Sections 4 and 7).

Overall, by reading this book, researchers and students will become aware of the current nanotechnology-based strategies intended to overcome all these external barriers. However, we should keep in mind that reaching the internal body fluids is only the starting point of the hazardous journey that drugs undergo before reaching their target. As shown in Figure 1.3, it is obvious that, given the difficulties and the lack of instructions for these molecules to go to their place of action, many of them will be lost and execute an unsuitable action at the wrong place. Indeed, once drug molecules are in the blood circulation, they will be exposed to the attack of degrading enzymes and/or get sequestered by plasmatic and tissular proteins, the pattern of this binding affinity being a major determinant of the final action (either no effect, the desirable or the unwanted effects).

The use of nanocarriers has been considered a way to protect drugs from degradation for a long time, however, preserving the stability of the nanocarrier itself has always been a concern that has not been sufficiently studied. On the other hand, changing the biodistribution of drugs is not, *per se*,

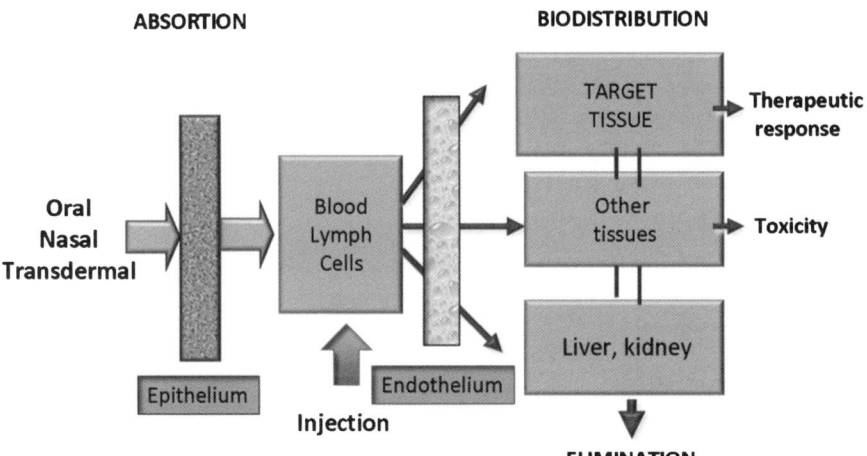

Figure 1.3 Schematic illustration of the barriers that could potentially be overcome through the use of nanocarriers.

necessarily good. For example, the nanocarriers developed at first were found to accumulate in the mononuclear phagocytic system (MPS) and related organs, a pattern that was of interest for the treatment of specific diseases, *i.e.* infectious diseases affecting these organs, but not for others. Fortunately, as described in the following sections, research conducted over the last decades has provided strategies for a greater control of carrier biodistribution. Nevertheless, the current achievements have not yet reached the level of perfect targeting, called by Paul Erlich the "magic bullet". Important advances towards this ambitious goal have been covered in Section 8.

1.3 The Key Milestones in the History of Nanocarriers and Drug Delivery

This section describes the early days when the "nano-drug delivery" field began, the key people who launched this exciting field, and the evolution of the field from its origins in the 1960s to the currently very active era of targeted nano-carriers. This analysis has taken into account a number of historical perspectives reported by key leaders in research.[1,2,5–7] Overall, this half a century period has been characterized by the concurrence of knowledge coming from different areas and a number of extraordinary provident discoveries. These discoveries, coming from the medical and pharmaceutical sectors, occurred as a consequence of the search for medical solutions and the increasing understanding of physio-pathology at both the cell and tissue levels. In the following lines we describe the critical milestones in the development of nanocarriers. These milestones are presented in a flow diagram (Figure 1.4).

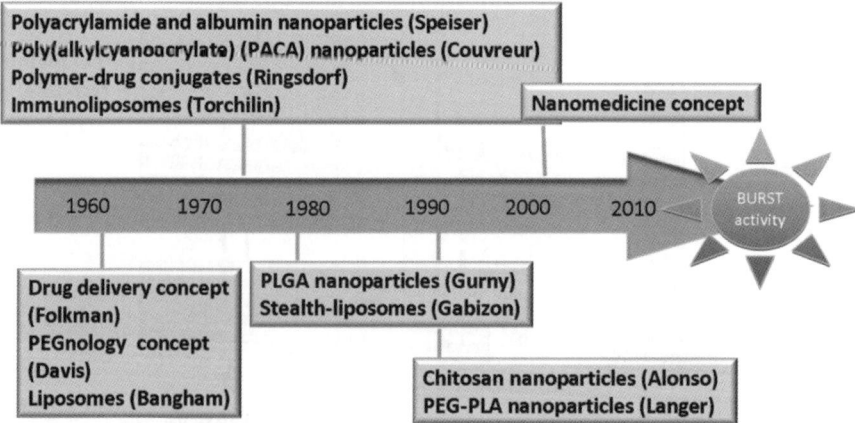

Figure 1.4   Flow diagram showing some critical milestones in the development of nanocarriers.

1.3.1 Milestone: The Concept of Controlled Drug Delivery

In the 1960s, probably motivated by the progress in biopharmaceutics and pharmacokinetics, the idea of prolonging the residence of drugs in the body by controlling their delivery became a major focus of attention. The pioneer of this concept was Judah Folkman, a surgeon who in the mid-1960s discovered that implanting a Silastic® (silicone rubber) tubing exposed to anaesthetic gases into rabbits resulted in a prolonged sleep.[8,9] He proposed that segments of such tubing containing a drug could be implanted in order to achieve a controlled drug delivery. Interestingly, at the same time, Alejandro Zaffaroni, an outstanding biochemist and entrepreneur, had also been thinking about the concept of controlled drug delivery. He heard about Folkman's work and decided to share his vision of founding a company focused on the concept of controlled drug delivery with him. This idea was soon realized and the company Alza started with the help of Folkman and others as the first company specialized in drug delivery.[10]

During the 70s there was a great development of the controlled release idea and a number of devices started entering the market. It was in this active period when Bob Langer developed with Judah Folkman the idea of controlling the release of proteins, using polymer matrices such as polyvyinilacetate.[11] This was the first example of successful delivery of a complex hydrosoluble molecule from a hydrophobic, non-degradable polymer matrix, thus launching the field of controlled delivery of macromolecules. The impact of this discovery has been great, as it stimulated the immense activity and clinical success of micro- and nano-therapeutics to the present day.[1,2,5,6,12] One of the various targets of application of macromolecular drug delivery promoted by Bob Langer is vaccination.[13] He proposed the single-dose vaccine strategy, consisting of associating an antigen to a polymer device, as a way of improving and simplifying vaccination campaigns. Later on this idea was adopted by the WHO and the "Bill and Melinda Gates Foundation", where it was considered a Grand Challenge in Global Health.

1.3.2 Milestone: The Discovery of Liposomes

Interestingly, in the same decade that the physicist Feynman (1918–1988) was awarded with the Nobel Prize in Physics (1965) and introduced the concept of nanotechnology, Bangham discovered the first nanostructured drug carrier by chance. Alec Bangham (1922–2010), currently known as the father of liposomes, was a leading hematologist whose interest was studying the behavior of blood cellular membranes. In the 1960s, when he was trying to figure out the organization of lipids, he observed the possibility of forming a kind of vesicular structure made of artificial membranes. He thought of this simple and original structure as a way of transporting and delivering drugs into cells.

Following the discovery of liposomes there has been thousands of papers and articles coming from all kinds of backgrounds and aimed at studying a number of aspects including their composition, preparation techniques, physicochemical properties, biological behavior, ability to target and deliver different types of drugs as well as their pharmacological and toxicological behavior.[14,15] A very large number of authors have contributed to the development of a great variety of liposomal compositions, loaded with many different drugs, and studied their stability, biodistribution and biodegradation.[16–18] From these contributions and others, little by little, the field has been continuously dipping into knowledge and novelty, with specific landmarks driving this process.

Throughout this productive process, there have been particularly relevant ideas, which have had a great impact on the development of nanomedicines; these are the design and development of immunoliposomes and "stealth" liposomes.[19–23] Even though these conceptual ideas were proposed separately, both were intended to modify the biodistribution pattern of drugs and to target them to specific cell populations. In a sense, the design of liposomes was about bringing to life Ehrlich's imagined concept of the "magic bullet". Although Paul Ehrlich was mainly thinking of killing bacteria, this targeting concept has always been in the minds of drug delivery nanotechnologists. In fact, a few years after the discovery of liposomes, in the 1970s, Vladimir Torchilin presented the original concept of immunoliposomes.[24] His idea was to target liposomes to specific cells using monoclonal antibodies as targeting ligands and gave the name of immunoliposomes to the resulting targeted nanostructures. Over the last few decades, this targeting concept has attracted a great deal of the attention and has been applied to a wide range of nanostructures and ligands. The evolution of this idea has also been motivated by the important advances in the field of cellular and molecular biology, which have led to the identification of specific targets.[25–27] As a consequence, nowadays there is a variety of targeting ligands starting from the monoclonal antibodies (*e.g.* anti-Her2; anti-CD19) but including, as well, a variety of large (proteins, peptides, aptamers) and small molecules (mannose, folic acid, transferrin among others).

A great help in the design of targeted liposomes has come from the concept of stealth nanocarriers, equipped with a protective shell that prevents their immediate uptake by the MPS and permits their circulation in the blood stream for extended periods of time. This long-circulating concept was taken from the protein chemistry field, where this approach was initially aimed at reducing the immunogenicity of proteins, as will be later described. Its application to the formulation of liposomes has led to successful developments[28–30] and, in particular, to the development of Doxil/Caelix®.[31] This formulation containing doxorubicine was initially approved for the treatment of Kaposi Sarcoma in AIDS patients and is being currently used for the treatment of recurrent breast cancer and ovarian cancer.[32]

The advances made in liposomes over the last decades have led to a number of marketed formulations (see Table 1.1) and a significant number of formulations under clinical development. Overall, the majority of systems have been approved for the treatment of various types of cancer. Moreover,

Table 1.1   Marketed liposomal formulations.

Trade name	Drug	Indication
Myocet®	Doxorubicine	Breast methastatic cancer
Doxyl®/Caelix®	Doxorubicine	Ovarian methastatic cancer
Daunosome®	Daunorubicine	Kaposi sarcoma
Onco-TCS®/Marquibo®	Vincristine	No Hodgkin lymphoma
Depocit®	Citarabine	Lymphomatose meningitis
Mepact®	Mipaphurtide	Osteosarcome
Albecet®	Amphotericin	Fungal infections
AmBisome®	Amphotericin	Fungal infections
Visudyne®	Verteporfrine	Age macular degeneration
Depodur®	Morphine	Pain relief
Octocog alfa®	Factor VIII	Haemophilia

there are currently a number of clinical trials evaluating the potential of these technologies for the simultaneous delivery of different anticancer drugs.[33,34]

1.3.3   Milestone: The Origin of PEGnology and Polymer Therapeutics

Concomitantly with the incipient revolution of biotechnology and the evidence of the necessity of improving the delivery of proteins to the body, another interesting idea arose in the mind of Frank Davis in the very productive 1960s. This was what he called the origin of PEGnology otherwise named as the "PEGylation concept".[35] This concept started as a result of his interest in developing a procedure to reduce the adverse immunological responses caused by selected bioactive proteins, which could be utilized for human therapy. What he found was that PEG-proteins showed not only greatly reduced immunogenicities upon intravenous injection in animals, but of equal interest, much longer circulating lives than their unpegylated counterparts. Interestingly, this concept has influenced major advances in drug delivery throughout the following decades. In particular, the subsequent work developed by Veronese and others[36] has extended the application of PEGnology to a variety of proteins, a result of which has been the clinical development and marketing in the 1990s of a number of therapeutic proteins, namely enzymes (such as L-asparaginase) and cytokines (including interferon and granulocyte colony-stimulating factor).[37] Moreover, the same approach has been applied to all kinds of nanocarriers, a result of which has been a great improvement in their ability to overcome biological barriers.

A few years after the inception of PEGnology, in the mid 1970s, polymer conjugation was extended by Ringsdorff to the linking of a variety of water-soluble polymers to chemotherapeutic agents.[38] This innovative approach was also reported by Maeda *et al.* for styrene-maleic acid copolymer-conjugated neocarzinostatin (SMANCS), and it eventually led to the concept of the "enhanced permeability and retention" (EPR) effect of solid tumors in 1986.[39,40] Subsequent research by Duncan and Kopececk led to the design of the first synthetic polymer–drug conjugates to progress to clinical trials.[41–44] Following these pioneering works, a number of polymer-drug conjugates arose at various places around the world, leading to the concept of polymer therapeutics.[45] So far, polymer therapeutics have been designed to improve drug therapeutic efficacy and reduce side effects of antineoplasic drugs, in which the prolonged circulation time plays a crucial role. This has been the result for a number of drugs already in routine clinical use (for example, SMANCS, PEG-adenosine deaminase, PEG-interferon alfa-2a and 2b, PEG-human G-CSF, PEG-HGH antagonist, PEG-antiTNF Fab) (see Table 8.2.1 in Chapter 8.2, for more information). Moreover, conjugation to hydrophilic polymers was also found to improve the water solubility of hydrophobic drugs such as paclitaxel, enabling easier formulation and patient administration.[46] Among synthetic polymer-drug conjugates, poly(L-glutamic acid) (PG)-paclitaxel (PG-TXL) (CT-2103, Xyotax®) has advanced to Phase III clinical trials and is positioned to be the first of its class to reach the market.[47,48]

An idea associated to the polymer conjugation chemistry proposed by Ringsdorf and coworkers for the delivery of drugs was the synthesis of block copolymers able to form polymeric micelles. These are nanosized (typically in the range of 10–100 nm) constructions formed from the self-assembly of amphiphilic block copolymers in aqueous environments. In water, the hydrophobic portion of the block copolymer self-associates into a semi-solid core, whereas the hydrophilic segment of the copolymer becomes exposed towards the external aqueous medium forming a shell. The resulting core-shell architecture is important for drug delivery purposes, because the hydrophobic core can act as a reservoir for water insoluble drugs, while the outer shell protects the micelle from rapid clearance. This original idea of using copolymer micelles for the formulation of hydrophobic drugs started in the in the 1980s,[49] and it has received a great deal of attention from many scientists who have contributed to the synthesis and formation of different block copolymers as well as in the formulation of several active compounds. Among the block copolymers, the most attractive ones have been those composed of polyethylene glycol (PEG) linked to biodegradable fragments such as poly(propylene oxide) (PPO), poly(D,L-lactic acid) (PDLLA), poly(ε-caprolactone) (PCL), and polyaminoacids such as poly glutamic acid, poly-L-aspartic acid.[50] In this area, Kataoka[51–53] has certainly been one of the most brilliant pioneers in the field.

Different types of micelles are those proposed by Torchilin, which are formed by conjugates of soluble copolymers with lipids.[54–56] The great variety of lipids conjugated with hydrophilic polymers, *i.e.* PEG, PVA and others, have led to an important accumulation of knowledge on their potential for the solubilization of hydrophobic drugs, drug targeting and intracellular drug delivery. One of the more studied conjugates is polyethylene glycol-phosphatidyl ethanolamine conjugate (PEGYPE).

The so-called micellization approach has led to the successful development of new formulations of anticancer drugs, which are currently under clinical evaluation. These are polyaminocid-PEG and PLA-PEG copolymer micelles.[57–59]

As with many other nanocarriers the late stage development of these nanostructures relies on the functionalization of the outer surface of the polymeric micelle in order to modify its physicochemical and biological properties. This can be accomplished in a regulated fashion by synthesizing end-functionalized block copolymers.

1.3.4 Milestone: The Origin of Nanoparticles

The search for the historic path of the growing field of nanoparticles is complicated due to the confusing terminology. The terms "nanoparticles", "nanocapsules", "nanospheres", "microcapsules", "microspheres", "colloidal carriers", and "latices" have all been used to refer to particles in the nanometre size range for the delivery of bioactive compounds. Currently, there is a definition that has been accepted for pharmaceutical purposes: "Nanoparticles may be defined as being submicronic (<1 μm) colloidal systems generally, but not necessarily, made of polymers".[60] Assuming this definition, in the following paragraphs we will refer to the origin and evolution of nanoparticles made of different biomaterials, starting with those made of acrylic polymers as they represent the land-mark in the field (Figure 1.5).

1.3.4.1 Poly(alkyl)-Acrylate-based Nanoparticles and Nanocapsules

Interestingly, at the same time that Judah Folkman and Alejandro Zafaroni were envisioning the potential of controlled drug delivery using polymer materials in the US, in Europe Peter Speiser was also exploring the same idea using miniaturised delivery systems. Indeed, it was in the 1960s when he developed polyacrylic beads for oral administration[61,62] and started the idea of

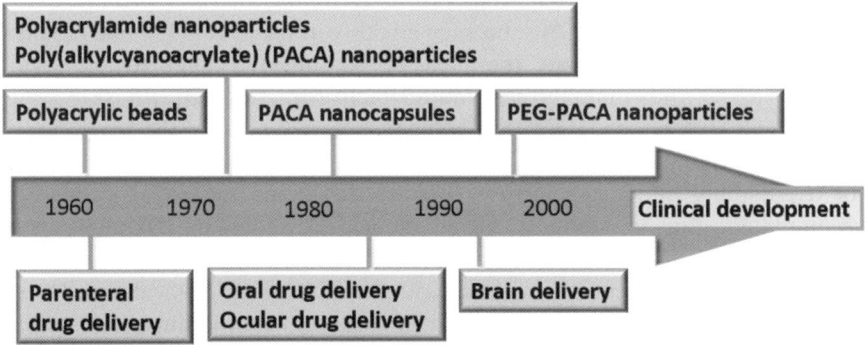

Figure 1.5 Milestones in the development of poly(acrylate) nanoparticles.

preparing polyacrylamide nanoparticles for vaccination purposes. This idea soon became a reality through the adequate adaptation of a "micelle polymerisation technique".[63] Subsequently, Patrick Couvreur made a contribution to this discovery by investigating the lysosomotropic character of nanoparticles.[64] Simultaneously and, more importantly, he produced the first rapidly biodegradable acrylic nanoparticles made of poly(alkyl cyanoacrylate) through the use of a controlled anionic polymerization technique.[65] This has proved to be a critical development that has attracted the attention of many researchers who have investigated the potential of these nanoparticles for a variety of applications (Figure 1.5).

A few years after the initial development of the nanoparticles, Al Khouri Fallouh *et al.*[66] applied the anionic polymerization technique in an oil-in-water system leading to the production, for the first time, of PACA nanocapsules. This represented an interesting alternative as it allows the efficient encapsulation of small lipophilic drugs and also macromolecules, *i.e.* insulin.[67,68]

As in the case of most nanocarriers, an important therapeutic application of nanoparticles has been in the area of cancer therapy. Couvreur and co-workers have been responsible for the majority of contributions in this field, going from the study of toxicity, mechanistic, pharmacokinetic and pharmacodynamic issues to the pharmaceutical development of a significant number of formulations.[69,70] Another application of PACA-based nanocarriers has been for improving the oral bioavailability of drugs. The evidence of the potential of PACA nanoparticles for increasing the oral absorption of drugs was first reported by Maincent *et al.*[71,72] This idea was soon applied to the challenging molecule of insulin, providing the first evidence of the potential of nanocarriers for the successful oral administration of insulin.[67,68] This report has become a landmark as it represents the origin of the use of polymeric nanocarriers for oral peptide delivery. In the same decade (1980s), PLCA nanoparticles were also proposed for the ocular drug delivery area, with the final goal of improving the transcorneal penetration of drugs. The proof-of-concept was first shown for the antiglaucoma drug pilocarpine[73-75] and later on for other compounds, such as aminoglucosides and beta-blockers.[76,77] More recently, the idea of using PACA nanoparticles for overcoming the BBB, preconized by Peter Speiser in the early 1980s, become a reality in the 1990s. This was thanks to Jörg Kreuter, who had the idea of coating the nanoparticles with polysorbate 80, thus making them more effective for this critical task.[78,79]

Briefly, PACA nanoparticles can be now considered one of the keystones in the exciting and initial tortuous path of polymer nanoparticles as nanocarriers for overcoming biological barriers. Besides the great amount of knowledge accumulated, this keystone is facing clinical development. Namely, poly (isohexyl cyanoacrylate) nanoparticles loaded with doxorubicine (Doxorubicin-Transdrug®) are currently being investigated in Phase II/III clinical trials by the company BioAlliance for the treatment of resistant hepatocellular carcinomas (see Chapter 11, for a detailed review). It is particularly noteworthy that the first clinical trial using this polymer as nanoparticles was initiated in 1992 at the Institut Gustave Roussy, France.[80]

1.3.4.2   Albumin-based nanoparticles

The development of albumin and gelatin nanoparticles in 1974 is also, in part, attributed to Peter Speiser and co-workers. They originally adapted a process that allowed them to form well-defined nanoparticles using a desolvating agent.[81–83] There was a precedent in the literature of albumin nanoparticles obtained by heating a W/O emulsion containing albumin, however because of the large size distribution (between 300 and 1000 nm) the authors preferred to call them microspheres.[84] These particles were initially proposed as radio-pharmaceuticals and, thus, the first report on their application presented the biodistribution after their labelling with 99mTc.[85] This work was followed by that of other groups using these particles as carriers for drugs[86–88] and also magnetite nanoparticles.[89,90] This latter report represents the origin of the targeting concept using an external magnetic field.

Despite the early development of albumin nanoparticles, as can be noted in Figure 1.6, for a couple of decades there was almost no activity related to their specific development and application. This was mostly due to the erroneous pre-conception that these denaturalized protein nanoparticles could lead to important immunological reactions and toxicity problems. The activity was accelerated in the 2000s when the use of these nanoparticles was proposed for the delivery of nucleic acid based compounds among others.[91,92] More recently, the focus has also been on the functionalization of these nanoparticles with antibodies and proteins, *i.e.* apolipoprotein E (apo E), either for an improved targeting[93,94] or a facilitated transport of drugs across the BBB.

Interestingly, in 2005, after more than 30 years of experience but with very modest activity in this specific subject, albumin nanoparticles have emerged as the first commercial drug-containing nanoparticle product.[95] This nanomedicine product, named as Abraxane[TM], consisting of human serum albumin nanoparticles loaded with paclitaxel, is currently indicated for the treatment of breast carcinoma.

1.3.4.3   PLGA-based Nanoparticles

Despite the early development of microspheres made of biodegradable poly(lactide-*co*-glycolide) (PLGA) copolymers, the presentation of this successful biomaterial in the form of nanoparticles did not occur until 1981. It was Robert Gurny in collaboration with Gil Banker and Nicholas Peppas, who reported for the first time the preparation of poly(lactic acid) nanoparticles for drug delivery purposes.[96] At that time, these researchers described the particles as "pseudolattices" because they were manufactured from previously polymerised material (PLGA), in contrast to lattices that are produced during polymerisation *in situ* (the case of acrylates).

As in the case of albumin nanoparticles, the activity regarding the development and application of these nanoparticles was scarce until the late 1990s, however, it speeded up in the 2000s with thousands of articles published in this field. The independent contributions of Ruxandra Gref[97] and Didier Bazile[98] to the development of PEGylated PLA and PLGA nanoparticles are particularly worth mentioning. In line with the achievements made for liposomes, the PEGylation approach was conceived as a strategy to prolong the blood circulation of the nanoparticles and avoid their accumulation in the MPS and related organs.

An additional remarkable development was the adaptation of the W/O/W emulsification technique for the nanoencapsulation of delicate macromolecules, *i.e.* proteins.[99] Using this technique it is possible to obtain well-defined nanoparticles, with a narrow size distribution (mean size normally around 200 nm) and with a great capacity for the association of model proteins as well as therapeutic enzymes.[100] A few years later, being conscious of the risk of protein damage during encapsulation and release from PLGA matrices, Alonso and co-workers defined new formulation strategies consisting of the incorporation of PEG and poloxamers into these matrices. Such strategies turned out to be particularly efficient for preserving the stability of interferon alpha,[101] proangiogenic factors[102] antigens[103,104] and plasmid DNA.[105,106]

A critical milestone in the evolution of PLGA nanoparticles has been the discovery of the positive effect of a PEG coating on their ability to overcome mucosal barriers. In 1998 Tobio *et al.* [103] reported for the first time that the presence of a PEG around PLA nanoparticles favored the transport of the associated protein (tetanus toxoid) across the nasal membrane. This finding caused a shift in the classical paradigm claiming the necessity of a relatively large size (1–10 microns) and a hydrophobic surface to facilitate the transport of antigens across mucosal barriers. Indeed, Alonso and co-workers have reported a number of studies illustrating that the small size of the nanoparticles (less than 500 nm) and their hydrophilic PEG surface were relevant factors for preserving their stability in mucosal fluids and facilitating their penetration into mucosal barriers, such as the nasal,[107,108] intestinal[109] and ocular barriers.[110] Accordingly, the proof of principle of nasal vaccination using tetanus toxoid-loaded PLA-PEG nanoparticles has led to very promising results.[111]

The positive evolution of PLGA-PEG nanoparticles with regard to their ability to overcome biological barriers, has not led so far to their introduction into clinical development. However, it is worth mentioning the presence of low molecular weight PLA-PEG block copolymer micelles in phase II clinical trials.[112] The earlier development of micelles compared to nanoparticles could be related to the simplicity of the methodology for producing these micelles together with their rapid degradation and elimination profile.

1.3.4.4   Chitosan Nanoparticles and Nanocapsules

In the late 1990s, Calvo *et al.* reported for the first time the production of chitosan nanoparticles using a very mild and simple ionic crosslinking technique.[113] Amazingly, as can be noted in Figure 1.6, the consequences of this initial report are illustrated in thousands of articles published to date. The driving force behind the design and development of these nanoparticles was the need for a nanocarrier that would easily associate and deliver delicate

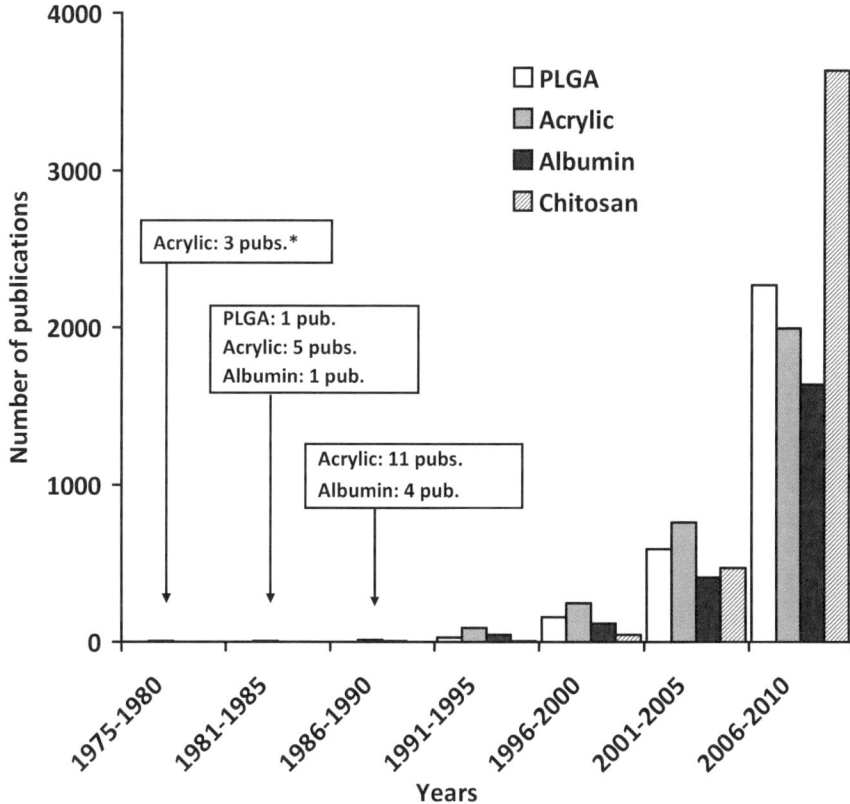

*pub. = publication

Figure 1.6   Evolution in the number of reports of some types of nanoparticles (PLGA, acrylic, chitosan, albumin), which are currently on the market or undergoing clinical development. Search done in the ISI Web of Knowledge using the following keywords: **PLGA nanoparticles**: (PLGA AND nano*) OR (PLA and nano*) OR ("polylactic acid" AND nano*) OR ("polylactide-co-glycolide" AND nano*) OR ("polylactic-co-glyco-lic acid" AND nano*) OR ("PLGA-PEG" AND nano*) OR ("PLA-PEG" AND nano*). **Acrylic nanoparticles**: (polyacrylate AND nano*) OR (polycyanoacrylate AND nano*) OR (polyalkylcyanoacrylate AND nano*) OR ("polyacrylic acid" AND nano*) OR (acrylate AND nano*). **Albumin nanoparticles**: albumin AND (nanop* OR nanos* OR nanoc*). **Chitosan nanoparticles**: (nanop* OR nanos* OR nanoc*).

macromolecules and exhibit and affinity for mucosal surfaces. A wide variety of macromolecules including gluthatione,[114] insulin,[115] heparin,[116] growth factors,[117] antigens (tetanus and diphtheria toxoids and the recombinant hepatitis B surface (rHBs) antigen),[118,119] plasmid DNA and siRNA[120–123] have already been associated to these nanoparticles. Simultaneously, the same authors developed alternative chitosan-based nanocarriers consisting of a lipophilic oily core surrounded by a coating made of a polyester and chitosan.[124] This nanostructure, named a nanocapsule, later evolved towards a single chitosan or chitosan-PEG coating and has been adapted for the encapsulation of either lipophilic[125] or hydrophilic peptides.[126]

Soon after their initial development, it was found that, as expected, chitosan nanoparticles could interact with the nasal mucosa and facilitate the transport of the associated macromolecule. For example, the reported work has shown that insulin-loaded chitosan nanoparticles can significantly improve insulin absorption as compared to the control solution of chitosan containing insulin.[115] Moreover, in the area of nasal vaccination it was observed that the nasal delivery of antigens, *i.e.* tetanus toxoid, associated to chitosan nanoparticles resulted in high and long-lasting immune responses.[127,128] A more advanced version of these nanoparticles, consisting of a hybrid mixture of chitosan and cyclodextrins, has also been found to further improve this absorption enhancing behavior.[129]

The use of chitosan-based nanoparticles for improving the oral absorption of macromolecules, *i.e.* insulin, has been a subject of intensive investigation. There are more than one hundred articles describing different types of chitosan-based nanoparticles, generally prepared according to the original ionic gelation technique,[113] which illustrate different levels of success in terms of oral insulin absorption.[130–135] Particularly promising are the recent data by Sonaje *et al.*, who showed that nanoparticles consisting of chitosan in combination with alginate, dextran or polyaminoacids were able to increase the oral absorption of insulin up to more than 10% absolute bioavailability.[136,137] On the other hand, as shown in Figure 1.7, chitosan nanocapsules have been found to enhance and prolong the absorption of the peptide salmon calcitonin (sCT). The mechanistic issues behind this interesting behavior have not been totally validated yet. However, the results of the *in vitro* and *in vivo* evaluation suggest that these nanocarriers have the ability to interact and diffuse through the mucus layer and finally remain associated to the underlying epithelium.[138]

Another area of potential application of chitosan nanoparticles and nanocapsules is ophthalmics. Confocal microscopy studies have shown that these nanostructures are able to adhere to the ocular mucosa and reach the superficial layers of the corneal epithelium, thus providing a way to shuttle bioactive compounds to different eye compartments.[139] A particularly interesting ophthalmic indication for chitosan-based nanoparticles has been identified in the area of ocular gene therapy. Work by De la Fuente[140,141] has clearly shown that nanoparticles consisting of chitosan and hyaluronic acid are very efficient vehicles for the delivery of pDNA to the conjunctival cells. Their

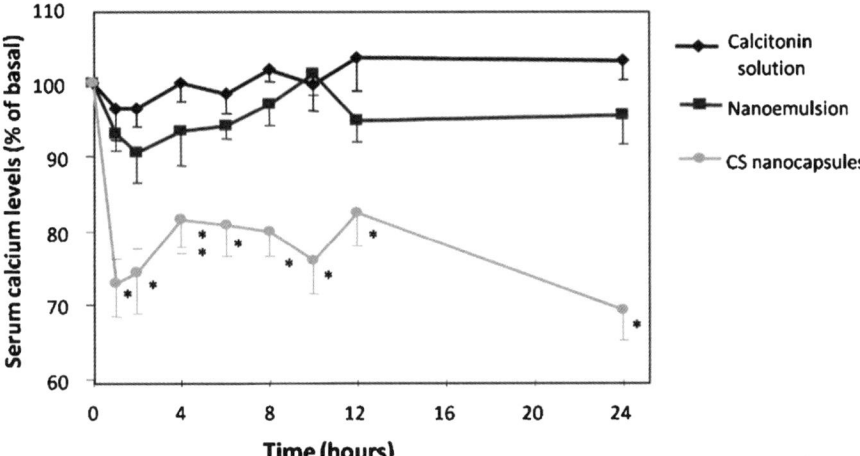

Figure 1.7 *In vivo* efficacy of chitosan nanocapsules for enhancing the intestinal
absorption of salmon calcitonin (sCT). (Reproduced with permission
from Springer Science + Business Media: C. Prego *et al.*, *Pharmacol. Res.*,
2006, **23**, 549.)

efficiency was explained by the conjunctival targeting properties of hyaluronic
acid combined with the bioadhesive behavior of chitosan.

A few authors have also explored the specific interest of chitosan nanoparticles
for particularly challenging biological barriers, *e.g.* pulmonary and blood-brain
barriers. For pulmonary delivery applications, the nanoparticle's aqueous
suspensions have been converted into aerosolized and inhalable powders[142]
and have led to positive results on the delivery of macromolecules, *e.g.* insulin and
pDNA.[143] On the other hand, regarding the BBB, there has been some interesting
academic work revealing the ability of chitosan-PEG nanoparticles functionalized with the monoclonal antibody OX26 to facilitate the transport of drugs
across the BBB and to treat experimental brain ischemia.[144]

Finally, after more than 15 years of research activity in chitosan-based
nanostructures, there is one prototype of chitosan nanocapsules undergoing
clinical evaluation for the treatment of psoriasis. As indicated, there are
thousands of articles and hundreds of patents showing the great potential of
these nanostructures for a variety of treatments, therefore, it could be
speculated that overcoming the regulatory hurdles associated with the first
development may open a door for an extensive exploitation of this
accumulated knowledge.

1.4 Current State of the Art in Nanocarriers Development

The section above presents the key milestones of nanocarriers and drug
delivery from the pharmaceutical technology perspective of the authors and

with particular emphasis on the efforts leading to clinical developments. The purpose of this section is to briefly summarize the current tendencies derived from these initial efforts, with a specific focus on the technology and the biomaterials used to produce them. The authors are aware that there are other approaches, *i.e.* solid lipid nanoparticles, dendrimers technology and others, which have not been discussed here.

Currently, the variety of nanostructures available is great and the number of review articles describing their nature are also important.[59,145–149] A very broad classification is the one dividing them into two categories based on the nature of the biomaterials: organic or inorganic materials (Figure 1.8). The most advanced studies making use of inorganic materials are those related to metallic nanoparticles,[150] which have been mostly used for diagnostic purposes. In fact, the first nanoparticles reaching the market have been those made of magnetite, indicated for MRI studies.[151] However, over the last decades, there has been a growing interest in using them for the treatment of cancer.[152] For example, hybrid organic–inorganic iron-based nanoconstructs were recently proposed by the groups of Couvreur and Ferey (so-called nano metal oxide frameworks, NanoMOFs), with tunable structures and porosities for better drug interactions and high loadings.[153] The specific composition and advances made in this specific field of iron nanoparticles has been extensively covered in previous reports and is beyond the scope of this chapter.[154,155] On the other hand, with regard to the nanocarriers made of organic materials, there is a tendency to classify them into different sub-categories, such as those made of lipids, proteins and polymer, of either synthetic or natural origin. This classification is appropriate from the conceptual point of view, however, current approaches are attempting to incorporate a variety of materials within the same nanostructure and form hybrid complex nanostructures. Finally, taking into account the structural organization of the biomaterials and the way of loading the active compounds, a classification can be made as described in the following lines. An illustration of these structures is presented in Figure 1.9.

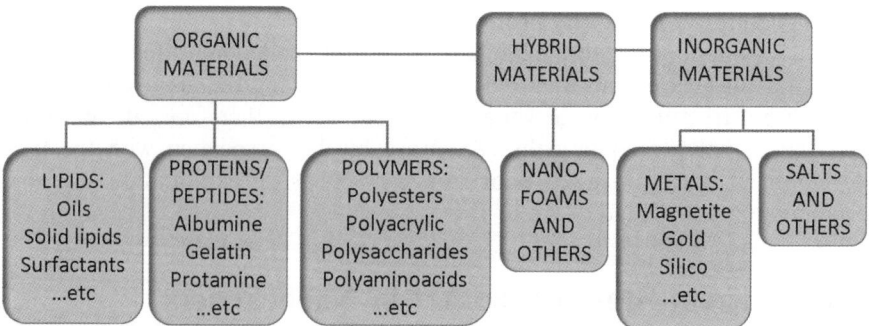

Figure 1.8   Classification of the materials currently used for the formation of nanostructures. The present chapter focuses on the use of organic materials.

Liposomes: the current approaches on liposomes are mainly projected towards the definition of complex nanostructures consisting of phospholipid bilayers, which include polymer-lipid conjugates (*e.g.* PEG-lecithin; PEG-cholesterol, *etc.*), polymers (*e.g.* polysaccharides such as chitosan and hyaluronic acid) and a whole range of macromolecules (proteins, peptides, monoclonal antibodies, *etc.*). These additional compounds may be well inserted in the lipid bilayer or simply forming a coating around it.[156]

Nanoparticles: a vast range of nanostructures could be grouped under the name of nanoparticles. Although the nanoparticles currently on the market or under clinical development are made of a single biomaterial (*i.e.* albumin or PIBCA), as in the case of liposomes, the actual tendency is the design of complex nanoparticles consisting of different polymers combined together or associated to other biomaterials, such as lipids or proteins. For example, there has been an interest in producing nanoparticles made of polymers containing cyclodextrin units linked to hydrophobic segments[157] as well as nanoparticles consisting of mixtures of polymers, *i.e.* chitosan, with cyclodextrins.[158] Another approach has been the formation of core-coated nanoparticles in which the core and the coating are made of different materials (different polymers, lipid core coated with polymers, polymer core coated with lipids or proteins, *etc.*).[159,160] Generally, in such nanostructures, the core has the role of accommodating and protecting the active ingredient, whereas the coating is adapted to confront specific biological barriers. This different structural organization is currently leading to the specific engineering of nanoparticles combining the structural elements in the appropriate form.

Nanocapsules: although nanocapsules might fall into the category of core-coated nanostructures, their specific denomination has come from the liquid nature of their core, which in most cases consist of an oil or mixture of oils and other components but it can also have an aqueous nature.[161] The composition of the core is dictated by the type of drug and/or the additional bioactive materials to be loaded into it. For example, there is a possibility of combining different oils with polymers and other lipids, which may help solubilize the drug or the complementary formulation ingredients, such as penetration enhancers, adjuvants, *etc.* On the other hand, the coating or shell can be simply made of hydrophilic protective polymers, such as PEG, or a variety of functional polymers and proteins. For example, we have recently reported a new class of nanocapsules with a shell made of polyaminoacids (polyarginine) and polysaccharides (chitosan, hyaluronic acid). Such nanocapsules have been found to be useful for the association of plasmid DNA and also antigens onto their surface, while the oily core can allocate additional adjuvant ingredients.[162]

Micelles and nanoassemblies: the initial micelles made of amphiphilic surfactants have paved the way to a great number of nanostructures consisting of amphiphilic diblock copolymers or polymers linked to a specific hydrophilic or hydrophobic domain. Probably, the most studied ones are those containing polyethylene oxide units, either in the form of conjugates with lipids, *i.e.*

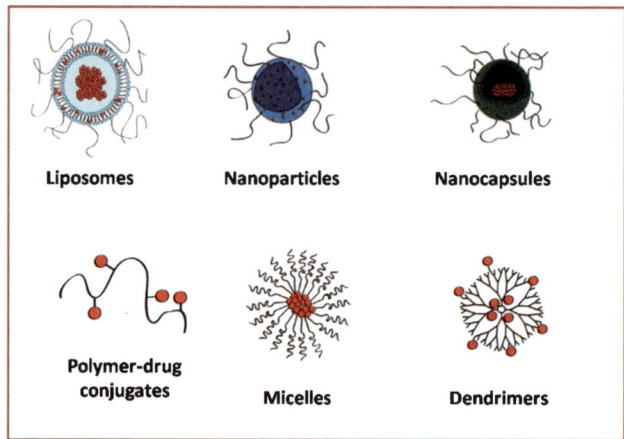

Liposomes   Nanoparticles   Nanocapsules

Polymer-drug
conjugates   Micelles   Dendrimers

Figure 1.9   Schematic illustration of a variety of current drug nanocarriers (courtesy
of Victoria Lozano).

phospholipids or cholesterol, or in the form of block copolymers, such as poloxamers of PLA-PEG. However, the current strategies go beyond the use of PEG and involve other hydrosoluble polymers for example chitosan.[163] In addition, micelles were originally conceived for the entrapment of lipophilic drugs, whereas nowadays there are possibilities of entrapping any kind of active molecule simply by adjusting the interaction forces between the cargo and the carrier or producing lipophilic derivatives of the parent compounds. Examples of these approaches are the entrapment of genetic material into hydrophobic chitosan derivatives.[164] Therefore, the possibilities of forming new micelles and assemblies are great and only determined by the specific advantages and safety of a particular biomaterial.

Polymer conjugates: thanks to the terrific activity in the polymer chemistry field the current display of polymers with potentialities of linking all kinds of bioactive molecules is enormous.[165] Particularly attractive are conjugates made with dendritic molecules (dendrons or dendrimers) as they offer the chance to associate a significant pay-load of different drug molecules, and finely and specifically modulate their delivery and interaction with the biological environment.[166] Additionally, these polymer conjugates can be presented in a simple solution or in the form of nanoassemblies, which might be of interest for modulating their interaction with biological compounds.

Self-assembled squalenoylated prodrugs: there are many prodrugs which have been used for drug delivery purposes. The squalenoylation technology discussed below represents an exception as it leads to the spontaneous formation of nanoparticles. In fact, squalene is a triterpene widely distributed in nature that is an intermediate in the cholesterol biosynthesis pathway. Advantage has been taken of the remarkable dynamically folded conformation of squalene to chemically conjugate this lipid with various therapeutic molecules in order to construct nanoassemblies of 100–300 nm.[167] This breakthrough concept of "squalenoylation" has led to impressive drug loading capacities (up to 50% drug loading, *i.e.* weight of drug/weight of drug transporter) and found applications in cancer (*i.e.* Gemcitabine, paclitaxel, *etc.*)[168] or antiviral therapy[169] as well as in nucleic acid delivery.[170] In a Lego-type approach, it is also possible to construct squalene-based multifunctional nanoparticles with additional imaging functionalities.[171] This new nanotechnology platform is expected to have important applications in pharmacology.

As indicated above, the classification presented here is logically related to the historical perspective of nanocarriers, and the advances made in their conceptual design and development are a consequence of long-term research efforts in this direction.

Irrespective of the specific nature of the nanocarriers described up until now, there are some common conceptual and experimental approaches, which have generally been applied to all kinds of nanocarriers. For example, in the field of intravenous drug delivery there is no doubt of the necessity to provide them with a hydrophilic polymer shell to facilitate their access to the target organs. Additionally, the use of hydrophilic coatings has been found to be of interest for protecting nanostructures in mucosal fluids, *i.e.* nasal and gastrointestinal fluids, and facilitating their contact with mucosal surfaces. This protection has been classically provided by PEG, however currently there is a search for other hydrophilic polymers, such as polysaccharides (dextran, hyaluronic acid and others), that may offer additional advantages.

Another generic approach, broadly applied to all kinds of nanostructures, is their functionalization with specific targeting ligands. While the first step of the approach was made with monoclonal antibodies—the concept of immunoliposomes—the number of targeting ligands has been gradually increasing. The most common functionalization approaches that are currently under study involve the use of monoclonal antibodies, peptides, aptamers and polysaccharides.[26,93,94] The actual evolution of this field is being determined by the identification of the targeting molecules in conjunction with adequate nanocarrier design and development of low-cost and reproducible technologies. The design and engineering of nanocarriers involves a number of features that are crucial for their final success. These features include an adequate stability, an appropriate internalization of the nanocarrier into the target cell (in the case of intracellular pharmacological targets) and controlled intracellular drug delivery.[12]

It is worth mentioning at this stage an advanced generic concept which aims at designing nanosystems intended to have multiple functions including the diagnostic, treatment and monitoring of diseases. These so called "nanotheragnostics" must be able to reach the appropriate biological target, to make it visible with the appropriate labeling (fluorescent, radioactive, metallic compound, *etc.*), to deliver the associated drug according to a predefined schedule and to follow-up the biochanges (using a biosensor), which should enable the cure or the adjustment of the biological functions.[172] Most of the nanostructures developed up until now have been explored for both the treatment and the diagnosis of diseases.[173] However, this attractive concept, which might open the door to personalized medicine, needs additional investigation, especially concerning the safety issue, before the start of clinical trials.

Overall, the current line of design and engineering of nanostructures aims at providing them with multiple functionalities not only for the purpose of diagnostic and treatment but also for achieving a more advanced control of their interaction with the human body. For example, in the area of gene therapy and also in vaccination, there is an interest in designing biomimetic structures, *i.e.* virus like particles, which simulate the behavior of a virus. It is obvious that the growth and evolution of this tendency is determined by the limited knowledge of our biological barriers and also the mechanistic issues underlying the behavior of natural specimens. Moreover, a major obstacle to be confronted in the definition of this new era of nanocarriers is represented by the safety concerns and regulatory hurdlers.

1.5   The Therapeutic Expectations for the Current and Future Nanocarriers

The application of nanotechnology to drug delivery is expected to change the landscape of pharmaceutical and biotechnology industries in the foreseeable future. The development pipelines of pharmaceutical companies are believed to be drying up, and a number of blockbuster drugs will come off patent in the near future. There is a conviction that the use of nanocarriers may provide patients with more efficacious and less toxic drugs. However, equally important is the fact that they can facilitate the administration of drugs, *i.e.* avoiding injection and making it practicable or at least more patient-friendly. For example, there is an increasing number of biopharmaceuticals and chemically synthesized peptides in industry pipelines which have not found their way to the market simply because they need to be administered by injection. The indications of these molecules are often chronic diseases, *i.e.* obesity and diabetes, as well as frequent disease-related events as in the case of pain treatment. The availability of a nanocarrier that would enable the transport of these complex molecules across external barriers, *i.e.* intestinal, nasal, pulmonary or skin, would be of great benefit for the development of new medicines. In addition, from a pharmacoeconomic perspective, the availability of a simple form of administration, *i.e.* oral, for these nanopacked macromolecules could also result in a reduction in the cost of the corresponding treatments. Therefore, from the perspective of overcoming these external barriers, the treatment of any disease could be substantially improved by the use of nanocarriers. However, as shown in the historical evolution of nanocarriers, there are specific therapeutic domains in which the expectations on the use of nanotechnologies are great. Several of these indications are briefly described below.

1.5.1 Cancer

This is by far the disease indication most frequently explored in nanomedicine. In fact, cancer treatment has always been behind the concept of targeted drug delivery. Within this broad frame, the use of nanocarriers may provide a number of advantages, which go from simply solubilizing and/or preventing the premature degradation of anticancer compounds to prolonging their residence time in the body and modifying the biodistribution pattern. As indicated in the historical view of nanocarriers, all these goals have already been achieved either for low molecular weight hydrophobic compounds, *e.g.* doxorubicine and taxanes, or large delicate proteins, *e.g.* interferon alfa or synthetic fragments of nucleic acids (antisens oligonucleotides or siRNA). Additionally, at present the use of these nanocarriers for cancer gene therapy is being extensively explored.[174–176] However, the level of active targeting and controlled delivery that is necessary in order to make these therapies effective is still a matter of debate. The future tendencies regarding this application, either with small or large bioactive molecules, will rely on their accurate and appropriate targeting and delivery. In this sense, it is important to keep in mind that the best target for most types of cancer are not the tumor cells but the circulating cancer cells. Reaching these cells requires concentrating the nanomedicine in the lymphatic circulation and specifically target the cancer cells using the appropriate signaling. Therefore, future development in this field will clearly depend on the advances made on molecular oncology.

1.5.2 Cardiovascular Diseases

The localization or prolonged retention of drugs in the cardiovascular system is not an easy task due to their natural tendency to associate to tissue compartments. The use of nanostructures makes it possible for the drug to remain in the vascular compartment for extended periods of time because of the shielding effect. In addition to this, there have been significant efforts oriented towards localizing the drugs in specific regions of the vascular system, in most cases, associated to a stent. Such approaches have mainly been oriented to the delivery of anti-clotting agents, which are intended to help prevent cardiac hypoxia and failure.[177]

1.5.3 CNS Diseases

The brain is one of the regions of the body with the most difficult access due to the presence of the brain-blood barrier (BBB) whose role is to prevent the entry of any potentially harmful molecule. The very powerful nature of this barrier makes the treatment of devastating diseases, either psychiatric diseases or degenerative diseases (*e.g.* multiple sclerosis, Parkinson, Alzheimer, *etc.*) very difficult.[178,179] In many instances, the difficulties arise from the high doses of drugs required in order to ensure a minimum access to the brain. This restrictive access leads to the accumulation of the drug in other non-relevant organs and, as a result, to the manifestation of important side effects. This situation, frequently observed for anti-psychotic drugs, finally leads to the drop-out of treatments and severe complications of the disease. As indicated in the historical view, the use of nanocarriers has been conceived since the mid 1990s as a strategy to overcome the BBB. The knowledge accumulated so far clearly indicates that in order to achieve a transport from the blood compartment to the brain tissue it is necessary for the cargo-containing nanocarrier to have, firstly, a prolonged circulation in the blood stream and, secondly, to be able to get across this barrier. This latter and complex step has been explored by a number of means including the use of penetration enhancers, such as cell penetrating peptides (CPP) and surfactants, or targeting compounds.[180,181] Overall, the possibilities appear to be great, in the sense that a variety of nanocarriers could do their work as long as they are appropriately decorated. However, despite the advances made up until now, there is not a clearly visible path for these nanocarriers towards the market, mainly because there are insufficient data concerning the elimination of the nanocarriers from the brain tissue. Therefore, for this specific indication there is a clear need to make significant progress in the accumulation of knowledge on the barrier itself, but mainly on the mechanistic issues underlying the interaction of biomaterials with the BBB.

1.5.4   Infectious Diseases

Interestingly, as indicated in the historical analysis, nanoparticles were developed at first in order to be used as adjuvants for the prevention of infectious transmittable diseases.[182,183] While this specific application has evolved quite productively, there has also been significant activity oriented towards the idea of antigen controlled delivery (single-dose vaccination) and to the potential use of nanocarriers for overcoming mucosal barriers, such as the nasal barrier (needle-free vaccination).[184] So far the most successful adjuvants recently approved or under clinical evaluation are based on nanoemulsions containing different kinds of surfactants and immunostimulants. In addition, significant progress has been made on the understanding of antigen–biomaterial interaction, which is thought to be valuable for the engineering of appropriate nanocarriers. The current tendency in this area aims at assembling antigens, immunostimulants and any other material that may facilitate the presentation to the immune system. For this purpose, multi-compartment systems with the possibility of allocating different molecules are particularly attractive. An example of this type of structure is illustrated in Figure 1.10. In addition, the physical parameters (size, geometry) and structural organization of all the components are known to be of relevance for achieving the goal. With regard to this specific application, it is important to keep in mind that the development of new vaccines intended to protect against devastating diseases (*e.g.* malaria, tuberculosis and aids) is critically

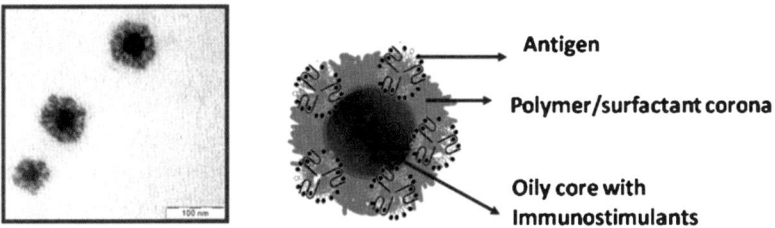

Figure 1.10 Schematic illustration of nanostructures accommodating antigens and immunostimulants (courtesy of Sara Vicente).

dependent on the design and development of appropriate adjuvants. In fact, currently, there is a variety of antigens for fighting these diseases under preclinical and clinical evaluation whose success is known to be absolutely dependent on the identification of the appropriate adjuvant vehicle.

In addition to the vaccine delivery area, nanocarriers offer interesting potential for the delivery of anti-infective drugs. Since the pioneering work from the late 1980s,[185,186] there has been modest activity in searching for improvement of the treatment of these diseases.[187] This modest activity is largely down to the fact that, because of good preventive medicine, these diseases are not having a great impact in the developed world, but only in the poorest regions. However, there are currently a number of initiatives promoted by the WHO, Gates Foundation, Wellcome Trust, government organizations and several pharmaceutical companies, which are seeking to confront these so-called neglected diseases. These are infectious diseases whose treatment could benefit from the use of appropriate drug-nanocarrier combinations. On the other hand, the general abuse of the use of anti-infective drugs in the wealthy countries has led to the phenomenon of drug resistance and, thus, to the search for more powerful anti-infective drugs. Many of these drugs are either very hydrophobic or large delicate macromolecules, which will certainly require innovative formulation technologies. Nanocarriers are expected to play a role in this demanding arena.

Even though the above-indicated diseases are those concentrating most of the research activity in nanocarriers for drug delivery, as mentioned in the beginning of this section, any disease may benefit from these new technologies. Indeed, among the greatest expectations for these nanocarriers are those related to their role in terms of overcoming the external barriers of the body, such as the skin and mucosal surfaces. Overcoming the skin barrier would not only facilitate the treatment of skin diseases but also vaccination and the treatment of certain systemic diseases, especially those related to the immune system. Alternatively, overcoming the ocular mucosal barrier would be of great benefit for the treatment of ocular diseases. Finally, overcoming the nasal and gastro-intestinal barriers is expected to have a terrific impact for the treatment of any disease based on the use of large and/or polar molecules, *i.e.* peptides, proteins, nucleic acid-based drugs, monoclonal antibodies, *etc.*

## 1.6	Conclusions

Overall, despite the skepticism of certain researchers and developers in the area of drug discovery and delivery with regard to the potential of drug nanocarriers, a clear message from this chapter is that, half a century after the origin of the idea of using nanocarriers for transporting drugs across biological barriers, we do have a significant number—close to 30— nanomedicines on the market. More importantly, the chapter illustrates the continuous growth of this field and the accumulation of multidisciplinary knowledge and experience. Nowadays, these advances are expected to lead to a burst in the production of nanomedicine products.

Acknowledgements

The authors wish to thank Sara Vicente, Adam McGlone and Noemi Csaba for their help in the preparation of this chapter.

References

1. O. C. Farokhzad and R. Langer, *ACS Nano*, 2009, **3**, 16.
2. A. S. Hoffman, *J. Controlled Release,* 2008, **132**, 153.
3. A. L. Porter and J. Youtie, *Nat. Nanotechnol.*, 2009, **4**, 534.
4. P. Ehrlich, Nobel Prize lecture, www.nobelprize.org/nobel_prizes/medicine/laureates/1908/ehrlich-lecture.pdf.
5. J. Kreuter, *Int. J. Pharm.*, 2007, **331**, 1.
6. C. Omid, B. Farokhzada and R. Langer, *Adv. Drug Delivery Rev.*, 2006, **58**, 1456.
7. R. Duncan, *Nat. Rev. Drug Discovery,* 2003, **2**, 347.
8. J. Folkman and D. M. Long, *J. Surg. Res.,* 1964, **4**, 139.
9. J. Folkman, D. M. Long and R. Rosenbau, *Science,* 1966, **154**, 148.
10. A. Zaffaroni, US Patent, 1971, 3598122.
11. R. Langer and J. Folkman, *Nature,* 1976, **263**, 797.
12. M. A. Phillips, M. L. Granb and N. A. Peppas, *Nano Today*, 2010, **5**, 143.
13. I. Preis and R. Langer, *J. Immunol. Methods,* 1979, **28**,193.
14. L. P. Kasi, G. Lopez-Berenstein, K. Mehta, M. Rosenblum, H. J. Glenn, T. P. Haynie, G. Mavligit and E. M. Hersh, *Int. J. Nucl. Med. Biol.*, 1984, **11**, 35.
15. G. Gregoriadis, *Liposomes as Drug Carriers,* John Wiley & Sons, Chichester, 1988.
16. D. J. Crommelin and G. Storm, *J. Liposome Res.*, 2003, **13**, 33.
17. V. P. Torchilin, *Nat. Rev. Drug Discovery*, 2005, **4**, 145.
18. A. Gabizon, *J. Liposome Res.*, 2003, **13**, 17.
19. A. L. Klibanov, K. Maruyama, V. P.Torchilin and L. Huang, *FEBS Lett.,* 1990, **268**, 235.

20. D. Papahadjopoulos, T. M. Allen, A. Gabizon, E. Mayhew, K. Matthay, S. K. Huang, K. D. Lee, M. C. Woodle, D. D. Lasic, C. Redemann and F. J. Martin, *Proc. Natl. Acad. Sci. U. S. A.*, 1991, **88**, 11460.
21. E. Mastrobattistaa, A. Gerben, B. Koninga and G. Storma, *Adv. Drug Delivery Rev.*, 1999, **40**, 103.
22. V.P. Torchilin, J. Narula, E. Halpern and B. A. Khaw, *Biochim. Biophys. Acta*, 1996, **1279**, 75.
23. A. Gabizon, H. Shmeeda, A. T. Horowitz and S. Zalipsky, *Adv. Drug Delivery Rev.*, 2004, **56**, 1177.
24. V. P. Torchilin, B. A. Khaw, V. N. Smirnov and E. Haber, *Biochem. Biophys. Res. Commun.*, 1979, **89**, 1114.
25. R. Langer, *Science,* 2001, **293**, 58.
26. F. X. Gu, R. Karnik, A. Z. Wang, F. Alexis, E. Levy-Nissenbaum, S. Hong1, R. S. Langer and O. C. Farokhzad, *NanoToday*, 2007, **40**, 14.
27. A. H. Faraji and P. Wipt, *Bioorg. Med. Chem.*, 2009, **17**, 2950.
28. M. T. Allen, C. B. Hansen and J. Rutledge, *Biochem. Biophys. Acta*, 1989, **981**, 27.
29. A. Gabizon and D. Papahadjopoulos, *Proc. Nat. Acad. Sci. U. S. A.*, 1988, **85**, 6949.
30. A. L. Klibanov, K. Maruyama, A. M. Beckerleg, V. P. Torchilin and L. Huang, *FEBS Lett.,* 1990, **268**, 235
31. M. L. Immordino, F. Dosio and L. Cattel, *Int. J. Nanomed.*, 2006, **1**, 297.
32. A. N. Gordon, J. T. Fleagle, D. Guthrie, D. E. Parkin, M. E. Gore and A. J. Lacave, *J. Clin. Oncol.*, 2001, **19**, 3312.
33. S. Chan, N. Davidson, E. Juozaityte, F. Erdkamp and A. Pluzansk, *Ann. Oncol.,* 2004, **15**, 1527.
34. M. Rosati, C. Raimondi, G. Baciarello, P. Grassi, S. Giovanni, E. Petrelli, M. Basile, M. Girolami, M. Di Seri and L. Frati, *Ann. Oncol.*, 2011, **22**, 315
35. F. F. Davis, *Adv. Drug Delivery Rev.*, 2002, **54**, 457.
36. F. M. Veronese and G. Paust, *Drug Discovery Today*, 2005, **10**, 1451.
37. R. Duncan, *Nat. Rev. Cancer*, 2006, **6**, 688.
38. H. Ringsdorf, *J. Polym. Sci. Symp.*, 1975, **51**, 135.
39. Y. Matsumura and H. Maeda, *Cancer Res.,* 1986, **46**, 6387.
40. H. Maeda, G. Y. Bharate and J. Daruwalla, *Eur. J. Pharm. Biopharm.*, 2009, **71**, 409.
41. R. Duncan, J. Kopecek, P. Rejmanova and J. B. Lloyd, *Biochim. Biophys. Acta,* 1983, **755**, 518.
42. R. Duncan, *J. Controlled Release*, 1989, **10**, 51.
43. R. Duncan L. W. Seymour, K. B. O'Hare, P. A. Flanagan, S. Wedge, I. C. Hume, K. Ulbrich, J. Strohalm, V. Subr, F. Spreafico, M. Grandi, M. Ripamonti, M. Farao and A. Suarato, *J. Controlled Release*, 1992, **19**, 331.
44. L. W. Seymour, D. R. Ferry, D. Anderson, S. Hesslewood, P. J. Julyan, R. Poyner, J. Doran, A. M. Young, S. Burtles and D. J. Kerr, *J. Clin. Oncol.*, 2002, **20**, 1668.

45. R. Duncan, *Nat. Rev. Drug Discovery*, 2003, **2**, 347.
46. P. Bailon and C. Y. Won, *Expert Opin. Drug Delivery*, 2009, **6**, 1.
47. C. Li, D.-F. Yu, R. A. Newman, F. Cabral, L. C. Stephens, N. Hunter, L. Milas and S. Wal, *Cancer Res.*, 1998, **58**, 2404.
48. C. Li and S. Wallace, *Adv. Drug Delivery Rev.*, 2008, **60**, 886.
49. L.Gros, H. Ringsdorf and H. Schupp, *Angew. Chem., Int. Ed.*, 1981, **20**, 305.
50. M. Jones and J. C. Leroux, *Eur. J. Pharm. Biopharm.*, 1999, **48**, 101.
51. K. Kataoka, G. S. Kwon, M. Y. Yama, T. Okano and Y. Sakurai, *J. Controlled Release*, 1993, **24**, 119.
52. G. S. Kwon, M. Naito, M. Yokoyama, T. Okano, Y. Sakurai and K. Kataoka. *Pharm. Res.*, 1995, **12**, 192.
53. A. Harada and K. Kataoka., *Macromolecules*, 1998, **31**, 288.
54. Z. Gao, A. Lukyanov, A. Singhal and V. Torchilin, *Nano Lett.*, 2002, **2**, 979.
55. V. P. Torchilin, A. N. Lukyanov, Z. Gao and B. Papahadjopoulos-Sternberg, *Proc. Natl. Acad. Sci. U. S. A.*, 2003, **100**, 6039.
56. V.P. Torchilin, *Pharm. Res.*, 2007, **24**, 1.
57. Y. Matsumura, T. Hamagichi, T. Ura, K. Muro, Y. Yamada, Y. Shimada, K. Shirao, T. Okusaka, H. Ueno, M. Ikeda and N. Watanabe, *Brit. J. Cancer*, 2004, **91**, 1775.
58. T. Y. Kim, D.-W. Kim, J.-Y. Chung , S. G. Shin, S. C. Kim, D. S. Heo, N. K. Kim and Y. J. Bang, *Clin. Cancer Res.*, 2004, **10**, 3708.
59. H. Chen, C. Khemtong, X. Yang, X. Chang and J. Gao, *Drug Discovery Today*, 2011, **16**, 354.
60. P. Couvreur, G. Barratt, E. Fattal, P. Legrand and C. Vauthier, *Crit. Rev. Ther. Drug Carrier Syst.*, 2002, **19**, 99.
61. S.C. Khanna and P. Speiser, *J. Pharm. Sci.*, 1969, **58**, 1114.
62. S.C. Khanna, T. Jecklin and P. Speiser, *J. Pharm. Sci.*, 1970, **59**, 614.
63. G. Birrenbach and P. P. Speiser, *J. Pharm. Sci.*, 1976, **65**, 1763.
64. P. Couvreur, P. Tulkens, M. Roland, A. Trouet and P. Speiser, *FEBS Lett.*, 1977, **84**, 323.
65. P. Couvreur, B. Kante, M. Roland, P. Guiot, P. Baudhin and P. Speiser, *J. Pharm. Pharmacol.*, 1979, **31**, 331.
66. N. K. Fallouh, L. Roblot-Treupel, H. Fessi, J. P. Devissaguet and F. Pusieux, *Int. J. Pharm.*, 1986, **28**, 125.
67. C. Damgé, C. Michel, M. Aprahamian and P. Couvreur, *Biomater. Clin. Appl.*, 1987, **17**, 643.
68. C. Damgé, C. Michel, M. Aprahamian and P. Couvreur, *Diabetes*, 1988, **37**, 246.
69. F. Brasseur, C. Verdun, P. Couvreur, C. Deckers and M. Roland, *Chim. Oggi*, 1987, **9**, 17.
70. C. Verdun, F. Brasseur, H. Vranckx, P. Couvreur and M. Roland, *Cancer Chemother. Pharmacol.*, 1990, **26**, 13.

71. P. Maincent, J. P. Devissaguet, R. L. Verge, P. A. Sado and P. Couvreur, *Appl. Biochem. Biotechnol.*, 1984, **10**, 263.
72. P. Maincent, R. L. Verge, P. A. Sado, P. Couvreur and J. P. Devissaguet, *J. Pharm. Sci.*, 1986, **75**, 955.
73. T. Harmia, J. Kreuter, P. P. Speiser, T. Boye, R. Gurny and A. Kubis, *Int. J. Pharm.*, 1986, **33**, 187.
74. R. Diepold, J. Kreuter, J. Himber, R. L. Gurny, V. H. L., J. R. Robinson, M. F. Saettone and O. E. Schnaudigel, *Graefe's Arch. Clin. Exp. Ophthalmol.*, 1989, **227**, 188.
75. R. Gurny, *Pharm. Acta Helv.*, 1981, **56**, 130.
76. C. Losa, P. Calvo, E. Castro, J. L. Vila Jato and M. J. Alonso, *J. Pharm. Pharmacol.*, 1991, **43**, 548.
77. C. M. Losa, L. Marchal-Heussler, F. Orallo, J. L. Vila Jato and M. J. Alonso, *Pharm. Res.*, 1993, **10**, 80.
78. R. Alyautdin, D. Gothier, V. Petrov, D. Kharkevich and J. Kreuter, *Eur. J. Pharm. Biopharm.*, 1995, **41**, 44.
79. R. N. Alyautdin, V. E. Petrov, K. Langer, A. Berthold, D. A. Kharkevich and J. Kreuter, *Pharm. Res.*, 1997, **14**, 325.
80. J. Kattan, J. P. Droz, P. Couvreur, J. P. Marino, A. Boutan-Laroze, P. Rougier, P. Brault, H. Vranckx, J. M. Grognet, X. Morge and H. Sancho-Garnier, *Invest. New Drugs*, 1992, **10**, 191.
81. J. J. Marty and R. C. Oppenheim, *Aust. J. Pharm. Sci.*, 1977, **6**, 65.
82. J. J. Marty, R. C. Oppenheim and P. Speiser, *Pharm. Acta Helv.*, 1978, **53**, 17.
83. R. C. Oppenheim, J. J. Marty and N. F. Steward, *Aust. J. Pharm. Sci.*, 1978, **6**, 65.
84. I. Zolle, F. Hosein, B. A. Rhodes, T. K. Natarajan and H. N. Wagner Jr., *J. Nucl. Med.*, 1973, **11**, 379.
85. U. Scheffel, B. A. Rhodes, T. K. Natarajan and H. N. Wagner Jr., *J. Nucl. Med.*, 1973, **13**, 498.
86. P. A. Kramer, *J. Pharm. Sci.*, 1974, 1646.
87. K. Sugibayashi, Y. Morimoto, T. Nadai and Y. Kato, *Chem. Pharm. Bull.*, 1977, **25**, 3433.
88. K. Sugibayashi, Y. Morimoto, T. Nadai, Y. Kato, A. Hasegawa and T. Arita, *Chem. Pharm. Bull.*, 1979, **27**, 204.
89. K. J.Widder, G. Flouret and A. E. Senyei, *Pharm. Sci.*, 1979, **68**, 79.
90. K. J. Widder, A. E. Senyei and D. F. Ranney, *Adv. Pharmacol. Chemother.*, 1979, **16**, 213.
91. K. Langer, C. Coester, C. Weber, H. Von Briesen and J. Kreuter, *Eur. J. Pharm. Biopharm.*, 2000, **17**, 303.
92. H. Wartlick, B. Spankuch-Schmitt, K. Strebhardt, J. Kreuter and K. Langer, *J. Controlled Release*, 2006, **96**, 483.
93. H. Wartlick, K. Michaelis, S. Balthasar, K. Strebhardt, J. Kreuter and K. Langer, *J. Drug Target.*, 2004, **12**, 461.

94. S. Balthasar, K. Michaelis, N. Dinauer, H. Von Briesen, J. Kreuter and K. Langer, *Biomaterials*, 2005, **26**, 2723.

95. W. Gradishar; S. Tjulandin, N. Davidson, H. Shaw; N. Deasy, P. Bhar, M. Hawkins and J. O'Shaughnessy, *J. Clin. Oncol.*, 2005, **23**, 7794.

96. R. Gurny, N. A. Peppas, D. D. Harrington and G. S. Banker, *Drug Dev. Ind. Pharm.*, 1981, **7**, 1.

97. R. Gref, Y. Minamitake, M. Peracchia, V. Trubetshoy, V. Torchillin and R. Langer, *Science,* 1994, **263**, 1600.

98. T. Verrecchia, G. Spenlehauer , D. V. Bazile, A. Murry-Brelier, Y. Archimbaud and M. Veillard, *J. Controlled Release*, 1995, **36**, 49.

99. D. Blanco and M. J. Alonso, *Eur. J. Pharm. Biopharm.*, 1997, **43**, 287.

100. M. Gaspar, D. Blanco, M. E. M. Cruz and M. J. Alonso, *J. Controlled Release*, 1998, **52**, 53.

101. A. Sánchez, M. Tobío, L. González, A. Fabra, M. J. Alonso, *Eur. J. Pharm. Sci.*, 2003, **18**, 221.

102. I. D'Angelo, M. Garcia-Fuentes, Y. Parajó, A. Welle, T. Vántus, A. Horváth, G. Bökönyi, G. Kéri and M. J. Alonso, *Mol. Pharm.*, 2010, **7**, 1724.

103. M. Tobío, R. Gref, A. Sánchez, R. Langer and M. J. Alonso, *Pharm. Res.*, 1998, **15**, 270.

104. P. Paolicelli, C. Prego, A. Sanchez and M. J. Alonso, *Nanomedicine*, 2010, **5**, 843.

105. C. Pérez, A. Sánchez, D. Putnam, D. Ting, R. Langer and M. J. Alonso, *J. Controlled Release*, 2001, **75**, 211.

106. N. Csaba, P. Caamaño, A. Sánchez, F. Domínguez and M. J. Alonso, *Biomacromolecules*, 2005, **6**, 271.

107. A. Vila, A. Sánchez, C. Évora, I. Soriano and M. J. Alonso, *Int. J. Pharm.*, 2005, **292**, 43.

108. A. Vila, H. Gill, O. McCallion and M. J. Alonso, *J. Controlled Release*, 2004, **98**, 231.

109. M. Tobío, A. Sánchez, A. Vila, I. Soriano, C. Évora, J. L. Vila Jato and M. J. Alonso, *Colloid Surf. B-Biointerfaces*, 2000, **18**, 315.

110. A. De Campos, A. Sánchez, R. Gref, P. Calvo and M. J. Alonso, *Eur. J. Pharm. Sci.,* 2003, **20**, 73.

111. N. Csaba, M.García Fuentes and M. J. Alonso, *Expert Opin. Drug Delivery*, 2006, **3**, 463.

112. H. Chen, C. Khemtong, X. Yang, X. Chang and J. Gao, *Drug Discovery Today*, 2011, **16**, 354.

113. P. Calvo, C. Remuñán-López, J. L. Vila-Jato and M. J. Alonso, *J. Appl. Polymer Sci.*, 1997, **63**, 125.

114. A. Trapani, A. Lopedota, M. Franco, N. Cioffi, E. Ieva, M. Garcia-Fuentes and M. J. Alonso, *Eur. J. Pharm. Biopharm.*, 2010, **75**, 26.

115. R. Fernández Urrusuno, P. Calvo, C. Remuñán-López, J. L. Vila Jato and M. J. Alonso, *Pharm. Res.*, 1999, **16**, 1576.

116. F.A. Oyarzun-Ampuero, J. Brea, M.I. Loza, D. Torres and M. J. Alonso, *Int. J. Pharm.*, 2009, **381**, 122.

117. Y. Parajó, I. d'Angelo, A. Welle, M. Garcia-Fuentes and M. J. Alonso, *Drug Delivery*, 2010, **17**, 596.

118. P. Calvo, C. Remuñán-López, J. L.Vila-Jato and M. J. Alonso, *Pharm. Res.*, 1997, **14**, 1431.

119. C. Prego, P. Paolicelli, B. Díaz, S. Vicente, A. Sánchez, A. González-Fernández and M. J. Alonso, *Vaccine*, 2010, **28**, 2607.

120. N. Csaba, A. Sánchez, E. Fernández-Megia, R. Novoa-Carballal and M. J. Alonso, *Eur. J. Pharm. Sci.*, 2004, **23**, 55.

121. N. Csaba, M. Köping-Höggård, E. Fernandez-Megia, R. Novoa-Carballal, R. Riguera and M. J. Alonso, *J. Biomed. Nanotechnol.*, 2009, **5**, 162.

122. N. Csaba, M. Köping-Höggård and M. J. Alonso, *Int. J. Pharm,* 2009, **382**, 205.

123. M. Raviña, E. Cubillo, D. Olmeda, R. Novoa-Carballal, E. Fernandez-Megia, R. Riguera, A. Sánchez, A. Cano and M. J. Alonso, *Pharm. Res.*, 2010, **27**, 2544.

124. P. Calvo, C. Remuñán-López, J. L. Vila-Jato and M. J. Alonso, *Colloid. Polymer Sci.*, 1997, **275**, 46.

125. M.V. Lozano, D. Torrecilla, D. Torres, A. Vidal, F. Domínguez and M. J. Alonso, *Biomacromolecules*, 2008, **9**, 2186.

126. C. Prego, D. Torres and M. J. Alonso, *J. Nanosci. Nanotechnol.*, 2006, **6**, 2921.

127. A. Vila, A. Sánchez, K.A. Janes, I. Behrens, T. Kissel, J. L. Vila Jato and M. J. Alonso, *Eur. J. Pharm. Biopharm.*, 2004, **57**, 123.

128. N. Csaba, M. García Fuentes and M. J. Alonso, *Adv. Drug Delivery Rev.*, 2009, **61**, 140.

129. D. Teijeiro-Osorio, C. Remuñán-López and M. J. Alonso, *Biomacromolecules*, 2009, **10**, 243.

130. C. Prego, D. Torres and M. J. Alonso, *Expert Opin. Drug Delivery*, 2005, **2**, 843.

131. B. Sarmento, A. Ribeiro, F. Veiga, D. Ferreira and R. Neufeld, *Biomacromolecules,* 2007, **8**, 3054.

132. B. Sarmento, A. Ribeiro, F. Veiga, P. Sampaio, R. Neufeld and D. Ferreira, *Pharm. Res.*, 2007, **24**, 2198.

133. C.P. Reis, A.J. Ribeiro, S. Houng, F. Veiga and R. J. Neufeld, *Eur. J. Pharm. Sci.*, 2007, **30**, 392.

134. C.P. Reis, F. Veiga, A.J. Ribeiro, R.J. Neufeld and C. Damgé, *J. Pharm. Sci.*, 2008, **97**, 5290.

135. S. Shu, X. Zhang, D. Teng, Z. Wang and C. Li, *Carbohydr. Res.*, 2009, **344**, 1197.

136. K. Sonaje, K. J. Lin, S. P. Wey, C. K. Lin, T. H. Yeh, H. N. Nguyen, C. W. Hsu, T. Ch. Yen, J. H. Juang and H. W. Sung, *Biomaterials*, 2010, **31**, 6849.

137. K. Sonaje, Y. J. Chen, H. L. Chen, S. P. Wey, J. H. Juang, H. N. Nguyen, C. W. Hsu, J. Lin and H. W. Sung, *Biomaterials*, 2010, **31**, 3384.

138. C. Prego, M. Fabre, D. Torres and M. J. Alonso, *Pharm. Res.*, 2006, **23**, 549.

139. A. M. De Campos, Y. Diebold, E. S. Carvahlo, A. Sánchez and M.J. Alonso, *Pharm. Res.*, 2004, **21**, 803.

140. M. De la Fuente, B. Seijo and M. J. Alonso, *Gene Ther.*, 2008, **15**, 668.

141. M. De la Fuente, M. Raviña, P. Paolicelli, A. Sanchez, B. Seijo and M. J. Alonso, *Adv. Drug Delivery Rev.*, 2010, **62**, 100.

142. R. O. Williams III, M. K. Baca, M. J. Alonso and C. Remuñán-López, *Int. J. Pharm.*, 1998, **174**, 209.

143. S. Al-Qadi, A. Grenha, D. Carrión-Recio, B. Seijo and C. Remuñán-López, *J. Controlled Release,* 2012, **157**, 383.

144. Y. Aktas, M. Yemisci, K. Andrieux, R. N. Gursoy, M. J. Alonso, E. Fernandez-Megía, R. Novoa-Carballal, E. Quiñoá, R. Riguera, M. F. Sargon, H. Hamdi Celik, A. S. Demir, A. Atilla Hincal, T. Dalkara, Y. Capan and P. Couvreur, *Bioconjugate Chem.*, 2005, **16**, 1503.

145. P. Hervella, V. Lozano, M. Garcia-Fuentes and M. J. Alonso, *J. Biomed. Nanotechnol.*, 2008, **4**, 276.

146. M. De la Fuente, N. Csaba, M. García-Fuentes and M. J. Alonso, *Nanomedicine,* 2008, **3**, 845.

147. C. Vauthier and P. Couvreur, *J. Biomed. Nanotechnol.,* 2007, **3**, 223.

148. O. Farokhzad and R. Langer, *Adv. Drug Delivery Rev.*, 2006, **58**, 1456.

149. O. C. Farokhzad and R. Langer, *ACS Nano*, 2009, **3**, 16.

150. F. Sonvico, C. Dubernet, P. Colombo and P. Couvreur, *Curr. Pharm. Design*, 2005, **11**, 2095.

151. B. Bonnemain, *J. Drug Targ.*, 1998, **6**, 167.

152. A. S. Lubbe, C. Bergemann, H. Riess, F. Schriever, P. Reichardt, K. Possinger, M. Matthias, B. Dorken, F. Herrmann, R. Gurtler, P. Hohenberger, N. Haas, R. Sohr, B. Sander, A. J. Lemke, D. Ohlendorf, W. Huhnt and D. Huhn, *Cancer Res.*, 1996, **56**, 4686.

153. P. Horcajada, T. Chalati, C. Serre, B. Gillet, C. Sebrie, T. Baati, J. F. Eubank, D. Heurtaux, P. Clayette, C. Kreuz, J.-S. Chang, Y. Hwang, V. Marsaud, P. Bories, L. Cynober, S. Gil, G. Férey, P. Couvreur and R. Gref, *Nat. Mater.*, 2010, **9**, 172.

154. J. R. McCarthy and R. Weissleder, *Adv. Drug Delivery Rev.*, 2008, **60**, 1241.

155. P. Ghosh, G. Han, M. De, C. K. Kim and V. M. Rotello, *Adv. Drug Delivery Rev.*, 2008, **60**, 1307.

156. T. Musacchio, V. P. Torchilin and P. Vladimir, *Front. Biosci., Landmark Ed.*, 2011, **16**, 1388.

157. S. Daoud-Mahammed, C. Ringard-Lefebvre, N. Razzouq, V. Rosilio, B. Gillet, P.Couvreur, C. Amiel and R. Gref, *J. Colloid Interface Sci.*, 2007, **307**, 83.

158. F. Maestrelli, M. Garcia-Fuentes, P. Mura and M. J. Alonso, *Eur. J. Pharm. Biopharm.*, 2006, **63**, 79.

159. A. L. Villemson, P. Couvreur, R. Gref and N. I. Larionova, *Polymer Sci. Series*, 2007, **49**, 708.

160. F. A. Oyarzun-Ampuero, G. Rivera-Rodríguez, M. J. Alonso, D. Torres, 2011, submitted.

161. G. Lambert, E. Fattal, H. Pinto-Alphandary, A. Gulik, P. Couvreur, *Pharm. Res.*, 2000, **17**, 707.

162. M. V. Lozano, G. Lollo, J. Brea, A. Vidal, M. I. Loza, D. Torres, M. J. Alonso, 2011, submitted.

163. X. Qu, V. V. Khutoryanskiy, A. Stewart, S. Rahman, B. Papahadjopoulos-Sternberg, C. Dufes, D. McCarthy, C. G. Wilson, R. Lyons, K. C. Carter, A. Schatzlein and I. F. Uchegbu, *Biomacromolecules*, 2006, **7**, 3452.

164. I. F. Uchegbu, L. Sadiq, A. Pardakhty, M. El-Hamadi, A. I. Gray, L. Tetley, W. Wang, B. H. Zinselmeyer and A. G. Schatzelein, *J. Drug Targeting*, 2004, **12**, 527.

165. R. Gaspar and R. Duncan, *Adv. Drug Delivery Rev.*, 2009, **61**, 1220.

166. M. A. Mintzer and M. W. Grinstaff , *Chem. Soc. Rev.*, 2011, **40**, 173.

167. P. Couvreur, B. Stella, L. H. Reddy, H. Hillaireau, C. Dubernet, D. Desmaële, S. Lepêtre-Mouelhi, F. Rocco, N. Dereuddre-Bosquet, P. Clayette, V. Rosilio, V. Marsaud, J. M. Renoir and L. Cattel, *Nano Lett.*, 2006, **6**, 2544.

168. L. H. Reddy, P. E. Marque, C. Dubernet, S. L. Mouelhi, D. Desmaële and P. Couvreur, *J. Pharmacol. Exp. Ther.*, 2008, **325**, 484.

169. R. Bekkara-Aounallah, M. Gref, L. H. Othman, B. Reddy, V. Pili, C. Allain, H. Bourgaux, S. Hillaireau, D. Lepêtre-Mouelhi, J. Desmaële, N. Nicolas, P. Chafi and P. Couvreur, *Adv. Funct. Mater.*, 2008, **18**, 3715.

170. M. Raouane, D. Desmaele, M. Gilbert-Sirieix, C. Gueutin, F. Zouhiri, C. Bourgaux, E. Lepeltier, R. Gref, R. Ben Salah, G. Clayman, L. Massaad-Massade and P. Couvreur, *J. Med. Chem.*, 2011, **54**, 4067.

171. J. L. Arias, L. Harivardhan Reddy, M. Othman, B. Gillet, D. Desmaële, F. Zouhiri, F. Dosio, R. Gref and P. Couvreur, *ACS Nano*, 2011, **13**, 1513.

172. T. Lammers, F. Kiessling, W. E. Hennink and G. Storm, *Mol. Pharm.*, 2010, **7**, 1899.

173. M. E. Gindy and R. K. Prud'homme, *Drug Delivery*, 2009, **6**, 865.

174. Y. Gao, X.-L. Liu and X.-R. Li, *Int. J. Nanomedicine*, 2011, **6**, 1017.

175. M. E. Davis, J. E. Zuckerman, C. Hang, J. Choi, D. Seligson, A. Tolcher, C. A. Alabi, Y. Yen, J. D. Heidel and A. Ribas, *Nature*, 2010, **464**, 1067.

176. A. K. Whitehead, R. Langer and D. G. Anderson, *Nat. Rev.*, 2009, **8**, 129.

177. L. Lei, S. R. Guo , W. L. Chen, H. J. Rong and F. Lu, *Expert Opin. Drug Delivery*, 2011, **8**, 813.

178. W. M. Pardridge, *NeuroRx*, 2005, **2**, 3.

179. A. G. de Boer and P. J. Gaillard, *Annu. Rev. Pharmacol.*, 2007, **47**, 323.

180. E. Garcia-Garcia, K. Andrieux, S. Gil and P. Couvreur, *Int. J. Pharm.*, 2005, **298**, 274.
181. A. Zensi, D. Begley, C. Pontikis, C. Legros, L. Mihoreanu, S. Wagner, C. Buchel, H. von Briesen and J. Kreuter, *J. Controlled Release*, 2009, **137**, 78.
182. J. Kreuter and P.P. Speiser, *Infect. Immun.*, 1976, **13**, 204.
183. J. Kreuter and P.P. Speiser, *J. Pharm. Sci.*, 1976, **65**, 1624.
184. S. Vicente, C. Prego, N. Csaba and M. J. Alonso, *J. Drug Delivery Sci. Technol.*, 2010, **20**, 267
185. S. Henry-Michelland, M. J. Alonso, A. Andremont, P. Maincent, J. Saucières and P. Couvreur, *Int. J. Pharm.*, 1987, **35**, 121.
186. M. Youssef, E. Fattal, M. J. Alonso, L. Roblot-Treupel, J. Sauzieres, C. Tancrede, A. Omnes, P. Couvreur and A. Andremont, *Antimicrob. Agents Chemother.*, 1988, **32**, 1204.
187. E. Taylor and R. Webster, *Int. J. Nanomed.*, 2011, **6**, 1463

Section 2
Nanostructures Overcoming the Intestinal Barrier

CHAPTER 2.1

Nanostructures Overcoming the Intestinal Barrier: Physiological Considerations and Mechanistic Issues

SAM MAHER[a], KATIE B. RYAN[b], TAUSEEF AHMAD[a], CAITRIONA M. O'DRISCOLL[b] AND DAVID J. BRAYDEN*[a]

[a] Conway Institute, University College Dublin, Belfield, Dublin 4, Ireland;
[b] School of Pharmacy, University College Cork, Cork, Ireland
*E-mail: david.brayden@ucd.ie

2.1.1 Oral Peptide and Protein Delivery Challenges

Only two peptides have been licensed to date for oral delivery, desmopressin and cyclosporine (CsA). The synthetic V_2 receptor agonist, desmopressin, is exceptionally potent, stable and relatively cheap to synthesize, and thus its oral bioavailability (F) of 0.16% is still commercially acceptable.[1] The undecapeptide, cyclosporine A, is cyclic, stable, relatively hydrophobic and, with an F of approximately 30% in an enteric coated microemulsion (Sandimmun Neoral®, Novartis, Switzerland), qualifies as a Class II molecule with high permeability and low solubility as defined by the Biopharmaceutical Classification System (BCS).[2] However, peptides and proteins generally fall into Class III of the BCS, where poor permeability is the predominant issue, and this has led to pursuance of absorption-promoting technologies as the primary and now most advanced strategy for oral peptide delivery (Table 2.1.1).

RSC Drug Discovery Series No. 22
Nanostructured Biomaterials for Overcoming Biological Barriers
Edited by Maria Jose Alonso and Noemi S. Csaba
© The Royal Society of Chemistry 2012
Published by the Royal Society of Chemistry, www.rsc.org

Table 2.1.1   Examples of absorption promoters in the clinic for oral peptide delivery with potential for ultimate inclusion as part of oral NP peptide formulations.

Excipient	Description	Clinical Phase	Source
Sodium cholate	Cholic acid, FDA-GRAS-list	Phase III (insulin)	Biocon Ltd (India)
Labrasol®	US Pharmacopoeia	Marketed (CsA, generic)	Sidmak Laboratories (USA)
Bile salts/carnitines/ surfactants	GRAS-listed claim	Phase III completed (calcitonin)	Unigene/Tarsa, (USA)
Eligen®, e.g. salcaprozate sodium (SNAC), and 5-CNAC	Novel excipient; provisional GRAS status claim (SNAC)	Phase I (GLP-1, insulin) with SNAC; Phase III (calcitonin) with 5-CNAC	Emisphere Technologies (USA)
Sodium caprate (C_{10}), solid dosage form	Present in foods (capric acid)	Phase I (insulin/GLP-1 analogue)	Merrion Pharma (Ireland)
Axcess™ aromatic alcohol technology: phenoxyethanol, phenyl ethanol, benzyl alcohol	GRAS-listed claims, Pharmacopoeia, solvents and plasticizers	Phase II (calcitonin)	Bone Medical (Australia)
Alkyl maltosides, Intravail® technology	Personal care and food products	Phase I (leptin)	Aegis Therapeutics (USA)

Abbreviations: SNAC: (n-(8-[2-hydroxybenzoyl]amino)caprylic acid) (N-(5-chlorosalicyloyl)-8-aminocaprylic acid); 5-CNAC: (N-(5-chlorosalicyloyl)-8-aminocaprylic acid); GLP-1: Glucagon-like Peptide-1.

Some of these promoters are being incorporated as components of NP peptide-, protein- and siRNA- formulations, so it is pertinent to discuss their current status both in NP and non-NP formulations. While it can be easily demonstrated that promoters ranging from medium chain fatty acids to large polysaccharides can boost peptide permeation in admixtures across *in vitro* models of the intestinal epithelium (*e.g.* Caco-2 monolayers[3]) and isolated intestinal mucosae,[4] as well as in preclinical oral studies,[5] few have made it to the clinic. There is considerable debate over the possible reasons for attrition and lack of translation of such drug delivery platforms. From a physiological standpoint, *in vitro* and *in situ* models offer limited predictability as they reveal genuine enhancer potential only if the delivery agent can reach the gut wall in sufficient concentration, with an assumption that the cargo also reaches it contemporaneously.[6] This intimate association of high concentrations of cargo with the enhancer is an important reason why some NP technologies may improve outcomes in oral peptide research, in addition to the obvious capacity to protect cargo. Reductionist models nonetheless ignore the significant impact of dilution in the lumen and the effects of GI tract transit time, and there is particular concern relating to physiological differences when attempting to correlate animal and human oral delivery data.[7] In addition, recent data suggests that overlying intestinal mucus presents a barrier to penetration of particulates,[8] again poorly modeled in *vitro*. PEGylated poly(sebacic) acid co-polymeric NPs however, seem to be able to traverse mucus well, dependent on the selected molecular weight and hydrophilic features of PEG, resulting in avoidance of entrapment in glycoprotein meshes.[9] Finally, in the rare occasion that an enhancer demonstrates efficacy for peptide delivery in animal models, translation to man is not guaranteed. For example, there was a relatively poor correlation ($R = 0.51$) in oral F for 43 drugs in a human *versus* dog comparative predictability study,[10] although it is recognized that the dog model can still provide useful guidance for clinical testing of some oral controlled release formulations with, for example, pH-dependent coatings on matrix tablets. We and others have calculated that the concentration of the enhancer, sodium caprate (C_{10}), requires at least a 10–15 fold increase when moving from *in vitro* to *in vivo* oral drug delivery models,[11] and this raises the specter of unavoidable intestinal tissue toxicity and the likelihood of different mechanisms of action pertaining at different concentrations, while at the same time presenting huge formulation challenges in terms of adequate enhancer loading.

The issue of enhancer toxicity as a barrier to translation may be somewhat exaggerated, though clearly it requires case-by-case analysis. In our extensive experience of working with C_{10} in a range of drug delivery models, we have noted the tremendous capacity of the epithelium for repair following histopathological changes *in vivo*, and it is clear that *in vitro* models tend to overestimate irreversible damage. C_{10} is an essential component of the Gastro-Intestinal Permeation Enhancement Technology[12,13] (GIPET^{-TM}, Merrion Pharmaceuticals Ltd, Ireland) and is present within coated matrix tablets at concentrations predicted to be greater than 100 mM for human delivery *in vivo*. While over 300 human subjects have been administered this formulation across 7 Phase I trials, without reported evidence of any intestinal side-effects, just 10 mM (1.95 mg ml^{-1}) C_{10} damages the intestinal epithelium *in vitro* following a 15 min exposure.[14] Endogenous surfactants and detergents, including bile salts, irreversibly damage intestinal cell cultures and are endogenous lipid solubilisers *in vivo*, but sodium cholate has been used without major toxicity problems in hexyl insulin PEGylated conjugate clinical trials.[15] Moreover, the use of chenodeoxycholate to dissolve gallstones has also been in Phase I trials without suggestion of overt toxicity.[16] Recently, the cytotoxicity-generating capacity of over 150 enhancers from different classes was measured by the methylthiazolyldiphenyl-tetrazolium bromide (MTT) conversion assay and correlated to ability to lower transepithelial electrical resistance (TEER) (a surrogate for increased transepithelial flux) across Caco-2 monolayers. The data suggested that in some instances these parameters can be dissociated, thus permitting a concentration window of enhancement not associated with cell death.[17] Nevertheless, an over-reliance on such *in vitro* assays might lead to premature abandonment of promising enhancers for NP formulations, reaffirming that there is no substitute for *in vivo* data. A second area of controversy relating to an aspect of potential toxicity is whether there is any evidence that enhancers permit inadvertent permeation of bystander bacterial or viral pathogens or lipopolysaccharide. We could find very little, which is perhaps not surprising given the huge differences in molecular weight and radii between peptides and proteins *versus* such macromolecular structures. Besides, even if this was a real rather than perceived issue, there is still the argument that significant pathogen permeation would require that pathogens be very closely associated with the enhancer at the epithelium similar to the required relationship between enhancers and active cargoes,[6] a rather unlikely scenario. There is the more important issue of a requirement for extremely high enhancer concentrations *in vivo*, which suggests that the currently available range of permeability-promoting agents have quite low potency in enhancing peptide delivery, a feature that leads to high intra-individual variation. Nanotechnology may be able to address some of these issues, perhaps by potentially allowing lower levels of entrapped enhancers to be used. NPs also offer the potential for slow transit of the entrapped peptide/enhancer *via* mucoadhesive mechanisms and also to promote delivery to the colon, where for instance proteolysis may be significantly lower and the efficacy of absorption enhancers may be greater.

Can the metabolic stability issue for oral delivery of peptides and proteins be overcome? This appears to be a largely surmountable problem, which again has tended to be overemphasized as a hurdle. Many polymers in particulate-, conjugated- or even as admixture formats provide protection against serine proteases and endo- and exo-peptidases, intestinal washes and homogenates, liver homogenates and serum proteases. Examples include thiolated poly-carbophil conjugates[18] and amphiphilic comb-shaped poly(ethylene)glycol (PEG) polymeric conjugates.[19] In the early 1990s by contrast, the potential solution offered was to use cocktails of peptidase inhibitors including bestatin, apamin and soyabean trypsin inhibitor, many of which turned out to have unacceptable toxicology. Figure 2.1.1 shows the sequential problems peptides and proteins encounter in the intestine in their difficult and usually fruitless journey to the systemic circulation. It is worth pointing out that peptidase metabolism is designed to break food-derived proteins down into di- and tri-amino acids for uptake by the human peptide transporter 1 (hPepT1) on the apical membrane in the small intestine.[20]

Since transported small hydrophilic peptides will end up in the hepatic portal vein, one can make a stronger case for development of oral peptide formulations targeting hepatic function, (*e.g.* insulin and glucagon-like peptide-1 (GLP-1)), rather than for systemic targets where peptides are susceptible to dilution, lack of tissue targeting and have the potential for further degradation. Even if an enhancer greatly increases the permeability coefficient of an entrapped peptide across the epithelium, extensive liver metabolism and rapid liver clearance due to efficient blood perfusion could still negate any positive impact on oral F,[21] so these parameters also need to be established in advance.

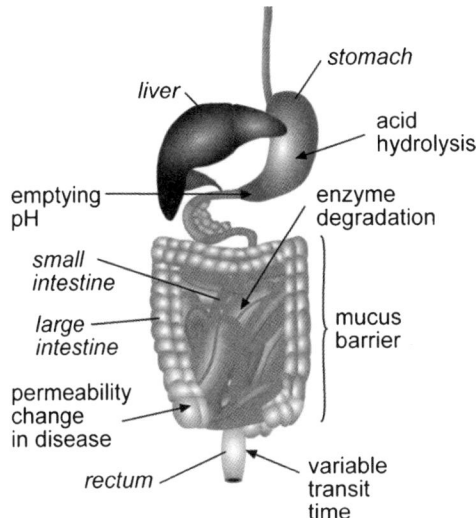

Figure 2.1.1 Features of the gastrointestinal tract that play a role in poor and variable oral F of peptides. Note susceptibility to stomach acid and proteolytic degradation in the small intestine. Diseases can also affect the extent of drug absorption, while excipients can induce altered permeability.

2.1.2 Physicochemical and Physiological Barriers to Drug Absorption

In the BCS, poor solubility is considered a physicochemical barrier, while poor permeability relates to intestinal epithelial physiology.[22] BCS class III drugs have high solubility but inadequate lipophilicity, which impedes passive permeability across the plasma membrane of intestinal epithelial cells, while BCS class IV drugs have poor solubility and poor permeability, and potential progression in the discovery pipeline generally depends on subsequent structural modification. Physicochemical criteria important for passive drug permeability can be revealed by various *in silico* models, notably Lipinski's rule of five, where molecular weight, sub-optimal partitioning between water and octanol, and the degree of ionization are used to predict intestinal absorption.[23] Considerable proteolytic action is present in pancreatic secretions, up to 40 g per day in adults, whilst proteolytic action from sloughed cells that have lysed is also high, given that the entire surface epithelium is rapidly regenerated.

To-date, prodrugs are the most innovative and successful approach to improving oral delivery of poorly absorbed drugs;[24] the drug is modified with either a moiety that either improves solubility for passive absorption (*e.g.* Sulindac, Clinoril®, Merck, UK) or directs it *via* carrier-mediated transporters (*e.g.* hPEPT1) that shuttle the drug across the epithelium (*e.g.* valacyclovir, Valtrex®, GlaxoSmithKline, UK). Drug solubility is often improved by selecting a more soluble salt, for instance the hydrochloride salt of tetracycline has six times more aqueous solubility.[22] Drug absorption can also be aided by excipients that promote better dissolution, many of which are listed in the United States Pharmacopoeia–National Formulary (USP-NF); (*e.g.* Labrasol[®], cyclosporin; Sidmak Laboratories, USA). On the other hand, development of excipients that promote better flux across the gut wall by modulating the intestinal barrier have not yet been successful in oral formulations, although various clinical trials are ongoing.[25] There have been attempts to deliver drugs in almost every segment of the gut, an example being ampicillin rectal suppositories (Doktacillin, Meda, Sweden[26]).

The oral cavity and oesophagus are the first parts of the GI tract encountered, but due to the short residence time, these regions are not very amenable to drug permeation from conventional oral formulations, although there are successful specialized mucoadhesive formulations permitting greater residence in the oral cavity and adequate buccal absorption (*e.g.* prochlorperazine, Buccastem[®], Reckitt Benckiser). After movement through the mucin-lubricated oesophagus, the dosage form enters the stomach, which is not designed to efficiently absorb dietary xenobiotics. pH disparities between the stomach and small intestinal fluids can have a considerable effect on solubility and resulting permeability of ionisable charged drugs. Gastric juice inactivates any drug that is acid labile and provides a source of proteolytic enzymes. The gastric mucosa is protected from gastric secretions by a carpet of mucus estimated to be 140 µm, which is an extrinsic barrier that can also impede gastric drug absorption. Fate of an oral dosage form in the stomach depends on residence time and pH, both of which can vary greatly depending on many factors, including diet, age, and disease state. In the fasted state, the pH in the human stomach is less than 3, but it can be as high as 7 in the fed state,[27] a pH which can dissolve enteric coatings thereby releasing acid labile drugs; this causes variability in efficacy and can lead to gastroesophageal irritation (*e.g.* bisphosphonates).

The small intestine is commonly the region of the gut where the highest fraction of a drug is absorbed. After leaving the stomach the formulation enters the duodenum, the shortest segment of the small intestine (~ 26 cm); here chime is neutralized by secretion of bicarbonate from Brunner's glands located in crypts of the duodenum. Pancreatic secretions and bile enter the duodenum *via* the pancreatic and bile ducts, respectively, providing further change in luminal pH, and considerable enzyme activity. Unlike the stomach, residence time in the small intestine is more consistent following a meal. However, intake of food can both positively and negatively impact drug absorption. Despite subtle changes in gross histology between the duodenum, jejunum and ileum that give rise to differences in absorptive and secretory characteristics, the epithelial cell composition at the mucosal surface is similar. Absorptive enterocytes are the major cell type in each of the segments of the small intestine, followed by mucus secretory goblets cells and the immune sampling Microfold cells (M-cells) which are organized individually or grouped together in the follicle associated epithelium of Peyer's patches (also see Chapter 2.3); thereafter enteroendocrine cells and paneth cells are less prevalent. The colon consists of the ascending, transverse, descending and sigmoid regions, and functions in fluid reabsorption. Although the surface area and blood flow to the colon is far less than the small intestine and it is not conventionally perceived as a drug absorption site, the potential of this region for targeted drug delivery is being studied. The greater residence time (2–30 h)[27] provides opportunity, and the solvent drag upon reabsorption of fluids often makes this region more sensitive to absorption promoters.[28] Furthermore, proteolytic action is considerably reduced in the large intestine.[29] In every region of the GI tract there are extrinsic and intrinsic barriers that can prevent absorption of BCS Class III and Class IV drugs (Figure 2.1.1). For instance, the mucus gel layer is thicker in the colon than the small intestine, while the distal colon and rectal absorption sites avoid the first pass effect. Upon efficient diffusion through mucus followed by passive or active permeation across enterocytes a drug molecule will then pass through the basement membrane to the endothelia of the capillaries supporting the hepatic portal vein.

2.1.3 Nanoparticle Carriers as a Potential Solution for Oral Peptide Delivery

Nanostructures with peptide cargoes formed by electrostatic complexation, ionic gelation or conjugation, should protect acid- or enzymatically- labile peptides and proteins from the hostile environment of the stomach and upper GI tract. They can promote optimal solubility, and provide sufficient co-localized concentrations of peptide and absorption enhancers, which in turn should enable better transport across the intestinal epithelium. It is important to point out that encapsulating a peptide or protein in NPs for oral delivery may not address the permeability issue *per se*; this is because NPs, while adequately protecting the cargo, may also be poorly absorbed. An alternative model is that payloads can be released in high concentration close to the gut wall. Further barriers to NP progress to the liver include the basement membrane, the sub-mucosal phagocytic immune cells, endothelia of capillaries and the hepatic portal vein as well as the Kupffer cells that line the walls of the liver sinusoids (Fig. 2.1.2). There are a number of preclinical drug delivery models that can be used to study each of these barriers and the influence of NP delivery platforms on the uptake of poorly permeable payloads (reviewed in ref. 30).

An understanding of particle composition and their physiological interaction with the cells that form the intestinal barrier is a prerequisite prior to *in vivo* studies. A decision to marry the use of NPs with promotion enhancers is a formative step in this understanding. For example, in a recent rat study examining delivery of salmon calcitonin (sCT) in NPs, an interesting comparison was made showing increased serum calcium reductions if the

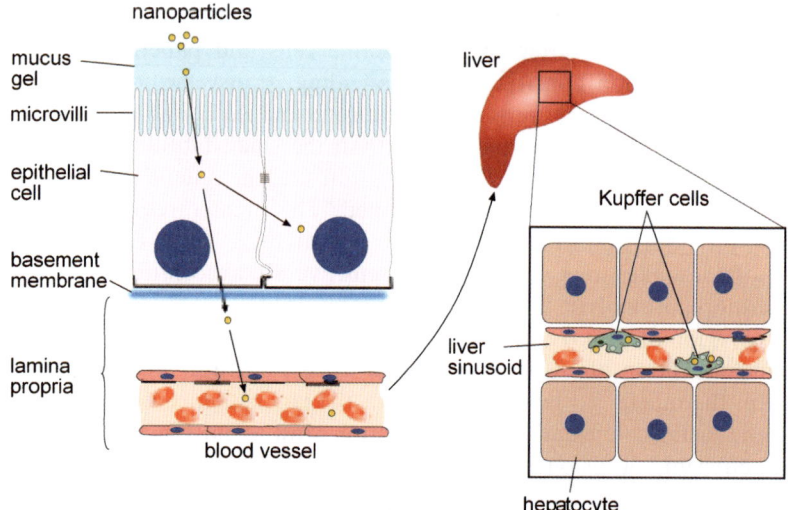

Figure 2.1.2 Barriers to oral F of NPs containing peptides from the intestinal epithelium to the systemic circulation. Provided NPs can penetrate the mucus glycoprotein mesh, the major barrier to absorption is the intestinal epithelium.[30,112] Upon uptake and translocation, NPs must also evade sub-mucosal immune cells that can phagocytose and digest NPs and their cargo *via* lysosomes. Further barriers are the basement membrane, the endothelium of capillaries, the hepatic portal vein and the Kupffer cells that line sinusoids.

enhancer, spermine, was incorporated with sCT in the same NP, compared to spermine formulated as an external mixture with the sCT-entrapped NPs.[31] Tailoring physicochemical and biological properties of NPs to safely and reproducibly increase delivery of cargoes is, however, exceptionally difficult.

Pharmaceutical scientists tend to use a rather liberal definition of 'nano', ranging from diameters of 100–1000 nm. Typically, polymeric and liposomal-based NPs require a minimum size of 100 nm in diameter in order to permit significant peptide and protein loading; however the US National Nanotechnology Initiative has set the range of 1–100 nm in diameter for NP constructs,[32] and the lower end of this scale is the traditional realm of inorganic chemistry (*e.g.* gold, titanium dioxide). Pharmaceutical scientists who are working in a NP size range around 100 nm tend to be working with 'nanocomplexes' (*e.g.* dendrimers, cyclodextrins, amphiphilic co-polymers); they do not necessarily fit the typical textbook image of a round polymeric or liposomal particle with associated cargo. Other NP formats of interest are recombinant NPs synthesised with exact dimensions for precise examination of the nature of NP-epithelial interactions. These include a variety of polymeric structures including elastin-like, silk-like, silk/elastin-like and artificial amino-based polymers.[33]

What is the actual evidence that a particulate format should promote peptide uptake by the intestinal epithelium? For over 20 years it has been repeatedly shown that gut-associated lymphoid tissue (GALT) (*i.e.* Peyer's patch) M-cells take up NPs and microparticles more avidly than either regular follicle-associated and non-associated epithelial enterocytes[34] and this led to interest in oral vaccine delivery of non-live subunit peptides and proteins using particulates to generate mucosal immunity (see Chapter 2.3, this volume). To summarize, there is reasonable consensus that for M-cell uptake, particle diameters should ideally be in the 200–500 nm range, and usually not larger than 3μm, but it is well known that particle polymer composition, surface charge, hydrophilicity and species differences may alter that profile.[35] For example, rabbit M-cells take up polystyrene particles more avidly than mouse or human counterparts, whereas interactions with poly(D,L-lactide-co-glycolic acid) (PLGA) particles are low compared to other species.[36] Much of this data was achieved using a controversial though now better accepted 'M-like' cell model comprising a co-culture of Caco-2 monolayers in various formats with human Raji B lymphocytes or Peyer's patch lymphocytes (reviewed in ref. 37). Thus, it was recently shown by us that novel glucan-based yeast derived particles of diameter 1–3 μm were preferentially taken up by the Caco-2/Raji co-cultures compared to Caco-2 mono-cultures (Figure 2.1.3),[38] an important finding since these same particles delivered a therapeutic siRNA that relieved inflammation in the dextran sulphate model of murine colitis following oral delivery.[39]

Data from isolated human ileal Peyer's patches corroborates the prevalence of M-cell uptake of Arg-Gly-Asp (RGD)-coated polystyrene NPs *via* integrin receptors and confirms an optimal particle diameter of 100–500 nm,[40] a size that may permit reasonable peptide or vaccine loading. It is worth bearing in mind that much of the intestinal particle uptake data refers to polystyrene,

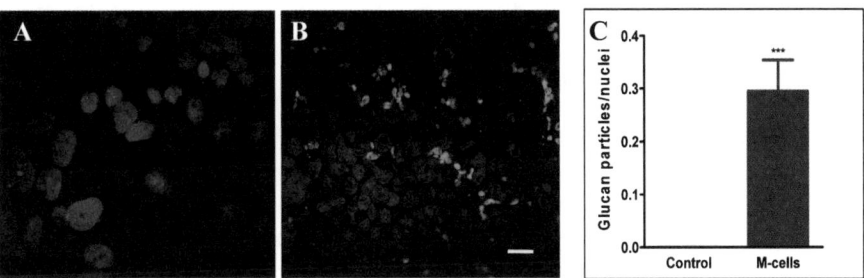

Figure 2.1.3   Confocal microscopy illustrating the uptake of fluorescently-labelled β-glucan particles (GPs) in (a) filter-grown human cultured Caco-2 monolayers and (b) 'M-like' cell co-cultures of Caco-2 converted by Raji B cells. Nuclei are stained blue (Hoechst®) and particles are green. Z-section slices taken from the middle of monolayers exposed to GPs show an abundance of internalized particles in the M-cell model (horizontal bars = 5 μm); (c) Quantification of GPs associated with Caco-2 and M-cell monolayers as determined using high content screening analysis. ***$P < 0.001$, unpaired Student's t-test. (T. Ahmad and D Brayden, unpublished data).

PLGA, chitosan and liposomes vesicular formats. No doubt data achieved with newer biomaterials of different shape, surface charge and hydrophilicity will alter our interpretation of the current M-cell and enterocyte uptake model. With novel fabrication technologies available to make protein-derived biomaterials to synthesize non-spherical NPs for injected implants,[41] efforts are underway to systematically assess cellular uptake and cytotoxicity of polymeric NPs of different shapes, initially by macrophages.[42] Dramatic improvements in NP targeting and internalization by intestinal epithelia as well as altered presentation by endocytosis can be expected with use of elliptical, cone- and rod-shaped particles. Increasingly, we will be describing biocompatible and biodegradable NPs for oral peptide delivery using more accurate measures of shape.

A novel tool to aid in toxicological assessment of new biomaterials is *in vitro* high content analysis (HCA), a multi-parametric spatio-temporal analysis with the power to predict *in vivo* toxicity that has superior sensitivity and specificity compared to conventional cytotoxicity methodology. For example, detection of cytotoxicity of poly(2-(dimethylamino ethyl)methacrylate) (pDMAEMA), a polymer being investigated as a gene delivery vector and anti-microbial agent, was more sensitive and quantitative when measured by HCA compared with MTT assay.[43] Through advanced imaging techniques, pharmaceutical scientists will also be providing more accurate quantitative measures describing the association of particle with the associated peptide, be it adsorbed to the surface or truly entrapped. At the same time, imaging techniques including positron emission tomography scanning are being used for the first time to track *in vivo* tissue distribution of insulin, for example when instilled with cell penetrating peptides, to rat intestinal loops.[44] These methodologies permit quantitative measures of organ distribution of oral NP formulations, and it may soon supersede current fluorescent imaging modalities due to more sensitive resolution.

The oral vaccine particle field has since moved on to using novel NP-conjugated peptide ligands arising from phage display against Peyer's patch tissue[45] or human 'M-like' cells,[46] while at the same time using gene arrays to source previously undiscovered human M-cell targets.[47] We still await convincing evidence that M-cell targeted NPs containing vaccines can generate increased mucosal immunity over untargeted counterparts when orally administered, although there is some recent suggestive data for PLGA-entrapped hepatitis B surface antigen in mice when the α-l-fucose specific lectin, *Lotus tetragonolobus* from Winged- or Asparagus pea, was used to target M-cells.[48] Studies investigating NP pathways into the immune system using ligated intestinal loops from mice confirm that fluorescent PLGA-NPs are held in mucus, before being delivered to M-cells, and are then subsequently distributed to both dendritic and B cell lymphocyte populations.[49] In further support of the mucosal immunity hypothesis for oral vaccine delivery, the work highlights the potential added value in targeting of antigen-presenting cells within the Peyer's patch dome, downstream from surface M-cells in the

FAE, although it remains to be seen if such formulation complexity may still be practical from a manufacturing perspective. Historically, the initial clinical trials of oral enterotoxigenic *Escherichia coli* (ETEC) vaccines using untargeted PLG-NPs in the μm diameter range were unsuccessful,[50] and this drove interest in M-cell-targeted NPs as well as in NP systems incorporating adjuvants. For peptides however, the premise is that, although enterocytes take up far less NPs than M-cells, this is far outweighed by their high density in the small intestine and therefore, they are the preferred target.

2.1.4 Dissecting the NP Uptake Pathways in the Intestine

If the NP can penetrate mucus overlying the small intestinal epithelium, imaging research aimed at tracking their pathways across epithelial cells confirms the dominance of transcellular pathways over paracellular,[21,51] both for M-cells and villous enterocytes. An estimate of the normal tight junction (TJ) diameter in the human small intestine is 0.5–2.0 nm,[52] so even in the presence of efficient TJ openers or even in inflammatory conditions where junction diameter may be increased by 10-fold, this is still expected to be too small for significant paracellular flux of 100–500 nm diameter NPs. The more relevant internalizing pathways for NP trafficking though epithelial cells comprise four potentially exploitable, but still relatively poorly understood, non-phagocyte, endocytotic pathways: clathrin-mediated pits, caveolae, macropinocytosis and clathrin/caveolae-independent endocytosis (reviewed in ref. 53) (Figure 2.1.4). Of these, it would appear that clathrin-mediated endocytosis (CME) is important in intestinal epithelia. CME can be divided into receptor-mediated and non-receptor (fluid-phase) mediated endocytosis and it is the former that has been more widely studied. Both processes involve formation of a clathrin-coated protein lattice, which forms a pit. With the help of the GTPase dynamin, vesicles of diameter 100–120 nm are prepared for invagination, followed by internalization and presentation to mildly-acidified early endosomes for sorting. From these, receptors can be recycled back to the apical membrane (*e.g.* transferrin) or they can move across the basolateral membrane to complete transcytosis. Should the vesicles continue to reside in endosomes, they are moved to late endosomes, which have a higher pH and prepare material for transfer to pre-lysosomal vesicles. Of importance for peptide-containing NPs, the lysosomal system contains numerous peptidases including cathepsin B, in addition to a very acidic milieu, so strategies are required to evade such a destructive fate, and to favour movement of NPs out of the early endosomes and direct them to the basolateral membrane.

The caveolae-dependent mechanism is also present in intestinal epithelia, but it is less dominant than CME and tends to take up particles in the lower 50–80 nm range and at a slower rate.[54] Like clathrin, caveolin is also a protein, of which sphingolipids and cholesterol comprise major components. Again, dynamin facilitates invagination and caveolar vesicle formation. In parallel

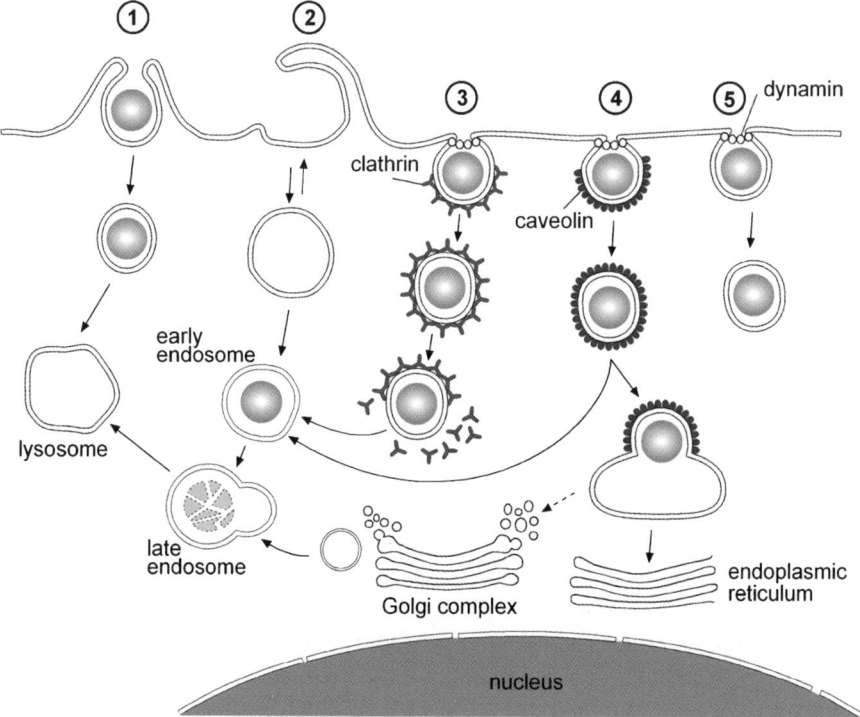

Figure 2.1.4   Schematic of transcellular pathways of NP cell uptake by intestinal
epithelia (modified from ref. 54 and 113). (1) Phagocytosis, (2)
macropinocytosis, (3) clathrin-dependent endocytosis. These paths
comprise formation of early endosomes, which merge with pre-
lysosomal vesicles containing peptidases. (4) Caveolin-dependent
endocytosis and (5) clathrin/caveolin-independent uptake pathways
lead to lysosomal-independent NP processing.

with the CME pathway, early escape of peptide-containing NPs from the
caveolar vesicle construct is essential to avoid default presentation to the
destructive lysosomal pathway. Cholesterol, albumin and folate appear to use
the caveoli-dependent pathway. Macropinocytosis is responsible for non-
selective internalization of larger 1–5 μm particles. It is likely therefore, that
this is the dominant process involved in uptake of the liposomes, PLGA and
polystyrene microparticles reported in the majority of Caco-2 monolayers and
intestinal epithelial studies to date. An important study using human intestinal
mucosae in Ussing chambers and Caco-2 monolayers revealed that the flux of
fluorescent polystryrene NPs, 200–500 nm in diameter, were blocked by a
selective inhibitor of macropinocytosis, *N*-isopropyl amiloride,[40] in preference
to the clathrin- or caveolae-dependent routes, although some overlap *via*
absorption of cholesterol-dominant membrane domains was not ruled out. A
recent study by Cartiera *et al.*[55] examined PLGA internalization by Caco-2
monolayers. Compared to other cell lines, Caco-2 monolayers took up

fluorescently-labeled 100 nm NPs slowly and sparsely and *via* endocytosis, but evidence for specific organelle association was lacking. Considerable attention is also focused on NP uptake *via* cholesterol-rich lipid rafts, a recent example being a DNA-associated NP entering airway epithelia associated with the plasma membrane surface raft protein, nuceolin.[56] This type of pathway is neither clathrin nor caveolae-dependent, but seems to require microtubule assistance once internalized.

In relation to M-cell particle uptake, the process of internalization of larger micron-sized particles appears to be predominantly phagocytosis, similar to the process used by M-cells for viruses and bacteria.[57] With the availability of more selective inhibitors, along with advances in cell imaging that permit intracellular tracking and counting of NPs in individual live cells, we are starting to see more complex dissection of NP uptake pathways and subsequent tracking of intracellular and transcytosis processing. As an example of the latter, Golberg *et al.*[58] examined uptake of G3.5 dendrimer NPs by Caco-2 cells using confocal microscopy and flow cytometry and, with the help of endocytotic inhibitors, were able to show that transepithelial transport by fully differentiated monolayers was dominated by clathrin- and dynamin-dependent processes, and that these subsequently interacted with TJs downstream to cause openings. Caveolin-dependent uptake processes for dendrimers seemed, however, to play a more important role in undifferentiated Caco-2 cells. Accepting the limitations of Caco-2 monolayers in having unrepresentatively high TEER values, and a lack of goblet cells, perhaps use of mucus-covered intestinal epithelial monolayers[59] in parallel with better access to fresh human intestinal tissue[11,40] will reveal a more accurate picture of NP uptake and transcytotic pathways.

2.1.5 Examples of NPs for Oral Peptide Delivery

If we exclude technologies that may have particulate features as a physicochemical by-product rather than through specifically engineered NP design (*e.g.* non-covalently linked particulate carriers, Eligen®, Emisphere, USA; supramolecular vesicles of C_{10} mixed with pay-load, GIPET®, Merrion Pharmaceuticals, Ireland; amphiphilic PEGylated conjugates, Biocon Ltd, India; micelles with enhancers for buccal delivery, Oralyn®, Generex, Canada), no specifically designed NP technology for oral peptides, proteins or genes has yet reached the clinic. If we further exclude M-cell targeting of vaccine-loaded microparticles and NPs (Chapter 2.3, this volume), we can capture key progress through examination of the oral insulin and, to a lesser extent, the oral calcitonin particulate literature. Many polymers and co-polymers have been used to synthesize NPs capable of oral peptide delivery. Damgé *et al.*[60] summarized the most interesting oral insulin polymeric NP formulations in detailed tables under the criteria of method of preparation, size, surface charge, biocompatibility, biodegradability, release profile stability, and oral F. Pharmacology of the insulin-loaded NPs has been investigated in normal and diabetic rats, but data is somewhat confounded by large variations in insulin dose levels between studies, along with rather inconsistent methodology ranging from ileal loop instillations of mixed solutions to oral gavage with mini-tablets. The main formulations cited are dominated by core polymers or co-polymers of PLG, chitosan, poly(ε-caprolactone) (PCL) and poly(isobutylcyanoacrylate) and to a lesser extent by alginate, solid-lipid, calcium phosphate, and gold NPs.[60] Variations include incorporation of surface PEG coatings or use of PEG in block co-polymer formats with PLG, chitosan, poly(sebacic) acid and methacrylate, as well as co-polymers of chitosan with dextran or poly(glutamic acid). Common features of these particles are diameters of 100–500 nm and relative F values ranging from 3–15%, bearing in mind the substantially higher oral doses compared to those injected. These NPs are delivered in formulations of considerable complexity and tend to use pH-sensitive coatings (*e.g.* Eudragit®) to allow stomach bypass and presentation in the small intestine. Other variations include embedding insulin-entrapped NPs in mucoadhesive gels and use of mucoadhesive or bioadhesive coatings (reviewed in ref. 61). It remains to be seen, however, if any of these approaches will ever make it to Phase I trials.

Perhaps the most concentrated efforts in oral insulin delivery using NPs involve chitosan derivatives, achievable due to the availability of free amino and hydroxyl groups for modification and attachment. Furthermore, since unmodified chitosan has limited solubility at pH values of the small intestinal regions, use of tri-methylated (TMC) NP versions of thiolated chitosan, with improved solubility at pH 5–7, may have considerable potential both for oral delivery and for colon-specific targeting.[62] Thiolated versions of chitosan are also being used in NP formats for peptide delivery with the enhancer, reduced glutathione. The chitosan-4-thiobutylamidine conjugate has bridged sulfur groups that permit increased attachment to mucus and to small intestinal regions. The rationale is improved peptidase resistance, mucoadhesion and permeation enhancement compared to unmodified chitosan.[63] Convincing *in vivo* data for systemic peptide delivery is awaited in order to prove that there is genuine advantage for thiomeric chitosan NPs over chitosan and other competing chitosan-derivatized polymeric NPs.[64] Poly(allyl)amine (PAA) is also a key material in novel amphiphilic quaternized nanostructures that is currently being assessed for oral insulin[65] and calcitonin[66] delivery. PAA grafted with palmitoyl groups in a quaternized format can be complexed with peptides to yield 200 nm diameter NPs, which retain *in vitro* and *in vivo* bioactivity and are peptidase resistant. Prototype examples of insulin-entrapped NPs for oral delivery on the basis of novel design features, polymers and current status are described in this volume (Chapter 2.2) and elsewhere.[60]

Targeted NPs to receptors on the small intestinal epithelia are based on the premise that untargeted formats will not deliver enough peptide systemically. Again, excluding the M-cell literature, evidence for proof of concept is sparse.

A relatively advanced system appears to be vitamin B12 conjugated to insulin-loaded dextran NPs, where increased oral F over untargeted NP formats was achieved in diabetic rats.[67] The B12 uptake pathway at first glance seems especially attractive since the vitamin is non-toxic and very soluble. The uptake pathway is in turn selective and the B12-intrinsic factor receptor is well expressed on small intestinal enterocytes. Attempts to make a successful oral peptide B12 receptor-targeting NP are, unfortunately, hampered by low B12 uptake capacity of 1–2 μg per day, low enzymatic protection for the peptide, as well as a relatively low number of B12 molecules that can be conjugated to the available particle surface, not to mention a slow oral uptake rate.[68] Improvements in using linker technology to reduce steric hindrance, and better conjugation chemistry suggest that a higher density of B12 per NP might be achievable and this may ultimately improve oral peptide F. A vitamin-B12 NP oral delivery system (CobOral™, Access Pharmaceuticals, USA), is also in preclinical research with insulin and human growth hormone as payloads. Convincing evidence is required to demonstrate that improved pharmacological outcomes are genuinely vitamin B12 receptor-mediated.

Lectins bind specifically to carbohydrates moieties of glycoproteins, soluble carbohydrates or glycolipids that result in agglutination of cells and polysaccharides.[69,70] It has been proposed that lectin–sugar interactions can also be used to trigger vesicular uptake.[69] While their exact physiologically role is not fully understood, they are implicated in many cell recognition and adhesion processes and may be of plant, microbial or animal origin.[71,72] Lectins have also been shown to facilitate endocytosis in enterocytes,[73] and the ability of lectins to bind to, and recognize, specific docking molecules on the intestinal epithelial surface has been exploited in targeted drug delivery of NPs and even prodrugs.[74] Certain lectins have been shown to interact directly with epithelial cells, through cytoadhesion,[75] whilst others exhibit mucoadhesive properties.[76] Targeted delivery with lectins may be possible because of the varied carbohydrate composition of different cell types and also due to the variability that exists between normal and diseased states.[69] In addition to targeting specific carbohydrate moieties on cells or within mucus, another way of exploiting lectin-mediated delivery is through endogenous lectins on the surface of epithelial cells; in this case, the encapsulation system contains a complimentary sugar moiety.[77,78] A number of lectins mediate drug delivery, including the plant lectins L*ycopersicon esculentum* agglutinin (LEA), wheat germ agglutinin (WGA), succinylated WGAs, and *Urtica dioica* agglutinin (UDA), which may be sourced from tomatoes, wheat germ and stinging nettle.[69] Initial studies focused on polystyrene particles modified with tomato lectin.[79] The intestinal enterocyte receptor, *N*-acetyl glucosamine has also received much attention. It is well expressed and binds to the WGA and tomato lectins, and studies using Caco-2 monolayers showed that PLG particles with WGA attached were bioadhesive and could be internalized.[80] On a note of caution, species differences can hamper development of targeted-NPs. An example is the binding of the lectin *Ulex europaeus agglutinin*-1

(UAE-1) which binds specifically to α-L-fucose, a sugar that is located on M-cells of mice but not of humans.[35]

A number of studies have observed a correlation between surface modification of particulates with targeting molecules and improved oral absorption. Historical data showed that targeting *Yersinia* bacterial receptors using *invasin* as a ligand on polystyrene NPs seemed to increase enterocyte particle uptake.[81] More recent studies have utilized carriers that are biocompatible. For example, PEG-PLA NPs functionalized with WGA through a non-covalent biotin–avidin interaction increased NP association with Caco-2 cells to 8.5%, compared to less than 1.5% for untargeted NPs.[82] Other receptors of note that have been targeted using a range of configurations in the small intestine include hPepT1 (the prodrug, valacyclovir[83]), transferrin (insulin conjugate in a nanoplex hydrogel[84]), biotin (PEGylated CT conjugates[85]), and the intestinal bile acid transporter, IBAT (conjugate and prodrug design[86]). Since many vitamin/nutrient-type receptors are well-expressed on tumors, there are intense efforts to develop parenteral targeted NPs to aid in tumor diagnostic imaging and to enable targeted delivery of chemotherapeutics in parallel.[87] Alternative approaches to increase cargo penetration across goblet cells include the targeting peptide (CSKSSDYQC), discovered through *in vivo* random phage display.[88] In theory, this peptide can be conjugated to peptide-encapsulated, bioadhesive-NPs to increase particle uptake, thereby overcoming the issue of entrapment in mucous. This approach mirrored a study to examine the increased M-cell uptake of latex particles studded with the M-cell targeting peptides, LETTCASLCYPS and VPPHPMTYSCQY,[45] discovered using the same panning methodology.

Cell penetrating peptides (CPPs) based on poly-cationic sequences (*e.g.* L-penetratin) have also been used in ad-mixtures to deliver insulin across the gut in rodents,[44] while octa-arginine has been incorporated in solid lipid NPs containing insulin and a 10% relative F was determined in rats.[89] CPPs have the advantage of not requiring complicated chemistry for attachment to particles and they may work as a surface coating or an entrapped component of an NP formulation. The downside is the non-specific nature of their action on cells as well as their susceptibility to proteolysis and the current heavy reliance on *in vitro* data. Overall, convincing evidence for oral delivery of peptides *in vivo via* ligand-coated NPs is lacking, likely due to variable receptor expression, insufficient increases in particle uptake by enterocytes, large variations in F, and also by an inability to reproduce and scale up these complicated formulations. One of the lessons of the last 20 years in oral peptide delivery is that simple formulations using scaleable robust formulation processes with well-known excipients and enhancers have a greater success as complex targeted particle concepts. There are some parallels with injected targeted NPs, where again, the potential compared to untargeted NPs has yet to be fully realized. Farokhzad and Langer argue that a renewed focus on self-assembling materials will make synthesis of scaleable targeted NPs easier, while use of particles with optimized biophysicochemical properties will improve cell uptake and overall PK.[90]

2.1.6 Overcoming the Intestinal Barrier with Novel Polymeric NP Constructs

There are interesting novel NP designs using novel biomaterials and chemistries that have warranted preclinical research. Noted earlier were the use of PEG-coatings to penetrate mucus, and the ability of short-chain low molecular weight PEG-coated NPs to reach target cells, which seem to be a feature of both injected and orally-delivered NPs. Other hydrophilic agents may perhaps substitute for PEG in this regard, and potential candidates for mucus penetration and improved enterocyte bioadhesion include poly (sialic) acid[91] and manosamine-coated NPs.[92] A recent construct includes a 50 nm PEG-coated calcium phosphate NP loaded with insulin.[93] PEG conjugation improved the insulin loading of the NP due to increased hydrophilicity, and sustained release was predicted in the small intestine. Apart from protection for the peptides against metabolism, an added advantage for PEG is that targeting ligands can be more easily conjugated, thereby enabling synthesis of targeted NPs. An example is the synthesis of a GLP-1-dibiotin-PEG conjugate, which proved to be stable to dipeptidyl peptidase IV and was orally-active in rats, likely due to biotin receptor-mediated transport.[85] Developing this concept into a NP format by which biotin could be conjugated to a PEG coating might be feasible. Recent advances in PEG chemistry reveal the potential of comb-shaped PEGs in protecting sCT from serine proteases. Poly(PEG) methacrylate was specifically conjugated to the cysteine-1 of sCT to form a stable NP complex, which provided a longer half life upon injection, with a direct relationship to overall conjugate molecular weight.[94] For such a large nanostructure to be developed adequately for oral delivery, improvements in degradable linker technology are required so that sCT can disassemble from the polymer close to the epithelial cell and then perhaps have its permeation facilitated by the use of enhancers. One could equally argue that it could be preferable to prioritize low molecular weight PEG-peptide amphiphilic nano-complexed conjugates for oral delivery, as they may be able to permeate intact without a requirement for linkers.[95]

Other NPs that have been considered for oral peptide delivery include surface-modified liposomes,[96] and biodegradable polymeric particles.[97] Coating of tripalmitin lipid NPs with PEG or chitosan did not affect permeability of Caco-2 monolayers and interaction of these NPs with Caco-2 monolayers was independent of coating type.[98] Chitosan NPs loaded with elcatonin and analogues thereof caused a significant and prolonged reduction in serum calcium levels when administered intra-gastrically to rats compared to PEG coated NPs or peptide solutions, suggesting that chitosan might impact on the mechanism of uptake, likely due to its mucoadhesive properties.[99–101] In support, an increased association of chitosan NPs with the mucus-secreting enterocyte/goblet human cell line HT29-MTX-E12 compared to Caco-2 monolayers was noted.[59] As an alternative to coating NPs with PEG or chitosan,[102] encapsulation of insulin in alginate–dextran NPs was effective in releasing the peptide at the higher pH values of the small intestine.[103] Further modification of the alginate–dextran core matrix with albumin–chitosan provided acid and protease protection, as well as enabling better uptake across intestinal epithelial cells.[104] Since albumin can be absorbed from the endothelia though caveolae mediated endocytosis,[105] it is possible this is the route used for passage from epithelia to the hepatic portal vein. When orally administered, insulin-loaded albumin–chitosan–alginate–dextran NPs reduced hyperglyce-mia in diabetic rats.[104]

The field of dendrimer nanostructures for oral peptide delivery has proceeded in parallel with that of co-polymer chemistry and novel PEGylation. Perhaps the most promising group are anionic poly(amido amine) (PAMAM) dendrimers. Focus is currently on use of low generation number dendrimers, co-polymerised with PEG and/or other biocompatible polymers; plus attachment of a range of fatty acid acyl chains, all with the goal of eventually increasing permeation of dendrimer-peptide nanostructures without the accompanying cytotoxicity (see Chapter 2.2, this volume). Delivery of nucleic acids to their nuclear targets to a sufficient level and in an intact form is a difficult challenge, and a number of NP-based platforms have been described.[30] Regarding materials being used for oral gene delivery using particles, this research strand largely evolved from initial studies using micron-sized PLGA to deliver rotavirus DNA vaccines across Peyer's patch M-cells.[106] While human DNA vaccination research using PLGA micro-particles seems to have stalled somewhat, DNA vaccine research with orally-delivered PLGA, chitosan and alginate particles continues in veterinary science, especially in fish vaccination.[107] Others have attempted to orally deliver chemically-modified stable phosphorothioate antisense oligonucleo-tides (ODNs) formulated in solid-dose multi-particulate mini-tablets with very high concentrations of C_{10}. In a Phase I trial, average F was < 10% and intra-subject variation was a feature of this study.[108] While it is unclear if that data was sufficient to encourage continuation of the program at Isis Pharmaceuticals, the same team has also demonstrated that micellar solutions of the bile salt chenodeoxycholate could aid permeation of ODNs in rat gut perfusions.[109] Oral siRNA delivery using NPs seems a considerably more achievable goal than for full genes due to an ability to more easily modify the smaller construct and a requirement for cytosolic as against nuclear delivery (see Chapter 2.2, this volume). Current intestinally-delivered siRNA constructs are designed for local targets including the colonic epithelial cells and enteric immune cells that are involved in colitis. Thanks to the use of NPs and to the lesser hurdle of cytosolic delivery for siRNA, perhaps the dose levels that might eventually be considered for oral human studies may be within more reasonable levels than those used previously for failed attempts at polymeric DNA vaccines or gene delivery.

2.1.7 Conclusions

Oral peptide delivery using NPs or any other permeation enhancing technology remains elusive. Of the oral peptides on the market, CsA is formulated as a self-emulsifying drug delivery system that, with the aid of surfactants, forms biodegradable emulsion nanostructures as small as 100 nm, which leads to improvement in drug solubility and concurrent reductions in intra-subject variation.[110] NPs may yet prove successful if they can be formulated with enhancers, provide protection against peptidases and can enable co-release in appropriate absorbing regions of the intestine. If intact nanostructures containing peptides are to be delivered across the intestinal epithelium, downstream endothelia and hepatocytes must also be systematically examined to visualize transport and safety. Modern biocompatible biomaterials, including modified glucans and thioketals with and without PEG coatings, in addition to providing better entrapment of peptides or siRNA in a NP format, could promote better transport across the intestinal mucosa. Self-assembled amphiphilic polymeric NP complexes may be suitable for some peptides due mainly to electrostatic interaction *via* a simple and scalable process, and the challenge is to activate the dissociation close to or beyond the epithelium. Dendrimer-type nanostructures have low polydispersity and have functionalized groups that allow for relatively easy conjugation of cargoes and targeting ligands, but *in vivo* safety and efficacy by the oral route is lacking. Targeted NPs have demonstrated limited progress in oral peptide delivery and the only formulation that seems to be anywhere close to the clinic is a NP conjugated with vitamin B12, noting that it has been in preclinical studies in various formats for 20 years. Finally, it is ironic that the formulations that were not actually designed as NPs, but have particulate nature (e.g. Eligen®, GIPET®, and Oralyn®), are the most advanced in clinical trials for oral peptides. These formulations use high concentrations of enhancers, where surfactant-generating micelles/vesicles are likely formed. Non-covalent interactions between *N*-acylated amino acids and peptide cargoes in the Eligen® technology also form detergent-like nanostructures, from which the trigger and timing of the dissociation steps for peptides are still not well understood.[111] Finally, although interest in oral PLG particles for vaccines and DNA vaccines has abated, there is exciting preclinical data on oral siRNA in NPs platforms for treatment of IBD.

Acknowledgements

This work was financially supported by Science Foundation Ireland (Strategic Research Cluster Grant 07/SRC/B1154). The authors would like to acknowledge the assistance of Eamonn Fitzpatrick of University College Dublin with the artwork.

References

1. A. Rembratt, C. Graugaard-Jensen, T. Senderovitz, J. P. Norgaard and J. C. Djurhuus, *Eur. J. Clin. Pharmacol.*, 2004, **60**, 397.
2. S. G. Yang, *Arch. Pharm. Res.*, 2010, **33**, 1835.
3. T. Lindmark, Y. Kimura and P. Artursson, *J. Pharmacol. Exp. Ther.*, 1998, **284**, 362.
4. A. Yamamoto, T. Okagawa, A. Kotani, T. Uchiyama, T. Shimura, S. Tabata, S. Kondo and S. Muranishi, *J. Pharm. Pharmacol.*, 1997, **49**, 1057.
5. M. Thanou, J. C. Verhoef, J. H. Verheijden and H. E. Junginger, *Pharm Res.*, 2001, **18**, 823.
6. M. Baluom, M. Friedman and A. Rubinstein, *J. Controlled Release*, 2001, **70**, 139.
7. T. T. Kararli, *Biopharm. Drug Dispos.*, 1995, **16**, 351.
8. S. K. Lai, Y. Y. Wang and J. Hanes, *Adv. Drug Delivery Rev.*, 2009, **61**, 158.
9. B. C. Tang, M. Dawson, S. K. Lai, Y. Y. Wang, J. S. Suk, M. Yang, P. Zeitlin, M. P. Boyle, J. Fu and J. Hanes, *Proc. Natl. Acad. Sci. U. S. A.*, 2009, **106**, 19268.
10. W. L. Chiou, H. Y. Jeong, S. M. Chung and T. C. Wu, *Pharm. Res.*, 2000, **17**, 135.
11. S. Maher, R. Kennelly, V. A. Bzik, A. W. Baird, X. Wang, D. Winter and D. J. Brayden, *Eur. J. Pharm. Sci.*, 2009, **38**, 291.
12. S. Maher, T. W. Leonard, J. Jacobsen and D. J. Brayden, *Adv. Drug Delivery Rev.*, 2009, **61**, 1427.
13. T. W. Leonard, J. Lynch, M. J. McKenna and D. J. Brayden, *Expert Opin. Drug Delivery*, 2006, **3**, 685.
14. X. Wang, S. Maher and D. J. Brayden, *Ther. Delivery*, 2010, **1**, 75.
15. H. Iyer, A. Khedkar and M. Verma, *Diabetes, Obes. Metab.*, 2010, **12**, 179.
16. M. L. Petroni, R. P. Jazrawi, P. Pazzi, A. Lanzini, M. Zuin, M. G. Pigozzi, M. Fracchia, G. Galatola, V. Alvisi, K. W. Heaton, M. Podda and T. C. Northfield, *Aliment. Pharmacol. Ther.*, 2001, **15**, 123.
17. K. Whitehead, N. Karr and S. Mitragotri, *Pharm. Res.*, 2008, **25**, 1782.
18. A. Bernkop-Schnurch, H. Zarti and G. F. Walker, *J. Pharmacol. Sci.*, 2001, **90**, 1907.
19. S. M. Ryan, X. Wang, G. Mantovani, C. T. Sayers, D. M. Haddleton and D. J. Brayden, *J. Controlled Release*, 2009, **135**, 51.
20. Y. Hu, D. E. Smith, K. Ma, D. Jappar, W. Thomas and K. M. Hillgren, *Mol. Pharm.*, 2008, **5**, 1122.
21. A.B. Ungell, in *Drug transport mechanisms across the intestinal epithelium*, ed. J. B. Dressman and C. Reppas, Informa USA, 2nd edn, 2009, pp. 21.

22. F. Gabor, C. Fillafer, L. Neutsch, G. Ratzinger, M. Wirth, *Improving oral delivery*, in *Handbook of Experimental Pharmacology*, ed. M Scgafer-Korting, Springer-Verlag, Berlin, 2010, vol. 197, pp. 345.

23. C. A. Lipinski, F. Lombardo, B. W. Dominy and P. J. Feeney, *Adv. Drug Delivery Rev.*, 2001, **46**, 3.

24. J. Rautio, H. Kumpulainen, T. Heimbach, R. Oliyai, D. Oh, T. Jarvinen and J. Savolainen, *Nat. Rev. Drug Discov.*, 2008, **7**, 255.

25. D. J. Brayden and S. Maher, *Ther. Delivery*, 2010, **1**, 5.

26. T. Lindmark, J. D. Soderholm, G. Olaison, G. Alvan, G. Ocklind and P. Artursson, *Pharm. Res.*, 1997, **14**, 930.

27. G. Van den Mooter, *Expert Opin. Drug Delivery*, 2006, **3**, 111.

28. S. Maher, X. Wang, V. Bzik, S. McClean and D. J. Brayden, *Eur. J. Pharm. Sci.*, 2009, **38**, 301.

29. J. F. Woodley, *Crit. Rev. Ther. Drug Carrier Syst.*, 1994, **11**, 61.

30. M. J. O'Neill, L. Bourre, S. Melgar and C. M. O'Driscoll, *Drug Discov. Today*, 2010, **16**, 203.

31. A. Makhlof, M. Werle, Y. Tozuka and H. Takeuchi, *J. Controlled Release*, 2011, **149**, 81.

32. N. Sadrieh and K. M. Tyner, *Ther. Delivery*, 2010, **1**, 83.

33. H. Ghandehari and A. Hatefi, *Adv. Drug Delivery Rev.*, 2010, **62**, 1403.

34. A. Azizi, A. Kumar, F. Diaz-Mitoma and J. Mestecky, *PLoS Pathog.*, 2010, **6**, e1001147.

35. A. des Rieux, V. Fievez, M. Garinot, Y. J. Schneider and V. Preat, *J. Controlled Release*, 2006, **116**, 1.

36. M. A. Clark, M. A. Jepson and B. H. Hirst, *Adv. Drug Delivery Rev.*, 2001, **50**, 81.

37. D. J. Brayden, M. A. Jepson and A. W. Baird, *Drug Discov. Today*, 2005, **10**, 1145.

38. T. Ahmad, M. Wang, M. P. Czech and D. J. Brayden, *Trans. 37th Int. Soc. Controlled Release Society*, Portland, 2010, Abstract 251.

39. M. Aouadi, G. J. Tesz, S. M. Nicoloro, M. Wang, M. Chouinard, E. Soto, G. R. Ostroff and M. P. Czech, *Nature*, 2009, **458**, 1180.

40. E. Gullberg, A. V. Keita, S. Y. Salim, M. Andersson, K. D. Caldwell, J. D. Soderholm and P. Artursson, *J. Pharmacol. Exp. Ther.*, 2006, **319**, 632.

41. W. Kim and E. L. Chaikof, *Adv. Drug. Delivery Rev.*, 2010, **62**, 1468.

42. G. Sharma, D. T. Valenta, Y. Altman, S. Harvey, H. Xie, S. Mitragotri and J. W. Smith, *J. Controlled Release*, 2010, **147**, 408.

43. L. A. Rawlinson, P. J. O'Brien and D. J. Brayden, *J. Controlled Release*, 2010, **146**, 84.

44. N. Kamei, M. Morishita, Y. Kanayama, K. Hasegawa, M. Nishimura, E. Hayashinaka, Y. Wada, Y. Watanabe and K. Takayama, *J. Controlled Release*, 2010, **146**, 16.

45. L. M. Higgins, I. Lambkin, G. Donnelly, D. Byrne, C. Wilson, J. Dee, M. Smith and D. J. O'Mahony, *Pharm. Res.*, 2004, **21**, 695.

46. V. Fievez, L. Plapied, C. Plaideau, D. Legendre, A. des Rieux, V. Pourcelle, H. Freichels, C. Jerome, J. Marchand, V. Preat and Y. J. Schneider, *Int. J. Pharm.*, 2010, **394**, 35.

47. D. Lo, B. Hilbush, S. Mah, D. Brayden, D. Byrne, L. Higgins and D. J. O'Mahony, *Adv. Drug Delivery Rev.*, 2002, **54**, 1213.

48. N. Mishra, S. Tiwari, B. Vaidya, G. P. Agrawal and S. P. Vyas, *J. Drug Targeting*, 2011, **19**, 67.

49. C. Primard, N. Rochereau, E. Luciani, C. Genin, T. Delair, S. Paul and B. Verrier, *Biomaterials*, 2010, **31**, 6060.

50. D. E. Katz, A. J. DeLorimier, M. K. Wolf, E. R. Hall, F. J. Cassels, J. E. van Hamont, R. L. Newcomer, M. A. Davachi, D. N. Taylor and C. E. McQueen, *Vaccine*, 2003, **21**, 341.

51. E. G. Ragnarsson, I. Schoultz, E. Gullberg, A. H. Carlsson, F. Tafazoli, M. Lerm, K. E. Magnusson, J. D. Soderholm and P. Artursson, *Lab. Invest.*, 2008, **88**, 1215.

52. K. D. Fine, C. A. Santa Ana, J. L. Porter and J. S. Fordtran, *Gastroenterology*, 1995, **108**, 983.

53. P. L. Tuma and A. L. Hubbard, *Physiol. Rev.*, 2003, **83**, 871.

54. H. Hillaireau and P. Couvreur, *Cell. Mol. Life Sci.*, 2009, **66**, 2873.

55. M. S. Cartiera, K. M. Johnson, V. Rajendran, M. J. Caplan and W. M. Saltzman, *Biomaterials*, 2009, **30**, 2790.

56. X. Chen, S. Shank, P. B. Davis and A. G. Ziady, *Mol. Ther.*, 2011, **19**, 93.

57. A. Torres-Medina, *Am. J. Vet. Res.*, 1984, **45**, 652.

58. D. S. Goldberg, H. Ghandehari and P. W. Swaan, *Pharm. Res.*, 2010, **27**, 1547.

59. I. Behrens, A. I. Pena, M. J. Alonso and T. Kissel, *Pharm. Res.*, 2002, **19**, 1185.

60. C. Damge, C. P. Reis and P. Maincent, *Expert Opin. Drug Delivery*, 2008, **5**, 45.

61. T. W. Wong, *J. Drug Target*, 2009, **18**, 79.

62. L. Yin, J. Ding, C. He, L. Cui, C. Tang and C. Yin, *Biomaterials*, 2009, **30**, 5691.

63. M. Werle, H. Takeuchi and A. Bernkop-Schnurch, *J. Pharmacol. Sci.*, 2009, **98**, 1643.

64. A. Chaudhury and S. Das, *AAPS PharmSciTech*, 2011, **12**, 10.

65. C. Thompson, W. P. Cheng, P. Gadad, K. Skene, M. Smith, G. Smith, A. McKinnon and R. Knott, *Pharmacol. Res.*, 2011, **28**, 886.

66. W. P. Cheng, C. Thompson, S. M. Ryan, T. Aguirre, L. Tetley and D. J. Brayden, *J. Controlled Release*, 2010, **147**, 289.

67. K. B. Chalasani, G. J. Russell-Jones, A. K. Jain, P. V. Diwan and S. K. Jain, *J. Controlled Release*, 2007, **122**, 141.

68. S. M. Clardy, D. G. Allis, T. J. Fairchild and R. P. Doyle, *Expert Opin. Drug Delivery*, 2011, **8**, 127.

69. C. Bies, C.-M. Lehr and J. F. Woodley, *Adv. Drug Delivery Rev.*, 2004, **56**, 425.

70. I. J. Goldstein, R. C. Hughes, M. Monsigny, T. Osawa and N. Sharon, *Nature*, 1980, **285**, 66.
71. J. D. Smart, *Adv. Drug Delivery Rev.*, 2004, **56**, 481.
72. J. P. Zanetta, S. Kuchler, S. Lehmann, A. Badache, S. Maschke, D. Thomas, P. Dufourcq and G. Vincendon, *Histochem. J.*, 1992, **24**, 791.
73. B. Naisbett and J. Woodley, *Int. J. Pharm.*, 1994, **110**, 127.
74. M. A. Robinson, S. T. Charlton, P. Garnier, X. T. Wang, S. S. Davis, A. C. Perkins, M. Frier, R. Duncan, T. J. Savage, D. A. Wyatt, S. A. Watson and B. G. Davis, *Proc. Natl. Acad. Sci. U. S. A.*, 2004, **101**, 14527.
75. C.-M. Lehr, *J. Controlled Release*, 2000, **65**, 19.
76. F. Gabor, E. Bogner, A. Weissenboeck and M. Wirth, *Adv. Drug Del. Rev.*, 2004, **56**, 459.
77. E. Palomino, *Adv. Drug Delivery Rev.*, 1994, **13**, 311.
78. J. D. Smart, *Adv. Drug Delivery Rev*, 2004, **56**, 481.
79. B. Carreno-Gómez, J. F. Woodley and A. T. Florence, *Int. J. Pharm.*, 1999, **183**, 7.
80. A. Weissenboeck, E. Bogner, M. Wirth and F. Gabor, *Pharm. Res.*, 2004, **21**, 1917.
81. N. Hussain and A. T. Florence, *Pharm. Res.*, 1998, **15**, 153.
82. R. Gref, P. Couvreur, G. Barratt and E. Mysiakine, *Biomaterials*, 2003, **24**, 4529.
83. A. E. Thomsen, M. S. Christensen, M. A. Bagger and B. Steffansen, *Eur. J. Pharm. Sci.*, 2004, **23**, 319.
84. J. P. Shofner, M. A. Phillips and N. A. Peppas, *Macromol. Biosci.*, 2010, **10**, 299.
85. M. Cetin, Y. S. Youn, Y. Capan and K. C. Lee, *AAPS PharmSciTech*, 2008, **9**, 1191.
86. A. Balakrishnan and J. E. Polli, *Mol. Pharm.*, 2006, **3**, 223.
87. S. A. Kularatne and P. S. Low, *Methods Mol. Biol.*, 2010, **624**, 249.
88. S. K. Kang, J. H. Woo, M. K. Kim, S. S. Woo, J. H. Choi, H. G. Lee, N. K. Lee and Y. J. Choi, *J. Biotechnol.*, 2008, **135**, 210.
89. Z. Zhang, H. Lv and J. Zhou, *Pharmazie*, 2009, **64**, 574.
90. O. C. Farokhzad and R. Langer, *ACS Nano*, 2009, **3**, 16.
91. G. Gregoriadis, S. Jain, I. Papaioannou and P. Laing, *Int. J. Pharm.*, 2005, **300**, 125.
92. H. H. Salman, C. Gamazo, M. A. Campanero and J. M. Irache, *J. Nanosci. Nanotechnol.*, 2006, **6**, 3203.
93. R. Ramachandran, W. Paul and C. P. Sharma, *J. Biomed. Mater. Res., Part B*, 2009, **88B**, 41.
94. S. M. Ryan, J. M. Frias, X. Wang, C. T. Sayers, D. M. Haddleton and D. J. Brayden, *J. Controlled Release*, 2011, **149**, 126.
95. P. Hazra, L. Adhikary, N. Dave, A. Khedkar, H. S. Manjunath, R. Anantharaman and H. Iyer, *Biotechnol. Prog.*, 2010, **26**, 1695.
96. H. Li, J. H. Song, J. S. Park and K. Han, *Int. J. Pharm.*, 2003, **258**, 11.

97. A. Vila, A. Sánchez, M. Tobio, P. Calvo and M. J. Alonso, *J. Controlled Release*, 2002, **78**, 15.
98. M. Garcia-Fuentes, C. Prego, D. Torres and M. J. Alonso, *Eur. J. Pharm. Sci.*, 2005, **25**, 133.
99. Y. Kawashima, H. Yamamoto, H. Takeuchi and Y. Kuno, *Pharm. Dev. Technol.*, 2000, **5**, 77.
100. C. Prego, M. Garcia, D. Torres and M. J. Alonso, *J. Controlled Release*, 2005, **101**, 151.
101. S. K. Lai, D. E. O'Hanlon, S. Harrold, S. T. Man, Y.-Y. Wang, R. Cone and J. Hanes, *Proc. Natl. Acad. Sci. U. S. A.*, 2007, **104**, 1482.
102. C. P. Reis, A. J. Ribeiro, S. Houng, F. Veiga and R. J. Neufeld, *Eur. J. Pharm. Sci.*, 2007, **30**, 392.
103. C. P. Reis, A. n. J. Ribeiro, F. Veiga, R. J. Neufeld and C. Damgé, *Drug Delivery*, 2008, **15**, 127.
104. D. Mehta and A. B. Malik, *Physiol. Rev.*, 2006, **86**, 279.
105. R. Langer, *Science*, 2001, **293**, 58.
106. S. C. Chen, D. H. Jones, E. F. Fynan, G. H. Farrar, J. C. Clegg, H. B. Greenberg and J. E. Herrmann, *J. Virol.*, 1998, **72**, 5757.
107. J. Tian, X. Sun, X. Chen, J. Yu, L. Qu and L. Wang, *Int. Immunopharmacol.*, 2008, **8**, 900.
108. L. G. Tillman, R. S. Geary and G. E. Hardee, *J. Pharm. Sci.*, 2008, **97**, 225.
109. K. Tsutsumi, S. K. Li, R. V. Hymas, C. L. Teng, L. G. Tillman, G. E. Hardee, W. I. Higuchi and N. F. Ho, *J. Pharm. Sci.*, 2008, **97**, 350.
110. K. Kohli, S. Chopra, D. Dhar, S. Arora and R. K. Khar, *Drug Discov. Today*, 2010, **15**, 958.
111. K. Henriksen, A. C. Bay-Jensen, C. Christiansen and M. A. Karsdal, *Expert Opin. Biol. Ther.*, 2010, **10**, 1617.
112. S. Maher, D. J. Brayden, L. Feighery and S. McClean, *Crit. Rev. Ther. Drug Carrier Syst.*, 2008, **25**, 117
113. S. Mayor and R. E. Pagano, *Nat. Rev. Mol. Cell Biol.*, 2007, **8**, 603.

CHAPTER 2.2

Nanostructures Overcoming the Intestinal Barrier: Drug Delivery Strategies

KATIE B. RYAN[a], SAM MAHER[b], DAVID J. BRAYDEN[b] AND CAITRIONA M. O'DRISCOLL*[a]

[a] School of Pharmacy, University College Cork, Cavanagh Pharmacy Building, College Road, Cork, Ireland; [b] Conway Institute, University College Dublin, Belfield, Dublin 4, Ireland
*E-mail: caitriona.odriscoll@ucc.ie

2.2.1 Introduction

The oral route is the preferred route for drug administration especially in the case of chronic therapies because of its convenience and wide patient acceptance. There are many challenges associated with orally delivering drugs that have sub-optimal biophysicochemical properties. Two major physiological obstacles include (i) poor stability associated with the relatively harsh environment in the gastrointestinal tract and (ii) inadequate epithelial transport of Bio-pharmaceutical Classification System (BCS) Class III drugs (high solubility and low permeability) and BCS Class IV drugs (low solubility and low permeability), due in part to extrinsic and intrinsic barriers to the systemic circulation, both of which considerably affect oral bioavailability (F).[1] It partly explains the increased research interest in alternative non-injected routes of delivery for biotech cargoes (nasal, pulmonary and transdermal), in an effort to increase F and patient compliance, and to possibly reduce side effects commonly

RSC Drug Discovery Series No. 22
Nanostructured Biomaterials for Overcoming Biological Barriers
Edited by Maria Jose Alonso and Noemi S. Csaba
© The Royal Society of Chemistry 2012
Published by the Royal Society of Chemistry, www.rsc.org

seen with injectables. There are additional biopharmaceutical challenges associated with the physicochemical and biological properties of the therapeutic cargo to be delivered; such as structure, molecular weight, cLogP, solubility, degree of ionisation, as well as gastrointestinal (GI) metabolism, stability in the biological fluids and permeation across the epithelium,[2] all of which have consequences for *F.* Lipinski *et al.* have outlined the interplay between drug properties and permeability in the context of oral absorption and the generation of lead molecules in drug discovery.[3,4] Nanocarriers, including nanovesicles and nanoparticles (NPs), are important technology platforms that may play a role in overcoming these challenges. They are classified as sub-micron sized particles that may be employed to modify the distribution profile of a drug at the cellular and tissue level by entrapment or adsorption of the therapeutic cargo.[5,6] It is generally accepted that nanotechnology is the understanding and control of matter in the loosely defined 1–100 nm dimension range,[7] and is proposed in the context of drug delivery to confer more favourable therapeutic properties in terms of drug safety or desirable pharmacokinetics through modification of drug absorption, distribution, metabolism and elimination.[2] Nanomaterials have unique physicochemical properties including small size, large surface area-to-mass ratio and high reactivity, all of which differentiate them from bulk materials of the same composition.[7] In addition, intestinal epithelial uptake of nanocarriers tends to be more efficient than microparticles (MPs) in animal models.[8,9] Rational selection of materials and formulation of novel technologies to enable (i) efficient adsorption or encapsulation, (ii) stabilisation of the therapeutic cargo, (iii) enhancement of the physicochemical properties of the candidate drug, (iv) targeted and controlled drug release (in a temporal or spatial manner), and (v) enhanced permeability and mucosal transport/uptake at the target site, have been the focus of much research.[10–12]

The application of a 'nano' approach for oral drug delivery is not without significant challenges. NP absorption varies considerably according to study design and lack of correlation between *in vitro* and animal models, and is influenced by the physicochemical properties of the technology employed (*i.e.* hydrophilicity or hydrophobicity, surface charge, particle size, shape and colloidal stability), as well as the dose administered and the species of animal used.[8,13,14] An important consideration of the 'nano' approach is the decreased drug carrying capacity in volume terms that is associated with the decrease in particle size compared to the equivalent micron-sized particles, hence the predominance of literature referring to carriers in the 100–200 nm diameter range. It can be further complicated *in vivo* by the presence of mucus, which acts as a barrier by slowing or inhibiting epithelial uptake of nanocarriers. In some instances, attraction to mucus can be exploited to increase transit time and contact with the mucosal surface, such as those based on mucoadhesive polymers.[14] The potential exists for nanocarriers to target delivery of the cargo regionally in the intestine for local treatments or to a target in the systemic circulation, although to-date, the clinical progression of either approach to promote the oral delivery has been disappointing. If the target is beyond the GI

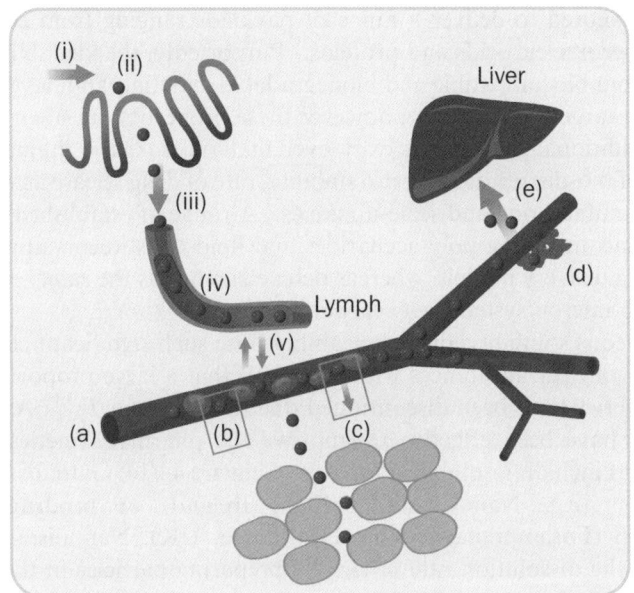

Figure 2.2.1 Diagram depicting the likely potential pathways taken by oral nanocarriers (not to scale). (i) Transport in the GI tract, (ii) absorption from the GI tract by M-cells or enterocytes, (iii) lymphatic uptake, (iv) transport through the lymph vessels and (v) passage into blood vessels; where it can be (a) carried by the blood flow; (b) undergo adhesion to capillary walls; (c) extravasate into tissue, or be taken up at different sites including (d) leaky tumoural tissue and (e) the liver. (Modified from ref. 14 with permission from Elsevier Ltd.)

tract, nanocarrier movement through Peyer's patch M-cells or non-Peyer's patch epithelial cells is only the first step. Subsequent passage through the mesenteric lymph, filtration in lymph nodes, transfer into blood and perhaps extravasation may also be necessary (Figure 2.2.1).[14] Particle uptake and distribution are affected by factors such as nanocarrier type, size, charge, concentration and time.[13,15,16] The complexities of particle uptake and distribution are reviewed in Chapter 2.1. This diversity of requirements not only to encapsulate and stabilise the payload, but also to achieve targeted delivery in therapeutic quantities represents significant obstacles that need to be systematically addressed.

2.2.2 Nanosystems – Materials and Formulation

A variety of nanosystems are being investigated for oral drug delivery including those based on polymers (*e.g.* nanocapsules, nanospheres based on synthetic or natural polymers, polymeric micelles), lipids (*e.g.* liposomes, micro/nano-emulsions, self-(micro)emulsifying drug delivery systems (S(M)EDDS), solid lipid nanoparticles (SLNs)) and inorganic (*e.g.* magnetite, gold and calcium phosphate (CaP)) NPs. Each has particular merits and drawbacks, and some have been tailored to deliver a range of payloads ranging from conventional drugs to large nucleic acids and proteins.[1] Purportedly, the ideal NP should be prepared from biocompatible and biodegradable materials and have good drug loading and targeting efficiency; however these properties are often difficult to achieve. Additionally, concerns exist over material toxicity, immunogenicity, specificity of bio-distribution, *in vivo* stability, rate of drug release as well as cost-effective manufacturing and scale-up issues.[7] A range of established biocompatible materials including polysaccharides and lipid constituents appear safe in various drug delivery models, whereas debate surrounds the safety and toxicity of other sub-micron systems, *e.g.* carbon nanotubes.[17,18]

Poor aqueous solubility and permeability pose such significant challenges in the delivery of drug substances with the result that a large proportion of new drug discoveries have been discontinued due to their poor F.[3,19] A number of approaches have been effective to improve the pharmacokinetics of poorly soluble drugs including solubilisation (*e.g.* Gelucire 44/14; Gattefosse, France), nanonisation (*e.g.* NanoCrystal®; Élan, Ireland) or prodrug chemical modification (Fosamprenavir; ViiV Healthcare, UK). Nanonisation is used to enhance the dissolution rate *in vivo* by preparing particles in the nano size range using bottom-up techniques such as precipitation, or top-down approaches which typically employ wet-milling techniques to achieve particle diminution.[20,21] A number of technologies have been developed including DissoCubes® (SkyePharma, UK), which uses high-pressure piston-gap homogenisation in water, and Nanopure® (Abbott, USA), which employs homogenisation in non-aqueous or water-reduced media.[20,21] NanoCrystal® (Élan, Ireland), which is the best known proprietary technology, uses a bead/pearl mill to reduce the size of the drug with subsequent dispersion in stabilising agents to prevent aggregation. This approach led to several products including aprepitant (Emend®, Merck & Co, USA), fenofibrate (Tricor®, Abbott Laboratories, USA) and sirolimus/rapamycin (Rapamune®, Pfizer (formerly Wyeth), USA). In addition, other drugs substances including 2-methoxyestradiol for the oral treatment of cancer, are at the clinical trial stage.[7] Baxter employed a combination approach in their NANOEDGE™ technology. Following a precipitation step, the drug is subject to high energy annealing, a step intended to prevent growth of precipitated nanocrystals.[21]

2.2.2.1 Polymeric Approaches

Polymer based NPs are one category that has been widely investigated to improve the F of poorly absorbed drugs including proteins, peptides, DNA as well as low molecular weight conventional drugs.[5,22] They have a number of functions in mediating oral drug delivery including protecting the payload, increasing GI epithelial permeability to the drug or drug-carrier complex and/or inhibition of drug efflux transporters.[23,24] Polymeric NPs may be sub-divided into 'nanocapsules', where the polymer surrounds the drug core, and 'nanospheres', in which the drug is distributed throughout the polymer matrix.[25]

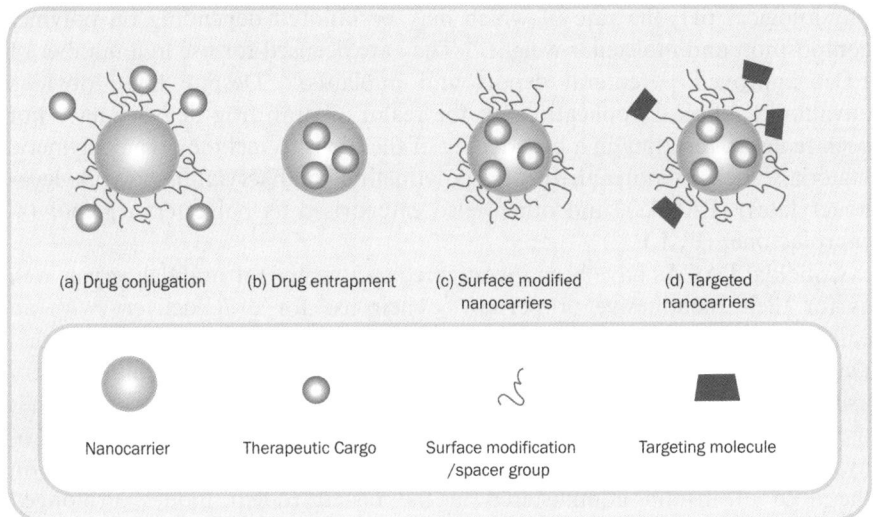

(a) Drug conjugation   (b) Drug entrapment   (c) Surface modified nanocarriers   (d) Targeted nanocarriers

Nanocarrier   Therapeutic Cargo   Surface modification /spacer group   Targeting molecule

Figure 2.2.2   Schematic illustration of the loading of therapeutic cargoes using oral 'generic' nanocarriers and their subsequent modification to create targeted delivery systems.

Alternatively, the drug may be adsorbed or conjugated to the exterior surface (Figure 2.2.2). Physicochemical properties of the polymer that influence selection include molecular weight, crystal property, co-polymer ratio, solubility, biodegradation and compatibility. These can directly influence features of the NP including size, charge, drug loading as well as the rate and extent of drug release. There are a number of advantages associated with polymeric NPs, firstly the versatility associated with the materials employed (natural, semi-synthetic or synthetic); and secondly the variety of methods by which they are processed (spray-drying, emulsion techniques, precipitation, solvent extraction–evaporation).[26] These factors suggest that polymeric NPs are suitable for the delivery of a wide range of cargoes. Thirdly, the architecture of the resultant NP may be modified to enhance targeting and cellular uptake (Figure 2.2.2). Finally, the stability associated with the carrier and encapsulated cargo, especially in comparison to other systems such as liposomes, seems to justify the current popularity of polymers in nanomedicine.[27]

2.2.2.1.1   Synthetic Polymers

Polyesters alone or in combination with other polymers are the most extensively investigated synthetic polymers for NP formation and drug delivery; in particular poly(L-lactic-co-glycolic acid) (PLGA) and poly(L-lactic acid) (PLA).[22] The popularity of these biocompatible polymers with established biomedical and pharmaceutical applications is related to the controlled release of the payload as they degrade *via* bulk erosion at physiological pH; the rate of which may be tailored depending on polymer composition and molecular weight.[28] They are licensed for use in a number of FDA approved parenteral depots and implants.[29] Despite these obvious advantages, clinical applications in the realm of oral drug delivery have not been realised, resulting in a broadening of the focus to include other polymeric materials such as polyanhydrides, poly(methyl methacrylate), poly(akylcyanoacrylates) (PACAs), and others also categorised as polyesters *e.g.* poly(ε-caprolactone) (PCL).

Colloidal PACAs have been investigated as a vector for oral delivery as well as for their bioadhesive properties.[30] Their use for drug delivery with an emphasis on the oral delivery of proteins and peptides has been reviewed.[31] Damgé and co-workers carried out formative studies on oral insulin delivery using PACA NPs. *In vivo* evaluation in a diabetic rat model demonstrated that the nanoencapsulated insulin formulations resulted in a reduction of hyperglycaemia. This effect occurred from day 2 up to day 20, depending on the dose of insulin administered in the fasted rodent model. Prolonged reduction in hyperglycaemia was attributed to preservation of insulin bioactivity through nanoencapsulation, protection of the payload from proteolytic enzymes and enhanced permeation across intestinal mucosa.[32,33]

Additional peptides, including the somatostatin analogue octreotide, have also been encapsulated in oil-containing PACA nanocapsules. *In vivo* studies of the nanocapsule formulation in a rat model demonstrated a significant reduction of prolactin secretion compared with free octreotide and a slight increase in plasma octreotide levels.[34] Polyisobutylcyanoacrylate (PIBCA) nanocapsules have also been investigated as a carrier for calcitonin (CT).[35] Nanoencapsulation increased intestinal stability of CT, but did not facilitate any significant overall enhancement of absorption.

Other polymeric approaches have also been considered to stabilise insulin delivery including the formulation of NPs using PCL in combination with the polycationic and biocompatible acrylic polymer, Eudragit® RS.[36,37] Oral administration of these insulin-NPs in a diabetic rat model decreased fasted glycaemia in a dose-dependent manner. This was attributed to the preservation of biological activity of insulin in the NP and also to the mucoadhesive properties of the Eudragit® RS, both of which facilitated better intestinal uptake of insulin.[37] Other NP strategies to enhance oral insulin delivery have been reviewed in detail by Damgé and colleagues.[38] Table 2.2.1 lists prototype examples of insulin-entrapped NPs for oral delivery on the basis of novel design features, polymers and current status; the table is an update on tables presented in the review by Damgé,[38] and therefore only includes more recent literature.

Polymeric Micelles

Polymeric micelles (PMs) possess a distinctive core-shell architecture, and are formed from amphiphilic copolymers which have a solubility difference

Table 2.2.1   Recent prototype insulin-NP formulations for oral delivery (preclinical).

NP Composition	Description	Stage (reference)
Chitosan-6-mercaptonicotinic acid	Thiolated 200 nm chitosan cationic NPs made by ionic gelation	Reduced plasma glucose in rats better than chitosan NPs.[175]
Zirconium phosphate	Ion exchange process without pre-intercalators; 200–300 nm NPs	*In vitro* release of stable intact insulin, no cytotoxicity.[176]
Solid-lipid NPs	Double emulsion process, 350 nm NPs,	Oral F of up to 4.5% in rats with glyceryl palmitostearate the key lipid.[177]
Alginate /dextran around calcium, bound to poloxamer and chitosan, coated with albumin.	Ionotropic gelation and polyelectrolyte complexation to form 400 nm NPs	13% oral *F* in diabetic rats compared to solutions.[178]
PCL and Eudragit RS® NPs	700 nm NPs, 70% *in vitro* release of aspart-insulin in 24 hours at pH 7.4	Sustained suppression of plasma glucose in diabetic rats following oral delivery.[37]
Quaternary Poly(allyl)amine palmitoyl nanoplexes	Amphiphilic polyelectrolyte 200 nm complexes *via* electrostatic interaction	Effects determined on Caco-2 cell tight junctions.[179]

between the hydrophilic and hydrophobic monomer units (Figure 2.2.3). PMs self-assemble in aqueous environments to form a hydrophobic core (which acts as a reservoir for lipophilic molecules) surrounded by a hydrophilic corona exposed to the aqueous environment. They have been investigated as carrier vehicles for a number of poorly water soluble drugs[39] including griseofulvin,[40] cyclosporin A[41] and efavirenz,[42] presumably by encapsulation in the hydrophobic core. Alternatively, PMs can be used as delivery platforms for water-soluble drugs by electrostatic interactions with the micellar corona.[43] Amphiphilic polymers have been investigated as an alternative to surfactants to prepare micelles because they form stable structures with low critical micelle concentrations (CMC), as well as structures with high kinetic stability that can resist dilution and subsequent loss of entrapped drug cargo in body fluids.[44] Materials commonly investigated to form the hydrophobic core of the copolymer micelles include biodegradable polymers, such as PLA, poly(-glycolic acid) (PGA) and PCL, and non-degradable polymers including poly(propylene oxide) PPO, whereas poly(ethylene oxide) (PEO), poly(acrylic acid) (PAA), poly(aspartic acid), poly(L-lysine) (PLL), poly(ethyleneimine) (PEI), and chitosan derivatives have been investigated as corona-forming segments enabling micellar stability.[43] Linear and branched copolymers based on PEO–PPO are commonly investigated in PM formulations.

PMs may be prepared by direct dissolution methods, emulsion and spray-drying techniques,[24] whilst the drug may be loaded by chemical conjugation,

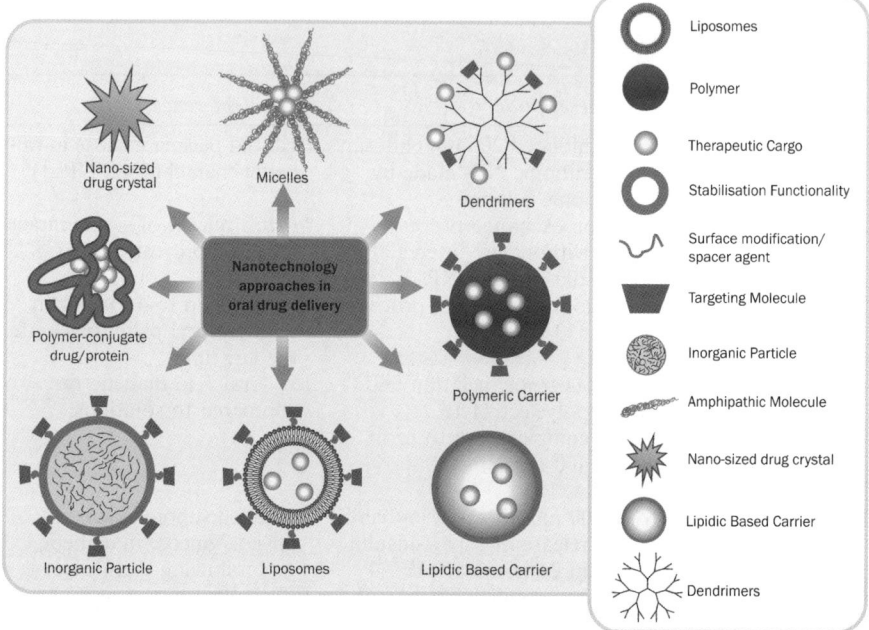

Figure 2.2.3 Schematic representation of selected nanotechnology approaches used in oral drug delivery. The illustration shows a variety of nanocarriers where the drug may be entrapped or bound to the nanocarrier. The nanocarriers may be further modified by conjugation of targeting moieties. (Modified from ref. 6, with permission from Nature Publishing Group).

physical entrapment or poly-ionic complexation.[45] Design, formulation and applications of PMs are broadly reviewed elsewhere,[46,47] and specifically for delivery by the oral route.[24] The application of poly-anionic, (PEO–PPO–PEO)–PAA (Pluronic®-PAA) copolymer micelles for the oral delivery of chemotherapeutic drugs has been reviewed.[43] The mucoadhesive micelles contain dehydrated PPO in the core and a multi-layered corona composed of hydrophilic PEO and partially ionised PAA segments which impart a pH-sensitive character to the NPs. *In vivo* studies using radiolabelled Pluronic-PAA copolymers found that following oral administration these substances were excreted rather than absorbed into the systemic circulation.[43]

Park's group, investigated the capacity of hydrotropic polymer micelles (HPMs) to overcome challenges of poor aqueous stability and encapsulation efficiency.[48] HPMs differ from conventional PMs based on the presence of water soluble molecules within the hydrophobic core (hydrotropic agents). These include nicotinamide and its derivatives including, *N*-picolylnicotinamide (PNA)[48] and *N,N*-diethylnicotinamide (DENA).[39] The solubility of poorly soluble agents is increased through hydrotropic interactions; a mechanism due to interactions including π–π stacking.[49] The same group compared solubilisation of paclitaxel in HPMs, which consisted of a micellar shell-forming poly(ethylene glycol) (PEG) block and a core-forming poly(2-(4-vinylbenzyloxy)-*N*,*N*-diethylnicotinamide) (PVBODENA) block, to that achieved using PMs composed of poly(ethylene glycol)-b-P(D,L-lactide) (PEG-b-PLA). The HPMs were deemed to be a superior carrier due to improved drug loading (37.4% *versus* 27.6%) and enhanced stability on storage. In addition, they were more effective than PEG-PLA micelle formulations in inhibiting the proliferation of human cancer cells.[50] However, *in vitro* release studies showed that complete paclitaxel release from the HPMs took about 24 h. In order to tailor release to coincide with GI transit, pH-sensitive HPMs (PEG-b-PVBODENA-*co*-AA) were prepared by including acrylic acid (AA) to destabilise the HPM structure at elevated pH levels, thereby facilitating an increase in oral *F* by ensuring release was concomitant with GI transit time and pH. *In vitro* release in simulated intestinal fluid (SIF) demonstrated that paclitaxel release was complete within 12 h from HPMs with more than 20% (mol) AA.[39]

A number of studies have been undertaken to investigate the factors affecting PM intestinal absorption. Despite the presence of the hydrophilic outer layer which could potentially limit interaction with the apical membrane, Mathot and co-workers,[51,52] reported that [monomethylether poly(oxyethylene glycol)$_{750}$-poly(caprolactone-co-trimethylene carbonate)] [mmePEG$_{750}$P(CL-co-TMC)] PMs were absorbed by fluid-phase endocytosis and that polymeric unimers diffused passively across the membrane. van Hasselt *et al.*[53] concluded that the *F* of lipophilic cargoes such as vitamin K depends on the presence of bile to extract the drug from the micelles. Despite the promise attributed to PMs as nanocarriers, more studies are warranted to elucidate their interaction, uptake and fate thereafter.

Dendrimers

Dendrimer nanostructures have shown promise as platform technologies not only for drug and gene delivery, but also for imaging and radiotherapy applications. The merits of such an approach are associated with their defined architecture and low polydispersity which arises from their step-wise synthesis. Additionally these uniform structures, which are truly nanosized (2.5 to 10 nm) offer the potential for selective functionalisation at the terminal groups (Figure 2.2.3).[54] These charged nanostructures appear to cross the intestinal epithelium to an appreciable extent, taking advantage of both paracellular and endocytotic transcellular routes (reviewed in ref. 55). Anionic poly(amidoamine) (PAMAM) dendrimers also have lower *in vitro* cytotoxicity compared to cationic or neutral formats. Cytotoxicity is still an issue, however, for high generation number G3 and above; it can be partially offset by conjugation to medium chain fatty acids, which confers the additional benefits of improved solubility for BCS Class II and IV drugs, as well as helping drugs to avoid P-glycoprotein (P-gp) efflux.[56] Most evidence to date has shown that dendrimers can promote solubilisation and epithelial cell permeation of hydrophobic small molecules in NP formats including naproxen,[57] doxorubicin[58] and camptothecin.[59] It is likely that the properties of dendrimers as both solubilizers and inhibitors of P-gp efflux have contributed to these datasets as much as any direct enhancement action on the epithelium. It is important to note that the majority of evidence that dendrimers can act as oral NP carriers comes from Caco-2 monolayers, with some limited data from rat intestinal everted sacs and *in situ* closed intestinal loops, but to our knowledge no evidence has been presented to date from the more challenging hurdle presented by *in situ* single pass gut perfusion methods. Recent evidence from a rat *in situ* small intestine closed loop instillation study, however, suggested that G2 PAMAM dendrimers could facilitate salmon calcitonin (sCT) but not insulin permeability,[60] suggesting a molecular weight cut off for the given vector. In addition, no permeation enhancement for these peptides was seen in closed colonic loops in the same study. With this exception, *in vivo* evidence of their potential for oral delivery of peptides and proteins is lacking.

pH Sensitive Systems

pH responsive polymers change from 'insoluble' to 'soluble' as the pH increases from acidic to neutral or slightly alkaline as they pass along different regions of the GI tract, and they can be effectively used to (i) delay drug release from delivery platforms, (ii) protect an acid or protease labile cargo from degradation or (iii) target delivery to the lower GI tract; all of which depend on the specific properties of the polymer employed. The most commonly used polymers are methacrylic acid copolymers, the Eudragit® family. Eudragit® L and Eudragit® S are soluble at pH $\geq$ 6 and 7, respectively, and are used commercially for enteric coating and for colon targeting applications. Additionally, pH dependent polymers based on hydroxylpropylmethylcellulose acetate phthalate and cellulose acetate phthalate have been developed.[61,62] Peppas and co-workers investigated pH-sensitive hydrogel nanospheres composed of poly(methacrylic acid-grafted poly(ethylene glycol)) P(MAA-g-PEG) for delivery of sCT.[63] The methacrylic acid moiety acts as a potential binding site for calcium (carboxylic acid pendant groups) and imparts pH sensitive characteristics to the nanospheres, whereas the PEG moiety is chosen to stabilise and protect the selected cargo.[64] These pH sensitive hydrogel nanospheres protect proteins from the acidic environment by remaining in a collapsed state at low pH but swell at the higher pH values of the duodenum and jejunum leading to release of the entrapped cargo. *In vitro* studies demonstrated that transport of sCT was enhanced in the presence of the P(MAA-g-PEG) nanospheres and it was suggested that the effect was mediated predominantly by a paracellular transport mechanism.[63] The same group also investigated nanospheres composed of copolymers of acrylic acid (AA) grafted with poly(ethylene glycol) P(AA-g-PEG), in addition to the P(MAA-g-PEG) nanospheres for pH-controlled insulin delivery.[11] The size of these NPs increased from 200 nm at pH 2 to 2 mm at pH $\sim$ 6, which corresponded to pH-controlled insulin release from the NP as the pH rose above the pK_a of the network.

In vivo studies in diabetic rats demonstrated that lower serum glucose levels were achieved with insulin-loaded copolymers compared to free insulin over 6 h. Furthermore, the P(AA-g-PEG) gel nanosphere formulation prepared with a molar monomer feed ratio of AA:EG (2:1) had the greatest effect on the serum glucose levels in the diabetic animal models.[11] The enhanced permeabilizing effect of the P(AA-g-PEG) compared to P(MAA-g-PEG) NPs was attributed to their faster swelling dynamics and calcium binding properties.[64]

It is important to note that the application of pH-sensitive technologies in oral drug delivery is not without difficulty. There are issues of high inter- and intra-subject variability in both fasted and fed states. pH values can also differ by up to 2 pH units in the same region between different individuals[65] and dramatic changes in local pH can exist in inflammatory bowel disease (IBD). For example, a proximal colonic pH value of 2.3 was detected in a patient suffering from ulcerative colitis, where normal pH is close to neutral.[66] Such variability must be considered when designing targeted formulations based on pH-sensitive technologies for medical conditions such as IBD. In addition, other pertinent factors include inter- and intra-individual variation in residence time and luminal fluid composition.

2.2.2.1.2 Biopolymers

Chitosan is a versatile, cationic natural biomaterial used to make NP carriers or to modify the surface properties of existing NPs.[67] In addition to useful properties such as biodegradability and biocompatibility, the polymer has muco/bioadhesive properties[23,68] and has been shown to reversibly alter paracellular permeability *in vitro*. Adherence of chitosan to the intestinal mucosal surface increases residence time, localising the payload at the epithelial surface.[69] In addition, chitosan can protect the drug from degradation and/or dilution which could permit better vectorial flux along an increased concentration gradient.[70] Another distinctive feature includes the potential for chitosan particle constructs to enhance penetration of small molecules and peptides across nasal[71] and intestinal mucosa.[72,73] Evidence from *in vitro* studies suggests that chitosan alters the structural organisation of TJ proteins.[72,74–76] The use of chitosan NPs as vehicles for oral peptide delivery has been previously reported.[77–79] Pan *et al.*[77] noted that chitosan NPs entrapping insulin enhanced intestinal absorption to a greater extent than chitosan solutions ad-mixed with insulin, with the former resulting in a prolonged reduction of plasma glucose levels in a diabetic rat model.[77] Chitosan has also been investigated for the oral delivery of DNA through the formation of polyelectrolyte complexes, as well as for delivery of vaccines and clotting factors.[80–84] Its low cell specificity and transfection efficiency represents a significant challenge that must be addressed before it could be considered for further progression as a gene delivery vector.[84] Additional considerations in the design of formulations for efficient delivery of DNA include the requirements for endosomal escape and nuclear localisation,[85] and this has not yet been optimized for chitosan-based NPs.

Chitosan has also been investigated for site-specific delivery to the colon because it can be digested by the colonic microflora.[86] However, it has been reported that specific colonic targeting requires cross-linking of chitosan to avoid degradation in the small intestine.[87] Nanocomplexes of chitosan with anionic polymers (*e.g.* alginate) or peptides (poly-γ-glutamic acid) have also demonstrated enhanced oral delivery of insulin in animal models.[88,89] The latter system was developed to improve the stability of chitosan NPs over a broader pH range.[89,90] Other examples of mucoadhesive polymers investigated for oral NP formulations include poly(acrylates).[91] There are practical drawbacks associated with mucoadhesives, which include physiological limitations associated with the high turnover rate of the intestinal mucus layer.[92]

Despite the initial promise associated with chitosan NPs for peptide delivery, a number of drawbacks exist including poor solubility at the slightly acid-to-neutral pH values found in the lumen of the small intestine, reducing its capacity to act as a penetration enhancer.[93] In an attempt to address such problems, a number of derivatives have been prepared including *N*-trimethylated chitosan (TMC) and thiolated chitosan derivatives to increase solubility at physiological pH[86,94,95] and improve mucoadhesion,[96] respectively. Thiolated chitosan improved absorption of buserelin and octreotide when delivered in admixtures.[97,98] However, studies that focus on solid forms of the chitosan and its derivatives may be more practical given the relative instability of peptides and proteins in solution.[95] The penetration and mucoadhesive properties of NPs prepared using thiolated derivatives of chitosan were compared to unmodified NPs.[96] *Ex vivo* analysis in a rat model demonstrated that the thiolated chitosan resulted in enhanced mucoadhesion and localisation within the mucus layer. However, in isolated porcine intestinal mucosa, the unmodified chitosan diffused better, leading the authors to suggest that the therapeutic advantage of these modified thiolated NPs is associated with protecting the drug from degradation or entrapment in the mucous gel rather than improved NP translocation.

2.2.2.2   Lipid Systems

Lipid-based formulation is a successful strategy for delivering drug candidates primarily with sub-optimal solubility in order to increase oral F.[99] They have been reported to improve drug absorption by a number of mechanisms, including increasing solubilisation, reducing drug precipitation in aqueous luminal fluids, enhancing membrane permeability, inhibiting efflux transporters and cytochrome P450 enzymes, enhancing chylomicron formation and/or improving lymphatic transport.[99,100] Some of the more commonly investigated lipid formulations for oral delivery include liposomes, SLNs and microemulsions (*e.g.* (S(M)EDDS)). Liposomes have been utilised successfully in a number of parenteral formulations, however this success has not been translated to the oral route primarily because of concerns over their stability under the harsh mechanical and biochemical processing within the intestinal lumen.[27] This limitation is due in part to the susceptibility of the phospholipid membrane to degradation by bile salts and phospholipases, resulting in complex deformation and exposure of the cargo to the intestinal contents.[29,101] Attempts to increase the liposomal stability, and their feasibility as oral peptide delivery vehicles, have focused on modifying vesicle components,[102] coating the liposome vesicle[103] or preparing cross-linked stable networks in the liposomal membrane ("polymerised" liposomes).[101] The latter approach was attempted in the Orasome[TM] delivery system (Endorex Corporation, USA) to promote the oral *F* of a number of hormone polypeptides, including insulin and human growth hormone, by targeting their delivery to the lower intestine in a stable format.[104] This formulation was difficult to reproduce, epithelial cell uptake levels were low and controlling subsequent release was problematic. The use of liposomes in drug delivery has been reviewed in depth elsewhere.[105]

SEDDS are physically stable, isotropic mixtures of oil, surfactant, co-surfactant and solubilized drug substance that are used to improve oral drug delivery in soft and hard gelatin capsules.[106] These have been investigated for their potential to enhance drug solubilisation, dissolution and to promote intestinal absorption.[107] Following dilution in aqueous fluids in the GI tract, these systems spontaneously emulsify to form micro- and/or nano-emulsions,[108] generating a high surface area of interaction between the formulation and GI fluids.[109] The characteristics of the resultant emulsions are dependent on excipient choice and the relative composition of formulation constituents.[106] Excipients and formulation strategies employed in lipid-based formulations has been comprehensively reviewed elsewhere.[110] The Lipid Formulation Classification System (LFCS) categorises lipid formulations according to those with common constituents.[111,112] This framework has been proposed to identify how factors such as formulation are likely to affect performance *in vivo*; however performance criteria need to be established to facilitate *in vitro–in vivo* correlation studies.[110,111] SEDDS and SMEDDS are categorised as Type II and Type III, respectively, according to the LFCS, with particle size above 100 nm (SEDDS) and below 100 nm (SMEDDS) being a distinguishing feature. The latter contains higher concentrations of hydrophilic surfactants and co-solvents. These drug delivery platforms have been successful clinically, *e.g.* cyclosporine A (Sandimmun Neoral®; Novartis, Switzerland). Cyclosporine A was solubilised in the nanosized, oily disperse phase (labrafil M 2125 CS (polyoxyethylated glycolized glycerides; Gattefosse; France).[14] Its increased absorption *in vivo* compared to the original Sandimmun® has been attributed to the improved dispersion characteristics.[113] Water-in-oil microemulsion technology for oral insulin delivery is also being investigated in clinical trials: Macrulin[TM] (Provalis PLC, UK), which contains insulin in the aqueous phase and both cholesterol and lecithin in the oil phase.[104] Porter and colleagues[109] have reviewed the effect of lipid formulations (lipid solutions, suspensions and self-emulsifying formulations) on the oral F of selected drug candidates.[109]

Many formulations have been examined to promote better oral delivery of the HIV-protease inhibitor, saquinavir. It has been proposed that novel nano-emulsions based on edible oils rich in polyunsaturated fatty acids (flax-seed, safflower oil) may be used to enhance the oral F and brain delivery of saquinavir.[114] We have previously investigated targeting saquinavir to the lymphatics in orally-administered lipid formulations.[115] It seems that some highly lipophilic drugs can associate with lymph lipoproteins in enterocytes, facilitating transport/uptake into the intestinal lymph nodes prior to transport to the systemic circulation. This would result in a number of benefits including a reduction of the first-pass effect for saquinavir[115] and/or increased lymphatic exposure,[116] which would be particularly useful in targeting diseases such as HIV known to spread *via* the lymphatic system.[117]

Lipid nanocapsules (LNCs) have been shown to promote the oral F of paclitaxel.[118] These LNCs are composed of medium-chain triglycerides and surfactants, which facilitate drug delivery by solubilisation and encapsulation of the drug in the liquid triglyceride core. Another primary constituent of the LNC includes Solutol® HS15 (a mixture of free polyethylene glycol 660 and polyethylene glycol 660 hydroxystearate) which has been shown to inhibit P-gp.[119] *In vivo* evaluation of paclitaxel-LNC in a rat model demonstrated an approximate three-fold increase in the plasma concentration compared to a commercially available formulation that contains Cremophor EL (Taxol®; Bristol-Myers Squibb, USA). A similar increase was observed for the group treated with Taxol® in conjunction with the P-gp inhibitor, verapamil.[118] In other studies, LNCs improved paclitaxel permeability across Caco-2 mono-layers by 3.5-fold compared to paclitaxel alone, an effect that was predominantly mediated by clathrin-dependent and caveolae-dependent endocytosis.[120]

A number of studies have been conducted using lipid delivery systems containing solid cores; examples include lipospheres and SLN. The use of SLNs for protein and peptide delivery has been reviewed.[121] In one case, Alonso and co-workers investigated SLNs composed of the solid triglyceride tripalmitin for macromolecular delivery. The same group also investigated modification of the inner structure of NPs made of tripalmitin, lecithin, and poly(ethylene glycol) (PEG)-stearate by co-encapsulating the liquid lipid Miglyol® 812 to facilitate better diffusion of encapsulated protein and peptide cargoes from the matrix.[122,123] For sCT, addition of Miglyol® oil to the core did not seem to affect encapsulation efficiency or the release profile due to the strong electrostatic interaction between the positively charged peptide and the negatively charged lipids.[123]

2.2.3 Surface Modification and Targeting

Debate exists surrounding the ideal surface properties of NPs that can facilitate intestinal epithelial transport. There is some consensus that physicochemical properties of NPs including size, hydrophobicity and zeta potential are important design features in the context of NP stability and mucosal uptake.[10] To achieve the overall aim of efficacious delivery, two primary strategies have been assessed: (i) modifying the physicochemical properties of the drugs and/or NP surfaces, and/or (ii) by coupling a ligand moiety on the NP surface for effective targeting to an area of recognition (Figures 2.2.2 and 2.2.3).[10] Surface engineering of injectable nanocarriers using hydrophilic polymers, most notably PEG, has been effective in creating "stealth" properties in technologies delivered by the parenteral route, as a means of reducing opsonisation and the premature elimination of the NPs by the reticuloendothelial system (RES), thereby promoting circulation and increasing serum half-life.[124,125] In the context of oral drug delivery, PEG coating has been examined for its ability to stabilise nanocarriers in physiological fluids,[126,127] to improve mucosal uptake,[128] to promote better diffusion through mucus,[129,130] and as a component of drug nanocarriers.[126,131] Naloxol is an example of a small drug molecule which has been PEGylated (NKTR-118), and is currently undergoing clinical trials for the treatment of opioid-induced constipation.[7] Surface modification of various nanocarriers including liposomes,[103] polymeric NPs[127] and SLN using hydrophilic polymers, such as PEG and chitosan, have been reported.[132,133] A novel approach has been investigated based on BioOral™ (BioSante, USA) technology that utilises CaP NPs to protect proteins including insulin from acidic degradation. Interestingly, modification of a proprietary formulation of CaP-PEG-insulin by encasing it in the protein, casein (CAPIC) has been suggested to further protect the protein from degradation due to the non-degradability of casein in acid.[104]

2.2.3.1 Targeting Ligands

Nanotechnologies for oral drug delivery have been effective in encapsulating and stabilising various payloads. The main bottleneck for many micro- and nano-systems is poor intestinal permeation, which is the significant obstacle in achieving efficacy by the oral route.[29] For those carriers that require epithelial cell uptake, they must persist for sufficient time in the intestinal lumen to come into contact with the apical surface in order to initiate uptake by transcytosis.[10] Strategies to improve residence time and epithelial cell uptake have been investigated. These include optimising particle size and surface properties and the use of mucoadhesive systems to increase GI residence as previously outlined; however, the effectiveness of mucoadhesive polymers can be impeded by (i) lack of specificity at the target site, (ii) high rate of mucus turnover and (iii) secretory processes.[134] As an alternative approach, "active" targeting of nanocarrier systems through functionalisation with epithelial cell receptor ligands, such as lectins and peptide transporters (*e.g.* human peptide transporter-1 (hPEPT1)) can potentially increase the specificity and efficacy of delivery. Many targeting molecules have been investigated including: bacterial surface proteins, *e.g.* invasin, which targets β-integrin,[135] the M-cell-homing peptide, CKS9, which was selected by phage display for targeting of chitosan NPs,[136] the arginine–glycine–aspartic acid (RGD) peptide for targeting NPs to the apical surface of M-cells *via* β integrin sub-types[137] and the c-terminal targeting peptide, *Clostridium perfringens enterotoxin* (the terminal 30 amino acids of CPE) (CPE30), for targeting PLGA NPs to the claudin-4 protein which is associated with TJs.[138]

Like many other oral formulation approaches, targeting nanocarriers to epithelia *in vivo* is difficult and they have additional hurdles to overcome compared to NPs that are passively absorbed. Formulation considerations in their design include mechanism of NP attachment, density and spatial arrangement on the surface, and maintenance of specific receptor binding activity. Physiological considerations include adequate receptor expression in the appropriate intestinal region in the average human population (see Chapter 2.1). With regard to putative ligands such as lectins, there are also concerns over species differences in expression, toxicity, immunogenicity and stability.[139] Given the high number of variables requiring optimisation and the practical limitations of low-throughput approaches, novel high-throughput screening methods have been investigated to increase the efficiency of the process.[125] Finally, it is also important to prove that ligand-grafted nanocarriers can actually penetrate the mucus layer to interact with the target.

2.2.4   Novel Technologies for Nucleic Acid Delivery

There are few studies describing delivery of siRNA by the oral route. Most focus on application of nanotechnology to deliver drugs locally, such as those developed to alleviate the symptoms of IBDs. Strategies used to improve non-viral delivery of gene therapeutics have recently been reviewed.[140] Also outlined in that review are preclinical models that are currently used to evaluate the stability and/or efficacy of nucleic acid therapies. Nanoparticles in-microsphere oral systems (NiMOS) have been designed to improve gene delivery and transfection in the GI tract.[141] These NiMOS delivery platforms were formulated using a double-emulsion like technique where the therapeutic agents are encapsulated in type B gelatin NPs (~200 nm) through controlled precipitation, prior to formulation as microspheres composed of a PCL matrix. This approach has the potential to enhance entrapment efficiency using non-immunogenic polymers (gelatin) to nanoencapsulate the therapeutic. PCL is resistant to acidic pH and is preferentially digested by lipases, which are present in the small and large intestine. PCL was incorporated to increase stability in the physiological fluids of the upper GI tract.[12,141,142] NiMOS microparticles with a diameter of less than 10 μm may in theory be suitable for uptake by the Peyer's patches, whereas particles > 10 μm could be suitable for controlled release in the GI tract.[12] The transfection potential of the delivery platform was confirmed qualitatively by encapsulating reporter plasmids expressing β-galactosidase or enhanced green fluorescent protein (EGFP-N1). Detection of EFGP-N1 expression (which is not intrinsically present in rats) in the small and large intestine five days post-administration was attributed to the localisation of transfected siRNA in the intestinal enterocytes. This was in contrast to gelatin NPs and naked DNA which did not result in transgene expression.[141] NiMOSs have also been investigated for the treatment of IBD using the dextran sulphate sodium (DSS)-induced colitis mouse model.[143] Orally-administered NiMOS loaded with RNA interference (i) down-regulated colonic levels of tumor necrosis factor-α (TNF-α), and decreased expression of other pro-inflammatory markers: interleukin-1β (IL-1β), interferon-γ (IFN-γ) and monocyte chemoattractant protein-1 (MCP-1).[143]

Another noteworthy recent advance includes the development of micron size β-1,3-D-glucan particles encapsulating siRNA particles (GeRPs) which have achieved efficient gene-silencing in mouse macrophages.[144] GeRPs were produced by purifying porous β-1,3-D-glucan particles (2–4 μm) from baker's yeast by a series of alkaline and solvent extractions. The anionic siRNA is complexed and neutralised *via* electrostatic interactions with cationic polyethyleneimine (PEI) layers in the core. After oral administration in a mouse model, it was postulated that these GeRPs were ultimately phagocytosed by macrophages and dendritic cells in the gut-associated lymphoid tissue (GALT), with subsequent release of siRNA through the porous GeRP shell due to the acidic pH in phagosomes. See Chapter 2.1 for discussion of the potential role of M-cell uptake as part of the mechanism. *In vivo* analysis in a mouse model demonstrated that silencing of the mitogen-activated protein kinase kinase kinase kinase 4 (Map4k4) in macrophages protected the rodents from the effects of lipopolysaccharide-induced lethality through inhibition of TNF-α and IL1-β production.[144] Accepting the perils of predicting oral absorption in man on the basis of rodent studies, it could mean that the dose levels of siRNA that might be considered for oral human studies may eventually be within reasonable parameters, however further studies are warranted.

Specific pathophysiological features of IBDs can be exploited to prepare a targeted delivery platform. This is highlighted in the design and formulation of thioketal NPs (TKNs) that selectively release their siRNA cargo in response to abnormally high levels of reactive oxygen species (ROS), which are found at the site of intestinal inflammation. TKNs are formulated from poly-(1,4-phenyleneacetone dimethylene thioketal), a polymer composed of ROS-sensitive thioketal linkages that are stable to acid and protease degradation.[145] *In vivo* evaluation of TKNs (~600 nm) loaded with siRNA against TNF-α in the mouse DSS model revealed a reduction in TNF-α mRNA levels in the colon and resulted in a protective effect from ulcerative colitis.[146]

2.2.5 Intestinal Site-Specific Delivery Using Nanosystems

Targeting the colon has been reasonably effective in the treatment of local intestinal diseases, especially IBD, but not yet for systemic delivery. Factors influencing both local and systemic delivery include physiological barriers (*e.g.* lower surface area, variable fluid volume and viscosity), transit time (colonic residence has been reported to vary from one hour to several days) and the remote location of the colon.[147,148] Challenges in the design and formulation of appropriate delivery platforms that efficiently deliver sufficient quantities of the payload to the target need to be addressed. Pertuit *et al.*[149] examined covalent binding of 5-aminosalicylic (5-ASA) to PCL prior to NP formulation in an attempt to overcome the problem of 5-ASA leakage in the upper GI tract. This approach caused drug retention in the matrix and slower drug release *in vitro* and alleviated the symptoms of experimental colitis in mice. It was, however, noted that there was limited scope for attachment to the PCL because there is only one carboxyl group per molecule accessible for binding. The colon is a valid target for both local and systemic drug delivery due to its distinct characteristics, such as lower levels of efflux transporters and of peptidases, and with microflora-based enzmes to be potentially exploited.[150,151] Colonic targeting is particularly desirable in the case of peptides and proteins due to significantly lower metabolic levels compared with the small intestine,[152] and where delayed release could be advantageous[153] as well as having good absorption efficiency for some drugs.[148] For example, Tubic-Grozdanis *et al.*[154] showed that the *F* of simvastatin was three times higher following delivery to the distal gut compared to delivery in the upper GI tract.[154]

To date, GI targeting systems have focused on region-specific luminal pH, time-dependent formulations or colonic enzymatic activity for activation of drug release in the colon. In inflammation, novel strategies based on exploiting pathophysiological changes in the tissue or an altered cellular immune response in inflamed tissues have been investigated. An increased presence of neutrophils, natural killer cells, mast cells and regulatory T-cells play an important role in IBD.[155,156] It has been established that particle uptake into colon-associated immune cells[157] or across the leaky intestine allows the selective accumulation in the targeted area, which has the benefit of enhancing efficacy while reducing adverse effects in non-target tissue. Higher particulate uptake in colitis tissue has been attributed to increased mucus levels and enhanced permeability.[149] Lamprecht *et al.*[158] investigated particle uptake in the trinitrobenzenesulfonic acid rat model of colitis and saw increased particle adherence at the thicker mucus layer and in ulcerated regions. More efficient deposition occurred in the case of smaller NPs (0.1 μm) compared to larger MPs (1–10 μm).[158] The same group incorporated rolipram in PLGA NPs to target the inflamed intestinal tissues in the same rat model. NPs were equally as effective as solutions of rolipram in mitigating colitis, however subsequent evaluation over a 5 day drug-free period revealed that relapse occurred in the solution-treated group along with a higher adverse-effect index compared to the NP formulation. The continued reduction in inflammation in the group treated with NPs was attributed to increased accumulation in inflamed tissues compared to rolipram solution.[159]

The lectin, wheat germ agglutinin (WGA), has been investigated for colon targeting due to its high binding efficiency to human colonic carcinomas and to human colonocytes.[160] Wang *et al.*[161] investigated the attachment of WGA to paclitaxel-loaded PLGA NPs for the potential treatment of colon cancer. Attachment of WGA resulted in greater anti-proliferative activity against human colonic Caco-2 and HT-29 cells when compared to the unmodified PLGA NPs. This was attributed to the over-expression of the WGA receptor target, *N*-acetyl-D-glucosamine-containing glycoprotein receptor.[161]

The large quantity and diversity of bacteria present in the colon can be exploited to release drugs in this region.[162–164] The concentrations of bacteria found at the mucosal surface, in the mucus gel layer and in the intestinal lumen[165] are exponentially higher than those found in the upper GI tract.[147] The colonic flora plays an important role in digestion of a range of indigestible xenobiotics including polysaccharides and drug substances.[165] Their metabolic and fermentative capability has already been exploited for targeted colonic drug delivery with a number of marketed small molecule formulations to treat colitic inflammation containing the following prodrugs; sulfasalazine, olsalazine and mesalazine. Activation of the inactive prodrugs is mediated by azoreductases produced by colonic microbiota, which cleave azo-bonds.[166] Polymeric excipients containing azo-bonds have also been investigated as matrices and coatings to protect drugs against absorption or degradation in the upper GI tract,[61] examples include azopolymers prepared with hydroxyethylmethacrylate (HEMA) cross-linked with styrene[167] or methylmethacrylate (MMA).[168] Drugs have also been conjugated to polymeric carriers including *N*-(2-hydroxypropyl)methacrylamide (HPMA) polymer by means of azo-bonds for targeted colonic delivery.[169,170] The degree of cross-linking and the hydrophilicity of azopolymers are important design parameters.[171] Combinations of the use of aromatic azo-bond functionalities in pH sensitive hydrogels,[172] or hydrogels systems which release mucoadhesive structures after degradation of azo networks have been investigated.[173]

Alternative materials to those activated by azoreductases including cyclodextrins and dextrans which are relatively stable in the upper intestine, but subject to enzymatic hydrolysis in the lower intestine are also of interest.[166] Basit and co-workers proposed that polysaccharide carriers appear to be promising in colon delivery because of the high level of polysaccharidase producing bacteria,[163] which can selectively digest these materials. Amylose, a resistant starch, has proven to be more popular than guar gum and pectin and is the only form in current clinical development.[87] These materials have the additional advantage of being cheap, non-toxic and biodegradable.[165] A colon-specific coating based on the starch amylose in combination with ethylcellulose (EC) (COLAL®), has been applied to tablets, granules and pellets to achieve colon targeting. EC is used to control the swelling of the amylose and forms a water impermeable barrier that is resistant to colonic bacterial digestion whereas digestion of amylose by the colonic bacteria creates pores in the EC film, thus enabling sustained drug release. One pellet preparation containing the corticosteroid, prednisolone sodium metasulphobenzoate, encased in the COLAL® coat (COLAL-PRED)®, was evaluated in Phase III clinical trials for the treatment of ulcerative colitis but did not demonstrate improved efficacy relative to oral prednisolone; although advantages in terms of toxicity profiles were suggested.[165]

McConnell *et al.*[162] compared the relative effectiveness of two physiological approaches to promote colon delivery, intestinal pH and microflora metabolism, in order to achieve systemic delivery of theophylline in human subjects. In that study, colon targeting was enhanced to a greater extent with the bacterial COLAL® platform compared to Eudragit® S coating. The latter resulted in premature drug release in the small intestine, shorter T_{max} and greater variability. This was attributed to the higher pH in the distal ileum compared to the colon which resulted in dissolution of the Eudragit® S coating in the small intestine. Studies in human subjects found that fasted and fed states had no impact on microflora induced drug release using a polysaccharide-based colon-targeting system.[147] Despite the potential of colon targeting, there is currently insufficient evidence to suggest that drugs can be better delivered to systemic targets by this route.[165]

2.2.6 Conclusions

Oral F is dependent on the biological and physicochemical properties of the drug carrier and the physiology of the delivery route, all of which have influenced the development of a number of distinctive oral NP formulations. Nanocarrier design, including the choice of material, and properties such as size, surface character including hydrophobicity and charge have important implications for drug loading efficiency, rate of particle degradation, drug release rate, stability, toxicity, efficacy, uptake, biodistribution and pharmacokinetics. Despite extensive research in the area of oral drug delivery, debate still exists regarding the optimal features required to maximise therapeutic outcomes. Many of the same nanotechnology platforms investigated for parenteral, topical and pulmonary delivery outlined in other chapters of this volume, including polymeric and lipid systems, have also been investigated for oral drug delivery. What differentiates their use in the context of oral delivery is the formulation design to achieve stability and region- or disease-specific targeting. Formulation strategies have been employed to trigger drug release in the GI tract regions in response to changes in pH and enzymatic activity, whereas time-controlled systems have been designed to exploit transit time in the GI tract. However, these systems in isolation have suffered from poor specificity and this has stimulated interest in combination approaches (*e.g.* pH sensitive technologies in combination with enzymatically triggered systems). *In vivo* studies have demonstrated that novel approaches for siRNA delivery have shown potential in treating local inflammation. Notwithstanding the promise offered by nanocarriers in oral drug delivery, considerable challenges remain that include the need to produce more effective, robust and biocompatible oral

NP platform technologies, some of which are quite sophisticated and complex, in a scalable and cost-effective manner. Nanocarrier uptake by cells lining the intestinal mucosa has been demonstrated by a number of studies; however, further studies are warranted to demonstrate unequivocally that this uptake translates to levels that are significant enough to reproducibly achieve therapeutic efficacy for the selected macromolecule. Additionally, research needs to focus on the development of new biomaterials that are specifically designed to facilitate oral delivery, and also on targeting ligands that recognise well-expressed region-specific receptors. The long-term safety of these and other delivery platforms must be evaluated in preclinical animal models that accurately predict efficacy and potential toxicity, assuming that is possible.

It is clear that increasing knowledge in the area of disease pathology will highlight potential opportunities for targeted drug delivery and enhanced therapeutic outcomes. The use of ROS (a signature of local inflammation) to target siRNA in the treatment of IBD is a salient example of such an approach. In addition, nano approaches will undoubtedly play an important role in the development of personalised medicines with the evolution of pharmacogenomics, and also in the field of theragnostics, which offer the opportunity to dramatically improve the diagnosis and treatment of both acute and chronic conditions and patient outcomes.

References

1. M. José Alonso, *Biomed. Pharmacother.*, 2004, **58**, 168.
2. H. Devalapally, A. Chakilam and M. M. Amiji, *J. Pharm. Sci.*, 2007, **96**, 2547.
3. C. A. Lipinski, *J. Pharmacol. Toxicol. Methods*, 2000, **44**, 235.
4. C. A. Lipinski, F. Lombardo, B. W. Dominy and P. J. Feeney, *Adv. Drug Del. Rev.*, 1997, **23**, 3.
5. P. Couvreur, C. Dubernet and F. Puisieux, *Eur. J. Pharm. Biopharm.*, 1995, **41**, 2.
6. D. Peer, J. M. Karp, S. Hong, O. C. Farokhzad, R. Margalit and R. Langer, *Nat. Nano*, 2007, **2**, 751.
7. L. Zhang, F. X. Gu, J. M. Chan, A. Z. Wang, R. S. Langer and O. C. Farokhzad, *Clin Pharmacol. Ther.*, 2007, **83**, 761.
8. M. P. Desai, V. Labhasetwar, G. L. Amidon and R. J. Levy, *Pharm. Res.*, 1996, **13**, 1838.
9. S. McClean, E. Prosser, E. Meehan, D. O'Malley, N. Clarke, Z. Ramtoola and D. Brayden, *Eur. J. Pharm. Sci.*, 1998, **6**, 153.
10. A. des Rieux, V. Fievez, M. Garinot, Y.-J. Schneider and V. Préat, *J. Controlled Release*, 2006, **116**, 1.
11. A. C. Foss, T. Goto, M. Morishita and N. A. Peppas, *Eur. J. Pharm. Biopharm.*, 2004, **57**, 163.
12. M. D. Bhavsar, S. B. Tiwari and M. M. Amiji, *J. Controlled Release*, 2006, **110**, 422.

13. M. P. Desai, V. Labhasetwar, E. Walter, R. J. Levy and G. L. Amidon, *Pharm. Res.*, 1997, **14**, 1568.
14. A. T. Florence, *Drug Discovery Today: Technol.*, 2005, **2**, 75.
15. D. A. Norris, N. Puri and P. J. Sinko, *Adv. Drug Del. Rev.*, 1998, **34**, 135.
16. M. S. Cartiera, K. M. Johnson, V. Rajendran, M. J. Caplan and W. M. Saltzman, *Biomaterials*, 2009, **30**, 2790.
17. W. Cheung, F. Pontoriero, O. Taratula, A. M. Chen and H. He, *Adv. Drug Delivery Rev.*, 2010, **62**, 633.
18. C. A. Poland, R. Duffin, I. Kinloch, A. Maynard, W. A. H. Wallace, A. Seaton, V. Stone, S. Brown, W. MacNee and K. Donaldson, *Nat. Nano*, 2008, **3**, 423.
19. S. Stegemann, F. Leveiller, D. Franchi, H. de Jong and H. Lindén, *Eur. J. Pharm. Sci.*, 2007, **31**, 249.
20. C. M. Keck and R. H. Müller, *Eur. J. Pharm. Biopharm.*, 2006, **62**, 3.
21. R. Shegokar and R. H. Müller, *Int. J. Pharm.*, 2010, **399**, 129.
22. J. Panyam and V. Labhasetwar, *Adv. Drug Delivery Rev.*, 2003, **55**, 329.
23. C. Prego, D. Torres, E. Fernandez-Megia, R. Novoa-Carballal, E. Quiñoá and M. J. Alonso, *J. Controlled Release*, 2006, **111**, 299.
24. G. Gaucher, P. Satturwar, M.-C. Jones, A. Furtos and J.-C. Leroux, *Eur. J. Pharm. Biopharm.*, 2010, **76**, 147.
25. C. Pinto Reis, R. J. Neufeld, A. J. Ribeiro and F. Veiga, *Nanomed. Nanotechnol. Biol. Med.*, 2006, **2**, 53.
26. K. Yin Win and S.-S. Feng, *Biomaterials*, 2005, **26**, 2713.
27. K. S. Soppimath, T. M. Aminabhavi, A. R. Kulkarni and W. E. Rudzinski, *J. Controlled Release*, 2001, **70**, 1.
28. J. Wang, B. M. Wang and S. P. Schwendeman, *J. Controlled Release*, 2002, **82**, 289.
29. H. Chen and R. Langer, *Adv. Drug Delivery Rev.*, 1998, **34**, 339.
30. C. Vauthier, C. Dubernet, E. Fattal, H. Pinto-Alphandary and P. Couvreur, *Adv. Drug Delivery Rev.*, 2003, **55**, 519.
31. A. Graf, A. McDowell and T. Rades, *Expert Opin. Drug Deliv.*, 2009, **6**, 371.
32. C. Damgé, C. Michel, M. Aprahamian and P. Couvreur, *Diabetes*, 1988, **37**, 246.
33. C. Damgé, C. Michel, M. Aprahamian, P. Couvreur and J. P. Devissaguet, *J. Controlled Release*, 1990, **13**, 233.
34. C. Damgé, J. Vonderscher, P. Marbach and M. Pinget, *J. Pharm. Pharmacol.*, 1997, **49**, 949.
35. P. J. Lowe and C. S. Temple, *J. Pharm. Pharmacol.*, 1994, **46**, 547.
36. C. Damgé, P. Maincent and N. Ubrich, *J. Controlled Release*, 2007, **117**, 163.
37. C. Damgé, M. Socha, N. Ubrich and P. Maincent, *J. Pharm. Sci.*, 2010, **99**, 879.
38. C. Damgé, C. P. Reis and P. Maincent, *Expert Opin. Drug Deliv.*, 2008, **5**, 45.

39. S. Kim, J. Y. Kim, K. M. Huh, G. Acharya and K. Park, *J. Controlled Release*, 2008, **132**, 222.
40. E. Pierri and K. Avgoustakis, *J. Biomed. Mater. Res. Part A*, 2005, **75A**, 639.
41. M. F. Francis, M. Cristea, Y. Yang and F. M. Winnik, *Pharm. Res.*, 2005, **22**, 209.
42. D. A. Chiappetta, C. Hocht, C. Taira and A. Sosnik, *Biomaterials*, 2011, **32**, 2379.
43. L. Bromberg, *J. Controlled Release*, 2008, **128**, 99.
44. M. Yokoyama, *Crit. Rev. Ther. Drug Carrier Syst.*, 1992, **9**, 213.
45. E. Batrakova, T. Bronich, J. Vetro and A. Kabanov, in *Nanoparticulates as drug carriers*, ed. V. P. Torchilin, Imperial College Press, London, 2006.
46. V. P. Torchilin, *Cell. Mol. Life Sci.*, 2004, **61**, 2549.
47. V. Torchilin, *Pharm. Res.*, 2007, **24**, 2333.
48. K. M. Huh, S. C. Lee, Y. W. Cho, J. Lee, J. H. Jeong and K. Park, *J. Controlled Release*, 2005, **101**, 59.
49. V. S. Trubetskoy, *Adv. Drug Delivery Rev.*, 1999, **37**, 81.
50. S. C. Lee, K. M. Huh, J. Lee, Y. W. Cho, R. E. Galinsky and K. Park, *Biomacromolecules*, 2007, **8**, 202.
51. F. Mathot, L. van Beijsterveldt, V. Préat, M. Brewster and A. Ariën, *J. Controlled Release*, 2006, **111**, 47.
52. F. Mathot, A. des Rieux, A. Ari n, Y. J. Schneider, M. Brewster and V. Preat, *J. Controlled Release*, 2007, **124**, 134.
53. P. M. van Hasselt, G. E. P. J. Janssens, C. J. F. Rijcken and C. F. van Nostrum, *J. Controlled Release*, 2008, **132**, e29.
54. A. T. Florence, *Adv. Drug Delivery Rev.*, 2005, **57**, 2104.
55. V. K. Yellepeddi, A. Kumar and S. Palakurthi, *Expert Opin. Drug Deliv.*, 2009, **6**, 835.
56. A. Saovapakhiran, A. D'Emanuele, D. Attwood and J. Penny, *Bioconj. Chem.*, 2009, **20**, 693.
57. M. Najlah, S. Freeman, D. Attwood and A. D'Emanuele, *Int. J. Pharm.*, 2007, **336**, 183.
58. W. Ke, Y. Zhao, R. Huang, C. Jiang and Y. Pei, *J. Pharm. Sci.*, 2008, **97**, 2208.
59. R. Kolhatkar, P. Swaan and H. Ghandehari, *Pharm. Res.*, 2008, **25**, 1723.
60. Y. Lin, T. Fujimori, N. Kawaguchi, Y. Tsujimoto, M. Nishimi, Z. Dong, H. Katsumi, T. Sakane and A. Yamamoto, *J. Controlled Release*, 2011, **149**, 21.
61. G. Van den Mooter, *Expert Opin. Drug Deliv.*, 2006, **3**, 111.
62. A. Gazzaniga, A. Maroni, M. E. Sangalli and L. Zema, *Expert Opin. Drug Deliv.*, 2006, **3**, 583.
63. M. Torres-Lugo, M. García, R. Record and N. A. Peppas, *Biotechnol. Prog.*, 2002, **18**, 612.
64. A. C. Foss and N. A. Peppas, *Eur. J. Pharm. Biopharm.*, 2004, **57**, 447.

65. J. Fallingborg, L. A. Christensen, M. Ingeman-Nielsen, B. A. Jacobsen, K. Abildgaard and H. H. Rasmussen, *Aliment. Pharmacol. Ther.*, 1989, **3**, 605.
66. J. Fallingborg, L. A. Christensen, B. A. Jacobsen and S. N. Rasmussen, *Dig. Dis. Sci.*, 1993, **38**, 1989.
67. P. A. Sandford, ACS, Polym. Preprints, Division of Polymer Chemistry, 1990.
68. C.-M. Lehr, J. A. Bouwstra, E. H. Schacht and H. E. Junginger, *Int. J. Pharm.*, 1992, **78**, 43.
69. E. E. Hassan and J. M. Gallo, *Pharm. Res.*, 1990, **7**, 491.
70. A. Bernkop-Schnürch, *Drug Discovery Today: Technologies*, 2005, **2**, 83.
71. L. Illum, N. F. Farraj and S. S. Davis, *Pharm. Res.*, 1994, **11**, 1186.
72. I. M. van der Lubben, J. C. Verhoef, G. Borchard and H. E. Junginger, *Eur. J. Pharm. Sci.*, 2001, **14**, 201.
73. H. L. Lueßen, C. O. Rentel, A. F. Kotzé, C. M. Lehr, A. G. de Boer, J. C. Verhoef and H. E. Junginger, *J. Controlled Release*, 1997, **45**, 15.
74. N. G. M. Schipper, S. Olsson, J. A. Hoogstraate, A. G. deBoer, K. M. Vårum and P. Artursson, *Pharm. Res.*, 1997, **14**, 923.
75. G. Borchard, H. L. Lueen, A. G. de Boer, J. C. Verhoef, C.-M. Lehr and H. E. Junginger, *J. Controlled Release*, 1996, **39**, 131.
76. J. Smith, E. Wood and M. Dornish, *Pharm. Res.*, 2004, **21**, 43.
77. Y. Pan, Y. J. Li, H. Y. Zhao, J. M. Zheng, H. Xu, G. Wei, J. S. Hao and F. D. Cui, *Int. J. Pharm.*, 2002, **249**, 139.
78. Z. Ma, T. M. Lim and L.-Y. Lim, *Int. J. Pharm.*, 2005, **293**, 271.
79. M. H. El-Shabouri, *Int. J. Pharm.*, 2002, **249**, 101.
80. J. L. Chew, C. B. Wolfowicz, H.-Q. Mao, K. W. Leong and K. Y. Chua, *Vaccine*, 2003, **21**, 2720.
81. W. Liu, S. Sun, Z. Cao, X. Zhang, K. Yao, W. W. Lu and K. D. K. Luk, *Biomaterials*, 2005, **26**, 2705.
82. J. M. Dang and K. W. Leong, *Adv. Drug Delivery Rev.*, 2006, **58**, 487.
83. K. Bowman, R. Sarkar, S. Raut and K. W. Leong, *J. Controlled Release*, 2008, **132**, 252.
84. S. Mao, W. Sun and T. Kissel, *Adv. Drug Delivery Rev.*, 2010, **62**, 12.
85. K. Bowman and K. W. Leong, *Int. J. Nanomed.*, 2006, **1**, 117.
86. R. Hejazi and M. Amiji, *J. Controlled Release*, 2003, **89**, 151.
87. E. L. McConnell, S. Murdan and A. W. Basit, *J. Pharm. Sci.*, 2008, **97**, 3820.
88. B. Sarmento, A. Ribeiro, F. Veiga, P. Sampaio, R. Neufeld and D. Ferreira, *Pharm. Res.*, 2007, **24**, 2198.
89. K. Sonaje, Y.-H. Lin, J.-H. Juang, S.-P. Wey, C.-T. Chen and H.-W. Sung, *Biomaterials*, 2009, **30**, 2329.
90. K. Sonaje, K.-J. Lin, S.-P. Wey, C.-K. Lin, T.-H. Yeh, H.-N. Nguyen, C.-W. Hsu, T.-C. Yen, J.-H. Juang and H.-W. Sung, *Biomaterials*, 2010, **31**, 6849.
91. M. Greindl and A. Bernkop-Schnürch, *Pharm. Res.*, 2006, **23**, 2183.

92. C. M. Lehr, J. A. Bouwstra, W. Kok, A. G. De Boer, J. J. Tukker, J. C. Verhoef, D. D. Breimer and H. E. Junginger, *J. Pharm. Pharmacol.*, 1992, **44**, 402.
93. C. K. S. Pillai, W. Paul and C. P. Sharma, *Progr. Polym. Sci.*, 2009, **34**, 641.
94. M. Thanou, J. C. Verhoef and H. E. Junginger, *Adv. Drug Delivery Rev.*, 2001, **50**, S91.
95. S. M. van der Merwe, J. C. Verhoef, J. H. M. Verheijden, A. F. Kotzé and H. E. Junginger, *Eur. J. Pharm. Biopharm.*, 2004, **58**, 225.
96. S. Dünnhaupt, J. Barthelmes, J. Hombach, D. Sakloetsakun, V. Arkhipova and A. Bernkop-Schnürch, *Int. J. Pharm.*, 2011, **408**, 191.
97. M. Thanou, B. I. Florea, M. W. E. Langemeÿer, J. C. Verhoef and H. E. Junginger, *Pharm. Res.*, 2000, **17**, 27.
98. M. Thanou, J. C. Verhoef, J. H. M. Verheijden and H. E. Junginger, *Pharm. Res.*, 2001, **18**, 823.
99. C. M. O'Driscoll and B. T. Griffin, *Adv. Drug Delivery Rev.*, 2008, **60**, 617.
100. C. J. H. Porter, N. L. Trevaskis and W. N. Charman, *Nat. Rev. Drug Discovery*, 2007, **6**, 231.
101. J. i. Okada, S. Cohen and R. Langer, *Pharm. Res.*, 1995, **12**, 576.
102. M. C. Taira, N. S. Chiaramoni, K. M. Pecuch and S. Alonso-Romanowski, *Drug Deliv.*, 2004, **11**, 123.
103. H. Li, J. H. Song, J. S. Park and K. Han, *Int. J. Pharm.*, 2003, **258**, 11.
104. K. Park, I. C. Kwon and K. Park, *React. Funct. Polym.*, 2011, **71**, 280.
105. V. P. Torchilin, *Nat. Rev. Drug Discov.*, 2005, **4**, 145.
106. D. J. Hauss, *Adv. Drug Delivery Rev.*, 2007, **59**, 667.
107. P. P. Constantinides, *Pharm. Res.*, 1995, **12**, 1561.
108. T. Gershanik and S. Benita, *Eur. J. Pharm. Biopharm.*, 2000, **50**, 179.
109. C. J. H. Porter, C. W. Pouton, J. F. Cuine and W. N. Charman, *Adv. Drug Delivery Rev.*, 2008, **60**, 673.
110. C. W. Pouton and C. J. H. Porter, *Adv. Drug Delivery Rev.*, 2008, **60**, 625.
111. C. W. Pouton, *Eur. J. Pharm. Sci.*, 2006, **29**, 278.
112. C. W. Pouton, *Eur. J. Pharm. Sci.*, 2000, **11**, S93.
113. E. A. Mueller, J. M. Kovarik, J. B. van Bree, W. Tetzloff, J. Grevel and K. Kutz, *Pharm. Res.*, 1994, **11**, 301.
114. T. K. Vyas, A. Shahiwala and M. M. Amiji, *Int. J. Pharm.*, 2008, **347**, 93.
115. B. T. Griffin and C. M. O'Driscoll, *J. Pharm. Pharmacol.*, 2006, **58**, 917.
116. N. L. Trevaskis, W. N. Charman and C. J. H. Porter, *Adv. Drug Delivery Rev.*, 2008, **60**, 702.
117. C. M. O'Driscoll, *Eur. J. Pharm. Sci.*, 2002, **15**, 405.
118. S. Peltier, J.-M. Oger, F. Lagarce, W. Couet and J.-P. Benoît, *Pharm. Res.*, 2006, **23**, 1243.
119. J. S. Coon, W. Knudson, K. Clodfelter, B. Lu and R. S. Weinstein, *Cancer Res.*, 1991, **51**, 897.

120. E. Roger, F. Lagarce, E. Garcion and J. P. Benoit, *J. Controlled Release*, 2009, **140**, 174.
121. A. J. Almeida and E. Souto, *Adv. Drug Delivery Rev.*, 2007, **59**, 478.
122. M. Garcia-Fuentes, M. J. Alonso and D. Torres, *J. Colloid Interface Sci.*, 2005, **285**, 590.
123. M. Garcia-Fuentes, D. Torres and M. J. Alonso, *Int. J. Pharm.*, 2005, **296**, 122.
124. R. Gref, Y. Minamitake, M. T. Peracchia, V. Trubetskoy, V. Torchilin and R. Langer, *Science*, 1994, **263**, 1600.
125. O. C. Farokhzad and R. Langer, *Adv. Drug Delivery Rev.*, 2006, **58**, 1456.
126. M. Tobío, A. Sánchez, A. Vila, I. Soriano, C. Evora, J. L. Vila-Jato and M. J. Alonso, *Colloids Surf., B.*, 2000, **18**, 315.
127. A. Vila, A. Sánchez, M. TobIo, P. Calvo and M. J. Alonso, *J. Controlled Release*, 2002, **78**, 15.
128. A. M. De Campos, A. Sánchez, R. Gref, P. Calvo and M. J. Alonso, *Eur. J. Pharm. Sci.*, 2003, **20**, 73.
129. S. K. Lai, Y.-Y. Wang and J. Hanes, *Adv. Drug Delivery Rev.*, 2009, **61**, 158.
130. B. C. Tang, M. Dawson, S. K. Lai, Y.-Y. Wang, J. S. Suk, M. Yang, P. Zeitlin, M. P. Boyle, J. Fu and J. Hanes, *Proc. Natl. Acad. Sci. U. S. A.*, 2009, **106**, 19268.
131. C. Prego, M. Fabre, D. Torres and M. Alonso, *Pharm. Res.*, 2006, **23**, 549.
132. M. Garcia-Fuentes, C. Prego, D. Torres and M. J. Alonso, *Eur. J. Pharm. Sci.*, 2005, **25**, 133.
133. Y. Kawashima, H. Yamamoto, H. Takeuchi and Y. Kuno, *Pharm. Dev. Technol.*, 2000, **5**, 77.
134. C.-M. Lehr, *J. Controlled Release*, 2000, **65**, 19.
135. N. Hussain and A. T. Florence, *Pharm. Res.*, 1998, **15**, 153.
136. M.-K. Yoo, S.-K. Kang, J.-H. Choi, I.-K. Park, H.-S. Na, H.-C. Lee, E.-B. Kim, N.-K. Lee, J.-W. Nah, Y.-J. Choi and C.-S. Cho, *Biomaterials*, 2010, **31**, 7738.
137. V. Fievez, L. Plapied, A. des Rieux, V. Pourcelle, H. Freichels, V. Wascotte, M.-L. Vanderhaeghen, C. Jerôme, A. Vanderplasschen, J. Marchand-Brynaert, Y.-J. Schneider and V. Préat, *Eur. J. Pharm. Biopharm.*, 2009, **73**, 16.
138. T. E. Rajapaksa, M. Stover-Hamer, X. Fernandez, H. A. Eckelhoefer and D. D. Lo, *J. Controlled Release*, 2010, **142**, 196.
139. F. Gabor, E. Bogner, A. Weissenboeck and M. Wirth, *Adv. Drug Delivery Rev.*, 2004, **56**, 459.
140. M. J. O'Neill, L. Bourre, S. Melgar and C. M. O'Driscoll, *Drug Discov. Today*, 2011, **16**, 203.
141. M. D. Bhavsar and M. M. Amiji, *J. Controlled Release*, 2007, **119**, 339.
142. M. A. Benoit, B. Baras and J. Gillard, *Int. J. Pharm.*, 1999, **184**, 73.
143. C. Kriegel and M. Amiji, *J. Controlled Release*, 2011, **150**, 77.

144. M. Aouadi, G. J. Tesz, S. M. Nicoloro, M. Wang, M. Chouinard, E. Soto, G. R. Ostroff and M. P. Czech, *Nature*, 2009, **458**, 1180.
145. S. Colonna, N. Gaggero, G. Carrea and P. Pasta, *Tetrahedron: Asymmetry*, 1996, **7**, 565.
146. D. S. Wilson, G. Dalmasso, L. Wang, S. V. Sitaraman, D. Merlin and N. Murthy, *Nat. Mater.*, 2010, **9**, 923.
147. A. W. Basit, M. D. Short and E. L. McConnell, *J. Drug Targeting*, 2009, **17**, 64.
148. A. Rubinstein, *Drug Discovery Today: Technol.*, 2005, **2**, 33.
149. D. Pertuit, B. Moulari, T. Betz, A. Nadaradjane, D. Neumann, L. Ismaïli, B. Refouvelet, Y. Pellequer and A. Lamprecht, *J. Controlled Release*, 2007, **123**, 211.
150. S. Berggren, C. Gall, N. Wollnitz, M. Ekelund, U. Karlbom, J. Hoogstraate, D. Schrenk and H. Lennernäs, *Mol. Pharm.*, 2007, **4**, 252.
151. I. Bièche, C. l. Narjoz, T. Asselah, S. Vacher, P. Marcellin, R. Lidereau, P. Beaune and I. de Waziers, *Pharmacogenetics Genomics*, 2007, **17**, 731.
152. M. Mackay, J. Phillips and J. Hastewell, *Adv. Drug Delivery Rev.*, 1997, **28**, 253.
153. V. C. Ibekwe, F. Liu, H. M. Fadda, M. K. Khela, D. F. Evans, G. E. Parsons and A. W. Basit, *J. Pharm. Sci.*, 2006, **95**, 2760.
154. M. Tubic-Grozdanis, J. Hilfinger, G. Amidon, J. Kim, P. Kijek, P. Staubach and P. Langguth, *Pharm. Res.*, 2008, **25**, 1591.
155. M. C. Allison, S. Cornwall, L. W. Poulter, A. P. Dhillon and R. E. Pounder, *Gut*, 1988, **29**, 1531.
156. A. Lamprecht, H. Yamamoto, H. Takeuchi and Y. Kawashima, *J. Pharmacol. Exp. Ther.*, 2005, **315**, 196.
157. Y. Tabata, Y. Inoue and Y. Ikada, *Vaccine*, 1996, **14**, 1677.
158. A. Lamprecht, U. Schäfer and C.-M. Lehr, *Pharm. Res.*, 2001, **18**, 788.
159. A. Lamprecht, N. Ubrich, H. Yamamoto, U. Schäefer, H. Takeuchi, P. Maincent, Y. Kawashima and C.-M. Lehr, *J. Pharmacol. Exp. Ther.*, 2001, **299**, 775.
160. J. Calderó, E. Campo, C. Ascaso, J. Ramos, M. J. Panadés and J. M. Reñé, *Virchows Arch.*, 1989, **415**, 347.
161. C. Wang, P. C. Ho and L. Y. Lim, *Int. J. Pharm.*, 2010, **400**, 201.
162. E. L. McConnell, M. D. Short and A. W. Basit, *J. Controlled Release*, 2008, **130**, 154.
163. A. W. Basit, *Drugs*, 2005, **65**, 1991.
164. T. Sousa, R. Paterson, V. Moore, A. Carlsson, B. Abrahamsson and A. W. Basit, *Int. J. Pharm.*, 2008, **363**, 1.
165. E. L. McConnell, F. Liu and A. W. Basit, *J. Drug Targeting*, 2009, **17**, 335.
166. D. R. Friend, *Adv. Drug Delivery Rev.*, 2005, **57**, 247.
167. M. Saffran, G. S. Kumar, C. Savariar, J. C. Burnham, F. Williams and D. C. Neckers, *Science*, 1986, **233**, 1081.

168. G. Van den Mooter, C. Samyn and R. Kinget, *Int. J. Pharm.*, 1992, **87**, 37.

169. S.-Q. Gao, Z.-R. Lu, B. Petri, P. Kopecková and J. Kopecek, *J. Controlled Release*, 2006, **110**, 323.

170. J. Kopecek and P. Kopecková, *Adv. Drug Delivery Rev.*, 2010, **62**, 122.

171. M. Roldo, E. Barbu, J. F. Brown, D. W. Laight, J. D. Smart and J. Tsibouklis, *Expert Opin. Drug Deliv.*, 2007, **4**, 547.

172. P. Chivukula, K. Dusek, D. Wang, M. Dusková-Smrcková, P. Kopecková and J. Kopecek, *Biomaterials*, 2006, **27**, 1140.

173. E. P. Kakoulides, J. D. Smart and J. Tsibouklis, *J. Controlled Release*, 1998, **54**, 95.

174. G. Millotti, G. Perera, C. Vigl, K. Pickl, F. M. Sinner and A. Bernkop-Schnürch, *Drug Deliv.*, 2011, **18**, 190.

175. A. Díaz, A. David, R. Pérez, M. L. González, A. Báez, S. E. Wark, P. Zhang, A. Clearfield and J. L. Colón, *Biomacromolecules*, 2010, **11**, 2465.

176. R. Yang, R. Gao, F. Li, H. He and X. Tang, *Drug Dev. Ind. Pharm.*, 2011, **37**, 139.

177. C. B. Woitiski, R. J. Neufeld, F. Veiga, R. A. Carvalho and I. V. Figueiredo, *Eur. J. Pharm. Sci.*, 2010, **41**, 556.

178. C. Thompson, W. Cheng, P. Gadad, K. Skene, M. Smith, G. Smith, A. McKinnon and R. Knott, *Pharm. Res.*, 2011, **28**, 886.

CHAPTER 2.3

Nanostructures for Oral Vaccine Delivery

CARLOS GAMAZO[a] AND JUAN M. IRACHE*[b]

[a] Department of Microbiology, University of Navarra, 31080 Pamplona, Spain; [b] Department of Pharmacy and Pharmaceutical Technology, University of Navarra, C/ Irunlarrea, 31080 Pamplona, Spain
*E-mail: jmirache@unav.es

2.3.1 Introduction

This chapter deals with mucosal adjuvants but also deals with health, safety, and confidence. Vaccination is the most effective means for disease prevention, however there are many concerns about this practice. Currently, few vaccines have been approved for human use and, surprisingly, most contain live microorganisms and are administered parenterally. Why do we use live vaccines if dead or fractionated ones are safer?[1] Why do we use needles for vaccine inoculation if the mucosal routes offer security and convenience?[2] Furthermore, since most viral and bacterial pathogens use mucosal epithelia, might it be possible to induce a protective mucosal immune response? The answer to this critical question is yes, and its rationale comes from adaptive evolution, as usual in biology. Animals have evolved a sophisticated mucosal immune system to protect us from that myriad of mucosal pathogens that try to gain access through our extensive mucosal epithelia. Two convincing bits of data to help think about the importance of mucosal immunity, and hence, of interest to the study of immunization by mucosal routes: most T-cells are found at mucosal surfaces, and the sIgA isotype comprises at least 70% of all Ig produced in mammals.[3] Furthermore, we will analyze later in this chapter

RSC Drug Discovery Series No. 22
Nanostructured Biomaterials for Overcoming Biological Barriers
Edited by Maria Jose Alonso and Noemi S. Csaba
© The Royal Society of Chemistry 2012
Published by the Royal Society of Chemistry, www.rsc.org

the fact that the activation of mucosal immunity induces both mucosal and systemic immune responses. In contrast, parenteral immunization usually fails to stimulate mucosal lymphatic tissues to generate protective immunity at these sites. Paradoxically, as said before, the few vaccines approved for human use are administered parenterally and do not induce satisfactory mucosal immunity.[4]

We are scientifically well prepared now to accept the logic of taking advantage of the mucosal immune system. Nevertheless, empirically this concept is not new, and was "assumed" centuries ago. It is well documented that to prevent the devastating plague of smallpox, the Chinese (medical manuscripts dated to 1122 B.C.) discovered the effectiveness of nose immunization with dried crusts of smallpox scabs.[5] For the same purpose, but passively, in India healthy people wore infected peoples' clothes and stayed with smallpox victims in order to get immunized. The oral route was also used in Turkey, where healthy people swallowed smallpox scabs from the infected ones. Later the skin was used to inoculate the scabs, and this was the historical origin of the current practices for vaccination. This new practice, named variolation, was introduced in England and Western Europe in the XVIII century. However, this was not a safe method as live viruses were used, just slightly attenuated after drying the scabs, and, in some cases, the host become highly infected and even die. Jenner (1796), introduced the use of the antigenic-related virus from cows (lat. *Vacca*), the virus vaccinia; by far less virulent that the human version smallpox. In the late XIX century, Pasteur, Koch and others developed the first laboratory-created vaccines. However, the mucosal routes for immunization were largely replaced by the parenteral ones. All these vaccines were attenuated-live ones, strong activators of the immune system, but unsafe, showing drawbacks, such as: problems related with residual virulence after attenuation that hamper their administration to immunocompromized or pregnant patients; mutations leading to reversion to virulence; lability during conservation. The use of safer killed or subunit vaccines are a big motivation for many researchers, however, these forms induced inadequate immune responses. Ramon, Glenny and others in the 1920–30s observed that immunizations with toxins could be improved by adsorption to inert substances, such as charcoal, collodion particles, dextran or aluminium potassium sulphate.[6] These substances were called adjuvants (*adyuvare* (Latin): "to help"). Following Glenny's discovery, aluminium salts were used in vaccine preparations with tetanus and diphtheria toxoids. Adjuvants have been used empirically for decades and, in spite of the fact that many adjuvants have been developed since then, insoluble aluminium salts are still used worldwide as the principle adjuvants in clinical vaccines. Ironically, aluminium is toxic. Recently, it was suggested that the intrinsic cytotoxicity of aluminium salts leads to the activation of NLRP3 inflammasome involved in innate immunity and citoquine response, crucial components of the host defence.[7,8] Many other receptors and signalization pathways of the innate immune system are being defined as mediators of the adjuvant effects, opening a new rational way for the development of new adjuvants.

Vaccinology is a fast moving science that requires the interdisciplinary collaboration of other sciences, such as pharmaceutical technology. In the 1950s, the pharmacist talked to the vaccinologist on emulsions (MF59 adjuvant is a good example)[9] today the "cross-talk" is on nanoparticles, Toll-like receptors (TLR)-agonist polymers and bioadhesion.[10,11] The new technologies to produce purified antigens require the intervention of pharmaceutical technology to produce new adjuvants that are more convenient to use, safer and easy to administer, even through mucosal routes. In this way, in the near future it is possible that the mucosal administration of vaccines, such as by oral, intranasal, or conjunctival routes, will be the method of choice for vaccine delivery.[12]

In the following section we will present an overview on the mucosal immune system. By understanding immune activation, we can rationally design adjuvants to re-direct the immune response to a protective one. Later, we will focus on the mechanisms of nanoparticle adjuvants for induction and modulation of the immune response.

2.3.2 Organization of the Mucosal Immune System

It has already been mentioned that the mucosal surface is the obvious easy portal of entry of many pathogens, and therefore a significant network of effector molecules and cells have evolved to be strategically situated at this area that covers around 400 m^2 in the human host. In order to simplify, we will introduce the mucosal immune system by considering three key functional properties: (i) "inter-network" organization; (ii) major regulation to maintain the adequate tolerance/inflammatory balance; (iii) role of the pattern-recognition-receptors (PRR) in mucosa to discriminate aggressions from the outside-world.

2.3.2.1 Common Mucosal Immune System

About 60% of the total body lymphocytes are in organized follicles along mucosae called broadly the mucosa-associated lymphoid tissue (MALT).[13] Depending on their physical situation they are designated as gut-associated lymphoid tissue (GALT), nasopharynx-associated lymphoid tissue (NALT), bronchus-associated lymphoid tissue (BALT), or conjunctiva-associated lymphoid tissue (CALT), among other major systems. The anatomic distribution is a direct evolutionary consequence of the function required, since they have to confront different challenges. What's more, the special design of the mucosal immune system means that the entry of an antigen in a discrete anatomical site renders a response both in that one and also in other distal mucosal areas. This is analogous to the antigens in blood that are filtered-trapped, processed and presented in the lymph nodes and the spleen, resulting in systemic immune responses. Thus, though MALT sites are anatomically separated, they are functionally connected in what has been named as the "common mucosal immune system". As a consequence, after the administration of a vaccine through the oral route, we can expect to find specific memory cells and effector responses, such as specific IgA at the intestinal, vaginal and other mucosal tracts.

2.3.2.2   Tolerance

Another preliminary and fundamental property of MALT is the natural predisposition to induce tolerance. This effect is as necessary as expected to avoid a continuous inflammatory response during our quotidian acts for everyday living, like food ingestion (antigens in food and water) or breathing (inhalation of foreign substances).[14] Experimentally, it is observed that the mucosal administration of a single high dose or repeated low doses of antigens induce systemic unresponsiveness (characterized by regulatory T-cells), exemplified by Tr1/Th3. Repeated high doses of antigens, like those coming from the symbiotic microbiota, may elicit a mucosal IgA-response (Th2/Th3 response), and only after a big challenge, such as the one produced by virulent pathogens that make a breach in the epithelium, a potent inflammatory response is induced (Th1/Th2/Th17) (Figure 2.3.1).

Epithelial cells play an important role in the decision of the type of immune response induced. Multiple mediators released from enterocytes establish a continuous cross-talk with the immune cells.[15] In turn, the overlying biofilm of intestinal microbiota (commonly and erroneously known as "microflora") have evolved several mechanisms for programming enterocytes and dendritic cells (DCs) to induce a regulatory response. In fact, the intestinal symbiotic microbiota has contributed largely to the organization of the inductive sites of the immune system, and to its natural predisposition to tolerance, as we will discuss later.[16,17]

2.3.2.3   PRR Recognition

An essential feature for any host is the discrimination between foreign and the body's own antigens, and even more finely tuned, between potential pathogens from the non-pathogenic, symbiotic microbiota. This property resides in a series of pattern-recognition receptors (PRR), proteins able to recognize conserved molecular markers that are shared by pathogens but absent in the host.[18] The most important of them are, probably, the Toll-like receptors (TLR). TLRs are expressed on/in many cells, including intestinal epithelial cells.[19] Currently, 13 TLRs have been described in mammals, with different capacities involved in the elicited immune responses. Thus, the specific recognition of the corresponding pathogen-associated molecular patterns (PAMPs) will direct differential innate and, further, adaptive immune responses. TLR3 (recognizes double-stranded RNA from viruses), TLR4 (bacterial lipopolysaccharide), TLR5 (bacterial flagellin), TLR7 (single-stranded RNA from virus) and TLR9 (bacterial CpG-containing DNA) are related with a preferential Th1 response. In contrast, TLR2 (recognizes

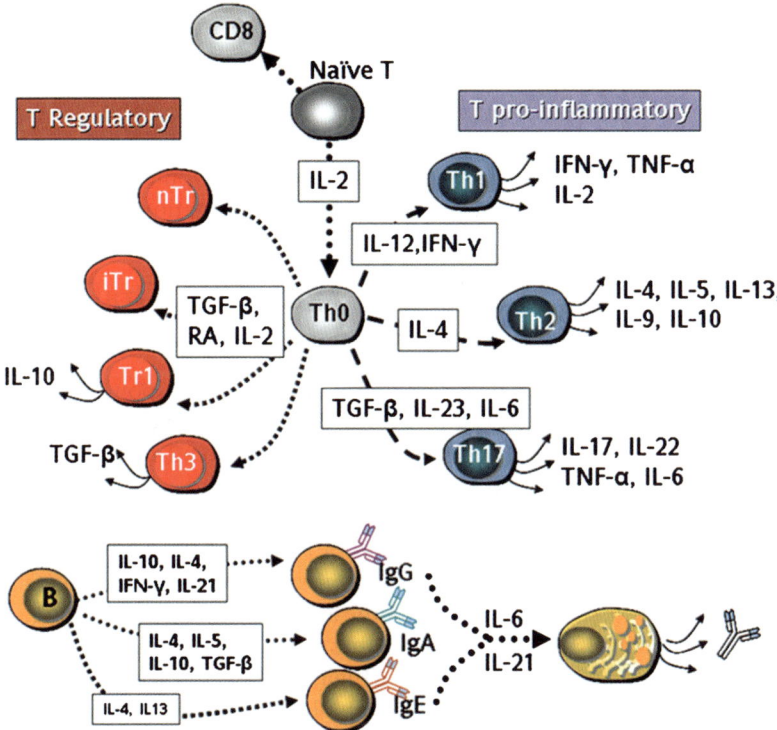

Figure 2.3.1 Organization and cellular traffic at gut-associated lymphoid tissues (GALT). GALT contains the largest pool of immunocompetent cells. The relative concentration of CD4$^+$ T subsets in MALT influences the outcome of the response. However, by vaccination, the natural predisposed polarization to Treg/Th2 may be influenced by the adjuvant and the route of administration.

bacterial peptidoglycane and lipopeptides) elicits a Th2 response, although in combination with TLR6 elicits a regulatory response (Th3/Treg).[20]

Interestingly, although TLRs are usually linked with DC antigen recognition, intestinal mucosa also recognizes the antigenic challenge by TLR, and this fact is of vital importance for the host.[21] Actually, the expression of TLRs in the intestinal epithelium is greater than that of other major organs, like the liver. Intracellular PRRs, such as NOD-like receptors (NLRs), also play an important role in the intestinal epithelium. Both TLRs and NLRs are expressed differentially along the intestine. This is again a critical evolving consequence to cover the expected pathogen signals along the epithelial intestinal surfaces, and has a direct effect in vaccinology, since each intestinal section may respond differently to the same antigenic stimulus.[22] For example, TLR4 expression is regulated by IFN-γ and TNF-α in intestinal epithelial cells and oral mucosal epithelial cells.[23] Ligation of TLR initiates a signalling cascade that results in the activation of the transcription factor NF-κB and subsequent up-regulation of co-stimulatory molecules as well as inflammatory cytokines and chemokines and hormonal factors. The characteristic recruitment of immune cells to damaged areas is in part dependent on the chemotactic IL-8 released by epithelial cells.

In addition to natural ligands, non-microbial, xenobiotic or artificial ligands for these TLRs are being described. This is an important point that is opening new dimensions in the adjuvant field for mucosal vaccination.[24,25]

2.3.3   Anatomy of GALT and Fate of the Orally Administered Antigens

The equilibrium between homeostasis and inflammatory response at mucosal level is the result of multifactorial processes, dependent in the first instance on the peculiar anatomy of GALT. The intestine (see Chapter 2.1) is considered to be the largest lymphoid organ in mammals and contains more immune cells than any other organ including the spleen and liver, with more than 10^{12} lymphocytes and the largest concentration of antibodies in the body.[26]

Antigens may follow different routes to cross the epithelial cell barrier. In any case, DCs act as sentinels, acquiring the antigens and then migrating to the subepithelial dome (SED) in Peyer's patches (PP) or to the mesenteric lymph nodes (MLN) to activate native T-cells (Figure 2.3.2). In fact, intestinal DCs have been recognized as the critical cells involved in the decision of tolerance *vs.* inflammation.[22]

There are a large variety of subsets of DCs that differ in their degree of "natural" or inducible predisposition to express and release a determined repertoire of cytokines and, subsequently, to mount a specific immune response. We can find DCs at lamina propria (LP) and PP in different degrees of maturation. Some are "resident" whereas others are "immigrant". In general terms, resident mucosal-DCs are microenvironmentally conditioned to favour tolerogenic responses, whereas newly recruited inflammatory DC may avoid it and drive inflammatory responses (Figure 2.3.2). This is an important item to consider in mucosal vaccination.

Basically, immature DCs are present in peripheral tissues and are mainly phagocytic cells; mature DCs are found in PP and MLN, and are specialized in antigen presentation, as they are characterized by the high surface expression of co-stimulatory molecules that is required for T-cell activation. Mature DCs derive from immature cells after a maturation process that is initiated by inflammatory stimuli and that leads to a massive migration of these DCs to the draining lymph nodes. In this case, an inflammatory response is induced. However, when immature or semimature DCs arrive from peripheral tissues to draining lymph nodes, tolerance induction is expected (Figures 2.3.2 and 2.3.3).

Different subsets of DCs have been described in mouse PPs.[27–29] These subsets can be distinguished by the expression of different surface markers and present specialized functions. For instance, the phenotype

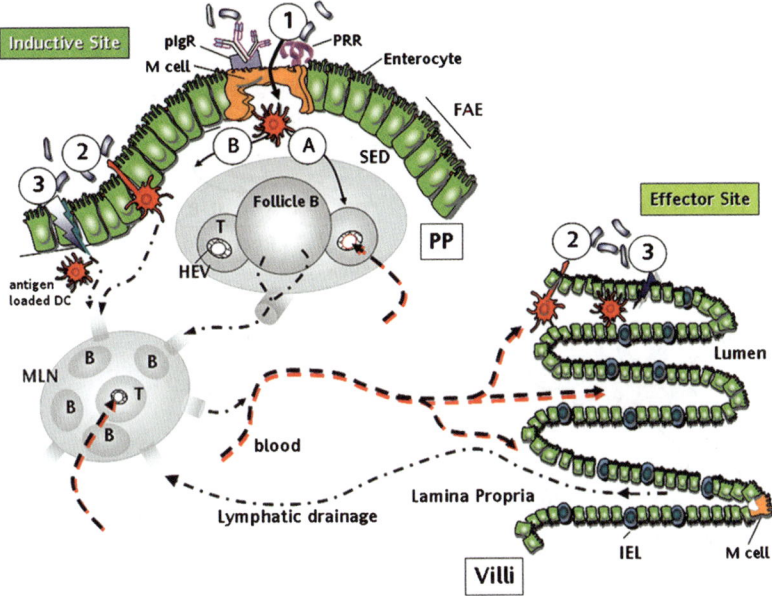

Figure 2.3.2 Antigen sampling in Peyer's patches and lamina propria. The inductive sites of GALT comprise the appendix, isolated lymphoid follicles, Peyer's patches (PP) (small intestine) and the lymphoglandular complexes (large intestine). Separating the GALT from the lumen is the follicle-associated epithelium (FAE) as a monolayer containing enterocytes (intestinal epithelial cells) and within them are the microfold cells (M cells). M-cells, which act as "antigen sampling" agents, conform like an open pocket where B- and T-cells and APCs are located in order to facilitate the capture of the transcytosed antigens. Underneath, a net of DCs and the subepithelial dome containing B-cell follicles separated by T-cell areas are positioned, known as interfollicular regions, also enriched in APCs. The lamina propria is considered the effector site within the GALT. Resident mucosal DCs form an extensive network that constantly survey the luminal microenvironment. The three known routes by which inert particles may gain access into the lamina propria from the lumen are represented: 1) through M cells; 2) by DCs expanding dendrites to the lumen; 3) through disrupted or injured epithelium.

CD11b⁺CD8α⁻CCR6⁺ is found primarily in the follicle-associated epithelium in the SED, and therefore are the first to uptake of luminal antigens transported by M-cells. Therefore, and as was discussed before, they will preferentially induce a Th2 differentiation and IgA-plasmablasts.[30]

It may be inferred that the route used by the antigen to cross the intestinal epithelium will condition the sort of the immune response elicited, since a different repertoire of conditioned or not-conditioned DC subsets will run to its encounter. We will now briefly discuss this action following one of the four main pathways: (i) through specialized M-cells; (ii) through dendrites extended

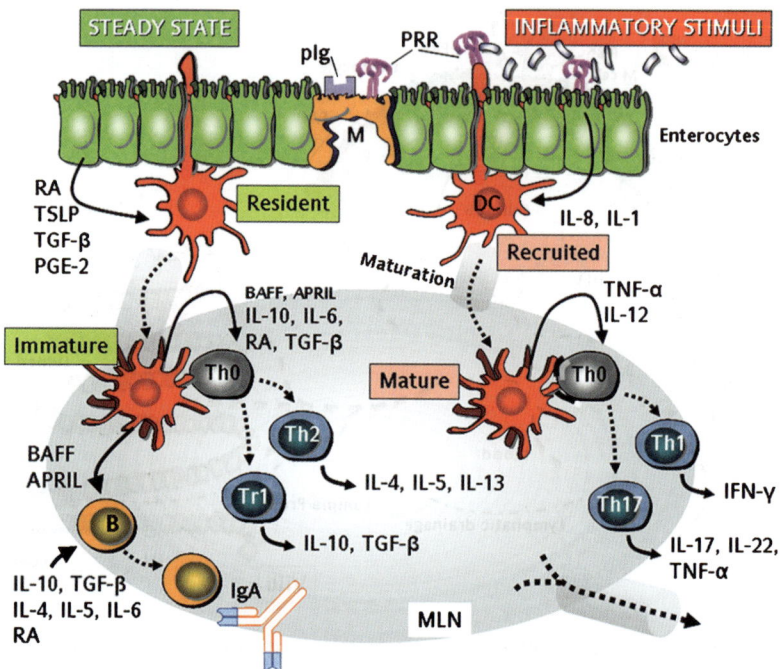

Figure 2.3.3 Dendritic cells and enterocytes condition the differentiation and maturation processes at mucosal level. Basically, intestinal epithelial mucosa may be found in two main states: steady state-conditioned and inflammatory condition. Both situations are dependent on the microenvironment dictated by the enterocytes and the type of subset and activation state of the DCs.

from special DCs; (iii) through the use invasion strategies such as the Type III secretion system employed by certain bacterial pathogens.

2.3.3.1 Transcytosis through M-Cells

M-cells occupy 10% of the intestinal lymphoid follicle surface area in humans and mice, and represent the primary route by which mucosal antigens are transported, in an unmodified form, through epithelial barrier (Figure 2.3.2.; see also Chapter 2.1). M-cells are mainly located in PP but also along the intestinal villi. In fact, villous M-cells are used for many pathogens to penetrate the epithelial barrier, and may be a significant portal of entry for vaccinal antigens.[31]

Antigens may take as little as 10 minutes to pass through M-cells. Several mechanisms/receptors are implicated, such as TLR-4.[32–34] However, apart from the mechanisms for entry, the status of the DCs that encounter the transcytosed antigen will mark the elicited immune response (Figure 2.3.3). DCs may be conditioned to rest in an immature state by factors released from the epithelial cells [chemokines: like TGF-beta, PGE-2, retinoic acid; and, in human, hormonal factors: like thymic stromal-derived lymhopoietin (TSLP)]. Under this state of low activation the DCs capture the transcytosed antigen and then migrate to the MLN where the presentation to naïve specific T-cells takes place. These immature DCs express a low level of co-stimulatory molecules (CD80, CD86, CD40, CD54), and under this state, the presentation of the antigenic determinant to T-cells will induce the release of IL-10 and TGF-β, that will differentiate the naïve T-cells into Th2 or Treg cells.

In some circumstances (Figure 2.3.2), the antigen passing through M-cells is taken directly by B-cells that, as professional antigen presenting cells (APCs), process and present an antigenic determinant to Th cells, and again, TGF-β and IL-10 are expressed, eliciting the clonal expansion and transformation to predominantly IgA-committed cells.[35] These plasmablasts can migrate *via* blood to local and distant mucosal and secretory tissues.

Another route through M-cells involves the pIgR, a specific receptor for IgA (containing or not the secretory component) but not for IgG or IgM. Through this receptor, the antigen linked to the luminal IgA travel through the M-cell to DCs present in the SED in PP. DCs are then transported to the T-cell areas in PP and/or in MLN. These DCs belong to the "conditioned" immature DCs, meaning that under reduced or absent inflammatory conditions, they will render a Th2 and/or Treg polarization (Figure 2.3.2).

It has been described that when antigens use the M-cells in villi, a T-independent generation of IgA may be generated. DCs located beneath M-cells in LP contribute to this archaic class-switching mediated by TGF-β, BAFF and APRIL chemoquines.[36]

2.3.3.2 *Via* Dendrites from Dendritic Cells

LP and PP contain special DCs able to project "dendrites" to sample luminal antigens, penetrating the epithelium without destroying its integrity.[37] This process to form transepithelial dendrites is dependent on the chemokine receptor CX_3CR1. Although this phenomenon occurs in steady-state conditions, it increases largely during a strong antigenic stimulus (microbial challenge or strong vaccinal stimulus). Let's consider two different situations: steady-state *vs.* inflammatory induced stimulus (Figure 2.3.3). In steady-state conditions, retinoic acid, thymic stromal lymphopoietin (TSLP) and TGF-β released by enterocytes will "condition" non-activated resident DCs to elicit a Th2 or regulatory responses, as described above.[38] However, following an inflammatory stimulus, a recruitment of CD expressing CX3CR1 to the LP or PP is observed, increasing the number of DCs extending dendrites into intestinal lumen. Most of these inflammatory stimuli come after PRR signalling by their own enterocytes, including IL-8. Recruited DCs also express a large number of surface PRRs in order to detect the luminal antigenic challenges. Under this state of high activation, DC-expressing massively co-stimulatory molecules, present the antigenic determinant to the specific T native cells in the T area of PP or MLN. The substantial distinctive release of IL-12 from those DCs will also contribute to the further differentiation of native cells to Th1/Th2/Th17, linked to an inflammatory response. It is then inferred that in both situations, enterocytes "sense" the presence of the antigenic challenge and transmit this information to DCs. Once again, resident DCs are conditioned to be tolerogenic, whereas DCs that are recruited to the gut in response to an aggressive challenge activate the inflammatory response.[39]

2.3.3.3   *Via* Epithelium Damage

Some microbial pathogens (such as a number of serovars of *Yersinia, Shigella, Salmonella, or Escherichia coli,* among others) use virulence factors to invade the intestinal epithelial cells inflicting a physical damage,[40] leading to the activation of cytosolic PRRs and enhanced production of chemokines and proinflammatory cytokines from the enterocytes (Figure 2.3.2). Neutrophils, macrophages and different subsets of DCs from MLN are markedly recruited to the site and become activated, and are capable of driving TH1/Th17 responses. IL-1 and IL-8 released by activated enterocytes may act as chemotactic agents to attract circulating DCs to the invaded epithelium. These sort of recruited DCs encounter the antigen before "environment-regulatory conditioning" can take place (see above), whereas some other incoming DCs are already genetically refractory to conditioning (such as CD11b-). In both cases, mature DCs release high levels of IL-12 to drive the differentiation to Th1 cells (Figures 2.3.1 and 2.3.2)

2.3.4   Mucosal Vaccine Adjuvants

Intestinal epithelium, traditionally identified as a physical barrier against luminal bacteria, is also an important source of immunomodulators which "educate" or "condition" resident DCs and lead to the recruitment of migratory DCs to initiate both innate and adaptive immune responses. Together, the default responses in the gut after mucosal immunization is to drive a tolerogenic or a non-inflammatory response; just a direct consequence of the evolving nature of the mucosal immune system. However, through the use of delivery systems such as mucosal adjuvants containing TLR agonists it is possible to abrogate it, eliciting the often desired balanced Th1/Th2/Th17 responses in both mucosal and systemic sites.

The efficient delivery of antigens to DCs and further activation are deeply involved in the success of an effective vaccine. Using appropriate adjuvants, DCs may be activated in the proper manner. Modified bacterial enterotoxins, such as detoxified cholera toxin (CT) and *Escherichia coli* heat labile toxin, are widely used mucosal adjuvants. However, they present important drawbacks: after oral administration they induce diarrhoea in humans; after nasal administration they may target olfactory neurons and gain access to olfactory bulbs and deeper structures in the brain parenchyma, causing facial paresis. This has induced major research on the development of new mucosal adjuvants.[41,42]

Particulate delivery systems belong to the category of adjuvants that facilitate the antigen uptake by APCs or by increasing the influx of professional APCs into the injection site.[43] Among the different types of particulated delivery systems, polymer nanoparticles are a group of delivery systems with interesting abilities as adjuvants for both conventional and mucosal vaccination.[44] In particular, poly(anhydride) nanoparticulate systems made by the copolymer of methyl vinyl ether and maleic anhydride have demonstrated their efficacy as adjuvants to induce Th1 immune responses.[45–48]

Furthermore, these nanoparticles may be exploited as mucosal adjuvants since they can enhance the delivery of the loaded antigen to the gut lymphoid cells due to their ability to be captured and internalized by cells of the GALT.

Interactions between particles and DCs depend on particle characteristics such as size and shape, charge and hydrophobicity, but the mechanisms responsible for DC maturation may be mostly related to TLR recognition in DCs. Many reports show how the stimulation with TLR-agonists induces the surface expression of co-stimulatory molecules, and hence the phenotypic modulation to the typical feature of a mature DC. Recently, it has been shown that poly(anhydride)-nanoparticles induced innate immune responses mediated by a TLR-2 and TLR-4 dependent manner.[24,25] These nanoparticles induced maturation of DCs with a significant up-regulation of CD40, CD80, and CD86 and a biased Th1 present response in animal models. This is an important finding since it has been recently shown that the use of multiple TLR-agonists carried by nanoparticles influence the induction of long-term memory cells, the ultimate goal for any vaccine being the stimulation of long-lasting protective immunological memory.[49,50]

We will discuss in more detail the use of nanoparticles as oral adjuvants in the following sections.

2.3.4.1 Factors Influencing the Efficacy of Nanoparticles as Oral Adjuvants

Recently, it has been pointed out that factors that may contribute to the efficacy of particulates as mucosal adjuvants are the following: (i) the materials used to prepare the particles, (ii) the nature of antigens used, (iii) the methods of antigen loading, and (iv) the routes of administration.[51] The three first factors would influence the physico-chemical properties of the resulting particulates, determining their size and surface characteristics.

2.3.4.1.1 Influence of the Size

Concerning the size of particulates, data from studies that attempted to correlate the size of the particles and their adjuvant activity have been controversial. Some researchers showed that larger particles were more potent than smaller ones,[52,53] while others showed the exact opposite.[54] There were also data showing that the size of particles did not affect the resultant immune responses,[55] and data from some studies suggested that there may be an ideal particle size with the most potent vaccine adjuvant activity.[56,57] Interestingly, it also becomes evident that the size of the particles influences the type of immune responses induced as well,[58,59] but it again remains controversial as to whether small or large particles favor T-cell subset or, widely speaking, antibody *vs.* cellular immune responses.[53,59,60] In any cases, particles in the nanoscale size rather than microscale size appear to be more adapted for cellular uptake and immune stimulation in the GI tract.[46,61]

Moreover, the homogeneity or dispersion of the size of particulates is another important factor influencing the immune response induced by particulates and, indirectly, this aspect has contributed to the controversial results. For example, in a previous study, bilosome particles (lipid vesicles containing bile salts) of 400–2500 nm and 60–350 nm were compared.[60] It is likely that in the 400–2500 nm range, the 400 nm particles and the 2500 nm particles had different adjuvant activities because data from another study showed that antigen-loaded particles of 200–600 nm and 2000–8000 nm induced different immune responses.[53] In another interesting study, ovalbumin (OVA)-coated nanoparticles of 230 and 700 nm were evaluated.[62] It was clear that the smaller nanoparticles induced stronger OVA-specific antibody and cellular immune responses than the largest ones.

2.3.4.1.2   *Influence of the Polymer Nature*

Another important factor influencing the ability of nanoparticles as adjuvants is related to the nature of the polymer used to prepare these antigen delivery systems. The polymer determines the stability of the resulting particles in the gastrointestinal tract as well their interaction with components of the mucosa. It is important to note that the stability may be an important factor influencing the release of the loaded antigen. Furthermore, the adequate selection of the polymer may also determine the antigen loading in the resulting nanoparticles. In any case, little information is available concerning how stable particles really need to be, how the antigen has to be released from the nanoparticles in the gastrointestinal tract, or as to whether M-cells really take up more than just a few particles *in vivo*. Among others, PLGA, chitosan, lipids and Gantrez AN (polyanhydride) have been proposed as materials for antigen encapsulation.

A great number of antigens have been successfully encapsulated in poly(D,L, lactide-co-glycolide) (PLGA) particles while fully maintaining their structural and antigenic integrity. In general, ovalbumin, peptides, bacterial toxoids, inactivated bacteria and, more recently, DNA plasmids entrapped in PLGA particles have demonstrated to significantly enhance the systemic and/or mucosal immune responses after oral administration.[63,64] Furthermore, PLGA have already been investigated in clinical trials with oral vaccines incorporated in PLGA microparticles.[65] Chitosan may also have an immunomodulatory effect as it has been shown to stimulate production of cytokines from immune cells *in vitro*[66] and enhance a naturally Th2/Th3-biased microenvironment at the mucosal level in absence of antigen.[67] Recently, the copolymers between methyl vinyl ether and maleic anhydride (PVM/MA, Gantrez®) have also been proposed as a material to prepare nanoparticles for oral antigen delivery.[68] These copolymers are widely employed as thickening and suspending agents, denture adhesives, and excipients for the preparation of transdermal patches. Finally, solid lipid nanoparticles[69] and nanoparticles made from cationic cross-linked polysaccharides[70] have also been proposed as antigen delivery systems for oral vaccination.

2.3.4.1.3   Influence of the Surface Properties

Last but not least, the modification of the surface properties of nanoparticles may also dramatically affect their stability, distribution within the gut and interaction with the GALT. In general, nanoparticles are relatively rapidly eliminated from the mucosa due to the continuous mucus turnover and peristaltism.[71] Thus, the immune response elicited with these antigen carriers is not as high as necessary to offer the adequate degree of protection to the host, and consequently, high and multiple oral doses are required.

In order to increase the residence of nanoparticles within the mucosa and to improve their capability to reach specific immunocompetent cells and the GALT, different strategies have been developed, including the improvement of the bioadhesive properties of nanoparticles. This approach can be obtained by modifying the surface properties of nanoparticles or by coupling a targeting molecule at their surface. These modifications mainly change the nanoparticle's zeta potential and their hydrophobicity, thus influencing the stability of the resulting nanoparticles and their ability to reach specific areas within the gut.

One of the most popular techniques for modification of the nanoparticle surface is pegylation. The adequate coating of nanoparticles with poly(ethylene glycol) (PEG) decreases the interaction of nanoparticles with the mucins and allows them to "diffuse" through the mucus layer, reaching the surface of the mucosa.[72,73] In addition, it was demonstrated that pegylation of poly(anhydride) nanoparticles yielded carriers with a high ability to avoid adhesive interactions within the stomach and to concentrate them in the small intestine mucosa of animals.[74]

Another strategy to render nanoparticles more efficient as adjuvants for vaccination relies on their association with compounds or molecules involved in either the colonisation process of microorganisms or the activation of the host immune system (see section 2.3.4). The colonization process of microorganisms to host tissues involves adherence to the surface of a cell, invasion or internalisation in the cell, multiplication and production of extracellular proteins that damage the host cells and facilitate the growth and spread of the pathogen.[75] Microorganisms can invade and colonize the host tissue by using a number of different specific adherence factors including lipoteichoic acids, outer membrane proteins,[76] flagellum,[77] fimbriae and pili,[78] lectins,[79] and glycoproteins (*i.e.* mannose proteins).[80] Most of these adhesive factors are also considered as immunomodulators and they are included in the generic denomination of PAMPs (see section 2.3.2.3).

In this biomimetic approach, a good microorganism to be imitated is *Salmonella* spp. This pathogen has the natural ability to invade both non-phagocytic host cells and Peyer's patches.[81] The real process by which *Salmonella* is able to colonize the gut mucosa is still unknown; although, different bacterial appendages, such as flagella, play an important role in the gut mucosa colonization and invasion. In this context, "Salmonella-like" nanoparticles have been obtained by the association of *Salmonella* Enteritidis flagellin (the main component of flagellar filament) to Gantrez AN nanoparticles.[82] These carriers displayed an important tropism for the ileum and their distribution within the gut correlated well with the described colonisation profile for *Salmonella* Enteritidis, including a broad concentration in Peyer's patches.[82] Using ovalbumin as a model antigen, "Salmonella-like" nanoparticles induced a strong and balanced secretion of both IgG2a (Th1) and IgG1 (Th2) specific antibodies. In addition, these nanoparticles were able to induce a much more strong mucosal IgA response than control nanoparticles.[83]

Glycoconjugates enriched in mannose residues have been also found to promote the interaction of a number of microorganisms, such as *Candida albicans*, *Listeria monocytogenes*, HIV, *Enterobacteriaceae* or *Bifidobacterium*, with different tissues and substrates.[84–86] This adhesive mechanism may be related to the high binding affinity of mannose residues to the so-called mannose-binding lectins which are expressed on the lymphoid and non-lymphoid cells of different mucosa.[87,88] Recently, mannose coating nanoparticles have been proposed for oral vaccination. Using ovalbumin as antigen model, these nanoparticles elicited a higher and more balanced specific antibody response (IgG1 and IgG2a) compared with conventional nanoparticles, which elicited a more typical Th2 response. In addition, mannose-coated nanoparticles were able to elicit a significant intestinal secretory IgA for at least 6 weeks after immunization.[43] These immunogenic responses appear to be related to the ability of mannose-coated nanoparticles to specifically target the ileum of animals. In fact, the delivery of antigens to the distal small intestine region facilitates the induction of Th1 responses (see section 2.3.3).

2.3.5 Efficacy of Nanoparticles as Mucosal Adjuvants for Vaccination

Mucosal vaccination using antigens loaded or encapsulated in particles appears to have a sound scientific rationale based on the protection of an antigen from exposure to extreme pH conditions, bile and pancreatic secretions[68,89] and provide a depot-effect.[90] At the same time, advantage is taken of the inherent inclination of particulates to be naturally captured by mucosal APCs as part of its duty as a sentinel in triggering of mucosal immunity against pathogens[91–93] (see section 2.3.2.3). In fact, it has been clearly demonstrated that nanoparticles can interact with different components of the mucosa. In addition, the use of nanoparticles offers another advantage, such as the possibility to load an immunoadjuvant (*i.e.*, cholera toxin B subunit, CTB, or lipopolysaccharide, LPS) with the antigen in order to increase and induce the more appropriate immune response.[94,95]

Table 2.3.1 summarises some of the research works related to the development of nanoparticles as adjuvants for mucosal vaccination.

Table 2.3.1 Examples of oral immunizations using antigen-loaded nanoparticles.

Antigen	Polymer	Observations	Refs
Tetanus toxoid	SbutPVA–PLGA NPs	• After 3 doses, high levels of specific IgG and IgA • Antibody response greatly influenced by the NP size	54
Heliocobacter pylori lysates	PLGA NPs	• Significantly specific IgA and IgG responses for at least 9 weeks	64
Pertussis toxoid and filamentous haemagglutinin from *Bordetella pertussis*	PLGA NPs	• Three doses conferred protection against *B. pertussis* challenge • Induction of a Th2 response (IL-5, and mucosal IgA)	94
Plasmodium falciparum synthetic peptide SPf66	PLGA NPs	• Systemic IgG antibody responses • Predominant Th1 response	95
Salmonella enteritidis outer membrane complex	Gantrez NPs	• Elicited protection in mice • Th1 response	96
DNA encoding for *T. gondii* GRA1 protein	Chitosan NPs	• High antibody (IgG2a/IgG1) levels	97
Rotavirus antigens DNA	PLGA NPs	• Intestinal IgA antibodies protected mice against rotavirus challenge	98
HBsAg	UEA-1 coated PLGA NPs	• Lectin coated nanoparticles elicited secretion of IgA and high levels of Il-2 and IFN-γ	99
HbsAg	Alginate-coated chitosan NPs	• NPs elicited the generation of antibodies detected in serum and in intestinal washings (IgA)	100
pDNA for IL4/IL6	Chitosan NPs	• High levels of IL-2, IL-4 and IL-6 • Full protection in mice	101

PLGA: poly(D,L-lactide-co-glycolide); SbutPVA–PLGA: sulfobutylated poly(viny-lalcohol)-graft-poly(lactide-co-glycolide) PLA: poly (D,L-lactic acid); BSA: bovine serum albumin; NP: nanoparticles; UEA: *Ulex Europaeus* 1 lectin; HBsAg: hepatitis B antigen; pDNA (IL4/IL6): fusion gene encoding porcine interleukin 4 and interleukin 6.

One of the first works related with the efficacy of nanoparticles as mucosal adjuvants for vaccination was performed by Kim and collaborators.[64] In this work, the oral immunization of BALB/c mice with PLGA nanoparticles containing *Helicobacter pylori* lysates induced strong mucosal and systemic immune responses. A specific mucosal IgA response was observed as well as serum IgG1 and IgG2b responses, which were significantly higher than for animals immunized with free lysates. PLGA particles have also been proposed for vaccination against malaria and tetanus. In the former, Carcaboso and co-workers proposed the use of PLGA particles for the encapsulation of Spf66 (peptide from merozoite stage-specific protein fragments of *Plasmodium falciparum*).[95] In this work a mixture of two different PLGA particles were administered by the oral route, and when animals were boosted, 3 weeks later significant systemic IgG antibody responses were elicited. These responses were comparable to those obtained with a traditional adjuvant (alum triple shot) and superior to the aqueous vaccine given by the oral route.[95] In the latter, PLGA nanoparticles modified with sulfobutylated poly(vinylalcohol) were used for the oral delivery of tetanus toxoid (TT). Again, these nanoparticles orally administered induced reproducible levels of higher specific IgG and IgA titers than control mice when immunized by the intraperitoneal route.[54]

AIDS prevention through vaccination has also merited special attention. For this purpose, one of the most popular antigens used is Tat protein due to its role in HIV-1 replication and T-cell immunosuppression.[102] The oral administration in mice of Tat-loaded chitosan nanoparticles induced a cell-mediated immunity able to inhibit the activity of the protein, which is a prerequisite for the development of an anti-AIDS protective vaccine.[102] In a more recent study, Patel and collaborators proposed the use of anionic nano-particles to enhance both cellular and humoral immune responses to Tat. These nanoparticles were found to be effective to generate high antibody titers.[103]

A recombinant hepatitis B surface antigen (HBsAg) was encapsulated alone or in combination with immunostimulatory CpG motifs in alginate-coated chitosan nanoparticles. After two oral immunizations in mice, both types of nanoparticles vaccines showed enhanced immune responses characterised by higher values of the CD69 expression in CD4+ and CD8+ T-lymphocytes and lower values of this marker in B-lymphocytes. However, the generation of anti-HBsAg antibodies in serum (IgG) and in the intestinal washings (sIgA) was significantly higher in animals treated with chitosan-coated nanoparticles than with uncoated ones.[100] In another interesting approach, Gupta and collaborators proposed the use of HBsAg-loaded PLGA nanoparticles coated with *Ulex europaeus* 1. These "lectinized" nanoparticles demonstrated a significantly higher interaction with M-cells, enhanced immune responses and higher secretion of IL-2 and IFN-γ than conventional nanoparticles.[99]

The oral administration of DNA plasmids encapsulated in nanoparticles can also induce high systemic and mucosal antibody responses to the encoded protein. In this context, Bivas-Benita and collaborators evaluated the potential of chitosan nanoparticles loaded with *Toxoplasma gondii* GRA1 encoding

DNA plasmid to elicit GRA-1-specific immune responses after oral administration using different prime/boost regimen.[98] Boosting with GRA1 DNA vaccine resulted in high anti-GRA1 antibody levels, characterized by similar serum antibody titers of IgG2a and IgG1.[98] In a recent elegant work, a novel nanoparticles-in-microsphere oral system (NiMOS) has been proposed for transfection in specific regions of the gut. In this device, pDNA is firstly encapsulated in gelatine nanoparticles, which are then incorporated in poly(ε-caprolactone) microparticles of less than 5.0 micrometers. The transfection efficiency of this system may be of interest for vaccination purposes.[104]

Another interesting approach for oral vaccination against viral diseases is the direct coating of pathogenic virus with polymers in order to obtain nanoparticles capable of protecting the microorganism against inactivation and destruction during their passage along the stomach. In this way, different pH-dependent polymers (*i.e.* polyacrylic acid copolymers, polyacrylic acid–polyvinyl pirrolidone, sodium alginate–chitosan and sodium alginate–spermidine) were proposed for the coating and encapsulation of live measles virus.[105] Oral immunization of guinea pigs with those nanoparticles induced high and lasted titers of protective antibody titers for up 12 months after the immunization.[106] In a similar way, poly(ester) nanoparticles have been also proposed to entrap whole rotavirus (strain SA11). Thus, poly(lactic acid) (PLA) nanoparticles elicited improved and long-lasting IgA and IgG antibody titers in comparison to the soluble antigen.[107]

Nevertheless, up to now and despite the increased interest in the development of nanoparticles as oral adjuvants for antigen delivery, few studies address their potential efficacy against a challenge. One of the first works was based on the oral delivery of a *Bordetella pertussis* antigen entrapped in PLGA nanoparticles against a murine respiratory challenge model. In this model, nanoencapsulated antigens elicited both systemic and mucosal responses, leading to a protection against this pathogen.[94] More recently, an antigenic complex from the outer membrane of *Salmonella* Enteritidis was loaded in poly(anhydride) nanoparticles. In mice, three weeks after challenge with a lethal dose of this microorganism, the protection conferred by immunization was 80%. By contrast, the control formulation conferred only 20% of protection and all the mice died 6 days after the challenge.[108]

Concerning DNA vaccines, some interesting results have been also reported. Thus, the oral administration of a single dose of pDNA (encoding for the outer capsid proteins of rotavirus) loaded in PLGA nanoparticles was sufficient to elicit rotavirus-specific serum IgG, IgM, and IgA as well as intestinal IgA in mice.[109] Moreover, after challenge with a homologous rotavirus, fecal rotavirus antigen was significantly reduced compared with controls.[97,109] More recently, chitosan nanoparticles were also used for the oral delivery (by feeding) of a DNA vaccine against *Vibrio anguillarum* in sea bass. After being challenged by intramuscular injection, a survival rate of 46% was recorded.[110] In a similar work, these chitosan nanoparticles were used to incorporate a

DNA construct containing the VP28 gene of the white spot syndrome virus (WSSV). These vaccines were investigated in black tiger shrimp (*Penaeus monodon*). The results showed significant survival rates after WSSV-challenged shrimp at 7, 15 and 30 days post-treatment (relative survival, 85%, 65% and 50%, respectively) whereas 100% mortality was observed in the control groups.[111] In another elegant study, the potential of calcium phosphate nanoparticles as vehicle to deliver a DNA vaccine for foot and mouth disease was investigated. This plasmid encoded for all the proteins required for FMD viral capsid assembly. The resulting nanoparticles were orally administered and evaluated in both mice and guinea pigs. In all cases, these nanoparticles induced significant cell and humoral mediated immune responses. More important, all the immunised animals with these nanoparticles were protected against the challenge and displayed a similar degree of protection as observed with the control vaccine administered.[112]

Polymer nanoparticles can also be used to enhance or potentiate the immunity conferred by a traditional vaccine. In this way, the association of a fusion gene of porcine IL-4 and IL-6 with chitosan nanoparticles was evaluated in mice previously vaccinated with inactivated *Escherichia coli* vaccine. After the challenge, mice treated with nanoparticles displayed a content of immunoglobulins and specific antibodies to *E. coli* significantly higher than controls. In addition, all the animals treated with nanoparticles survived the challenge and did not show any symptoms or lesions, whereas the control mice manifested serious clinical symptoms.[101] In another approach, oligonucleotides enriched in CpG motifs were entrapped in chitosan nanoparticles and co-administered with a *Salmonella* Paratyphi vaccine in mice. The mice were orally challenged with virulent bacteria 35 days after inoculation. A synergistic effect between CpG and nanoparticles was found. In fact, animals treated with these nanoparticles displayed increased levels of immunoglobulins and specific antibodies to paratyphoid vaccine. More important, all the immunized animals treated with the nanoparticles survived, while the control mice fell ill with evident lesions provoked by *Salmonella*.[113]

2.3.6   Conclusions

In summary, intestinal epithelium, traditionally identified as a physical barrier against luminal bacteria, is also equipped with a repertoire of cells geared towards a tolerance status, meaning an active induction of T-cells secreting cytokines such as TGF-β, IL-4 or IL-10, known as anti-inflammatory cytokines. Thus, the default response in the gut after mucosal immunization is to drive a tolerogenic or a non-inflammatory response. This is an evolving consequence of food ingestion. Food antigens, that lack co-stimulatory "danger" signals, are programmed for mucosal tolerance. The continuous sampling of the intestinal contents by immature DCs, which do not have many co-stimulatory molecules, will render an anti-inflammatory response. However, pathogens present "danger molecules" (PAMPs) that bind mucosal epithelium and cause epithelial cells to release cytokines and chemokines that attract immune cells for induction of an inflammatory response. Therefore, modifying the vaccines with immunological adjuvants that carry PAMPs or other PRR-agonists may be a way to face the natural mucosal tolerogenic tendency. Thus, nanoparticles have been revealed as good oral adjuvants, being able to induce simultaneously a peripheral and a mucosal immunity. Through the use of nanoparticulated delivery systems it is possible to abrogate tolerance and elicit the desired balanced Th1/Th2/Th17 type responses in both mucosal and systemic sites.

References

1. S. H. E. Kaufmann, *Nat. Rev. Microbiol.*, 2007, **5**, 491.
2. S. Mitragotri, *Nat. Rev. Microbiol.*, 2005, **5**, 905.
3. A. J. Macpherson, K. D. McCoy, F. E. Johansen and P. Brandtzaeg, *Mucosal Immunol.*, 2008, **1**, 11.
4. Centers for Disease Control and Prevention, Department of Health and Human Services, http://www.cdc.gov/vaccines/recs/schedules/default.htm.
5. D. R. Hopkins, *The Greatest Killer: Smallpox in History*, University of Chicago Press, Chicago, 2002.
6. K. Landsteiner and J. Jacobs, *J. Exp. Med.*, 1935, **61**, 643.
7. S. C. Eisenbarth, O. R. Colegio, W. O'Connor, F. S. Sutterwala and R.A. Flavell, *Nature*, 2008, **453**, 1122.
8. P. Marrack, A. S. McKee and M. W. Munks, *Nat. Rev. Immunol.*, 2009, **9**, 287.
9. A. Podda and G. Del Giudice, *Expert Rev. Vaccines*, 2003, **2**, 197.
10. A. S. McKee, M. W. Munks and P. Marrack, *Immunity*, 2007, **27**, 687.
11. L. R. Coffman, A. Sher and R. A. Seder, *Immunity*, 2010, **33**, 492.
12. G.T. Nepom, *Nat. Rev. Immunol.*, 2008, **8**, 409.
13. D. Guy-Grand and P. Vassalli, *Curr. Opin. Immunol.*, 1993, **5**, 247.
14. C. F. Anderson, M. Oukka, V. J. Kuchroo and D. Sacks, *J. Exp. Med.*, 2007, **204**, 285.
15. M. Swamy, C. Jamora, W. Havran and A. Hayday, *Nat. Immunol.*, 2010, **11**, 656.
16. A. J. Macpherson and N. L. Harris, *Nat. Rev. Immunol.*, 2004, **4**, 478.
17. A. M. Mowat, *Nat. Rev. Immunol.*, 2003, **3**, 331.
18. R. Janeway and C. A. Medzhitov, *Annu. Rev. Immunol.*, 2002, **20**, 197.
19. K. Takeda, T. Kaisho and S. Akira, *Annu. Rev. Immunol.*, 2003, **21**, 335.
20. O. Takeuchi, S. Sato, T. Horiuchi, K. Hoshino, K. Takeda, Z. Dong, R. L. Modlin and S. Akira, *J. Immunol.*, 2002, **169**, 10.
21. A. T. Gewirtz, *Curr. Pharm. Des.*, 2003, **9**, 1.
22. S. C. Ng, M. A. Kamm, A. J. Stagg and S. C. Knight, *Inflammatory Bowel Dis.*, 2010, **16**, 1787.
23. S. R. Krutzik, M. T. Ochoa, P. A. Sielinq, S. Uematsu, Y. W. Ng, A. Legaspi, P. T. Liu, S. T. Cole, P. J. Godowski, Y. Maeda, E. N. Sarno,

M. V. Norgard, P. J. Brennan, S. Akira, T. H. Rea and R. L. Modlin, *Nat. Med.*, 2003, **9**, 525.
24. I. Tamayo, J. M. Irache, C. Mansilla, J. Ochoa-Repáraz, J. J. Lasarte and C. Gamazo, *Clin. Vaccine Immunol.*, 2010, **17**, 1356.
25. A. I. Camacho, R. Da Costa Martins, I. Tamayo, J. de Souza, J. J. Lasarte, C. Mansilla, I. Esparza, J. M. Irache and C. Gamazo, *Vaccine*, 2011, **29**, 7130.
26. L. Mayer, *J. Pediatr. Gastroenterol. Nutr.*, 2000, **30**, S4.
27. R. M. Salazar-Gonzalez, J. H. Niess, D. J. Zammit, R. Ravindran, A. Srinivasan, J. R. Maxwell, T. Stoklasek, R. Yadav, I. R. Williams, X. Gu, B. A. McCormick, M. A. Pazos, A. T. Vella, L. Lefrancois, H. C. Reinecker and S. J. McSorley, *Immunity*, 2006, **24**, 623.
28. K. E. Persson, E. Jaensson and W. W. Agace, *Immunobiology*, 2010, **215**, 692.
29. M. Rescigno and A. Di Sabatino, *J. Clin. Invest.*, 2009, **119**, 2441.
30. A Sato, M. Hashiguchi, E. Toda, A. Iwasaki, S. Hachimura and S. Kaminogawa, *J. Immunol.*, 2003, **171**, 3684.
31. M. H. Jang, M. N. Kweon, K. Iwatani, M. Yamamoto, K. Terahara, C. Sasakawa, T. Suzuki, T. Nochi, Y. Yokota, P. D. Rennert, T. Hiroi, H. Tamagawa, H. Iijima, J. Kunisawa, Y. Yuki and H. Kiyono, *Proc. Natl. Acad. Sci. U. S. A.*, 2004, **101**, 6110.
32. P. Tyrer, A. R. Foxwell, A. W. Cripps, M. A. Apicella and J. M. Kyd, *Infect. Immun.*, 2006, **74**, 625.
33. C. J. Rey, N. Garin, F. Spertini and B. Corthésy, *J. Immunol.*, 2004, **172**, 3026.
34. N. J. Mantis, M. C. Cheung, K. R. Chintalacharuvu, J.Rey, B. Corthesy and M.R. Neutra, *J. Immunol.*, 2002, **169**, 1844.
35. A. Cerutti, *Nat. Rev. Immunol.*, 2008, **8**, 421.
36. S. Fagarasan, S. Kawamotom, O. Kanagawa and K. Suzuki, *Annu. Rev. Immunol.*, 2010, **28**, 243.
37. M. Rescigno, M. Urbano, B. Valzasina, M. Francolini, G. Rotta, R. Bonasio, F. Granucci, J. P. Kraehenbuhl and P. Ricciardi-Castagnoli, *Nat. Immunol.*, 2001, **2**, 361.
38. M. Rimoldi, M. Chieppa, V. Salucci, F. Avogadri, A. Sonzogni, G. M. Sampietro, A. Nespoli, G. Viale, P. Allavena and M. Rescigno, *Nat. Immunol.*, 2005, **6**, 507.
39. D. I. Iliev, G. Matteoli and M. Rescigno, *J. Exp. Med.*, 2007, **204**, 2253.
40. P. Cossart and P. J. Sansonetti, *Science,* 2004, **304**, 242.
41. W. Chen, G. B. Patel, H. Yan and J. Zhang, *Hum. Vaccines*, 2010, **17**, 6.
42. J. Holmgren, C. Czerkinsky C, K. Eriksson and A. M. Harandi, *Vaccine*, 2003, **21**, S89.
43. J. M. Irache, H. H. Salman, S. Gomez, S. Espuelas and C. Gamazo, *Front. Biosci., Scholar Ed.*, 2010, **2**, 876.
44. J. F. Mann, R. Acevedo, J. D. Campo, O. Perez and V. A. Ferro, *Expert Rev. Vaccines*, 2009, **8**, 103.

45. J. Ochoa, J. M. Irache, I. Tamayo, A. Walz, V. G. Del Vecchio and C. Gamazo, *Vaccine*, 2007, **25**, 4410.
46. M. Estevan, J. M. Irache, M. J. Grilló, J. M. Blasco and C. Gamazo, *Vet. Microbiol.*, 2006, **118**, 124.
47. S. Gomez, C. Gamazo, B. San Román, M. Ferrer, M. L. Sanz and J. M. Irache, *Vaccine*, 2007, **25**, 5263.
48. S. Gomez, C. Gamazo, B. San Román, C. Vauthier, M. Ferrer and J. M. Irache, *J. Nanosci. Nanotechnol.*, 2006, **6**, 3283.
49. N. Lycke and M. Bemark, *Mucosal Immunol.*, 2010, **3**, 556.
50. S. P. Kasturi, I. Skountzou, R. A. Albrecht, D. Koutsonanos, T. Hua, H. I. Nakaya, R. Ravindran, S. Stewart, M. Alam, M. Kwissa, F. Villinger, N. Murthy, J. Steel, J. Jacob, R. J. Hogan, A. García-Sastre, R. Compans and B. Pulendran, *Nature*, 2011, **470**, 543.
51. M. O. Oyewumi, A. Kumar and Z. Cui, *Expert. Rev. Vaccines*, 2010, **9**, 1095.
52. I. Gutierro, R. M. Hernandez, M. Igartua, A. R. Gascon and J. L. Pedraz, *Vaccine*, 2002, **21**, 67.
53. V. Kanchan and A. K. Panda, *Biomaterials*, 2007, **28**, 5344.
54. T. Jung, W. Kamm, A. Breitenbach, K. D. Hungerer, E. Hundt and T. Kissel, *Pharm. Res.*, 2001, **18**, 352.
55. J. Wendorf, J. Chesko, J. Kazzaz, M. Ugozzoli, M. Vajdy, D. O'Hagan and M. Singh, *Hum. Vaccines*, 2008, **4**, 44.
56. T. Fifis, A. Gamvrellis, B. Crimeen-Irwin, G. A. Pietersz, J. Li, P. L. Mottram, I. F. McKenzie and M. Plebanski, *J. Immunol.*, 2004, **173**, 3148.
57. M. Kalkanidis, G. A. Pietersz, S. D. Xiang, P. L. Mottram, B. Crimeen-Irwin, K. Ardipradja and M. Plebanski, *Methods*, 2006, **40**, 20.
58. P. L. Mottram, D. Leong, B. Crimeen-Irwin, S. Gloster, S. D. Xiang, J. Meanger, R. Ghildyal, N. Vardaxis and M. Plebanski, *Mol. Pharm.*, 2007, **4**, 73.
59. A. Caputo, A. Castaldello, E. Brocca-Cofano, R. Voltan, F. Bortolazzi, G. Altavilla, K. Sparnacci, M. Laus, L. Tondelli, R. Gavioli and B. Ensoli, *Vaccine*, 2009, **27**, 3605.
60. J. F. Mann, E. Shakir, K.C. Carter, A. B. Mullen, J. Alexander and V. A. Ferro, *Vaccine*, 2009, **27**, 3643.
61. M. Shakweh, G. Ponchel and E. Fattal, *Expert. Opin. Drug Delivery*, 2004, **1**, 141.
62. X. Li, B. R. Sloat, N. Yanasarn and Z. Cui, *Eur. J. Pharm. Biopharm.*, 2011, **78**, 107.
63. K. J., Maloy, A. M., Donachie, D. T. O'Hagan, A. M. Mowat, *Immunology*, 1994, **81**, 661.
64. S. Y. Kim, H. J. Doh, M. H. Jang, Y. J. Ha, S. I. Chung and H. J. Park, *Helicobacter*, 1999, **4**, 33.
65. D. E. Katz, A. J. DeLorimier, M. K. Wolf, E. R. Hall, F. J. Cassels, J. E. van Hamont, R. L. Newcomer, M. A. Davachi, D. N. Taylor and C. E. McQueen, *Vaccine*, 2003, **21**, 341.
66. M. Otterlei, K. M. Varum, L. Ryan and T. Espevik, *Vaccine*, 1994, **12**, 825.

67. C. Porporatto, I. D. Bianco and S. G. Correa, *J. Leukocyte Biol.*, 2005, **78**, 62.
68. P. Arbós, M. A. Campanero, M. A. Arangoa, M. J. Renedo and J. M. Irache, *J. Controlled Release*, 2003, **89**, 19.
69. C. Olbrich, R. H. Muller, K. Tabatt, O. Kayser, C. Schulze and R. Schade, ATLA, *Altern. Lab. Anim.*, 2002, **30**, 443.
70. S. Peppoloni, P. Ruggiero, M. Contorni, M. Morandi, M. Pizza, R. Rappuoli, A. Podda and G. Del Giudice, *Expert Rev. Vaccines*, 2003, **2**, 285.
71. G. Ponchel and J. M. Irache, *Adv. Drug Del. Rev.*, 1998, **34**, 191.
72. A. Vila, A. Sanchez, M. Tobio, P. Calvo and M. J. Alonso, *J. Controlled Release*, 2002, **78**, 15.
73. K. Yoncheva, S. Gomez, M. A. Campanero, C. Gamazo and J. M. Irache, *Expert Opin. Drug Deliv.*, 2005, **2**, 205.
74. K. Yoncheva, E. Lizarraga and J. M. Irache, *Eur. J. Pharm. Sci.*, 2005, **24**, 411.
75. P. J. Sansonetti, *Med. Microbiol. Immunol.*, 1993, **182**, 223.
76. A. A. Fadl, K. S. Venkitanarayanan and M. I. Khan, *J. Appl. Microbiol.*, 2002, **92**, 180.
77. K. H. Darwin and V. L. Miller, *Clin. Microbiol. Rev.*, 1999, **12**, 405.
78. A. D. Humphries, S. M. Townsend, R. A. Kingsley, T. L. Nicholson, R. M. Tsolis and A. J. Baumler, *FEMS Microbiol. Lett.*, 2001, **201**, 121.
79. H. Kaltner and B. Stierstorfer, *Acta Anat.*, 1998, **161**, 162.
80. D. H. Lloyd, J. Viac, D. Werling, C. A. Reme and H. Gatto, *Vet. Dermatol.*, 2007, **18**, 197.
81. J. C. Sirard, F. Niedergang and J.P. Kraehenbuhl, *Immunol. Rev.* 1999, **171**, 5.
82. H. Salman, C. Gamazo, M. A. Campanero and J. M. Irache, *J. Controlled Release*, 2005, **106**, 1.
83. H. Salman, J. M. Irache and C. Gamazo, *Vaccine*, 2009, **27**, 4784.
84. F. Dalle, T. Jouault, P. A. Trinel, J. Esnault, J. M. Mallet, P. d'Athis, D. Poulain and A. Bonnin, *Infect. Immun.*, 2003, **71**, 7061.
85. J. M. Irache, H. Salman, C. Gamazo and S. Espuelas, *Expert Opin. Drug Deliv.*, 2008, **5**, 703.
86. D. L. Jack and M. W. Turner, *Biochem. Soc. Trans.*, 2003, **31**, 753.
87. K. Uemura, M. Ska, T. Nakagawa, N. Kawasaki, S. Thiel, J. C. Jensenius and T. Kawasaki, *J. Immunol.*, 2002, **169**, 6945.
88. S. Wagner, N. J. Lynch, W. Walter, W. J. Schwaeble and M. Loos, *J. Immunol.*, 2003, **170**, 1462.
89. M. Desai, J. Hilfinger, G. Amidon, R. J. Levy and V. Labhastwar, *J. Microencapsul.*, 1999, **17**, 215.
90. T. Storni, T. M. Kundig, G. Senti and P. Johansen, *Adv. Drug Delivery Rev.*, 2005, **57**, 333.
91. J. P. Krahenbuhl and M. R. Neutra, *Annu. Rev. Cell Dev. Biol.*, 2000, **16**, 301.
92. S. Espuelas, J. M. Irache and C. Gamazo, *Inmunologia*, 2005, **24**, 207.

93. D. T. O'Hagan, J. P. McGee, M. Lindblad and J. Holmgren, *Int. J. Pharm.*, 1995, **119**, 251.
94. M. A. Conway, L. Madrigal-Estebas, S. McClean, D. J. Brayden and K. H. Mills, *Vaccine*, 2001, **19**, 1940.
95. A. M. Carcaboso, R. M. Hernandez, M. Igartua, A. R. Gascon, J. E. Rosas, M. E. Patarroyo and J. L. Pedraz, *Int. J. Pharm.*, 2003, **260**, 273.
96. C. Gamazo, J. Ochoa-Reparaz and J. M. Irache, *Clin. Microbiol. Infect.*, 2004, **10**, 332.
97. M. Bivas-Benita, M. Laloup, S. Versteyhe, J. Dewit, B. J. De, E. Jongert and G. Borchard, *Int. J. Pharm.*, 2003, **266**, 17.
98. J. E. Herrmann, S. C. Chen, D. H. Jones, A. Tinsley-Bown, E. F. Fynan, H. B. Greenberg and G. H. Farrar, *Virology*, 1999, **259**, 148.
99. P. N. Gupta, K. Khatri, A. K. Goyal, N. Mishra and S. P. Vyas, *J. Drug Targeting* 2007, **15**, 701.
100. O. Borges, J. Tavares, A. de Sousa, G. Borchard, H. E. Junginger and A. Cordeiro-da-Silva, *Eur. J. Pharm. Sci.*, 2007, **32**, 278.
101. H. Zhang, C. Cheng, M. Zheng, J. L. Chen, M. J. Meng, Z. Z. Zhao, Q. Chen, Z. Xie, J. L. Li, Y. Yang, Y. Shen, H. N. Wang, Z. Z. Wang and R. Gao, *Vaccine*, 2007, **25**, 7094.
102. H. Le Buanec, C. Vetu, A. Lachgar, M. A. Benoit, J. Gillard, S. Paturance, J. Aucoututier, V. Gane, D. Zagury and B. Bizzini, *Biomed. Pharmacother.*, 2001, **55**, 316.
103. J. Patel, D. Galey, J. Jones, P. Ray, J. G. Woodward, A. Nath and R. J. Mumper, *Vaccine*, 2006, **24**, 3564.
104. M. D. Bhavsar and M. M. Amiji, *AAPS PharmSciTech.* 2008, **9**, 288.
105. E. A. Nechaeva, N. Varaksin, T. Ryabicheva, M. Smolina, T. Kolokoltsova, A. Vilesov, N. Aksenova, R. Stankevich and R. Isidorov, *Ann. N. Y. Acad. Sci.*, 2001, **944**, 180.
106. E. Nechaeva, *Exp. Rev. Vaccines*, 2002, **1**, 385.
107. B. Nayak, A. K. Panda, P. Ray and A. R. Ray, *J. Microencapsulation*, 2009, **26**, 154.
108. J. Ochoa-Reparaz, B. Sesma, M. Alvarez, M. J. Renedo, J. M. Irache and C. Gamazo, *Vet. Res.*, 2004, **35**, 291.
109. S. C. Chen, D. H. Jones, E. F. Fynan, G. H. Farrar, J. C. Clegg, H. E. Greenberg and J. E. Herrmann, *J. Virol.*, 1998, **72**, 5757.
110. S. Rajesh Kumar, V. P. Ishaq Ahmed, V. Parameswaran, R. Sudhakaran, V. Sarath Babu and A. S. Sahul Hameed, *Fish Shellfish Immunol.*, 2008, **25**, 47.
111. S. Rajesh Kumar, C. Venkatesan, M. Sarathi, V. Sarathbabu, J. Thomas, K. Anver Basha and A. S. Sahul Hameed, *Fish Shellfish Immunol.*, 2009, **26**, 429.
112. D. H. Joyappa, C. A. Kumar, N. Banumathi, G. R. Reddy and V. V. Suryanarayana, *Vet Microbiol.*, 2009, **20**, 58.
113. M. L. Fu, S. C. Ying, M. Wu, H. Li, K. Y. Wu, Y. Yang, H. Zhang, C. Cheng, Z. Z. Wang, X. Y. Wang, X. B. Lv, Y. Z. Zhang and R. Gao, *Biomed. Environ. Sci.*, 2006, **19**, 315.

Section 3
Nanocarriers Overcoming the Nasal Barriers

CHAPTER 3.1

Nanocarriers Overcoming the Nasal Barriers: Physiological Considerations and Mechanistic Issues

ANTÓNIO J. ALMEIDA* AND HELENA F. FLORINDO

Research Institute for Medicines and Pharmaceutical Sciences (iMed.UL), Faculty of Pharmacy, University of Lisbon. Av. Prof. Gama Pinto, 1649-003 Lisbon, Portugal
*E-mail: aalmeida@ff.ul.pt

3.1.1 Introduction

Mucosal absorption has been rather neglected in the advanced drug delivery market, perhaps because of the obstacles that still have to be overcome in order for mucosal routes to become commercially viable alternatives for the delivery of a large number of proteins and other large biomolecules.[1] The mucous surfaces of the body (mouth, eye, nose, rectum and vagina) offer less of a barrier than the skin or the GI tract to the systemic absorption of drugs and the advantage of bypassing the hepato-gastrointestinal first pass elimination associated with the oral route. They are ideal for rapid absorption but practical difficulties include the fact that most mucosal sites are not suitable for dosage forms that must remain in place for an extended period. Among the mucosal sites, the nasal route of administration is assuming greater importance as an alternative non-injectable route for local and systemic therapy or immunisation. Moreover, in

RSC Drug Discovery Series No. 22
Nanostructured Biomaterials for Overcoming Biological Barriers
Edited by Maria Jose Alonso and Noemi S. Csaba
© The Royal Society of Chemistry 2012
Published by the Royal Society of Chemistry, www.rsc.org

recent years it has emerged as a promising approach for brain delivery of soluble and nanoparticulate drugs, *via* the olfactory neuroepithelium, involving paracellular, transcellular and/or neuronal transport.

3.1.2   The Nasal Route of Administration

The nasal route of administration is a non-invasive, painless and easily accessible route that has widespread researchers' interest as an alternative for the local and systemic delivery of drugs and/or induction of systemic and mucosal antibody responses in saliva, respiratory and female genital tract secretions.[2,3] Moreover, it does not require formulation sterilization, hospitalization and specialized professionals for the administration of drugs and therefore has attracted the attention of the scientific community as an economic option to have in consideration, especially in the development of medicines to be used in underdeveloped countries.[2,4]

The nasal permeability, low proteolytic activity and epithelium high vascularization, more specifically the presence of venous sinusoids and arteriovenous anastomosis, allow a higher rate and extent of drug absorption, being able to overcome the enzymatic or acid degradation and the first-pass hepatic mechanism associated to the oral administration route, as the blood goes directly from the nose to the systemic circulation.[4–7] In addition, nasal large surface area (~ 150–160 cm^2 in humans) due to microvilli (~ 400 per cell) and porous endothelial membrane contribute to higher and rapid permeation of drugs when administered through this route.[2,8–10]

It has been shown that the intranasal administration of drugs as anti-histamines and corticosteroids allows the rapid development of the desired therapeutic effects, without causing adverse reactions as a consequence of their systemic delivery, being therefore preferable to their oral administration.[11] Imigran® and Miacalcic® are examples of formulations to be administered by the nasal route, being nasal sprays used to treat migraine and osteoporosis, respectively.[7] In addition, it has been demonstrated that the nasal adminis-tration of several drugs resulted in blood levels similar to those obtained using the intravenous route of administration.[2,7]

It is possible to target the central nervous system (CNS) by the intranasal delivery of substances with low molecular weight or even peptides or proteins.[7,12,13] However, it seems that CNS targeting through the nasal mucosa is very dependent on drug physicochemical properties, as in some studies the presence of the intranasal administered drugs was not detected in the brain area.[14,15]

Within the vaccination field, the stimulation of local nasal immune tissues, more specifically the nasopharinx-associated lymphoid tissue (NALT) and antigen-presenting cells (APC), by the targeted delivery of an antigen has been able to induce both local and systemic immune responses.[7,16–19] In fact, infections by potential pathogens have been prevented, including some infection-related diseases such as cervical cancer, due to a long-term antigen-specific antibody production and cellular immune responses.[20,21,22] Nevertheless, there are differences between human and mice nasal mucosa that need to be taken into consideration during preclinical studies involving the nasal administration of drugs and antigens. In fact, NALT as organized and diffuse lymphoid aggregates, as observed in laboratory animals such as mice, is replaced in superior animals by the Waldeyer's ring, which in humans consists of the adenoid or nasopharyngeal tonsil, the bilateral pharyngeal lymphoid bands, the bilateral tubal and palatine tonsils and the bilateral lingual tonsils, where macrophages and dendritic cells form a dense network.[23,24] Similar tissues have been found in several other animal species, including monkeys, horses, sheep, cattle, mice and hamsters.[24,25]

On the other hand, the intrinsic characteristics of this route of delivery also limits its use by high molecular weight compounds and large volumes ($> 200 \mu l$), these being solubility and possible dose limitations.[7,26] In fact, the physiological barriers that protect the organism by the infection of potential pathogens that use the nasal cavity as portal of entry, can also compromise drug permeability.[27] Those normal defence mechanisms and other physiological barriers are discussed below.

Despite the importance of the intrinsic characteristics of the nasal cavity, the dosage form properties also have an important effect on the epithelium permeability, especially when biodegradable and biocompatible polymers, *e.g.* hydroxypropylcellulose, polylactide (PLA), polylactide-co-glycolide (PLGA), poly-ε-caprolactone (PCL), hyaluronic acid, polyacrylic acid, poloxamers, polyethylene oxide, chitosan, alginate, lipids and other appropriate excipients,[17,27–29] are used to prepare drug/antigen carriers as solutions,[30] powders,[7] micro and nanoparticles,[31,32] liposomes,[33,34] niosomes[35] and hydrogels.[36–38] In fact, the prolonged time of contact of mucoadhesive delivery systems with the nasal mucosal surface results in enhanced drug permeability and extended periods of drug release when compared to conventional dosage forms.[7]

3.1.3 Physiological Barriers Associated to the Nasal Cavity

Besides being an effective route of administration, the nasal cavity has physiological barriers that prevent potential pathogens to overcome this portal of entry and thus develop local and systemic infections, but can also compromise the permeability of drugs through the nasal mucosa.[2]

The nasal cavity is composed by anterior and posterior vestibules, an atrium, an olfactory region, the nasopharynx and a respiratory area, which presents the highest permeability due to its rich vasculature and large surface area and great amount of nasal secretions. On the other hand, the nasal vestibule and the atrium present a low permeability mainly because of keratinized cells present on the first and the stratified cells and low surface of the latter.[2,39] As a result, the respiratory area is the one usually targeted for the permeation of compounds.

This nasal anatomical formation has a pseudostratified columnar epithelium that includes nonciliated cuboidal cells, ciliated cells at the apical surface, basal cells and goblet cells, being also found in trachea, bronchi and bronchioles. Goblet cells are glandular columnar epithelium cells that produce the secretions that will constitute the mucus layer that is transported towards nasopharynx by cilia movement, while basal cells are stem cells that will replace other necessary epithelial cells.[9] The relative high permeability of nasal epitheliums is due to the two cell layers that separate nasal lumen from the lamina propria and therefore the vasculature. A loose connective tissue forms the lamina propria present under this epithelium and contains mucous glands. The M-cells overlay the organized and diffuse lymphoid aggregates, which are interspersed with the nasal mucosal surface, and are mostly responsible for the uptake of nanoparticulate systems at this cavity.[27] In fact, the mucosal immune system is responsible for the induction of an effective immune response at mucosal tissues.

The neuronal regulation of nasal cycles of relaxation, a consequence of sympathetic stimulation, and congestion due to parasympathetic stimulation, regulates the amounts of drugs that actually cross the nasal epithelium, being improved by the latter.[40]

The main barriers that can be identified at the nasal cavity are: (1) the enzymatic activity, (2) the mucus layer, and (3) the mucociliary clearance and ciliary beating present at the nasal epithelium. All these need to be taken into consideration when developing a drug or antigen delivery approach.[2,27] The mucus is a complex secretion composed by water (90%), mucins (1%), salts (1%), lipids, mucopolysaccharides and proteins (1%), as albumin and lysozyme. Mucus rheological properties are consequence of the non-covalent bonds, such as hydrogen, electrostatic and hydrophobic bonds, formed between mucin glycoproteins of high molecular weight, which are also responsible for its negative charge at neutral pH.[9,41]

Approximately 1.5–2 mL of mucus is produced every day, constituting a thick layer (5 μm) and an enzymatic barrier as it contains a large number of peptidases, proteases and oxidative and conjugative enzymes that degrade compounds that reach nasal mucosa and thus reduce the permeation of drugs such peptides (insulin, LHRH and calcitonin).[7,42] As a consequence, the use of peptidase and protease inhibitors, such as amastatin, puromycin, have improved the nasal absorption of human growth hormones, leucin-enkephalin and LHRH growth hormones.[2,43]

It is important to mention that another critical factor for drug absorption is the mucus turnover at nasal epithelium as it will determine the excretion of a highly mucoadhesive system and thus limit the release of the drug substance. The viscosity of the layer is also fundamental, as a thick layer will prevent the interpenetration of the drug but a low viscosity will also make the adhesion of the drug delivery system difficult.[44]

The mucociliary clearance and ciliary beating remove mucus and prevents substance adherence to nasal mucosa, which will then be drained to the nasopharynx and discharged by the gastrointestinal tract. Therefore, the drug time of contact with nasal mucosa is reduced and thus so is their permeability.

The modification of the nasal mucosa structure, for example by the use of absorption enhancers such as spermine, chitosan or oleic acid, to change the physiology of nasal cavity has been explored to enhance drug permeability.[31,32] In fact, those absorption enhancers can act through a single or a combination of several mechanisms, such as the increase of membrane fluidity, decrease of mucous layer viscosity, disruption of tight junctions, inhibition of proteolytic enzymes, increase of blood flow, dissociation of protein aggregates and increase of paracellular or transcellular transport. However, these alterations have to be characterized and controlled as they can compromise the normal function of those barriers and cause adverse effects or even allow the development of certain infections.[2]

Thus, the use of water-soluble prodrugs improves nasal drug absorption, as these compounds need to be dissolved in the nasal secretions in order to permeate through the epithelium.[9,41] In addition, mucoadhesive hydrophilic polymers have been associated to active pharmaceutical ingredients in order to specifically target and deliver the drug at the nasal region, where the formulation will be retained and release the drug near the site of action, increasing therefore its bioavailability. This target delivery and higher bioavailability will allow the use of lower doses of the active substance and simultaneously the reduction of the drug–dose dependent side-effects that may occur due to its localization at non-disease sites.[45]

In addition, the pH of a dosage form formulation can be changed by nasal cavity pH, which is around 5.5–6.5 in adults and 5.0–7.0 in children. In fact, variation in the pH of nasal secretions can alter drug ionization and therefore modify its permeation through the epithelium.[2,7]

3.1.4 Mechanisms of Particle Uptake and Translocation at the Nasal Cavity

One of the characteristics influencing the increased interest in the nasal cavity as a site for systemic drug and antigen delivery is the ability of large molecules to permeate through the nasal mucosa into the systemic circulation. Particulate uptake also occurs in the nasal mucosa, and particles up to approximately 1 μm have been shown to rapidly enter the bloodstream following intranasal administration.[46] Therefore, the use of the nasal route for administration of nano/microparticulates has been extensively studied. Intranasal vaccination experiments against *Streptococcus pneumoniae* were conducted in the early 20th century resulting in the protection of rabbits immunized with a single dose of a suspension of killed *pneumococci*.[47] Nevertheless, evidence that the nasal administration of antigens can be a useful route of immunisation has been provided by Peters and Allison (in 1929), who investigated the possibility of inducing immunity to scarlet fever by repeated applications of erythrogenic toxin to the nasal mucosa.[48] From the group of 62 humans involved in the study, 46 showed immunity to some extent. A revival of interest in intranasal immunization started with the work of Waldman *et al.* who detected influenza antibodies in human respiratory secretions after intranasal administration of inactivated virus.[49,50] An attempt to immunise male volunteers against an inactivated strain of rhinovirus resulted in protection of the individuals given the vaccine nasally, whereas the frequency with which illness developed in the volunteers administered intramuscular vaccine did not differ from that of the seronegative controls.[51] These had developed only serum antibodies, while the volunteers who were immunised intranasally developed both serum and secretory antibodies. Finally, the penetration of mucosal surfaces by pathogenic bacteria and viruses, including experimental infections, indicate the existence of a particle uptake pathway at the nasal mucosa.[52–54]

As for the uptake and translocation of intact particulates in 1960 the absorption of 1–5 μm diameter particles by the tonsils in calves was described,[55] and in 1989 carbon particles were reported to be taken up at the respiratory mucosa associated lymphoid tissue in sheep,[25,56] whereas polystyrene particles were found in blood circulation after intranasal administration to rats.[57]

As aforementioned, the mucosal surface of the nasal cavity is covered by an epithelium, which constitutes an efficient physical barrier that protects the host from environmental pathogens. A controlled transepithelial transport of antigens or microorganisms must occur, however, in order to trigger immune responses or tolerance necessary for the protection and integrity of mucosal surfaces. Epithelial barriers at different mucosal sites of the body differ in their cellular organisation, and the antigen-sampling mechanisms at these regions are adapted accordingly. There is, however, a common feature in all these mucosal surfaces: the mucosa-associated lymphoid tissue (MALT), present in all of them, although with a different organisation, depending on the body region and the species.

Although the structure of epithelia may somehow vary according to the anatomical site, the routes and the mechanisms of drug absorption are generally applicable to most epithelial types. The route of diffusion through the membrane follows the Fick's law, which states that the rate of diffusion across a membrane is proportional to the concentration gradient across the membrane (eqn (3.1.1)).[58]

$$\frac{dm}{dt} = \frac{Dk}{h} A.C = PA.C \tag{3.1.1}$$

Where dm/dt is the rate of diffusion across the membrane, D is the diffusion coefficient of the drug in the membrane, k is the partition coefficient of the drug into the membrane, h is the membrane thickness, A is the available surface area, ΔC is the concentration gradient and P is the permeability constant (defined as Dk/h). Thus, drug absorption is mainly driven by a concentration gradient.

However, for particulates, a chemical potential (ΔC) no longer exists but is replaced by a convective kinetic energy, which causes movement and displacement of the particle through a medium (mucus in the present case). Absorption from mucosal sites can be described also by the Fick's diffusion law, but instead of a concentration gradient we may consider that Brownian motion for particles <1 µm provides the force for random motion and diffusion through a medium.[59,60] Translocation permeability is thus given by the Einstein–Stokes equation (eqn (3.1.2)), where k is the Boltzman constant, T is the temperature, η is the viscosity, r is the radius of the particle and h is the thickness of the mucus layer.

$$P = \frac{kT}{6\eta\pi rh} \qquad (3.1.2)$$

It is not solubility but other factors that control solid–liquid interfaces and interactions between particles and the mucus, such as surface tension, wettability, hydrophobic interactions, ionisable surfaces (zeta potential) and particle size. Thus, mucoadhesion phenomena play a crucial role in particle uptake at the nasal mucosa. Once close to the cell the absorption of particles will depend on cell–surface interactions.

Although many scientists have studied the phenomenon, the mucosal uptake and translocation of solid particles across the epithelial layer at the nasal cavity was for a long time controversial and is still not fully understood.[4,46,61] Three possible modes of uptake of particles have been proposed, some less efficient than others (Figure 3.1.1).

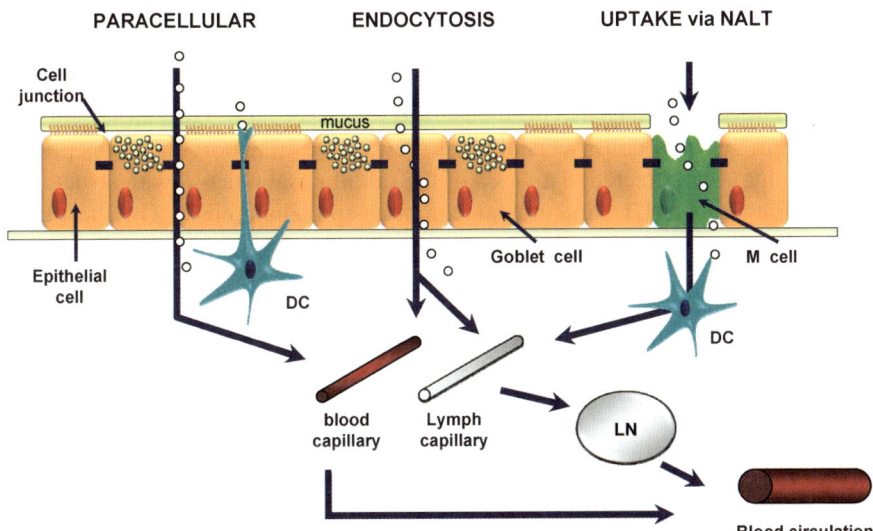

Figure 3.1.1 Routes and mechanisms of particle uptake and transport across the nasal epithelium.

3.1.4.1   Transport *via* the M-cells of NALT

The roles of the NALT and, in superior animals, of the Waldeyer's ring, have long been demonstrated. These structures are able to take up and transport particulate and soluble antigens.[23,25,56,62–65] The uptake of polystyrene particles, together with the penetration of mucosal surfaces by pathogenic bacteria, showed the existence of a particle uptake pathway at the nasal mucosa.[4,46,57] A mechanism for the mode of entry and processing of intranasally administered particulate antigens has been proposed by Kuper *et al.* and is related to that described for other MALT components.[24] This mechanism was later confirmed by Eyles *et al.* using FITC-labelled polystyrene particles of 1.1 μm diameter.[66] Briefly, nasal administration of particulate antigens results in uptake mainly by the M-cells, while soluble antigens are mainly absorbed at the nasal epithelium. Antigens of the first type will be processed at the NALT and preferentially drained to the posterior cervical lymph nodes (PCLN). After uptake at the nasal epithelium, soluble antigens will be carried by antigen presenting cells to the superficial cervical lymph nodes (SCLN), which in turn drain to the PCLN. At the SCLN, soluble antigens may induce a systemic immune response or a status of specific tolerance. On the other hand, as PCLN are involved in the enhancement of a secretory immune response, these antigens can also induce this type of immunity. The final result of NALT stimulation will depend on the balance between the activation in the posterior or superficial cervical lymph nodes. Therefore, given the sites of preferential uptake, the nature of the antigen plays an important role in the ultimate response.[4]

3.1.4.2   Transcellular Transport *via* the Epithelial Cells

Most small, lipophilic drug molecules are absorbed by passive diffusion. Active transport mediated by carriers for di- and tri-peptides has been demonstrated, whereas endocytosis has also been observed for larger molecules, although this mechanism remains controversial.[67] In fact, active transport of polypeptides ($\approx$ 4000 Da) was detected in an area of the rabbit nasal mucosa with structures resembling the MALT, and not at common epithelial cells.[68] On the other hand, the transport of 10–500 nm polystyrene particles across nasal epithelial membranes has been clearly demonstrated, although in minimal amounts.[69]

3.1.4.3   Paracellular Transport Between Epithelial Cells

The paracellular uptake of 10–200 nm polystyrene particles across excised rabbit nasal mucosa has been reported.[69] However, the paracellular pathway is dependent on the opening of the tight junction between epithelial cells. Protein kinase C is strongly involved in the complicated signal transduction process that regulates the tight junctions. It has been shown that chitosan and other polycationic materials can open tight junctions.[70,71] Therefore, this is the main mode of entry of the chitosan-based drug delivery systems, increasing paracellular transport of polar drugs. Furthermore, chitosan polymers are mucoadhesive, increasing the contact time for the transport of the drug across the membrane and the absorption of drugs and antigens, presenting immunoadjuvant properties. Consequently, chitosan is used in several intranasal pharmaceutical forms, including powders, liquids, gels, microparticles and nanoparticles.[72] More recently positively and negatively charged ultraflexible liposomes containing salmon calcitonin were reported to decrease the calcium level in a rate and extent almost equivalent to that obtained with a subcutaneous injection of the protein.[73] Although the authors did not provide a discussion of the possible mode of transport of the liposomes, a pathway similar to that of ultradeformable lipid vesicles used for transdermal delivery, *i.e.* a paracellular route, is implicit.[74]

Irrespective of the mechanism, it is certain is that the onset of detection of particles on blood is really short, even considering that they may first be taken by the lymphatic circulation. After nasal administration, fluorescent carboxylated polystyrene particles of 0.51 μm and 0.83 μm in diameter were detected in peripheral blood after 10 min and 12 min, respectively.[57,75] The data produced by several research groups in the last 25 years show that both the extent of absorption and the mechanism of particle uptake at the nasal mucosa, as in other mucosal sites, are dependent on particle size, surface characteristics, presence of specific permeation enhancers and other factors, such as vehicle volume.

It would appear that particles of certain compositions are capable of uptake by the NALT and subsequent translocation through the lymphatics. Nevertheless, there is a clear dependence on particle size. Smaller particles are taken up and translocate to a higher degree than the larger particles.[46,76]

Particle absorption improves with a decrease in particle size, as demonstrated by the tissue levels of radiolabelled tetanus toxoid following nasal administration of the antigen encapsulated in PLA-PEG particles of different size, being higher for PLA-PEG 200 nm particles than for PLA-PEG microparticles of 1 μm, 5 μm or 10 μm,[19] Particle size must also be taken into account when formulations are intended for nasal immunization, since smaller particles with a with a size range of 10–200 nm enter the initial lymphatic vessels from the interstitial space by directly diffusing through lymphatic endothelial cell junctions, whereas larger particles 200 nm to 1 μm are too large to pass through endothelial cell junctions and are transported into the lymph vessel following uptake by dendritic cells.[77]

Concerning surface characteristics, as a general rule, greater uptake is observed for more hydrophobic particles, which is consistent with the particle surface properties that affect phagocytosis.[78] Nevertheless, it has been shown that the more hydrophobic the particle surface, the lower the percentage transported regardless of size of the particle.[46] Hydrophilic PLA-PEG nanoparticles showed higher nasal uptake than the more hydrophobic PLA nanoparticles of the same particle size.[79] According to Lai *et al.*[80] these results seem to be related to the interaction of PEG with the mucus. As mentioned above, interaction between particles and the mucus is crucial for particle uptake, and PEGylation promotes particle diffusion through mucus. On the other hand, PLA-PEG nanoparticles show increased stability, which correlated well with the enhanced mucosal transport and prolonged blood circulation time of encapsulated protein antigen (tetanus toxoid) following nasal administration in rats, whereas the corresponding non-PEGylated PLA particles strongly aggregated upon contact with biological fluids.[79,81] This has been particularly well reviewed by Csaba *et al.*[81] We think both explanations are valid and these phenomena may act simultaneously, contributing to an increased uptake of PLA-PEG nanoparticles.

The presence of specific targeting ligands is also an important feature of nasally administered particulate drug or vaccine carriers. Monoclonal antibodies with specificity for M-cells, or lectins, which bind to specific carbohydrate residues found on M-cells, can increase the uptake of micro/ nanoparticulate delivery systems.[82]

Another common approach to promote the uptake of particulate drug delivery systems at the nasal cavity is the already mentioned use of permeation enhancers, most of which have been studied to improve the absorption of insulin.[43] These compounds include surfactants, cyclodextrins, bile salts (sodium deoxycholate, sodium dihydrofusidate), cationic polymers (chitosan, sperminated gelatin, aminated gelatin), and fatty acids (oleic acid). In fact, the efficiency of nanoparticulate nasal vaccine formulations based on PLGA or PCL, either with the antigens entrapped or adsorbed, is significantly increased by adding permeation enhancers to the formulations, such as chitosan, oleic acid and spermine (Figure 3.1.2.).[31,32,83–85]

Besides, particle uptake at the nasal mucosa appears to depend also on vehicle properties and volume. The *in vivo* distribution of 1.1 μm fluorescent polystyrene carboxylate microspheres is clearly dependent on volume. Mice treated with 10 μL volumes of showed little evidence of particle translocation to liver and spleen, although significant particle transfer to PCLN (which drain NALT) occurred. In contrast, appreciable lung deposition of the microspheres occurred when the particles were nasally instilled in 50 μL volumes.[66]

Despite the amount of data available demonstrating that particle uptake is a reality and that polymeric micro- and nanoparticles are useful delivery systems, no set criteria are available for the design of a good particulate carrier for nasal delivery of drugs or antigens. This is partly due to the publication of conflicting and confusing data, including variation in methodology, mode of evaluation and animal species.

3.1.4.4 Particle Uptake and Translocation to the CNS

Although the phenomenon has been known for decades, in recent years the nasal cavity has emerged as a promising site for brain delivery of soluble and nanoparticulate drugs *via* the olfactory neuroepithelium. Early studies carried out with poliovirus instilled intranasally into chimpanzees and Rhesus monkeys revealed that the olfactory nerve and olfactory bulbs are, indeed, portals of entry to the CNS for intranasally instilled virus particles. The

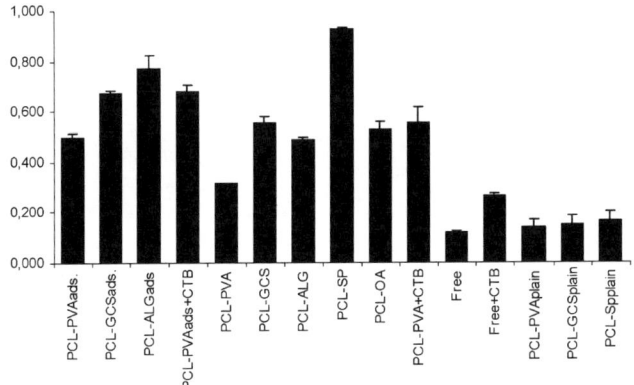

Figure 3.1.2   Secretory IgA levels in lung washes of mice intranasally immunised with different PCL nanoparticulate formulations containing *Streptococcus equi* antigens, after 12 weeks of booster dose (mean $\pm$ s.d.; n = 4). Legend: PCL-PVA – polycaprolactone–polyvinyl alcohol nanoparticles; PCL-GCS – polycaprolactone–glycol-chitosan nanoparticles; PCL-ALG – polycaprolactone–alginate nanoparticles; CTB – cholera toxin B subunit; PCL-SP – polycaprolactone–spermine nanoparticles; PCL-OA – polycaprolactone–oleic acid nanoparticles; Free – free antigens; Free+CTB – free antigens admixed with cholera toxin B subunit; ads – formulations containing the antigens loaded by adsorption onto the nanoparticle surface; plain – empty nanoparticles. In the non-specified nanoparticulate formulations the antigens were loaded by encapsulation within the particles. Reproduced from ref. 31 with permission from Elsevier Ltd.

authors even determined the transport velocity of the virus in the axoplasm of axons to be 2.4 mm h^{-1}.[86,87] In fact, the close proximity of nasal olfactory mucosa and olfactory bulb requires only a short distance to be covered by neuronal transport (Figure 3.1.3).

Studies carried out in animal models have shown that drug transport into the CNS is dependent on drug molecular weight, hydrophobicity and degree of ionization, as well as the mucociliary clearance and the P-glycoprotein efflux mechanism. Therefore, the use of nanoparticulate carrier systems has been investigated as an alternative approach to nose-to-brain drug delivery.

The mechanisms of particle uptake are not fully understood but may involve paracellular, transcellular and/or neuronal transport.[89] In a study conducted in mice, it was shown that nanoparticle transport of chitosan-coated polystyrene or polysorbate-coated polystyrene nanoparticles (100 nm or 200 nm in diameter) was exclusively transcellular.[90] None of the nanoparticle formulations showed preference for uptake into olfactory axons over other nasal epithelial cells and no 100 nm particles could be found in the olfactory bulbs. The authors speculated that an optimal nanoparticle diameter for axonal transport is <100 nm in mice.[90] Nevertheless, particle size is only one of the parameters governing this process, surface characteristics of nanoparticles

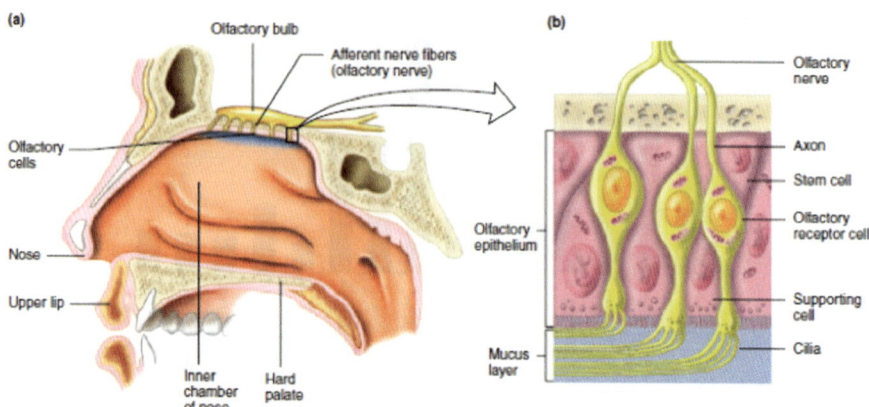

Figure 3.1.3 Location (a) and enlargement (b) of a portion of the olfactory epithelium showing the structure of the olfactory receptor cells. Reproduced from ref. 88, with permission from McGraw-Hill Companies, Inc.

being also essential determinants. Although the transport pathways of particulates from the nasal cavity to the CNS, and its toxicological implications have been reviewed and discussed,[87] their therapeutic applications still raise relevant questions, such as the intended therapeutic effect and the amount of particles that are effectively translocated, as well as the fate of nanoparticles after their translocation to specific cell types or to subcellular structures in the brain. Therefore, further studies in suitable animal models and humans are required, particularly those concerning the safety aspect involved in the nose-to-brain delivery of drugs.

3.1.5 Conclusions

Mucosal delivery and uptake of particulate systems is a subject which is now starting to be understood. Among the several mucosal sites, the nasal cavity is a potential alternative to the parenteral route due to its easier accessibility. Drugs and antigens may be attached to a suitable particulate carrier for therapeutic purposes upon uptake at the nasal cavity. Several local or systemic therapeutic applications may be foreseen, such as immunisation with protein antigens or DNA, infectious disease treatment and cancer. Nasal administration of particulate antigens may produce effective vaccines both for protection of the upper respiratory tract and for induction of long lasting protective systemic immunity.

Nevertheless, a more functional control of the uptake and translocation process must be achieved through the full investigation of the factors that affect particulate absorption. Indeed, tailoring particulate carries in terms of particle size, surface characteristics and the presence of specific permeation enhancers would allow selection of the translocation route and control the *in vivo* fate and efficacy of the formulations.

References

1. N. M. K. Ghilzai and A. Desai, in *Pharma Tech 2004*. Business Briefing World Market Series, London, 2004, (http://www.touchbriefings.com/pdf/890/ACF1A06.pdf).
2. H. R. Costantino, L. Illum, G. Brandt, P. H. Johnson and S. C. Quay, *Int. J. Pharm.*, 2007, **337**, 1.
3. P. A. Kozlowski, S. B. Williams, R. M. Lynch, T. P. Flanigan, R. R. Patterson, S. Cu-Uvin and M. R. Neutra, *J. Immunol.*, 2002, **169**, 566.
4. A. J. Almeida and H. O. Alpar, *J. Drug Targeting*, 1996, **3**, 455.
5. S. S. Davis, *Adv. Drug Delivery Rev.*, 2001, **51**, 21.
6. R. J. Soane, M. Frier, A. C. Perkins, N. S. Jones, S. S. Davis and L. Illum, *Int. J. Pharm.*, 1999, **178**, 55.
7. P. J. Watts and A. Smith, *Drug Discovery Today*, 2011, **16**, 4.
8. X. F. Liu, J. R. Fawcett, R. G. Thorne, T. A. Defor and W. H. Frey, *J. Neurol. Sci.*, 2001, **187**, 91.
9. G. P. Andrews, T. P. Laverty and D. S. Jones, *Eur. J. Pharm. Biopharm.*, 2009, **71**, 505.
10. E. Marttin, N. G. M. Schipper, J. C. Verhoef and F. W. H. M. Merkus, *Adv. Drug Delivery Rev.*, 1998, **29**, 13.
11. R. J. Salib and P. H. Howarth, *Drug Safety*, 2003, **26**, 863.
12. M. Kumar, A. Misra, A. K. Babbar, A. K. Mishra, P. Mishra and K. Pathak, *Int. J. Pharm.*, 2008, **358**, 285.
13. W. A. Banks, *Eur. J. Pharmacol.*, 2004, **490**, 5.
14. Z. Yang, Y. Huang, G. F. Gan and R. J. Sawchuk, *J. Pharm. Sci.*, 2005, **94**, 1577.
15. M. P. van den Berg, P. Merkus, S. G. Romeijn, J. C. Verhoef and F. W. H. M. Merkus, *Pharm. Res.*, 2004, **21**, 799.
16. A. W. Zuercher, S. E. Coffin, M. C. Thurnheer, P. Fundova and J. J. Cebra, *J. Immunol.*, 2002, **168**, 1796.
17. A. H. Krauland, D. Guggi and A. Bernkop-Schnurch, *Int. J. Pharm.*, 2006, **307**, 270.
18. I. M. van der Lubben, G. Kersten, M. M. Fretz, C. Beuvery, J. C. Verhoef and H. E. Junginger, *Vaccine*, 2003, **21**, 1400.
19. A. Vila, A. Sanchez, C. Evora, I. Soriano, O. McCallion and M. J. Alonso, *Int. J. Pharm.*, 2005, **292**, 43.
20. K. R. Van Kampen, Z. K. Shi, P. Gao, J. F. Zhang, K. W. Foster, D. T. Chen, D. Marks, C. A. Elmets and D. C. C. Tang, *Vaccine*, 2005, **23**, 1029.
21. H. O. Alpar, J. E. Eyles, E. D. Williamson, S. Somavarapu, *Adv. Drug Delivery Rev.*, 2001, **51**, 173.
22. R. B. Belshe, *J. Infect. Dis.*, 2004, **190**, 2096.

23. C. Balmelli, S. Demotz, H. Acha-Orbea, P. De Grandi and D. Nardelli-Haefliger, *J. Virol.*, 2002, **76**, 12596.
24. C. F. Kuper, P. J. Koornstra, D. M. H. Hameleers, J. Biewenga, B. J. Spit, A. M. Duijvestijn, P. J. C. V. Vriesman and T. Sminia, *Immunol. Today*, 1992, **13**, 219.
25. W. Chen, M. R. Alley, B. W. Manktelow, D. Hopcroft and R. Bennett, *J. Comp. Pathol.*, 1991, **104**, 47.
26. P. Dondeti, H. Zia and T. E. Needham, *Int. J. Pharm.*, 1996, **127**, 115.
27. H. O. Alpar, S. Somavarapu, K. N. Atuah and V. Bramwell, *Adv. Drug Delivery Rev.*, 2005, **57**, 411.
28. L. Illum, I. Jabbal-Gill, M. Hinchcliffe, A. N. Fisher and S. S. Davis, *Adv. Drug Delivery Rev.*, 2001, **51**, 81.
29. C. Tas, C. K. Ozkan, A. Savaser, Y. Ozkan, U. Tasdemir and H. Altunay, *Eur. J. Pharm. Biopharm.*, 2006, **64**, 246.
30. S. Y. Yu, Y. Zhao, F. L. Wu, X. Zhang, W. L. Lu, H. Zhang and Q. Zhang, *Int. J. Pharm.*, 2004, **281**, 11.
31. H. F. Florindo, S. Pandit, L. Lacerda, L. M. D. Goncalves, H. O. Alpar and A. J. Almeida, *Biomaterials*, 2009, **30**, 879.
32. H. F. Florindo, S. Pandit, L. M. D. Goncalves, M. Videira, O. Alpar and A. J. Almeida, *Biomaterials*, 2009, **30**, 5161.
33. A. Shahiwala and A. Misra, *J. Pharm. Pharmacol.*, 2006, **58**, 19.
34. J. W. Wiseman, C. A. Goddard and W. H. Colledge, *Gene Ther.*, 2001, **8**, 1562.
35. S. A. Dsouza, J. Ray, S. Pandey and N. Udupa, *J. Pharm. Pharmacol.*, 1997, **49**, 145.
36. J. Wu, W. Wei, L. Y. Wang, Z. G. Su and G. H. Ma, *Biomaterials*, 2007, **28**, 2220.
37. M. Koping-Hoggard, A. Sanchez and M. J. Alonso, *Expert Rev. Vaccines*, 2005, **4**, 185.
38. R. C. Read, S. C. Naylor, C. W. Potter, J. Bond, I. Jabbal-Gill, A. Fisher, L. Illum and R. Jennings, *Vaccine*, 2005, **23**, 4367.
39. N. Jones, *Adv. Drug Delivery Rev.*, 2001, **51**, 5.
40. G. Philip, F. M. Baroody, D. Proud, R. M. Naclerio and A. G. Togias, *J. Allergy Clin. Immun.*, 1994, **94**, 1035.
41. M. I. Ugwoke, R. U. Agu, N. Verbeke and R. Kinget, *Adv. Drug Delivery Rev.*, 2005, **57**, 1640.
42. G. S. Tirucherai, C. Yang and A. K. Mitra, *Expert Opin. Biol. Ther.*, 2001, **1**, 49.
43. X. P. Duan and S. R. Mao, *Drug Discovery Today*, 2010, **15**, 416.
44. H. Hagerstrom, M. Paulsson and K. Edsman, *Eur. J. Pharm. Sci.*, 2000, **9**, 301.
45. J. Woodley, *Clin. Pharmacokinet.*, 2001, **40**, 77.
46. M. D. Donovan and Y. Huang, *Adv. Drug Delivery Rev.*, 1998, **29**, 147.
47. C. G. Bull and C. M. Mckee, *Am. J. Hyg.*, 1929, **9**, 490.
48. B. A. Peters and S. F. Allison, *Lancet*, 1929, **1**, 1035.

49. R. H. Waldman, J. A. Kasel, R. V. Fulk, Y. Togo, R. B. Hornick, G. G. Heiner, A. T. Dawkins and J. J. Mann, *Nature*, 1968, **218**, 594.
50. J. Mestecky, J. Bienenstock, J. R. McGhee, M. E. Lamm, W. Strober, J. J. Cebra, L. Mayer and P. L. Ogra in *Mucosal Immunology*, ed. J. Mestecky, M. E. Lamm, W. Strober, J. Bienenstock, J. R. McGhee and L. Mayer, Elsevier, Amsterdam, 2005, p xxxiii.
51. J. C. Perkins, D. N. Tucker, H. L. S. Knopf, R. P. Wenzel, A. Z. Kapikian and R. M. Chanock, *Am. J. Epidemiol.*, 1969, **90**, 519.
52. E. R. Moxon, A. L. Smith, D. R. Averill and D. H. Smith, *J. Infect. Dis.*, 1974, **129**, 154.
53. I. E. Salit, E. Vanmelle and L. Tomalty, *Can. J. Microbiol.*, 1984, **30**, 1022.
54. M. Novak, Z. Moldoveanu, D. P. Schafer, J. Mestecky and R. W. Compans, *Vaccine*, 1993, **11**, 55.
55. J. M. Payne, B. F. Sansom, R. J. Garner, A. R. Thomson and B. J. Miles, *Nature*, 1960, **188**, 586.
56. W. Chen, M. R. Alley and B. W. Manktelow, *J. Comp. Pathol.*, 1989, **101**, 327.
57. H. O. Alpar, A. J. Almeida and M. R. W. Brown, *J. Drug Targeting*, 1994, **2**, 147.
58. A. M. Hillery in *Drug Delivery and Targeting for Pharmacists and Pharmaceutical Scientists*, ed. A. M. Hillery, A. W. Lloyd and J. Swarbrick, Taylor & Francis, London, 2001, p 1.
59. D. A. Norris and P. J. Sinko, *J. Appl. Polym. Sci.*, 1997, **63**, 1481.
60. D. A. Norris, N. Puri and P. J. Sinko, *Adv. Drug Delivery Rev.*, 1998, **34**, 135.
61. T. H. Ermak and P. J. Giannasca, *Adv. Drug Delivery Rev.*, 1998, **34**, 261.
62. P. Brandtzaeg, P. Joahnsen, I. Farstad and G. Haraldsen, *Ann. N. Y. Acad. Sci.*, 1997, **820**, 1.
63. Y. Fujimura, *Virchows Arch.*, 2000, **436**, 560.
64. Y. Fujimura, M. Takaeda, H. Ikai, K. Haruma, T. Akisada, T. Harada, T. Sakai and M. Ohuchi, *Virchows Arch.*, 2004, **444**, 36.
65. Y. Fujimura, T. Akisada, T. Harada and K. Haruma, *Med. Mol. Morphol.*, 2006, **39**, 181.
66. J. E. Eyles, V. W. Bramwell, E. D. Williamson and H. O. Alpar, *Vaccine*, 2001, **19**, 4732.
67. K. Ohtake, H. Natsume, H. Ueda and Y. Morimoto, *J. Controlled Release*, 2002, **82**, 263.
68. D. Cremaschi, C. Rossetti, M. T. Draghetti, C. Manzoni and V. Aliverti, *Pflugers Archiv – Eur. J. Physiol.*, 1991, **419**, 425.
69. Y. Huang and M. Donovan, *Int. J. Pharm. Adv.*, 1996, **1**, 298.
70. J. M. Smith, M. Dornish and E. J. Wood, *Biomaterials*, 2005, **26**, 3269.
71. L. Illum, *J. Pharm. Sci.*, 2007, **96**, 473.
72. M. Amidi, E. Mastrobattista, W. Jiskoot and W. E. Hennink, *Adv. Drug Delivery Rev.*, 2010, **62**, 59.

73. M. Chen, X. R. Li, Y. X. Zhou, K. W. Yang, X. W. Chen, Q. Deng, Y. Liu and L. J. Ren, *Peptides*, 2009, **30**, 1288.
74. G. Cevc, A. Schatzlein and H. Richardsen, *Bichim. Biophys. Acta, Biomembranes*, 2002, **1564**, 21.
75. A. J. Almeida, H. O. Alpar and M. R. W. Brown, *J. Pharm. Pharmacol.*, 1993, **45**, 198.
76. J. Ali, M. Ali, S. Baboota, J. K. Sahni, C. Ramassamy, L. Dao and Bhavna, *Curr. Pharm. Design*, 2010, **16**, 1644.
77. M. Bachmann and G. Jennings, *Nat. Rev. Immunol.*, 2010, **10**, 787.
78. H. O. Alpar and A. J. Almeida, *Eur. J. Pharm. Biopharm.*, 1994, **40**, 198.
79. M. Tobio, R. Gref, A. Sanchez, R. Langer and M. J. Alonso, *Pharm. Res.*, 1998, **15**, 270.
80. K. Lai, D. O'Hanlon, S. Harrold, S. Man, Y. Wang, R. Cone and J. Hanes, *Proc. Natl. Acad. Sci. U. S. A.*, 2007, **104**, 1482.
81. N. Csaba, M. Garcia-Fuentes and M. J. Alonso, *Adv. Drug Delivery Rev.*, 2009, **61**, 140.
82. P. J. Giannasca, J. A. Boden and T. P. Monath, *Infect. Immun.*, 1997, **65**, 4288.
83. H. F. Florindo, S. Pandit, L. M. D. Goncalves, H. O. Alpar and A. J. Almeida, *Vaccine*, 2008, **26**, 4168.
84. H. F. Florindo, S. Pandit, L. M. D. Goncalves, H. O. Alpar and A. J. Almeida, *Vaccine*, 2009, **27**, 1230.
85. F. Oyarzun-Ampuero, M. Garcia-Fuentes, D. Torres and M. Alonso, *J. Drug Delivery Sci. Technol.*, 2010, **20**, 277.
86. D. Bodian and H. A. Howe, *Bull. Johns Hopkins Hosp.*, 1941, **69**, 79.
87. G. Oberdörster, E. Oberdörster and J. Oberdörster, *Environ. Health Perspect.*, 2005, **113**, 823.
88. E. P. Widmaier, H. Raff and K. T. Strang, *Vander, Sherman & Luciano's Human Physiology: The Mechanisms of Body Functions*, McGraw Hill, New York, 2004, p. 240.
89. A. Mistry, S. Stolnik and L. Illum, *Int. J. Pharm.*, 2009, **379**, 146.
90. A. Mistry, S. Z. Glud, J. Kjems, J. Randel, K. A. Howard, S. Stolnik and L. Illum, *J. Drug Targeting*, 2009, **17**, 543.

CHAPTER 3.2

Nanostructures Overcoming the Nasal Barrier: Protein and Peptide Delivery Strategies

CECILIA PREGO[a]† AND FRANCISCO M. GOYCOOLEA[*b]

[a] Department of Pharmaceutical Technology, University of Santiago de Compostela, 15782, Santiago de Compostela, Spain; †Current address: CZ Veterinaria S.A., Poligono La Relva, Torneiros s/n, 36400 Porriño, Pontevedra, Spain; [b] Institute of Plant Biology and Biotechnology, Westfalian Wilehlms University of Münster, Schlossgarten 3, 48149 Münster, Germany
*E-mail: goycoole@uni-muenster.de

3.2.1 Introduction

Biotechnology has developed a number of therapeutic peptides and proteins for treating incurable diseases. The approval of biopharmaceuticals represents a significant and growing activity of the overall pharmaceutical market.[1,2] From 2005 to 2010, a total of 58 biopharmaceuticals were approved within the European Union and/or the United States. Among them, 40% represented genuinely new biological entities; however, around 50% of them were versions of previously approved substances. The most prominent blockbusters are monoclonal antibodies-based products for the treatment of cancer, followed by insulin and its analogs. Other products include hormones, growth factors, blood-related proteins, and subunit proteins.

In spite of these advances, the administration of such biopharmaceuticals usually requires a parenteral route. Over the past decade or so, the quest for

RSC Drug Discovery Series No. 22
Nanostructured Biomaterials for Overcoming Biological Barriers
Edited by Maria Jose Alonso and Noemi S. Csaba
© The Royal Society of Chemistry 2012
Published by the Royal Society of Chemistry, www.rsc.org

effective methods to deliver peptide and proteins to the systemic circulation avoiding parenteral administration has been at the forefront of research. The nasal route has attracted much attention as it is easier and more comfortable to the patient and it avoids enterohepatic circulation and gut enzymes. This naturally makes it a very attractive alternative for the delivery of peptides and recombinant proteins. In addition to the general well-advocated advantages of the nasal administration of peptide/protein drugs, the normal endogenous pulsatile secretory pattern characteristic of feedback-regulated hormones (*e.g.* insulin) may be mimicked more closely when administered intranasally than by s.c. injection due to a rapid absorption and corresponding high peak amplitude. On the other hand, over the past decade or so the potential of the nasal route has been considered as particularly promising for the administration of therapeutic compounds, such as neuropeptides and hormones (*e.g.* estradiol), directly to the brain whose therapeutic use has been hampered because, when administered systemically, these compounds do not readily pass the blood–brain barrier (BBB), and they evoke potent hormone-like side effects when circulating in the blood.[3–6] In this regards, a number of studies have addressed the suitability of the nasal route to target therapeutic peptides to the central nervous system (CNS) using olfactory and trigeminal neural pathways connecting the nasal passages to the brain.[7–11]

In spite of the great advantages of the nasal route, the bioavailability of large molecular weight (M_w) biopharmaceutical drugs from the nasal cavity is normally affected by both physiological barriers and by the poor permeability of these type of drugs. As a result, interest in the development of new dosage forms have increased over the past two decades. Nanotechnology approaches have firmly demonstrated a promising route in this regard. The focus of the present Chapter is centred on the various strategies to utilize nanocarriers for the systemic delivery of bioactive proteins through the nasal route. Only slight attention is paid to intranasal vaccine delivery strategies, as these are the subject of the following chapter by B. Slütter and W. Jiskoot on nasal vaccine delivery (Chapter 3.3).

3.2.2   Nasal Bioavailability of Therapeutic Peptides and Proteins

The delivery of therapeutic peptides or proteins across the nasal epithelium is limited by physiological and metabolic barriers, which will affect to a varying extent the pharmacological bioavailability. Tightly impermeable epithelial cell layers in the nasal cavity and also covering the nasopharyngeal associated lymphoid tissue (NALT) as well as the short residence time of formulations in the nasal cavity are severe physiological limitations for protein/vaccine delivery in the upper respiratory tract.[12] Mucociliary clearance and ciliary beating are the mechanisms responsible for the short residence time of any particulate formulation deposited in the nasal cavity. In addition, deposition site, environment (*e.g.* pH), membrane permeability, the presence of the mucus layer and immunological clearance, are other well-identified physiological barriers that need to be overcome to achieve an effective systemic response after a peptide or protein is administered intranasally. Additional factors that contribute to a low bioavailability include metabolizing enzymes that degrade the molecule prior or during absorption, efflux transporters and nasal congestion. Besides, the physicochemical properties inherent to the peptide/protein molecule are also of crucial importance, mainly the hydrophilicity and M_w are among the factors that limit their permeability across the nasal epithelium.

Normally, the nasal epithelium is considered to have a lower metabolic activity than the gastrointestinal tract. However, it has been shown that in the nasal mucosa there is also significant proteases activity.[13] Approximately 1.5–2 litres of mucus are secreted daily by goblet cells and serous glands within the nasal cavity, known to contain *ca.* 1% of proteins, including several proteases such as neutral endopeptidases (enkephalinase and catepsin B), lysozyme, aminopeptidase peroxidase and carboxipeptidase N[14]. Small peptides are relatively resistant to the action of endopeptidases, however large peptides are considerably degraded by these enzymes. These proteases occur as extra- and intracellular enzymes, hence the actions of intracellular enzymes will not be significant if the peptides are absorbed by both passive diffusion or actively *via* the transcellular pathway, thus never coming into contact with the inside of the cell. This basically depends on the lipophilicity of the active compound. However, for transcytotic transport, enzymatic degradation may be relevant. Hence, suitable strategies are needed to protect bioactivity and prevent the loss of bioavailability of metabolically liable polypeptides. Among the strategies to prevent enzymatic degradation, various compounds have been used such as specific and non-specific protease inhibitors including bacitracin, bestatin, diprotinin and aprotinin, which inhibit leucineaminopeptidase, dipeptidyl peptidase and trypsin, respectively.[14–20] Drug delivery nanocarriers are effective in this regard, as they are bound to protect the formation of the enzyme-substrate complex.

In general, low M_w drugs tend to be lost less easily by several-fold than are high M_w ones mostly due to clearance and membrane permeability. Early studies addressed the effects of the M_w of a large list of water soluble drugs on their nasal absorption. The nasal absorption was estimated by measuring the excretion of these compounds in bile and urine. Water soluble, low M_w, peptides enter *via* passive diffusion through aqueous channels of the membrane. Below $M_w \sim 300$ Da the permeation of drugs is not affected by the physicochemical properties. However, for peptides/proteins of $M_w > 300$ Da absorption rates fell drastically as the M_w increased.[21,22] Indeed, relatively small peptides such as pentapeptide metkephamide (M_w 660 Da), desmopressin (M_w 1069 Da) and buserelin (M_w 1000 Da) are absorbed to near 100%, whilst for gonadorelin (M_w 1300 Da), salmon calcitonin (M_w 3400 Da) or insulin (M_w 6000 Da) the absorption falls sharply to $< 3\%$.[21,22]

Nasal spray formulations of low M_w peptide drugs have been successfully developed, such as buserelin (Suprefact®), approved in Canada for the treatment of diabetes insipidus and primary nocturnal enuresis, prostatic cancer, estradiol dependent endometriosis and infertility. Other commercially available peptide nasal sprays include desmopresin acetate (Minirin®), prescribed also as an anti-diuretic; nafarelin acetate (Synarel®), an agonistic analog of gonadotropin-releasing hormone (GnRH) prescribed in control of precocious puberty, and salmon calcitonin (Miacalcin®) for the treatment of postmenopausal osteoporosis (Table 3.2.1). Even when in general the nasal bioavailability of these peptides is typically low ($< \sim 3$–10%), they are prescribed for non-parenteral application in light of the clinical advantage, practical application and cost of development. However, for proteins of larger size that currently are prescribed parenterally, *e.g.* insulin, human growth hormone, exenatide or glucagon, the development of strategies to promote drug absorption and increase their nasal bioavailability ($< 1\%$) have been at the focus of pharmaceutical research.

The early approaches to develop effective formulations for intranasal administration of proteins were based in the use of absorption or penetration enhancers.[23] Among these, bile salts (*e.g.* sodium deoxycholate, sodium glycocholate, sodium taurocholate or sodium taurodihydrofusidate) are known to decrease the viscosity of the mucus and create transient hydrophilic pores in the membrane bilayer. EDTA and fatty acid salts (*e.g.* sodium caprate and sodium laurate) increase the paracellular transport by removal of luminal calcium, thus increasing the permeability of the tight junctions. In turn, non-ionic detergents (*e.g.* Laureth-9) are known to alter membrane structure and permeability. However, the use of this type of penetration enhancers in chronic therapy entails the potential risk that other large molecules can enter the systemic circulation once the epithelial barrier is breached. Besides, some of these (*e.g.* bile salts) have reported to cause a deleterious effect to the epithelial tissue, either directly, by perturbing vital cell structures and/or functions, or indirectly, by permeabilizing the epithelium.[14] These features led to the search of alternative enhance penetration agents of lower toxicity. In this regard, cell-penetrating peptides have been evaluated as nasal absorption enhancers of

Table 3.2.1 Commercially available peptide drugs applied by the nasal route

Drugs	Molecular weight (Da)	Commercial name	Company	FDA Approval date
Desmopressin acetate	1183	Minirin®	Sanofi-Aventis	1978
Salmon calcitonin	3432	Miacalcin®	Novartis	1995
Buserelin acetate[a]	1239	Suprefact®	Sanofi-Aventis	—
Nafarelin acetate	1321	Synarel®	Pfizer	1990
Oxytocin	1007	Syntocinon®	Novartis	1995

[a]Approved in Canada

insulin,[24-26] with pronounced effects on the increased absorption of the peptide. Also, cyclodextrins have been used as solubilizers and absorption enhancers for peptide nasal drug delivery[27-29] as well as to increase the stability of peptides like insulin.[30] Cyclodextrins are circular oligosaccharides typically of five to seven glucose residues. It is known that cyclodextrins are able to suppress the aggregation of therapeutically relevant proteins, such as insulin, human growth hormone and salmon calcitonin, as well as several other proteins. This stems on their ability to bind to aromatic residues in Tyr, Phe and Trp which can lead to preferential stabilization of the unfolded state of the protein, thus effectively reducing the folding rates, as recently reviewed.[31] Indeed, the partial inclusion of hydrophobic chains of macromolecules in the lipophilic cavity of cyclodextrins has been suggested to lead to increase in the stability of peptides in nasal enzymatic systems.[30,32] Other intrinsic advantages of cyclodextrins include their ability to enhance the nasal absorption of peptides and proteins by inhibiting their enzymatic degradation, inhibit the function of P-glycoprotein and disrupt the epithelial membrane by extraction of phospholipids and proteins, and/or by opening cellular tight junctions.[27,30,33,34] Altogether, these properties make cyclodextrin compounds extremely attractive components of intranasal formulations.

Another key approach aiming to enhance intranasal drug absorption of high M_w protein drugs includes the use of mucoadhesive systems that reduce the mucociliary clearance of drug formulations and thereby increases the contact time between the drug and the absorption site.[35] The main strategies under this approach are use of microparticles, bioadhesive liquid formulations and liquid gelling formulations.

Solid microparticles based on bioadhesive polymers have been found to prolong the retention time and improve the bioavailability of insulin and desmopresin;[14,36] also liophylized microparticles comprising hydroxypropyl methyl cellulose (HPMC), hydroxypropyl-β-cyclodextrin and D-alpha-tocopheryl poly(ethylene glycol 1000) have been found to have the highest capacity to enhance the permeabilization of lysozyme in a cell monolayer of primary in primary human nasal epithelial (HNE) cells.[37] In other studies it has been found that starch dry powder microparticles are able to promote the intranasal absorption of octeotride as model drug in anaesthetised rats.[38] Meanwhile, the nasal absorption promoting activity of other polysaccharide microparticles, including alginate, pectin and cellulose derivative (Avicel PH10), was studied.

A general conclusion derived from these studies was that the capacity to enhance the intranasal absorption of octeotride is directly related to the Ca^{2+} binding capacity and Sephadex 125 (crosslinked dextran), alginate and pectin resulted with the highest bioavailability, while chitosan had hardly any effect. These results, however, are bound to be influenced by the water uptake behaviour of the various powder formulations and must be considered cautiously.

A more recent approach documents the use of a liquid formulation polymer conjugate in which oligoarginine, a cell penetrating peptide, is linked to the sodium salt of poly(*N*-vinylacetamide-co-acrylic acid) (PNVA-co-AA), a random co-polymer composed of *N*-vinylacetamide and acrylic acid as an absorption enhancer of insulin and exendin-4 after intranasal administration of a physical mixture of the polymer and the therapeutic peptides. The expectation was to obtain a polymer that would enhance the cellular uptake of the bioactive molecules that are physically mixed with them. Insulin, exendin-4, and oligoarginine-linked PNVA-co-AA were administered to rats. It was found that oligoarginines linked to PNVA-co-AA possessed a significant penetration-enhancing function.[39]

3.2.3 Nanocarriers for Intranasal Delivery of Peptide/ Proteins

Compared with other dosage forms such as powders or microspheres, nanocarriers have distinct advantages in improving the transport of peptide and protein macromolecules across the nasal mucosa. Studies are consistent with the notion that the efficacy of transport across the nasal mucosa is deeply influenced by the particle size. Hence, for micro- and nanoparticles of PLA-PEG the highest transport of tetanus toxoid was found for nanoparticles of 200 nm.[40] Hence, drug delivery systems with nanometric dimensions have been the focus of research and innovation over the past years. Other studies have confirmed that it is not only the size of the nanocarrier that determines its properties as an effective transporter across epithelial barriers, but also the nature of the coating material is critical in preserving the stability of the nanoparticles in the mucosal fluids and, as a consequence, in favouring their interaction with the nasal mucosa as well as in preventing the interaction of particles with blood proteins, and hence their recognition by mononuclear phagocyte system.[41,42] In this regards, nanocarriers which incorporate bioadhesive polymers at their surface have been found as particularly attractive for intranasal administration of therapeutic proteins, as is further discussed below.

Nanosystems that have been studied in the development of intranasal vehicles for therapeutic peptides and proteins include polymer-based nanospheres (or nanoparticles), polymer conjugates, liposomes, solid-lipid nanoparticles, nanocapsules and *in situ* gelling systems. According with their specific composition and structural complexity, distinctive terms have been coined to designate the various nanosystems, such as "aquasomes" (nanoparticulate carrier systems with three layered self-assembled structures), "archaesomes" (vesicles composed of glycerolipids of Archaea with potent adjuvant activity), "proteasomes" (high-molecular-weight multi-subunit enzyme complexes with catalytic activity that is specifically due to assembly pattern of enzymes), among others.[43] In Figure 3.2.1 a schematic representation of two different type of nanocarriers, namely, matrix nanoparticles and oil-core nanocapsules, is illustrated.

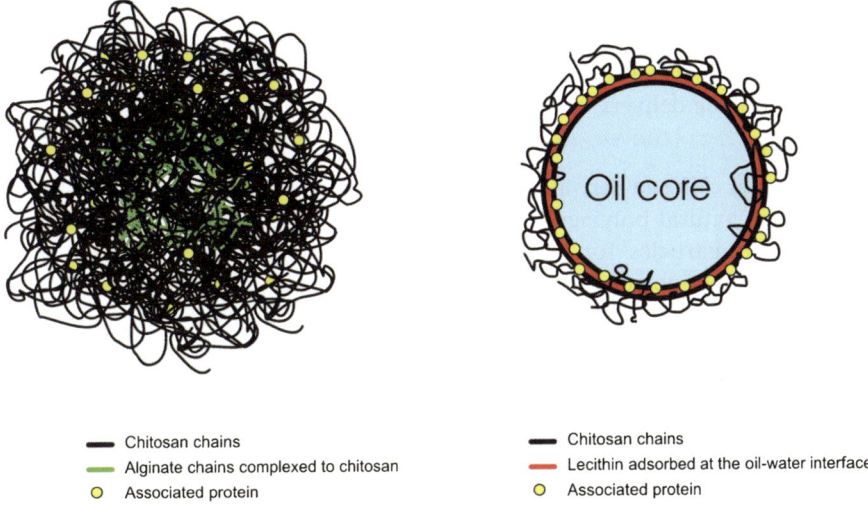

- ━━ Chitosan chains
- ━━ Alginate chains complexed to chitosan
- ○ Associated protein

- ━━ Chitosan chains
- ━━ Lecithin adsorbed at the oil-water interface
- ○ Associated protein

Figure 3.2.1 Schematic representation of protein-loaded chitosan–alginate hybrid nanoparticle (left) and a lecithin–chitosan oil-core nanocapsule (right).

3.2.3.1 Polymer Nanoparticles

Nanoparticles are sub-micron-size colloidal particles formed by a polymeric matrix in which the drug can be associated or encapsulated into their matrix, adsorbed or conjugated onto their surface. Polymers that have been used to harness drug delivery nanocarriers for peptides and proteins include natural or synthetic materials. The list of natural polymers includes, but is not limited to: chitosan, starch derivatives, alginate, dextran sulfate and cellulose derivatives, while among the synthetic ones, polyesters such as polylactic acid (PLA), poly(glycolic acid) (PGA), poly(lactic co-glycolic) (PLGA), along with poly-ε-caprolactone, poly(methyl methacrylate), poly(vinyl) alcohol (PVA), poly(-amino acids), are among the main ones. Regardless of their origin, these polymeric materials share as the key property, the fact that they are biocompatible and biodegradable. Although the repertoire of biocompatible polymers amenable for use in the preparation of nanoparticles suitable for peptide and protein delivery is large, only a few of them have so far proved to be effective in the design of adequate systems for transmucosal nasal delivery. This is so because it is not easy to combine the needed physicochemical properties, suitable technology and safety to harness adequate colloidal nanoparticles with the needed biopharmaceutical performance (*e.g.* permeabilizing, bioadhesive properties, *etc.*) into one single macromolecule.

In the following sections we address in detail the various types nanocarrier systems with proof-of-concept of their effectiveness as intranasal carriers of peptide/protein, with attention placed on their physicochemical and biopharmaceutical properties. Table 3.2.2 summarizes a representative selection of proof-of-principle of intranasal administration of therapeutic peptides/proteins in different animal models by various nanoparticle systems. Although the main focus of this chapter is on non-vaccine peptide/proteins, for the sake of comparison, we have also included in the Table some examples of systems intended for the delivery of ovalbumin and tetanus toxoid.

Chitosan and chitosan derivatives-based nanoparticles

Among the natural polymers that have been more extensively investigated to harness nanoparticles for drug delivery and applications in nanomedicine, chitosan is undoubtedly at the top of the list, and particularly so for the development of nanocarrier systems for nasal, ocular, oral and pulmonary transmucosal peptide and protein.[44,45] Chitosan is a linear heteropolysaccharide that is comprised by $(1{\rightarrow}4)$-linked 2-amino-2-deoxy-β-D-glucose (D units) and 2-acetamido-2-deoxy-β-D-glucose (A units) sugar residues. The molar ratio of A units with respect to the total (A + D units) is regarded as the degree of acetylation (DA%). Together, the M_w and the DA% are the key properties that determine to a large extent physicochemical and biological properties of a given chitosan. Chitosan is soluble in acidic aqueous solution at pH below its intrinsic pK_o (~ 6.1). The polycationic character of chitosan in solution confers it with a high capacity for the electrostatic association and delivery of therapeutic hydrophilic macromolecules (*e.g.* hormones, pDNA, siRNA, antigens, among other), thus effectively, protecting their bioactivity against enzymatic and hydrolytic degradation. Chitosan is a relatively non-toxic, biocompatible material.[46] It is known to be biodegraded by human chitotriosidase,[47] lysozyme which is highly concentrated in mucosal surfaces,[48–50] and by bacterial chitosanases.[51] Other biological properties that make chitosan an ideal candidate for the development of intranasal nanocarriers include its high mucoadhesivity,[52,53] biodegradability, biocompatibility[54,55] and capacity to promote the absorption of poorly absorbable macromolecules across epithelial barriers by transient widening of cell tight junctions, thus modifying the paracellular transport.[56–60] This effect has been mainly attributed to a reorganization of the actin rings[56] and to a translocation of the ZO-1 and occludin tight junction proteins[58] from the membrane to the cytoskeleton. As a result of the biological and physicochemical properties of chitosan, it is considered an almost unique biopolymer. Hence, it has been utilized extensively in pharmaceutical research, including the development of novel nanocarriers where its biological properties are fully exploited, particularly in the development of drug delivery formulations suitable for administration by non-invasive routes, including the intranasal administration of peptides and proteins.[59–65]

The group led by Prof. Maria J. Alonso at Universidad de Santiago de Compostela (Spain) were pioneers in the development of a nanoparticle drug delivery platform based on nanoparticles obtained by ionotropic gelation of chitosan with pentasodium tripolyphosphate (TPP).[66] Among the main advantages of this type of hydrophilic nanoparticle system is that the process to prepare them is simple and does not need the use of high temperatures,

Table 3.2.2 Examples of protein-loaded nanoparticle systems and proof-of-concept for their use as intranasal carriers.

Constituent polymer[a]	Other components[b]	Active molecule[c]	Size (d. nm)	Zeta potential (mV)	Assoc. efficiency (%)	Dose[d] (IU kg^{-1}) or (µg kg^{-1})	Animal model[e]	Ref.
Chitosan-Cl	TPP	INS	352 ± 11	+33 ± 0.2	92.5 ± 0.5	5	Rabbit	65
Chitosan-Cl	TPP, ALG	INS	275 ± 1	+43 ± 4.0	55.8 ± 1.2	5	Rabbit	79
Chitosan-Cl	TPP, SBβCD	INS	327 ± 27	+32 ± 0.1	94.9 ± 0.1	5	Rabbit	74
Chitosan	Na$_2$SO$_4$	OVA	385 ± 8	+25 ± 0.6	84.8 ± 0.3	2.5	Rat	71
Chitosan	TPP, GEL	TT	354 ± 27	+37 ± 0.3	55.1 ± 3.4	40	Rat	63
Chitosan-*g*-PEG	TPP	INS	218 ± 12	+16 ± 0.4	18.6 ± 2.4	5	Rabbit	69
Chitosan-Au		INS	100	+65	53	10	Dia_Rat	84
Chitosan-PLGA		^{125}I TT	500 ± 29	+22 ± 1.1	90.8 ± 3.8	n.r.	BALB/c mice	42
Potatoe starch	GLCh; LPC	INS	218 ± 6	n.r.	n.r.	10	Dia_Rat	85
PLA	GEL	^{125}I TT	153 ± 3	n.r.	36.7 ± 0.3	n.r.	Rat	41
PLA-PEG			137 ± 1	n.r.	35.6 ± 0.6			
Amine-modified		INS	250 ± 39	+18 ± 3.3	n.r.	4	Rat	89
PVA-*g*-PLA			290 ± 62	+23 ± 2.5	n.r.	10	Dia_Rat	

[a]PEG = polyethylenglycol; PLA = polylactic acid; PLGA = polylactic co-glycolic acid; PVA-*g*-PLA = polyvinyl alcohol-graft-polylactic acid. [b]Gelling and co-gelling compounds and/or permeation enhancers: TPP = pentasodium triphosphate; SBβCD = sulfobutyl ester β-cyclodextrin; GLCh = Na-glycocholate; LPC = lysophospatidylcholine; GEL = gelatin. [c]Active molecules: INS = insulin; OVA = ovalbumin; TT = tetanous toxoid. [d]Dose of insulin in IU kg^{-1}; dose of other proteins in µg kg^{-1}. [e]Unless otherwise indicated in column, assays were conducted on healthy animals. Dia_Rat = stroptozotocin-induced diabetic rat; n.r. = not reported values.

vigorous stirring, use of organic solvents nor sonication. Instead, the nanoparticles are formed spontaneously upon mixing chitosan and TPP in aqueous phase, at room temperature and under gentle stirring. Such mild conditions, along with the polycationic nature of chitosan, confer to the colloidal particles the capacity to effectively associate and preserve the stability and bioactivity of therapeutic macromolecules against enzymatic and hydrolytic degradation. Indeed, this system has shown high loading capacity for hydrophilic proteins such as insulin,[65,67–69] bovine serum albumin[70,71] and tetanus toxoid.[42,63] The formation of chitosan-based nanoparticles by ionotropic gelation with TPP has served as the basis for the development of drug delivery platforms for the transmucosal delivery of a number of different peptides and proteins (*e.g.* insulin, tetanus toxoid, ovalbumin) and other biological macromolecules (*e.g.* pDNA, siRNA, heparin). The chitosan–TPP nanoparticle platform constitutes the basis of a number of new systems that have been developed over the past decade by incorporating other co-gelled polysaccharides, such as hyaluronic acid, konjac glucomannan, alginate and cyclodextrins,[72,73] as well as for nanoparticles obtained from chemical derivatives of chitosan that make use of ionotropic gelation of chitosan.

Effectiveness of chitosan-based nanoparticles
Several studies focused on the efficacy of chitosan-based nanoparticles for improving the nasal bioavailability of peptide and protein macromolecules have addressed the delivery of insulin, due to the ease to measure its systemic absorption in various animal models.[65,74] But also other proteins, such as tetanus toxoid[42,63] and ovalbumin,[70,71] have been successfully loaded to chitosan-based nanoparticles and tested *in vivo*.

In studies of intranasal administration of insulin, the efficacy of chitosan-based nanoparticles[75] at increasing the hypoglycaemic response in healthy rabbits has consistently been demonstrated.[65,69,74,75] Instillation of insulin-loaded chitosan–TPP nanoparticles led to a significant decrease of the glycaemia levels ($\sim 40\%$ reduction) compared to insulin–chitosan in acetate buffer that did not show any significant effect from insulin alone solution in the same buffer ($\sim 85\%$ reduction in blood glucose levels occurring 30 min post-administration)[65] and in both cases the blood glucose nadir occurred within the same timespan (~ 45–60 min). The mechanisms whereby chitosan-based nanoparticles improved the nasal transport of insulin across the nasal epithelium were thought to be related to the capacity of chitosan to open the tight junctions, in line with previous reports.[56,59] In addition, based on the positive data obtained for chitosan nanoparticles as compared to chitosan solution, it was proposed that the underlying action mechanisms of nanoparticles may differ to those of chitosan solutions. Two possibilities have been considered in this regard: (a) chitosan nanoparticles could intensify the contact of insulin with the absorptive epithelium as compared to chitosan solutions; (b) some nanoparticles could enter the nasal epithelium, thus working as true peptide carriers from the mucosal fluids to the nasal epithelium, thus exploiting the endocytotic/transcytotic pathways. This last interpretation agrees well with the ability of these particles to be internalized by epithelial cells *in vivo*, as shown in cross sections epithelium excised from rats by probing the presence of nanoparticles prepared with fluorescein labelled chitosan (Fl-chitosan) using confocal laser scanning microscopy (CLSM).[74] In addition, it may be pointed out that another explanation to the improved bioavailability of peptide drugs incorporated in particulate formulations in comparison to nasal solutions can be seen in the steeper concentration gradient of the drug that can be achieved by utilizing nano or microparticles.[76] Consequently, it seems likely that the suggested mechanisms may operate simultaneously. Additional advantages associated with the use of chitosan nanoparticles encompass entrapment of the peptide/protein drug within the particle matrix, consequently ensuring protection against enzymatic degradation and the possibility of controlled drug release.

Hybrid chitosan-based nanoparticles

A recently developed class of chitosan-based hybrid nanoparticles intended for intranasal delivery of peptides, have been harnessed by incorporating oligosaccharides such as cyclodextrin derivatives[74,77,78] or other polysaccharides such as alginate[79] into the chitosan–TPP system. In a recent study conducted in our group, a similar hypoglycaemic minimum as that observed following instillation of insulin-loaded chitosan–TPP nanoparticles was also observed for insulin-loaded chitosan–TPP–alginate ($M_w \sim 18$ or ~ 32 kDa) nanoparticles when tested in the same experiment. Interestingly enough, however, the hypoglycaemic effect corresponding to the one of the chitosan–TPP–alginate formulations (comprising alginate of $M_w \sim 18$ kDa), was significantly prolonged by up to 5 h with respect to the control (Figure 3.2.2). With regard to the bioadhesive behaviour of chitosan–TPP–alginate nanoparticles, it is noteworthy that the presence of alginate could further contribute to this mechanism due to its high affinity for Ca^{2+} known to be present in nasal secretions.[80] This leads us to think on the possibility that the insulin is delivered more slowly *in vivo* due to a stronger interaction of the peptide with the alginate-containing nanocarrier. This effect is observed despite the fact that alginate was present only in minor amounts in these formulations (*i.e.* <10%).

In turn, systems comprising cyclodextrins have been found to be able to load efficiently insulin, as well as other therapeutic macromolecules, and to exhibit some specific advantages when compared with nanoparticles that comprise only chitosan and TPP.[74] In general, the nanoparticles that comprise cyclodextrins exhibit less cytotoxicity in Calu-3 cells in a wide range of doses. Insulin-loaded chitosan–cyclodextrin–TPP nanoparticles were administered intranasally to healthy rabbits and resulted in a 35% decrease in the glucose plasmatic concentration with respect to the basal levels, as compared with 14% induced by the administration of insulin in solution (5 IU kg^{-1}). The mechanisms whereby these nanoparticles are able to enhance the systemic

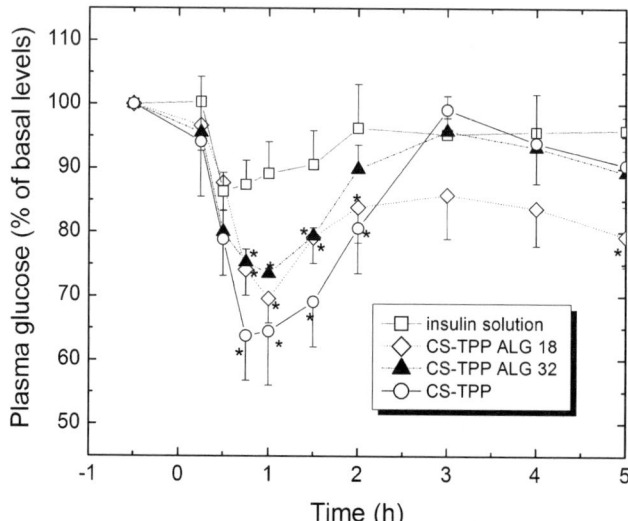

Figure 3.2.2   Plasma glucose levels achieved in rabbits after nasal instillation (at pH 4.3; mean ± SD, *n* = 6) of: insulin acetate solution pH 4.3(◇); insulin-loaded CS–TPP nanoparticles suspended in acetate buffer pH 4.3 (O); insulin-loaded chitosan–TPP–alginate M_w ~ 18 kDa nanoparticles suspended in acetate buffer pH 4.3 (◇); insulin-loaded chitosan–TPP–alginate M_w ~ 32 kDa nanoparticles suspended in acetate buffer pH 4.3 (△). *Statistical significant differences (p ≤ 0.05) between insulin-loaded nanoparticle and insulin acetate solution control group. Reproduced with permission of American Chemical Society.[79]

absorption of insulin are suggested to be similar to those operating in chitosan–TPP nanoparticles.

Chitosan derivatives-based nanoparticles

Chemical derivatives of chitosan have also shown very interesting potential as building blocks of nanoparticles for peptide nasal delivery. Nanoparticles obtained from chitosan derivatives have been prepared by ionotropic gelation using TPP. Pegylated chitosan was synthesized by reacting PEG-aldehyde with amine groups of chitosan, aiming to improve the solubility and stability of chitosan in biological media.[81] Nanoparticles of PEG–chitosan were prepared by ionotropic crosslinking with TPP.[75] These particles exhibited a very high association efficiency of insulin (> 78%). Pegylated chitosan nanoparticle formulations showed faster insulin release as compared to insulin chitosan nanoparticles. With increasing M_w of the PEG grafts, the insulin release was retarded. The pegylated formulations with fastest insulin release *in vitro*, showed also the highest insulin plasma concentrations and the fastest reduction in blood glucose levels after intranasal administration to healthy rabbits. These formulations lowered blood glucose down to 54% of baseline values at 120 min post-administration and the peak blood insulin levels reached 218.1($\pm$33.8) IU ml^{-1} at 30 min. It is proposed that the PEG chains surface coverage cause the nanoparticles to diffuse into the mucus layer, thus effectively prolonging the nasal residence time and sustaining the release of insulin, as hypothesized previously.[81] In turn, nanoparticles produced with quaternized trimethyl chitosan (TMC) harnessed by ionotropic gelation with TPP were loaded with ovalbumin and their transport across the nasal epithelium of rat was studied by loading FTIC-ovalbumin.[82] In agreement with previous studies in unmodified chitosan, it was shown that only when the protein was administered in the nanoparticles, their presence was observed by CLSM, whereas for soluble FITC-ovalbumin no fluorescence was detected, which indicates that uptake of soluble FITC-albumin was negligible. According to these authors, the results are consistent with the idea that the transport of nanoparticles is unlikely to occur *via* paracellular permeabilization, and is likely due to intracellular uptake by epithelial and NALT cells. As an explanation to why TMC nanoparticles do not transport themselves *via* the paracellular pathway, it has been argued that the TMC in solution is unable to open cellular tight junctions.[83]

Chitosan-coated gold nanoparticles

A different type of system has been described in which chitosan was used as a reducing agent in the formation of metallic gold nanoparticles.[84] Insulin was associated at the surface of these particles with reasonably high efficiency of 53%. A significant reduction in blood glucose levels following nasal administration of insulin-loaded chitosan-reduced gold nanoparticles was found as compared to insulin solution (P < 0.01). At a relatively lower dose of 10 IU kg^{-1}, nasal administration led to 20.27 and 34.12% reduction at the end of 2 and 3 h, respectively. These were considered as very promising results. The novelty of this approach lies in the use of a hybrid nanocarrier system comprising a hard metallic core coated by a shell of hydrophilic biocompatible polymer, where a large amount of protein can be loaded, and it underlies the importance of the composition of the surface of the particles in increasing the pharmacodynamic activity of the administered insulin.

Other polysaccharide-based nanoparticles

In a different series of studies, insulin-loaded nanoparticles were prepared by cross-linking epichlorohydrin with starch in the presence of a permeation enhancer such as sodium glycocholate or lysophosphatidylcholine.[85] These particles were nasally administered intranasally to healthy rats. The particles, containing sodium glycocholate, increased the plasma insulin levels significantly. Additionally, this system produced a higher hypoglycemic effect as compared to nanoparticles that contain only lysophosphatidylcholine. It is worth pointing out that other polysaccharides with well documented bioadhesion and capacity to enhance the intranasal absorption of peptides (*e.g.* cellulose derivatives, pectins, hyaluronan)[36] so far have hardly been exploited in the design of new nanoparticulate systems for nasal delivery of peptides. This clearly offers a field of great opportunity for the future development of innovative drug delivery nanocarrier systems.

Synthetic polymer-based nanocarriers

Nanoparticles comprising synthetic polymers intended for intranasal delivery of proteins have mostly been based on polyesters such as PLA, poly(lactic-co-glycolic acid) (PLGA) and their copolymers of variable molecular weight and composition. These polymers have been approved by the Food and Drug Agency of the USA for a number of clinical applications.

Polymeric nanoparticles intended for intranasal administration of tetanus toxoid were obtained by the double emulsion technique and comprised PLA or PLA–polyethylenglycol (PEG) loaded with tetanus toxoid TT and a trace of [127]I (1 µCi).[41,86,87] These particles had a size in the range of 136–151 nm and a zeta potential of −32 to −48 mV and drug association efficiency of 31–37%. The nanoparticles comprising PLA were more hydrophobic than those containing PLA–PEG. The absorption behaviour and biodistribution after intranasal administration to rats were studied from the concentration of [125]TT in several tissues related to the mononuclear phagocyte system (MPS). It was found that 24 h after intranasal instillation, the percentage of radioactivity detected in lymph nodes, lungs, liver and spleen was 3–6 fold higher in PLA–PEG than in PLA nanoparticles. Based on the results of this work, the hypothesis is that the PLA–PEG nanoparticles are partially taken up by the M-cells of the NALT, but could also be transported by a transcellular or paracellular pathway to the submucosa layer and be drained to the lymphatics and blood. In the preceding study,[87] it was found that PLA–PEG nanoparticles had greater stability against enzymatic degradation in simulated gastric fluids than PLA systems, and were internalized. The PEG coating is believed to reduce the interaction of the nanoparticles with the enzymes and to favour the transport of the particles across the enterocytes when administered orally to rats.

An alternative strategy to the use of PLA–PEG-based nanoparticles is to attach PLGA to a hydrophilic central chain of polyviny alcohol (PVA). The properties of this polymer can be modulated by substituting part of the PVA chain with negatively (sulfobutyl-PVA) or positively (diethylaminoethyl-PVA) charged derivatives. These studies revealed that branched copolymers based on sulfobutylated-PVA were promising for the nasal delivery of the model antigen tetanus toxoid, resulting in enhanced IgG and IgA antibody production as compared to a control TT solution in mice.[88]

A different approach proposes to combine different enhancer principles, amphiphilicity, cationic charges, and bioadhesion into a single macromolecule. To this end, a water-soluble amine-modified polyester was synthesized, namely, poly(vinyl 3-(diethyl)propylcarbamate *co*-vinyl acetate-*co*-vinyl alcohol) as cationic backbone, and the corresponding graft-polyesters with short (L-lactic acid) side chains, for their potential to form insulin nanocomplexes by spontaneous self-assembly.[89] The formed nanocomplexes had an average size of 250–290 nm and a zeta potential of ~+20 mV and a high insulin loading efficiency (insulin/polymer mass ratio 1.0 : 1.7). The insulin nanocomplexes were tested in both healthy and stroptozotozin-induced diabetic rat, in both cases the animals were anaesthetized. The nanocomplexes reduced the glucose plasmatic levels in fasted healthy rats by 20% after 50–80 min, and in streptozotocin induced diabetic rats by 30% within 75–95 min, compared to basal levels. The insulin nanocomplex obtained with the more hydrophobic amphiphilic comb-like lactic acid grafted polyester derivative, INS/P(26)-2$_{LL}$ (bearing 26 amino groups per PVA molecule, DP = 300, and 48 short lactide chains per backbone with an average chain length of about three lactic acid units), were more effective at a threefold higher polymer concentration, increasing the relative bioavailability of a 5 IU kg^{-1} dose from 2.8 to 5.7%. The results of this study are opposite to those observed in TT-loaded PLA–PEG nanoparticles that showed enhanced transport compared to the more hydrophobic counterparts comprised only by PLA. Possibly, different mechanisms operate in the cellular internalization of both type of nanoparticles, which may in turn be influenced not only by the surface charge and hydrophilicity, but also by the molecular architecture of the nanocarrier surface.

Interestingly enough, the difference in surface electrical charge between PLA–PEG nanoparticles, with a negative zeta potential (−32 to −48 mV), and INS/P(26)-2$_{LL}$ nanocomplexes, with a zeta potential of +20 mV, does not seem to determine the capacity to transport the peptides across the nasal epithelial barriers and their subsequent systemic absorption.

Intranasal delivery of therapeutic peptides directly to specific central nervous system (CNS) targets using polymeric nanoparticles represents a major challenge in drug delivery. Indeed, numerous attempts have been made to investigate the feasibility of the nasal route to achieve delivery of a therapeutic peptides and proteins to the brain by way of transport through the olfactory neuroepithelium, thus effectively circumventing the blood-to-brain barrier (BBB). To this end, poly(L-lactic acid–D-lactic acid) nanoparticles (D,LPLA NPs) (of size ~ 100 nm) loaded with thyrotropin-releasing hormone (TRH; Protirelin), an anticonvulsant used in certain intractable epileptic patients, were developed and their efficacy was tested intranasally in a rat kindling model of temporal lobe epilepsy.[90] The results showed that the NPs treatment resulted in a significant reduction in the seizures and related parameters. The results of this work provide proof-of-principle that intranasal delivery of sustained-release TRH-NPs may be neuroprotective and can be utilized to suppress a number of seizure characteristics *in vivo*.

In a different study, a nanoparticle system comprised by a core of PLA overcoated with polysorbate 80 was loaded with neurotoxin-1 (NT), an analgesic peptide (M_w ~ 6952.19 Da), which was separated from the venom of *Naja naja atra*. The results showed that nanoparticles could exert enhanced delivery of NT into the brain significantly after i.n. administration.[91]

As a remark to these studies, it can be said that even though intranasal delivery to the brain has been an area of active research over the past years, with only marginal success, it has been pinpointed that the use of new polymer nanocarrier systems may render this approach more successful in enhancing neuropeptide bioavailability.[11]

3.2.3.2   Lipidic Nanosystems

Liposomes

Beyond the extended use of liposomes as drug delivery systems in cancer therapy, their utilization as carriers for antigenic proteins in intranasal immunization has been investigated almost since their discovery.[92,93] The various studies are consistent in concluding that, in general, the liposomal formulation are not able to improve the systemic responses to antigens administered intranasally up to the level required for effective immunization.[64]

On the other hand, a multivesicular liposomic (MVL) formulation has been documented for the nasal administration of insulin. The MVL was coated either with chitosan (CS-MVL) or carbopol (CP-MVL) so as to achieve a sustained release protein delivery profile *via* the nasal and ocular mucosae.[94] As a result, insulin-loaded coated MVLs of sizes in the range 26–34 µm and a high protein loading capacity (up to 62%) were obtained. Their pharmacological performance was tested in STZ-induced diabetic rats and compared to that of uncoated MVLs conventional liposomes and free insulin solution. In diabetic STZ-induced diabetic rats, CS-MVLs effectively reduced the blood glucose levels by 35% up to 48 h after nasal administration, compared to a marginal reduction of 22% by non-coated MVLs up to 12 h after administration. Meanwhile, the comparison between CS-MVLs and CP-MVLs revealed maximum reductions of 65% in plasmatic glucose levels and a 55% reduction, respectively, attained at 8 h in both cases. The response observed in the nasal administration has been mainly attributed to the prolonged residence of the mucoadhesive MVLs in the mucosa as well as the protective effect of the MVLs against enzymatic attacks of the peptide drug.

Nanocapsules

Nanocapsules are comprised of an oily core, a surfactant and hydrophilic polymer coating. Nanocapsules have shown promising potential as an effective drug delivery platform for transmucosal administration of peptides, lipophilic drugs and vaccines.

Nanocapsules are characterized by a vesicular system composed by an inner core containing an oil stabilized by lecithin and a surface polymer coating. For the effective intranasal delivery of peptides and proteins, the polymer of choice to form the nanocapsules has been in most cases chitosan. The combination of hydrophobic and hydrophilic materials for peptide and protein delivery into a nanocarrier offer the advantages of the lipids protecting the macromolecules from degradation and those of the polysaccharide at the surface providing a close interaction of the nanocarrier with the mucosal surface.

On the other hand, chitosan-based nanocapsules were first developed in mid 90s also in the same group. The technology used for their preparation was based on the solvent displacement technique with an important modification: the incorporation of chitosan into the external aqueous phase where the formation of nanocapsules takes place.[76] Chitosan nanocapsules were initially designed for the encapsulation of lipophilic drugs such as indomethacin,[95] cyclosporine[96] and taxanes[97] in which the drug is encapsulated in the oil. However, these nanocarriers have also demonstrated the capacity to load hydrophilic macromolecules, peptides and vaccines,[98–102] with the aim of challenging transmucosal delivery. In this case, the association of the drug to chitosan nanocapsules was either at the oil–water interface or adsorbed to the surface of the nanocarrier. The hypothesis behind the use of these nanocarriers for peptide and protein delivery was based on the ability of the oily core to protect the associated macromolecules from degradation in the mucosal fluids.[103,104] Additionally, the surface coated with chitosan will improve the interaction of the nanocarrier with the epithelium thanks to the previously mentioned properties.

Chitosan nanocapsules have been initially explored for ocular drug delivery. Results demonstrated that this type of nanocarrier significantly enhanced the drug concentration in ocular tissues and improved the pharmacokinetic parameters as compared to uncoated nanocapsules or the commercial formulations.[105] From this work, the authors concluded that a polymeric coating with the cationic polysaccharide chitosan adds benefits to the nanocarrier for ocular drug delivery. These promising results together with the low ocular toxicity of chitosan nanocapsules led to the investigation of other mucosal surfaces such as nasal and even oral administration.[62,98] Salmon calcitonin was selected as a peptidic macromolecule to evaluate the potential of chitosan nanocapsules for nasal peptide delivery.[62] The association of the salmon calcitonin to the nanocarrier took place through the interaction of the hydrophobic regions of the peptide with the negatively charged lecithin.[106] The incorporation of salmon calcitonin to chitosan nanocapsules was affected by the chitosan coating. In fact, the positive character of both the peptide and the polymer led to a competition for the anionic binding sites in the oily core. In spite of this competition, a significant association of salmon calcitonin to chitosan nanocapsules was achieved, being higher than 40%.

These nanocapsules were tested *in vivo* after nasal administration to rats, with different controls: the uncoated nanoemulsion, an aqueous solution of salmon calcitonin and an aqueous solution of salmon calcitonin with or without chitosan (Figure 3.2.3).[62] The hypocalcemic effect elicited by the chitosan nanocapsules was significantly greater than the pharmacological profile obtained by the controls. It is remarkable that the administration of a colloidal carrier (chitosan nanocapsules or the uncoated nanoemulsion) led to a better response than the solutions of the peptide with or without chitosan. It

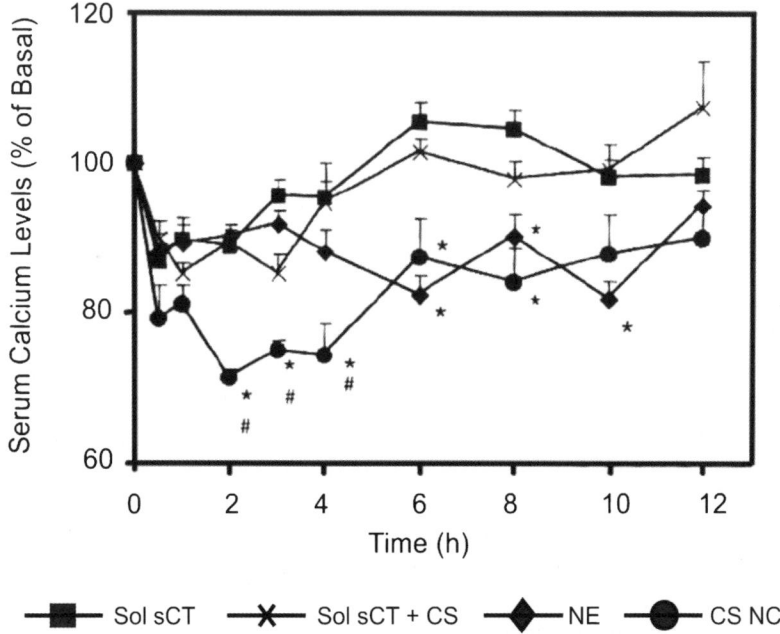

Figure 3.2.3 Serum calcium levels after nasal administration in rats of salmon calcitonin (sCT, dose: 15 UI kg^{-1}) in aqueous solution with or without chitosan (sol sCT and sol sCT + CS, respectively) or encapsulated in the control nanoemulsion (NE) or in chitosan nanocapsules (CS NC); (mean ± SE; n = 6). *Significantly different from sCT solutions ($p < 0.05$). #Significantly different from nanoemulsion ($p < 0.05$). Reproduced with permission of Association de Pharmacie Galénique Industrielle.[62]

was speculated that the effect obtained with the nanoemulsion was related to the lipids' absorption-enhancing effects[107] and/or to their ability to protect the associated drug.[104,108] Consequently, these results confirmed that the development of the nanocarrier was crucial to obtain a response, and the hypocalcemic effect was greatly improved with a chitosan coating.

Chitosan nanocapsules have also recently been tested as a carrier for intranasal vaccination,[71,102] using ovalbumin and hepatitis B surface antigen as model antigens. The authors found that the immune response generated was significantly higher for the nanocarrier with a chitosan coating than for the uncoated nanocarrier, highlighting the importance of a chitosan layer coating the nanocarrier.

3.2.4 Conclusions

The design of nanocarriers for transmucosal delivery of peptides and proteins has generated interest because of the well-known advantages of a needle-free administration. Nanocarriers offer great potential for macromolecular delivery through the mucosal surfaces, evidenced by the number of nanocarriers which have demonstrated *in vivo* effectiveness. Their versatility in terms of structure, composition and physico-chemical properties offer a myriad of opportunities for enhancing the nasal peptide/protein delivery. Focusing on nanocarriers including chitosan on their composition, it was observed that the properties of the cationic polysaccharide are critical for the efficacy of the nanocarrier. The presence of chitosan was responsible for the close interaction between the nanocarrier and the mucosal surface, improving the residence time of the nanocarriers and thus facilitating the transport of the associated macro-molecule across the epithelium.

At present, there is the proof-of-concept that chitosan-based nanocarriers can enhance the absorption of different peptides in small-scale animal models. It is expected that this accumulated information will lead to the evaluation of the efficacy of these nanocarriers in large-scale animals, giving an indication of the potential of nanocarriers for clinical use.

References

1. G. Walsh, *Nat. Biotechnol.*, 2010, **28**, 917.
2. Z. Antosova, M. Mackova, V. Kral and T. Macek, *Trends Biotechnol.*, 2009, **27**, 628.
3. W. M. Pardridge, *J. Neurovirol.*, 1999, **5**, 556.
4. J. Born, T. Lange, W. Kern, G. P. McGregor, U. Bickel and H. L. Fehm, *Nat. Neurosci.* 2002, **5**, 514.
5. X. Wang, N. Chi and X. Tang, *Eur. J. Pharm. Biopharm.*, 2008, **70**, 735.
6. A. Mistry, S. Stolnik and L. Illum, *Int. J. Pharm.*, 2009, **379**, 146.
7. M. J. Kubek, I. Ringel and A. J. Domb, *Issues related to intranasal delivery of neuropeptides to temporal lobe targets. Blood-brain barrier: Drug delivery and brain pathology*, ed. D. Kobiler, D. Lustig and S. Saphira, New York, Kluwer Academic/Plennum Publishers, 2001, p. 323.
8. C. Dufes, J. Olivier, F. Gaillard, A. Gaillard, W. Couet and J. Muller, *Int. J. Pharm.*, 2003, **255**, 87.
9. T. K. Vyas, I. Salphati and L. Z. Benet, *Curr. Drug Delivery*, 2005, **2**, 165.
10. M. J. Kubek, M. Yard, D. K. Lahiri and A. J. Domb, in *Nanoparticles for pharmaceutical applications*, ed. A. J. Domb, Y.Tabata and M. N. Ravi Kumar, New York, American Scientific Publishers, 2007, p. 73.
11. M. J. Kubek, Y. Domb and M. C. Veronesi, *Neurotherapeutics*, 2009, **6**, 359–371.
12. P. Arora, S. Sharma, S. Garg, *Drug Discovery Today*, 2002, **7**, 967.
13. V. H. Lee, *Int. FIP Satellite Symp. Disposition Delivery Peptide Drugs*, Leiden, 1987.
14. A. B. Lansley and G. P. Martin, in *Drug Delivery and Targeting for Pharmacists and Pharmaceutical Scientists*, ed. A.M. Hilklery, A.W. Lloyd and J. Swarbrick, Taylor & Francis, London, 2001, p. 238.

15. V. H. Lee, *J. Controlled Release*, 1990, **13**, 213.
16. V. D. Hoang, A. R. Uchenna, J. Mark, K. Renaat and V. Norbert, *Int. J. Pharm.*, 2002, **238**, 247.
17. M. A. Hussain, A. B. Shenvi, S. M. Rowe and E. Shefter, *Pharm. Res.*, 1989, **6**, 186.
18. H. S. Gwak, Y. M. Cho and I. K. Chun, *J. Pharm. Pharmacol.*, 2003, **55**, 1207.
19. R. U. Agu, H. Vu Dang, M. Jorissen, R. Kinget and N. Verbeke, *Peptides*, 2004, **25**, 563.
20. A. Yamamoto, A. M. Luo and V. H. L. Lee, *Pharm. Res.*, 1988, **5**, S107.
21. C. McMartin, L. E. F. Hutchinson, R. Hyde and G. E. Peters, *J. Pharm. Sci.*, 1987, **76**, 535.
22. A. Fisher, K. Brown, S. Davis, G. Parr and D. A. Smith, *J. Pharm. Pharmacol.*, 1987, **39**, 357.
23. L. Hedin, B. Olsson, M. Diczfalusy, C. Flyg, A. S. Petersson, S. Rosberg and K. Albertsson-Wikland, *J. Clin. Endocrinol. Metab.*, 1993, **76**, 962.
24. E. S. Khafagy, M. Morishita, K. Isowa, J. Imai and K. Takayama, *J. Controlled Release*, 2009, **133**, 103.
25. B. Gupta, T. S. Levchenko and V. P. Torchilin, *Adv. Drug Deliv. Rev.*, 2005, **57**, 637.
26. M. Morishita, N. Kamei, J. Ehara, K. Isowa and K. Takayama, *J. Controlled Release*, 2007, **118**, 177.
27. F. W. H. M. Merkus, J. C. Verhoef, E. Marttin, S. G. Romeijn, P. H. M. van der Kuy, W. A. J. J. Hermens and N. G. M. Schipper, *Adv. Drug Deliv. Rev.*, 1999, **36**, 41.
28. K. Matsubara, K. Abe, T. Irie, K. Uekama, *J. Pharm. Sci.*, 1995, **84**, 1295.
29. K. Matsubara, Y. Ando, T. Irie and K. Uekama, *Pharm. Res.*, 1997, **14**, 1401.
30. Y. Dotsikas and Y. L. Loukas, *J. Pharm. Biomed. Anal.*, 2002, **29**, 487.
31. S. Frokjaer and D. E. Otzen, *Nat. Rev.*, 2005, **4**, 298.
32. T. Irie and M. E. Brewester, *Adv. Drug Deliv. Rev.*, 1999, **36**, 101.
33. E. Marttin, J. C. Verhoef, S. G. Romeijn and F. W. H. M. Merkus, *Pharm. Res.*, 1995, **12**, 1151.
34. E. Marttin, J. C. Verhoef and F. W. H. M. Merkus, *J. Drug Target.*, 1998, **6**, 17.
35. M. Zhou and D. Donovan, *Int. J. Pharm.*, 1996, **135**, 115.
36. C. R. Oechslein, G. Fricker and T. Kissel, *Int. J. Pharm.*, 1996, **139**, 25.
37. H.-J. Cho, P. Balakrishnan, S. J. Chung, C. K. Shim and D. D. Kim, *Int. J. Pharm.*, 2011, **416**, 77.
38. P. Edman, E. Bjork and L. Ryden, *J. Controlled Release*, 1992, **21**, 165.
39. S. Sakuma, M. Suita, Y. Masaoka, M. Kataoka, N. Nakajima, N. Shinkai, H. Yamauchi, K. Hiwatari, H. Tachikawa, R. Kimura and S. Yamashita, *J. Controlled Release*, 2010, **148**, 187–196

40. A. Vila, H. Gill, O. McCallion and M. J. Alonso, *J. Controlled Release*, 2004, **98**, 231.
41. M. Tobío, R. Gref, A. Sánchez, R. Langer and M. J. Alonso, *Pharm. Res.*, 1998, **15**, 270.
42. A. Vila, A. Sanchez, M. Tobío, P. Calvo and M. J. Alonso, *J. Controlled Release*, 2002, **78**, 15.
43. S., Marakanam, M. S. Umashankar, R. K. Sachdeva and M. Gulati, *Nanomed.: Nanotechnol., Biol. Med.*, 2010, **6**, 419.
44. J. J. Wang, Z. W. Zeng, R. Z. Xiao, T. Xie, G. L. Zhou, L. Guang, X. R. Zhan and S. L. Wang, *Int. J. Nanomed.*, 2011, **6**, 765.
45. H. Peniche and C. Peniche, *Polym. Int.*, 2011, **60**, 883.
46. T. Kean and M. Thanou, *Adv. Drug Del. Rev.*, 2010, **62**, 3.
47. C. Gorzelanny, B. Pöppelmann, K. Pappelbaum, B. M. Moerschbacher and S. W. Schneider, *Biomaterials*, 2010, **31**, 8556.
48. S. H. Pangburn, P. V. Trescony and J. Heller, *Biomaterials*, 1982, **3**, 105
49. S. Aiba-i, *Int. J. Biol. Macromol.*, 1992, **14**, 225.
50. S. Hirano, H. Tsuchida and N. Nagao, *Biomaterials*, 1989, **10**, 574.
51. T. Fukamizo, Y. Honda, S. Goto, I. Boucher and R. Brzezinski, *J. Biochem.* **1995**, 311, 377.
52. C. M. Lehr, J. A. Bouwstra, E. H. Schacht and H. E. Junginger, *Int. J. Pharm.*, 1992, **78**, 43.
53. C. Shruti, M. Saiqa, K. Jasjeet, I. Zeemat and T. Sushma, *J. Pharm. Pharmacol.*, 2006, **58**, 1021.
54. C. Chatelet, O. Damour and A. Domard, *Biomaterials*, 2001, **22**, 261.
55. P. J. Vandevord, H. W. T. Matthew, S. P. Desilva, L. Mayton, B. Wu and P. H. Wooley, *J. Biomed. Mater. Res.*, 2002, **59**, 585.
56. P. Artursson, T. Lindmark, S. S. Davis and L. Illum, *Pharm. Res.*, 1994, **11**, 1358.
57. N. G. M. Schipper, S. Olsson, A. J. Hoostraat, A. G. De Boer, K. M. Varum and P. Artursson, *Pharm. Res.*, 1997, **14**, 923.
58. J. Smith, E. Wood and M. Dornish, *Pharm. Res.*, 2004, **21**, 43.
59. D. Vllasaliu, R. Exposito-Harris, A. Heras, L. Casettaria, M. Garnetta, L. Illum and S. Stolnik, *Int. J. Pharm.*, 2010, **400**, 183.
60. V. Dodane, M. A. Khan and J. R. Merwin, *Int. J. Pharm.*, 1999, **182**, 21.
61. K. A. Janes, P. Calvo and M. J. Alonso, *Adv. Drug Delivery Rev.*, 2001, **47**, 83.
62. C. Prego, D. Torres and M. J. Alonso, *J. Drug Delivery Sci. Technol.*, 2006, **16**, 331.
63. A. Vila, A. Sánchez, K. Janes, I. Behrens, T. Kissel, J. L. Vila-Jato and M. J. Alonso, *Eur. J. Pharm. Biopharm.*, 2004, **57**, 123.
64. N. Csaba, M. Garcia-Fuentes and M. J. Alonso, *Adv. Drug Delivery Rev.*, 2009, **61**, 140.
65. R. Fernandez-Urrusuno, P. Calvo, C. Remuñán-López, J. L. Vila-Jato and M. J. Alonso, *Pharm. Res.*, 1999, **16**, 1576.

66. P. Calvo, C. Remuñán-López, J. L. Vila-Jato and M. J. Alonso, *J. Appl. Polym. Sci.*, 1997, **63**, 125.
67. Y. Pan, Y. J. Li, H. Y. Zhao, J. M. Zheng, H. Xu, G. Wei, J. S. Hao and F. D. Cui, *Int. J. Pharm.*, 2002, **249**, 139.
68. Z. S. Ma, H. H. Yeoh and L. Y. Lim, *J. Pharm. Sci.*, 2002, **91**, 1396.
69. X. Zhang, H. Zhang, Z. Wu, Z. Wang, H. Niu and C. Li, *Eur. J. Pharm. Biopharm.*, 2008, **68**, 526
70. Y. Xu and Y. Du, *Int. J. Pharm.*, 2003, **250**, 215.
71. T. Nagamoto, Y. Hattori, K. Takayama and Y. Maitani, *Pharm. Res.*, 2004, **21**, 671.
72. F. M. Goycoolea, I. Higuera-Ciapara and M. J. Alonso, in *Handbook of Natural-based Polymers for Biomedical Applications*, ed. R.L. Reis, N. M. Neves, J. F.Mano, M. E. Gomes, A.P. Marques and H.S. Azevedo, Woodhead Publishing Ltd, 2008, p. 644.
73. M. J. Alonso, C. Prego, M. García-Fuentes, in *Nanoparticles for Pharmaceutical Application*, ed. J. Domb, Y. Tabata, M. N. Ravi Kumar and S. Farber, American Scientific Publishers, 2007, p. 135.
74. D. Teijeiro-Osorio, C. Remuñán-López and M. J. Alonso, *Biomacromolecules*, 2009, **10**, 243.
75. P. Calvo, C. Remunán-López, J. L. Vila-Jato and M. J. Alonso, *Colloid Polym. Sci.*, 1997, **275**, 46.
76. V. M. Leitner, D. Guggi, A. H. Krauland and A. Bernkop-Schnürch, *J. Controlled Release*, 2004, **100**, 87
77. A. H. Krauland and M. J. Alonso, *Int. J. Pharm.*, 2007, **340**, 134.
78. A. Trapani, M. Garcia-Fuentes and M. J. Alonso, *Nanotechnology*, 2008, **19**, 185101
79. F. M. Goycoolea, G. Lollo, C. Remuñán-López, F. Quaglia and M. J. Alonso, *Biomacromolecules*, 2009, **10**, 1736
80. G. T. Grant, E. R. Morris, D. A. Rees, P. J. Smith and D. Thom, *FEBS Letters*, 1973, **32**, 195.
81. Serra L., Doménech J., Peppas N.A. *Eur. J. Pharm. Biopharm.*, 2006, **63**, 11.
82. M. Amidi, S. G. Romeijn, G. Borchard, H. E. Junginger, W. E. Hennink and W. Jiskoot, *J. Controlled Release*, 2006, **111**, 107
83. A. F. Kotze, H. L. Lueben, B. J. De Leeuw, B. G. De Boer, J. Coos Verhoef and H. E. Junginger, *J. Controlled Release*, 1998, **51**, 35
84. D. R. Bhumkar, H. M. Joshi, M. Sastry and V. B. Pokharkar, *Pharm. Res.*, 2007, **24**, 1415.
85. A. K. Jain, R. K. Khar, F. J. Ahmed and P. V. Diwan, *Eur. J. Pharm. Biopharm.*, 2008, **69**, 426
86. A. Vila, A. Sanchez, C. Evora, I. Soriano, O. McCallion and M. J. Alonso, *Int. J. Pharm.* 2005, **292**, 43.
87. M. Tobío, A. Sánchez, A. Vila, I. Soriano, C. Evora, J. L. Vila-Jato and M. J. Alonso, *Colloids Surf., B*, 2000, **18**, 315.

88. T. Jung, W. Kamm, A. Breitenbach, K.D. Hungerer, E. Hundt and T. Kissel, *Pharm. Res.*, 2003, **18,** 352

89. M. Simon, M. Wittmar, T. Kissel and T. Linn, *Pharm. Res.*, 2005, **22**, 1879.

90. M. C. Veronesi, Y. Aldouby, Y. Domb and M. J. Kubek, *Brain Res.,* 2009, 1303, 151–160.

91. Y. Ruana, L. Yaoa, B. Zhang, B. Zhang and J. Guo, *Peptides*, 2011, **32**, 1526.

92. A. D. Bangham, M. M. Standish and J. C. Watkins, *J. Mol. Biol.* 1963, **13**, 238.

93. A. C. Allison and G. Gregoriadis, *Nature*, 1974, **252**, 252.

94. A. K. Jain, K. B. Chalasani, R. K. Khar, F. J. Ahmed and P. V. Diwan, *J. Drug Targeting*, 2007, **15**, 417.

95. P. Calvo, M. J. Alonso, J. L. Vila-Jato and J. R. Robinson, *J. Pharm. Pharmacol.*, 1996, **48**, 1147.

96. P. Calvo, J. L. Vila-Jato and M. J. Alonso, *J. Pharm. Sci.*, 1996, **85**, 530.

97. M. V. Lozano, D. Torrecilla, D. Torres, A. Vidal, F. Domínguez and M. J. Alonso, *Biomacromolecules*, 2008, **9**, 2186.

98. C. Prego, M. Garcia, D. Torres and M. J. Alonso, *J. Controlled Release,* 2005, **101**, 151.

99. C. Prego, M. Fabre, D. Torres and M. J. Alonso, *Pharm. Res.*, 2006, **23**, 549.

100. C. Prego, D. Torres, E. Fernandez-Megia, R. Novoa-Carballal, E. Quiñoa and M. J. Alonso, *J. Controlled Release*, 2006, **111**, 299.

101. C. Prego, D. Torres and M. J. Alonso, *J. Nanosci. Nanotechnol.*, 2006, **6**, 2921.

102. S. Vicente, B. Diaz, A. Sanchez, A. Gonzalez-Fernández and M. J. Alonso, *2nd Pharm. Sci. Fair*, Nice, France, 2010.

103. C. Damge, C. Michel, M. Aprahamia, P. Couvreur and J. P. Devissaguet, *J. Controlled Release*, 1990, **13**, 233.

104. P. J. Lowe and C. S. Temple, *J. Pharm. Pharmacol.*, 1994, **46**, 547.

105. P. Calvo, J. L. Vila-Jato and M. J. Alonso, *Int. J. Pharm.*, 1997, **153**, 41.

106. R. M. Epand, R. F.Epand, R. C. Orlowski, R. J. Schlueter, L. T. Boni and S. W. Hui, *Biochemistry*, 1983, **22**, 5074.

107. S. Muranishi, *Crit. Rev. Ther. Drug Carrier Syst.*, 1990, **7**, 1.

108. Y. H. Sang and P. T. Gwan, *J. Pharm. Sci.*, 2004, **93**, 488.

CHAPTER 3.3

Nanostructures for Nasal Vaccine Delivery

BRAM SLÜTTER[a] AND WIM JISKOOT*[b]

[a] Department of Microbiology, University of Iowa, 51 Newton Rd, Iowa City,
IA 52242, USA. E-mail: bernard-slutter@uiowa.edu; [b] Leiden/Amsterdam
Center for Drug Research, Leiden University, Einsteinweg 55, 2333 CC
Leiden, The Netherlands
*E-mail: w.jiskoot@lacdr.leidenuniv.nl

3.3.1 Introduction

The nasal mucosa is an attractive site for vaccination and consequently has
gained increased attention over the last years. The nasal cavity is easily
accessible, low on proteolytic enzymes compared to the oral route and is lined
with high numbers of immune competent cells, including those in the
nasopharyngeal-associated lymphoid tissue (NALT). These unique features
potentially allow the induction of protective immune responses with increased
patient comfort and would abrogate the need for administration by trained
personnel.[1]

Nasal vaccination, however, has its limitations. Poor absorption of the
antigen through epithelium, combined with a limited residence time in the
nasal cavity, requires the administration of a very high dose compared to
parenteral administration.[2] Moreover, nasal administration of antigen can lead
to tolerance or unresponsiveness instead of immunity.[3,4] Therefore, despite the
considerable efforts on research and development, only one nasal vaccine is
currently on the market (an influenza vaccine, Flumist®). This vaccine consists
of a highly immunogenic cold-adapted live attenuated influenza virus and is

RSC Drug Discovery Series No. 22
Nanostructured Biomaterials for Overcoming Biological Barriers
Edited by Maria Jose Alonso and Noemi S. Csaba
© The Royal Society of Chemistry 2012
Published by the Royal Society of Chemistry, www.rsc.org

therefore apt to meet the challenges of absorption and tolerance. However because of its live-attenuated nature it is indicated only for people between the ages of 2–49, and not for infants and elderly which are important target populations for an influenza vaccine. Development of subunit vaccines for nasal vaccination would therefore be desirable.

Nanostructures have been described as interesting tools to enhance the delivery and immunogenicity of nasally administered vaccines. This chapter will review the physiological hurdles to be taken by a nasal vaccine and describes the progress of designing nasal subunit vaccines using nanoparticulate delivery systems.

3.3.2 The Nasal Cavity as Application Site for Vaccination

Although the concept of nasal vaccination may sound innovative, the practice of nasal vaccine delivery has been age old. Before the introduction of the hollow needle at the end of the 19th century, vaccines were generally administered intradermally (by scratching the skin) or by nasal application.[6] Intramuscular (i.m.) and subcutaneous (s.c.) injection (also referred to as the "parenteral" route, which in fact does not discriminate these routes from nasal, as nasal administration is also a parenteral route) were quickly adopted as the preferred routes of administration. Injectable vaccines allow accurate dosing and have proven to be an effective way of inducing systemic antibody responses (Table 3.3.1).

Despite the successes of injectable vaccines (the eradication of smallpox being its most referred to feat), these strategies may need to be revised if they are to meet the demands of the 21st century. The use of needles requires trained personal, and improper (or re-)use of needles has had serious impact on the spread of blood borne pathogens like HIV and hepatitis B/C. Emphasizing this fact, the WHO estimated in 2005 that from the 1 billion injections administered in the course of childhood programs in 14 different developing countries about 50% were administered unsafely.[7] Moreover, vaccination *via* the needle can cause local side effects ranging from pain at the injection site to stiffness and severe swelling, reducing patient compliance. Importantly, s.c. and i.m. administration can only result in systemic antibody and/or T-cell responses, but does not induce local or mucosal immunity. Inducing mucosal immunity, in particular the excretion of secretory IgA (sIgA) onto the mucosal surface, could be very instrumental in clearing pathogens before they cross (*e.g.* in case of HIV) or infect (*e.g.* influenza A) the mucosal epithelium. Moreover, sIgA is thought to mediate heterosubtypic immunity.[8,9]

In this respect, nasal vaccination is very promising. The nose is easily accessible allowing non invasive administration without the use of a complicated delivery device. As the vaccine is applied on the outside of the body absolute sterility is not required and local side effects are restricted to "a funny feeling" or a running nose.[10] Compared to the oral route, where these

Table 3.3.1   Advantages and disadvantages of different routes of immunization (adapted from Slütter *et al.*[5]).

Immunization route	Advantages	Disadvantages
Parenteral	Powerful systemic immune response	Invasive
	Accurate dosing	Sterile formulation required
		Local inflammation
		Limited (no) mucosal immune response
Nasal	Non invasive	Mucociliary clearance
	Mucosal and systemic immune response	Inefficient uptake of soluble antigens
	Easily accessible	Possible delivery through olfactory nerve
	Little degradation (compared to oral)	Application device needed
Oral	Non invasive	Vaccine digestion in stomach and gut.
	Mucosal and systemic immune response	Inefficient uptake of soluble antigens
	Large surface area	Mucosal tolerance
Pulmonary	Non invasive	Delivery of antigen highly variable from person to person
	Mucosal and systemic immune response	Dry powder inhaler or nebulizer needed
	Little degradation (compared to oral)	Clearance from lungs
Dermal	Non or minimally invasive	May require (minimally) invasive technology (*e.g.* tattooing, microneedles)
	Large, easily accessible application area	
	High density of immune cells in skin	Patch or application device needed
	Mucosal immune response possible	Less established technology

advantages may also apply, the enzymatic activity in the nose is relatively low, which is favourable for antigen stability at the administration site. Importantly, the nasal sub-epithelium is equipped with a high density of dendritic cells (DCs). These versatile professional antigen presenting cells (APCs) are believed to be key cell types that link innate and adaptive immunity, due to their ability to present epitopes of digested pathogens *via* both MHCI and MHCII molecules on the cell surfaces. Therefore, DCs that have encountered and endocytosed pathogens at the nasal epithelium can prime both CD4+ T-cells (required to help B-cells produce antibodies) as well as CD8+ T-cells (cytotoxic T-lymphocytes, CTLs) in the draining lymph nodes. Finally, local immunity in the upper airways, as well as systemic immunity, could be mediated by the lymphoid tissue referred to as nasopharyngeal-associated lymphoid tissue (NALT). NALT is most pronounced in the nasopharynx and the Waldeyer's ring, which includes the nasopharyngeal, tubal, palatine and lingual tonsils. NALT comprises agglomerates of cells involved in the initiation and execution of an immune response, like DCs, T-cells and B-cells,[11] situated underneath the nasal epithelium. In mice, germinal centers (places where plasma cells are located) have been reported in the NALT after challenge with a reovirus,[12] and Shimoda *et al.* showed that B-cells in the sub-epithelial region of the nose are prone to switch from IgM to IgA,[13] suggesting a prominent role of the NALT in producing mucosal immune responses. In addition, epithelium covering NALT comprises a high number of microfold cells (M-cells). These epithelial cells differ from ordinary epithelial cells as they do not contain cilia and have relatively high concentrations of cytoskeleton protein vimentin, making them capable of transcytosing matter from the luminal side towards the basolateral side. M-cells are found in very low numbers in regular human respiratory epithelium, however NALT is densely covered with M-cells. *In vitro* studies suggest an important role for both B-cells and T-cells in the differentiation and maintenance of M-cells, confirming the close relation of M-cells with NALT and making M-cells an important possible portal for antigen entrance.[14]

Mucosal immunity after nasal vaccination is, however, not restricted to the upper airways. *Via* a system called the common mucosal immune system, after nasal immunization sIgA antibodies can be detected also in other mucosal secretions.[15] This may prove an important advantage to reduce the risk of, for instance, sexually transmitted diseases.

3.3.3 Challenges in Nasal Vaccination

3.3.3.1 Nasal Subunit Vaccine Design is Tough

Despite the considerable advantages nasal vaccination offers, i.m. and s.c. injection is still the most common way of applying vaccines. Designing nasal vaccines has proven to be challenging and the considerable efforts that have been put in their development have only yielded 1 licensed vaccine.

Why is it that difficult to design an effective and safe nasal vaccine? The major challenge seems to be the physiological hurdles an antigen has to overcome after it has been applied nasally. In the following paragraphs, we will discuss the main physiological hurdles involved.

3.3.3.2 Mucus

The nasal cavity has primarily evolved to keep substances out rather than letting them in. Like every mucosal lining the nasal epithelium is protected with a layer of mucus. Although its composition can be different in several diseased states, in general it consists of 95% water, 2% mucins, 1% salts, <1% lipids and 1% other proteins.[16] The mucin fibers form a porous network structure that can trap particulates by entanglement or electrostatic interaction. Although the nasal mucus layer is relatively thin (<5 μm) compared to other mucosa, it forms a very effective barrier. Not only because of its high viscosity, but also because of rapid ciliary movement towards the esophagus,[17] effectively replacing the entire mucus layer every 20 min.[5] This greatly limits the time for the antigen to be taken up by the epithelium.

3.3.3.3 Epithelium

If an antigen successfully makes it through the mucus layer, passage through the nasal epithelium is its next obstacle. Although it has been reported that DCs can sample antigen by sticking their dendrites through the epithelium into the lumen,[18] it is generally believed that the bulk of the antigen actually has to cross the epithelium in order to be endocytosed by DCs. The epithelium consists of a layer of epithelial cells that are closely stacked together by tight junctions.[19] The epithelium changes from pseudo-stratified to semi-columnar from the anterior to the posterior end of the nasal cavity, which impacts the permeability of the epithelium (Table 3.3.2). Nonetheless, even in the most permeable area the diameter of the tight junction does not exceed 1 Å (0.1 nm), leaving little to no space for intercellular transport of molecules, especially not large proteins and even bulkier vaccine delivery systems.[20] Tight junctions can open under certain circumstances and widening up to 10 Å has been documented, thereby increasing the permeability of the epithelium for small molecules. Even so the epithelium will remain a formidable barrier as the size of antigens generally exceeds 10 Å.

3.3.3.4 Nasal Tolerance

The nasal cavity is exposed to a wide range of environmental elements on a daily basis. Only a few of these entries need to be dealt with *via* an (innate or adaptive) immune response, as most of these elements do not pose a threat to the individual or are rapidly removed *via* mucociliary clearance. Mucosal surfaces in general are therefore renowned for induction of tolerance rather than immunity.[21] Oral tolerance is the best described phenomenon,[22] but nasal tolerance has also been described. Various studies have reported the induction of regulatory T-cells (T_{reg}) and anti-inflammatory cytokines like IL-10 and TGF-ß after nasal application of antigens.[23–25] This may offer interesting therapeutic approaches to treat autoimmune diseases like type I diabetes,

Table 3.3.2 Characteristics of human nasal epithelium (adapted from Pires *et al.*[19]).

	Surface Area	Vascularization	Permeability
Vestibule	≈0.6 cm²	Low	Poor
Atrium	<10 cm²	Low	Medium
Respiratory (turbinate)	≈150–160 cm²	High	Good

rheumatoid arthritis[26] and Crohn's disease,[27] but is undesirable when vaccination against pathogens is concerned.

3.3.4 Nasal Vaccine Design

The challenges mentioned in the former paragraphs need to be addressed to effectuate an effective and safe nasal vaccine. In fact, the only licensed nasal vaccine so far (Flumist®) is perfectly adapted. It prolongs the nasal residence time of the virus due to its capacity to colonize the nasal cavity, crosses the mucus and epithelium due to the unique structure of the influenza virion, and provides several danger signals due to the presence of viral RNA and apoptotic epithelial cells.[28,29] For non-replicating or subunit vaccines, however, the vaccine formulation has to provide these characteristics. The next paragraphs will address the various approaches presented in the literature to overcome the major hurdles in nasal vaccination and will review the role of nanostructures in optimizing the immune response against nasally applied antigens.

3.3.4.1 Penetrating Mucus and Prolonging Nasal Residence Time

Nature seems to be well aware of the restrictions the mucus layer imposes on the accessibility of the epithelium. Pathogens therefore developed strategies to effectively penetrate the mucus layer before being cleared from the nasal cavity towards the digestive system. For instance, the excretion of mucinases and sialidases can degrade the mucins, thereby acting as mucolytic agents.[30] The neuraminidase protein of influenza is a well-known example of a mucinase. In fact, treatment with neuraminidase inhibitors prevents the mucus from becoming fluidic[31] and is an effective treatment against influenza infection.[32] Neuraminidase may therefore be an interesting additive for a nasal vaccine formulation as it could improve the diffusion of the antigen through the mucus. Liposomes containing neuraminidase and hemagglutinin (referred to as virosomes) have shown to be an excellent nasal carrier for the F-protein of human respiratory syncytial virus[33] and DNA constructs encoding the F-protein of mumps virus.[34]

A second strategy to penetrate the mucus is to adapt the delivery system to the properties of mucus rather than to change the mucus structure itself. Although the mucus layer seems almost impregnable to particulate matter, recent studies have shown that particles up to hundreds of nanometres in diameter can diffuse through mucus in a matter of minutes.[35] As referred to earlier, mucin fibers in the mucus can trap fibrous material by entangling it and can bind charged species by electrostatic interactions. Not coincidentally, many viruses have a compact form and little surface charges, allowing them to diffuse through mucus with relative easy. Using this approach Saltzman and coworkers coated poly(lactic-co-glycolic acid) (PLGA) nanoparticles with polyethylene glycol (PEG) molecules (M ~2 kDa) to shield PLGA's negative surface charge.[36] They showed that a PEG coating increased the rate of diffusion through a mucus layer tenfold. However, when longer (M $\sim$ 10 kDa) PEG chains were used diffusion was impaired, indicating that entanglement of PEG molecules with mucin fibers can make PLGA particles become trapped in the mucus.

Interestingly however, generally the use of charged species rather than uncharged ones is advocated. The idea behind this is that many of these charged polymers are so-called "mucoadhesives". These substances interact with the mucus and epithelium and slow down the clearance of co-administered antigens, thereby increasing the antigen's nasal residence time and the probability of antigen uptake by the epithelium. The mechanism of mucoadhesion differs between substances.[37] Positively charged polymers like chitosan, N-trimethyl chitosan (TMC) and protamine interact with the negatively charged mucus and cell membranes of the nasal epithelium. Negatively charged polymers like alginate and poly(acrylic acid) seem to deploy hydrogen bonding[38] and thiolated polymers can bind to mucins through disulfide bonds.[39] Particles prepared with mucoadhesive agents like alginate,[40] chitosan,[41] TMC,[2,42] poly(acrylic acid)[43] and thiomers[44] have shown to increase the nasal residence time of the antigen and improve the resulting immune response.

In conclusion, nasal vaccine formulation can be designed to either influence the diffusion through or adhesion to the mucus. Higher antibody titres have been obtained with both approaches, showing that if it comes to designing nasal vaccine formulations multiple options could be considered.

3.3.4.2 Crossing Epithelium and M-cell Targeting

Application of "tight junction openers" like chitosan and its water soluble derivative N-trimethyl chitosan (TMC) have been shown to improve the transport of both small molecular weight compounds (*e.g.* mannitol) as well larger protein like insulin or ovalbumin *in vitro* as well *in vivo*.[45–47] An important feature of a tight junction opener is the ability to temporarily increase the permeability of epithelium without destroying it. Positively charged polymers like poly ethylene imine (PEI) have also been shown to improve the transport rate of various proteins, but histological examination has reported damage to the respiratory tract,[48] which obviously is highly undesirable.

Tight junction opening can only increase the permeability of the epithelium to a certain degree. Naturally, the encapsulation of antigen into nanostructures will almost completely impede diffusion through these tight junctions. It is therefore believed that the trafficking of particulates over the nasal epithelium is a result of active rather than passive transport. Consequently, improving transport *via* the intracellular route should be promoted. Epithelial cells can be infected by microorganisms like rhinovirus[49] and may take up particulates. However, as epithelial cells have not been cited as a cell type capable of transcytosing antigens to the subepithelial space, antigens are unlikely to reach B-cells and DCs *via* this route.

A more pertinent approach would be the targeting of M-cells. The combination of their transcytotic capacity and convenient location makes these cells an attractive target to get antigens over the epithelial barrier and present them to both DCs and IgA prone B-cells (Figure 3.3.1).

M-cells have been reported to efficiently take up antigens with a particulate nature and deliver them to the NALT[50] rather than to the general afferent lymphatics (Figure 3.3.2). Their bias toward transcytosis of particulate matter could be explained as an evolutionary advantage, as pathogens are of particulate nature. However, the mechanism of its preference is poorly understood and it is unclear what the ideal size or surface characteristics of the particulate should be.

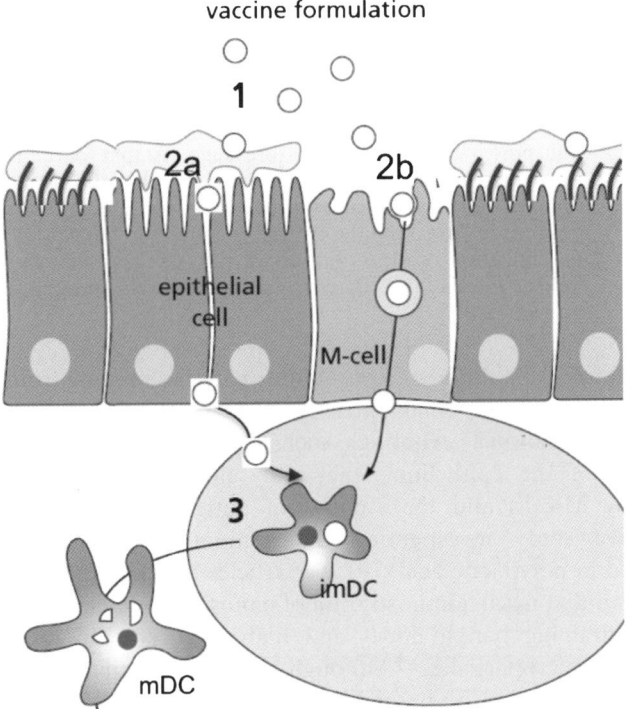

Figure 3.3.1 The 3 basic challenges in nasal vaccine design. Antigen is trapped in the mucus (1) and readily removed from the nasal cavity by mucociliary clearance leading to a short nasal residence time of the antigen. Transport over the nasal epithelium is limited and has to occur through the intercellular (2a) or transcellular (2b) route. Finally, immature dendritic cells (imDC) will have to mature (mDC) after uptake of the antigen (3) to initiate an adaptive immune response, a process impaired due to the tolerogenic environment of the nasal epithelium.

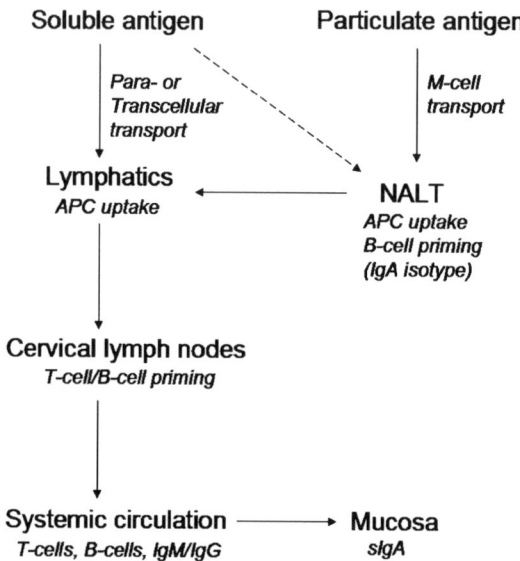

Figure 3.3.2   Nasally applied antigens can reach APCs in the afferent lymphatics either *via* diffusion through the nasal epithelium or through active transport *via* M-cells. Literature suggests that particles are mainly transported through M-cells, whereas soluble antigens may take either route. B-cells associated with the NALT can quickly isotype switch to produce IgA, making the antigen encounter in the NALT pertinent for the induction of sIgA. Either way, it is in the cervical lymph nodes where T-cells and B-cells will be actively primed for systemic protection.

In vivo studies using different sizes of latex particles ranging from 20 nm to 2 μm showed that especially small particles (20 nm) had reached the bloodstream after nasal application.[51] Although such studies may inform us about the permeability of the epithelium, they may not be representative for the transport by M-cells and the amount of antigen reaching DCs. A more representative study investigating the amount of tetanus toxoid (TT) encapsulated in poly(lactic acid) (PLA) particles reaching the draining lymph nodes showed that nasal administration of nanosized particles resulted in more TT in the draining lymph nodes and higher TT specific IgG titers than application of microparticles.[52] Although this indicates that nanoparticles are more easily transported by nasal M-cells than microparticles, no differences between TT loaded sulfobutylated PLGA particles with a diameter of 200 nm and 500 nm were found,[53] indicating that M-cell transport is not further enhanced by reducing the nanoparticle size under 500 nm. A recent study, however, showed that the delivery of the model antigen ovalbumin to the cervical lymph nodes in a nanoconjugated form with TMC (hydrodynamic diameter *ca.* 30 nm), was twice as effective as the nasal application of ovalbumin-containing TMC nanoparticles (diameter *ca.* 300 nm).[45] Moreover, the ovalbumin specific IgG titers were higher after nasal administration of the nanoconjugate compared to the nanoparticles. Unfortunately, the authors were unable to show whether the delivery advantage was M-cell dependent, but it does suggest that a small sized nasal vaccine delivery system is beneficial.

In recent years, several M-cell surface markers have been identified and used to target nanostructures containing antigens to M-cells.[54] Specific carbohydrate residues (mainly sialic acid and galactose) on nasal M-cells can be recognized by lectins (isolectin B_4, *Maackia amurensis I* lectin).[55] Nasal application of the model antigen horse radish peroxidase (HRP) with isolectin B_4 significantly enhanced the antibody (IgG and sIgA) response to HRP in comparison with administration of HRP alone. Also. targeting integrin receptor $\alpha 5\beta 1$, which is located on the apical surface of M-cells, by attachment of an RGD peptide on PLGA particles has shown improved transport by M-cell *in vitro*,[56] although only a limited increase in IgG titers after oral application was observed. IgG and IgA antibodies themselves may also improve uptake by Fc-receptor mediated M-cell targeting.[57] The availability of M-cell culturing systems will certainly reveal new M-cell targets in the near future that could aid the transport of (particulate) antigens over the epithelial barrier.

3.3.4.3 Overcoming Nasal Tolerance and Use of Adjuvants

Key to the induction of a pro-inflammatory response is providing DCs, that have taken up antigen, with danger signals in order for them to mature.[58] DC maturation leads to increased expression of co-stimulatory molecules like CD40, CD80, CD86 and the production of pro-inflammatory cytokines. Co-stimulation and exposure to pro-inflammatory cytokines are essential to the proper priming of T-cells and lack of these factors may induce anergic T-cells. As purified antigens often lack the necessary danger signals to induce DC maturation, co-stimulatory agents or immune potentiators (adjuvants) could be added to the formulation.

The arsenal of adjuvants at our disposal is steadily growing. Whereas for more than a century alum was the only approved adjuvant for human use, recently new adjuvants like squalene emulsions (*e.g.*, MF-59) and non toxic variants of lipopolysaccharide (LPS) (*e.g.*, MPL) were licensed for parenteral administration. Increasing knowledge on the activation of DCs has speeded up this process, as it has explained the mode of action of several adjuvants for which the mechanism was unknown until recently (*e.g.*, alum, MDP and LPS), which is a prerequisite for approval by the American Food and Drug Administration. Moreover, the observation that DC maturation can be triggered *via* specific pathogen recognition receptors (PRRs, *e.g.* Toll-like receptors and NOD-like receptors) has led to the identification of new ligands that could act as adjuvants. As the signaling cascade resulting from activation of PRRs is becoming clearer, in the near future it may be possible to select the proper adjuvant according to the nature of the vaccine and the type of immune response required.

The use of adjuvants can be very beneficial for nasal vaccination, but not every adjuvant is a good nasal adjuvant. In fact, all currently used adjuvants are approved for injectable vaccines only. Whether or not an adjuvant is a good choice for nasal use will depend on the type of response required (*e.g.* antibodies or cytotoxic T-lymphocytes), the dose of adjuvant required and whether the physical characteristics of adjuvant allow effective uptake through the epithelium. In concurrence with the "danger model", adjuvants are supposed to be dangerous goods and consequently there is only a small margin between immune potentiation and toxicity. Therefore, not all adjuvants are suitable because their effective dose is a toxic one. For instance, heat labile enterotoxin (LT) was successfully used as an adjuvant with virosomes, but after nasal administration LT was presumably taken up by olfactory neurons, eventually leading to Bell's palsy.[59] Whether an adjuvant will have a strong positive effect on the immune response raised by a nasal vaccine likely depends in part on how easily the adjuvant is absorbed into the nasal epithelium and on the expression of its complementary receptor on nasal DCs. For instance, alum is a poor nasal adjuvant as it will not diffuse through the epithelium (and at the same time may cause local irritation). TLR ligands, however, could be powerful adjuvants, especially if their transport over the nasal epithelium is enhanced using a delivery system. Encapsulating adjuvants into carrier systems may kill two birds with one stone, because it enhances the delivery of the adjuvant over the epithelium and enables the simultaneous delivery of the antigen and the adjuvant to the same DC. This last process is crucial for efficient activation of both CD4+ and CD8+ T-cells.[60–62] Indeed synergistic effects of the TLR9 ligand CpG in TMC particles,[63] the NOD like receptor ligand muramyl dipeptide (MDP) in chitosan particles and TLR-4 ligand MPL in liposomes have been reported after nasal vaccination.[64]

3.3.4.4 Further Clinical and Formulation Challenges

The former paragraphs described how nanostructures could improve the efficacy of nasal vaccines in animals, mainly mice. Mice have been instrumental in studying mechanistic aspects of nasal vaccination, but will never be able to fully predict the immune response in humans. For instance, apart from differences in the immune system and architecture of the nose, a very obvious difference between mice and man is the size of the nasal epithelium. Relatively to its bodyweight, the surface area of nasal epithelium of mice is 4 times larger than the human epithelium, which could cause an overestimation of the absorption of the antigen in mice compared to humans. Moreover, many pathogens do not cause disease in mice. Mouse strains susceptible to these pathogens are being developed, but still these models have their limitations.

Furthermore, the pharmaceutical form will be very important. Distribution patterns after application of nasal sprays or nasal drops is very different as the size of aerosol determine whether the antigen is deposited.[65] Taking into account the localization of NALT, a nasal spray would therefore seem an obvious choice, however this would still require a liquid vaccine formulation. Liquid formulations are generally unstable, require cold storage and imply a relatively short shelf life. A product in dry powder form is generally more stable and could therefore be more easily distributed; also to countries where maintaining the cold chain is not self-evident. Lyophilization of vaccines without damaging antigenic epitopes has been shown on a pharmaceutical scale. If a simple, robust and cheap delivery device can be developed to apply the vaccine as a powder or to reconstitute the vaccine just before application, nasal vaccination may become the new standard in vaccination world-wide.

3.3.5   Conclusion

Nasal vaccination has a great potential to improve both vaccine efficiency and patient comfort, but nasal vaccines will require specific formulations to be effective.

Nanostructures can be very instrumental in enhancing immunogenicity of nasal vaccines, but the major challenge is to manage the interplay between antigen, delivery vehicle and adjuvant in such a way that the optimal immune response is obtained. Nanoparticles should most likely be as small as possible to optimally diffuse through the mucus and penetrate the mucosal epithelium. A mucoadhesive and tight junction-opening agent could be applied to create the ideal environment for the nanostructures to reach the subepithelial space. Finally, the nanostructures should contain both antigen and adjuvant to be endocytosed by the same DC and effectively start an adaptive immune response.

References

1. L. Illum and S. S. Davis, *Adv. Drug Delivery Rev.,* 2001, **51**, 1.
2. B. Slütter, S. Bal, C. Keijzer, R. Mallants, N. Hagenaars, I. Que, E. Kaijzel, W. van Eden, P. Augustijns, C. Lowik, J. Bouwstra, F. Broere and W. Jiskoot, *Vaccine*, 2010, **28**, 6282.
3. J. Mestecky, Z. Moldoveanu and C. O. Elson, *Vaccine*, 2005, **23**, 1800.
4. W. W. Unger, F. Hauet-Broere, W. Jansen, L. A. van Berkel, G. Kraal and J. N. Samsom, *J. Immunol.* 2003, **171**, 4592.
5. B. Slütter, N. Hagenaars and W. Jiskoot, *J. Drug Targeting*, 2008, **16**, 1.
6. P. Brandtzaeg, *Ann. N. Y. Acad. Sci.*, 1996, **778**, 1.
7. E. L. Giudice and J. D. Campbell, *Adv. Drug Delivery Rev.*, 2006, **58**, 68.
8. P. Brandtzaeg, *Vaccine*, 2007, **25**, 5467.
9. S. I. Tamura, H. Asanuma, Y. Ito, Y. Hirabayashi, Y. Suzuki, T. Nagamine, C. Aizawa, T. Kurata and A. Oya, *Eur. J. Immunol.*, 1992, **22**, 477.
10. E. M. Flood, M. D. Rousculp, K. J. Ryan, K. M. Beusterien, V. M. Divino, S. L. Toback, M. Sasane, S. L. Block, M. C. Hall and P. J. Mahadevia, *Clin. Ther.*, 2010, **32**, 1448.

11. K. I. Jeong, H. Suzuki, H. Nakayama and K. Doi, *J. Anat.*, 2000, **196**, 443.
12. A. W. Zuercher, S. E. Coffin, M. C. Thurnheer, P. Fundova and J. J. Cebra, *J. Immunol.*, 2002, **168**, 1796.
13. M. Shimoda, T. Nakamura, Y. Takahashi, H. Asanuma, S. Tamura, T. Kurata, T. Mizuochi, N. Azuma, C. Kanno and T. Takemori, *J. Exp. Med.*, 2001, **194**, 1597.
14. H. S. Park, K. P. Francis, J. Yu and P. P. Cleary, *J. Immunol.* 2003, **171**, 2532.
15. A. Hasegawa, Y. Fu and K. Koyama, *Am. J. Reprod. Immunol.*, 2002, **48**, 305.
16. B. Heurtault, B. Frisch and F. Pons, *Expert. Opin. Drug Delivery*, 2010, **7**, 829.
17. H. Y. Reynolds, *Curr. Opin. Pulm. Med.*, 2002, **8**, 154.
18. M. R. Neutra, E. Pringault and J. P. Kraehenbuhl, *Annu. Rev. Immunol.*, 1996, **14**, 275.
19. A. Pires, A. Fortuna, G. Alves and A. Falcao, *J. Pharm. Pharm. Sci.*, 2009, **12**, 288.
20. L. Illum, *J. Pharm. Sci.* 2007, **96**, 473.
21. R. A. Langlois and K. L. Legge, *Immunol. Res.*, 2007, **39**, 128.
22. A. M. Faria and H. L. Weiner, *Immunol. Rev.*, 2005, **206**, 232.
23. D. S. Donaldson, K. K. Tong and N. A. Williams, *Mucosal Immunol.*, 2011, **4**, 227.
24. S. Till, L. Jopling, P. Wachholz, R. Robson, S. Qin, D. Andrew, L. Wu, J. van Neerven, T. Williams, S. Durham and I. Sabroe, *J. Immunol.*, 2001, **166**, 2303.
25. J. B. Sun, C. Czerkinsky and J. Holmgren, *Scand. J. Immunol.*, 2010, **71**, 1.
26. F. Broere, L. Wieten, E. I. Klein Koerkamp, J. A. van Roon, T. Guichelaar, F. P. Lafeber and W. van Eden, *J. Immunol.*, 2008, **181**, 899.
27. A. M. Faria and H. L. Weiner, *Clin. Dev. Immunol.*, 2006, **13**, 143.
28. S. Koyama, C. Coban, T. Aoshi, T. Horii, S. Akira and K. J. Ishii, *Expert Rev. Vaccines*, 2009, **8**, 1099.
29. T. Ichinohe, H. K. Lee, Y. Ogura, R. Flavell and A. Iwasaki, *J. Exp. Med.*, 2009, **206**, 79.
30. R. Wiggins, S. J. Hicks, P. W. Soothill, M. R. Millar and A. P. Corfield, *Sex. Transm. Infect.*, 2001, **77**, 402.
31. D. A. Boltz, J. R. Aldridge Jr., R. G. Webster and E. A. Govorkova, *Drugs*, 2010, **70**, 1349.
32. V. A. Jagannath, G. V. Asokan, Z. Fedorowicz, J. S. Singaram and T. W. Lee, *Cochrane Db. Syst. Rev.*, 2010, CD008139.
33. M. G. Cusi, *Hum. Vaccin.*, 2006, **2**, 1.
34. M. G. Cusi, C. Terrosi, G. G. Savellini, G. Di Genova, R. Zurbriggen and P. Correale, *Vaccine*, 2004, **22**, 735.
35. Y. Cu and W. M. Saltzman, *Nat. Mater.* 2009, **8**, 11.
36. Y. Cu and W. M. Saltzman, *Mol. Pharm.* 2009, **6**, 173.
37. J. D. Smart, *Adv. Drug Delivery Rev.*, 2005, **57**, 1556.

38. M. N. Patel, J. D. Smart, T. G. Nevell, R. J. Ewen, P. J. Eaton and J. Tsibouklis, *Biomacromolecules*, 2003, **4**, 1184.
39. A. Bernkop-Schnurch, A. Weithaler, K, Albrecht and A. Greimel, *Int. J. Pharm.*, 2006, **317**, 76.
40. M. Tafaghodi, S. A. Sajadi Tabassi and M. R. Jaafari, *Int. J. Pharm.*, 2006, **319**, 37.
41. T. Nagamoto, Y. Hattori, K. Takayama and Y. Maitani, *Pharm. Res.*, 2004, **21**, 671.
42. M. Amidi, S. G. Romeijn, J. C. Verhoef, H. E. Junginger, L. Bungener, A. Huckriede, D. J. Crommelin and W. Jiskoot, *Vaccine*, 2007, **25**, 144.
43. D. Coucke, M. Schotsaert, C. Libert, E. Pringels, C. Vervaet, P. Foreman, X. Saelens and J. P. Remon, *Vaccine*, 2009, **27**, 1279.
44. V. M. Leitner, D. Guggi, A. H. Krauland and A. Bernkop-Schnurch, *J. Controlled Release*, 2004, **100**, 87.
45. B. Slütter, S. M. Bal, I. Que, E. Kaijzel, C. Lowik, J. Bouwstra and W. Jiskoot, *Mol. Pharm.*, 2010, **7**, 2207.
46. S. Yu, Y. Zhao, F. Wu, X. Zhang, W. Lu, H. Zhang and Q. Zhang, *Int. J. Pharm.*, 2004, **281**, 11.
47. E. Pringels, C. Vervaet, R. Verbeeck, P. Foreman and J. P. Remon, *Eur. J. Pharm. Biopharm.*, 2008, **68**, 201.
48. S. Castellani, S. Di Gioia, T. Trotta, A. B. Maffione and M. Conese, *J. Biomed. Biotechnol.*, 2010, 103976.
49. N. Lopez-Souza, S. Favoreto, H. Wong, T. Ward, S. Yagi, D. Schnurr, W. E. Finkbeiner, G. M. Dolganov, J. H. Widdicombe, H. A. Boushey and P. C. Avila, *J. Allergy Clin. Immunol.*, 2009, **123**, 1384.
50. Y. Fujimura, M. Takeda, H. Ikai, K. Haruma, T. Akisada, T. Harada, T. Sakai and M. Ohuchi, *Virchows Arch.*, 2004, **444**, 36.
51. J. Brooking, S. S. Davis and L. Illum, *J. Drug Target.*, 2001, **9**, 267.
52. A. Vila, A. Sanchez, C. Evora, I. Soriano, O. McCallion and M. J. Alonso, *Int. J. Pharm.*, 2005, **292**, 43.
53. T. Jung, W. Kamm, A. Breitenbach, K. D. Hungerer, E. Hundt and T. Kissel, *Pharm. Res.*, 2001, **18**, 352.
54. P. J. Giannasca, J. A. Boden and T. P. Monath, *Infect. Immun.*, 1997, **65**, 4288.
55. S. Takata, O. Ohtani and Y. Watanabe, *Arch. Histol. Cytol.*, 2000, **63**, 305.
56. M. Garinot, V. Fievez, V. Pourcelle, F. Stoffelbach, A. des Rieux, L. Plapied, I. Theate, H. Freichels, C. Jerome, J. Marchand-Brynaert, Y. J. Schneider and V. Preat, *J. Controlled Release.*, 2007, **120**, 195.
57. L. Favre, F. Spertini and B. Corthesy, *J. Immunol.*, 2005, **175**, 2793.
58. P. Matzinger, *Science*, 2002, **296**, 301.
59. M. Mutsch, W. Zhou, P. Rhodes, M. Bopp, R. T. Chen, T. Linder, C. Spyr and R. Steffen, *N. Engl. J. Med.*, 2004, **350**, 896.
60. D. T. O'Hagan, M. Singh and J. B. Ulmer, *Methods*, 2006, **40**, 10.
61. S. Fischer, E. Schlosser, M. Mueller, N. Csaba, H. P. Merkle, M. Groettrup and B. Gander, *J. Drug Targeting*, 2009, **17**, 652.

62. J. M. Blander and R. Medzhitov, *Nature*, 2006, **440**, 808.
63. B. Slütter and W. Jiskoot, *J. Controlled Release*, 2010, **148**, 117.
64. S. A. Moschos, V. W. Bramwell, S. Somavarapu and H. O. Alpar, *Immunol. Cell Biol.*, 2004, **82**, 628.
65. Y. Cheng, T. D. Holmes, J. Gao, R. A. Guilmette, S. Li, Y. Surakitbanharn and C. Rowlings, *J. Aerosol Med.*, 2001, **14**, 267.

Section 4
Nanostructures Overcoming the Ocular Barrier

CHAPTER 4.1

Nanostructures Overcoming the Ocular Barrier: Physiological Considerations and Mechanistic Issues

CLIVE G. WILSON*[a] AND LAY EAN TAN[b]

[a] Strathclyde Institute of Pharmacy and Biomedical Sciences, University of Strathclyde, 161 Cathedral Street, Glasgow, G4 0RE, UK; [b] Ferring Controlled Therapeutics, 1 Redwood Place, Peel Park Campus, East Kilbride, G74 5PB, UK
*E-mail: c.g.wilson@strath.ac.uk

4.1.1 Introduction

The evolution of the human eye into a compact and efficient system that translates images into a structured, three-dimensional view of the world has provided the key sense for man. Unlike most systems encountered in the human body, the eye forms a closed space whose geometry is preserved by hydraulic forces and uses internal muscles to focus an image onto the retina by positioning and stretching the lens appropriately. The analogy of the eye as a camera is often made since the key elements of lenses, transduction of light into stored data by the retina, and control of light intensity by the iris is immediately familiar. The system is, however, orders of magnitude more complicated and more sophisticated than its electronic counterpart.

The importance of the visual sense set against the possible physical and biochemical insults necessitates the establishment of a number of barriers

RSC Drug Discovery Series No. 22
Nanostructured Biomaterials for Overcoming Biological Barriers
Edited by Maria Jose Alonso and Noemi S. Csaba
© The Royal Society of Chemistry 2012
Published by the Royal Society of Chemistry, www.rsc.org

which isolate the environment of the eye from the rest of the body. In addition, the metabolic demands of the eye are huge. Faced with the possibility of photochemical reactions due to concentration of light by the lens, the retina is under constant potential threat and must be served by an efficient microvasculature to remove waste products and supply nutrients and oxygen at high fluxes. Thus, from a drug delivery point of view, the eye poses a number of challenges for effective therapy based on these physiological and anatomical constraints. These restrictions can be described in terms of a number of sequential barriers. First, the provision of a clear focused image requires the outer surface to be kept optically clear, which is physically accomplished by tear secretion and blinking, and biochemically and immunologically supported by the conjunctival lining. This will limit the residence time of materials administered topically. The connective tissue barriers of sclera and cornea and the third by the blood-retinal barriers, which behave similarly to the blood-brain barrier, provide the second limitation. In addition to these structural barriers outside and on the periphery of the retina, other factors within the eye need to be considered including transporter expression, melanin binding and the physical state of the vitreous humour which contribute to the loss of drug from reservoir compartments and limit the exposure of target tissues to formulation.

4.1.2 Topical Ocular Barriers

The smallest permanent resident of the external ocular environment are skin-associated microbes and the manner in which these potential threats are kept from colonizing the tissue might provide a useful insight into possible therapeutic approaches in particulate delivery. Tears are stabilized by goblet cell secretions, which provide mucins as an anti-evaporative component of the structure. Lipidic components from the Meibomian glands modify the outer zones of the tear film to sustain tear film integrity during full eye opening. A decrease in goblet cell density is associated with disease[1] and at the surface of the eye, bacterial infections occur *via* disturbances in the surface mucin barrier (*e.g.* organisms such as *Staphylococcus* aureus or encapsulated *Streptococcus* pneumoniae) or by the microbes traversing the mucin layer without apparently damaging it (non-encapsulated *Streptococcus* pneumoniae). Investigators have proposed the use of biodegradable PSA (polysebacic acid)–PEG (polyethylene glycol) as a shielding for nanoparticulate systems, which apparently cross the mucin layer[2] due to the lack of interactions with the hydrophilic and uncharged surface of PEG.

If drug preparations are applied externally, they will interact with the tear film immediately. The temperature of the liquid is sensed by thermoreceptors in the cornea and sclera, which are also triggered by menthol.[3] Tears are secreted at a rate of 0.5–2.2 µl min^{-1} and normally have pH values in a range between 7.14 to 7.82.[3] Solutions which have high or low pH are not tolerated by the eye and cause the conjunctival vessels to engorge and to become leaky.

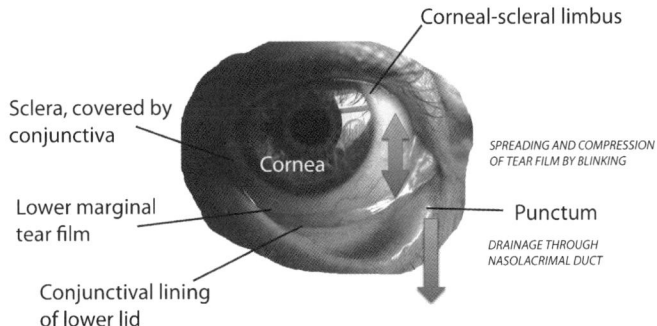

Figure 4.1.1   The external features of the eye.

Irrespective of the nature of the solution, the refractive change following instillation causes blinking and excessive lacrimation, during which the synchronized movement of the eyelids spreads the precorneal tear film across the cornea and pushes it towards the nasolacrimal duct (Figure 4.1.1).

Precorneal drainage is efficient. An aqueous instilled dose leaves the precorneal area within five minutes of instillation in humans and excess liquid wells out of the eye and down the cheek. It is generally assumed that drugs are absorbed through the cornea, but as the drug reservoir in the lower marginal strip is thickened, less mixing occurs and contact is longest below the corneal limbus. Slowing down the rate of drainage by using mucoadhesive polymers, such as chitosan-based derivatives and conjugates,[5] has been shown to increase the topical delivery of ofloxacin, and direct gelation in the lower marginal strip in gellan gum preparations allows the reduction in the number of doses of timolol maleate.[6]

4.1.2.1   Corneal Barriers

The cornea is the transparent window of the frontal globe and is formed from several layers: an external hydrophobic epithelium about 50 µm thick bounded by Bowman's layer, followed by the hydrophilic stromal layer and finally the thin Descemet's membrane and endothelium which controls the transparency of the cornea when alive. The corneal epithelium is composed of five to six layers of cells, increasing to eight to ten at the periphery with a total thickness of 50–100 µm. The cells at the base are columnar, but as new cells squeeze these forward, they become flatter so that three groups of cells are usually identified: basal cells, an intermediate zone of 2–3 layers of polygonal cells (wing shaped) and squamous cells. The multi-layer cell structure is keratinized and forms a relatively impermeable barrier to both hydrophobic and hydrophilic drug molecules until the outermost layer is damaged, suggesting that tight junctions exist between the cells of the outer layer. Thus, unless the epithelium is removed, nanoparticles cannot access the stroma.[7] The reflex blinking, rapid drainage by nasolacrimal apparatus and clearance by the conjunctival lymphatic system contribute to a low degree of drug penetration. In recent years, novel pharmaceutical approaches have attempted to optimize drug penetration across the cornea epithelium, including the use of absorption enhancers[8] and prodrugs.[9] Calvo and colleagues demonstrated that the corneal penetration of encapsulated indomethacin in the form of nanoparticles, nanocapsules and nanoemulsions was 3-fold higher than Indocollyre® commercial eye drops in an *ex vivo* study.[10] No differences were observed in penetration rate between the three nanocarrier systems, leading to the conclusion that the colloidal nature was the key factor, unrelated to the individual chemical composition of the systems. Similarly, De and co-workers illustrated that polycarboxylic acid nanoparticles achieved greater transcorneal drug release compared to the commercial formulation Alphagan™, attributed to a sustained delivery.[11] Although these data are encouraging, *in vitro* and *ex vivo* experimental protocols commonly neglect drainage and non-corneal absorption, which contribute to the main elimination pathways.[12] The experiments involved incubation periods of a few hours, and thus do not reflect the actual scenario of drug-corneal interaction and experimental artefacts cannot be avoided. The corneal uptake and retention of novel PLGA–chitosan nanoplexes in rabbit eyes resulted in a higher and sustained corneal concentration of rhodamine in eyes compared to eye drops.[13] The confocal image analysis suggested nanoplexes penetrated into the corneal epithelium through the intercellular and intracellular pathways. The latter finding is surprising considering the estimated size of the corneal epithelial gap junction is only 0.8 nm.[14] The intercellular route in this case is unlikely since the nanoplexes were much larger.

A complete permeation of corneal epithelium by nanocarriers has not been reported, but partial penetration has been commonly noted to be influenced by size and charge. For example, nanoparticles of hydrolysable dye attained higher penetration rate into the cornea than that of micron-sized particles.[15] Since the cornea is negatively charged at physiological pH, it should attract particle surfaces with opposite charges at short ranges. Positively charged colloids are reported to show greater corneal uptake than negatively or neutral charged particles.[16] The wettability study performed by Klang and co-workers demonstrated a greater spreading coefficient and lower contact angle between an excised corneal tissue and a drop of cationic emulsion compared to anionic emulsion.[17] This suggests improved surface interaction with positively charged emulsions facilitated nanoparticle uptake by the cornea. Law and colleagues demonstrated that the cornea surface was completely coated by the positively charged liposomes, which bound intimately; whereas the debris of the negatively charged liposomes was found to deposit on the cornea surface.[18] The authors highlighted the importance of charge interaction in prolonging the residence time of nanocarriers on the cornea surface for higher probability of drug absorption and nanoparticle uptake. Jani and colleagues have used direct corneal injection of albumin nanoparticles containing the plasmid pCMV.Fit23K which persisted for four weeks, expressed the intraceptor levels for 5 weeks and resisted subsequent mechanico-chemical trauma.[19]

4.1.2.2 The Scleral and Conjunctival Barriers

The sclera is the outer white tough part of the eye, which is an important structural element, with the site of insertion for extra-ocular muscles. It covers 80% of the exterior surface and is white and non-transparent. It borders the transparent cornea at the pars planar and is divided into three layers: episclera, stroma and lamina fusca. Only a limited number of blood vessels, originating from arteriolar branches of the anterior ciliary vessels are found and superficial vessels are mainly confined to the loose outer episclera. Scleral permeability approximates that of the corneal stroma and has been shown to be permeable to solutes up to 70 kDa.[20] The sclera is overlaid by a thin membranous covering of conjunctival tissue which is composed of squamous non-keratinised cells, which form a thin mucous membrane overlying a connective tissue layer with many fine blood vessels. The tissue, visible as a pink flat mass covering the anterior sclera, is the bulbar conjunctiva and is loosely attached; that lining the inside surface of the eyelids (palpebral conjunctiva) is more firmly anchored. The conjunctiva covers the frontal part of the eye, apart from the cornea, and is connected to the epidermis of the eyelids. It is important for tear film stability and maintaining anti-inflammatory status. The conjunctiva shows diversity: that covering the lid margin and bulbar conjunctiva is composed of non-keratinised, stratified squamous epithelium whereas the tarsal and fornix conjunctiva is composed of cuboidal and columnar epithelium of various thickness and has a microvillus surface architecture. Goblet cells are scarce near the lid margin and corneal limbus but abundant elsewhere. The tissue shows greater permeability (sixteen fold higher pore density and larger intercellular pore sizes) and the total area exceeds that of the cornea by approximately 400%. Bulbar and palpebral conjunctiva had comparable permeability for polyethylene glycol (PEG) oligomers of molecular weights below 1000 g mol^{-1}, which suggests neutral and hydrophilic therapeutic compounds should have equal degree of absorption through conjunctiva tissue,[21] although the authors commented that the flux would be expected to vary *in vivo* since other factors such as surface area, metabolic activity and vascularization may affect drug absorption. It has been shown that more than 20–50% of an instilled low molecular weight drug was absorbed through palpebral conjunctiva into systemic circulation.[22,23] The conjunctivae serve as a secondary reservoir for drugs and instillation of one preparation within 5 minutes of the first usually results in a reduction of the concentration of drug from the primary instillation.

Several workers have calculated that polar solutes of up to 40 kDa cross the conjunctival epithelium by restricted diffusion through intercelluar pores equivalent size of 5.5 nm.[24] It is not surprising therefore that absorption of drug compounds into the conjunctiva is favoured over the cornea. Figure 4.1.2 illustrates the possible routes of nanoparticulate uptake into the conjunctival tissue. Kompella and co-workers showed nanoparticle uptake was detected an hour earlier in the conjunctiva compared to the cornea.[7] Larger solutes and particulate systems were unable to cross the conjunctiva epithelia barrier and have to adapt non-diffusional pathways, suggesting direct tissue entry is size-dependant.[24] This phenomenon was later illustrated using PLGA particles of varying sizes. Smaller 100 nm particles exhibited the highest uptake into the rabbit conjunctival epithelial cells compared to larger 800 nm and 10 μm particles.[25] In the same study, the authors established that the uptake mechanism was mediated through an adsorptive-type endocytosis and was inhibited by the vesicle formation blocker cytochalasin D. The uptake process was however saturable, at a concentration of nanoparticulates of 4 mg mL^{-1}.

Subconjunctival injection bypasses the epithelium layer and allows direct access to the tissue space. Injected particles of 20 nm can passively diffuse through the sclera-choroid layer and are simultaneously cleared through capillary fenestrations present at the episcleral, conjunctival and periocular blood vessels into systemic and lymphatic circulations, leading to unsuccessful transmission to the retina.[26] The importance of lymphatic drainage was further confirmed by confocal microscopy showing accumulation of nanoparticles in the cervical, axillary and the mesenteric lymph nodes. Phagocytosis by macrophages of the reticulo-endothelial system located in the spleen and liver will eventually clear the particles from both circulations.[27] Conversely, particles of > 200 nm exceeding the cut-off permeability of the sclera and gaps between the capillaries, were found to be retained at the injection site for at least 2 months. Prolonged residence time enables sustained drug release that will become available to the retina. Therefore, larger particulate systems that are unable to cross both the static and dynamic barriers are more efficient drug carrier for drug delivery to the retina.[27]

4.1.3   Barriers of the Inner Globe

Once drug has entered or been introduced into the eye, it is distributed by diffusion and convective flow processes and eliminated through the retina or drains through the canal of Schlemm and to a lesser extent through the uveal tract. To inflate the globe and nourish the internal tissue, secretions of the ciliary body distribute liquid, but the intervening structures between anterior and posterior poles create potentially complex distribution patterns, especially for larger particulate structures. If there is no charge interaction, nanoparticles will be treated like large solutes of the size of macromolecules.

4.1.3.1   Aqueous Humour

For modelling drug dispersion in the eye following direct injection, the relative volumes of the aqueous spaces and the rates of turnover are important determinants of kinetics. Two spaces are clearly discernable as illustrated in Figure 4.1.3.

The aqueous humour can be aspirated during eye surgery and about 150 μl can be routinely obtained. Photographic measurements using a slit lamp showed that the volumes of the anterior chamber measured in both eyes of 39

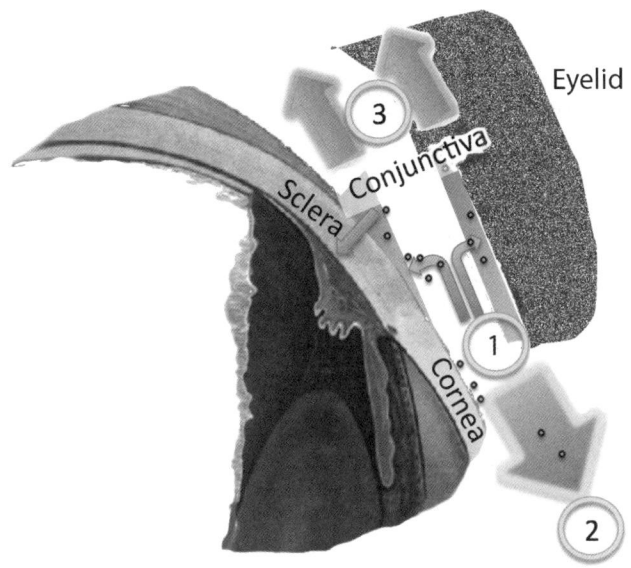

Figure 4.1.2   The deposition of nanoparticulates administered topically. The uptake of materials into the conjunctival tissue is assisted by high surface area and porosity (1), although the majority of material will be cleared by drainage (2). Once in the conjunctival tissue, nanoparticulates will be cleared through the lymphatic system (3).

normal subjects (mean age, 28 years; range, 19–56 years) were 209 ± 37 μL. The aqueous humour is similar in composition to and isosmolar with plasma but contains much less protein (30 mg mL^{-1} *vs.* 80 mg mL^{-1}), has higher ascorbate, pyruvate and lactate than plasma, and less glucose and urea. The aqueous humour is formed by the ciliary epithelium and flows into the posterior chamber providing nourishment for the lens and the epithelia. It flows through the pupil and slowly circles in the anterior chamber entering the canal of Schlemm and also drains through the low-pressure pathway of the episcleral veins. This pathway, uveoscleral outflow, is thought to account for about 35% of drainage in monkeys; older estimates for man are 5–20%.[28] From here it enters the systemic circulation. Drugs administered to the cornea will be prevented from further inward diffusion by this outflow. The resistance to outflow generates a small intra-ocular pressure of about 16 mm Hg; abnormally high pressures (< 25 mm Hg) are strongly associated with retinal anoxia (glaucomatous damage). The rate of production by the ciliary body is around 2–3 μL min^{-1},[29] yielding estimates of turnover of 1.5% min^{-1} or an equivalent half-life of 47 minutes. The same volume is drawn off *via* Schlemm's canal from where it is conducted into veins.[30] The epithelium of the iris and of the ciliary body pump anionic drugs from the aqueous into the blood: this limits entry into the aqueous from the blood, a phenomenon which is termed the blood-aqueous barrier.

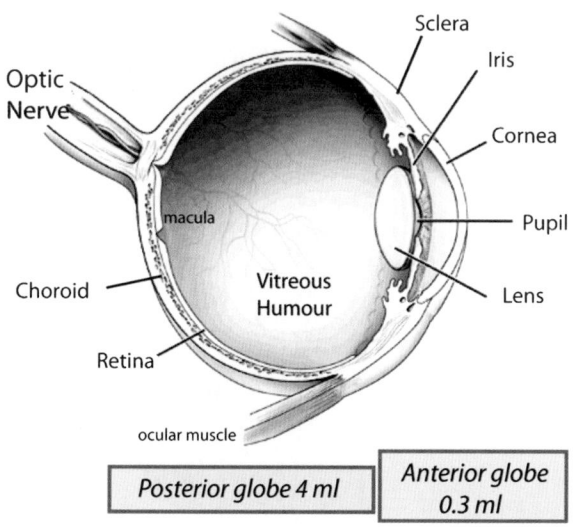

Figure 4.1.3   The general anatomical features of the eye, illustrated in cross section. Image adapted from National Eye Institute bank and used with permission.

4.1.3.2   Uveal Tract

The uveal tract of the eye is composed of the iris, ciliary body and the choroid. Though the epithelium and stroma of the iris are both pigmented, differences in the amount of melanocytes in the stroma and thus light scattering determine one person's eye colour. In the ciliary body melanin is located only in the outer pigmented epithelium.[31] Many drugs are known to bind to melanin. It has been shown that the pigmentation of the iris has a significant effect on the ocular residence time and free concentrations of some drugs.

The ciliary body, situated posterior to the iris, performs several functions. It connects the anterior part of the choroid to the circumference of the iris and contains the ciliary muscles necessary for accommodation.

4.1.3.3   The Lens

The lens is suspended in the proximal quarter of the eye just behind the iris. The tissue is made up of a single cell type, which is precisely aligned and encased in a capsular bag that has tendon attachments to the ciliary muscles allowing changes in dimensions. Since there is no regeneration of lens fibre, any damage may be retained throughout life. This makes the lens more susceptible to metabolic changes associated with disease states leading to permanent opacity.[32] In modelling, the lens is seen as an important barrier confining the conduit of molecules mainly to extreme edges of the structure.

4.1.3.4 Vitreous Humour

The vitreous occupies 80% of the globe and has a volume and wet weight of approximately 4 mL and 4 g, respectively. The tissue density of approximately 1.0 g mL^{-1},[33] reflects a highly hydrated matrix of the vitreous as water constitutes almost 98% of its content.[34,35] It has the outward appearance of a transparent, viscoelastic gel, with a refractive index of 1.33.

The whole structure is anchored rather like a barrage balloon to various structures: anteriorly to the ora serrata, ciliary epithelium, zonular fibres and posteriorly to the optic nerve head, macula and the retina vessels, unless there is a posterior vitreous detachment (PVD). In juvenile vitreous, pseudoplastic hyaluronan that is trapped within the elastic collageneous network contributes to the stiffness of the vitreous gel by imposing an internal tension against the collagen fibrils as it swells to achieve an equilibrium hydration state.[36] However, owing to the unequal distribution of the collagen and hyaluronan,[37–39] the gel is in essence heterogeneous and two zones can be distinguished: a cortical zone, characterised by more densely arranged collagen fibrils, and a more liquid central vitreous region.

The vitreous is largely an unstirred fluid body, externally isolated from the surrounding tissue by its boundaries: the cortex and outer hyaloid lamina. Within the vitreous cortex, collagen and hyaluronan intertwined with each other, provides a passive barrier (without cellular interference) to molecular transport, assuming to sustain internal composition and maintain homeostasis.[40] When a 50 μL of 10 μm particle suspension was injected into an *ex vivo* Miyake-Apple eye preparation, particle movement was impeded by pockets in the vitreous structure leading to localization at the injection site. Particle interaction with collagen fibres was clearly seen which suggests that the steric hindrance is partly attributable to the fibrillar network. When treated with collagenase, the fibrous structures are destroyed and particles move freely throughout the cuvette under small thermal gradients (provided by the blue LED used during the process of imaging) (Figure 4.1.4). Jongebloed and Worst illustrated the presence of individual spaces within the vitreous structure, which they described as cisterns.[41] The dimension of the cisterns varies regionally and can be demonstrated by the injection of red and white ink mixture into various sections of a dissected vitreous. The larger white ink particles (0.25–0.5 μm) adhered to the superficial layers of the cisternal wall, whereas the smaller red ink particles (0.1–0.25 μm) manage to penetrate deeper where the meshwork was tighter.[40]

The flow processes within the vitreous are mainly passive diffusion and convective flow. The vitreous behaves isotropically towards diffusing molecules and the diffusion coefficient of small molecules in the vitreous is only marginally less than the diffusion coefficient of the molecule in free solution. As the animal gets larger, convective effects become more important. Acid orange 8 diffusivity within bovine vitreous humour is about half that in free solution.[42] Convective flow has been suggested to be responsible for no more than 30% of the intravitreal transport in man – in the mouse it is

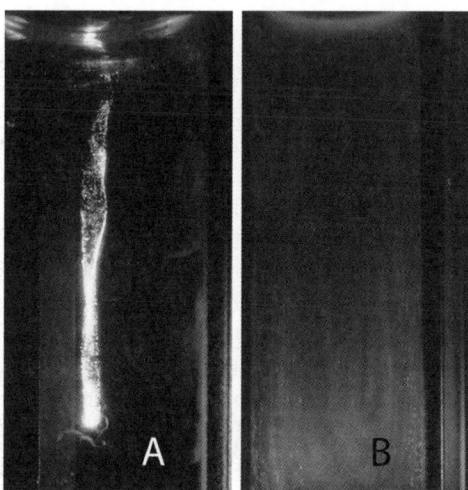

Figure 4.1.4 Particles (10 μm in phosphate buffered saline) added to sheep vitreous humour that has been decanted into a cuvette. The needle tract fortuitously enters a cistern and delineates the pocket with settling microparticles (A). Treatment of the tissue with collagenase prior to adding the particles destroys the cisternal structure (B).

recognised as being unimportant. Net in-flow of liquid moves from the posterior anterior chamber to the vitreous, and drug elimination from an intra-vitreal injection is either anteriorly to the aqueous chamber or posteriorly to the retina.[43-44] Compounds that can cross the blood-retinal barrier (BRB) will be cleared across the retina and have relatively short half-lives. Compounds that cannot traverse the BRB are normally hydrophilic and large. They must diffuse to the retrozonular space and are cleared through the aqueous chamber. These compounds typically have longer intravitreal half-lives dependent on molecular weight and steric factors.

After adolescence, portions of the vitreous become more fluid as the gel volume decreases from ~80% in young human adults to ~50% in the elderly population of age above 60 years. Our group illustrated the impact of vitreous liquefaction, where sodium fluorescein (MW ~376 Da) and FD 150 kDa were cleared faster from the partially liquefied vitreous compared to the control. The intravitreal movement of 1 μm particle that should not be affected by simple diffusion processes was found to show greater dispersion and accelerated sedimentation in the liquefied eye.[45] Spielberg and Leys, in a clinical study, have documented that elderly populations required more frequent injections of bevacizumab (mean age: 68.5 years; 3.75 injections) than younger individuals (mean age: 39.5 years; 1.75 injections). This suggests that the loss of vitreous structure with age reduces its capability in sustaining injected substances leading to shorter intravitreal half-lives available for therapeutic treatment.[46] The correlation is that when the vitreous body is partially removed, access from anterior to posterior parts of the eye is faster, especially if the lens has been removed and an anterior vitrectomy is performed.[47]

4.1.3.5 Posterior Vitreous Detachment (PVD)

As the vitreous degenerates, the vitreous body starts to collapse causing a traction which results in the pulling away of the vitreous cortex from posterior pole. This is a normal physiological consequence of ageing which leads to the pathophysiological condition PVD. Although the condition has few clinical consequences, it involves a breakdown of the blood retinal barrier, which normally has a low passive permeability. Macular oedema is associated with marked changes at the vitreoretinal margins, which would be predicted to be associated with changes in active transport and solute flow.

4.1.3.6 The Choroid and Retinal Blood Supply

The retinal pigment epithelium and photoreceptors are the most metabolically active tissues in the body and is supplied by two separate systems, which do not overlap; the inner retina is supplied by retinal vascular system and the outer retina by the choroid below the sclera. The choroid, supplying nutrition to the retina, is a highly pigmented loose connective tissue having an apparent thickness of approximately 220 μm at the posterior pole and 100 μm anteriorly. The choroid supplies nutrition to the retina. It also contains pigment cells melanocytes that produce melanin.

The choriocapillaris is worth special consideration as it forms a critical barrier between the sclera and retina. The inner vessels of the choroid form a diffuse monolayer, the choriocapillaris, which are fenestrated and therefore permeable. The fenestrations, approximately 60 nm in diameter, allow nutrient waste product exchange but the adjacent retinal pigment epithelial cells (RPE) have tight junctions and form an outer blood-retinal barrier which prevents access to the sub-retinal space above the RPE. The choroidal blood flow is anteriorly directed. Together with the fenestrated nature of the choriocapillaris, this poses a large hurdle to retinal drug delivery. Systemically or peri-ocularly administered drugs must establish a concentration gradient across the RPE for productive absorption while fighting anterior clearance through the choriocapillaris; however, actively targeted nanoparticles, having a surface motif of a linear peptide arginine–glycine–aspartic acid and transferring, are reported to cross the choroidal circulation.[48]

Retinal vessels from the branching of the ophthalmic artery supply the inner retina; these vessels are blind ended and obstruction results in loss of supply to a complete area (capillary infarction). The vessels are not fenestrated and like the vessels of the brain, do not leak small molecules such as fluorescein unless they are ruptured. This is in contrast to capillary beds elsewhere in the body that are leaky. This structure is sometimes referred to as the retinal capillary endothelial barrier. There is outward pumping of anions across the retina by the RPE and the endothelial cells of the retinal vessels.

The suprachoroid is approximately 30 μm in thickness and comprises thin interconnected lamellae of melanocytes, fibroblasts and connective tissue fibres. These are separated by a thin space known as the suprachoroidal space. In this space hydrostatic pressure is a few mm Hg lower than the intraocular pressure (IOP). The small gradient allows drainage of aqueous humour through the tissue spaces of the ciliary muscles into the suprachoroidal space. Venous drainage takes place through a series of large vortex veins; each of these drains a sector of the choroid into the superior and inferior ophthalmic veins of the orbit.

4.1.3.7 The Retina

The retina extends forward as a globe shaped wineglass to the sclera almost external to the skull. That part of the sclera devoid of retina is the pars planar which is used as an access point for injection or for close delivery to the iris-ciliary body (ICB). When stripped from its basement membrane and opened out, the collapsed retina is a circular disk approximately 42 mm in diameter and 0.5 mm in thickness. The organisation of the retina is based on a three-neuron chain, (photoreceptor cell – bipolar cell – ganglion cell) and accompanying cells (horizontal, amacrine and Müller cells), stratified into ten layers that can be identified histologically. The Müller cells extend into the inner limiting membrane and the structures remain associated with the vitreous until vitreous detachment occurs (Figure 4.1.5).

The blood-retinal barrier, or BRB is composed of two components: the retinal vascular epithelium and the retinal pigment epithelium (RPE) and is therefore separated into an inner (iBRB) and an outer barrier respectively (oBRB). The barriers are not distinct membrane elements as might be inferred from the tissue section, but dark staining zones identified are due to the presence of high density occludens junctions.

In diabetic eye disease, the failure of normal endothelial function is an initial pathophysiological change, an alteration of the iBRB that leads to diabetic macular oedema, the most frequent cause of vision loss due to diabetes. In age-related macular degeneration (AMD), the RPE begins to atrophy, leading to the yellow crystalline deposits that may slowly form more obvious deposits or drusen. The integrity of the oBRB keeps the initial pathophysiological response of the choroid to this insult from growing new vasculature and invading the retina, precipitating the more serious and vision threatening wet AMD.[49]

4.1.3.8 Retinal Pigment Epithelium

The RPE cells are joined laterally at their apical surface by junctional complexes, electrically coupled by gap junctions and secured by zonulae

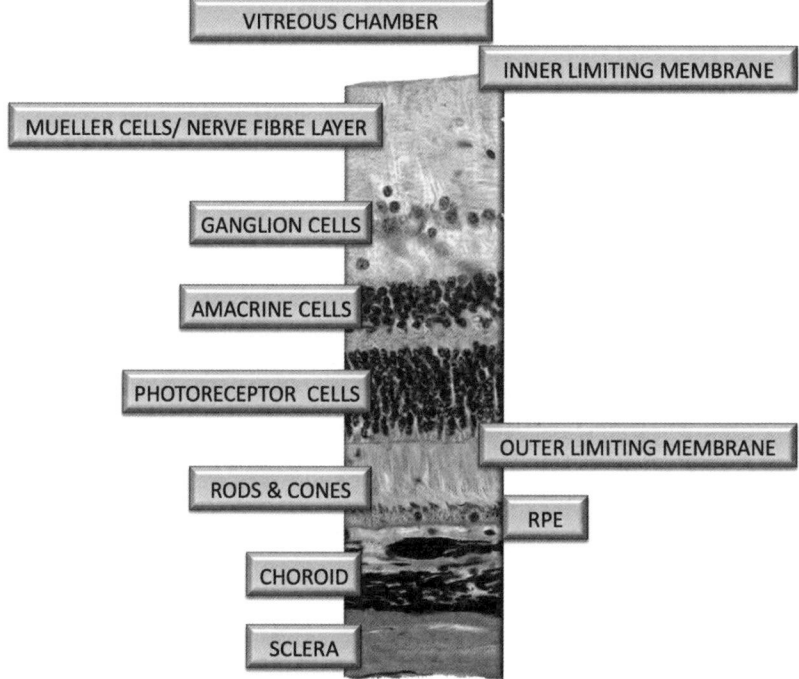

Figure 4.1.5 The layers of the retina, illustrating outer and inner limiting membranes.

occludens and zonulae adherens. On the basal side there are no tight junctions. Melanin is abundant in this layer and is of a different type to that seen deposited in the choroidal space. The innermost retinal pigment epithelium separates the outer photoreceptor rods from the choroidal blood supply. Separated from the retina by Bruch's membrane, the RPE serves to regulate nutrients to the retina, phagocytise retinal debris, remove metabolic end products and control the visual cycle. The basal membrane is highly infolded, extending 1 μm into the cytoplasm, suggesting that it is involved in transport functions.[50] The inner surface facing the photoreceptors has a surface with microvilli, 5 to 7 mm long which fold around a third of the photoreceptors.[51]

The functions of retinal pigment epithelium cells not only include phagocytosis of photoreceptor outer segments and active transport of materials from the blood supply but also production of the inter-photoreceptor matrix (IPM) which contributes to the adhesion of the retina to the RPE. RPE cells support the vitamin A metabolism in the visual cycle by moving retinoids between photoreceptor and pigment epithelium cells *via* the aqueous space of the IPM. A recent review illustrates the many important functions of the RPE in metabolic support of the eye.[52] The RPE is a "tight" ion transporting barrier and paracellular transport of polar solutes across the RPE from the choroid is restricted. This is reflected by the trans-epithelial electrical resistance (TEER) of the cell layer(s). It has been reported that the choroidal TEER ($\sim 9 \, \Omega \, cm^2$) is less than 10% of the total resistance of isolated bovine RPE-choroid (100–150 $\Omega \, cm^2$). Passive RPE diffusion has been shown to be a function of lipophilicity. The endothelium of the retinal vessels represents the inner blood-retinal barrier and offers considerable resistance to systemic penetration of drugs. The RPE has a high capacity to absorb water and lactate in the direction retina to choroid. The monocarboxylate transporter (MCT1) moves lactate and H^+, and water is co-transported with high efficiency. The role appears to be to maintain adhesion of the retina and to regulate the pH within the retina.

Carrier-mediated membrane transport proteins on the RPE selectively transport nutrients, metabolites, and xenobiotics between the choriocapillaris and cells of the distal retina and include: amino acid,[53–55] peptide,[56] di-carboxylate glucose,[57] mono-carboxylic acid,[58,59] nucleoside,[60] organic anion and organic cation[61] transporters. Membrane barriers such as the efflux pumps including multi-drug resistance protein (P-gp) and multi-drug resistance-associated protein (MRP) pumps have also been identified on the RPE. Exploitation of these transport systems may be the key to circumventing the outer blood-retinal barrier.

4.1.3.9   The Retinal Vascular Barrier

The inner BRB functions to isolate the retina by due to retinal vascular endothelial cells whose tightness is improved by the extensions of the Müller cells surrounding the retinal blood vessels. Through aquaporins, particularly AQP4, a process called K^+ siphoning dehydrates the retina. It has been suggested that experimental autoimmune uveitis, induced in mice by injection of proteins associated with photoreceptors, is caused by disruption of the AQP4 function.[62]

4.1.4   Conclusion

The physiological barriers to transport of large solutes and nanoparticles markedly restrict access of the posterior eye from anterior sites, especially when administered topically. Although transport may be noted, there must be enough flux of a therapeutic agent through the tissue to be useful. In ageing and in disease, both access and clearance are altered which may lead to the need to adjust the delivery concept accordingly.

References

1. J. D. Nelson and J. C. Wright, *Arch. Ophthalmol.*, 1984, **102**, 1049.
2. B. C. Tang, M. Dawson, S. K. Lai, Y-Y Wang, J. S. Suk, M. Yang, P. Zeitlin, M. P. Boyle, J. Fu and J. Hanes, *Proc. Natl. Acad. Sci. U. S. A.*, 2009, **106**, 19268.

3. E. de la Peña, A. Mälkiä, H. Cabedo, C. Belmonte and F. Viana, *J. Physiol.*, 2005, **567**, 415.

4. R. D. Schoenwald, Ocular pharmacokinetics and pharmacodynamics, in: *Ophthalmic drug delivery system*, ed. A. K. Mitra, 2nd edn, Drugs and the pharmaceutical sciences, series ed. J. Swarbrick, Marcel Dekker, 2003, p. 135–179.

5. Y. Zambito and G. Di Colo, *J. Drug Delivery Sci. Technol.*, 2010, **20**, 45.

6. J. L. Greaves and C. G. Wilson (with A. Rozier, J. Grove and B. Plazonnet), *Curr. Eye Res.*, 1990, **9**, 415.

7. U. B. Kompella, S. Sundaram, S. Raghava and E. R. Escobar, *Mol. Vision*, 2006, **12**, 1185.

8. R. Liu, Z. Liu, C. Zhang and B. Zhang, *J. Pharm. Sci.*, 2011, DOI: 10.1002/jps.22540.

9. M. A. Babizhayev, *Drug Test Anal.*, 2011, DOI: 10.1002/dta.265.

10. P. Calvo, J. L. Vila-Jato and M. J. Alonso, *J. Pharm. Sci.,*1995, **85**, 530.

11. T. K. De, E. J. Bergey, S. J. Chung, D. J. Rodman, D. J. Bharali and P. N. Prasad, *J. Microencapsulation*, 2004, **21**, 841.

12. L. Rabinovich-Guilatt, P. Couvreur, G. Lambert and C. Dubernet, *J. Drug Targeting*, 2004, **12**, 623.

13. G. K. Jain, S. A. Pathan, S. Akhter, N. Jayabalan, S. Talegaonkar, R. K. Khar and F. J. Ahmad, *Colloids Surf., B*, 2011, **82**, 397.

14. A. Edwards and M. R. Prausnitz, *Pharm. Res.*, 2001, **18**, 1497.

15. K. Baba, Y. Tanaka, A. Kubota, H. Kasai, S. Yokokura, H. Nakanishiand K. Nishida, *J. Controlled Release*, 2011, DOI: 10.1016/j.jconrel.2011.04.019.

16. H. E. Schaeffer and D. L. Krohn, *Invest. Ophthalmol. Visual Sci.*, 1982, **22**, 220.

17. S. Klang, M. Abdulrazik and S. Benita, *Pharm. Dev. Technol.*, 2000, **5**, 521.

18. S. L. Law, K. J. Huang and C. H. Chiang, *J. Controlled Release,* 2000, **63**, 135.

19. P. D. Jani, N. Singh, C. Jenkins, S. Raghava, Y. Mo, S. Amin, U. B. Kompella and B. K. Ambati, *Invest. Ophthalmol. Visual Sci.*, 2007, **48**, 2030.

20. J. Ambati, C. S. Canakis, J. W. Miller, E. S. Gragoudas, A. Edwards, D. J. Weissgold, I. Kin, F. C. Delori and A. P. Adamis, *Invest. Ophthalmol. Visual Sci.*, 2000, **41**, 1181.

21. K. M. Hämäläinen, K. Kananen, S. Auriola, K. Kontturi and A. Urtti, *Invest. Ophthalmol. Visual Sci.*, 1997, **38**, 627.

22. A. Urtti, L. Salminen and O. Miinalainen, *Int. J. Pharm.*, 1985, **23**, 147.

23. S. C. Chang and V. H. L. Lee, *J. Ocul. Pharmacol.*, 1987, **3**, 159.

24. Y. Horibe, K.-I. Hosoya, K.-J. Kim, T. Ogiso and V. H. L. Lee, *Pharm. Res.*, 1997, **14**, 1246.

25. M. G. Qaddoumi, H. Ueda, J. Yang, J. Davda, V. Labhassetwar and V. H. L. Lee, *Pharm. Res.*, 2004, **21**, 641.

26. A. C. Amrite and U. B. Kompella, *J. Pharm. Pharmacol.*, 2005, **57**, 1555.

27. A. C. Amrite, H. F. Edelhauser, S. R. Singh and U. B. Kompella, *Mol. Vis*ion, 2008, **14**, 150.
28. A. Bill and C. I. Phillips, *Exp. Eye Res.*, 1971, **12**, 275.
29. R. Brubaker, The physiology of the aqueous humour formation, in *Glaucoma Applied Pharmacology in Medical Treatment*, ed. S. M. Drance and A. H. Neufield, Grune & Stratton, Orlando, Florida, 1984, p. 35.
30. A. Faller, *Der korper des Menschen, Einfurhung in Bau und Funktion*, Thieme Verlag, Stuttgart, 1988.
31. G. Raviola, *Exp. Eye Res.*, 1977, **25**, 27.
32. J. V. Forrester, A. D. Dick, P. McMenamin and W. R. Lee, *The eye, basic science in practice*, W.B. Saunders, London, 1996.
33. N. Soman and R. Banerjee, *Biomed. Mater. Eng.*, 2003, **13**, 59.
34. P. N. Bishop, *Prog. Retinal Eye Res.*, 2000, **19**, 323.
35. J. Sebag, *The Vitreous*, Sptringer-Verlag, New York, 1989.
36. C. S. Nickerson, H. L. Karageozian, J. Park and J. A. Kornfield, *Macromol. Symp.*, 2005, **227**, 183.
37. E. A. Balaz, Physiology of the vitreous body, in *Vitreous body in retina surgery: special emphasis on reoperations*, C.V. Mosby, St. Louis, 1960, p. 29.
38. B. Lee, M. Litt and G. Buchsbaum, *Biorheology*, 1994, **31**, 339.
39. T. Matsuura, Y. Hara, S. Maruoka, S. Kawasaki, S. Sasaki and M. Annaka, *Macromolecules*, 2004, **37**, 7784.
40. J. G. F. Worst and L. I. Los, Chapter 3: Functional anatomy of the vitreous, in *Cisternal anatomy of the vitreous*, Kugler Publications, Amsterdam, Netherlands, 1995, p. 33.
41. W. L. Jongebloed and J. F. G. Worst, *Doc. Ophthalmol.*, 1987, **67**, 183.
42. J. Xu, J. J. Heys, V. H. Barocas and T. W. Randolph, *Pharm. Res.*, 2000, **17**, 664.
43. D. M. Maurice, S. Mishima, Ocular pharmacokinetics, in *Pharmacology of the eye*, ed. M. L. Sears, Springer-Verlag, New York, 1984, p 19–116.
44. D. M. Maurice, *J. Ocul. Pharmacol. Ther.*, 2001, **17**, 393.
45. L. E. Tan, W. Orilla, P. M. Hughes, S. Tsai, J. A. Burke and C. G. Wilson, *Invest. Ophthalmol. Visual Sci.*, 2011, **52**, 1111.
46. L. Spielberg and A. Leys, *Bull. Soc. Belge Ophthalmol.*, 2009, **312**, 17.
47. H. Lund-Anderson, B. Sander, The vitreous, in *Adler's Physiology of the Eye* ed. P. L. Kaufman and A. Alm, 10th edn, Mosby, St Louis, Missouri,, 2003, p. 293–316.
48. S. R. Singh, H. E. Grossniklaus, S. J. Kang, H. F. Edelhauser, B. K. Ambati and U. B. Kompella, *Gene Ther.*, 2009, **16**, 645.
49. J. Cunha-Vaz, *Eur. Ophthal. Rev.*, 2009, **3**, 105.
50. M. La Cour, The Retinal Pigment Epithelium, in *Adlers Physiology of the Eye*, ed. P.L. Kaufman and A. Alm, 10th edn, Mosby, St Louis, Missouri, 2003, p 348–357.

51. R. K. Sharma, B. E. J. Ehinger, Development and structure of the retina, in *Adlers Physiology of the Eye*, ed. P. L. Kaufman and A. Alm, 10th edn, Mosby, St Louis, Missouri, 2003, p 319–347.
52. R. Simó, M. Villarroel, L. Corraliza, C., Hernández and M. Garcia-Ramírez, *J. Biomed. Biotechnol.*, 2010, **2010**, 190724.
53. Y. Miyamoto, P. Kulanthaivel, F. H. Leibach and V. Ganapathy, *Invest. Ophthalmol. Visual Sci.*, 1991, **32**, 2542.
54. J. W. Leibach, D. R. Cool, M. A. Del Monte, V. Ganapathy, F. H. Leibach and Y. Miyamoto, *Curr. Eye Res.*, 1993, **12**, 29.
55. D. V. Pow, *Neurochem. Int.*, 2001, **121**, 89.
56. S. C. Ocheltree, R. F. Keep, H. Shen, D. Yang, B. A. Hughes and D. E. Smith, *Pharm. Res.*, 2003, **20**, 1364.
57. G. J. Mantych, G. S. Hageman and S. U. Devaskar, *Endocrinology*, 1993, **133**, 600.
58. R. M. Knott, M. Robertson, E. Muckersie, V. A. Folefac, F. E. Fairhurst, S. M. Wileman and J. V. Forrester, *Diabetologia*, 1999, **42**, 870.
59. N. J. Philp, D. Wang, H. Yoon and L. M. Hjelmeland, *Invest. Ophthalmol. Visual Sci.*, 2003, **44**, 1716.
60. E. F. Williams, I. Ezeonu and K. Dutt, *Curr. Eye Res.*, 1994, **13**, 109.
61. Y. H. Han, D. H. Sweet, D. N. Hu and J. B. Pritchard, *J. Pharmacol. Exp. Ther.*, 2001, **296**, 450.
62. E. Motulsky, P. Koch, S. Janssens, M. Liénart, A. M. Vanbellinghen, N. Bolaky, C. C. Chan, L. Caspers, M. D. Martin-Martinez, H. Xu, C. Delporte and F. Willermain, *Mol. Vision*, 2010, **16**, 602.

CHAPTER 4.2

Nanostructures Overcoming the Ocular Barrier: Drug Delivery Strategies

ARTO URTTI

Centre for Drug Research, Faculty of Pharmacy, University of Helsinki, Viikinkaari 5 E, 00014 University of Helsinki, Finland
E-mail: arto.urtti@helsinki.fi

4.2.1 Introduction

Various cellular and tissue targets in ocular drug treatment have distinct challenges of drug delivery. Anterior targets are involved in the treatment of infections, inflammatory conditions, and glaucoma. In these cases the drug should reach cornea, conjunctiva, iris, trabecular meshwork or ciliary body at adequate concentrations. These targets are also relevant in the case of local anesthesia and dilatation of pupils. These targets and conditions are treated most frequently with topical ocular eye drops. The disease conditions of the posterior eye segment cannot be treated using topical ocular eye drops due to the inadequate drug distribution to the posterior tissues. Posterior tissues, such as neural retina, retinal pigment epithelium, and choroid are involved in the retinal degenerations (like age-related macular degeneration, retinitis pigmentosa, diabetic retinopathy), posterior segment infections and inflammatory conditions and neovascularisation affecting choroidal or retinal vessels (Figure 4.2.1). In these cases, the route of administration is usually intravitreal, but periocular, subretinal and systemic administration may be possible in some cases. In general, only a negligible fraction of the systemically administered

RSC Drug Discovery Series No. 22
Nanostructured Biomaterials for Overcoming Biological Barriers
Edited by Maria Jose Alonso and Noemi S. Csaba
© The Royal Society of Chemistry 2012
Published by the Royal Society of Chemistry, www.rsc.org

STRUCTURE OF EYE

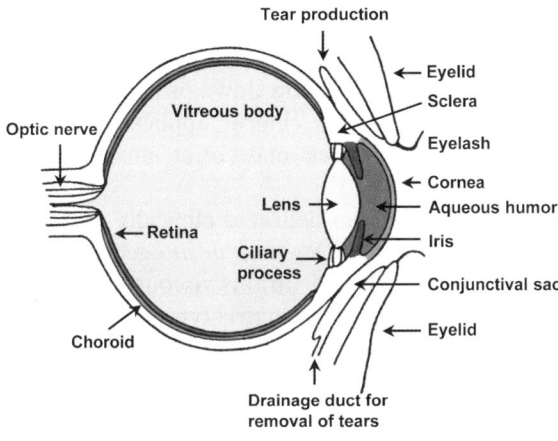

Figure 4.2.1 Anatomy of the eye. Anterior segment comprises of the cornea, conjunctiva, aqueous humour, iris, ciliary processes, and lens. The posterior segment includes vitreous body, retina, choroid, and optic nerve. Sclera is partly in the anterior and partly in the posterior segment.

drug reaches the ocular tissues. The eye is protected by blood-retinal barrier and blood – anterior chamber barriers. These barriers further limit drug access to the ocular tissues. Therefore, the major attention in the field, and in this chapter, is directed to the local routes of drug administration to the eye.

Even though the local delivery allows one to administer the drug to the close proximity of the target tissue, still the bioavailability is quite poor.[1] Only less than 5% of the topically administered drug enters the anterior chamber, in many cases the bioavailability is less than 1%.[2] The fraction of the drug that reaches the posterior segment is much lower, and even for fairly lipophilic small molecular weight drugs the concentration in the retina is 10^{-6}–10^{-5} times that in the tear fluid.[3] Thus, increasing the bioavailability of topical ocular drug administration, the only option for out-patients, is an important challenge for ophthalmic nanomedicine. The second challenge of topical ocular administration is to prolong the duration of drug activity, thereby improving the patient compliance that is problem in the glaucoma treatment. For posterior segment diseases the challenges are somewhat different. Invasive intravitreal injections have become widely used after launching of the anti-angiogenesis antibodies (Avastin, Lucentis) in the treatment of wet age-related macular degeneration. The injections are given monthly for the rest of the patients' lifetime. Long-acting drug delivery systems would improve the treatment of posterior segment diseases. Current systems are implantable matrices, but more elegant and less invasive long-acting systems would be preferable.[4,5] Cell based targeting is another challenge for nanomedicine in the posterior eye segment. Compared to some other fields, like oncology, cellular drug targeting in the eye has been sparsely studied. Cellular delivery is the key issue in the use of larger molecules with intracellular sites of action, such as transcription factors and many other proteins, miRNA, siRNA, and DNA. Inadequate delivery properties may slow down or hamper the progress of these novel therapeutic options towards clinical applications. Overall, improved drug delivery with nanostructures may offer answers to major clinical ophthalmology problems.

Topical ocular drugs have been delivered clinically using polymer matrices (Lacrisert) and reservoir systems (Ocusert), *in situ* gelling eye drops (Timoptic XE), microspheres (Betoptic S), and various viscous eye drops with polymeric excipients. Interestingly, several ocular matrix type inserts were used clinically in the Soviet Union in the 1960's.[5] These systems can be classified based on the therapeutic advantages in two categories: prolonged action systems and formulations for reduced ocular discomfort. Ocusert that released pilocarpine over one week offers a major prolongation of activity, but this device was withdrawn from the market, because it drops from the patients' eye occasionally. All other systems provide modest control of drug release and prolongation of residence time on the ocular surface. Major improvements surpassing these levels should be possible with future materials and formulations.

Current clinical intravitreal systems deliver ganciclovir (Vitrasert) and fluocinolone acetonide (Retisert) over several months. These systems are implants that must be surgically placed in the eye. Even though microparticles and nanosystems have been investigated for intravitreal administration the basic science background is mostly missing, and these systems have not entered the clinics. There are, however, good opportunities to achieve major therapeutic gains with new innovative intravitreal drug formulations.

Ocular pharmacokinetics depends on the molecular drug properties, ocular tissue and cell level barriers, compartment volumes, and fluid flow features in the eye.[1,5] These aspects should be taken into account in the design of nanostructured drug delivery systems. In this chapter, the ocular drug administration routes are described in pharmacokinetic context. The delivery system requirements are then discussed from this perspective.

4.2.2   Topical Ocular Drug Delivery

In topical ocular drug delivery the dug formulation is instilled or applied on the ocular surface. After application, the topical drug should absorb to the cornea or conjunctiva (for example for corneal and conjunctival infections and inflammations) or across the cornea into the anterior chamber from where the drug easily reaches the iris, trabecular meshwork and ciliary body (*i.e.* targets of glaucoma medications, mydriatic drugs).[6] The drug is eliminated from the anterior chamber *via* aqueous humour outflow and distribution to the veins in

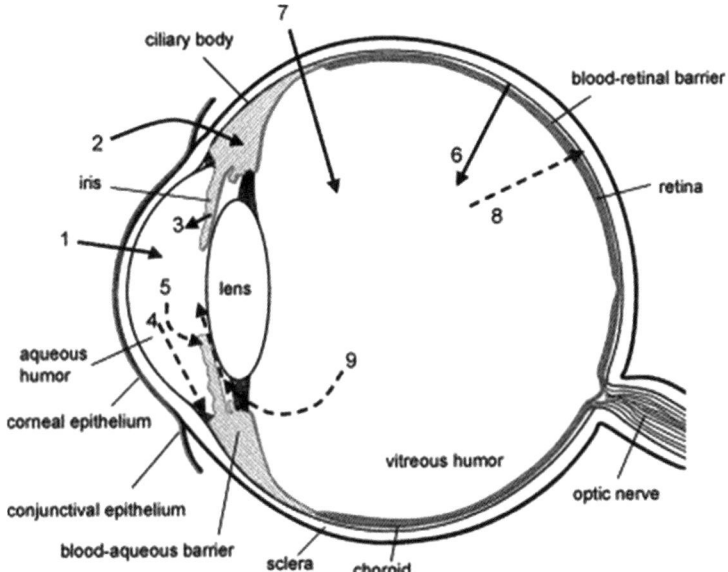

Figure 4.2.2 Ocular anatomy and the routes of drug transfer. The numbers refer to: (1) trans-corneal permeation from the lacrimal fluid into the anterior chamber, (2) non-corneal drug permeation across the conjunctiva and sclera into the anterior uvea, (3) drug distribution from the blood stream *via* blood–aqueous barrier into the anterior chamber, (4) elimination of drug from the anterior chamber by the aqueous humor turnover to the trabecular meshwork and Schlemm's canal, (5) drug elimination from the aqueous humor into the systemic uveoscleral circulation, (6) drug distribution from the blood into the posterior eye across the blood–retina barrier, (7) intravitreal drug administration, (8) drug elimination from the vitreous *via* posterior route across the blood retina barrier, and (9) drug elimination from the vitreous *via* anterior route to the posterior chamber. From ref. 4.

the anterior uvea (iris and ciliary body).[1,6] The routes of drug absorption, distribution and elimination in the eye are shown in Figure 4.2.2.

Eye drops are the most commonly used formulation. After instillation the eye drop spreads on the corneal and conjunctival surface and then drains to the nanolacrimal duct in a few minutes.[7] Therefore, the contact time on the ocular surface is limited. Furthermore, as described in the previous chapter, the corneal epithelium is a barrier that limits ocular drug absorption.[8] The third limiting factor is the rapid loss of drugs to the systemic circulation *via* conjunctival vasculature.[9] For example, pilocarpine and timolol show ocular absorption of 1–4%,[3,8] and systemic absorption *via* conjunctiva constitutes about 20–30% of the dose.[9] The rest of the dose is drained with lacrimal fluid to the nose where further systemic absorption may take place. Next, a quantitative approach is taken to describe the pharmacokinetics on the ocular surface.

4.2.2.1 Pharmacokinetics of Topical Ocular Drugs

After eye drop administration on the ocular surface the drug absorption must compete with several parallel processes of drug removal (Figure 4.2.3). These processes can be described in terms of drug clearances. Drug clearance by tear turnover (Cl_{tf}) is simply the rate of tear turnover (about 1 μl min^{-1}). Volume of tear fluid is 7 μL and typical eye drop volume is 40 μL (thus, volume of distribution in tear fluid would be $V_d = 47$ μL). Thus, the elimination rate constant by tear turnover alone would be $k_{tf} = Cl_{tf}/V_d = 1/47 = 0.02$ min^{-1}. Then, the half-life of elimination from tear fluid should be $t_{1/2} = \ln2/k_{tf} = 0.69/0.02 = 34$ min. In reality, the half-life of pilocarpine in the tear fluid is about 1 min.[7,8] It is obvious that the normal tear turnover is a minor factor in drug elimination from tear fluid. The major clearance factors are the drainage of extra fluid volume of eye drop and conjunctival clearance. The volume reaches the normal levels of 7 μL in a few minutes. Conjunctival clearance for a drug can be calculated as follows: $Cl_{cj} = P_{cj} \times S_{cj}$, where P_{cj} is the permeability of the conjunctiva and S_{cj} is the surface area of the conjunctiva. Conjunctival drug permeabilities are in the range of 0.5–5.0 × 10^{-6} cm s^{-1},[10] and the surface area is about 20 cm^2. In the same manner, we can calculate the clearance of drug from the lacrimal fluid through the corneal epithelium: $Cl_{co} = P_{co} \times S_{co}$, where subset "co" stands for the cornea. Small molecular weight drugs have corneal permeabilities of (0.14–75) × 10^{-6} cm s^{-1},[11] and the surface area of the cornea is 1 cm^2 leading to $Cl_{co} = 0.01$–4.5 μL min^{-1}. Then, we can consider a drug delivery system that is placed in the conjunctival sac where it releases drug at constant rate but the delivery system is retained, not being removed by the lacrimal flow or drainage of the system. In this case the drug bioavailability (assuming there is no corneal metabolism) is $F_{co} = Cl_{co} / (Cl_{tf} + Cl_{cj})$. For timolol, $F_{co} = 0.09$ meaning that the the maximal achievable fraction of drug absorption is about 9%. If the delivery system is drained from the conjunctival sac, the bioavailability will be lower than 9%. This analysis

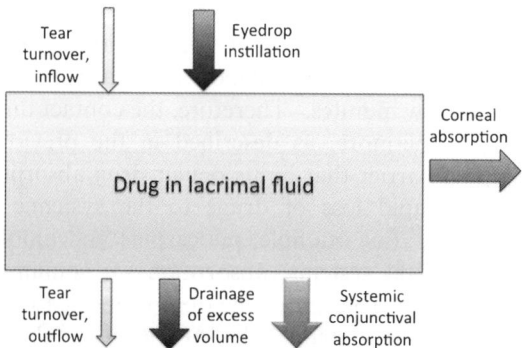

Figure 4.2.3   Precorneal drug elimination processes.

informs us that for each drug pharmacokinetic constraints set the upper limit for topical ocular drug bioavailability.

4.2.2.2 Drug Delivery Approaches

Using pharmacokinetic approach we can divide the drug delivery approaches into three categories.

Firstly, let us consider decreased drug removal from the corneal and conjunctival surface. In this case, the eye drop is formulated so that it remains on the ocular surface, but the drug is freely diffusing in the system. Viscous water-soluble polymer based solutions, gels and *in situ* forming gels with high diffusivity would fall into this category. Loss factor of solution drainage is not operating in this case, but other clearance factors are functional. The maximal bioavailability is determined as described above. In this case, the half-life of drug concentration in the preocular area (*e.g.* timolol) is $t_{1/2} = V_d/(Cl_{tf} + Cl_{co} + Cl_{cj}) = 3$ min. In summary, bioavailability is expected to improve modestly (perhaps 2 fold) and pre-ocular retention is prolonged slightly. However, the duration of drug activity is expected to increase only slightly because the overall drug retention on the ocular surface is extended only by about 10 min. The rate of systemic drug absorption is reduced and this is reflected in several fold decrease in systemic peak drug concentrations if the drug has very rapid systemic absorption and short half-life in the systemic absorption. Otherwise, this effect may be modest.

Secondly, we consider a controlled release formulation. This system is retained in the conjunctival sac and the drug is released in controlled manner from the device (*e.g.* polymer matrices, reservoir devices, controlled release semi-solid formulations). The upper limit of bioavailability is set as described in 4.2.2.1, but the benefit of prolonged duration of action can be achieved. The drug concentration in tear fluid at steady state will be $C_{ss,tf} = R_{ss}/(Cl_{tf} + Cl_{co} + Cl_{cj})$, where R_{ss} is the rate of drug release from the device. Interestingly, the conjunctival sac acts like a sink condition environment. This is due to the rapid drug clearance to the conjunctival vasculature. Therefore, in most cases *in vitro* release rate can be used in the steady state equation above. Drug input *via* cornea into the aqueous humour will be $J_{co} = C_{ss,tf}P_{co}S_{co}$ and the steady state concentration in the aqueous humour will be $C_{ss,aq} = J_{co}/Cl_{aq}$ where Cl_{aq} is the drug clearance from the aqueous humour. The values of drug clearance in this compartment range from 2 μL min^{-1} (macromolecules that are eliminated only *via* trabecular meshwork in the aqueous humour) to 20–30 μL min^{-1} (small lipophilic compounds that are eliminated *via* aqueous humour flow and vasculature of anterior uvea). A topical ocular delivery system can maintain the steady state concentration in the aqueous humour over long periods (weeks, months) if the delivery system is capable of long release periods and is not lost from the conjunctival sac. This kind of system can also decrease the peak drug concentrations in the systemic circulation, but probably not affect the extent systemic absorption. This approach offers the benefit of long action and potential for improved patient compliance.

The third option is the specific drug delivery across the cornea. Contact lenses are placed on the corneal surface and this gives an advantage of more direct drug delivery to the cornea. This approach is described elsewhere in this book. If the drug is targeted to the corneal epithelium, it will not be subject to the loss processes of free drug (*i.e.* Cl_{tf}, Cl_{cj}). In principle, this approach should provide opportunity to increase the bioavailability further for small molecular weight drugs. For larger molecules, not at all capable of permeating the corneal epithelium, this is the only option for topical ocular administration. These drugs must be shuttled with nanosystems to the corneal epithelium or across this tissue; the free drug has too low a permeability in the corneal epithelium and, in the case of protein drugs, the drug may be also metabolized to inactive fragments during the permeation).

Polymeric nanoparticles and liposomes have been investigated already for several decades as topical ocular drug delivery systems.[12–16] Usually these formulations extend the topical ocular retention time modestly, and they release the drug over period of a few hours. This combination leads typically to a situation where drug release is relatively slow compared to its retention time, and benefits to ocular drug delivery are not substantial. Similar modest improvements are achievable with simpler means, such as optimisation of eye drop pH and increased viscosity with water-soluble polymeric excipients (*e.g.* Carbopol, HPMC, PVA).[17,18]

Regular liposomes (phosphatidyl cholines, phosphatidyl glycerol, PEGylated derivatives) or nanoparticles (polycyanoacrylates, polylactides and polyglyco-lides and their co-polymers) are not able to get across the corneal epithelium, because the intercellular space in the tissue is limited (porosity about 10^{-7}, pore diameter about 2 nm).[19] In order to achieve nanoparticle permeation to the corneal epithelial cells, the nanosystem should bind on the corneal epithelium with electrostatic forces (*i.e.* cationic charges on the nanoparticle) or ligand–receptor interactions. The success of cationic charges is shown in the work of Toropainen *et al.*[20] in which cationic liposomes (DOTAP/DOPE) were complexed with protamine sulphate and plasmid DNA. These nanocomplexes were able to deliver pDNA into the corneal epithelial cells in culture and *in vivo*. Efficacy *in vivo* was several orders of magnitude lower than *in vitro*, but yet able to deliver DNA into the corneal epithelium. This resulted in one-week secretion of the encoded protein to the lacrimal fluid and protein delivery to the aqueous humour of rabbits. This illustrates how nanostructures can overcome the barrier of corneal epithelium and convert it to a protein secreting platform. This approach may be useful for protein drug delivery into the lacrimal fluid and anterior chamber. The marker protein (secreted alkaline phosphatase) as such had zero permeability in the corneal epithelium.

Several groups have investigated hyaluronic acid derivatized nanoparticles in topical ocular drug delivery.[21–24] The corneal epithelium expresses CD44 receptors that are capable of internalizing hyaluronic acid and DNA polyplexes with hyaluronic acid coating.[21] Overall, drug targeting and subsequent transcytosis in the corneal epithelium have been only sparsely studied. These approaches will require sub-cellular mechanistic investigations and realistic cell models. Unfortunately, the widely used HCE model of the corneal epithelium turned out to deviate substantially from its *in vivo* counterpart.[25]

Ahmed and Patton[26] showed that the corneal epithelium is the main route of drug entry into the aqueous humour. However, some larger compounds, like insulin, may have preferable entry *via* conjunctiva. Small compounds are rapidly removed by the conjunctival blood flow, but this loss is slower in the case of large molecules. This route may be viable for targeted nanostructures, *i.e.*, formulations homing to the conjunctival epithelium and gradually releasing the drug in these cells.

In summary, nanostructures may provide major improvements in ocular drug delivery. For prolonged duration of activity, the system must be capable of releasing the drug for a long time and be retained in the conjunctival sac. Corneal epithelial targeting may lead to significant improvements in bioavailability, including for larger molecules, such as oligonucleotides, DNA and proteins, but the formulation must have high affinity on the corneal epithelium and be internalized to these cells. Expression and functions of drug transporters in the corneal epithelium were recently summarized, but interplay of transporters with nanoparticle drug delivery systems have not been studied.

4.2.3 Intravitreal Drug Delivery

Intravitreal drug delivery means injection or implantation of the drug directly to the vitreal cavity. Invasive intravitreal drug administration is practiced in order to reach adequate drug concentrations in the retina and choroid. Intravitreal injections gained wide clinical use a few years ago when antibody injections changed the treatment of age related macular degeneration. Antibodies are injected repetitively on a monthly basis for years, possibly for decades. Antibodies have a long half-life of about 100 hours in the vitreous,[27,28] while small molecules have half-lives of a few hours.[29,30] The injection interval of one month means about 7 half lives of the antibody, *i.e.*, the drug concentration undergoes more than 100 fold changes in the course of the treatment. For a small molecular weight drug with a half-life of 3 hours this approach (100 fold concentration fluctuation and injections in simple solutions) would lead to daily intravitreal injections. It should be noted that 100-fold fluctuation is safe for antibodies, but this is not the case for all drugs.

Intracellular therapeutic compounds, such as miRNA, siRNA, DNA and transcription factors, must be delivered efficiently into the intracellular compartments. Otherwise, adequate activity is not reached. In the context of intravitreal drug delivery, this would mean nanosystems capable of targeting and internalisation of the correct cell type in the retina or choroid.

Clearly, the drug delivery challenges of intravitreal drug delivery are: (1) prolongation the duration of activity, (2) reduction of the concentration fluctuation, and (3) drug targeting to the retinal and choroidal cells.

4.2.3.1   Pharmacokinetics of Intravitreal Drugs

After intravitreal injection the drugs distribute in the vitreous by diffusion and convection (see Chapter 4.1). Vitreal drug elimination takes place *via* anterior and posterior routes.[1].The anterior route means drug diffusion to the posterior chamber and flow in the aqueous humour flow to the anterior chamber and elimination though trabecular meshwork to the Schlemm's canal and further to blood circulation. Drugs capable of crossing the blood-retina barrier (BRB) may escape the vitreous humour from the back of the eye.[1] This barrier is composed of the inner BRB (retinal capillary walls) and outer BRB (retinal pigment epithelium). Since small and lipophilic compounds are able to pass the BRB easier, their half-life in vitreous is much shorter (few hours) than that of large molecules, such as proteins (about 100 h). The volume of the vitreous (V_{vitr}) is about 1.8 mL. Therefore, $Cl_{vitr} = V_{vitr}k_{vitr} = V_{vitr}\ln2/t_{\frac{1}{2},vitr}$. The values for a small molecule with half-life of 3 h and a protein with a half-life of 100 h are about 6 μL min^{-1} and 0.2 μL min^{-1}, respectively. Figure 4.2.4 illustrates the routes of drug elimination from the vitreal cavity. It should be noted that diffusion in the vitreous is restricted leading to concentration gradients within the tissue. The local concentration gradients have an impact on drug elimination, because the rate of elimination across the eliminating membrane (*e.g.* retinal pigment epithelium, RPE) $J_{rpe} = CP_{rpe}S_{rpe}$, where C is the concentration next to the RPE on the vitreal side, P_{rpe} is the permeability in the RPE, and S_{rpe} is the surface area of the RPE. *In vivo* vitreal clearances (see above) are therefore smaller than the clearances based on the membrane properties, and aqueous humour flow rate. Subsequently, the injection site has

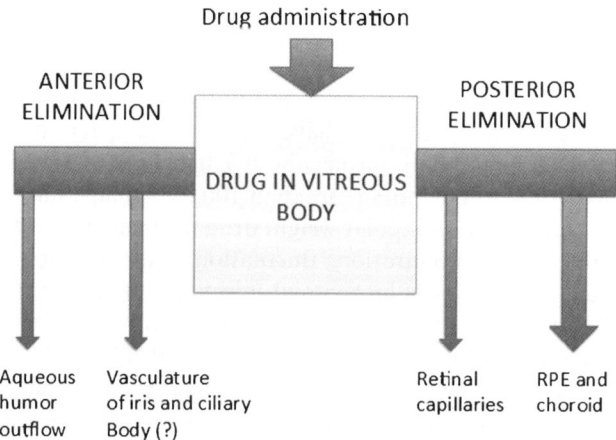

Figure 4.2.4   Elimination routes of drugs from the vitreous body.

an influence on drug concentration distribution in the vitreous, retina and choroid.

4.2.3.2 Drug delivery approaches

Intravitreal drug delivery approaches were summarized in more detail recently.[4] Current clinically accepted delivery systems aim at prolonged action and reduced concentration fluctuation. The systems are delivering ganciclovir (Vitrasert) and fluocinolone acetonide (Retisert) over months. The ganciclovir system is non-degradable polyvinyl acetate insert – ethylene vinyl acetate implant. Either the system must be removed or a new system is sutured to the eye. Ozurdex, a dexamethasone releasing matrix for treatment of macular oedema is biodegradable and, therefore, it does not have to be removed.[4] The system uses FDA approved material PLGA. An injectable system would be less invasive and clinically easier to use.

Nanoparticles may be feasible for prolonged drug delivery and cell based targeting. After intravitreal injection the particles may either diffuse in the vitreous or be retained at the injection site. Pitkänen *et al.*[31,32] investigated the mobility of the nanocomplexes and macromolecules in the isolated vitreous and in the eyecups with vitreal body. The complexes had limited diffusion in the vitreous, and neural retina further decreased their transport. Other studies have indicated that nanoparticles with poly(ethylene glycol) coating have improved mobility in the vitreous and may reach the retinal layers.[33] Several studies have dealt with the cell level delivery mechanisms. Particularly, RPE is efficient in internalizing particles. Due to its phagocytic capacity even microparticles were internalised *in vitro*. These cells seem also to express the transgene over a prolonged period (2 months) after nanocomplex mediated gene transfer.[34] Gene transfer in an active form at the nuclear level appears to be the key feature. Polyethylene imine DNA complexes showed about 100 times higher levels than poly-L-lysine complexes when assessed based on the expression per plasmid copy delivered into the nucleus.[35]

Intravitreal delivery properties of drugs and gene medicines may be improved with nanostructures, but systematic mechanistic research is needed in the field. Structure activity properties of nanoparticles in terms of mobility in the vitreous, delivery in the different retinal layers, interactions with vitreous components, and cellular targeting are still poorly known. Delivery system technologies, such as targeted and triggered release nanoparticles, may be useful as intravitreal systems.

4.2.4 Periocular Drug Delivery

Periocular drug delivery is practiced as injection or implantation of the drug formulation to the subconjunctival space (between conjunctiva and sclera), deeper next to the sclera (parabulbar, retrobulbar injections) or in the sclera (injection or implantation under scleral flap).[1] This method is not used as often as topical ocular or intravitreal administrations. Sometimes it is used, for example, to achieve improved local anesthetic action in the anterior part of the eye. Periocular injection is an attractive alternative, since it is less invasive than the intravitreal injection, and it may give opportunity to deliver adequate drug concentrations to the posterior segment of the eye. The relative roles of the periocular barriers are just emerging.

4.2.4.1 Pharmacokinetics of periocular drug administration

Pharmacokinetics of periocular injections have been discussed in more detail in two recent publications.[36,37] The barriers to periocular drug administration are illustrated in Figure 4.2.5. The scheme and associated model of periocular drug delivery was built based on *in vivo* and *in vitro* results, anatomical and physiological factors, and molecular properties. The subconjunctivally injected drug is eliminated by blood and lymphatic flow processes. Part of the injected drug is able to permeate to the sclera and further diffuse to suprachoroidal space. In the choroid, a large part of the drug is eliminated *via* choroidal blood flow, and a smaller fraction may permeate across the RPE into the retina.

Periocular drug delivery is treated in pharmacokinetics as a series of barriers and loss processes. At each step the fraction of drug entering the next step is defined as the fraction $F = J_{12}/(J_{10} + J_{12})$, where J_{12} is the clearance from donor compartment to the next step (*e.g.* from subconjunctival space to sclera) and $J_{10} + J_{12}$ designates the total clearance from the compartment (*e.g.* eliminating loss by the subconjunctical flow processes plus clearance for drug entry to the sclera). This analysis reveals that (1) most of the drug dose is cleared from the subconjunctival space *via* loss processes, (2) the sclera is a relatively unimportant barrier, (3) choroidal blood flow removes most of the drug reaching the tissue, (4) RPE is a more important barrier than sclera, and (5) large molecules are delivered better than small molecules to the retina and choroid. The last surprising point is explained as follows. Periocular delivery process involves major loss processes (subconjunctival and choroidal flows)

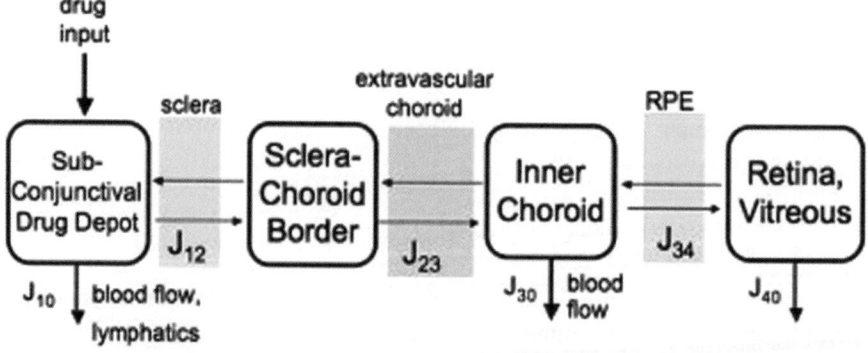

Figure 4.2.5 Scheme of periocular drug delivery.

where macromolecules are eliminated at much lower efficacy than the small molecules. These factors compensate for the slower permeation of macromolecules in the membranes (sclera, RPE).

4.2.4.2 Drug Delivery Approaches

Periocular drug delivery would definitely benefit from advanced drug formulations. So far, many polymeric systems have been tested, including controlled release poly(orthoesters) that can be injected in semi-solid form and thereafter they undergo bioerosion.[38]

Considering the complex periocular drug delivery process there are several points of interest for investigators of nanostructures (see ref. 36). Firstly, in the subconjunctival space the drug escape to the systemic blood circulation could be slowed down using nanoparticles. It is known that albumin is cleared 15–60 times slower from the injection site than small molecules. Secondly, sclera is a fairly leaky tissue where some nanostructures may be able to permeate. It is known that at least molecules with a diameter of 13 nm are able to traverse across sclera. Thirdly, in the choroid, macromolecules are eliminated 20 times slower than small molecules, and 60 nm is the cut-off size of the vascular fenestrations. Finally, the RPE has steep size dependence in permeability, but on the other hand it may endocytose nanoparticles, as has been shown in many *in vitro* studies. So far, nanoparticle permeation has been studied in the sclera,[39–42] but the complexity of the overall permeation process should not be underestimated. In conclusion, the periocular route of administration offers opportunities to build nanostructures for improved bioavailability and prolonged activity.

4.2.5 Other routes of ocular drug delivery

4.2.5.1 Systemic delivery

Systemic drug delivery is used clinically in some cases. For example, massive doses of intravenous antibiotics (*e.g.* cephalosporins) are given to treat serious ocular infections, intravenous mannitol has been used to lower critically high intraocular pressure, and carbonic anhydrase inhibitors, like acetazolamide, have been used in glaucoma treatment. This is feasible in some selected cases, particularly with drugs having a broad therapeutic index, like antibiotics. Systemic carbonic anhydrase inhibitors cause serious adverse effects and local administration of these drugs has been developed. The weight of the eye is about 1/10 000 of the whole body. Accordingly, the fraction of ocular blood flow is negligible compared to the blood flow in the liver and spleen, the organs that clear nanoparticles from the systemic blood circulation. Thus, a successful systemically given ocular drug targeting system should have exceptionally high selectivity for the eye and efficient surface coating to evade the reticuloendothelial system in the liver and spleen.

4.2.5.2   Sub-retinal administration

Subretinal injections are given between the neural retina and RPE. This is a surgical technique that requires special training and expertise. It is potentially dangerous, since failed injection might cause retinal detachment. This method has been used successfully to inject AAV-2 viral vectors for gene therapy of retinal degeneration (congenital amaunorosis) in humans.[42] This experiment was highly successful resulting in the recovery of vision in some patients. However, it is not certain how feasible this technique of drug delivery is in wider patient populations.

4.2.6   Conclusion

Ocular drug delivery involves many routes of drug administration each having unique pharmacokinetic constraints. It is obvious that nanostructure based drug formulations represent a major potential for improved ocular drug delivery. So far nanostructure studies have not resulted in systematic understanding of their behaviour in the eye. Mechanistic understanding of the nanostructures in the eye is needed and will emerge in the future. This will lead to improved pharmaceutical treatments of the ocular diseases.

References

1. D. M. Maurice and S. Mishima, Ocular pharmacokinetics, in *Handbook of experimental pharmacology*, ed. M. L. Sears, vol. 69, Springer Verlag, Berlin-Heidelberg, 1984, pp. 16–119.
2. K. Järvinen, T. Järvinen and A. Urtti, *Adv. Drug Delivery Rev.*, 1995, **16**, 3.
3. A. Urtti, J. D. Pipkin, G. S. Rork, T. Sendo, U. Finne and A. J. Repta, *Int. J. Pharm.*, 1990, **61**, 241.
4. E. Del Amo and A. Urtti, *Drug Discovery Today*, 2008, **13**, 135.
5. V. H. Lee and J. R. Robinson, *J. Ocul. Pharmacol.*, 1986, **2**, 67.
6. A. Urtti, *Adv. Drug Delivery Rev.*, 2006, **58**, 1131.
7. V. H. Lee and J. R. Robinson, *J. Pharm. Sci.*, 1979, **68**, 673.
8. J. W. Sieg and J. R. Robinson, *J. Pharm. Sci.*, 1976, **65**, 1816.
9. A. Urtti, L. Salminen, O. Miinalainen, *Int. J. Pharm.*, 1985, **23**, 147.
10. M.R. Prausnitz and J.S. Noonan, *J. Pharm. Sci.*, 1998, **87**, 1479.
11. H. Kidron, E. del Amo, K.S. Vellonen, A. Tissari and A. Urtti. *Pharm. Res.*, 2010, **27**, 1398.
12. C. Losa, P. Calvo, E. Castro, J.L.Vila-Jato and M.J. Alonso, *J. Pharm. Pharmacol.*, 1991, **43**, 548.
13. R. Cavalli, M.R. Gasco, P. Chetoni, S. Burgalassi and M. F. Saettone, *Int. J. Pharm.*, 2002, **15**, 241.
14. C. Giannavola, C. Bucolo, A. Maltese, D. Paolino, M. A. Vandelli, G. Puglisi, V. H. Lee and M. Fresta, *Pharm. Res.*, 2003, **20**, 584.

15. K. Yoncheva, S. Gómez, M. A. Campanero, C. Gamazo and J. M. Irache, *Expert Opin. Drug Delivery*, 2005, **2**, 205.

16. E. B. Souto, S. Doktorovova, E. Gonzalez-Mira, M. A. Egea and M. L. Garcia, *Curr. Eye Res.*, 2010, **35**, 537.

17. T. F. Patton and J. R. Robinson, *J. Pharm. Sci.*, 1975, **64**, 1312.

18. M.F. Saettone, B. Giannaccini, A. Teneggi, P. Savigni and N.Tellini, *J. Pharm. Pharmacol.*, 1982, **34**, 464.

19. K.M. Hämäläinen K. Kontturi, L. Murtomäki, S. Auriola and A. Urtti, *J. Controlled Release*, 1997, **49**, 97.

20. E. Toropainen, M. Hornof, K. Kaarniranta and A. Urtti, *J. Gene. Med.*, 2007, **9**, 208.

21. M. Hornof, M. de la Fuente, M. Hallikainen, R. H.Tammi and A. Urtti, *J. Gene. Med.*, 2008, **10**, 70.

22. M. de la Fuente, B. Seijo and M. J. Alonso, *Invest. Ophthalmol. Visual Sci.*, 2008, **49**, 2016.

23. I. Yenice, M. C. Mocan, E. Palaska, A. Bochot, E. Bilensoy, I. Vural, M. Irkeç and A. A. Hincal, *Exp. Eye Res.*, 2008, **87**, 162.

24. S. Wadhwa, R. Paliwal, S. R. Paliwal and S. P. Vyas, *J. Drug Targeting*, 2010, **18**, 292.

25. D. Greco, K. S. Vellonen, H. Turner, M. Häkli, T. Tervo, P. Auvinen, J. M. Wolosin and A. Urtti, *Mol. Vision*, 2010, **16**, 2109.

26. I. Ahmed and T.F. Patton, *Invest. Ophthalmol. Visual Sci.*, 1985, **26**, 584.

27. S.J. Bakri, M.R. Snyder, J.M. Reid, J.S. Pulido, M.K. Ezzat and R.J. Singh, *Ophthalmology*, 2007, **114**, 2179.

28. H. Kim, K.G. Csaky, C.C. Chan, P.M. Bungay, R.J. Lutz, R.L. Dedrick, P. Yuan, J. Rosenberg, A.J. Grillo-Lopez, W.H. Wilson and M.R. Robinson, *Exp. Eye Res.*, 2006, **82**, 760.

29. W. Liu, Q.F. Liu, R. Perkins, G. Drusano, A. Louie, A. Madu, U. Mian, M. Mayers and M. H. Miller, *Antimicrob. Agents Chemother.*, 1998, **42**, 1417.

30. P.M. Hughes, R. Krishnamoorthy and A.K. Mitra, *J. Ocul. Pharmacol. Ther.*, 1996, **12**, 209.

31. L. Pitkänen, M. Ruponen, J. Nieminen and A. Urtti, *Pharm. Res.*, 2003, **20**, 576.

32. L. Pitkänen, J. Pelkonen, M. Ruponen, S. Rönkkö and A. Urtti, *AAPS J.*, 2004, **6**, article 25.

33. L. Peeters, N. N. Sanders, K. Braeckmans, K. Boussery, J. Van de Voorde, S. C. De Smedt and J. Demeester, *Invest Ophthalmol Visual Sci.*, 2005, **46**, 3553.

34. E. Mannermaa, S. Rönkkö and A. Urtti, *Curr. Eye Res.*, 2005, **30**, 345.

35. M. Männistö, M. Reinisalo, M. Ruponen, P. Honkakoski, M. Tammi and A. Urtti, *J. Gene. Med.*, 2007, **9**, 479.

36. V.P. Ranta, E. Mannermaa, K. Lummepuro, A. Subrizi, A. Laukkanen, M. Antopolsky, L. Murtomäki, M. Hornof and A. Urtti, *J. Controlled Release*, 2010, 148, 42.

37. V.P. Ranta and A. Urtti, *Adv. Drug Delivery Rev.*, 2006, **58,** 1164.
38. J. Heller, *Adv. Drug Delivery Rev.*, 2005, **57**, 2053.
39. N. N. Sanders, L. Peeters, I. Lentacker, J. Demeester and S. C. De Smedt, *J. Controlled Release*, 2007, **122**, 226.
40. A. C. Amrite, H. F. Edelhauser, S. R. Singh and U. B. Kompella, *Mol. Vis.*, 2008, **14**, 150.
41. E. S. Kim, C. Durairaj, R. S. Kadam, S. J. Lee, Y. Mo, D. H. Geroski, U. B. Kompella and H. F. Edelhauser, *Pharm. Res.*, 2009, **26**, 1155.
42. T. R. Thrimawithana, S. Young, C. R. Bunt, C. Green and R. G. Alany, *Drug Discovery Today*, 2011, **16**, 270.
43. J. W. Bainbridge, A. J. Smith, S. S. Barker, S. Robbie, R. Henderson, K. Balaggan, A. Viswanathan, G. E. Holder, A. Stockman, N. Tyler, S. Petersen-Jones, S. S. Bhattacharya, A. J. Thrasher, F. W. Fitzke, B. J. Carter, G. S. Rubin, A. T. Moore and R. R. Ali, *N. Engl. J. Med.*, 2008, **358**, 2231.

CHAPTER 4.3

Ocular Drug Delivery from Nanostructured Contact Lenses

CARMEN ALVAREZ-LORENZO* AND
ANGEL CONCHEIRO

Departamento de Farmacia y Tecnologia Farmaceutica, Facultad de
Farmacia, Universidad de Santiago de Compostela, 15782-Santiago de
Compostela, Spain
*E-mail: carmen.alvarez.lorenzo@usc.es

4.3.1 Contact Lens-Drug Combination Products

The numerous and efficient barriers to local and systemic drug delivery to the eye make the development of sustained release systems a highly challenging issue.[1,2] Up to now, contact lenses (CLs) can be considered as the devices that can stay on the eye for more time with good patient compliance. This fact together with their polymeric nature, which can be modulated up to certain extent to fit specific purposes, has pointed out CLs as promising devices that accomplish the primary function of correcting vision deficiencies while playing a role as drug depots suitable for management of ocular disorders. Neutral CLs could be also merely used as medicated bandages in patients with visual accuracy, performing only as drug delivery systems. From a regulatory perspective, CLs intended for both drug delivery and refractive correction would be considered drug-device combination products.[3–5] Although medicated CLs have not been commercialized yet, the intense research carried out in the last three decades has resulted in a better knowledge on the dynamics of post-lens ocular fluid and drug diffusion towards the cornea, improved

RSC Drug Discovery Series No. 22
Nanostructured Biomaterials for Overcoming Biological Barriers
Edited by Maria Jose Alonso and Noemi S. Csaba
© The Royal Society of Chemistry 2012
Published by the Royal Society of Chemistry, www.rsc.org

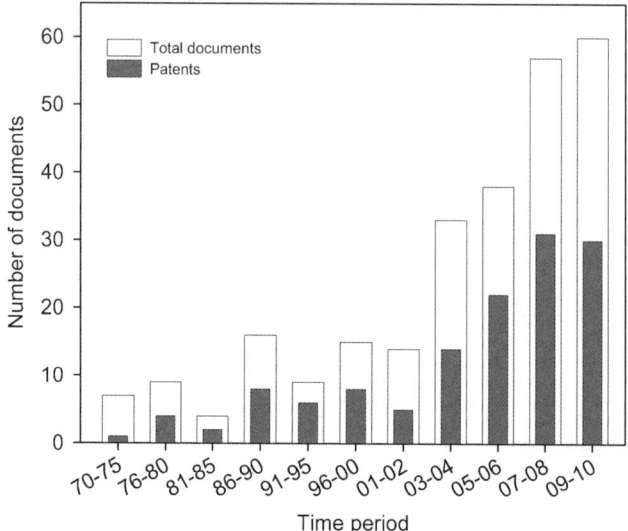

Figure 4.3.1 Evolution of the number of scientific documents on drug-eluting CLs from 1970 to 2010, according to the SciFinder® Scholar data base.

approaches to modulate CLs composition and drug release kinetics, and already promising *in vivo* results. The exponential increase in the scientific publications and patents appearing in the last few years about medicated CLs (Figure 4.3.1) helps to forecast that their use in the clinical arena is not far off. The next sections cover a brief overview of the advantages and limitations of CLs as drug carriers and then a more extensive analysis of the approaches made to modify the polymeric networks at the nanoscale in order to improve drug loading and release performance.

4.3.2 Drug-eluting Conventional Contact Lenses

Since the first rigid and soft CLs based on methyl methacrylate (MMA) and 2-hydroxyethyl methacrylate (HEMA) that appeared in 1936 and 1954, respectively,[6–8] a plethora of monomers has been synthesized for preparing CLs searching for adequate optical and mechanical features, low density, enough oxygen permeability, and more recently antibiofouling properties.[9–12] Depending on the nature of the monomers and the proportion of water in the network, CLs are currently classified in two main groups: hydrophilic and hydrophobic (Table 4.3.1). The oxygen permeability of hydrophilic soft CLs (SCLs) increases as the content in water raises[13] and some with a degree of swelling above 35% have been approved for 7 days wearing. By contrast, silicone-based CLs (which first appeared in 1979) are more gas-permeable than water and thus an increase in water content diminishes oxygen permeability.[14] Adequate combination of hydrophilic and gas-permeable components led to

Table 4.3.1 Classification of CLs according to the composition and hydrophilicity [FDA. Premarket Notification (510(k)) Guidance Document for Daily Wear Contact Lenses. http://www.fda.gov/cdrh/ode/conta.html, accessed February 2011].

Classification	Group	Description
Hydrophilic	I	Non ionic,[a] low water content
	II	Non-ionic,[a] high water content
	III	Ionic,[b] low water content
	IV	Ionic,[b] high water content
Hydrophobic	I	Without silicon and fluorine
	II	With silicon but not fluorine
	III	With silicon and fluorine
	IV	With fluorine but not silicon

[a]having an ionic content of $<1\%$ mole fraction at pH 7.2. [b]having an ionic content of $>1\%$ mole fraction at pH 7.2.

the approval of CLs for 30 days continuous wear in 2001.[15,16] The number of people who regularly wear CLs (*ca.* 125 million) is exponentially increasing and it is expected to supersede those wearing traditional pairs of glasses in the next decade. The market is currently split between the various types of CLs and has been recently estimated at 90% soft materials and 10% hard gas-permeable lenses; silicon-hydrogel representing 75% of SCLs sales.[17]

Although CLs are mainly intended for correction of ametropia problems, they have been also found suitable for therapeutical purposes, namely the management of a wide variety of ophthalmic disorders refractory to other treatment modalities.[18] According to the FDA, CLs can be classified in three cathegories: non-therapeutic, specialized use, and therapeutic (Table 4.3.2). The potential applications of therapeutic CLs are quite diverse, and comprehensive reviews can be found elsewhere.[19–24] Most therapeutic CLs

Table 4.3.2 Classification of CLs according to their use.[18,23]

Classification	Application
Non-therapeutic use	Correction of refractive ametropia, aphakia, and presbyopia
Specialized use	Treatment of keratoconus
Therapeutic use	Increased comfort and pain relief from exposed nerve endings;
	Promotion of corneal healing by protecting migrated and/or newly formed cells from the blinking action of the eyelids;
	Mechanical protection of the cornea;
	Maintenance of corneal epithelial hydration;
	Vision enhancement using plano or powered CLs to smooth an irregular corneal surface;
	Drug delivery for enhanced permeation and absorption

primarily work by inducing physical changes on the eye, providing mechanical protection and favoring the healing. Although drug delivery is one of the oldest pursued aims of therapeutic CLs, their use for this purpose is, however, still infrequent because of being the most challenging.[25–29] The therapeutic benefits that CLs can offer as drug release platforms are enormous, since controlled concentrations of drug on the cornea surface may greatly enhance ocular bioavailability while minimizing lost of drug by premature removal from eye surface or by unwanted systemic absorption, resulting in improved efficacy with minimized collateral effects. An additional objective is to simplify administration and improve compliance of therapeutic regimes. On the other hand, the risks of lesions and infections that involve the common uses of the lenses[15,21] gives additional interest to the medicated ones.

Conventional SCLs can be loaded with drugs by immersion in a drug solution or by dropping the ophthalmic solution in the concavity of the lens before insertion or on its surface after insertion (Figure 4.3.2).[30–39] The drug can be hosted in the SCL by remaining free in the aqueous phase or establishing non-specific interactions with the polymer network. Once on the eye, the drug is released to the post-lens tear film, between the cornea and the lens, where it can remain for long time. The motion of the SCL during blinking enhances the mass transfer in the post-lens tear film.[40] Furthermore, the renewal of the lachrymal fluid trapped between the lens and the cornea is much slower than in the uncovered surface. Additionally, during the time interval between successive blinks the external surface of the lens partially dehydrates.[41] As a consequence, the amount of drug that diffuses towards the corneal epithelium is estimated to be 5 times greater than the amount that is released to the lachrymal fluid that bathes its external surface.[40] The accumulation of drug on the corneal surface enables a more efficient penetration, resulting in remarkable improvements in both ocular bioavailability and pharmacological response, compared to conventional eye drops administration.[32,37,38,42] For example, in the case of

  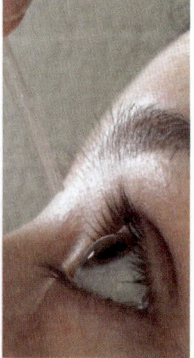

Figure 4.3.2   Common ways to load SCLs with drugs by immersion in a drug solution or by dropping the ophthalmic solution in the concavity of the lens before insertion or on its surface after insertion.

pilocarpine and timolol, delivery from SCL may enhance 20–35 times the fraction of drug that enters the cornea.[40,43] Medicated contact lenses may be particularly convenient for clinical conditions requiring a high intraocular concentration of drug, such as anterior (even posterior) segment inflammations, angle closure glaucoma or infections.[44] In some cases, drug release from SCLs can even lead to a decrease of the dose required to attain the desired therapeutic effect.[39,45,46] The research in this field is aimed at producing CLs with the ability to absorb high amounts of drugs and to deliver them in a controlled manner. It should be noted that the general use of CLs as drug release systems may not be suitable for all ophthalmic drugs since, in some cases, an excessive contact time between drug and cornea may exacerbate the topical side effects.[47] This aspect should be also taken into account when designing medicated CLs.

4.3.2.1 Key Factors for Drug Loading Yield and Release Rate

4.3.2.1.1 Loading

The total amount of drug loaded strongly determines the efficiency of SCLs as drug delivery systems. Most SCLs exhibit a limited loading, providing only one-tenth of the concentration in the aqueous humor that can be achieved when eye drops are used.[48–50] In the absence of specific interaction mechanisms, the loading only takes place in the aqueous phase of the hydrogels by a simple equilibrium with the drug solution, and the amount loaded can be estimated using the following equation:[51]

$$Loading\ (in\ aqueous\ phase) = (Vs/Wp){\cdot}C_0 \qquad (4.3.1)$$

where V_s is the volume of water sorbed by the hydrogel, W_p the dried hydrogel weight, and C_0 the concentration of drug in the loading solution. This explains that, for example, when tobramicin is instilled on high water content (71%) Permalens lenses, greater antibiotic concentrations in corneal tissue are achieved than that obtained with low water content (38.6%) Plano T therapeutic lenses.[38] Nevertheless, the sustaining of the release is in general more efficient from low water CLs[52] as will be explained below. Therefore, the feasibility of using drug-loaded SCLs depends on whether the drug and the hydrogel material can be matched so that the lens uptakes a sufficient quantity of drug and releases it in a controlled fashion.

Drug loading can be increased by raising the network/water partition coefficient, $K_{N/W}$, as predicted by the following expression:[51]

$$Loading\ (total) = [(Vs + K_{N/W}\ Vp)/Wp]{\cdot}C_0 \qquad (4.3.2)$$

where V_p is the volume of dried polymer and the other symbols have the same meaning as in the former equation. Nevertheless, irreversible binding of the drug to the network has to be avoided.[53] It has been recently reported that non-ionic hydrophilic drugs scarcely adsorb on SCLs. Cationic substances show some affinity for anionic and hydrated methacrylic acid-based SCLs (lens type IV, Table 4.3.1), anionic drugs mainly interact with nonionic and hydrated NVP-based SCLs (lens type II), and nonionic substances adsorb on hydrophobic silicone-based CLs (lens type III).[28]

On the other hand, in the few cases in which the loading can be enough for therapeutic purposes, the major limitation lies in the fact that the release occurs far too quickly to maintain therapeutic levels in the ocular structures for long enough periods of time.[7,54,55] Drug loading kinetics also affects the release behavior. Loading by immersion in drug solutions requires a time that depends on the mesh size of the network (determined by the cross-linking density and the degree of swelling), the molecular size of the drug, and the concentration of the drug in the loading solution.[25] Diffusion until equilibrium can take less than 30 min but, most commonly, up to several days may be needed.[33,56,57] The uniform distribution of the drug throughout the network is required for reproducible performance during release. As the drug molecules diffuse out, the layers near to the surface can be replenished from the deeper region of the lens, which acts as a reservoir making sustained release possible.[52,55,58–60] SCLs loaded until equilibrium with antimicrobial agents (antifungal, aminoglucosids or fluorquinolones) provide high ocular drug bioavailability, being an alternative to subconjunctival injections.[39,61–66] In a recent clinical study carried out with vasurfilcon A and etafilcon A lenses loaded with timolol maleate or birmonidine tartrate, 30 min wearing drug-loaded lenses enabled the control of intraocular pressure (IOP) in patients suffering glaucoma.[67]

4.3.2.1.2 Release

In absence of mechanisms of drug retention, a burst or dose-bumping release is generally observed. Drug release rate from SCLs obeys the diffusion laws that mainly take into account the thickness and degree of hydration of the lenses and the drug concentration in the network. The following equation can be used to predict, under sink conditions, the release of a drug by diffusion through hydrophilic lenses:[68]

$$\frac{dM}{dt} = \frac{8DM_\infty}{l^2} \exp\left(\frac{-\pi^2 Dt}{l^2}\right) \qquad (4.3.3)$$

In this expression, M_∞ represents the total amount of drug released, l is the lens thickness and D is the coefficient of drug diffusion, which is expected to remain constant if no changes in the swelling degree occur. The period of time over which a drug is released can be prolonged by either increasing l or by decreasing D.[69] For a given drug, the diffusion time decreases as the water content increases. For a given lens, the lower the molecular weight of the drug, the shorter the release time.[68] In this regard, the differences in the ability of silicon-containing (lotrafilcon and balafilcon) and PHEMA-containing (etafilcon, alphafilcon, polymacon, vifilcon, and omalfilcon) commercial CLs to absorb and release cromolyn sodium, ketorolac tromethamine, dexamethasone sodium, and ketotifen fumarate are very illustrative.[55] An *in vitro* rapid uptake and release (<60 min) was observed for the first three drugs regardless of the composition of the lens, due to the low molecular size of the drugs and the high content in water of the lenses. Cromolyn sodium, ciprofloxacin, idoxuridine, pilocarpine and prednisolone were also rapidly released from vifilcon, etafilcon and polymacon lenses when placed in saline fresh solution.[52] The more gradual uptake and release (for 2 h) of ketotifen fumarate, an amphiphilic histamine, is related to its low aqueous solubility (0.01 mg mL^{-1}) which affects the mass transfer rate to the surface of the lens. Table 4.3.3 summarizes the results of another comparative loading study carried out with six commercially available silicon-hydrogel CLs (balafilcon A, comfilcon A, galyfilcon A, lotrafilcon A, lotrafilcon B, and senofilcon) and three hydrophilic SCLs (alphafilcon A, etafilcon A, and polymacon). The loading was carried out by immersion in 0.3% ciprofloxacin-HCl[70] or 0.1% dexamethasone[71] solutions. Lotrafilcon A and etafilcon A exhibited, respectively, the lowest and the highest ability to load ciprofloxacin. The hydrophilic SCLs released greater amounts of drug compared to the silicone hydrogels (Table 4.3.3), being enough to meet the MIC$_{90}$ of ocular pathogens. However, no lens was able to sustain the release for more than 10 minutes. Similar release profiles were obtained for dexamethasone. Although alphafilcon A and lotrafilcon A loaded significantly greater amounts of dexamethasone, the former (a hydrophilic SCL) released the drug much faster (Table 4.3.3). The results obtained in these and other studies[72,73] suggest that an adequate balance between the silicon monomers and *N,N*-dimethylacrylamide (present in lotrafilcon A, lotrafilcon B, and senofilcon A) may lead to sustained drug release and that this feature could be a good starting point for developing medicated silicone-hydrogel CLs.

Just recently the use of biocompatible materials that can act as drug diffusion barriers on CLs has begun to be tested. Chauhan and coworkers have evaluated the usefulness of a vitamin E barrier within silicon-hydrogel CLs in such a way that the drug is forced to take a long tortuous path to diffuse from the lens, resulting in extended release.[69,74] Inmersion of drug-loaded lenses in vitamin E hydroalcoholic solutions led to the creation of a hydrophobic layer that notably increased the release time of timolol, dexamethasone 21-disodium phosphate and fluconazole from various commercially available contact lenses.[75] Vitamin E also has the beneficial effect of blocking UV radiation, but causes a decrease in oxygen and ion permeability. Therefore, the content in vitamin E should be the lowest as possible to modulate drug release rate without detrimental effects on other critical features. Similarly, surface treatment with octadecyl isocyanate or coating with poly(lactic-co-glycolic acid) films have been tested to control the release of norfloxacin and ciprofloxacin with promising antimicrobial results.[76,77]

4.3.3 Nanostructured Contact Lenses for Drug Delivery

Research with commercially available CLs has pointed out the difficulties to achieve sufficient drug loading and adequate control of drug release using

Table 4.3.3 Maximum amounts of ciprofloxacin and dexamethasone loaded and released from six commercially available silicon-hydrogel CLs (balafilcon A, comfilcon A, galyfilcon A, lotrafilcon A, lotrafilcon B, and senofilcon) and three acrylic SCL (alphafilcon A, etafilcon A, and polymacon). Data taken from Hui et al.[70] and Boone et al.[71]

Lens	Water content (%)	Ciprofloxacin loaded/mg per lens	Ciprofloxacin released (%)	Dexamethasone loaded/mg per lens	Dexamethasone released (%)
Alphafilcon A (SoftLens 66)	66 (group II)	0.68 (0.18)	56.2	0.118 (0.010)	56 (2)
Balafilcon A (PureVision)	36 (group III)	0.53 (0.28)	20.0	0.050 (0.003)	65 (4)
Comfilcon A (Biofinity)	48 (group I)	0.93 (0.04)	6.7	0.039 (0.004)	67 (9)
Etafilcon A (Acuvue2)	58 (group IV)	1.31 (0.11)	27.8	0.037 (0.004)	84 (9)
Galyfilcon A (Acuvue Advance)	47 (group I)	0.77 (0.13)	12.5	0.034 (0.006)	77 (8)
Lotrafilcon A (Night and Day)	24 (group I)	1.12 (0.01)	2.0	0.102 (0.011)	11 (1)
Lotrafilcon B (O$_2$Optix)	33 (group I)	1.01 (0.07)	6.3	0.038 (0.005)	59 (7)
Polymacon (SoftLens 38)	38 (group I)	0.71 (0.03)	37.5	0.067 (0.008)	42 (5)
Senofilcon A (Acuvue OASYS)	38 (group I)	1.11 (0.07)	4.4	0.056 (0.006)	44 (5)

networks that do not effectively interact with the target drug. Several approaches are being tested to overcome these limitations, while maintaining the oxygen permeability, the hydrophilicity, and the optical, morphological and mechanical properties required for the primary function of vision correction: (i) chemically-reversible immobilization of drugs in the hydrogel through labile bonds;[68,78] (ii) copolymerization with monomers able to interact directly with the drug;[46,53,79,80,81] (iii) incorporation of drug-loaded colloidal systems into or onto the lens;[82–86] (iv) grafting of cyclodextrins to preformed lenses;[87] and (v) molecular imprinting[26,27,88] (Figure 4.3.3). The last three approaches are based on modifications at the nano-scale level of the drug environment in the network and will be the aim of the next sections.

4.3.3.1 Colloidal Nanocarriers

Nanometric surfactant aggregates and lipid vesicles have been explored to modulate drug incorporation, partition and release kinetics from SCLs.[83,89] The nanometric structures, before or after being loaded with drug molecules, are incorporated into the lens newtork during polymerization. Empty nanostructures can be loaded by immersion of the lens in a drug solution. Hydrophobic drugs partition into the nanostructures, leading to enhanced loading. If the drug molecules are oppositely charged to those of the surfactant or lipid, they together can form part of the nanostructures.[90] If the proportion and the size of the colloidal particles dispersed in the matrix are sufficiently small, the lens stays transparent. Once the CL is placed on the eye, the

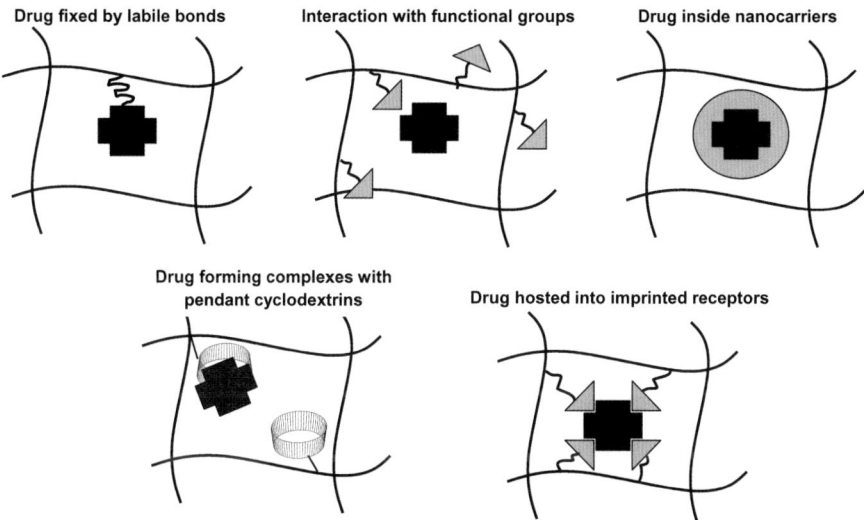

Figure 4.3.3 Different approaches to improve the ability of contact lenses to act as drug-eluting devices.

nanoparticles behave as depots that slowly release the drug to the lens matrix and then the drug has to diffuse to reach the postlens tear film.

4.3.3.1.1 Micelles and Microvesicles

Chauhan and coworkers[82,83] have developed PHEMA hydrogels containing oil-in-water microemulsion droplets (10–20 nm) bearing lidocaine dissolved in the oil phase. Microemulsions prepared with Tween 80 led to opaque hydrogels due to aggregation of the droplets caused by desorption of the surfactant. In contrast, microemulsions stabilized with Brij 97 and/or a silica shell did not affect the transparency of the hydrogels. However, the distribution of the droplets was irregular and domains with and without them were observed. This can be explained by the lack of interactions between the microemulsion droplets and the network. During polymerization, the droplets are pushed towards the bulk solution until the medium becomes too viscous and the droplets are trapped by the hydrogel. Lidocaine release profiles showed a burst of 30–50% dose, due to the drug adsorbed on the surface of the droplets and the amount freely dispersed on the hydrogel network, followed by a much slower release after some hours. No significant influence of the cross-linking density[82] nor of the lens thickness[83] was observed, which indicated that after the burst effect, the droplets inside the network controlled the release. Optimization of the microemulsion and the micelle-containing networks resulted in lenses able to sustain the release of cyclosporine A and that can stand autoclaving and packing in storage medium and still provide extended drug release at therapeutic rates.[85] The lenses were prepared by incorporating in the HEMA polymerization medium microemulsions (Figure 4.3.4) or micelles (of the same composition as the microemulsions but without ethyl butyrate) loaded with cyclosporine A.

Both microemulsion and micelle-containing hydrogels sustained cyclosporine A release providing therapeutic levels for 20 days (Figure 4.3.5). The higher the content in surfactant, the slower the release was. From the fabrication and scale-up point of view, micellar systems may be advantageous compared to the microemulsion-laden lenses, which require a two-step process. Furthermore, the oil phase can eventually diffuse out the lens causing ocular toxicity. The surfactants are expected to exist in the polymer network in three stages: (i) free; (ii) forming micellar aggregates; and (iii) interacting with the polymer chains. Similarly, the drug can be hosted in the network (i) as free molecules, (ii) inside the micelles, and (iii) interacting with the polymer chains (Figure 4.3.6). The hydrophobicity of the micellar core plays a relevant role in the drug partitioning. If the affinity of the drug for the surfactant is sufficiently large, the micelles can act as depots able to regulate drug release. Thus, the nature and chain length of the surfactant may affect the drug partitioning inside the surfactant domains of the hydrogel and the release kinetics.[91] This effect was shown in a study carried out with Brij surfactants ($C_{18}H_x(OCH_2CH_2)yOH$), namely Brij 78 ($x = 37$, $y = 20$), Brij 97 ($x = 35$, $y = 10$), Brij 98 ($x = 35$, $y = $

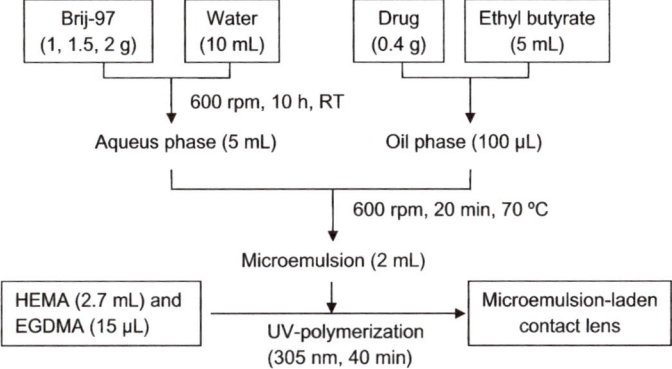

Figure 4.3.4 Scheme of the preparation procedure of SCLs containing microemulsion droplets loaded with cyclosporine A, as described by Kapoor et al.[85]

20) and Brij 700 ($x = 37$, $y = 100$). Cyclosporine A had a high partitioning inside the micelles of long chain (Brij 700) or saturated (Brij 78) surfactants and sustained release was achived for several days; by contrast, the micellar partition coefficients of dexamethasone and dexamethasone-21-acetate were smaller and the lenses could not effectively extend the release.

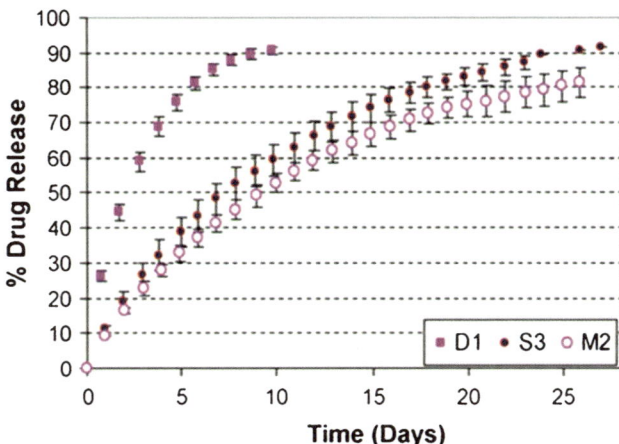

Figure 4.3.5 Cyclosporine A release profiles from Brij 97 surfactant laden (S3), microemulsion-laden (M2) and pure p-HEMA (D1) gels. All the gels were 200 μm thick in dry state and S3, M2 and D1 gels contained 49.2, 52.2 and 53.3 μg of drug, respectively. Reprinted from Kapoor et al.[85] with permission from Elsevier.

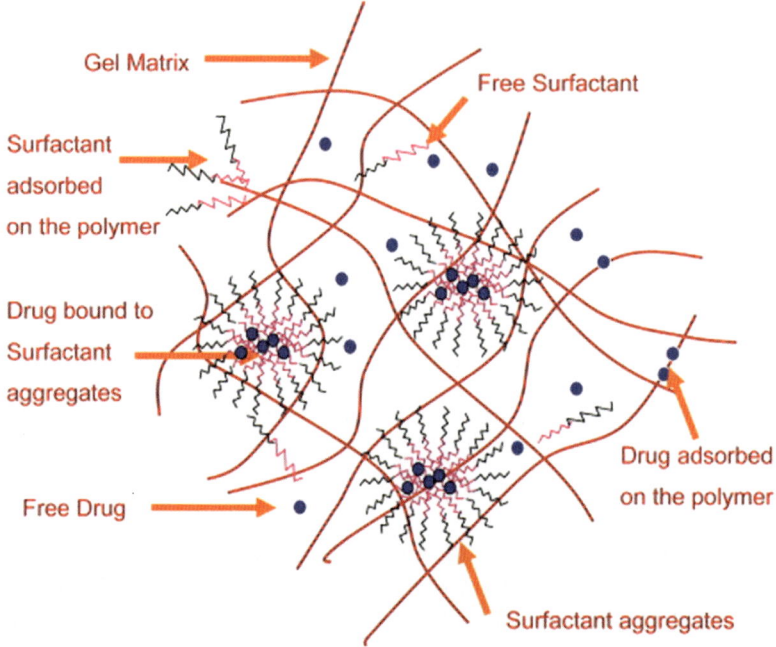

Figure 4.3.6   Schematic view of the microstructure of a surfactant-laden hydrogel. Reprinted from Kapoor *et al.*[91] with permission from Elsevier.

4.3.3.1.2   *Liposomes Onto or Into*

Liposomes have been also tested as nanocarriers to control drug release from inside or from the surface of CLs. Similarly to the microemulsion- and micelle-laden networks, dimyristoyl phosphatidylcholine (DMPC) liposomes have been dispersed in PHEMA hydrogels in order to load hydrophobic drugs in the lipid bilayers or hydrophilic ones in the aqueous cores.[84] The lipid fraction in the liposomes and the concentration of liposomes in the polymerization mixture can be adjusted to a certain extent to increase the amount of drug loaded into the lenses. After a burst (15–30% of the dose) in the first 24 h, the release of lidocaine was sustained for more than one week owing to the slow diffusion of the drug across the lipid layers of the vesicles. The liposome-loaded hydrogels showed decaying release rates and required storage in a drug saturated buffer to prevent unloading, which raised two important drawbacks.

SCLs bearing surface-immobilized liposomes have been also investigated as a way to achieve sustained release.[92] Several steps were required for the immobilization: (i) covalent binding of polyethylenimine onto the hydroxyl groups of SCL surface; (ii) binding of NHS-PEG-biotin to the amine groups *via* carbodiimide chemistry; (iii) immersion of the lenses in a Neutr-Avidin solution in order to link to the PEG-biotin layer; and (iv) incubation of the Neutr-Avidin coated lenses into a biotinylated-liposome suspension (Figure 4.3.7).[92,93] Carboxyfluorescein release profiles indicated that liposomes were

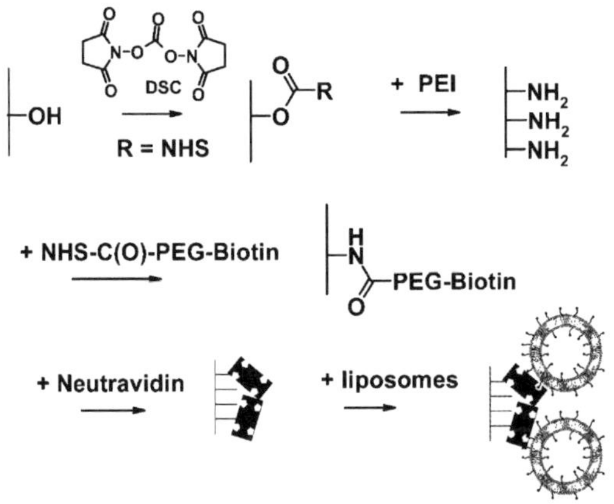

Figure 4.3.7 Chemical reactions leading to the attachment of liposomes onto the surfaces of soft contact lenses. Reprinted from Danion *et al.*[92] with permission from John Wiley and Sons.

not damaged during immobilization and that storage up to 1 month at 4 °C resulted in minimal leakage of their content.[92] Multilayers of liposomes could be formed by adding more NeutrAvidin in order to bind a second layer of biotinylated-liposomes, and so forth. CLs bearing up to 10 layers of levofloxacin-loaded liposomes were obtained in this way. Some CLs were then soaked overnight in a 5 mg mL^{-1} levofloxacin solution for combining a burst release of the antimicrobial agent with the sustained release provided by the liposome layers. Such a combination resulted in a complete inhibition of the growth of *S. aureus* within 2 hours and thus may be promising for *in vivo* treatment of the fast exponential growth at the beginning of the infection. Furthermore, the sustained release may complete the antibacterial effect.[94]

4.3.3.2 Pendant Cyclodextrins

The search of approaches for loading hydrophobic drugs by creating microenvironments suitable for hosting without altering the physical features of the lens has prompted the design of networks functionalized with pendant cyclodextrins (CD).[87,95] First attempts to copolymerize the lens monomers with CD monomers evidenced that CD units can host hydrophobic drugs totally or partially into their cavities through thermodynamically favourable interactions and provide an enhancement in drug affinity sufficient for increasing the loading and sustaining the release of acetazolamide, hydrocortisone, and pueranin.[96,97] However, copolymerization CD monomers usually results in an increase in the rigidity of the network and a decrease in water content.[96] Oppositely, the post-functionalization with CDs of already-preformed lenses avoids changes in the mechanical and optical features of the starting networks.[87] Pristine CD units (without any previous modification) have been anchored to networks made of HEMA and small proportions of glycidyl methacrylate (GMA). To do that, the hydroxyl groups of CD reacted with the epoxy groups of GMA under mild conditions. Pendant CDs improved the friction coefficient of the hydrogels, enhanced 15-fold the amount of diclofenac loaded, prevented a premature discharge of the drug in the multipurpose storage liquids and sustained the release for two weeks.[87] If loaded with miconazole, the hydrogels completely prevented *Candida albicans* biofilm formation *in vitro*.[95]

4.3.3.3 Imprinted Pockets

4.3.3.3.1 Molecular Imprinting Technology: Rigid vs. Swellable Networks

Incorporation of "functional" monomers possessing chemical groups able to interact with a given drug to the SCL network may remarkably raise the affinity for the drug. The interaction should be enough to increase the polymeric network/loading solution drug partition coefficient, thus, to drive the drug uptake.[79,80] Rational approaches for the selection of the functional monomers involves calorimetric analysis, computational modeling and, more recently biomimesis.[88,98,99] Nevertheless, the possibilities of using the most suitable monomer to interact with a certain drug at the adequate ratio are limited by biocompatibility issues and by the need of maintaining the network dimensions unaltered under the pH, temperature and osmotic pressure conditions which the lens has to face during handling and wearing. Molecular imprinting technology pursues the optimization of the spatial distribution of the functional monomers, in order to achieve the maximum efficiency of the interactions between the drug and the polymeric network. To do that, the drug molecules are used as templates during polymerization, by adding the drug to the monomers solution (including backbone and functional monomers and cross-linker) in order to induce their rearrangement as a function of the affinity for the drug. The conformation of the monomers is fixed during polymerization, and once the drug molecules that acted as templates are removed "imprinted pockets" with the size and the most suitable chemical groups to interact with the drug are obtained (Figure 4.3.8). It is expected that if the imprinted network enters again into contact with the drug, the loading can take place in the tailored pockets with a greater affinity than in the case of the networks synthesized in the absence of template (non-imprinted networks). This technology has been progressively developed over the last forty years as a tool for endowing rigid highly cross-linked polymeric systems with the ability to recognize target species.[100,101] Two modalities of regulation of the functional monomer/template association previous to the polymerization have been developed: (i) the pre-organized or covalent approach, which starts from template molecules bound to the functional monomers by reversible covalent bonds; and (ii) the self-assembly or non-covalent approach, which involves non-covalent or metal coordination multiple-point interaction

① Drug molecules are added to the monomers soup.

② Functional monomers arrange around drug molecules.

③ Polymerization fixes the position of the monomers.

④ The template is removed and imprinted cavities are revealed.

Figure 4.3.8 Schematic view of the preparation of hydrogels with high affinity for a template drug applying the molecular imprinting technology. Reprinted from Alvarez-Lorenzo *et al.*[105] with permission from Editions de Santé.

of the template with various functional monomers to form stable and soluble complexes of appropriate stoichiometry.[102,103] The non-covalent imprinting protocol enables more versatile combinations of templates and monomers, and provides faster bond association and dissociation kinetics than the covalent imprinting approach.[102,104] In either case, the preparation of conventional molecularly imprinted polymers (MIPs) requires the co-polymerization of the functional monomer-template complexes with high proportions of cross-linking agents in a porogen solvent, and the subsequent removal of the template molecules, in order to reveal the recognition cavities complementary in shape and functionality to the template molecules (*i.e.*, specific receptors).

The success of the imprinting strongly depends on the stability and solubility of the functional monomers-template assemblies during polymerization. Simple polymerization in the presence of a drug (chemically stable enough to endure the polymerization conditions) does not ensure an imprinting effect.

If the molar ratio in the complex is not appropriate or if the assemblies dissociate to some extent during polymerization, the functional monomers would be far apart from both the template and each other, resulting in a small difference between imprinted and non-imprinted (conventional) networks.[106,107] The polymerization should be carried out in a solvent medium that does not interfere or compete with the target molecules for the interactions with the functional monomers and at the lowest temperature possible to shift the equilibrium toward the formation of target-monomers complexes.[107] Highly cross-linked MIPs have been shown to be useful in selectively rebinding the target molecules from mixtures of chemical species, and thus have aroused interest for a wide range of applications, including solid phase extraction, environment remediation, chromatography, sensors, immunoassay or artificial catalyse.[108–110] In the last few years, the molecular imprinting technology has also received increasing attention in the drug delivery field as a way to obtain affinity-regulated and feed-back controlled drug release.[103,111–117]

Differently from the rigid polymer networks to which molecular imprinting is routinely applied, SCLs have much lower cross-linking density in order to be flexible enough to adapt to the ocular surface without causing mechanical damage. Swelling of the lens network in the preservation liquids or in the lachrymal fluid is likely to distort the structure of the imprinted pockets. Nevertheless, the compromised physical stability can be overcome by endowing the networks with high affinity binding pockets, *i.e.*, obtained at a minimal conformational energy.[118] Those pockets can memorize the structural features of the drug and undergo an "induced fit" in presence of the drug, recovering the same conformation as upon polymerization.[112,119] Loosely cross-linked imprinted hydrogels ressemble to a certain extent the recognition capacity of biomacromolecules (*e.g.* receptors, enzymes, antibodies). Natural evolution has determined the unique details of protein's native state, such as its shape and charge distribution, that enable it to recognize and interact with specific molecules.[118] Based on biomimetic principles, SCLs endowed with high affinity imprinted pockets are expected to be able to load the drug and, subsequently, to sustain the release. Therefore, a highly precise design of imprinted SCLs is mandatory to achieve an optimum balance between the mechanical properties required for the ocular application and the stability of the imprinted cavities in the lens structure required for drug affinity and selectivity. Research on drug-imprinted SCLs has focused so far on four therapeutic groups, namely β-adrenergic antagonists (timolol), antimicrobials (norfloxacin), antihistamines (ketotifen), and carbonic anhydrase inhibitors (acetazolamide), and two comfort ingredients (polyvinylpyrrolidone and hyaluronic acid).

4.3.3.3.2   Timolol-imprinted SCLs

The timolol structure is particularly suitable for providing imprinted systems since it offers multiple sites for interacting with functional monomers through ionic and hydrogen bonds.[120] A comprehensive study of the incidence of the nature of the functional monomer and the backbone monomer, the degree of cross-linking and the drug : functional monomers ratio on loading and release performance of the lenses has been carried out. Non-imprinted (conventional) and imprinted (synthesized in the presence of 23 mM drug) PHEMA hydrogels with particularly low cross-linking density (0.128 mol%) and 0.7 mm thickness were prepared using as functional monomers methacrylic acid (MAA) or methyl methacrylate (MMA) in a concentration range of 0 to 5.12 mol%.[120] After boiling in water for 3 minutes for removal of unreacted monomers, the imprinted hydrogels sustained the release of the remaining 70% drug for several hours in saline medium. Once the drug was released, the imprinted PHEMA hydrogels prepared with the lowest MAA proportion were able to reload 3 times more timolol than the non-imprinted ones from a pH 5.5 drug solution; the release rate being similar to that upon polymerization.

Further studies with imprinted lenses based on *N,N*-diethyl acrylamide (DEAA) and synthesized with different proportions of MAA (1.28–5.12 mol%) and cross-linker EGDMA (0.32–8.34 mol%), revealed that a minimum of 0.9 mol% EGDMA was required to achieve remarkably greater loading (one order of magnitude) compared to the non-imprinted lenses.[121] The imprinted lenses loaded a therapeutic dose of timolol, sustained its release in lachrymal fluid for more than 12 h, and reloaded another dose overnight to be ready for use the next day.[120,121] Keeping constant the proportions of the functional monomer (MAA, 100 mM) and cross-linker (EGDMA, 140 mM), it was shown that the backbone monomers used to prepare the SCLs also play a relevant role in the success of the imprinting and in the drug loading and release properties.[122] Four backbone compositions were tested: (i) *N,N*-diethylacrylamide (DEAA), (ii) 2-hydroxyethylmethacrylate (HEMA), (iii) 1-(tristrimethyl-siloxysilylpropyl)-methacrylate (SiMA) and *N,N*-dimethylacrylamide (DMAA) 50 : 50 v/v, and (iv) MMA : DMAA 50 : 50 v/v solutions. Lenses of 0.3 mm thickness were prepared by UV irradiation in the presence of timolol maleate (25 mM). Drug sorption isotherms in water at 37 °C revealed that the overall affinity of the lenses for timolol ranked in the order HEMA > SiMA-DMAA > MMA-DMAA > DEAA. The highest imprinting effect, *i.e.*, the greatest relative increase in overall affinity, with respect to non-imprinted systems, was obtained for the last two systems which contain the backbone monomers with lower affinity for timolol (Figure 4.3.9).[122] In the hydrated non-imprinted lenses, MAA groups are too far apart to form binding sites for timolol. By contrast, the imprinting procedure provides pockets with groups of MAA spaced close together and considerably increases the loading capacity. All lenses showed sustained release in 0.9% NaCl solution over 2–8 hours. Timolol diffusion coefficients through the imprinted lenses were $2.2 \cdot \times 10^{-9}$ cm^2 s^{-1} for DEAA-based lenses, $9.9 \cdot \times 10^{-9}$ cm^2 s^{-1} for HEMA-based lenses, $66.5 \cdot \times 10^{-9}$ cm^2 s^{-1} for MMA–DMAA-based lenses, and $71.3 \cdot \times 10^{-9}$ cm^2 s^{-1} for SiMA–DMAA-based lenses. These values confirm that timolol molecules move out easier from the hydrophilic networks that possess low affinity for the drug (*i.e.*, MMA-DMAA and SiMA-DMAA).

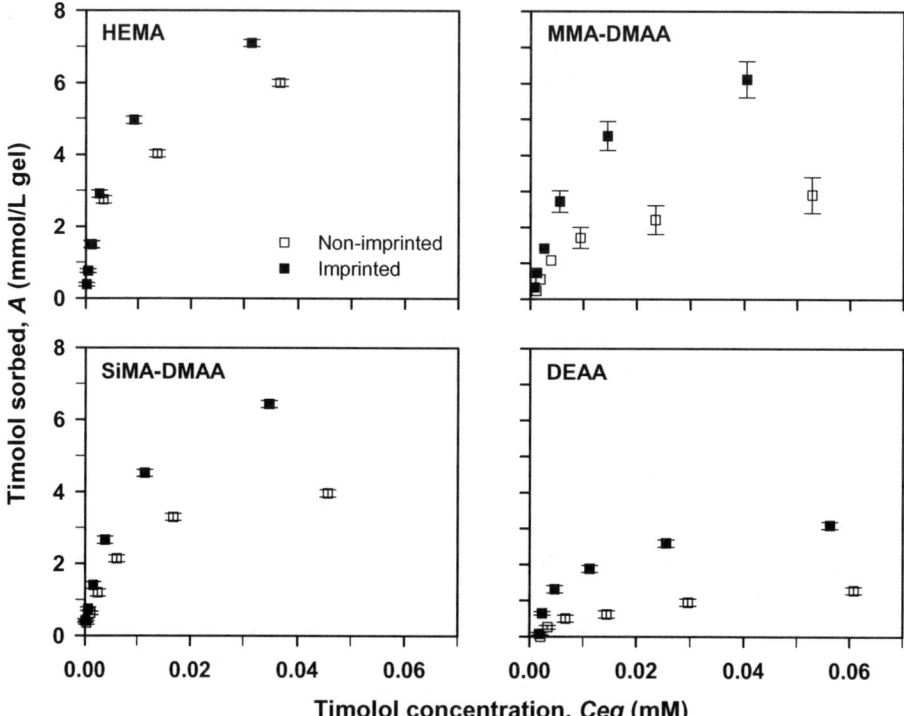

Figure 4.3.9   Timolol sorption isotherms of imprinted and non-imprinted lenses
made by photopolymerization of methacrylic acid (MAA, 100mM)
and cross-linker ethyleneglycol dimethacrylate (EGDMA, 140 mM)
with 2-hydroxyethyl methacrylate (HEMA), methyl methacrylate
(MMA) and *N,N*-dimethyl acrylamide (DMAA), 1-(tristrimethylsiloxy-
silylpropyl)-methacrylate (SiMA) and DMAA, or *N,N,*-diethyl
acrylamide (DEAA) as backbone monomers. Reprinted from
Hiratani and Alvarez-Lorenzo[122] with permission from Elsevier.

Ultrathin DEAA-based lenses (14 mm diameter and 0.08 mm center
thickness) evidenced the suitability of the imprinted networks for controlled
timolol delivery *in vivo*. Timolol levels in rabbits' lachrymal fluid were
monitored after the insertion of imprinted (34 µg drug per lens) and non-
imprinted lenses (20 µg drug per lens) or after the instillation of timolol eye-
drop solutions of 0.068% (total dose 34 µg) or of 0.25% (commercial solution,
total dose 125 µg) (Figure 4.3.10).[45] The imprinted lenses were able to maintain
the timolol release for 180 min, compared to the 90 min of the non-imprinted
ones. Both displayed the maximum ocular level at around 5 min, followed by
monoexponential decay. Irrespective of the initial concentration, timolol
applied as drops was flushed out of the eye in less than 60 min. Furthermore,
the area under the timolol concentration–time curve was 3.3-fold and 8.7-fold
greater for imprinted SCLs than for the non-imprinted ones and eyedrops,
respectively. Drug levels attained in the tear fluid were proportional to the

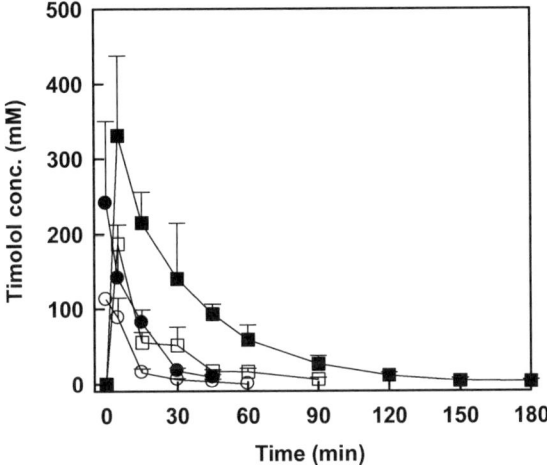

Figure 4.3.10   Timolol tear fluid concentration–time profiles after application of drug-loaded imprinted and non-imprinted contact lenses or instillation of eye drops. The doses were 34 μg for imprinted contact lens (solid squares), 21 μg for non-imprinted contact lens (open squares), and 34 μg (open circles) and 125 μg (solid circles) when 0.068% or 0.25% timolol eye drops were instilled. Each point represents the mean ± S.D. (*n* =3–5). Reprinted from Hiratani *et al.*[45] with permission from Elsevier.[45]

amounts loaded by the SCLs. These results indicate that imprinted SCLs reduce the precorneal elimination of the drug compared to the eyedrops and, in consequence, much smaller amount of drug is needed to achieve the desired therapeutic levels.

The influence of timolol/functional monomer molar ratio on the imprinting effect was analyzed using MAA as a functional monomer in hydrogels prepared with DMAA and tris(trimethylsiloxy)silylpropyl methacrylate (TRIS)[123] and using acrylic acid (AAc) as functional monomer in pHEMA hydrogels.[99] Isothermal titration calorimetry (ITC) revealed that each timolol molecule requires 6 to 8 AAc monomers to saturate the binding and that these ratios could be the most suitable for creating imprinted cavities. Imprinted poly(HEMA-co-AAc) hydrogels of 0.2 mm thickness prepared with timolol : AAc 1 : 6 or 1 : 8 molar ratio sustained the release better than the other hydrogels. In the case of DMAA-TRIS-MAA hydrogels, timolol release rate strongly decreased by increasing the MAA:timolol ratio in the gel recipe from 4 : 1 to 16 : 1. This considerable influence of the timolol:functional monomer ratio is related to changes in conformation of the imprinted cavities.[104,106] As the relative amount of timolol decreases, more functional monomers are available to gather during synthesis, forming efficient imprinted cavities with greater multiple-point binding constants. Each binding site is more perfectly constructed and its affinity for timolol delays the release process.

4.3.3.3.3 Norfloxacin-imprinted SCLs

Norfloxacin possesses carboxylic acid and amino groups that can electro-statically interact with ionized groups of other molecules, and an aromatic ring that can establish hydrophobic interactions. For a rational design of the imprinted networks, ITC was applied to evaluate the interactions of this drug with AAc and to elucidate the optimum template : functional monomer molar ratio, which was established in 1 : 4.[98] PHEMA-based hydrogels prepared with norfloxacin : AAc molar ratios ranging from 1 : 2 to 1 : 16 revealed that imprinted hydrogels synthesized using norfloxacin : AAc 1 : 4 molar ratio released the drug at a rate 3.5-times lower than non-imprinted hydrogels, providing a 24-hours controlled release (Figure 4.3.11). Imprinted hydrogels of various AAc contents and thicknesses exhibited similar loading/release behavior, evidencing the robustness of the imprinting technique.

4.3.3.3.4 Ketotifen-imprinted SCL

Byrne *et al.* have designed ketotifen-imprinted SCLs using as functional monomers those more similar to the chemical groups present in the physiological histamine H1-receptor.[124] HEMA, AA, NVP and acrylamide (AM) were selected as monomers taking into account that residues such as aspartic acid, lysine, arginine and tyrosine form the active site in the H_1-receptor. The guiding hypothesis was that if antihistamines have high affinity for the H_1-receptor, a hydrogel with similar chemical functionality would bind the antihistamine tightly, increasing the loading and delaying release kinetics. The hydrogels were prepared by photopolymerization starting from 5 mol% cross-linker (PEG200DMA), 92 mol% HEMA and 3 mol% functional monomers, namely AAc, NVP and acrylamide (AM).[125] The most biomimetic formulation, poly(AAc-co-AM-co-NVP-co-HEMA-co-PEG200DMA), exhibited a loading that was six times greater than that of the control network and three times higher than that of the networks containing one or two functional monomers. Release of therapeutically relevant amounts of drug in artificial lachrymal fluid containing lysozyme was sustained for 4 days, suggesting that the delivery *in vivo* might be also extended for several days (Figure 4.3.12). Under *sink* conditions, the release profiles evidenced a decreasing release rate that fitted to Fickian diffusion; the ketotifen fumarate diffusion coefficient from biomimetic networks being as low as $5.57 \cdot \times 10^{-10}$ cm^2 s^{-1}.[126] However, experiments carried out under finite conditions by imitating the volume and the flow rate (3 μL min^{-1}) of lachrymal fluid, evidenced even lower release rates, with profiles that fitted to zero-order kinetics for 3.5 days.[127] These studies suggest that *in vitro* finite turnover conditions may simulate the *in vivo* environment better than *sink* conditions do.

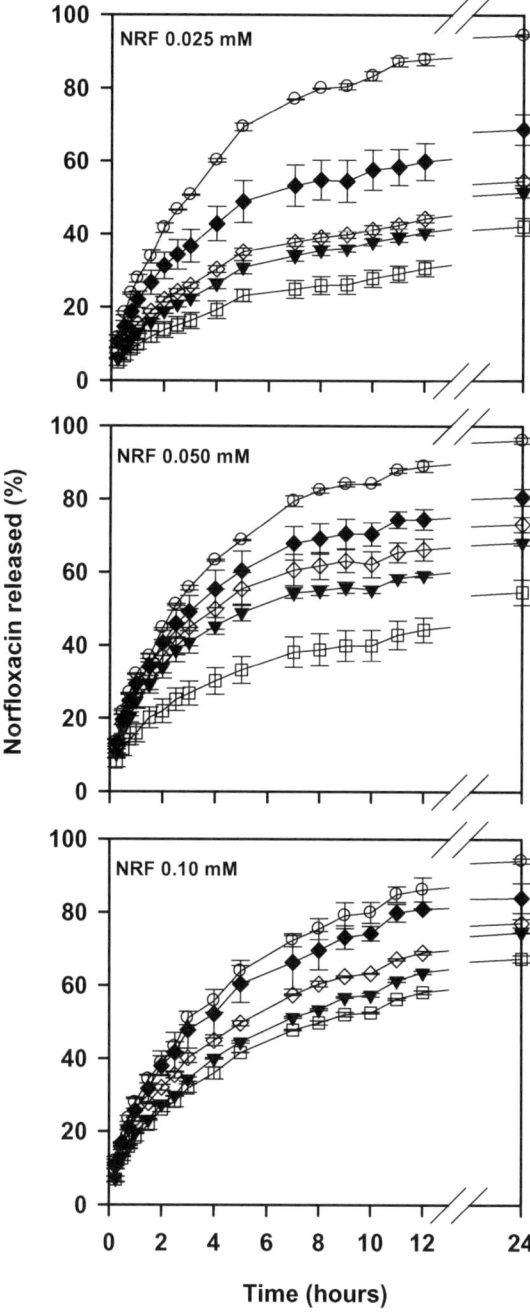

Figure 4.3.11 Norfloxacin (NRF) release profiles in lachrymal fluid from PHEMA hydrogels synthesized with AAc 200 mM and EGDMA 160 mM at different NRF : AAc molar ratios; namely zero, *i.e.* non-imprinted hydrogels (○), 1 : 16 (◆), 1 : 10 (◇), 1 : 6 (▼), and 1 : 4 (□). The hydrogels (thickness 0.4 mm) were previously loaded by immersion in 0.025 mM, 0.050 mM or 0.10 mM NRF solutions (*n* = 3). Reprinted from Alvarez-Lorenzo *et al.*[98] with permission from Elsevier.

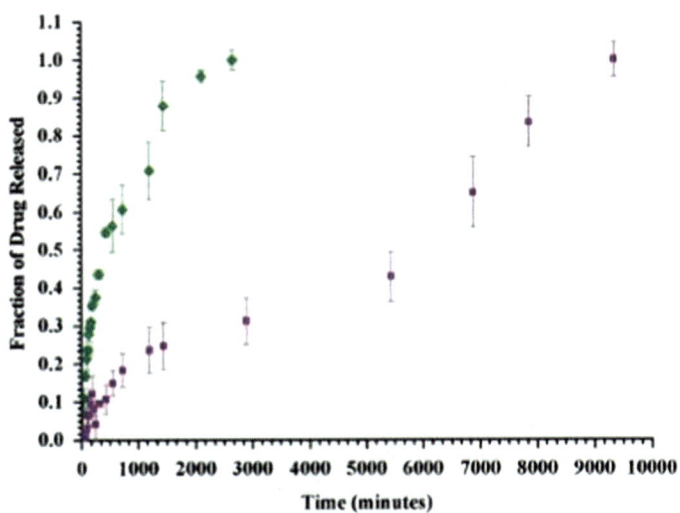

Figure 4.3.12   Ketotifen release profiles from a poly(AA-co-AM-co-HEMA-co-PEG200DMA) network, in artificial lachrymal fluid without (green diamonds) and with lysozyme (purple squares). Reprinted from Venkatesh *et al.*[127] with permission from Elsevier.

4.3.3.3.5   CAI- imprinted SCL

Carbonic anhydrase inhibitors (CAIs; *e.g.*, acetazolamide, ethoxzolamide) currently delivered by oral route have been shown useful to reduce the IOP. However the relevant side effects of CAIs at the doses required to achieve ocular therapeutic concentration prompted the search of topical formulations.[128] Hydrogels with ability to load and to sustain the release of acetazolamide have been recently developed combining biomimetic principles and molecular imprinting technology.[129] Although there are different isoforms of carbonic anhydrases, the active site of most of them consists of a cone-shaped cavity that contains a Zn^{2+} ion coordinated to three histidine residues in a tetrahedral geometry with a solvent molecule as the fourth ligand (Figure 4.3.13).[130] Acetazolamide inhibits the activity of the enzyme by binding to the Zn^{2+} ion. One oxygen atom of the sulfonamide interacts with the –NH group of treonine 199 and the other points toward the zinc ion, while the –NH function of the ionized sulfonamide group replaces the water molecule that binds to zinc and forms hydrogen bonds with the –OH group of threonine 99 (Figure 4.3.13).[131] Monomers bearing chemical groups similar to those of the amino acids involved in the active binding site were chosen to prepare biomimetic hydrogels: the zinc ions were introduced as methacrylate salt ($ZnMA_2$); the hydroxyl and amino groups can be supplied by 2-hydroxyethyl methacrylate (HEMA) and *N*-hydroxyethyl acrylamide (HEAA); and 4-vinylimidazole (4-VI) resembles histidine (Figure 4.3.13). Since 4-VI is not a common component of SCLs, the possibility of replacing it with

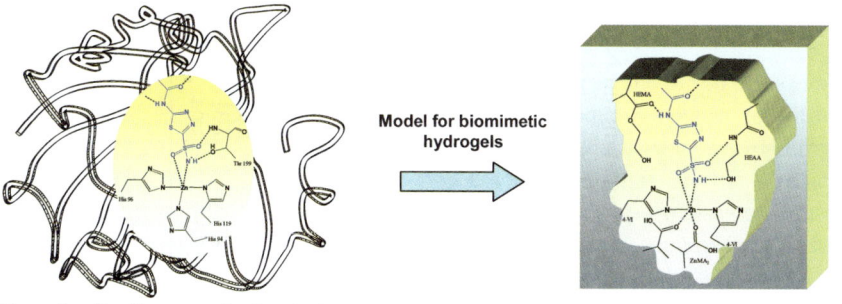

The active site of human carbonic anhydrase II, interacting with acetazolamide

Hydrogels with receptor-like binding sites uptake and control the release of carbonic anhydrase inhibitors

Figure 4.3.13 Schematic view of the biomimetic approach followed for the design of acetazolamide-imprinted networks. Reprinted from Ribeiro *et al.*[129] with permission from the American Chemical Society.

1-vinylimidazole (1-VI) was tested. Several sets of hydrogels were synthesized with biomimetic level increasing from A to J formulations (Table 4.3.4). After boiling and immersion in saline medium, zinc ions content remained above 70% only in hydrogels incorporating 4VI.

Hydrogels bearing 4VI moities showed network/water partition coefficient, $K_{N/W}$, values three times greater than those prepared with 1VI (Table 4.3.4). Minor improvement was achieved by adding acetazolamide as template during polymerization. This means that using the functional monomers that mimic the best the components of the carbonic anhydrase receptor, it is feasible to endow the hydrogels with high affinity for acetazolamide. Such a notable increase in affinity was also evidenced in a better control of drug release (Table 4.3.4).

4.3.3.3.6 Comfort Ingredient-imprinted SCL

Approaches to make SCLs more comfortable and to prevent dry eye syndrome[132] involve the incorporation of hydrophilic polymers to the SCL structure with the aim of wrapping the lens with a cushion-like layer, softening the contact with the eye and the lid.[133,134] Those hydrophilic polymers act as demulcents or as comfort ingredients, decreasing the friction coefficient and enhancing the ocular wetting.[135,136] Loading of the demulcent/comfort ingredient in SCLs could enable a sustained release for the whole period of time that most disposable lenses are used (one week). Compared to artificial tears that are intermittently applied, a sustained delivery from demulcent-semi interpenetrated SCLs may resemble better the lubrication role of natural tears.[136] Imprinted SCLs for hyaluronic acid, which consists of repeating units of glucuronic acid and *N*-acetylglucosamine, have been developed, mimicking the structure of the binding receptor in the human body, *i.e.*, the hyaluronic acid binding protein CD44 (Figure 4.3.14). To prepare the hydrogels, hyaluronic acid (6.5 mg) was added to nelfilcon A monomeric solution (1 g)

Table 4.3.4 Hydrogel compositions and acetazolamide network/water partition coefficients and release rate constants in NaCl 0.9% solution obtained after fitting to the square-root kinetics. Mean values and, in parenthesis, standard deviations ($n = 6$). Data taken from Ribeiro et al.[129]

Formulation	HEMA/ml	ZnMA$_2$/g	HEAA/g	1VI/g	4VI/g	ACT/g	K$_{N/W}$	K$_H$(% h$^{-0.5}$)
0	8	—	—	—	—	—	5.40 (0.18)	24.20
A	8	0.185	—	—	—	—	6.67 (1.46)	10.48
B	8	0.185	—	—	—	0.173	5.11 (1.77)	11.67
C	8	0.185	—	0.22	—	—	6.63 (0.96)	10.56
D	8	0.185	—	0.22	—	0.173	7.79 (0.58)	12.61
E	8	0.185	—	—	0.22	—	16.40 (0.16)	6.86
F	8	0.185	—	—	0.22	0.173	15.35 (1.26)	8.49
G	8	0.185	0.14	—	—	—	6.52 (0.48)	8.61
H	8	0.185	0.14	—	—	0.173	5.90 (0.50)	8.99
I	8	0.185	0.14	—	0.22	—	15.29 (1.47)	7.31
J	8	0.185	0.14	—	0.22	0.173	15.64 (0.72)	13.04

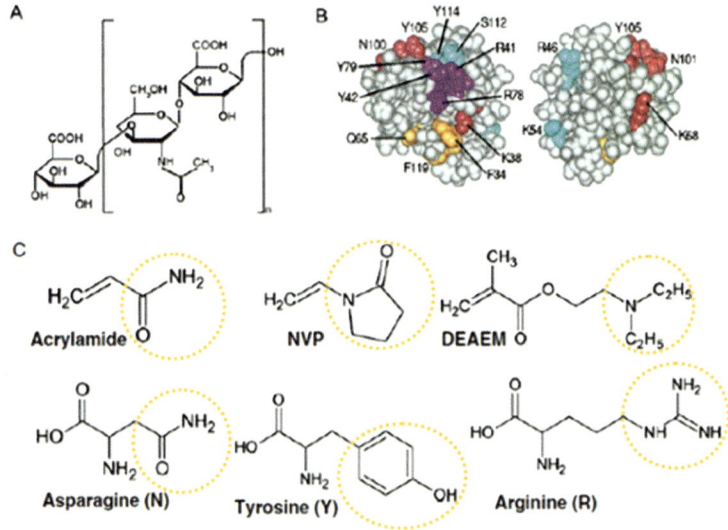

Figure 4.3.14 Hyaluronic acid (A), hyaluronic acid binding protein CD44 (B) and functional monomers chosen to mimic the amino acids found on the binding site CD44. Reprinted from Ali and Byrne[137] with permission from Spinger.

containing variable proportions of AM (its amide group resembles that of asparagines), NVP (hydrogen bonding capability similar to that of tyrosine) and 2-(diethylamino)ethyl methacrylate (DEAEM, with a positively charged group like arginine or lysine) in order to facilitate a multiple binding points interaction.[137] Hyaluronic acid release rate was precisely tuned by modifying the relative proportion of the functional monomers. Furthermore, beyond a certain proportion, the delivery was even completely prevented, which evidences the strength of the binding.

4.3.3.3.7 *Supercritical Fluid-Assisted Flurbiprofen Post-Imprinting*

Molecularly imprinted networks, as described above, are tailor-designed for a specific drug, because of the polymerization step, *i.e.*, each imprinted network is only valid (at least preferentially) for a given drug. Supercritical fluid-assisted imprinting is aimed to prepare on-demand tailored cavities into already preformed commercially available SCLs, *i.e.*, as a function of the drug required for each wearer. The hypothesis is that, after the SCL synthesis, there is still enough mobility and available free volume in the polymeric chains. Therefore, the chains can still be "reorganized" and even "fixed" when a template molecule is incorporated in the network and can establish specific interactions with certain regions of the polymer.[138] This process is known as post-imprinting. Impregnation of the network with the drug followed by extraction using supercritical fluids appears as a feasible way to accomplish such an aim. Supercritical carbon dioxide ($scCO_2$) is the most commonly used due to its abundance, low cost, non-flammability, chemical inertia and safety. It is miscible with water and can act as plasticizer of most polymeric materials. Furthermore, it has a low critical temperature (31.05 °C) that allows working at temperatures suitable for thermal-labile substances.[139] Compared to common liquid plasticizers, $scCO_2$ penetrates deeper in dense polymeric networks and can be easily removed after processing.[140] Supercritical solvent impregnation and extraction using $scCO_2$ already proved their advantages for the development of drug impregnated polymeric materials which can be used as DDSs for many biomedical applications.[141–143] Flurbiprofen load/release capability of daily-wear Hilafilcon B SCLs has been improved after sequential flurbiprofen impregnation and extraction steps in $scCO_2$ (Figure 4.3.15).[138] The processed SCLs showed a higher affinity for flurbiprofen in aqueous solution than for the structurally-related ibuprofen and dexamethasone, which suggests the creation of molecularly imprinted cavities driven by both physical (swelling/plasticization) and chemical (carbonyl groups in the network with C–F group in the drug) interactions. Furthermore, processing with $scCO_2$ did not alter the glass transition temperature, optical transparency, oxygen permeability, or contact angle of SCLs and permitted the preparation of hydrophobic drug-eluting SCLs in a much shorter time than conventional molecular imprinting methods.

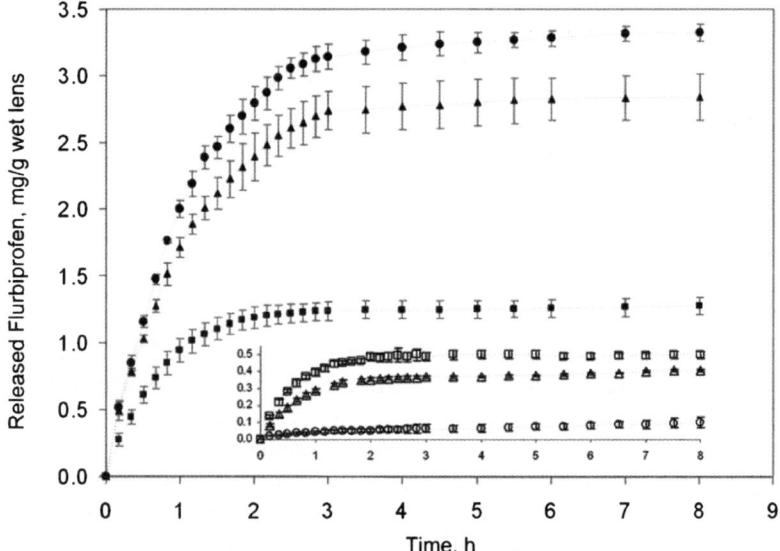

Figure 4.3.15   Flurbiprofen release in water from Hilafilcon B SCLs loaded by immersion in drug solution (small graph, empty symbols: (□) 1-cycle; (△) 2-cycles; and (○) 3-cycles) and by $scCO_2$ impregnation/extraction (filled symbols: (■) 1-cycle; (▲) 2-cycles; and (●) 3-cycles). Reprinted from Yañez *et al.*[138] with permission from Elsevier.

4.3.4 Conclusions

Simplicity of use, comfort, and prolonged drug release to the precorneal area make SCLs very attractive as ocular drug delivery systems able to overcome deficiences inherent in classical ophthalmic formulations. Although drug-eluting SCLs have been pursued for several decades, current technological advances that enable the incorporation of nanocarriers or the modification of the structure of the SCL at the nanoscale level, with minimal changes in composition, may lead the way to success. Progress in this field is occurring at various steps in the development process. Biomimetic principles that are commonly used for drug discovery are expected to be a very useful guide for endowing the networks with tailored drug binding sites, mimicking physiological drug-receptors. The optimization of nanocarriers in terms of composition, size and stability, the application of the molecular imprinting technology to achieve an adequate arrangement of the lens constituents, or the improvement of polymerization techniques that enable a precise control of the network formation, have already been demonstrated to be useful for tuning the nanostructure of the lenses in order to improve their performance as drug delivery systems.

Acknowledgements

This work was supported by MICINN (SAF2008-01679 and SAF2011-22771) and FEDER.

References

1. P. M. Hughes, O. Olejnik, J. E. Chang-Lin and C. G. Wilson, *Adv. Drug Deliv. Rev.*, 2005, **57**, 2010–2032.
2. V. R. Kearns and R. L. Williams, *Expert Rev. Med. Devices,* 2009, **6**, 277–290.
3. G. D. Novack, *Clin. Pharm. Ther.,* 2009, **85**, 539–543.
4. S. Leppard, in *Drug-device combination products. Delivery technologies and applications*, ed. A. Lewis, Woodhead Publishing Limited and CRC Press LLC, Boca Raton, 2010, p. 464–495.
5. J. Greenbaum, in *Drug-device combination products. Delivery technologies and applications*, ed. A. Lewis, Woodhead Publishing Limited and CRC Press LLC, Boca Raton, 2010, p. 496–529.
6. O. Witcherle and D. Lim, *Nature,* 1960, **185**, 117–118.
7. T. T. McMahon and K. Zadnik, *Cornea*, 2000, **19**, 730–740.
8. J. L. Munoa-Roiz and E. Aramendia-Salvador, http://www.oftalmo.com/publicaciones/lentes/cap2.htm, accessed February 2011.
9. P. C. Nicolson and J. Vogt, *Biomaterials*, 2001, **22**, 3273–3283.
10. A. Yamauchi, in *Gels Handbook*, ed. Y. Osada and K. Kajiwara, vol. 3., Academic Press, San Diego, 2001, p. 166–179.
11. T. Goda and K. Ishihara, *Expert Rev. Med. Devices,* 2006, **3**, 167–174.

12. J. Kopecek, *J. Polym. Sci., Part A: Polym. Chem.*, 2009, **47**, 5929–5946.
13. J. F. Kunzler and J. A. McGee, *Chem. Ind.*, 1995, **16**, 651–655.
14. A. W. Lloyd, R. G. A. Faragher and S. P. Denyer, *Biomaterials*, 2001, **22**, 769–785.
15. P. C. Donshik, *Ophtalmol. Clin. N. Am.*, 2003, **16**, 79–83.
16. H. Hiratani, T. Nakajima and K. J. Yamamoto, *Jpn. Soc. Biomater.*, 2002, **20**, 221–227.
17. J. Rajecki, http://www.modernmedicine.com/modernmedicine/Modern+Medicine+Now/Contact-lens-market-poised-for-continued-growth/ArticleStandard/Article/detail/679444, March 2010.
18. C. Shah, S. Raj and G. N. Foulks, *Ophtalmol. Clin. N. Am.*, 2003, **16**, 95–101.
19. L. Lim, D. T. H. Tan and W. K. Chan, *CLAO J.*, 2001, **27**, 179–185.
20. A. Kanpolat and O. Ucakhan, *Cornea*, 2003, **22**, 726–734.
21. R. Mely, *Ophthalmologica*, 2004, **218**, 33–38.
22. J. Bendoriene and U. Vogt, *Eye Contact Lens*, 2006, **32**, 104–108.
23. G. W. DeNaeyer, *New O.D.*, 2008, **December**, 11–15.
24. C. L. Schultz and D. W. Morck, *Clin. Exp. Opthalmol.*, 2010, **93**, 61–65.
25. M. F. Refojo, F. L. Leong, I. M. Chan and F. I. Tolentino, *Retina*, 1983, **3**, 45–49.
26. C. Alvarez-Lorenzo, H. Hiratani and A. Concheiro, *Am. J. Drug Delivery*, 2006, **4**, 131–151.
27. L. Xinming, C. Yingde, A. W. Lloyd, S. V. Mikhalovsky, S. R. Sandeman, C. A. Howel and L. Liewen, *Contact Lens Anterior Eye*, 2008, **31**, 57–64.
28. N. Tabuchi, T. Watanabe, M. Hattori, K. Sakai, H. Sakai, M. Abe, *J. Oleo Sci.*, 2009, **58**, 43–52.
29. C. Schultz, J. Breaux, J. Schentag and D. Morck, *Clin. Exp. Optom.*, 2011, **94**, 212–218.
30. J. Sedlacek, *Cs. Oftal.*, 1965, **21**, 509–512.
31. A. Gasset and H.E. Kaufman, *Am. J. Ophtalmol.*, 1970, **69**, 252–259.
32. S. R. Waltman and H. E. Kaufman, *Inv. Ophthalmol.*, 1970, **9**, 250–255.
33. J. S. Hillman, *Brit. J. Ophthalmol.*, 1974, **58**, 674–679.
34. M. Ruben and R. Watkins, *Brit. J. Ophthalmol.*, 1975, **59**, 455–458.
35. V. J. Marmion and M. R. Jain, *Trans. Ophthalmol. Soc. UK*, 1976, **96**, 319–321.
36. M. P. Rubinstein and J. E. Evans, *Contact Lens Anterior Eye*, 1997, **20**, 9–11.
37. D. S. Hull, H. F. Edelhauser and R. A. Hyndiuk, *Arch. Ophthalmol.*, 1974, **92**, 413–416.
38. A. Y. Matoba and J. P. McCulley, *Ophthalmology*, 1985, **92**, 97–99.
39. M. R. Jain, *Brit. J. Ophthalmol.*, 1988, **72**, 150–154.
40. C. C. Li and A. Chauhan, *Ind. Eng. Chem. Res.*, 2006, **45**, 3718–3734.
41. F. Fornaseiro, J. M. Prausnitz and C. J. Radke, *J. Membr. Sci.*, 2006, **275**, 229–243.
42. T. Vandorselaer, H. Youssfi, L. E. Caspers-Valu, P. Dumont and L. Vauthier, *J. Fr. Ophtalmol.*, 2001, **24**, 1025–1033.

43. C. C. Li and A. Chauhan, *J. Drug Delivery Sci. Technol.*, 2007, **17**, 69–79.
44. C. Schultz, J. Breaux, J. Schentag and D. Morck, *Clin. Exp. Optom.*, 2011, **94**, 212–218.
45. H. Hiratani, A. Fujiwara, Y. Tamiya, Y. Mizutani and C. Alvarez-Lorenzo, *Biomaterials*, 2005, **26**, 1293–1298.
46. J. Xu, X. Li and F. Sun, *Acta Biomater.*, 2010, **6**, 486–493.
47. J.A. Silbert, *J. Am. Optom. Assoc.*, 1996, **67**, 165–172.
48. T. P. Heyrman, M. L. McDermott, J. L. Ubels and H. F. Edelhauser, *J. Cataract. Refr. Surg.*, 1989, **15**, 169–175.
49. J. M. Chapman, L. Cheeks and K. Green, *J. Cataract. Refr. Surg.*, 1992, **18**, 456–459.
50. T. Momose, N. Ito, A. Kanai, Y. Watanabe and M. Shibata, *CLAO J.*, 1997, **23**, 96–99.
51. S. W. Kim, Y. H. Bae and T. Okano, *Pharm. Res.*, 1992, **9**, 283–290
52. G. A. Lesher and G. G. Gunderson, *Optometry Vision Sci.*, 1993, **70**, 1012–1018.
53. M. N. Miranda and S. Garcia-Castineiras, *CLAO J.*, 1983, **9**, 43–48.
54. L. Krejci, I. Brettschneider and R. Praus, *Acta Univ. Carol., Med.*, 1975, **21**, 387–396.
55. C. C. S. Karlgard, N. S. Wong, L. W. Jones and C. Moresoli, *Int. J. Pharm.*, 2003, **257**, 141–151.
56. S. M. Podos, B. Becker, C. F. Asseff and J. Harstein, *Am. J. Ophtalmol.*, 1972, **73**, 336–341.
57. G. H. Hsiue, J. A. Guu and C. C. Cheng, *Biomaterials*, 2001, **22**, 1763–1769.
58. J. J. Salz, A. L. Reader, L. J. Schwartz and K. Van Le, *J. Refract. Corneal Surg.*, 1994, **10**, 640–646
59. A. Dracopoulos, V. Bantseev and J. G. Sivak, *Invest. Opthalmol. Visc. Sci.*, 2005, **46S**, abstract 912.
60. L. C. Winterton, J. M. Lally, K. B. Sentell and L. L. Chapoy, *J. Biomed. Mater. Res. B: Appl. Biomater.*, 2007, **80 B**, 424–432.
61. E. M. Hehl, R. Beck, K. Luthard, R. Guthoff and B. Drewelow, *Eur. J. Clin. Pharmacol.*, 1999, **55**, 317–323.
62. X. Tian, M. Iwatsu and A. Kanai, *CLAO J.*, 2001, **27**, 209–215.
63. X. Tian, M. Iwatsu, K. Sado and A. Kanai, *CLAO J.*, 2001, **27**, 216–220.
64. I. Pinilla-Lozano, V. Polo-Llorens, J. M. Larrosa-Poves, S. Pérez Oliván, L. Izaguirre and J. Gorricho, *Rev. Esp. Contact.*, 1998; http://www.oftalmo.com/sec/98-tomo-2/Ind-98-2.htm, accessed February 2011.
65. I. Pinilla-Lozano, J.M. Larrosa-Poves, S. Pérez-Oliván, V. Polo Llorens, L. Izaguirre and J. Navascues, *Rev. Esp. Contact.*, 1998; http://www.oftalmo.com/sec/98-tomo-2/Ind-98-2.htm, accessed February 2011.
66. E. M. Anderson, M. L. Noble, S. Garty, H. Maa, J. D. Bryers, T. T. Shen and B. D. Ratner, *Biomaterials*, 2009, **30**, 5675–5681.
67. C. L. Schultz, T. R. Poling and J. O. Mint, *Clin. Exp. Optom.*, 2009, **92**, 343–348.

68. G. Wajs and J. C. Meslard, *Crit. Rev. Ther. Drug*, 1986, **2**, 275–289
69. C. C. Peng, J. Kim and A. Chauhan, *Biomaterials*, 2010, **31**, 4032–4047.
70. A. Hui, A. Boone and L. Jones, *Eye Contact Lens*, 2008, **34**, 266–271.
71. A. Boone, A. Hui and L. Jones, *Eye Contact Lens*, 2009, **35**, 260–267.
72. J. Kim, A. Conway and A. Chauhan, *Biomaterials*, 2008, **29**, 2259–2269.
73. J. Xu, X. Li and F. Sun, *Drug Delivery*, 2011, **18**, 150–158.
74. A. Chauhan and J. Kim, US Patent 20100330146 A1 20101230.
75. J. Kim, C. C. Peng and A. Chauhan, *J. Controlled Release*, 2010, **148**, 110–116.
76. E. M. Anderson, M. L. Noble, S. Garty, H. Ma, J .D. Bryers, T. T. Shen and B. D. Ratner, *Biomaterials*, 2008, **30**, 5675–5681.
77. J. B. Ciolino, T. R. Hoare, N. G. Iwata, I. Behlau, C. H. Dohlman, R. Langer and D.S. Kohane, *Invest. Ophth. Vis. Sci.*, 2009, **50**, 3346–3352.
78. J. P. Vairon, L. Yean, J. C. Meslard, F. Subira and C. Bunel, *Lactualité Chimique*, 1992, **Septembre–Octobre**, 330–335.
79. R. Uchida, T. Sato, H. Tanigawa and K. Uno, *J. Controlled Release*, 2003, **92**, 259–264.
80. T. Sato, R. Uchida, H. Tanigawa, K. Uno and A. Murakami, *J. Appl. Polym. Sci.*, 2005, **98**, 731–735.
81. P. Andrade-Vivero, E. Fernandez-Gabriel, C. Alvarez-Lorenzo and A. Concheiro, *J. Pharm. Sci.*, 2007, **96**, 802–813.
82. D. Gulsen and A. Chauhan, *Invest. Ophthalmol. Vis. Sci.*, 2004, **45**, 2342–2347.
83. D. Gulsen and A. Chauhan, *Int. J. Pharm.*, 2005, **292**, 95–117.
84. D. Gulsen, C. C. Li and A. Chauhan, *Current Eye Res.*, 2005, **30**, 1071–1080.
85. Y. Kapoor and A. Chauhan, *Int. J. Pharm.*, 2008, **361**, 222–229.
86. D. Anne, B. Heidi, M. Yves and V. Patrick, *J. Biomed. Mater. Res.*, 2007, **80A**, 41–51.
87. J. F. R. dos Santos, C. Alvarez-Lorenzo, M. Silva, L. Balsa, J. Couceiro, J. J. Torres-Labandeira and A. Concheiro, *Biomaterials*, 2009, **30**, 1348–1355.
88. C. J. White and M. E. Byrne, *Expert Opin. Drug Del.*, 2010, **7**, 765–780.
89. C. C. Li, M. Abrahamson, Y. Kapoor and A. Chauhan, *J. Colloid Interface Sci.*, 2007, **315**, 297–306.
90. Y. Kapoor and A. Chauhan, *J. Colloid Interface Sci.*, 2008, **322**, 624–633.
91. Y. Kapoor, J. C. Thomas, G. Tan, V. T. John and A. Chauhan, *Biomaterials*, 2009, **30**, 867–878.
92. A. Danion, H. Brochu, Y. Martin and P. Vermette, *J. Biomed. Mater. Res.*, 2007, **82A**, 41–51.
93. P. Vermette, H. J. Griesser, P. Kambouris and L. Meagher, *Biomacromolecules*, 2004, **5**, 1496–1502.
94. A. Danion and V. P. Arsenault, *J. Pharm. Sci.*, 2007, **96**, 2350–2363.
95. J. F. R. dos Santos, J. J. Torres-Labandeira, N. Matthijs, T. Coenye, A. Concheiro and C. Alvarez-Lorenzo, *Acta Biomater.*, 2010, **6**, 3919–3926.
96. J. F. R. dos Santos, R. Couceiro, A. Concheiro, J. J. Torres-Labandeira and C. Alvarez-Lorenzo, Acta Biomater., 2008, **4**, 745–755.

97. J. Xu, X. Li and F. Sun, *Acta Biomater.*, 2010, **6**, 486–493.
98. C. Alvarez-Lorenzo, F. Yañez, R. Barreiro-Iglesias and A. Concheiro, *J. Controlled Release*, 2006, **113**, 236–244.
99. F. Yañez, I. Chianella, S. A. Piletsky, A. Concheiro and C. Alvarez-Lorenzo, *Anal. Chim. Acta*, 2010, **659**, 178–185.
100. D. Cunliffe, A. Kirby and C. Alexander, *Adv. Drug Del. Rev.*, 2005, **57**, 1836–1853.
101. M. J. Whitcombe, I. Chianella, L. Larcombe, S. A. Piletsky, J. Noble, R. Porter and A. Horgan, *Chem. Soc. Rev.*, 2011, **40**, 1547–1571.
102. G. Wulff and A. Biffis, in *Molecularly Imprinted Polymers*, ed. B. Sellergren, Elsevier, Amsterdam, 2001, p. 71–111.
103. C. Alvarez-Lorenzo and A. Concheiro, in *Smart Nano- and Microparticles*, ed. R. Arshady and K. Kono, Kentus Books, London, 2006, p. 279–336.
104. A. G. Mayes and M. J. Whitcombe, *Adv. Drug Delivery Rev.*, 2005, **57**, 1742–1778.
105. C. Alvarez-Lorenzo, F. Yañez and A. Concheiro, *J. Drug Delivery Sci. Technol.*, 2010, **20**, 237–248.
106. H. S. Andersson, J. G. Karlsson, S. A. Piletsky, A. C. Koch-Schmidt, K. Mosbach and I. A. Nicholls, *J. Chromatogr., A*, 1999, **848**, 39–49.
107. O. Ramström and R. J. Ansell, *Chirality*, 1998, **10**, 195–209.
108. V. B. Kandimalla and H. X. Ju, *Anal. Bioanal. Chem.*, 2004, **380**, 587–605.
109. V. Pichon and F. Chapuis-Hugon, *Anal. Chim. Acta*, 2008, **622**, 48–61.
110. J. Haginaka, *J. Sep. Sci.*, 2009, **32**, 1548–1565.
111. M. E. Byrne, K. Park and N. A. Peppas, *Adv. Drug Delivery Rev.*, 2002, **54**, 149–161.
112. C. Alvarez-Lorenzo and A. Concheiro, *J. Chromatogr., B*, 2004, **804**, 231–245.
113. J. Z. Hilt and M. E. Byrne, *Adv. Drug Delivery Rev.*, 2004, **56**, 1599–1620.
114. D. Cunliffe, A. Kirby and C. Alexander, *Adv. Drug Delivery Rev.*, 2005, **57**, 1836–1853.
115. C. Alvarez-Lorenzo and A. Concheiro, in *Biotechnology Annual Review*, ed. M. R. El-Gewely, Vol. 12, Elsevier, Amsterdam, 2006, p. 225–268.
116. M. E. Byrne and V. Salian, *Int. J. Pharm.*, 2008, **364**, 188–212.
117. D. R. Kryscio and N. A. Peppas, *AICHE J.*, 2009, **55**, 1311–1324.
118. K. Ito, J. Chuang, C. Alvarez-Lorenzo, T. Watanabe, N. Ando, A. Yu. Grosberg, *Progr. Polym. Sci.*, 2003, **28**, 1489–1515.
119. C. Alvarez-Lorenzo, O. Guney, T. Oya, Y. Sakai, M. Kobayashi, T. Enoki, Y. Takeoka, T. Ishibashi, K. Kuroda, K. Tanaka, G. Wang, A. Yu. Grosberg, S. Masamune and T. Tanaka, *Macromolecules*, 2000, **33**, 8693–8697.
120. C. Alvarez-Lorenzo, H. Hiratani, J. L. Gómez-Amoza, R. Martinez-Pacheco, C. Souto and A. Concheiro, *J. Pharm. Sci.*, 2002, **91**, 2182–2192.
121. H. Hiratani and C. Alvarez-Lorenzo, *J. Controlled Release*, 2002, **83**, 223–230.

122. H. Hiratani and C. Alvarez-Lorenzo, *Biomaterials*, 2003, **25**, 1105–1113.
123. H. Hiratani, Y. Mizutani and C. Alvarez-Lorenzo, *Macromol. Biosci.*, 2005, **5**, 728–733.
124. S. Venkatesh, S. P. Sizemore and M. E. Byrne, *Biomaterials*, 2007, **28**, 717–724.
125. S. Venkatesh, S. P. Sizemore and M. E. Byrne, *Mater. Res. Soc. Symp. Proc.*, 2006, 897E.
126. S. Venkatesh, J. Saha, S. Pass and M. E. Byrne, *Eur. J. Pharm. Biopharm.*, 2008, **69**, 852–860.
127. M. Ali, S. Horikawa, S. Venkatesh, J. Saha, S. Pass, J.W. Hong and M.E. Byrne, *J. Controlled Release*, 2007, **124**, 154–162.
128. F. Mincione, A. Scozzafava and C. T. Supuran, *Curr. Pharm. Des.*, 2008, **14**, 649–654.
129. A. Ribeiro, F. Veiga, D. Santos, J. J. Torres-Labandeira, A. Concheiro and C. Alvarez-Lorenzo, *Biomacromolecules*, 2011, **12**, 701–709.
130. F. Abbate, A. Casini, A. Scozzafava and C. T. Supuran, *Bioorg. Med. Chem.*, 2004, **14**, 2357–2361.
131. S. Lindskog, *Pharmacol. Ther.*, 1997, **74**, 1–20.
132. K. Maruyama, N. Yokoi, Y. Takehisa and S. Kinoshita, *Invest. Ophthalmol. Visual Sci.*, 2004, **45**, 2563–2568.
133. S. Schwarz and J. Nick, *Optician*, 2006, **231**, 22–26.
134. R. C. Peterson, J. S. Wolffsohn, J. Nick, L. Winterton and J. Lally, *Contact Lens Anterior Eye*, 2006, **29**, 127–134.
135. L. C. Winterton, J. M. Lally, K. B. Sentell and L. L. Chapoy, *J. Biomed. Mater. Res., B*, 2007, **80 B**, 424–432.
136. F. Yañez, A. Concheiro and C. Alvarez-Lorenzo, *Eur. J. Pharm. Biopharm.*, 2008, **69**, 1094–1103.
137. M. Ali and M. E. Byrne, *Pharm. Res.*, 2009, **26**, 714–726.
138. F. Yañez, L. Martikainen, M. E. Braga, C. Alvarez-Lorenzo, A. Concheiro, C. M. Duarte, M. H. Gil and H. C. de Sousa, *Acta Biomater.*, 2011, **7**, 1019–1030.
139. E. Beckman, *J. Supercrit. Fluids*, 2004, **28**, 121–191.
140. O. R. Davies, A. L. Lewis, M. J. Whitaker, H. Tai, K. M. Shakesheff and S. M. Howdle, *Adv. Drug Delivery Rev.*, 2008, **60**, 373–387.
141. M. E. M. Braga, M. T. Vaz Pato, H. S. R. Costa Silva, E. I. Ferreira, M. H. Gil, C. M. M. Duarte, H. C. de Sousa, *J. Supercrit. Fluids,* 2008, **44**, 245–257.
142. M. V. Natu, M. H. Gil and H. C. de Sousa, *J. Supercrit. Fluids,* 2008, **47**, 93–1027.
143. V. P. Costa, M. E. Braga, C. M. Duarte, C. Alvarez-Lorenzo, A. Concheiro, M. H. Gil and H. C. de Sousa, *J. Supercrit. Fluids*, 2010, **53**, 165–173.

Section 5
Nanostructures for Overcoming the Pulmonary Barriers

CHAPTER 5.1

Nanostructures for Overcoming the Pulmonary Barriers: Physiological Considerations and Mechanistic Issues

JULIAN KIRCH†[a], CHRISTIAN A. RUGE†[a],
CRAIG SCHNEIDER†[b], JUSTIN HANES[b] AND
CLAUS-MICHAEL LEHR*[a,c]

[a] Department of Biopharmaceutics and Pharmaceutical Technology, Saarland University, Campus A4 1, D-66123 Saarbrücken, Germany; [b] Departments of Ophthalmology, Biomedical Engineering, Chemical & Biomolecular Engineering and Oncology, Center for Cancer Nanotechnology, Excellence Institute for NanoBioTechnology and Center for Nanomedicine, Johns Hopkins University School of Medicine 400 North Broadway, Baltimore, MD 21287, USA; [c] Helmholtz-Institute for Pharmaceutical Research Saarland (HIPS), Saarland University, Campus A4 1, D-66123 Saarbrücken, Germany
†These authors contributed equally.
*E-mail: lehr@mx.uni-saarland.de

RSC Drug Discovery Series No. 22
Nanostructured Biomaterials for Overcoming Biological Barriers
Edited by Maria Jose Alonso and Noemi S. Csaba
© The Royal Society of Chemistry 2012
Published by the Royal Society of Chemistry, www.rsc.org

5.1.1 The Lungs as a Route for Drug Delivery

5.1.1.1 Introduction: Why is the Lung a Good Place to Deliver Drugs?

Inhalation therapy and pulmonary drug delivery is, and has long been, a highly desired route for the administration of active pharmacological ingredients (API). Here, both systemic and local administration play an important role in acute and chronic therapy: similar to other locally constrained diseases, for pulmonary diseases, the local administration of potent API's is favored compared to parenteral administration due to side effects and issues of controlled release. This is particularly important in the case of diseases like COPD or Pulmonary Hypertension where frequent and long-lasting treatment is necessary. Here, the need for modern formulations with sustained, local release is paramount. Systemic delivery *via* the lungs holds several advantages over other forms of administration: pulmonary structures not only provide large and highly vascularized surfaces (about 140 m^2 of total surface area[1]) and thus high potential for rapid absorption, but also exhibit low catabolic activity, compared to the GI tract. Therefore, the lungs may be the place to deliver more unstable drugs with low bioavailability when orally administered. For example, biopharmaceuticals fit this description and several are already on the market as inhalable formulations.

Furthermore, the development of formulations suited for pulmonary delivery may contribute to faster and cheaper therapeutic modalities in public and private health care. This is of high significance in cases where intravenous (i.v.), intramuscular (i.m.), or oral administration is not desired or possible. An example is the treatment of diseases in third world countries, where repeated usage of one needle poses a high risk of infection, particularly with regard to HIV. In this context, however, it should be mentioned that the further development of administration technologies (dry powder inhalers, nebulizers *etc.*) goes hand in hand with galenics, and still holds room for optimization. Especially in terms of compliance and handling of inhalation devices, administration *via* the pulmonary route requires a high degree of coordination and training of patients, often limiting its acceptance in clinical practice.

5.1.1.2 Physiological Features Influencing Lung Deposition

Particles deposit in the various regions of the lung based on their physical and chemical properties and the properties of the airstreams in which they are entrained. Deposition is also dependent upon differences in the physiology and anatomy of the individual patient (healthy *vs.* diseased lungs and adult *vs.* child). Deposition occurs when a particle (or water droplet) collides with the airway surface liquid that lines the airway epithelium. Targeting deposition to a specific region of the lung is possible by controlling the physical and chemical properties of aerosolized particles.

Aerosol deposition in the lungs occurs by three major mechanisms: inertial impaction, sedimentation, and diffusion (Figure 5.1.1).[2] Deposition *via* inertial impaction and sedimentation depend upon the aerodynamic diameter of the aerosol particles. Particle aerodynamic diameter is defined as:

$$d_{aero} = \frac{d\sqrt{\frac{\rho}{\rho_w}}}{\gamma} \tag{5.1.1}$$

where d is the geometric diameter, ρ is the particle bulk density, ρ_w is the mass density of water, and γ is the shape factor (1 for a sphere).[4] Thus, large, dense particles will have large aerodynamic diameters, whereas small, low density particles will have small aerodynamic diameters.

Deposition *via* diffusion is typically limited to particles with a diameter less than 500 nm and is not dependent on particle density.[5] In general, particles with large aerodynamic diameters will deposit in the upper airways of the lungs, as they have sufficient mass to rapidly sediment out of the airstream onto the lung surfaces and/or sufficient inertia to maintain trajectory out of the airstream and impact on the airway surface during air stream bifurcation (see Figure 5.1.1). Particles with d_{aero} 8 μm and larger typically deposit in the upper airways, mouth, and throat, whereas bronchial deposition is typically observed for particles with d_{aero} between 4–10 μm (Figure 5.1.2). Small particles (20–50 nm $< d <$ 2–5 μm), including nanoparticles, deposit primarily in the alveolar region *via* diffusion and/or sedimentation, although a significant amount of particles between 100 nm and 1 μm fail to deposit and are exhaled (Figure 5.1.2).[5–8] Alveolar deposition of these small particles can be enhanced by

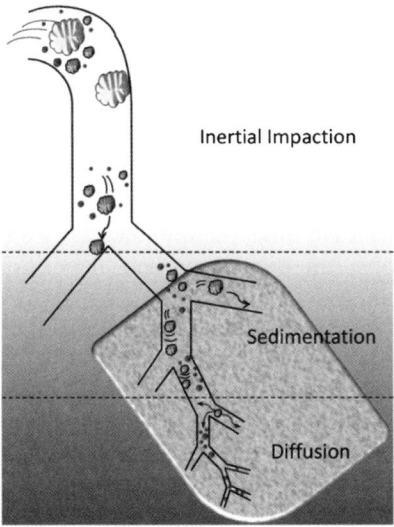

Figure 5.1.1 Primary deposition mechanisms of inhaled particles in the respiratory tract. Reprinted with permission from Elsevier, ref. 3.

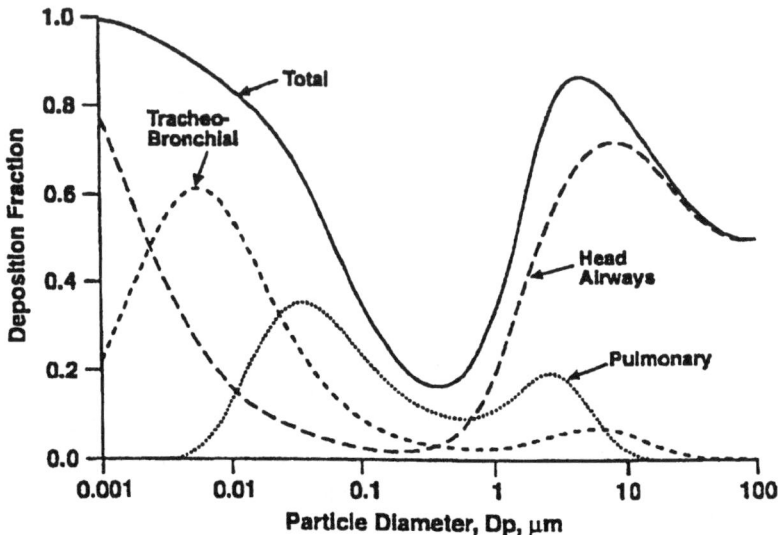

Figure 5.1.2   Particle deposition in the human respiratory tract as a function of particle aerodynamic diameter. Reprinted with permission from Medical Physics Publishing, ref. 9.

improving the patient's inhalation technique; specifically, breath holding increases the amount of time particles may diffuse and/or settle by gravity and deposit on the alveolar surface, resulting in improved alveolar deposition.[7]

The shape of a particle may also affect its deposition in the lung. For example, the aerodynamic diameter of a rod-shaped particle is approximately two to three times less than a spherical particle with a diameter equal to the width of the rod.[10–11] Due to the reduced aerodynamic diameter, long but thin particles may potentially deposit in the deep lung.[12–13] However, the deposition of rod shaped particles may also be affected by another mechanism, interception.[13–15]

The deposition of particles in the lung is also dependent upon variation in the anatomy of the individual patient, as well as inhalation technique.[16] To deposit in the lungs, particles must traverse the complex, highly branched lung airway structure. Any irregularities in this structure, such as airway surface irregularities and partial obstructions can disrupt particle deposition and create uneven particle distributions in the lung.[17–18] For example, cystic fibrosis, chronic obstructive pulmonary disease, and asthma can result in airway obstructions (such as mucus plugs in small airways), infection, and/or inflammation.[19–22] These obstructions typically result in increased deposition in the central airways and reduced deposition in the alveolar region of the lung.[23–26]

5.1.1.3   New Approaches

To deliver particles with large geometric diameters to the deep lung, Edwards *et al.* developed large porous particles (LPP; Figure 5.1.3).[27] By making

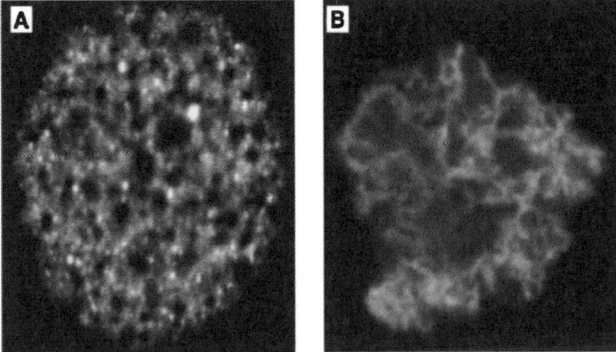

Figure 5.1.3 Confocal microscopy images of (A) porous PLGA and (B) porous PLAL-Lys particles. Fluorescein isothiocyanate–dextran was encapsulated in the PLGA particle to render the pore spaces of the particles visible in the fluorescent confocal image (A). The PLAL-Lys particles were fluorescently labeled through the reaction of rhodamine isothiocyanate with lysine amine groups on the surface of the particles (B). The PLGA and PLAL-Lys particles are highly porous as evidenced by the appearance of fluorescence throughout the particle structure. Figure reprinted with permission of AAAS, from ref. 27.

particles sufficiently porous, they significantly reduced particle density (from ~ 1 g cm^{-3} for nonporous particles to ~ 0.1 g cm^{-3} for porous particles) and made particles with geometric diameters up to 20 µm that possessed an aerodynamic diameter conducive to delivery to the alveolar region of the lung. The advantage of having these large particles in the alveoli is that the particles were large enough to avoid uptake by alveolar macrophages, which efficiently engulf particles in the size range of 1 µm to 5 µm.[27–29]

The LPP technology has been applied with a variety of biodegradable polymers and excipients,[27,30–33] and can provide sustained release of a variety of hydrophobic and hydrophilic drugs.[27,30–32] For example, Fu *et al.* showed that LPP could be made from copolymers of PEG, sebacic anhydride, and 1,3-bis-(carboxyphenoxy) propane (CPP).[32] These particles were capable of loading a model hydrophobic drug (rhodamine) and hydrophilic plasmid DNA and the release rate could be modulated by changing the composition of the monomers in the terpolymer. Dunbar *et al.* studied the aerosol characteristics of LPP's *in vitro* (in a multistage liquid impinger) and in human subjects via γ-scintigraphy.[34] They found LPP could be efficiently aerosolized with a mean *in vivo* lung deposition relative to the total metered dose of 59.0% and 37.3% for 3 and 5 µm mass mean aerodynamic diameter (MMAD) powders, respectively. Other studies in animal models with the LPP technology have shown that LPP can improve the efficacy of drugs and prolong the duration of action.[30,35]

5.1.2 What Happens After Landing? A Shift of Paradigm in Pulmonary Drug Delivery

The deposition and clearance of aerosol particles in the lungs has been under investigation for decades. However, the rate, extent, and various mechanisms of bio–nano interactions – *i.e.* mucociliary clearance, uptake by macrophages or translocation across the epithelium into the blood stream – of inhaled materials have often remained unclear. Generation of suitable aerosols to effectively deposit particles in the desired region of the lungs has long been thought to be the only pivotal factor for successful pulmonary drug absorption, and thus bioavailability. However, voices contesting this opinion are rising.[36–37] Acknowledging that efficient deposition of aerosol particles in the lungs is still a necessary, but not always sufficient condition for successful pulmonary drug delivery, there is an increasing need to understand and control what happens after a particle has landed.

With respect to the bio–nano interaction processes following deposition, the so-called air–blood barrier comprises both cellular and non-cellular elements. The cellular elements of the air–blood barrier are diverse and complex. Besides epithelial cells, which show dramatic differences in morphology and function between the upper (airways) and lower (alveolar) regions of the respiratory tract, there are macrophages and other cells of the immune system. Obviously, the interactions of drugs – either as particles or molecules – with epithelial cells and especially the macrophages deserve a closer consideration.

The non-cellular elements determine the solubility and rate of dissolution of particulate matter in the pulmonary lining fluids, its agglomeration or disintegration, water uptake, adsorption of biomolecules, as well as mucociliary clearance. The possible alteration of particle properties during inhalation, deposition and immersion in lung lining fluids play a distinct role in the interaction with both cellular and non-cellular barriers. However, the significance of such changes for particle clearance or translocation through the epithelium as well as the mechanisms of particle mobility and transport within the mucus blanket are still unclear.[40] Additionally in this context, the role of adsorbed biomolecules on particle clearance by macrophages, effects of particle shape and surface chemistry and synergistic effects, *e.g.* with mucociliary clearance, are being investigated but are still largely unexplored.[41]

Compared to oral drug delivery, it is barely known if and to what extent the composition and the structure of the lining fluids in the lung play a role in drug dissolution and absorption. Because of the relatively short distance from lumen to tissue (at least in the alveolar region), permeation of small molecule drugs (primarily BCS class II compounds) is typically not a rate-limiting step for drug delivery to the blood from particles deposited in the alveoli. Therefore, the dissolution rate of the drug from particles often determines the release of the drug and, therefore, the pharmacokinetics of the formulation. The two essential elements of the lung lining fluid are mucus and pulmonary surfactant, which vary in volume and thickness along the respiratory tract. The rather small total fluid volume in the lungs of 25–45 mL (see section 5.1.4) may be a challenge for new, often poorly soluble API's.[42] Thus, insoluble drugs can quickly reach their saturation concentration and prevent further dissolution. Making drugs in nanoparticle form increases the surface area to volume ratio, thereby increasing solubilization.[43–44]

Other aspects that are not clear yet are physical, chemical or metabolic changes of the carriers and drug materials following deposition. Concerning physical and chemical changes, surface properties, agglomeration status, density and water content are certainly indispensable, not merely for deposition but also for translocation and mobility of particles in mucus. With respect to metabolic changes, the lungs exhibit relatively low expression levels of metabolizing enzymes compared to the gastro-intestinal tract. However, premature drug and/or carrier degradation cannot be neglected, particularly because the interest in sustained release formulations for pulmonary application has been great and is still rising.

Mechanistic studies on the translocation of particulates across the pulmonary epithelium are rare and often do not allow generalized conclusions.[45–48] In particular, results obtained with bioresistent (nano-) materials cannot simply be transferred to particular drug carriers made of biodegradable excipients. When the drug and/or excipient is released relatively fast from the carrier by *e.g.* rapid dissolution or biodegradation – which is the case for all dry-powder inhalation aerosols currently on the market – mechanisms of molecular transport across the pulmonary cellular and non-cellular barriers have to be considered, as well as possible changes of these structures due to deposited nanomaterials. The role of membrane transporters and vesicular uptake *versus* paracellular transport is gaining momentum in pulmonary drug delivery, but still far from satisfyingly clarified.[49–50] This book chapter aims to highlight the essential challenges arising from such a shift of paradigm, and to address the resulting new problems as well as opportunities in pulmonary drug delivery with respect to cellular and non-cellular barriers.

5.1.3 Cellular Barriers

5.1.3.1 Epithelial Barriers

The barrier formed by the epithelium of the lungs is as large as it is diverse: according to its function, the epithelium in the conducting airways differs substantially from that in the peripheral lungs (Figure 5.1.4). In the peripheral lungs, the main function of the epi- and endothelium is to provide a large and thin surface to facilitate gas exchange. Therefore, the epithelium in this area is comprised of an extremely thin cell monolayer.[51] It is composed of two major cell types: alveolar type I and alveolar type II cells. Although type I cells only account for 10% of the alveolar cell number, they cover more than 90% of the surface in the peripheral lung, this is mainly due to their extremely outstretched morphology with protruding nuclei which facilitate the exchange of oxygen

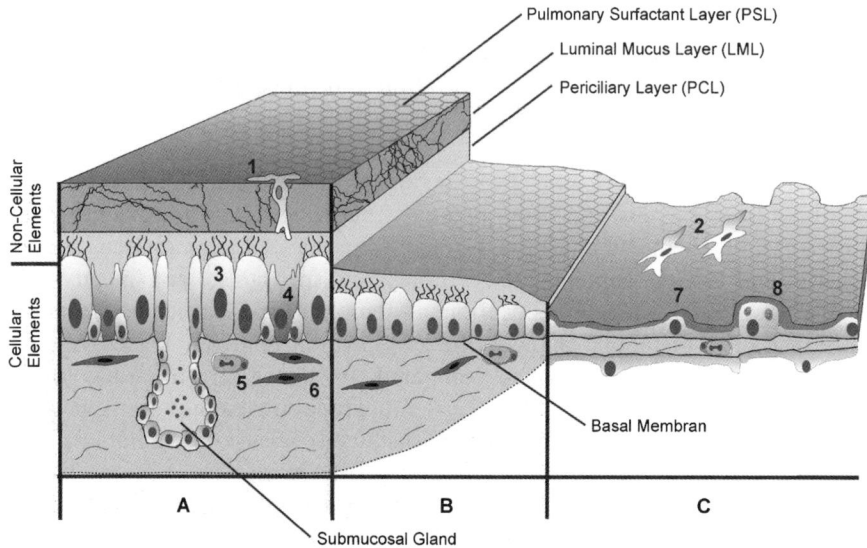

Figure 5.1.4   Cellular and non-cellular elements of the pulmonary air–blood–barrier. Trachea and bronchi (A) are lined with a thick fluid layer, composed of the luminal mucus layer (LML) and periciliary layer (PCL). The basal connective tissue is interfused with capillaries (5) and muscle fibers (6). Mucus is produced by goblet cells (4) and submucosal glands, and propelled by ciliated cells (3). This fluid layer decreases in thickness starting from the bronchioles (B), to a final value of 0.09–0.89 μm in the alveolar region (C). [38] Here, alveolar type I cells (7) cover the main part of the surface. Type II cells (8) serve secretory purposes and may be progenitive to type I cells. Airway macrophages patrol all pulmonary surfaces (1) and may cross the non-cellular barriers. Alveolar macrophages (2) can be characterized by close proximity to the air interface and form the first line of defense in the peripheral lungs (modified after Sturm *et al.*[39]).

and carbon dioxide.[52] Type II cells are more compact in shape and mainly serve as secretory cells for alveolar surfactant, which is discussed in section 5.1.4.2. It is believed that type II cells also act as progenitors to type I cells, although there is evidence that other cells can also proliferate into alveolar type I cells.[52] Further up in the airway region of the respiratory tract, the thickness of the epithelium successively increases to form a significant cellular barrier in the bronchi and the trachea. The dominating cell types in these parts of the lungs are ciliated cells, which account for 50% of the total cell population, secretory (mucous, goblet, serous or Clara) cells and basal cells.[52] The different composition and morphology of this part of the epithelium is highly adapted to its function: whereas in the alveolar region the epithelium is evolutionary optimized for gas exchange, the main purpose of the upper lungs is the conduction and filtering of the inhaled air. The pseudostratified epithelial layer of the upper and central lungs is lined with a fluid layer of far greater volume than the layer of surfactant in the peripheral lungs; this fluid layer is the mucus blanket.[53-54] Along with the thickness of the epithelium, the pulmonary fluid layer (*i.e.* non-cellular barriers) grows in thickness and surface coverage: forming single droplets on top of the scarcely ciliated cells of the lower bronchii, mucus is transported in larger flakes or even continuous streams in the trachea.[55] Foreign material deposited there is trapped within this layer and can be removed by constant propulsion of mucus by the coordinated beating of underlying cilia (see 5.1.4.1). Efficient clearance by coordinated ciliary beating has been under investigation for a long time and is the subject of several excellent reviews; however, it is not fully understood to date.[55-57]

Uptake of inhalable material (PM_{10}, particulate matter < 10 μm in diameter) into and its translocation across the epithelium of the lungs as well as further transport to other organs of the body has been under investigation for several decades.[58] However, significant scientific effort has been focused on toxicological effects of inhaled material. Availability of data regarding toxicological potential of particles with varying properties is comprehensive. Of this broad population of particles, so far only the nanosized fraction below a diameter of 100 nm ($PM_{0.1}$) is considered to be able to penetrate with high efficiency through the air–blood tissue barrier upon deposition.[50]

Regarding the mechanisms of particle uptake into and translocation through cells, various possibilities are being discussed. Here, we do not take into account inflammatory effects which alter barrier properties of the epithelium, instead this section focuses on more specific ways of particle translocation.[59] Not only common endocytic pathways like phagocytosis or Clathrin-, and Caveolin-mediated pinocytosis,[49] but also yet unknown mechanisms are believed to be relevant within this context. It has been shown that TiO_2 nanoparticles can penetrate the alveolar epithelium. As membrane association of the internalized nanoparticles could not be observed, it was concluded that the uptake may not involve endocytic mechanisms.[60] Instead, several active and passive processes such as diffusion through pores or adhesion-mediated uptake are currently considered. Further studies suggest a protein-mediated uptake, which involves specific receptor binding.[61] In recent years, an even more complex mechanism of particle translocation has been observed and described. *In vitro* results confirmed a possible mechanism of particle translocation which involves not only epithelial cells, but concerted interactions of epithelial cells with dendritic cells and macrophages.[62-63] In contrast, particles which already exceed the thickness of the relevant lung fluid layer with their diameter (> 5 μm) may be translocated into the epithelium according to another mechanism. Studies by Gehr *et al.* showed that the forces exerted on large polymeric particles by wetting with pulmonary surfactant not only submerged particles into alveolar fluids, but could be strong enough to push larger particles into the epithelium. According to the constraints regarding the size of the particles, this mechanism does not apply for particles in the nano-range. Uptake studies for nano- and submicron-sized particles in the region of the upper and central lungs are rare, as the dominant barrier for such particles is clearly the non-cellular mucus blanket (see 5.1.4.1).

The expanded understanding of the described transport processes is to a large extent due to the development of suitable cell culture models in recent years.[62,64] Although cell culture models cannot mimic all aspects of the pulmonary epithelium, they are valuable tools, especially for screening purposes. Major drawbacks compared to *in vivo* or *ex vivo* models are connected with issues of cell differentiation and functional characteristics (secretion of mucus and surfactant, transepithelial electrical resistance *etc.*). Here, well-established alveolar (A549) or bronchial (Calu-3, 16HBE14o-) models have been particularly useful in early stages of the development of inhalative medicines and toxicology. Dual- or even triple-cell co-cultures particularly facilitated the investigation of particle translocation and clearance, especially with respect to the interplay between epithelial and immune cells.[62,65] Although co-culture models are more physiologically relevant, they cannot depict all characteristics of the *in vivo* situation. Nevertheless, much progress in this area has been due to the efforts to improve such model systems. Potential interspecies incompatibilities in current co-culture models should be addressed in the future.

5.1.3.2 Macrophage Clearance

Besides mucociliary clearance, removal of particulate matter from the lungs is mainly mediated by surface macrophages. In the conducting zone of the lungs, airway macrophages cooperate with mucociliary clearance in terms of particle clearance. However, in the peripheral lungs, it is the alveolar macrophage (AM) that is discussed to be the most important phagocytes in particle uptake and clearance.[63]

AMs are resident mononuclear phagocytes that derive from hematopoietic stem cells in the bone narrow, and reach the alveolar tissue as monocytes *via* the blood where they are likely interstitial macrophages, before they move to the luminal side of the lung to become AMs.[63] AMs are the only macrophages in the body that are in close proximity to an air-interface and exposed to air. They are a major cellular component of the alveolar lining fluid – about 80% of cells recovered from a human bronchoalveolar lavage are AMs – and therefore, present in high numbers: about 5.99×10^6 AMs in a human lung,[66,67] which is about 12.4 AMs per alveolus.[63]

With the lungs being the largest epithelial tissue exposed to the surrounding environment, AMs fulfill a crucial function in pulmonary immune reactions and the host defense system,[68] and can be considered as the first line of defense against foreign material that reaches the alveolar environment.[69] Using their actin skeleton to spread out tiny filaments (pseudopodia), AMs are highly mobile cells with an area of movement of about 18.8 μm^2 per cell.[63] This allows AMs to patrol the alveolar space and quickly arrive on site in response to stimuli. Once they have reached the site of action, they can secrete a wide variety of mediators such as reactive oxygen species, TNF-alpha, chemokines and complement components. Overall, AMs are crucially involved in recovering the alveolar architecture and maintaining sterility in the peripheral lungs.[68]

With respect to their complex role in the lungs, AMs can also greatly influence the residence time of particle-based drug delivery systems. Once a particle is deposited in to the peripheral lung it first contacts the alveolar lining fluid, and encounters the pulmonary surfactant system. Consequently, it will be displaced into the subphase where interaction with AMs is most likely.[46,60] Particle phagocytosis by AMs and subsequent intracellular dissolution leads to removal of the API from the site of action and may prevent delivery of the drug to the target tissue, overall reducing the bioavailability of the drug.[37,70]

On the other hand, uptake of a (nano-) particulate carrier system by AMs can be a desired effect, especially in the case of infectious diseases such as tuberculosis, where the AM is the affected and therefore targeted cell-type. In this scenario, selective and controlled internalization of carrier systems loaded with anti-infective drugs by the AM could be a potential strategy for selective therapy.

The main process of particle uptake by AMs is phagocytosis. However, before this process occurs, it is very likely that deposited particles are opsonized with soluble components of the alveolar lining fluid (refer to section 4.3 of this chapter). Confrontation with AMs depends on the contingency of their presence at the site of deposition, but it is also possible that further AMs are directed towards the particulate matter *via* chemotaxis, until close proximity to the foreign material is reached.[69]

Phagocytosis in general can be defined as an actin-dependent uptake of particles larger than 500 nm by immune cells.[71] This energy-dependent process takes place upon polymerization of actin into organized structures, leading to membrane extensions that can engulf the particle.[63] Other cells with phagocytic activity besides macrophages are neutrophils and dendritic cells, where the latter show a lower activity for uptake of particles.[63,72]

On the other hand, there are several particle parameters that can influence the uptake by AMs. For instance, in terms of particle size it has been shown that active particle uptake *via* phagocytosis occurs primarily in the geometric diameter range between 1–5 μm,[29,73] whereas particles less than 500 nm are taken up sporadically and by non-specific mechanisms.[74–75] As outlined before, particles in the micrometer range are likely to be taken up mainly *via* active phagocytosis (or also macropinocytosis). However, this is unlikely for nanoparticles. The uptake here is probably also size dependent, using other pathways than phagocytosis: whereas nanoparticles bigger than 0.2 μm are probably taken up *via* pinocytosis, smaller particles (less than 150 nm) can be internalized *via* calveolae (50–100 nm) or clathrin-mediated (100–120 nm) uptake.[63]

Furthermore, particle shape can determine whether a particle is internalized or not. Champion *et al.* showed that the overall shape is not the primary factor influencing this effect, but the local shape, *i.e.* the shape of the particle at the position where initial cell contact is made.[76] Here, they successfully showed that aspherical particles with aspect ratios greater than 20 were internalized to an increased degree when the macrophage approached the particle at points with high curvature, whereas spherical particles were shown to be internalized from each side equally. Besides the particle shape, the material and material properties affect particle uptake by AM. For example, increased mechanical robustness and overall stiffness of particles leads to increased phagocytosis.[63] More importantly, the material composition of a particle can affect the adsorption of biomolecules with opsonin function, which is likely to influence the uptake by AM significantly.

Upon the interaction with (nano-) particles, AM can induce and trigger inflammatory reactions. Release of TNF-alpha, IL-1alpha, and IL-1beta can occur, leading to expression of adhesion molecules and release of other chemokines and growth factors.[69] Pozzi *et al.*[77] showed that exposure of macrophages to fine particulate matter (~ 40 nm in geometric diameter) lead to increased levels of TNF-alpha. $PM_{0.1}$ can induce generation of reactive oxygen species (ROS), causing oxidative stress in alveolar macrophages.[78]

On the other hand, AMs also control inflammation by release of chemokine inhibitors or TNF-alpha soluble receptors,[79] or can self-regulate inflammatory processes *via* production of IL-10, which can then reduce release of IL-1 or TNF-alpha.[80]

Furthermore, AMs can also interact with other immune cells during inflammatory situations induced by particulate matter. For instance, dendritic cells (DCs) are situated above and underneath the airway epithelium. They can extend their dendrites between epithelial cells to sample the luminal space for antigens and are able to report antigenic information to the pulmonary lymph nodes.[69] Thereby, they are able to interact with particle carrying AMs and can even receive particles from these cells. Blank *et al.* demonstrated that upon such a cell–cell interaction, particles even can be transferred from the AM to DCs, and that DCs, AMs, and epithelial cells essentially seem to form a network of cellular interaction.[81]

Once particle matter has been internalized by these professional phagocytes, the cells are able to leave the alveolar space and transport the particle cargo out of the lungs. The actual clearance, *i.e.* the removal of the particle-loaded AMs from the lungs and its further processing, is generally a short process (24–48 h).[63] The predominant clearance pathway from the lungs for AMs with ingested particles is probably *via* transport to the upper airways and the mucociliary escalator.[68] In case of particle transfer to DCs clearance probably occurs *via* the lymph. Redistribution of particles among AM (exocytosis and re-uptake by other AM) is also possible as a way to distribute the particle burden among the AMs.[82]

Interestingly, there are some differences among different species, as in mice, hamsters, and rats significant higher particle amounts are removed *via* AM clearance when compared to humans, probably a result from anatomical and structural differences within the lungs.[45] Nevertheless, to date, it is not yet fully understood how AMs find their way out of the lung.

Concerning cell culture models for AMs, primary cells isolated from various sources (mice, rat, pig, human) obtained by bronchoalveolar lavage (BAL) are commonly used. It is also possible to obtain alveolar tissue from lung surgeries, and to isolate AMs.[83] Besides usage of primary cells, there are also a few immortalized cell lines of AM available, such as murine (MH-S[84]), rat (NR 8383[85]) or porcine AM (3D4/21, 3D4/31[86]).

5.1.4 Non-Cellular Barriers – The Lung Surface Lining

As mentioned in the previous section, the non-cellular barriers of the airways pose a major challenge for drug and particle absorption.

The functional sectioning of the lungs into a conducting part (airways) and a gas-exchanging part (alveoli) is not only reflected by the varying cell types in the epithelium as we proceed from the throat to the alveoli, but is also reflected by changes in the non-cellular barrier of the lung surface lining (Figure 5.1.5).

In the conducting airways, the lung surface lining mainly consists of mucus, whereas in the deep lungs, pulmonary surfactant is most prevalent. In the lungs, the non-cellular matter can greatly influence the surface characteristics of the deposited delivery system. Compared to the alveolar region, issues of fluid characteristics, fluid dynamics, diffusional properties within bio-gels and interactions of particles with mucus and mucociliary clearance are of even greater importance in the upper lungs.

The main aspects important for the fate of the particle or the drug are dissolution, disintegration, biological modifications and clearance. After first contact with most outer (mostly non-cellular) pulmonary structures, *i.e.* mucus or pulmonary surfactant, an inhaled (nano-) particle is likely altered by its surrounding environment. Regardless of whether the conducting airways or the alveolar epithelium is the site of deposition, in both compartments of the lung soluble compounds (*i.e.* proteins, glycoproteins, lipids) are secreted by

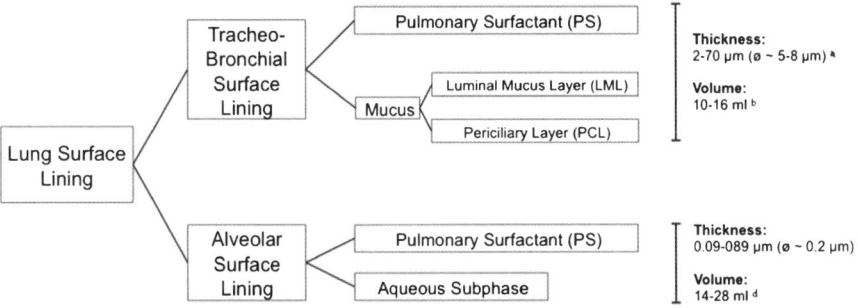

Figure 5.1.5 Nomenclature and dimensions of the pulmonary non-cellular barriers. [a]According to Sims *et al.*[87]. [b]Assuming an airway surface area of 2 m^2 and an average thickness of 5–8 μm[42,87]. [c]According to Bastacky *et al.*[38]. [d]Assuming an alveolar surface are of 70–140 m^2 and an average thickness of 0.2 μm.[1,38,88]

respective cells (*e.g.* goblet cells in the airway epithelium, or AT II cells in the alveolar tissue) and can bind to particulate delivery systems once they have landed.

Among these biomolecules, proteins are probably the dominant components, whereas lipids or salts may play a mediating role in such particle protein interactions. On the particle side, the binding is most likely influenced by general particle characteristics, such as size, shape or curvature, and surface properties, *i.e.* charge, roughness, and hydrophobicity.[89]

The principle phenomenon of protein adsorption to (nano-) particles and its relevance for further biological effects has recently been described in the so-called *protein corona theory*.[90,91] So far, this theory has been restricted to systems that are applied systemically. However, within the lungs, relatively little is known about how such biomolecules, as intrinsic parts of a functional lung, influence clearance and translocation of particulate drug delivery systems.

To maximize bioavailability of drugs or drug carriers, all clearance phenomena need to be circumvented or at least minimized without altering these biological barriers. This task has not yet been fully accomplished, but there are several approaches with promising outlook.

In this section, the aspect of particle clearance and the involvement of the non-cellular part of the pulmonary barrier herein, will be discussed. Here, interactions of nanomaterials with pulmonary surfactant and clearance by mucociliary activity have to be considered. In this context, the role of pulmonary surfactant and mucus in the fate of deposited material will be discussed.

5.1.4.1   Mucus and Mucociliary Clearance

Mucus is the primary defense mechanism of mucosal tissues and has evolved to efficiently trap and remove nano- and micron-sized objects, like viruses and bacteria.[92–93] Respiratory mucus lines the luminal side of the lung airways, from the trachea to the terminal bronchioles. Mucus functions as a selectively permeable barrier, allowing many nutrients, proteins, and other molecules to move through it, while excluding pathogens and other potentially dangerous materials from accessing the underlying cells.[93]

In the human lungs, the mucus layer is believed to be $\sim$2–70 μm thick,[87,94,95] and is composed of two distinct layers [94,96]: the periciliary (sol) layer ($\sim$5–10 μm thick)[97] and the luminal (gel) layer (varies throughout respiratory tract, increasing from distal to proximal airways).[97,98] The periciliary layer (PCL) is generally believed to be a watery, low viscosity layer[99] (although this view has recently been challenged[100]), located closer to the epithelial surface than the luminal gel layer, in which the cilia (located on the surface of certain airway epithelial cells) are constantly beating to propel mucus up and out of the lung.[93,96] The cilia beat in a coordinated fashion to propel the gel layer at speeds up to 5–10 mm min^{-1}.[93,101–103] The ciliary beating pushes the mucus gel blanket away from the epithelial surface, thus forming a low viscosity, water layer (the PCL).

The luminal mucus layer is a highly entangled, viscoelatic polymer network composed of secreted mucin proteins, cells, bacteria, nutrients, protective factors, and waste.[104–106] The major components of healthy mucus are water (90–98%) and mucins (2–5%).[96,107–110] Figure 5.1.6 shows the complex structure of a mature mucin protein. Mucins are highly glycosylated proteins with 40% to 80% of their weight coming from O-linked glycans found within PTS (proline, threonine, serine) rich domains.[93] These glycans provide mucus with a highly negatively charged, brush like structure. Mucins also contain cysteine rich domains where no glycosylation is present. These "naked" regions form into hydrophobic beads or globules that are stabilized by internal disulfide bonds.[111] The hydrophobic regions are a major contributor to the

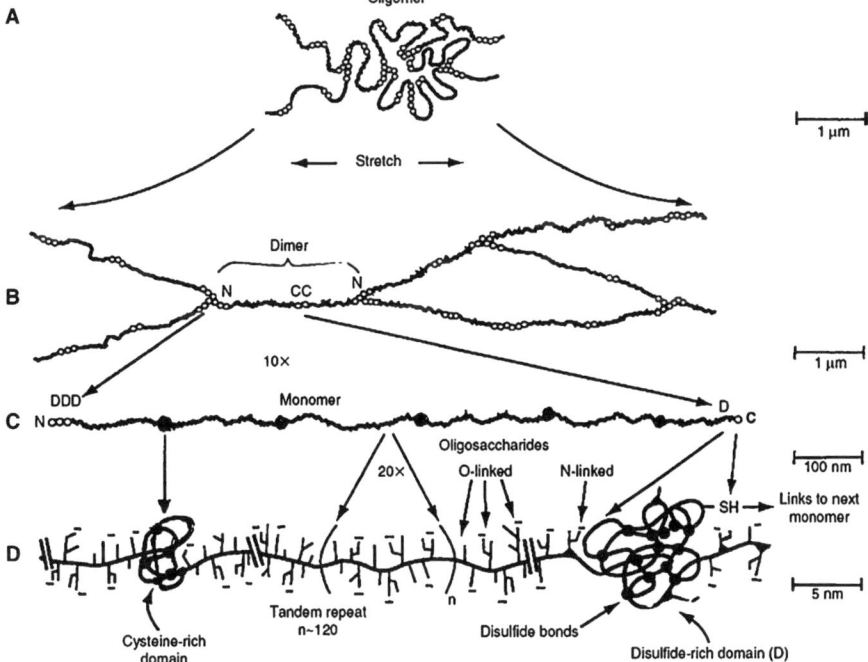

Figure 5.1.6 Major biochemical features of gel-forming mucins. (A) Several mucin monomers are shown linked together in an oligomeric gel. (B) Mucin monomers are crosslinked end-to-end *via* disulfide bonds between disulfide-rich domains (labeled "D") near the amino- and carboxyl-termini. (C) Interspersed along each fiber are "naked" globular protein regions with small exposed hydrophobic patches. These regions are stabilized by multiple disulfide bonds. (D) Individual mucin fibers are densely glycosylated with O- and N-linked glycans, most of which are negatively charged with sialic acid or sulfate groups. Reprinted with permission from Elsevier, ref. 121.

highly viscoelastic behavior of mucin gels, as they form dynamic, physical crosslinks with other mucins via hydrophobic interactions.

There are two major classes of mucin proteins: cell-associated mucins and secreted, gel-forming mucins.[112] Cell-associated mucins are much smaller than their secreted counterparts, varying in length from 10 to 500 nm. They have a transmembrane domain that anchors them to the surface of airway epithelial cells and they are believed to have roles in cell signaling.[113–118] Secreted, gel-forming mucins are much larger and polymerize together end to end *via* disulfide bonds to form mucin oligomers up to several microns long.[119,120] These large, highly glycosylated proteins come together to form a viscoelastic gel network *via* entanglements and hydrophobic interactions. In a 2% to 5% mucus gel, each mucin overlaps or entangles up 10 to 100 other mucins.[93]

A hallmark of mucus is its shear-dependent bulk viscosity that is roughly 1 000 to 10 000 times more viscous than water at low (physiological) shear rates.[54,93] The elasticity of mucus under the shearing of cilia allows it to be pushed effectively from the lungs (distal to proximal) to the trachea and then pharynx, where it then enters the digestive tract to inactivate and degrade any harmful organisms and toxins entrapped in the selectively permeable gel. When the components of mucus are out of balance, the viscoelasticity is altered, resulting in impaired mucociliary clearance. Several respiratory diseases such as asthma, chronic obstructive pulmonary disease (COPD), and cystic fibrosis (CF) are related to abnormal respiratory mucus and impaired mucociliary clearance.[122–126]

Mucus, as a polymer network, is inherently riddled with aqueous pores.[54,127] The size of these pores, described by the average mesh spacing, allows small molecules like proteins and nutrients to move through the mucus layer, while preventing bacteria and other large, potentially harmful particles from doing the same. EM estimates of the mucus mesh show mesh spacings ranging from 20 to 200 nm, although these results are the subject of controversy, as sample dehydration and fixation may change the structure from that of native, hydrated mucus.[128,129] Recently, muco-inert probe particles were used to determine the average mesh spacing of various mucus secretions.[127,130,131] Based on an obstruction scaling model, the average mesh spacing of human cervicovaginal mucus, human cystic fibrotic sputum, and human chronic rhinosinusitis mucus to be 340 ± 70 nm,[127], 145 ± 50 nm[131] and 150 ± 50 nm,[132] respectively. The results from these experiments revealed that mucus is riddled with pores that are larger than the diameter of many viruses. But if this is the case, how does mucus prevent viruses from infecting epithelial surfaces?

Mucus is not just a steric barrier to deposited particulates; it is also adhesive and thus able to stick to particles like flypaper. To most objects, mucus is an effective adhesive that can immobilize particles by hydrophobic and electrostatic interactions, as well as hydrogen bonding. The flexible nature of mucin fibers allows them to conform to the surface of an object and form multiple, low affinity bonds to trap it.[121] These bonds have short half-lives (thermal energy is sufficient to break individual bonds), but at any given time one or more bonds will keep the object sequestered and held in place by the elasticity of the gel.

Most antibodies made in the human body are found in mucosal secretions, where they protect the body from viruses and bacteria.[133,134] The F_c region of an antibody is capable of forming low affinity bonds with mucus and the F_{ab} portion can specifically recognize moieties on the surface of pathogens.[93,128] Thus, even viruses that are smaller than the average mucus mesh spacing and are not adherent to mucins can be specifically trapped with the help of antibody intermediates.[121]

5.1.4.2 Ways to Overcome Mucociliary Clearance

As discussed in the previous section, mucus efficiently traps a wide variety of particles by steric obstruction and adhesive interactions. In the lungs, objects trapped in the mucus gel are transported at rates up to 5–10 mm min^{-1} by beating cilia and are delivered to the GI tract for inactivation and digestion. The luminal gel layer of respiratory mucus is replaced as rapidly as every 10 to 20 minutes, resulting in extremely efficient clearance of inhaled particles. Trapping and rapid clearance is crucial to protect us from the onslaught of pathogens and environmental toxins we breathe in each day. However, the "mucociliary escalator" also serves as a major barrier to the delivery of therapeutic nanoparticles. Strategies to address this barrier and more efficiently deliver therapeutic nanoparticles to the lungs include: mucoadhesive particles, mucus-penetrating particles (MPP), and mucolytics.

Mucoadhesive particles
One strategy that has been tested to potentially improve the retention, pharmacokinetics, and efficacy of drugs delivered locally to the conducting airways has been to deliver therapeutics in mucoadhesive controlled release systems. Although this has been studied extensively in the gastrointestinal tract, the study of such systems in the lung has been relatively limited. However, there are several studies that have examined the effects of inhaled mucoadhesive nano- and microparticles on particle retention and drug PK.[135–141] Although some of these studies show improved performance of mucoadhesive particles, the mechanism by which mucoadhesion improves particle retention (and thereby PK) is not well understood. Evidence points to mucoadhesion resulting in changes in mucus rheology (increase in viscoelasticity), which has been shown to decrease mucociliary clearance in studies of natural and artificial mucus.[142–144]

A recent study by Henning *et al.*[145] investigating the mucociliary clearance (MCC) of various biodegradable polymer nanoparticles in a chicken trachea model of mucociliary clearance provided more insight into the mechanism by which mucoadhesive systems may improve the retention of particles in the lung. They found that particles composed of biodegradable polymers were cleared at different rates and clearance was related to their surface chemistry and mucoadhesion (Figure 5.1.7). PLGA-PEG particles were transported the fastest ($\sim$6 mm min^{-1}), approximately twice as fast as PLGA and PVA-g-PLGA particles ($\sim$3 mm min^{-1}). Incorporation of PEG into nanoparticles is known to reduce interactions with proteins and other biomolecules[146,147] (and mucus in particular[130,148,149]) and the clearance rate of PLGA-PEG particles was very similar to the normal MCC rate in the ICRP model (5.5 mm min^{-1}).[150] This suggests that PLGA-PEG did not adhere sufficiently to the tracheal mucus to alter the rheology and MCC. However, at this moment, it is not entirely clear whether the observed differences in mucociliary clearance of the different coated particles were indeed solely due to effects of mucoadhesion.

Although mucoadhesion is an interesting strategy to overcome the mucus barrier in the lungs, it could be dangerous in a clinical setting, especially for patients already prone to infection and impaired mucociliary clearance (COPD, asthma, CF patients, *etc.*), as the reduced clearance time could allow opportunistic pathogens more time to infect epithelial surfaces of the airways or increase mucus plugging of the airways.[151] Although mucociliary clearance may be somewhat decreased by this mechanism, long term retention of these particles is not likely, as the particles will presumably be bound to the outer, luminal mucus layer and thus will presumably be cleared from the lungs with

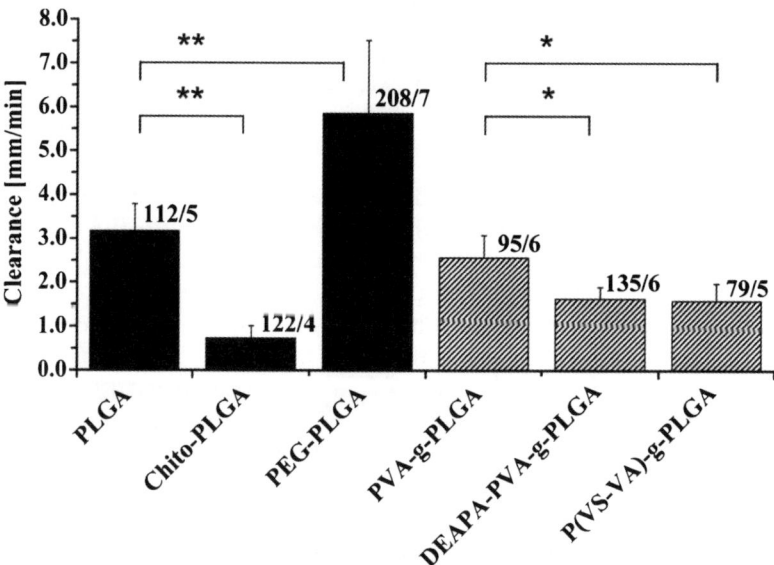

Figure 5.1.7   Mucociliary clearance of PLGA-based nanoparticles. Mucociliary transport rates (mean $\pm$ SD) were significantly different (*$p < 0.01$; **$p < 0.001$). Numbers beside the bars designate the numbers of particles/tracheas investigated, for example, "100/5" indicates 100 particles tracked on five tracheas. Reprinted with permission from ref. 145. The publisher for this copyrighted material is Mary Ann Liebert, Inc. Publishers.

this layer (minutes to hours). Further, gene delivery vehicles or other systems that require uptake into cells for effective treatment will be ineffective as they would not have access to airway epithelial cells.

Mucus-penetrating particles (MPP)

Although the luminal mucus layer (LML) is removed rapidly, the periciliary layer (PCL) is thought to be cleared much more slowly.[152–155] Ciliary stroke analysis predicts that the PCL is almost completely stationary.[156,157] Thus, particles that could diffuse through the LML into the PCL might be retained for prolonged periods of time and in the case of gene carriers, would allow more efficient uptake of particles into epithelial cells (Figure 5.1.8).

Cone and Saltzman investigated the diffusion of antibodies and viruses in human cervical mucus and found that monovalent antibodies and some viruses can move through mucus as rapidly as they move in water.[128,158] Upon studying the viruses capable of diffusing in mucus, they found that these viruses had highly charged surfaces with alternating positive and negative charges separated by only 5 Å.[93,128] This highly charged, yet net neutral surface prevents hydrophobic and electrostatic interactions between the viruses and mucus, allowing them to diffuse rapidly through the low viscosity pores (with viscosity of water).

Unfortunately, most conventional nanoparticles are adhesively trapped in mucus due to hydrophobic interactions (PLGA, PLA, and PCL) or electrostatic interactions (PEI, chitosan).[159–167] Further, it would be extremely difficult to engineer particles with evenly dispersed positive and negative charges (especially at spacing of 5 Å) that were not immunogenic. To engineer particles with similar surface properties to viruses, one approach was to use polymers that are net neutral, hydrophilic, and not likely to undergo extensive hydrogen bonding. Poly(ethylene glycol) (PEG) fit this criteria, but,

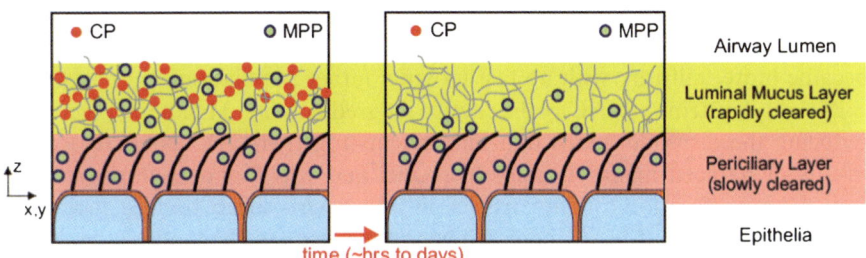

Figure 5.1.8 Fate of mucus-penetrating particles (MPP) and conventional mucoadhesive particles (CP) in the lung airways. MPP readily penetrate the luminal mucus layer (LML) and enter the underlying periciliary layer (PCL). In contrast, CP are immobilized in the LML. As the LML layer is cleared, CP are removed along with the LML whereas MPP in the PCL are retained, leading to prolonged residence time for MPP. Thus, there is very low drug dosing with CP, whereas MPP, because they are retained longer, will continue to release drug.

paradoxically, had previously been believed to be mucoadhesive.[163–164,168–176] However, extensive studies testing various coating densities and molecular weights of PEG on particle surfaces had not been performed.

Hanes and coworkers discovered that densely coating particles with low molecular weight PEG (2 kDa to 5 kDa) allowed the particles to transport rapidly in undiluted human cervicovaginal mucus (CVM).[130,148,177] Previous studies with PEG-coated particles had concluded that PEG was mucoadhesive but they used either high MW PEG (10 kDa or greater) or did not obtain a sufficiently dense coating with low MW PEG.[148] Particles coated with high MW PEG were likely stuck due to PEG entanglement in the mucus gel (due to the longer strands) and/or increased hydrogen bonding. Poorly coated particles likely left some of the hydrophobic or electrostatic surface of the particle exposed, allowing binding to mucin fibers.[148]

Mucolytics to Disrupt the Mucus Barrier
A variety of mucolytics have been used for the treatment of lung diseases, such as CF and COPD. Mucolytics act by chemically or physically degrading components of the mucus mesh, thereby reducing mucus bulk viscoelasticity and improving mucociliary clearance.[178]

Pulmozyme, a recombinant human DNAse, is commonly used in the treatment of cystic fibrosis. In CF patients, dead immune cells accumulate due to constant infection and inflammation, as well as impaired mucociliary clearance. The increased solids from DNA (and other cell debris) leads to futher entanglement of the mucus, increased viscoelasticity, and reduced mucociliary clearance. Pulmozyme works by degrading DNA into smaller pieces, reducing bulk viscoelasticity, and improving mucociliary clearance.[179,180] Sanders *et al.* saw only moderate improvement in the transport of polystyrene particles in CF sputum following treatment with pulmozyme.[101] Dawson *et al.* observed no difference in the diffusion rates of polystyrene particles in CF sputum before or after DNAse treatment, although the macroviscosity of the sputum was reduced.[181] Further, they noticed that the transport of particles became more uniform, with no fast moving outlier population.

N-Acetylcysteine (NAC), the active ingredient in Mucinex®, acts by reducing the disulfide bonds that link mucins together.[182,183] NAC treatment results in reduced macroviscosity of mucus and was found to improve gene transfection of cationic nanoparticles in an *ex vivo* sheep trachea model.[165] They also found that NAC increased gene expression in the mouse lung, but it did not improve nasal potential difference in CF null mice. Suk *et al.* showed, with muco-inert nanoparticle probes, that NAC treatment increased the average mesh spacing of CF sputum from 145 ± 50 to 230 ± 50 nm,[184] and showed that NAC improved gene transfection in an LPS-induced mouse model of mucus hyperplasia.[185]

A variety of other mucolytics (Nacystelyn, gelsolin, and thymosinB4) are in clinical use or in clinical trials that have not been evaluated as adjuvants to improve transport and efficacy of nanoparticles in the lungs.[186–189] However, as the different results from pulmozyme and NAC show, each of these mucolytics should be tested on a case by case basis to ascertain their usefulness in combination with nanoparticle therapeutics.

5.1.4.2 Pulmonary Surfactant

The first biological, essentially non-cellular barrier an inhaled nanoparticle or nanoparticle-containing formulation will encounter in the peripheral lung is the alveolar lining fluid,[190] sometimes also referred to as alveolar lining layer.[191] This ultra-thin liquid layer (thickness between 90–890 nm; area-weighted average about 200 nm)[38] covers the epithelial tissue in the alveolar region of the lungs.

Besides alveolar macrophages and loose dendritic cells and lymphocytes,[192] an integral part of the alveolar surface layer is pulmonary surfactant (PS). This surfactant layer is described to be continuous from the alveolar to the conducting airways.[38,193] Analysis of BAL revealed PS as a complex mixture composed by weight of about 90% lipids and 10% proteins.[194] Within the lipid fraction, the largest part comprises phospholipids with up to 70% dipalmi-toylphosphatidylcholine (DPPC)[194] Among the protein moiety, four surfactant proteins (SP) are known to date: SP-A, -B, -C and SP-D. The small SP-B and SP-C (17.4 kDa and 4.2 kDa, respectively) are extremely hydrophobic proteins and highly associate with lipids.[195] Together with the lipid fraction of PS, they form an entity with a biophysical function of utmost importance as they enable lipids secreted by alveolar type II cells to be promoted to the air–liquid interface of the alveolus. Here, they can spread to form a monolayer with their hydrophobic tail towards the air phase. By doing so, the surface tension in the lungs is reduced (below 2 mJ m^{-2}),[196] and alveoli are thereby prevented from collapsing.[194] Overall, the formation of highly complex lipid membranes from which lipids are exchanged and spread to the air-interface to allow normal breathing is crucially dependent on SP-B and SP-C.

In contrast, SP-A and SP-D are large proteins (630 kDa and 520 kDa, respectively) of a rather hydrophilic nature. They feature both collagen-rich and C-type lectin (calcium dependent) domains, whereupon they belong to the family of "collectins".[197] Besides some functions of these proteins that contribute to the biophysical functionality of PS, the major role of SP-A and SP-D assigns them to the host immune defense system. These two biomolecules can be considered as highly active opsonins, that bind to a large variety of biological patterns presented on bacteria, viruses or fungi, and can modulate numerous immuno-cellular effects to conduct the reaction of the immune system on pathogens towards a sterile deep lung. For excellent reviews on the immune-regulatory functions of SP-A and SP-D in the innate immune system, the reader is referred to works by Wright and others.[198–200]

Schürch *et al.* showed that upon first contact with PS, an inhaled particle is immediately displaced into the alveolar lining fluid, due to wetting with phospholipids and resulting high surface pressure.[46] Furthermore, particle displacement into the alveolar lining fluid was shown to be independent of the particle surface roughness and also the anatomical site of the lung's deposition in hamster lungs.[191] The effect of particle size on particle displacement is not well known, but it is likely not a major factor in this phenomena.

Concerning the interaction of inhaled particles – and especially nanoparticles – with PS, there are two compelling points of view. On the one hand, nanoparticles can have implications on the biophysical functionality of PS, which has been the subject of numerous studies in the recent past.[201–203] Overall, these studies showed that nanoparticulate matter can interfere with the PS function in the lung. Impediment of the surface tension reducing function or even disruption of the surfactant film are a potential risk in nanosafety of nanoparticle based delivery systems, making it a crucial parameter to be elucidated for each particle formulation that is intended for pulmonary drug delivery.[204]

On the other hand, components of PS – especially surfactant proteins and phospholipids – can adsorb to nanoparticles once submerged in the so-called aqueous subphase of PS. Adsorption of such biomolecules can lead to a "pulmonary surfactant" corona, that may influence the further biological fate of the nanoparticles. Gasser *et al.* demonstrated binding of phospholipids to carbon nanotubes, and that such a phospholipid coating also influenced the binding pattern of plasma proteins, when subsequently incubated in blood plasma.[205] These findings indicate that nanomaterials, which enter the body *via* the pulmonary route, may be altered in a way that "secondary protein adsorption" and thereby cellular effects or biodistribution *via* the blood stream may be altered through such a pulmonary pre-coating. Furthermore, concerning cellular interaction with AM, phospholipids have been shown to reduce the phagocytosis of microparticles by AM, when adsorbed to the particle surface.[206,207]

Of great interest are the interactions of nanostructures in the peripheral lungs with the pulmonary collectins SP-A and SP-D. Interestingly, there is only sporadic information available on such interactions, and the field of bio–nano interactions in the peripheral lung including these two proteins is rather in its infancy. So far, material dependent binding of SP-A and SP-D to nanostructured systems such as carbon nanotubes or metal oxide nanoparticles could be demonstrated in some studies.[208,209]

For instance, Schulze *et al.*[208] studied the adsorption of SP-A to different metal oxide nanomaterials after incubation in SP-A-containing bronchoalveolar lavage fluid. Among the tested particles, obvious differences were observed (Figure 5.1.9). Interestingly, when SP-A adsorption was selectively studied *via* immunostaining, some of the tested materials did bind SP-A to a high degree, whereas total protein adsorption was comparably lower than for other nanoparticles studied (*e.g.* TiO_2 A *versus* TiO_2 B, or CeO_2 A *versus* CeO_2 B). Overall, these results indicated that the adsorption of SP-A (the most abundant protein in PS) can be very different from the adsorption of other proteins. This points to the need for detailed protein binding studies, including

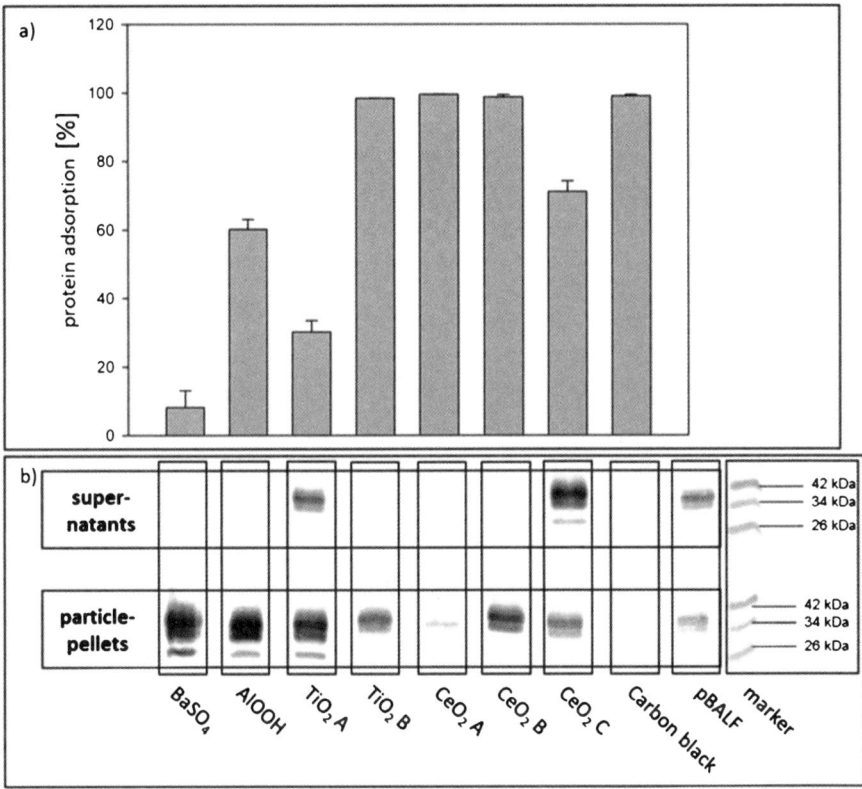

Figure 5.1.9 Adsorption of proteins from broncho-alveolar lavage fluid at a nanoparticle/protein ratio of 10 : 1. (a) Total protein adsorption: the particles show striking differences, even when derived from the same bulk material. (b) Immunoblot of SP-A from supernatant and pelleted nanoparticles after conditioning in porcine bronchoalveolar lavage fluid (SP-A monomer: 36 kDa). Reprinted with permission from ref. 208.

SP-A (and also SP-D), when new nanomaterials relevant to inhalation are evaluated in terms of characterization in biologically relevant fluids.

Consequently, adsorbed surfactant proteins can mediate the subsequent cellular response to such nano-structures. It has been shown that opsonization with pulmonary surfactant proteins can lead to increased particle uptake and clearance by phagocytes.[210–213] Furthermore, surfactant proteins can modulate the binding of particles to epithelial cells.[214] Concerning their role in receptor mediated recycling of PS components by AT II cells,[194,215] particle adsorbed surfactant proteins may even influence the translocation of particles across epithelial barriers. Regarding their great abundance, affinity to various structures,[216] and numerous functions the pulmonary collectins fulfill in the lung host defense system,[198] it is more than likely that these proteins are significantly involved in clearance or even translocation processes. Therefore, PS components, especially surfactant proteins, are essential to be included in *in vitro* experiments for evaluation of nanoparticle-based pulmonary drug delivery systems.[217]

Besides proteins and other PS components, there also other substances in BAL that can act as opsonins to enhance phagocytosis such as fibronectin, C-reactive protein, or IgG as the most prevalent immunoglobulin in the deep lung.[69]

Nevertheless, their position at the air-interface suggests pulmonary surfactant proteins play the leading role in bio-nano interactions in the peripheral lung.

To mimic the PS system *in vitro*, different approaches are possible with varying complexity in terms of compositions. Semisynthetic models of PS can be obtained using single lipid components such as dipalmitoyl-glycero-phosphocholine (DPPC) or dipalmitoyl-glycero-phosphoglycerol (DPPG) mixed in appropriate ratios with isolated surfactant proteins (mainly SP-B and SP-C). Such systems are mainly used in biophysical studies to investigate the influence of nanoparticles on the surface tension of PS.[202,218] Single surfactant proteins can be isolated from BAL performed with animal lungs (mostly rat and pig), or from BAL performed during bronchoscopy of alveolar proteinosis patients to isolate human proteins. Recombinant production of SP-A and SP-D is also possible, but restricted to monomeric products. More complex systems can be isolated from animal lungs after BAL. Obtained exogenous PS is clinically used to treat respiratory distress syndrome (RDS). Some products such as Alveofact® (bovine), Survanta® or Curosurf® (both porcine) are commercially available, and therefore, often used in studies addressing interactions of nanoparticles with PS.[201,204] The disadvantage of these systems in terms of usage as an *in vitro* model is that they are obtained by lipophilic extractions of BAL, and thus contain the lipids and the hydrophobic surfactant proteins, but lack the immunologically relevant hydrophilic SP-A.[219] Therefore, the most complex system is native surfactant, which also can be obtained from BAL from animal lungs with subsequent gradient centrifugation. It contains both the lipid fraction, but also SP-A, and is therefore the most suitable, but also most complex *in vitro* model for PS.[219]

5.1.5 Conclusion

Inhalation of therapeutic compounds, either for local treatment of lung-relevant diseases or as route of administration for increased systemic bioavailability, has been under investigation for several decades now. Here, research has mainly been focused on efficient deposition into the lungs and optimization of inhalation technologies.

However, with an increasing interest in nano-sized delivery systems for pulmonary application, there is also an increasing need for a thorough consideration of biological effects such systems by themselves exhibit on the sensitive physiological functions of the lungs. Therefore, there is an indispensable need to understand the various and complex mechanisms that occur at the bio–nano interface whenever a (nano-) particulate system has landed in the lungs, as well as the cellular responses resulting from such interactions.

In this chapter, we tried to address the main physiological considerations and resulting mechanistic issues that might evolve, or have already been shown to derive from such interactions. In this context, besides giving a brief overview on most important issues in particle lung deposition, we emphasized the role the diverse cellular and non-cellular barriers can play in particle lung interactions. Here, we pointed out that one has to consider those non-cellular components of the pulmonary air–blood-barrier, as these represent the most outer, and therefore first and foremost, biological barriers for inhaled (nano-) particles.

However, we are aware that in a short chapter like this, we had to be selective, and not all aspects could be addressed condignly.

Nevertheless, we are convinced that the presented chapter helps pharmaceutical scientists starting or already working in the field of pulmonary drug delivery to get an efficient and up-to-date overview on the important aspects in this discipline.

References

1. E. R. Weibel, *Physiol. Rev.*, 1973, **53** (2), 419–95.
2. H. Landahl, *Bull. Math. Biophys.*, 1950, **12**, 43–56.
3. T. C. Carvalho, J. I. Peters, and R. O. Williams, *Int. J. Pharm.*, 2011, **406** (1–2), 1–10.
4. T. M. Crowder, J. A. Rosati, J. D. Schroeter, A. J. Hickey, and T. B. Martonen, *Pharm. Res.*, 2002, **19** (3), 239–45.
5. H. Schulz, *Pharm. Sci. Technol. Today*, 1998, **1**, 336–344.
6. M. Lippmann, and R. E. Albert, *Am. Ind. Hyg. Assoc. J.*, 1969, **30** (3), 257–75.
7. J. Heyder, *Proc. Am. Thorac. Soc.*, 2004, **1** (4), 315–20.
8. C. Darquenne, P. Brand, J. Heyder, and M. Paiva, *J. Appl. Physiol.*, 1997, **83** (3), 966–74.
9. M. B. Snipes, Biokinetics of inhaled radionuclides, in *Internal Radiation Dosimetry*, ed. O. G. Raabe, Medical Physics, Madison, WI, 1994, pp 181–204.
10. W. C. Hinds, *Aerosol Technology: Properties, Behavior, and Measurement of Airborn Particles, 2nd edn*, Wiley, New York, 1999.
11. W. Stoeber, Dynamic shape factors of nonspherical aerosol particles in *Assessment of Airborne Particles*, ed. T. T. Mercer, P. E. Morrow, and W. Stoeber, Charles C Thomas, Springfield, 1972, pp 249–289.
12. V. Timbrell, *Ann. Occup. Hyg.*, 1982, **26**, 347–369.
13. B. Asgharian, and C. P. Yu, *J. Aerosol Med.*, 1988, **1**, 37–50.

14. V. Timbrell, *Ann. N. Y. Acad. Sci.*, 1965, **132** (1), 255–73.
15. I. Gonda, Targeting by deposition, in *Pharmaceutical Inhalation Aerosol Technology*, ed. A. J. Hickey, Marcel Dekker, New York, 1992, pp 61–82.
16. J. Heyder, J. Gebhart, W. Stahlhofen, and B. Stuck, *Ann. Occup. Hyg.*, 1982, **26** (1–4), 137–47.
17. C. S. Kim, and M. A. Eldridge, *J. Appl. Physiol.*, 1985, **59** (6), 1766–72.
18. G. C. Smaldone, and M. S. Messina, *J. Appl. Physiol.*, 1985, **59** (2), 509–14.
19. R. E. Kavanaugh, J. D. Unadkat, and A. L. Smith, Drug disposition in cystic fibrosis, in *Cystic Fibrosis*, ed. P. B. Davis, McGraw-Hill, New York, 1993, pp 91–136.
20. P. J. Anderson, J. D. Blanchard, J. D. Brain, H. A. Feldman, J. J. McNamara, and J. Heyder, *Am. Rev. Respir. Dis.*, 1989, **140** (5), 1317–24.
21. T. Martonen, I. Katz, and W. Cress, *Pharm. Res.*, 1995, **12** (1), 96–102.
22. P. Brand, T. Meyer, K. Sommerer, N. Weber, and G. Scheuch, *Exp. Lung Res.*, 2002, **28** (1), 39–54.
23. F. Gagnadoux, P. Diot, S. Marchand, R. Thompson, K. Dieckman, E. Lemarie, F. Varaigne, C. Maurage, J. L. Baulieu, and J. C. Rolland, *Rev. Mal. Respir.*, 1996, **13** (1), 55–60.
24. C. C. Kuni, J. R. Budd, W. E. Regelmann, R. P. Ducret, and R. J. Boudreau, *Clin. Nucl. Med.*, 1993, **18** (1), 15–8.
25. B. L. Laube, D. Y. Chang, A. N. Blask, and B. J. Rosenstein, *Chest*, 1992, **101** (5), 1302–8.
26. L. M. Marshall, P. W. Francis, and F. A. Khafagi, *J. Paediatr. Child Health*, 1994, **30** (1), 65–7.
27. D. A. Edwards, J. Hanes, G. Caponetti, J. Hrkach, A. Ben-Jebria, M. L. Eskew, J. Mintzes, D. Deaver, N. Lotan, and R. Langer, *Science*, 1997, **276** (5320), 1868–71.
28. D. A. Edwards, A. Ben-Jebria, and R. Langer, *J. Appl. Physiol.*, 1998, **85** (2), 379–85.
29. Y. Tabata, and Y. Ikada, *Biomaterials*, 1988, **9** (4), 356–62.
30. A. Ben-Jebria, D. Chen, M. L. Eskew, R. Vanbever, R. Langer, and D. A. Edwards, *Pharm. Res.*, 1999, **16** (4), 555–61.
31. J. Fu, J. Fiegel, E. Krauland, and J. Hanes, *Biomaterials*, 2002, **23** (22), 4425–4433.
32. J. Fu, J. Fiegel, and J. Hanes, *Macromolecules*, 2004, **37** (19), 7174–7180.
33. J. Fiegel, J. Fu, and J. Hanes, *J. Controlled Release*, 2004, **96** (3), 411–23.
34. C. Dunbar, G. Scheuch, K. Sommerer, M. DeLong, A. Verma, and R. Batycky, *Int. J. Pharm.*, 2002, **245** (1–2), 179–89.
35. L. Garcia-Contreras, J. Fiegel, M. J. Telko, K. Elbert, A. Hawi, M. Thomas, J. VerBerkmoes, W. A. Germishuizen, P. B. Fourie, A. J. Hickey, and D. Edwards, *Antimicrob. Agents Chemother.*, 2007, **51** (8), 2830–6.
36. J. S. Patton, J. D. Brain, L. A. Davies, J. Fiegel, M. Gumbleton, K.-J. Kim, M. Sakagami, R. Vanbever, and C. Ehrhardt, *J. Aerosol Med. Pulmonary Drug Delivery*, 2010, **23** (S2), S-71–S-87.

37. M. Bur, A. Henning, S. Hein, M. Schneider, and C.-M. Lehr, *Inhalation Toxicol.*, 2009, **21** (s1), 137–143.

38. J. Bastacky, C. Y. Lee, J. Goerke, H. Koushafar, D. Yager, L. Kenaga, T. P. Speed, Y. Chen, and J. A. Clements, *J. Appl. Physiol.*, 1995, **79** (5), 1615–28.

39. R. Sturm, and W. Hofmann, *J. Hazard Mater.*, 2009, **170** (1), 210–8.

40. J. Fiegel, C. Ehrhardt, U. F. Schaefer, C. M. Lehr, and J. Hanes, *Pharm. Res.*, 2003, **20** (5), 788–96.

41. J. A. Champion, and S. Mitragotri, *Proc. Natl. Acad. Sci. U. S. A.*, 2006, **103** (13), 4930–4.

42. J. S. Patton, *Adv. Drug Delivery Rev.*, 1996, **19** (1), 3–36.

43. J. Jinno, N. Kamada, M. Miyake, K. Yamada, T. Mukai, M. Odomi, H. Toguchi, G. G. Liversidge, K. Higaki, and T. Kimura, *J. Controlled Release*, 2006, **111** (1–2), 56–64.

44. J. Z. Yang, A. L. Young, P. C. Chiang, A. Thurston, and D. K. Pretzer, *J. Pharm. Sci.*, 2008, **97** (11), 4869–78.

45. W. Kreyling, *J. Aerosol Med.*, 1990, **3**, S-93-S-110.

46. S. Schürch, P. Gehr, V. Im Hof, M. Geiser, and F. Green, *Respir. Physiol.*, 1990, **80** (1), 17–32.

47. G. Oberdörster, Z. Sharp, V. Atudorei, A. Elder, R. Gelein, W. Kreyling, and C. Cox, *Inhalation Toxicol.*, 2004, **16** (6–7), 437–445.

48. P. Gehr, F. H. Green, M. Geiser, V. Im Hof, M. M. Lee, and S. Schurch, *J. Aerosol Med.*, 1996, **9** (2), 163–81.

49. S. D. Conner, and S. L. Schmid, *Nature*, 2003, **422** (6927), 37–44.

50. M. Geiser, B. Rothen-Rutishauser, N. Kapp, S. Schurch, W. Kreyling, H. Schulz, M. Semmler, V. Im Hof, J. Heyder, and P. Gehr, *Environ. Health Perspect.*, 2005, **113** (11), 1555–60.

51. G. Scheuch, M. J. Kohlhaeufl, P. Brand, and R. Siekmeier, *Adv. Drug Delivery Rev.*, 2006, **58** (9–10), 996–1008.

52. A. Steimer, E. Haltner, and C. M. Lehr, *J. Aerosol Med.*, 2005, **18** (2), 137–82.

53. C. M. Evans, and J. S. Koo, *Pharmacol. Ther.*, 2009, **121** (3), 332–48.

54. S. K. Lai, Y. Y. Wang, D. Wirtz, and J. Hanes, *Adv. Drug Delivery Rev.*, 2009, **61** (2), 86–100.

55. J. Iravani, and G. N. Melville, *Pharmacol. Ther. B*, 1976, **2** (3), 471–92.

56. M. B. Antunes, and N. A. Cohen, *Curr. Opin. Allergy Clin. Immunol.*, 2007, **7** (1), 5–10.

57. A. Braiman, and Z. Priel, *Respir. Physiol. Neurobiol.*, 2008, **163** (1–3), 202–7.

58. M. M. Bailey, and C. J. Berkland, *Med. Res. Rev.*, 2009, **29** (1), 196–212.

59. N. R. Yacobi, H. C. Phuleria, L. Demaio, C. H. Liang, C. A. Peng, C. Sioutas, Z. Borok, K. J. Kim, and E. D. Crandall, *Toxicol. In Vitro*, 2007, **21** (8), 1373–81.

60. M. Geiser, and W. G. Kreyling, *Particle Fibre Toxicol.*, 2010, **7**, 2.

61. O. Schmid, W. Moller, M. Semmler-Behnke, G. A. Ferron, E. Karg, J. Lipka, H. Schulz, W. G. Kreyling, and T. Stoeger, *Biomarkers*, 2009, **14** Suppl 1, 67–73.

62. B. Rothen-Rutishauser, F. Blank, C. Muhlfeld, and P. Gehr, *Expert Opin. Drug Metab. Toxicol.*, 2008, **4** (8), 1075–89.

63. M. Geiser, *J. Aerosol Med. Pulmonary Drug Delivery*, 2010, **23** (4), 207–17.

64. A. D. Lehmann, N. Daum, M. Bur, C. M. Lehr, P. Gehr, and B. M. Rothen-Rutishauser, *Eur. J. Pharm. Biopharm.*, **77** (3), 398–406.

65. B. M. Rothen-Rutishauser, S. G. Kiama and, P. Gehr, *Am. J. Respir. Cell Mol. Biol.*, 2005, **32** (4), 281–9.

66. J. R. Spurzem, C. Saltini, W. Rom, R. J. Winchester, and R. G. Crystal, *Am. Rev. Respir. Dis.*, 1987, **136** (2), 276–80.

67. K. C. Stone, R. R. Mercer, P. Gehr, B. Stockstill, and J. D. Crapo, *Am. J. Respir. Cell Mol. Biol.*, 1992, **6** (2), 235–43.

68. M. Dörger, and F. Krombach, *J. Aerosol Med.*, 2001, **13** (4), 369–80.

69. L. P. Nicod, *Eur. Respir. Rev.*, 2005, **14** (95), 45–50.

70. J. A. Champion, Y. K. Katare, and S. Mitragotri, *J. Controlled Release*, 2007, **121** (1–2), 3–9.

71. R. May, and L. Machesky, *J. Cell Sci.*, 2001, **114** (6), 1061–1077.

72. S. G. Kiama, L. Cochand, L. Karlsson, L. P. Nicod, P. Gehr, *J. Aerosol Med.*, 2001, **14** (3), 289–99.

73. D. Edwards, A. Ben-Jebria, and R. Langer, *J. Appl. Physiol.*, 1998, **85** (2), 379–385.

74. M. Geiser, L. M. Cruz-Orive, V. Im Hof, and P. Gehr, *J. Microsc.*, 1990, **160** (Pt 1), 75–88.

75. M. Geiser, *Microsc. Res. Technol.*, 2002, **57** (6), 512–22.

76. J. A. Champion, and S. Mitragotri, *Pharm. Res.*, 2009, **26** (1), 244–9.

77. R. Pozzi, B. De Berardis, L. Paoletti, and C. Guastadisegni, *Environ. Res.*, 2005, **99** (3), 344–54.

78 I. Beck-Speier, N. Dayal, E. Karg, K. L. Maier, G. Schumann, H. Schulz, M. Semmler, S. Takenaka, K. Stettmaier, W. Bors, A. Ghio, J. M. Samet, and J. Heyder, *Free Radic. Biol. Med.*, 2005, **38** (8), 1080–92.

79. B. Galve-de Rochemonteix, L. P. Nicod, and J. M. Dayer, *Am. J. Respir. Cell Mol. Biol.*, 1996, **14** (3), 279–87.

80. L. P. Nicod, F. el Habre, J. M. Dayer, and N. Boehringer, *Am. J. Respir. Cell Mol. Biol.*, 1995, **13** (1), 83–90.

81. F. Blank, B. Rothen-Rutishauser, and P. Gehr, *Am. J. Respir. Cell Mol. Biol.*, 2007, **36** (6), 669–77.

82. B. E. Lehnert, Y. E. Valdez, and G. L. Tietjen, *Am. J. Respir. Cell Mol. Biol.*, 1989, **1** (2), 145–54.

83. J. Q. Davies, and S. Gordon, *Methods Mol. Biol.*, 2005, **290**, 105–116.

84. I. N. Mbawuike, and H. B. Herscowitz, *J. Leukoc. Biol.*, 1989, **46** (2), 119–27.

85. R. J. Helmke, V. F. German, and J. A. Mangos, *In Vitro Cell. Dev. Biol.: Anim.*, 1989, **25** (1), 44–48.

86. H. M. Weingartl, M. Sabara, J. Pasick, E. Van Moorlehem, and L. Babiuk, *J. Virol. Methods*, 2002, **104** (2), 203–216.
87. D. E. Sims, and M. M. Horne, *Am. J. Physiol.*, 1997, **273** (5 Pt 1), L1036–41.
88. P. Gehr, M. Bachofen, and E. R. Weibel, *Resp. Physiol.*, 1978, **32** (2), 121–40.
89. A. E. Nel, L. Mädler, D. Velegol, T. Xia, E. M. V. Hoek, P. Somasundaran, F. Klaessig, V. Castranova, and M. Thompson, *Nature*, 2009, **8** (7), 543–557.
90. T. Cedervall, I. Lynch, S. Lindman, T. Berggard, E. Thulin, H. Nilsson, K. A. Dawson, and S. Linse, *Proc. Natl. Acad. Sci. U. S. A.*, 2007, **104** (7), 2050–5.
91. I. Lynch, and K. Dawson, *Nano Today*, 2008, **3** (1–2), 40–47.
92. M. R. Knowles, and R. C. Boucher, *J. Clin. Invest.*, 2002, **109** (5), 571–7.
93. R. Cone, Mucus in *Mucosal Immunlogy*, ed., Michael E. Lamm, Jerry R. McGhee, Lloyd Mayer, Jiri Mestecky and John Bienenstock, 3rd edn, Academic Press: San Diego, 1999, pp. 43–64.
94. B. K. Rubin, *Respir. Care*, 2002, **47** (7), 761–8.
95. H. Rahmoune, and K. L. Shephard, *J. Appl. Physiol.*, 1995, **78** (6), 2020–4.
96. M. S. Quraishi, N. S. Jones, and J. Mason, *Clinical Otolaryngology Allied Sci.*, 1998, **23** (5), 403–13.
97. M. A. Sleigh, J. R. Blake, and N. Liron, *Am. Rev. Respir. Dis.*, 1988, **137** (3), 726–41.
98. J. H. Widdicombe, S. J. Bastacky, D. X. Wu, and C. Y. Lee, *Eur. Respir. J.*, 1997, **10** (12), 2892–7.
99. D. J. Smith, E. A. Gaffney, and J. R. Blake, *Respir. Physiol. Neurobiol.*, 2008, **163** (1–3), 178–88.
100. R. C. Boucher, *Annu. Rev. Med.*, 2007, **58**, 157–70.
101. N. N. Sanders, S. C. De Smedt, E. Van Rompaey, P. Simoens, F. De Baets, and J. Demeester, *Am. J. Respir. Crit. Care Med.*, 2000, **162** (5), 1905–11.
102. F. Konrad, T. Schreiber, D. Brecht-Kraus, and M. Georgieff, *Chest*, 1994, **105** (1), 237–41.
103. W. M. Foster, E. G. Langenback, and E. H. Bergofsky, *Ann. Occup. Hyg.*, 1982, **26** (1–4), 227–44.
104. I. Carlstedt, and J. K. Sheehan, *Symp. Soc. Exp. Biol.*, 1989, **43**, 289–316.
105. D. J. Thornton, and J. K. Sheehan, *Proc. Am. Thorac. Soc.*, 2004, **1** (1), 54–61.
106. C. Wickstrom, J. R. Davies, G. V. Eriksen, E. C. Veerman, and I. Carlstedt, *Biochem. J.*, 1998, **334** (3), 685–693.
107. A. Allen, G. Flemstrom, A. Garner, and E. Kivilaakso, *Physiol. Rev.*, 1993, **73** (4), 823–857.
108. I. Carlstedt, H. Lindgren, J. K. Sheehan, U. Ulmsten, and L. Wingerup, *Biochem. J.*, 1983, **211** (1), 13–22.
109. C.-C. W. Chao, S. M. Butala, and A. Herp, *Exp. Eye Res.*, 1988, **47** (2), 185–196.

110. J. M. Samet, and P. W. Cheng, *Environ. Health Perspect.*, 1994, **102** (Suppl 2), 89–103.
111. J. K. Sheehan, K. Oates, and I. Carlstedt, *Biochem. J.*, 1986, **239** (1), 147–53.
112. J. Dekker, J. W. Rossen, H. A. Buller, and A. W. Einerhand, *Trends Biochem. Sci.*, 2002, **27** (3), 126–31.
113. M. E. Behrens, P. M. Grandgenett, J. M. Bailey, P. K. Singh, C. H. Yi, F. Yu, and M. A. Hollingsworth, *Oncogene*, 2010, **29** (42), 5667–77.
114. B. G. Bitler, A. Goverdhan, and J. A. Schroeder, *J. Cell Sci.*, 2010, **123** (Pt 10), 1716–23.
115. S. Bafna, S. Kaur, and S. K. Batra, *Oncogene*, 2010, **29** (20), 2893–904.
116. S. Jepson, M. Komatsu, B. Haq, M. E. Arango, D. Huang, C. A. Carraway, and K. L. Carraway, *Oncogene*, 2002, **21** (49), 7524–32.
117. K. L. Carraway, 3rd, E. A. Rossi, M. Komatsu, S. A. Price-Schiavi, D. Huang, P. M. Guy, M. E. Carvajal, N. Fregien, C. A. Carraway, and K. L. Carraway, *J. Biol. Chem.*, 1999, **274** (9), 5263–6.
118. T. A. Springer, *Nature*, 1990, **346** (6283), 425–34.
119. J. K. Sheehan, P. S. Richardson, D. C. Fung, M. Howard, and D. J. Thornton, *Am. J. Respir. Cell Mol. Biol.*, 1995, **13** (6), 748–56.
120. I. Carlstedt, and J. K. Sheehan, *Monogr. Allergy*, 1988, **24**, 16–24.
121. R. A. Cone, *Adv. Drug Delivery Rev.*, 2009, **61** (2), 75–85.
122. S. Girod, J. M. Zahm, C. Plotkowski, G. Beck, and E. Puchelle, *Eur. Respir. J.*, 1992, **5** (4), 477–87.
123. S. H. Randell, and R. C. Boucher, *Am. J. Respir. Cell Mol. Biol.*, 2006, **35** (1), 20–8.
124. B. L. Slomiany, and A. Slomiany, *J. Physiol. Pharmacol.*, 1991, **42** (2), 147–61.
125. M. J. Dulfano, K. Adler, and W. Philippoff, *Am. Rev. Respir. Dis.*, 1971, **104** (1), 88–98.
126. C. Galabert, J. Jacquot, J. M. Zahm, and E. Puchelle, *Clin. Chim. Acta*, 1987, **164** (2), 139–49.
127. S. K. Lai, Y. Y. Wang, K. Hida, R. Cone, J. Hanes, *Proc. Natl. Acad. Sci. U. S. A.*, 2010, **107** (2), 598–603.
128. S. S. Olmsted, J. L. Padgett, A. I. Yudin, K. J. Whaley, T. R. Moench, and R. A. Cone, *Biophys. J.*, 2001, **81** (4), 1930–7.
129. A. I. Yudin, F. W. Hanson, and D. F. Katz, *Biol. Reprod.*, 1989, **40** (3), 661–71.
130. S. K. Lai, D. E. O'Hanlon, S. Harrold, S. T. Man, Y. Y. Wang, R. Cone, and J. Hanes, *Proc. Natl. Acad. Sci. U. S. A.*, 2007, **104** (5), 1482–7.
131. J. S. Suk, S. K. Lai, Y. Y. Wang, L. M. Ensign, P. L. Zeitlin, M. P. Boyle, and J. Hanes, *Biomaterials*, 2009, **30** (13), 2591–7.
132. S. K. Lai, J. S. Suk, A. Pace, Y. Y. Wang, M. Yang, O. Mert, J. Chen, J. Kim, and J. Hanes, *Biomaterials*, 2011, **32** (26), 6285–90.
133. J. Mestecky, M. W. Russell, S. Jackson, and T. A. Brown, *Clin. Immunol. Immunopathol.*, 1986, **40** (1), 105–14.
134. M. E. Conley, and D. L. Delacroix, *Ann. Intern. Med.*, 1987, **106** (6), 892–9.

135. H. Yamamoto, Y. Kuno, S. Sugimoto, H. Takeuchi, and Y. Kawashima, *J Controlled Release*, 2005, **102** (2), 373–81.
136. K. Surendrakumar, G. P. Martyn, E. C. Hodgers, M. Jansen, and J. A. Blair, *J. Controlled Release*, 2003, **91** (3), 385–94.
137. M. Sakagami, W. Kinoshita, K. Sakon, J. Sato, and Y. Makino, *J. Controlled Release*, 2002, **80** (1–3), 207–18.
138. M. Sakagami, K. Sakon, W. Kinoshita, and Y. Makino, *J. Controlled Release*, 2001, **77** (1–2), 117–29.
139. X. B. Liu, J. X. Ye, L. H. Quan, C. Y. Liu, X. L. Deng, M. Yang, and Y. H. Liao, *Eur. J. Pharm. Biopharm.*, 2008, **70** (3), 845–52.
140. S. Bai, V. Gupta, and F. Ahsan, *J. Aerosol Med. Pulmonary Drug Delivery*, **23** (2), 97–104.
141. A. Makhlof, M. Werle, Y. Tozuka, and H. Takeuchi, *Int. J. Pharm.*, **397** (1–2), 92–5.
142. A. J. Shah, and M. D. Donovan, *AAPS PharmSciTech*, 2007, **8** (2), Article 33.
143. A. J. Shah, and M. D. Donovan, *AAPS PharmSciTech*, 2007, **8** (2), Article 32.
144. V. Gerber, P. Gehr, R. Straub, M. Frenz, M. King, and V. Im Hof, *Respir. Physiol.*, 1997, **107** (1), 67–74.
145. A. Henning, M. Schneider, N. Nafee, L. Muijs, E. Rytting, X. Wang, T. Kissel, D. Grafahrend, D. Klee, and C. M. Lehr, *J. Aerosol Med. Pulmonary Drug Delivery*, 2010, **23** (4), 233–41.
146. R. Gref, Y. Minamitake, M. T. Peracchia, V. Trubetskoy, V. Torchilin, and R. Langer, *Science*, 1994, **263** (5153), 1600–3.
147. M. Vittaz, D. Bazile, G. Spenlehauer, T. Verrecchia, M. Veillard, F. Puisieux, and D. Labarre, *Biomaterials*, 1996, **17** (16), 1575–81.
148. Y. Y. Wang, S. K. Lai, J. S. Suk, A. Pace, R. Cone, and J. Hanes, *Angew. Chem., Int. Ed. Engl.*, 2008, **47** (50), 9726–9.
149. B. C. Tang, M. Dawson, S. K. Lai, Y. Y. Wang, J. S. Suk, M. Yang, P. Zeitlin, M. P. Boyle, J. Fu, and J. Hanes, *Proc. Natl. Acad. Sci. U. S. A.*, 2009, **106** (46), 19268–73.
150. M. R. Bailey, E. Ansoborlo, R. A. Guilmette, and F. Paquet, *Radiation Protection Dosimetry*, 2007, **127**, 31–34.
151. A. B. Lansley, *Adv. Drug Delivery Rev.*, 1993, **11** (3), 299–327.
152. J. C. Lay, M. R. Stang, P. E. Fisher, J. R. Yankaskas, and W. D. Bennett, *J. Aerosol Med.*, 2003, **16** (2), 153–66.
153. M. A. Sleigh, The nature and action of respiratory tract cilia in *Respiratory Defense Mechanisms, Part I*, ed. J. D. Brain, ,D. F. Proctor, ,and L. M. Reid, Dekker, New York, 1977, pp 247–288.
154. J. R. Blake, and M. A. Sleigh, *Biol. Rev. Camb. Philos. Soc.*, 1974, **49** (1), 85–125.
155. G. R. Fulford, and J. R. Blake, *J. Theor. Biol.*, 1986, **121** (4), 381–402.
156. M. King, M. Agarwal, and J. B. Shukla, *Biorheology*, 1993, **30** (1), 49–61.
157. P. Satir, and M. A. Sleigh, *Annu. Rev. Physiol.*, 1990, **52**, 137–55.

158. W. M. Saltzman, M. L. Radomsky, K. J. Whaley, and R. A. Cone, *Biophys. J.*, 1994, **66** (2), 508–515.

159. M. Yang, S. K. Lai, Y. Wang, W. Zhong, C. Happe, M. Zhang, J. Fu, and J. Hanes, *Angew. Chem.*, 2011, **50**, 1–5.

160. A. T. Florence, *Pharm. Res.*, 1997, **14** (3), 259–66.

161. P. Arbos, M. A. Campanero, M. A. Arangoa, M. J. Renedo, and J. M. Irache, *J. Controlled Release*, 2003, **89** (1), 19–30.

162. I. Behrens, A. I. Pena, M. J. Alonso, and T. Kissel, *Pharm. Res.*, 2002, **19** (8), 1185–93.

163. K. Yoncheva, L. Guembe, M. A. Campanero, and J. M. Irache, *Int. J. Pharm.*, 2007, **334** (1–2), 156–65.

164. M. Fresta, G. Fontana, C. Bucolo, G. Cavallaro, G. Giammona, and G. Puglisi, *J. Pharm. Sci.*, 2001, **90** (3), 288–97.

165. S. Ferrari, C. Kitson, R. Farley, R. Steel, C. Marriott, D. A. Parkins, M. Scarpa, B. Wainwright, M. J. Evans, W. H. Colledge, D. M. Geddes, and E. W. Alton, *Gene Ther.*, 2001, **8** (18), 1380–6.

166. M. Dawson, E. Krauland, D. Wirtz, and J. Hanes, *Biotechnol. Prog.*, 2004, **20** (3), 851–7.

167. H. Takeuchi, H. Yamamoto, and Y. Kawashima, *Adv. Drug Delivery Rev.*, 2001, **47** (1), 39–54.

168. Y. Huang, W. Leobandung, A. Foss, and N. A. Peppas, *J. Controlled Release*, 2000, **65** (1–2), 63–71.

169. P. Bures, Y. Huang, E. Oral, and N. A. Peppas, *J. Controlled Release*, 2001, **72** (1–3), 25–33.

170. N. A. Peppas, K. B. Keys, M. Torres-Lugo, and A. M. Lowman, *J. Controlled Release*, 1999, **62** (1–2), 81–7.

171. A. G. Serrano, and J. Perez-Gil, *Chem. Phys. Lipids*, 2006, **141** (1–2), 105–18.

172. B. S. Lele, and A. S. Hoffman, *J. Controlled Release*, 2000, **69** (2), 237–48.

173. J. J. Sahlin, and N. A. Peppas, *J. Biomater. Sci. Polym. Ed.*, 1997, **8** (6), 421–36.

174. C. Giannavola, C. Bucolo, A. Maltese, D. Paolino, M. A. Vandelli, G. Puglisi, V. H. Lee, and M. Fresta, *Pharm. Res.*, 2003, **20** (4), 584–90.

175. Z. Sezgin, N. Yuksel, and T. Baykara, *Int. J. Pharm.*, 2007, **332** (1–2), 161–7.

176. K. Yoncheva, E. Lizarraga, and J. M. Irache, *Eur. J. Pharm. Sci.*, 2005, **24** (5), 411–9.

177. J. Hanes, M. Dawson, D. Wirtz, J. Fu, and E. Krauland, Drug and gene carrier particles that rapidly move through mucus barriers, 2005, PCT Publication Number: WO/2005/072710.

178. S. K. Lai, Y. Y. Wang, R. Cone, D. Wirtz and, J. Hanes, *PLoS One*, 2009, **4** (1), e4294.

179. S. Shak, D. J. Capon, R. Hellmiss, S. A. Marsters, and C. L. Baker, *Proc. Natl. Acad. Sci. U. S. A.*, 1990, **87** (23), 9188–92.

180. P. L. Shah, S. F. Scott, R. A. Knight, C. Marriott, C. Ranasinha, and M. E. Hodson, *Thorax*, 1996, **51** (2), 119–25.

181. M. Dawson, D. Wirtz, and J. Hanes, *J. Biol. Chem.*, 2003, **278** (50), 50393–50401.
182. A. L. Sheffner, E. M. Medler, L. W. Jacobs, and H. P. Sarett, *Am. Rev. Respir. Dis.*, 1964, **90**, 721 9.
183. M. O. Henke, and F. Ratjen, *Paediatr. Respir. Rev.*, 2007, **8** (1), 24–9.
184. J. S. Suk, S. K. Lai, N. J. Boylan, M. R. Dawson, M. P. Boyle, and J. Hanes, *Nanomedicine (Lond.)*, 2011, **6** (2), 365–75.
185. J. S. Suk, N. J. Boylan, K. Trehan, B. C. Tang, C. S. Schneider, J. Lin, M. Boyle, P. Zeitlin, S. K. Lai, M. Cooper, and J. Hanes, *Molecular Therapy*, in press, DOI:10.1038/mt.2011.160.
186. E. M. App, D. Baran, I. Dab, A. Malfroot, M. Coffiner, F. Vanderbist, and M. King, *Eur. Respir. J.*, 2002, **19** (2), 294–302.
187. R. P. Tomkiewicz, E. M. App, G. T. De Sanctis, M. Coffiner, P. Maes, B. K. Rubin, and M. King, *Pulmonary Pharmacol.*, 1995, **8** (6), 259–65.
188. C. A. Vasconcellos, P. G. Allen, M. E. Wohl, J. M. Drazen, P. A. Janmey, and T. P. Stossel, *Science*, 1994, **263** (5149), 969–71.
189. B. K. Rubin, A. P. Kater, and A. L. Goldstein, *Chest*, 2006, **130** (5), 1433–40.
190. U. Pison, R. Herold, and S. Schürch, *Colloids Surfaces, A*, 1996, **114**, 165–184.
191. M. Geiser, S. Schurch, and P. Gehr, *J. Appl. Physiol.*, 2003, **94** (5), 1793–801.
192. S. Sutinen, H. Riska, R. Backman, S. H. Sutinen, and B. Fröseth, *Respir. Med.*, 1995, **89** (2), 85–92.
193. M. Geiser, V. Im Hof, W. Siegenthaler, R. Grunder, and P. Gehr, *Microsc. Res. Technol.*, 1997, **36** (5), 428–37.
194. J. Goerke, *Biochim. Biophys. Acta*, 1998, **1408** (2–3), 79–89.
195. J. Perez-Gil, and K. M. Keough, *Biochim. Biophys. Acta*, 1998, **1408** (2–3), 203–17.
196. H. Bachofen, S. Schürch, M. Urbinelli, and E. R. Weibel, *J. Appl. Physiol.*, 1987, **62** (5), 1878–87.
197. U. Kishore, T. J. Greenhough, P. Waters, A. K. Shrive, R. Ghai, M. F. Kamran, A. L. Bernal, K. B. M. Reid, T. Madan, and T. Chakraborty, *Mol. Immunol.*, 2006, **43** (9), 1293–315.
198. J. R. Wright, *Nat. Rev. Immunol.*, 2005, **5** (1), 58–68.
199. F. X. McCormack, and J. A. Whitsett, *J. Clin. Invest.*, 2002, **109** (6), 707–12.
200. H. Sano, and Y. Kuroki, *Mol. Immunol.*, 2005, **42** (3), 279–287.
201. C. Schleh, C. Mühlfeld, K. Pulskamp, A. Schmiedl, M. Nassimi, H. D. Lauenstein, A. Braun, N. Krug, V. J. Erpenbeck, and J. M. Hohlfeld, *Respir. Res.*, 2009, **10**, 90.
202. R. K. Harishchandra, M. Saleem, and H.-J. Galla, *J. R. Soc. Interface*, 2010, **7** (Suppl_1), S15–S26.
203. M. S. Bakshi, L. Zhao, R. Smith, F. Possmayer, and N. O. Petersen, *Biophys. J.*, 2008, **94** (3), 855–868.

204. M. Beck-Broichsitter, C. Ruppert, T. Schmehl, A. Guenther, T. Betz, U. Bakowsky, W. Seeger, T. Kissel, and T. Gessler, *Nanomed.: Nanotechnol., Biol., Med.*, 2010.

205. M. Gasser, B. Rothen-Rutishauser, H. F. Krug, P. Gehr, M. Nelle, B. Yan, and P. Wick, *J. Nanobiotechnol.*, 2010, **8** (1), 31.

206. C. Evora, I. Soriano, R. A. Rogers, K. N. Shakesheff, J. Hanes, and R. Langer, *J. Controlled Release*, 1998, **51** (2–3), 143–52.

207. B. G. Jones, P. A. Dickinson, M. Gumbleton, and I. W. Kellaway, *J. Pharm. Pharmacol.*, 2002, **54** (8), 1065–72.

208. C. Schulze, U. F. Schaefer, C. A. Ruge, W. Wohlleben, and C.-M. Lehr, *Eur. J. Pharm. Biopharm.*, 2011, **77** (3), 376–83.

209. C. Salvador-Morales, P. Townsend, E. Flahaut, C. Venien-Bryan, A. Vlandas, M. L. H. Green, and R. B. Sim, *Carbon*, 2007, **45** (3), 607–617.

210. B. Stringer, and L. Kobzik, *Am. J. Respir. Cell Mol. Biol.*, 1996, **14** (2), 155–60.

211. H. Manz-Keinke, C. Egenhofer, H. Plattner, and J. Schlepper-Schäfer, *Exp. Cell Res.*, 1991, **192** (2), 597–603.

212. V. J. Erpenbeck, D. C. Malherbe, S. Sommer, A. Schmiedl, W. Steinhilber, A. J. Ghio, N. Krug, J. R. Wright, and J. M. Hohlfeld, *Am. J. Physiol. Lung Cell Mol. Physiol.*, 2005, **288** (4), L692–8.

213. C. Winkler, K. Hüper, A.-C. Wedekind, S. Rochlitzer, C. Hartwig, M. Müller, A. Braun, N. Krug, J. M. Hohlfeld, and V. J. Erpenbeck, *Exp. Lung Res.*, 2010, **36** (9), 522–30.

214. C. Schleh, V. J. Erpenbeck, C. Winkler, H. D. Lauenstein, M. Nassimi, A. Braun, N. Krug, and J. M. Hohlfeld, *Respir. Res.*, 2010, **11**, 83.

215. M. J. Tino, and J. R. Wright, *Biochim. Biophys. Acta*, 1998, **1408** (2–3), 241–63.

216. B. A. Seaton, E. C. Crouch, F. X. Mccormack, J. F. Head, K. L. Hartshorn, and R. Mendelsohn, *Innate Immunity*, 2010, **16** (3), 143–150.

217. C. Schleh, B. Rothen-Rutishauser, and W. G. Kreyling, *Eur. J. Pharm. Biopharm.*, 2010.

218. T. Ku, S. Gill, R. Löbenberg, S. Azarmi, W. Roa, and E. J. Prenner, *J. Nanosci. Nanotechnol.*, 2008, **8** (6), 2971–2978.

219. W. Bernhard, J. Mottaghian, A. Gebert, G. A. Rau, H. von Der Hardt, and C. F. Poets, *Am. J. Respir. Crit. Care Med.*, 2000, **162** (4 Pt 1), 1524–33.

CHAPTER 5.2

Nanostructures for Overcoming the Pulmonary Barrier: Drug Delivery Strategies

PAOLO COLOMBO*, FABIO SONVICO AND
FRANCESCA BUTTINI

Dipartimento Farmaceutico, Università degli Studi di Parma, Viale Usberti
27/a, 43125 Parma, Italy
*E-mail: farmac2@unipr.it

5.2.1 Pulmonary Drug Delivery Applications

The administration of active substances to lungs *via* aerosols has steadily attracted the interest of clinicians and pharmaceutical scientists thanks to the characteristics of this administration route. The large surface of the airways, the vascularization, the permeability of the alveolar epithelium, the low enzymatic activity and the avoidance of the first pass metabolism are all specificities of pulmonary route for drug delivery. Pulmonary drug delivery can be applied not only to the treatment of respiratory diseases, such as asthma, chronic obstructive pulmonary disease (COPD) and lung infections, but is also suitable for systemic treatments.[1,2] In particular, recent developments have evidenced that the lungs are promising routes for the efficient delivery of peptides, proteins or genetic material.[3,4] Insulin, human growth hormone, vaccines for influenza and measles and gene therapies for cystic fibrosis have been clinically administered to patients *via* the lungs.[5-9]

RSC Drug Discovery Series No. 22
Nanostructured Biomaterials for Overcoming Biological Barriers
Edited by Maria Jose Alonso and Noemi S. Csaba
© The Royal Society of Chemistry 2012
Published by the Royal Society of Chemistry, www.rsc.org

Until recently, the formulations developed for pulmonary delivery have been limited to the active substances and a restricted list of excipients essentially devoted to enabling and regulating the aerosolization process. As a consequence, the *in vivo* fate of the drug, *i.e.* interaction with lung epithelia, residence time, dissolution and therapeutic efficacy, was essentially determined by the physico-chemical properties of the active substance.

The application of pharmaceutical nanotechnologies to pulmonary delivery is expected to completely change this fate because nanoparticles are modifying the properties of the active substances masking their unfavorable properties and deeply affecting their interaction with the respiratory tissues.

When used as drug carriers, nanoparticles have shown enhanced drug water dispersion, sustained release, protection of labile molecules from chemical or enzymatic degradation and the possibility of internalization by phagocytic or epithelial cells. However, lung targeting of these delivery systems *via* intravenous administration was found to be ineffectual with low quantities of nanoparticles localized in the lung even when lung-targeted types of nanoparticles were tested.[10] Hence the most effective way to deliver nanopharmaceuticals to the lungs is by inhalation of aerosols. In addition, for pulmonary administration some other advantages are expected from the use of nanoparticles, such as improved formulation homogeneity, better delivery efficiency, increased residence time, reduction of the doses and side effects and improvement in patience compliance.[11]

However, both computational models and experimental data show that nano-sized materials, *i.e.* materials with distribution from 1 nm to 1 µm, are not efficiently deposited in the airways.[12] Hence when dealing with a nanoparticulate formulation, one option is to formulate nanoparticles in such a way that they are vehiculated or carried by droplets or particles in the range 1–3 µm, useful to deposit and release efficiently in the lungs. The alternative is to provide a formulation able to generate directly an aerosol in which particle distribution is in the low range of nanodimensions, *i.e.* below 50 nm. In fact, ultrafine particles with dimensions in the range of few tens of nanometres are able to deposit in the deep lungs chiefly by a diffusional effect. A further challenge to the task of delivering nanoparticles to the lung is the need of a device able to produce a respirable aerosol. Hence, the formulation containing the nanoparticles has to match with the device in order to provide the best performance in terms of delivered dose, physical and chemical stability and patient safety.

In this section the most recent approaches to delivering nano-pharmaceutics to the lung have been reviewed. In particular, nebulizers, pressurized metered dose inhalers and dry powder inhalers have been investigated for the administration of formulations containing nanoparticles. Those devices represent the gold standards on the market of respiratory medicines. A brief overview of the advantages and the hurdles posed by each of these technologies for nanoparticle delivery will be presented.

5.2.2 Nebulizers

Nebulization of a suspension of nanoparticles would appear to be the most straightforward method for their delivery. A variety of nebulizers are used for the administration of drugs in hospital settings or by patients at home (Figure 5.2.1). The two major types of nebulizers are the jet nebulizers and ultrasonic devices. Air jet nebulizers operate on the principle of the Venturi effect: pressurized air supplied by a compressor is forced through a nozzle at high speed over the end of a capillary tube dipping into the drug reservoir. Owing to the lower pressure created by the airstream at the nozzle, liquid may be drawn up through the feeding tube into the airstream where the high speed airflow and shear forces break the liquid into droplets to form an aerosol. An ultrasonic nebulizer uses a piezoelectric transducer placed at the bottom of the reservoir to induce mechanical waves into the liquid. Interference of these waves at the reservoir surface leads to the production of a fountain of liquid droplets at the surface. The dispersion of droplets in the atmosphere above the reservoir generates an aerosol. The design of the device and the presence of one or more baffles shatters or eliminates bigger droplets allowing the output only of droplets in size range ideal for inhalation (1–5 μm). Another type of nebulizer device, which have been introduced more recently, is based on perforated vibrating membranes, defined also as mesh technology. In this case, an electronic circuit generates a signal that causes a perforated membrane to vibrate in resonant bending mode. The vibration of the membrane creates mechanical pressure on the liquid causing it to squeeze through the holes of the membrane and creating an aerosol cloud. The size and density of the holes determine droplet size and delivery rate of the aerosol.[13–15]

All these nebulizers efficiently deliver drug solutions, especially for drugs requiring high dosage or substances that present formulation problems in pressurized metered dose inhalers, such as proteins. Nebulizers, however, were found to be less suitable for the aerosolization of suspensions. Several issues may affect the delivery of drug suspensions: in air jet nebulizers particles larger than the output droplets are not aerosolized due to the effect of impaction on

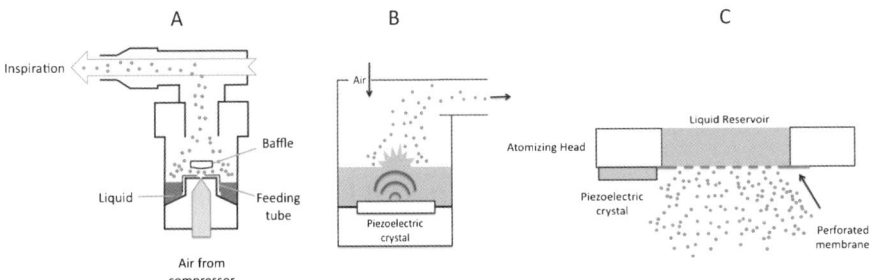

Figure 5.2.1 Different types of nebulizers: air jet (A), ultrasonic (B) and perforated vibrating membrane (C) nebulizers.

baffles. However, in ultrasonic nebulizers the aerosol formation occurs only on the liquid–air interface and thus few particles are effectively present in the droplets, the particles being mainly in the bulk of the liquid. Finally, the holes of the perforated membranes defining the droplet size in mesh technology nebulizers act as sieves delivering only particles two or three times smaller than mesh size.[16,17] Nanoparticle suspensions could constitute an approach to overcoming all these drawbacks.

Nanoparticles should be useful for the delivery of poorly soluble drugs, increasing their saturation solubility and dissolution rate. Furthermore, they could be more evenly distributed in the droplets produced by the nebulizers compared to micron-sized particles. Reduction in size produces a longer residence time at the deposition site as a consequence of the tendency of fine particles to stick to the mucosal surface, but also due to a reduced clearance by cilia and macrophages. The dispersing medium, the composition and surface properties of nanoparticles, the concentration of the suspension, and the quantity and nature of the stabilizing agents in the formulation are the factors expected to be critical for the efficient delivery with a nebulizer.

Nebulization properties of NanoCrystals[TM] of beclometasone dipropionate (BDP) obtained by milling (mean size 267 ± 84 nm) in an aqueous solution containing PVA 2.5% w/v were compared to those of a micronized suspension of the same drug with particle size around 10 μm. A commercial air jet nebulizer was used to study the formulations. The nanosuspension behaved similarly to water in terms of nebulization output and rate: indeed nanoparticles did not affect the aerosol formation process and aerosol deposition into a 7-stages cascade impactor. Compared to the micronized suspension, the NanoCrystals[TM] formulation was superior both in terms of BDP output fraction (0.40 *vs.* 0.14) and respirable BDP fraction (0.14 *vs.* 0.03). The nanoparticles emitted from the nebulizer, examined by scanning electron microscopy, were found to be unmodified in term of morphology and size. The major claim of this formulation was the stabilization of the nanoparticles against Ostwald ripening (dissolution of small crystals and redeposition on the surfaces of larger crystals) and aggregation phenomena.[18]

In another study, a similar BDP NanoCrystals[TM] formulation (mean particle size 164 nm, stabilized with 0.25% w/v Tyloxapol) was aerosolized with an ultrasonic nebulizer. The performance was compared to a commercial pMDI product (Vanceril®, Schering-Plough) delivering 42 μg of BDP per actuation. Deposition studies with Andersen Cascade Impactor comparing the actuation of the nebulizer against pMDI, evidenced that BDP respirable dose was higher (56–72% of the emitted dose *vs.* 36%) for the nebulizer delivering the nanoparticulate formulation. The emitted BDP dose from the two devices was comparable: 31.4–67.4 μg per two seconds actuation for the nebulizer with NanoCrystals[TM] formulation, 36.1 μg per actuation for the pMDI product, although the throat deposition was only 9–10% of the emitted dose for NanoCrystals[TM], whereas it reached 53% for the pMDI.[19]

Some studies directly investigated the effect of different nebulization technologies on the aerosolization of nanoparticle suspensions. A bupavarquone nanosuspension has been proposed for pulmonary treatment of fungal pneumonia caused by *Pneumocystis carinii* in immune-compromised patients. Stable bupavarquone nanosuspension (mean particle size 400–500 nm) was studied using four different nebulizers: two types of air jet nebulizers, one ultrasonic nebulizer and one vibrating mesh based device. The studies performed included emitted dose and mass median aerodynamic diameter (MMAD) of aerosol droplets produced by the different devices. The physical stability of nanoparticles, both in the emitted aerosol and in the residual formulation in the device at the end of the nebulization was also assessed. The results showed that buparvaquone nanoparticles did not affect MMAD of droplets produced by the four nebulizers, since the values obtained were equal to reference aqueous solutions containing only the surfactants. Jet and ultrasonic nebulizers led to the production of droplets with MMAD in the range 3.4–5.5 μm and respirable fraction between 44 and 72%. The vibrating mesh technology-based nebulizer produced larger droplets, 8.9–10.9 μm, with a consequently much lower respirable fraction (18–27%). Concerning nanoparticle stability, the mean particle size after nebulization remained almost the same for all the nebulizers. However, the formation of nanoparticle aggregates was more evident for both the emitted and residual suspensions in one of the jet nebulizers. The aggregation phenomena observed for this jet nebulizer have been attributed to the evaporation effect of the liquid formulation occurring during the aerosolization by Venturi process.[20]

In a similar study on budesonide submicron suspension (BSS, mean particle size 730 nm), the formulation was compared to the commercial suspension Pulmicort® Respules (0.5 mg mL^{-1} suspension, $d_{0.5}$ 4.43 μm; Astra Zeneca) using three different nebulizers: the air jet Pari LC Plus®, the ultrasonic Multisonic® (Otto Schill GmbH) and the perforate vibrating membrane inhaler e-Flow™ (PARI GmbH). Nebulizers were found to have similar performances in terms of output (0.9–1.6 g), MMAD (2.98–4.95 μm) and fine particle fraction (63–92%, highest for e-Flow™ and lowest for Pari LC Plus®). In contrast, it was noted that budesonide respirable dose was almost double for the BSS with respect to the Pulmicort® suspension whichever nebulizer was used. Nevertheless, it was evidenced that the ultrasonic nebulizer, despite it having similar aerosol characteristics to those of the others, was highly inefficient in delivering budesonide particles within the aerosol droplets, even with the submicron suspension. In fact, for the ultrasonic nebulizer the emitted and respirable doses were actually much lower than the doses measured for the other two nebulizers. These data were explained considering the typical aerosol formation mechanism of ultrasonic nebulizers. Particles of both submicron and micronized suspensions were less available at the surface of the liquid, and are less entrained in droplet formation. Usually, this effect less influences nanosuspension, but in this case, despite mean particle size in the order of 700 nm, the large particle size distribution of the budesonide submicron suspension led to the presence of a high fraction of particles in the micron size range.[21]

The pharmacokinetics after pulmonary administration of a nanosuspension of budesonide has been recently studied in a clinical trial and compared to the commercial budesonide suspension Pulmicort® Respules (Astra Zeneca). The nanosuspension aerosolized using a jet nebulizer showed a shorter nebulization time compared to the commercial suspension. The nanoparticle formulation was well tolerated, with no evidence of bronchospasm. A linear dose proportionality in pharmacokinetic parameters, such as C_{max} and AUC, was shown for the nanosuspension of budesonide in the dose range 0.5–1 mg. Concerning the comparison with the commercial suspension, budesonide nanoparticle suspension evidenced similar AUCs suggesting a similar lung absorption. However, significantly higher C_{max} and shorter t_{max} were registered for the nanosuspension, indicating higher rate of dissolution and absorption.[22]

Drug-loaded polymeric nanoparticles produced with polylactic-co-glycolic adic (PLGA) or other biocompatible, biodegradable polymers have been studied for pulmonary delivery using nebulizers. Dailey and coworkers made a study of PLGA nanoparticles (mean particle size 93–104 nm) and poly-vinylalcohol grafted PLGA nanoparticles (PVA-PLGA size 101–116 nm) loaded with coumarin-6 as model drug have been studied. The nanoparticle suspension's aerosolization was studied with three different nebulizers. Nanoparticles were produced by the solvent displacement method, using 0.0025% Alveofact® as surfactant. They differed for surface hydrophobicity and for different degradation rate. PLGA is in fact a slowly degrading polymer that may give rise to some concerns about lung safety in case of frequent administrations or prolonged therapies. The derivative obtained by grafting PLGA chain to a PVA backbone gives a comb-like polymer with more hydrophilic properties and customized characteristics, *i.e.* more rapid degradation profiles and release properties. Different nanoparticle formulations did not affect droplet MMAD and size distribution of nebulized suspensions, when compared with control solutions of NaCl 0.9% and glucose 5%. As shown also in other studies, the vibrating membrane nebulizer was the least efficient in terms of emitted aerosol and produced significantly larger droplets of MMAD (8 μm). MMADs close to 4 μm were obtained for the air jet and ultrasonic nebulizers. Concerning nanoparticle stability, data showed enhanced nanoparticle aggregation within the emitted aerosol and in the residual suspension left in the nebulizers. This phenomenon was evidenced for all the nebulization technologies, even if a dependence on the nebulizer type and on the surface properties of nanoparticles was present. The authors found that surface hydrophobicity plays a pivotal role in the tendency of nanoparticles to aggregate. Nanoparticles with a hydrophobic surface showed a higher tendency to form agglomerates in comparison to their more hydrophilic homologues. On the other side, the jet nebulizer technology was the one that showed the highest formation of agglomerates; agglomerates were found also in the emitted or residual formulation of the ultrasonic and the vibrating mesh technology based nebulizers, but contained less than the 10% of the total cumarin-6. It has been postulated that liquid evaporation as well as shear forces involved in aerosol formation contributed to promote a higher frequency of contacts between nanoparticles leading to agglomeration.[23]

In another study from the same group, nanoparticles with a positive or negative surface charge were obtained by solvent displacement method using diethylaminopropyl amine-poly(vinyl alcohol) grafted PLGA (DA-PVA-PLGA) alone and with increasing amounts of carboxymethyl cellulose (CMC) (particle size 76–256.6 nm and zeta potential between 58.9 and −46.6 mV). These particles do not require any surfactant for their production and hence have a definite advantage over crystal nanosuspensions that need high amounts of potentially toxic surfactants for their stabilization. Positive surface nanoparticles showed agglomeration when nebulized with air jet and ultrasonic nebulizers, with large agglomerates present both in the reservoir and in the aerosol droplets. Negatively charged nanoparticles, in contrast, did not show any sign of agglomeration. This evidence was explained with the complete condensation of the grafted cationic PLGA derivative in the presence of high amounts of CMC. Formulations containing free DA-PVA-PLGA polymer, as in the case of positively charged nanoparticles, showed high amounts of agglomeration due to the negative effect of the free polymer on suspension stability during nebulization.[24]

Aggregation of the formulation during the aerosolization process is not the only concern when dealing with nanoparticles. Stability of the nanoparticle suspension during storage conditions before the actual delivery *via* the nebulizer is one of the major issues of this type of formulations. Long term stability is reported for formulations with an optimized quantity of surfactants. Budesonide nanosuspensions stabilized with 0.5% w/w of lecithin and 0.2% w/w of tyloxapol were found to be physically stable for one year at room temperature with no sign of Ostwald ripening or agglomeration.[25] Another proposed solution is to freeze-dry the nanoparticles and reconstitute the colloidal suspension in the instant before delivery by inhalation. Hydrophilic polymers (dextrans, PEG, PVP), sugars (lactose, glucose, sucrose) and sugar alcohols (mannitol and sorbitol) are used as cryoprotectants and lyoprotectants. These substances are able to stabilize the freeze-dried nanosuspensions by means of molecular movement retardation and forming high viscosity amorphous phases.[26] Nanoparticles produced with the biodegradable comblike polymer DA-PVA-PLGA and CMC have been freeze-dried using various lyoprotectants, such as glucose, sucrose, lactose or mannitol. It was found that an increase in concentration of the lyoprotectant was generally beneficial for the stabilization of nanoparticles properties, such as size and surface charge. Concerning the different substances, lactose was the best protectant while mannitol and glucose led to more relevant agglomeration. The resuspended nanoparticles were aerosolized and no differences were found between different formulations in terms of aerosol output, MMAD of droplets and nanoparticle size and charge using either an air jet nebulizer or a vibrating perforated membrane nebulizer.[27]

Nebulizers for the delivery of nanosuspensions have been used at various levels of product development either for toxicological evaluation of nanoparticles or to test *in vivo* formulations for diagnostic, anti-tuberculosis or chemotherapy applications.

Solid lipid nanoparticles have been proposed as an alternative to liposomes and polymeric nanoparticles for drug delivery. The lung toxicity of solid lipid nanoparticles (SLN) produced by means of high-pressure homogenization using triglycerides (Softisan® 154), phospholipids (Phasopholipon® 90G) and stabilized with PEG-hydroxystearate (mean particle size 98.4 nm, surface charge -14.6 mV) has been evaluated on BALB/c mice. Mice were exposed on a daily basis in a closed Plexiglas® box to an aerosol of SLN obtained using a jet-driven aerosol generator system for 16 days. With deposited doses of SLN up to 200 µg per day no sign of inflammation, such as production of chemokine KC, Interleukine-6 or lactate dehydrogenase release, were evidenced with respect to baseline levels in bronchoalveolar lavage fluid.[28]

Aerosolization by means of nebulizers appears especially useful in studies carrying out proof-of-concept experiments in animal models. Insulin-loaded PLGA nanoparticles (mean particle size 400 nm, drug content 2.98%) were administered with a mesh nebulizer to artificially ventilated guinea pigs. Although nanoparticles showed a high burst release, *in vivo* studies evidenced that a dose of 3.9 IU kg^{-1} of insulin loaded nanoparticles induced a prolonged hypoglycemic effect lasting for 48 hours. The prolonged reduction in glucose blood levels obtained by the inhalation of insulin-loaded PLGA nanoparticles was attributed to the controlled release of a fraction of insulin dose.[29]

In another study, PLGA and chitosan surface-modified PLGA nanospheres were produced by emulsion solvent diffusion technique for pulmonary delivery of elcatonin (median particle size 650 nm, suface charge -3.6 mV and $+21.2$ mV for the PLGA and chitosan surface modified PLGA nanoparticles, respectively). Animal experiments performed on artificially ventilated guinea pigs, using a mesh type ultrasonic nebulizer, evidenced that by delivering the same amount of elcatonin (100 IU kg^{-1}) to the animals, a significantly lower and prolonged reduction in blood calcium level was obtained when using chitosan modified nanoparticles. This result was explained by a combined effect of prolonged release from nanoparticles, extended residence time on lung tissue and enhancement in drug absorption provided by the presence of chitosan on nanoparticle surface.[30]

Nanoparticles administered by nebulization have been proposed as carriers of radionuclides or anticancer drugs for the diagnosis and therapy of primary or metastatic lung cancers. The regional modality of treatment could increase exposure of the tumor to the drug, while minimizing systemic side effects.[31]

Celecoxib, a lipophilic inhibitor of cyclooxigenase-2 that showed synergistic activity with anticancer agents in the treatment of non-small cell lung cancer, has been proposed for pulmonary administration with nano-structured lipid carriers (NLC). Celecoxib-loaded nanoparticles were produced by high-pressure homogenization using Compritol, Miglyol and sodium thaurocholate as surfactant (mean particle size 217 nm, surface charge −25.3 mV, drug loading 4% w/w). A nanoparticle suspension was administered to BALB/c mice using a jet nebulizer coupled with the Inexpose[TM] system consisting of 12 ports located peripherally around a central delivery plenum. Lung concentration of celecoxib remained almost constant for 2.5 hours, and above the detection limit up to 12 hours after pulmonary administration of the nanoparticle suspension. Blood clearance was found to be much faster for the drug solution than for the nanosuspension (20.03 l kg h^{-1} *vs.* 0.93 l kg h^{-1} for nanoparticles). Plasma pharmacokinetics reflected this latter observation: the nanoparticle formulation showed significantly higher plasma levels at all time points and was detected for up to 24 hours. Peak plasma concentration was reached after 4 hours, while AUC was found to be 20 times higher than the values obtained for the drug solution.[32]

Solid lipid nanoparticles radiolabelled with ^{99m}Tc, using the lipophilic chelator hexamethylpropyleneamine oxime (mean diameter 200 nm, surface charge −17.2 mV, 20 mCi) have been administered to Wistar rats using a suitable mask and an ultrasonic nebulizer. Gamma scintigraphic images a few minutes after aerosolization evidenced a strong deposition in the whole lung. However, after 45–60 minutes SLN were detected mainly in lymphatic tissues, such as periaortic, axillary and inguinal lymph nodes, suggesting nanoparticle clearance through a lymphatic route. Radioactivity distribution 4 hours after inhalation evidenced a high retention of nanoparticles in lung tissue (23.3%) but a high relative particle concentration was achieved in lymph nodes (overall 16.7%). The results provide the possibility of using radiolabelled nanoparticles as lympho-scintigraphic agents for the diagnosis of lung cancer-derived metastases.[33]

One of the hot topics of nanoparticle administration to the lungs is the treatment of pulmonary infections. Inhalable alginate nanoparticles encapsulating isoniazid, rifampicin and pyrazinamide produced by cation-iduced gelification and stabilized with chitosan (mean particle size 235 nm, with drug encapsulation efficiencies between 70 and 90%) were administered to guinea pigs using an air-jet nebulizer. Using a single administration of therapeutic dose of the drugs, *i.e.* isoniazione 10 mg kg^{-1}, rifampicin 12 mg kg^{-1} and pyrazinamide 15 mg kg^{-1}, detectable levels of the three compounds were observed from 3 hours after aerosolization up to 14, 10 and 14 days, respectively. Tissue concentrations above MIC were found up to 15 days in lungs, liver and spleen. Plasma levels after nebulization of a solution of the three drugs fell to undetectable levels within 12–24 h. The nanoparticulate delivery system seems particularly promising because of being multi-drug and assuring a prolonged exposure to anti-tuberculosis drugs. The result would simplify dose regimen and improve patient compliance, which is the main cause of failure in the several-months-long multi-drug tuberculosis therapies.[34]

Inhalation of a nanosuspension of itraconazole has been proposed as innovative treatment of fungal lung infections due to *Aspergillus* in immunocompromised patients. Amorphous nanoparticles of itraconazole were prepared by the ultra-rapid feezing technique (mean diameter 0.25 μm) and compared to crystalline itraconazole nanoparticles obtained by wet ball milling (mean diameter 0.57 μm). The two formulations were administered to Sprague–Dawley rats using a vibrating mesh nebulizer equipped with a nose-only dosing apparatus. The delivered doses were in both cases equivalent to 100 mg of itraconazole and no differences were found in terms of nebulization performance and lung deposition. However, cumulative antifungal drug concentration absorbed into the blood was 3.8 times higher for amorphous nanoparticles than for crystalline ones. These results were attributed to the higher supersaturation level of the drug solution obtainable with amorphous nanoparticles. The lower level of lung clearance evidenced for the amorphous formulation was explained by the rapid dissolution, while nanocrystals are more prone to clearance of the lung immune-defense system. Amorphous nano-particles appear to be an ideal approach for the delivery to the lung of high doses of extremely poorly water-soluble drug both for local and systemic therapy.[35]

In conclusion, owing to their versatility, robustness and the different available aerosol delivery technologies, nebulizers have attracted renewed attention as systems to test and develop products based on nanoparticle suspensions. Studies conducted in this field evidenced that along with the optimization of the nanoparticle suspension characteristics (concentration, size distribution, stabilizers, surface properties, *etc.*) the choice of a suitable nebulizer (technology, producer, design, performance) is pivotal to obtain optimal aerosolization and a consistent delivery to the lungs.[36] Recent studies on new aerosol technologies, such as condensation aerosol generation, are opening the way for new opportunities to produce nanoparticulate aerosols directly from drug solutions in suitable solvents. With the direct production of nanoparticles from solutions, issues connected to the long term or in use stability of nanoparticle suspensions are avoided.[37] On the other hand, new portable and small dose nebulizer devices, called also soft mist inhalers, have been developed as alternative to the popular pMDI and DPI handheld devices. Recently, it has been shown that solid lipid nanoparticles obtained from emulsions using supercritical fluid extraction (mean volume diameter 30 nm) are able to be loaded with high content (10–20% w/w) of model drugs. Ketoprofen or indomethacin were found to be efficiently delivered using the innovative nebulizer system AERx. This portable multidose system delivered small volumes (50 μl) of the SLN suspension by mechanically pressurizing the liquid content of a single dose blister through a perforated nozzle sealed layer placed on the top of the blister. The emitted dose was between 50 and 60%, while the aerosol formed showed MMAD around 3 μm and fine particle fractions were found to be between 65 and 79%.[38]

In this section, we overviewed a number of studies in which nebulizers have been used in order to deliver nanoparticles to the lungs. Even though the aerosolization of liposomal preparations has been far more popular in the pharmaceutical field, we decided not to consider this topic in order to focus on the problems and pitfalls specifically related to the use of nanoparticles suspensions. Nevertheless, the delivery of liposomal formulations has been treated in the section of the chapter dedicated to dry powder inhalers, because of the specific interest and challenge related to the administration of liposomes as inhalable dry powder.

5.2.3   Pressurized Metered Dose Inhalers

Pressurized metered dose inhalers (pMDI) are devices in which precise doses of drug solution or suspension in a liquefied compressed gas is delivered by actuation of a metering valve (Figure 5.2.2). Introduced in the mid-1950's pMDIs have become the gold standard devices for delivering inhaled drugs. The major change in this technology in 50 years has been the phasing out of chlorofluorocarbons propellants as a consequence of the enacting of the Montreal Protocol on Substances that Deplete the Ozone Layer. In favour of more ozone-friendly devices, metered dose inhalers today use hydrofluoralkanes as propellants.[39] Despite their popularity, the application of nanotechnologies to these types of devices is only in its infancy, and very few studies have been published up to now.

Some authors suggested that significant bioactivity of inhaled drugs administered with some commercial HFA pMDIs, such as Qvar, may reside in the production of fine (< 1000 nm) and ultrafine particles (< 100 nm) during the aerosolization process.[40] At the same time nanosuspension of

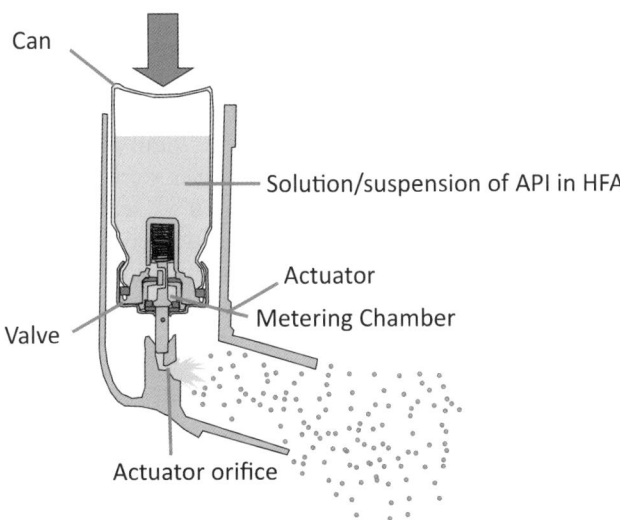

Figure 5.2.2   Pressurized metered dose inhaler.

propellant-insoluble drugs may improve formulation homogeneity, delivered dose consistency and fine particle dose fraction.

An approach proposed for the production of nanoparticles suitable for metered dose inhalers is the use of a reverse microemulsion. Nanoparticles of salbutamol sulphate or salbutamol sulphate and lactose were produced by snap freezing and subsequently freeze-drying a W/O microemulsion. The emulsion was obtained with water and iso-octane using lecithin and propan-2-ol as surface-active agents. Nanoparticles with diameters of around 35–70 nm were produced. In order to prevent agglomeration in the propellant, nanoparticles had to be dispersed in a suitable organic liquid (*n*-hexane) prior to filling with HFA-227 into MDI canisters. No sedimentation or creaming occurred in the pressurized formulation over several months. The actuation of such formulated MDIs produced an aerosol with MMAD around 1.5 µm and a fine particle fraction comprised between 60 and 65%.[41]

In another application, an emulsification process using lecithin as surfactant followed by freeze-drying was used to produce nanoparticles of lactose and insulin (mean particle size around 550 nm). Nanoparticles were suspended in HFA134a using essential oils, such as cineole and citral, as stabilizers. Studying the aerosolization properties with a multi-stage liquid impinger, the fine particle fraction of the aerosol was negatively affected by an important throat deposition (74%). The optimization of the nanoparticle production by the addition of glycerol monooleate as cosurfactant (mean particle size 350 nm) reduced throat deposition (35.5% of the emitted dose), greatly improving the fine particle fraction (FPF$_{<1.7\ \mu m}$ 45%).[42]

The encapsulation of nanoparticles in micron-sized particles whose surface was designed to be solvated by the propellant, provided physical stability to the formulation, and releasing of nanoparticles in aqueous fluid. Particles composed of water soluble HFA-philic oligo(lactide)-grafted-chitosan were produced by a emulsification solvent diffusion technique. They were used to encapsulate PLGA nanoparticles (mean particle size 241 nm) containing a fluorescent dye, coumarin-6. The microparticles obtained, with sizes ranging from 0.5 and 4 µm, could be dispersed in HFA-227 and produced a physically stable suspension. In contrast, PLGA nanoparticles were poorly stable in the propellant since they agglomerated within few seconds. The aerosol characteristics of this core–shell formulation evidenced a respirable fraction much higher than that obtained for the formulation containing PLGA nanoparticles alone (72.4% *vs.* 20.1% respectively). *In vitro,* microparticles were able to release the PLGA nanoparticles in a cell culture media and nanoparticles were then actively taken by Calu-3 cells infected with *Chlamydia pneumoniae.*[43]

A similar approach has been adopted for the delivery of curcumin-loaded PLGA nanoparticles encapsulated in PEG/chitosan graft copolymer microparticles. PLGA nanoparticles loaded with the drug were prepared by single emulsion solvent evaporation technique (particle size 243.4 ± 34.7 nm), while PEG/chitosan microspheres containing nanoparticles were produced by spray drying (mean volume diameter 3.5 $\pm$ 0.46 μm). Microparticles were able to swell in aqueous media up to 70 μm and that property was considered useful to escape pulmonary macrophage uptake and to prolong residence time into the lung. Microparticles could be dispersed in HFA 227 and were loaded in a pMDI crimp-sealed with 100 μl metering valves. A negative effect on microparticle concentration in the propellant was evidenced on evaluating the creaming time of the suspensions, which in fact was reduced from more than 10 minutes for 0.2 mg mL^{-1} to less than 5 minutes for 2 mg mL^{-1} suspensions. A similar effect was observed also on the fine particle fraction and throat deposition upon aerosolization: the percentage of curcumin deposited in the throat of the Next Generation Impactor increased from approximately 20% to 37% upon increasing the concentration of the microparticle suspension from 0.2 to 2 mg mL^{-1}. At the same time the fine particle fraction decreased from 65 to 40%. This effect was attributed to the presence of a greater number of aggregates for high microparticle concentrations as a consequence of the limited solvating capacity of the HFA 227 towards the microparticles.[44]

In a recent studies, low density flocs of anisotropic crystals such as rods, plates or needles with one nano-sized dimension, have been used for obtaining stable suspension in HFA propellants. Nanorods of bovine serum albumin and itraconazole produced by thin film freezing were subsequently freeze-dried and suspended in HFA 227. Very stable suspensions (stable up to 2 years in the case of itraconazole) were obtained because of the peculiar morphology of the crystals. The nanorods formed high volume weakly flocculated "open" flocs. The flocs were disrupted by the shear forces developed in the pMDI actuator and the suspension could be delivered at high dose per actuation (emitted dose 1–2 mg) obtaining satisfactory fine particle fractions (38–47% for BSA and 49–64% for itraconazole).[45,46]

The very recently emerging attention to the formulation of nanoparticles in pressurized metered dose inhalers demonstrate that this is a burgeoning field in which improvements are possible and eagerly expected, and opportunities are still to be taken.

5.2.4 Dry Powder Inhalers

Nanoparticles exhibit characteristics that make them ideal for pulmonary drug delivery and for treating lung-specific diseases such as lung cancer. Research has shown that nanoparticles avoid mucociliary clearance and in some cases phagocytic clearance by remaining in the lung lining fluid until dissolution or translocation by the epithelium cells. Nanoparticles can be incorporated into larger carrier particles to produce the appropriate size for pulmonary drug delivery.[47]

However, nanosize imparts to particles a poor possibility to deposit in the lung on account of their high tendency to be exhaled. Furthermore, devices for inhalation usually deliver particles having a size higher than 0.5 μm. Nanoparticles can be administered not only by their incorporation in solution for nebulizers or MDI products; several studies are focused on the possibility to dry nanoparticles and to administer them using dry powder inhalers.[48]

In this case nanoparticles have to be transformed into a micronized powder by carrier particles with an appropriate pulmonary size. Few methods have been employed to temporarily shift nanoparticles to micron-size particles to allow for dosage, aerosolization and deposition of drug. Once particles deposit on the pulmonary mucosa, nanoparticles should recover their primary size and perform their action.

In the development of novel DPIs, the spray drying technique offers a number of advantages over the lyophilization technique. Spray-dried powders have been studied containing nanoparticles dissolving in the lungs into polymeric nanoparticles with dimensions sufficiently small to avoid mucociliary and phagocytic clearance. The spray drying technique was utilized for the stabilization of nanoparticles and the development of uniform size particles with desired properties for pulmonary administration. The technique overcomes constraints associated with the lyophilization technique, such as the formation of hard cake, the need for further micronization, the addition of coarse carriers for aerosolization and a heterogeneous size distribution pattern.

The use of liposomes in aerosol has many potential advantages, including carrier suitability for most lipophilic drugs, aqueous compatibility, sustained release and intracellular delivery. Among pulmonary drug delivery systems, dry powder inhaler formulations (DPIs) stand out because of the stability of drugs and formulations. Liposomal drug DPIs have promising features for pulmonary administration, particularly with respect to controlled delivery, increased potency and reduced toxicity. In addition they uniformly deposit the drug locally, are propellant free, patient compliant, high dose carrying, stable and patent protected.

Nanoparticles via an aerosolized route were also studied for improving the clinical results in lung transplantation. Aerosolized immunosuppressant formulations are not commercially available at present. The administration of aerosolized cyclosporine after lung transplantation improves survival and decreases the risk of organ rejection thanks to the local immunosuppressive effect with minimal systemic toxicities.[47]

Tacrolimus, a primary immunosuppressant, is effective, potent, safe and superior to cyclosporine. The feasibility of pulmonary delivery of a liposomal encapsulated tacrolimus dry powder inhaler for prolonged drug retention in the lungs was evaluated as a rescue therapy to prevent refractory rejection of the lung after transplantation.[47]

Liposomes have been presented as an alternative for the administration of biomolecules through several mucosal surfaces, since they are versatile and tend to be relatively innocuous (produced with natural and biodegradable compounds), and also provide protection to the encapsulated material.

Tacrolimus encapsulated liposomes were prepared by a thin film evaporation technique and the liposomal dispersion was passed through high pressure homogenizer. Tacrolimus nano-liposomes (NLs) were separated by centrifugation and were dispersed in phosphate buffer saline containing different additives, such as lactose, sucrose, and trehalose and L-leucine, as antiadherents. The nano-dispersion was spray dried and the microparticles obtained characterized. In particular, this study demonstrates that trehalose-based powder had low density, good flowability, and an FPF value around 70%. *In vitro* release studies of the developed NLs dry powder were performed, with spray dried plain tacrolimus with lactose (STL) used as reference. *In vitro* drug release studies showed 90% drug release within 6 hours form STL and 18 hours from spray dried NLs, following a Higuchi model.

In vivo studies of NLs dry powder were conducted to test the possibility of obtaining a prolonged localized action of liposomal tacrolimus in lungs. The study was performed using an intratracheal instillations technique by administration of 200 µg of tacrolimus equivalent in STL or developed NLs dry powder in albino rats. Plain tacrolimus was rapidly absorbed from the lungs and then into the systemic circulation, or rapidly metabolized; in contrast, the liposomal-encapsulated tacrolimus showed prolonged residence of 24 hours, suggesting slow clearance from the airways. The investigation provides a practical approach for direct delivery of tacrolimus encapsulated in NLs for controlled and prolonged retention at the site of action. Long drug residence within the lungs after pulmonary administration of tacrolimus NLs dry powder is expected to necessitate less frequent administration or lower doses of tacrolimus, reducing associated systemic toxicities.[47]

In recent years, significant effort has been dedicated to increase nanotechnology applications for pulmonary delivery since it offers a potential means of improving the delivery of small molecule drugs, as well as macromolecules such as proteins, peptides, or genes to the tissue of interest. Furthermore, most of the new molecules being identified as active pharmaceutical ingredients are poorly water-soluble, drawing the research towards new approaches to the formulaton of these APIs. A promising approach is to use a particle engineering technology to overcome poor wetting and low dissolution rate of an API. Techniques that have been commonly used include mechanical milling, spray drying, solid dispersion, and supercritical CO_2 precipitation techniques like rapid expansion of supercritical solutions (RESS), solution enhanced dispersion by supercritical fluid (SEDS), precipitation with a compressed antisolvent (PCA) and supercritical antisolvent precipitation process (SAS). Nanoparticles, whether amorphous or crystalline, offer an additional interesting way of formulating drugs having poor water solubility. By presenting drugs at the nanoscale, dissolution can be rapid and as a result the bioavailability of poorly soluble drugs can be significantly improved.

Budesonide solubility was improved by the transformation of a drug nanosuspensions into dry powder formulations capable of effective deposition and rapid dissolution.[48] Budesonide is considered one of the most valuable therapeutic agents for the prophylactic treatment of asthma, despite its poor solubility in water (21.5 µg mL^{-1}). This drug is already applied through dry powder inhalers (Pulmicort), metered dose inhalers (Rhinocort) or capsules

(Entocort).[49] In this study different surfactants, including hydrophobic (cetyl alcohol and Span 85), hydrophilic (Pluronic, PVA, and PVP), and amphoteric (lecithin) ones, were used to create surface charge on the nanoparticles. Then, charge interactions were modified to flocculate nanoparticles into nanoparticle agglomerates exhibiting a particle size range of 2–4 μm.

Budesonide nanosuspensions were prepared using a precipitation technique: solutions of the drug in acetone were directly injected into a non-solvent (water). Nanosuspensions were lyophilized to obtain dry powders composed of micron-sized agglomerates, which showed a fine particle fraction around 90%. The aerodynamic characteristics were studied *in vitro* using a Tisch Ambient Cascade Impactor (Tisch Environmental, Inc.). The study was carried out by applying 20 mg powder manually into the orifice of the instrument at three airflow rates. The dissolution rate of prepared nanoparticles and nanoparticle agglomerates was measured and compared with the dissolution characteristics of the stock drug. The dissolution of budesonide particles was carried out at 37 °C in a 400 mL beaker. A known amount (10 mg) of lyophilized powder was suspended in 10 mL phosphate buffered saline (PBS, pH 7.4) in a dialysis bag membrane unit (M_w cut-off 10 000 Da) and the unit was allowed to float in a beaker containing 300 mL of PBS. The dissolution of budesonide nanoparticles was faster than stock budesonide due the increased surface area obtained by decreasing the particle size. In particular, nanoparticles prepared using hydrophobic and amphoteric surfactants (cetyl alcohol or lecithin) showed faster drug dissolution than nanoparticles containing the hydrophilic surfactant.

In another study, the objective was to extend the spray freezing into liquid (SFL) process to produce rapid dissolving, high potency powders with high surface areas and dissolution rate.[50] The spray freezing into liquid particle engineering process was developed to improve the wetting and enhance the dissolution rate of poorly water soluble APIs. Briefly, a feed solution containing poorly water soluble API and excipients is directly atomized into a cryogenic liquid to produce frozen particles. The frozen particles are then collected and lyophilized to obtain dry powders.[51] The atomization in conjunction with rapid freezing rates have led to nanostructured aggregates composed of amorphous API nanoparticles. Pharmaceutical powders produced by the current particle engineering technologies often show very low drug potency or drug/surfactant ratio: these SFL formulations had relatively low drug potency, typically 33%. In order to increase the potency to 50–90%, high concentrations of APIs were dissolved in pure or mixed organic solvents to prepare the feed solutions. Only small amounts of surfactant or polymer were sufficient to form SFL nanostructured aggregates with amorphous API, high surface areas and enhanced wettability; these properties enhanced dissolution rates.

Danazol was used as a model API and the dissolution rate was determined as a function of the danazol potency. High potency danazol/PVP K-15 powders were produced by the SFL process using acetonitrile.[52] The potency of the SFL danazol/PVP K-15 powders ranged from 33 to 91%. The sample with 91% potency was prepared from the acetonitrile/methylene chloride mixture. A relatively small amount of PVP K-15 formed SFL nanostructured aggregates with amorphous structure. These properties of the SFL powders led to fast dissolution rates. Bulk danazol showed a large crystalline plate shape with a fractured edge. SEM analysis revealed that the SFL danazol/PVP K-15 nanostructured aggregates had a porous morphology and were composed of many smooth primary nanoparticles with a diameter of about 100 nm. SFL danazol/PVP K-15 micro-aggregates had a geometric diameter of about 700 nm. In particular, SFL aggregates with 91% potency were about 3 μm in diameter, which were larger and less porous compared to the SFL aggregates with lower potency.

The dissolution rate is directly proportional to the wetted surface area of the API, which in turn increases with decreasing particle size. The dissolution rate of micronized bulk danazol was slow; only 30% of the danazol dissolved in 2 minutes. However, about 87% of the danazol in the SFL powder with 91% potency dissolved in 2 minutes and 100% of the danazol dissolved in 10 min. The increased dissolution rate of the SFL powders may be attributed to amorphous nature of danazol, reduction in particle size, enhanced surface area and increased wettability.

Nanoparticles could represent a novel strategy to control the pharmacokinetics of inhaled drugs and to create an effective prolonged release of drug deposited into lungs. Normally, particles deposited on the pulmonary mucosa are removed in a few hours due to the efficient clearance of API either through phagocytosis in the alveolar region or *via* the rapid absorption of delivered therapeutics. Therefore, development of appropriate drug-loaded nanoparticles with suitable aerodynamic characteristics that have prolonged release of drug once deposited in the lung, is considered one of the major challenges in inhalation therapy. In particular, the drug-loaded carrier particles, targeted to the deep lung, should be aerodynamically small enough to reach the lungs (0.5–5 μm) but able to escape from the alveolar macrophages.

A particular carrier system was developed for controlling pulmonary drug delivery and combines the benefits of both NPs and the respirable/swellable hydrogel microparticles. Drug-loaded hydrogel microparticles are particles with aerodynamic characteristics allowing them to be respirable when dry but attain large swollen sizes once deposited on moist lung surfaces to reduce macrophage uptake rates. These microparticles are based on PEG graft copolymerized onto chitosan in combination with Pluronic F-108 and were prepared *via* cryomilling.[53,54]

The ensuing carrier system comprised of semi-interpenetrating polymeric network (semi-IPN) microspheres. This semi-IPN was based on spray-dried ionotropically crosslinked alginate microspheres containing homogeneously entrapped self-assembled nanoparticles. The nanoparticles were prepared fr... the *N*-phthaloyl chitosan (NPHCs) graft copolymerized with poly(ethylene glycol) (PEG).[55] Semi-IPN hydrogel microspheres were loaded with bovine serum albumin (BSA), as a model for protein drugs, and the *in vitro* release profile was determined in PBS, pH 7.4 at 37 °C. The BSA release from the microspheres showed three different phases. Initially, there was a burst release within the first 2 h. This burst release was attributed to the fast initial swelling of the hydrogel microspheres. Then, the microspheres showed a slow sustained release of BSA up to 4 days. The drug released in this stage may be the drug that is physically entrapped in the microspheres. In the final phase, a slower release was observed which may be attributed to the release from the nanoparticles upon starting degradation of the microspheres.

The objective of a recent investigation was the preparation of surfactant-free composite microcarriers (MC) containing redispersable NPs suitable for dry-powder inhalation.[56] Nanoparticles in this case were prepared by a solvent displacement technique avoiding any surfactants. Microcarriers were prepared by spray-drying nanoparticle suspensions with lactose, mannitol or α-cyclodextrin as stabilizers. Stabilization of nanoparticle suspensions by spray-drying with α-cyclodextrin yielded redispersible particles smaller than 200 nm and was a more efficient than commonly used excipients.

Nanoparticle uptake into macrophages was studied in a cell culture model of promonocytic cell line U937 against PLGA reference microparticles at time points between 30 minuntes and 8 hours. The cell culture experiments with redispersed nanoparticles in RPMI-medium (1 mg mL^{-1}) seem to suggest less interaction and uptake with macrophages compared to polymeric micro-particles. Nanoparticles smaller than 200 nm were not actively taken up by macrophages, showed lower affinity to the macrophages and a smaller fraction was internalized.

Another study showed the development of carrier particles for respiratory drug delivery incorporating effervescent technology to increase the efficiency of nanoparticles release from DPI. In particular, the effervescence of the carrier particles generated the forces that helped the nanoparticles to disperse more efficiently, avoiding particle aggregation.[57]

In particular, polybutylcyanoacrylate nanoparticles used as model system for pulmonary delivery were included in formulations containing excipients for effervescence and lubricants. Butylcyanoacrylate nanoparticles were prepared by polymerization of the monomers in HCl 0.01 N solution, containing Dextran 70 kDa and fluorescein isothiocyanate-dextran 70 kDa (FITC). The particles were purified by centrifugation and resuspended in sterile water. The active release mechanism due to the effervescence increased drug dissolution and enhanced the dispersion of nanoparticles over the formed gas bubbles interface. Using lactose carrier particles that dissolve without effervescence or the new effervescent carrier particles, drug release and dispersion of nanoparticles were compared.

Butylcyanoacrylate nanoparticles were included in effervescent or non-effervescent lactose carrier particles using a spray-drying technique. In order to compare the different formulations and the effects of effervescence and excipients, four types of powders were produced. A suspension containing nanoparticles was added to either a 7% lactose solution, or to a 7% lactose solution containing PEG 6000 and L-leucine, or to an effervescent formulation solution, or to an effervescent formulation solution containing PEG 6000 and L-leucine. Sodium carbonate and citric acid were used to create the effervescence reaction. Ammonia was used to increase the pH of the solution to inhibit an effervescent reaction prior to spray drying. The pH was maintained at approximately 8.0.

Different amounts of lactose had a large impact on the size and morphology of the carrier particles. Increasing the amount of lactose led to smaller and denser particles and also produced particles with a more spherical shape. The MMAD was around 10 μm and the FPF was between 12.5 and 17.9%. A large improvement in respirability due to reduced particle size MMAD was achieved when 5 mL of both solutions containing 2.5% L-leucine and 2.5% PEG 6000 was added to the formulation. In this case the FPF value was equal to 46.5% and the MMAD to 2.17 μm. The nanoparticles were distributed continuously throughout the carrier particle matrix.

The effervescent properties of the microparticles were observed when they were exposed to water, aqueous surfaces or moist air.

Figure 5.2.3 shows the swelling and dissolution of the carrier particles after exposure to humid air. The lactose matrix of the particles dissolved, while a bubble of more than 30 μm was developed (Figure 5.2.3 (B)). Nanoparticles were distributed throughout the gas bubble.

The formulations with effervescent release mechanisms and the lactose formulations that contained L-leucine and PEG 6000 were able to release nanoparticles less agglomerated compared to the carrier particles made of lactose without an active release. This study demonstrated the promising effect of effervescent reaction as active release mechanism of nanoparticles included in DPI.

In another study, microspheres containing lipid/chitosan nanoparticle complexes, intended for the pulmonary administration of macromolecules, using a spray-drying technique, were reported. For this purpose, mannitol, known for its non-toxic and degradable properties, was chosen as a

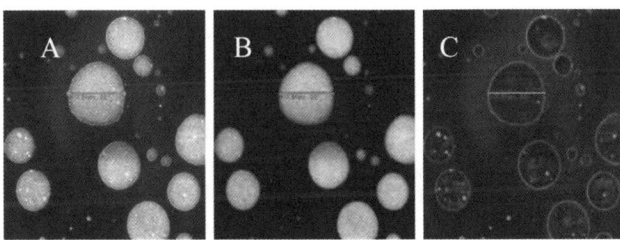

Figure 5.2.3 Confocal microscopy picture of effervescent particles exposed to humidity. (A) Super-imposed pictures B and C, showing gas bubbles of different diameters. (B) Nanoparticles distributed throughout the gas bubble. (C) Dissolved carrier matrix; the diameter of the largest circle is 31.94 μm. Reproduced with permission from ref. 57.

microencapsulation excipient and insulin as model protein. Chitosan, a very attractive polysaccharide due to its biodegradability and mucoadhesivity, has demonstrated low or absent toxicity in cell lines representative of the pulmonary route (16HBE14o- and Calu-3). These nanoparticle-loaded microspheres, obtained by spray-drying a suspension of nanoparticles in mannitol, exhibited adequate aerodynamic properties for lung delivery and proved to be biocompatible with respiratory epithelial cell layers (Calu-3 and A549).[58]

In this work, chitosan nanoparticles (CS-NP) were prepared using the ionotropic gelation of CS with pentasodium tripolyphosphate (TPP). CS and TPP were dissolved in purified water in order to obtain solutions of 1 mg mL^{-1} and 0.42 mg mL^{-1}, respectively. The spontaneous formation of nanoparticles occurred upon incorporation of 1.2 mL of the TPP solution in 3 mL of the CS solution.

The insulin loaded CS-NPs were obtained following protein dissolution in NaOH 0.01 M and its consequent incorporation in the TPP solution. The insulin concentration in the TPP solution was calculated in order to obtain CS-NPs with a theoretical content of 30% (w/w) insulin respective to CS. CS-NPs were concentrated by centrifugation and re-suspended in 100 µL of purified water.

Unloaded CS-NP presented a size around 430 nm and a positive zeta potential of approximately +44 mV. Insulin was associated to the nanosystem with an efficiency of 68%, achieving a particle drug loading of 36%.

The second step was the preparation of lipid chitosan nanoparticle (L/CS-NP) complexes by adding the previously prepared CS-NP suspension to a dry lipid film prepared using dipalmitoylphosphatidylcholine (DPPC) and dimyristoylphosphatidyl glycerol (DMPG). This film was hydrated for 30 min with a suspension of the CS-NP (unloaded or insulin loaded), forming L/CS-NP systems of ratio 3 : 1 (w/w). The two formulations of L/CS-NP display a similar size of around 2 µm.

Finally, dry powders containing the L/CS-NP complexes were obtained by spray-drying an aqueous solution of mannitol with the L/CS-NP suspension. The chosen carbohydrate/complexes ratio was the optimum used to prepare microspheres containing mannitol and nanoparticles. Previous work on the spray-drying of solid lipid nanoparticles demonstrated that the presence of carbohydrates like mannitol, lactose and trehalose provided an increased stability to the spray-dried product, because the sugar layer around particles prevented the lipids coalescence.[59]

The spray-drying technique led to the production of well-defined microspheres with spherical shape, and no aggregation. The Feret diameters were approximately 3 µm, apparent tap densities were low and significantly different for both formulations (containing DPPC and DPPC+DMPG), varying between 0.4 and 0.5 g cm^{-3} and real densities were around 1.4 g cm^{-3}. These properties contributed to aerodynamic diameters between 2.1 and 2.7 µm.

The release of insulin was determined by incubating the different formulations (CS-NP, L/CS-NP complexes and complex-loaded microspheres) in 5 ml of pH 7.4 phosphate buffer with mild horizontal shaking at 37 °C. A further release study using a physical mixture of insulin and unloaded L/CS-NP complexes was performed. This was used as a control in order to evaluate the occurrence of adsorption phenomena.

The insulin release from CS-NP was very rapid, exhibiting the typical initial burst effect, and at 15 min the maximum amount (80%) of insulin was already delivered. The authors say this release behaviour suggests that the interaction between CS and insulin is very weak, allowing the insulin release from the CS-NP by a dissociation mechanism. The insulin release profile displayed by the two L/CS-NP formulations assayed is significantly different to that of CS-NP (Figure 5.2.4). A slight initial burst effect, followed by a very slight increase in the protein release until 90 min was observed; the formulation containing only DPPC released 43% of the insulin content at 15 min, while the one containing both lipids delivered 30% in the same period of time.

The difference found in the release profile presented by both L/CS-NP complexes formulations was undoubtedly the result of different interactions between the CS-NP and the phospholipids of each formulation of complexes. Finally, concerning the L/CS-NP complex-loaded microspheres, it could be

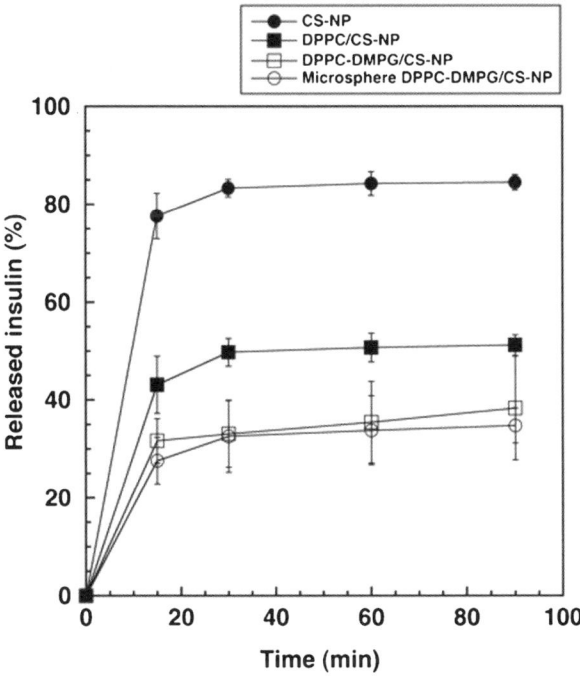

Figure 5.2.4   Release profile of insulin from (●) chitosan nanoparticles (CS-NP), (■) dipalmitoylphosphatidylcholine nanoparticle complexes (DPPC/CS-NP), (□) dipalmitoylphosphatidylcholine-dimyristoylphosphatidyl glycerol nanoparticles complexes (DPPC-DMPG/CS-NP), (○) microsphere containing DPPC-DMPG/CS-NP complexes. Modified from ref. 59.

confirmed that, as expected, mannitol did not influence the protein release profile, allowing the immediate delivery of the complexes. Therefore, mannitol would act only as an inert carrier of the L/CS-NP complexes.

The presence of phospholipids was a determinant in controlling the release of the encapsulated insulin, particularly when both lipids, DPPC and DMPG, were present. Moreover, the microencapsulation process did not have any effect on the insulin release profile.

A novel approach to sustained-release aerosol formulations involved the creation of a micron-scale porous particle composed of self-assembled biodegradable nanoparticles that delivers the nanoparticles effectively to the lungs. Nanoparticle formulations are also an interesting strategy to sustained delivery in the lungs given the avid uptake of nanoparticles by macrophages and other cells of the immune system. This is particularly interesting in the treatment of tuberculosis (TB) since the *Mycobacterium* is taken up by alveolar macrophages. Rifampicin, an anti-tuberculosis antibiotic, was formulated in a dry powder nanoparticle form suited for reducing the treatment dose of antibiotics for TB treatment and for the steady delivery of antibiotic to infected cellular tissue.[60]

Sung and co-authors developed "porous nanoparticle-aggregate particles" (PNAP) which released the nanoparticles from the aggregate once delivered into the body, acting as "Trojan" delivery systems for nanoparticles that would, otherwise, not deposit in the lungs as they are in dry form.[60] The PNAP could be composed entirely of nanoparticles or the nanoparticles can be dispersed throughout a matrix of an inert pharmaceutical excipient, such as leucine, a hydrophobic amino acid previously demonstrated to improve powder dispersibility and aerosolization properties.

Briefly, rifampicin-containing nanoparticles (NP) were prepared using a solvent evaporation method. An aliquot of solution containing PLGA and rifampicin in dichloromethane (DCM) was emulsified with 1% PVA aqueous solution. The emulsion was transferred to a 0.1% PVA solution and allowed to stir overnight for removal of DCM. The nanoparticles were collected by centrifugation and washed with fresh water to remove residual PVA and drug.

Then, porous particles containing 40% nanoparticles by weight (PNAP40) or 80% NP by weight (PNAP80) were prepared by spray drying a suspension of NPs in a L-leucine water solution. The volume median diameter of both PNAPs formulations was around 4 μm, as determined by laser diffraction. Both powders exhibited properties suitable for deposition in the respiratory tract with FPF 35.5 ± 2.1% and 44.7 ± 2.3% for PNAP40 and PNAP80, respectively. Scanning electron microscopy showed that the PNAP had a thin-walled structure composed of visible aggregated nanoparticles having a spherical structure and smooth surface, as shown in Figure 5.2.5.

In vitro release of rifampicin showed a burst release of approximately 80%, with the remainder available for release over a period beyond eight hours.

Guinea pigs were employed in pharmacokinetic studies. Animals were divided into five groups receiving rifampicin as follows: intravenous (IV) solution, oral suspension and dry powder *via* intratracheal insufflation in three forms: porous rifampicin powder without NPs, PNAPs containing 40% nanoparticles (PNAP40) and PNAPs containing 80% nanoparticles (PNAP80).

Plasma levels revealed statistical differences between insufflated groups and IV or oral treatment at all time points. No significant differences in average rifampicin plasma concentrations at the various time points was found between the groups treated by insufflation (with or without nanoparticles). PNAP delivered to guinea pigs by insufflation achieved systemic levels of rifampicin, detected for six to eight hours. Moreover, rifampicin concentrations remained detectable in lung tissue and cells up to and beyond eight hours.

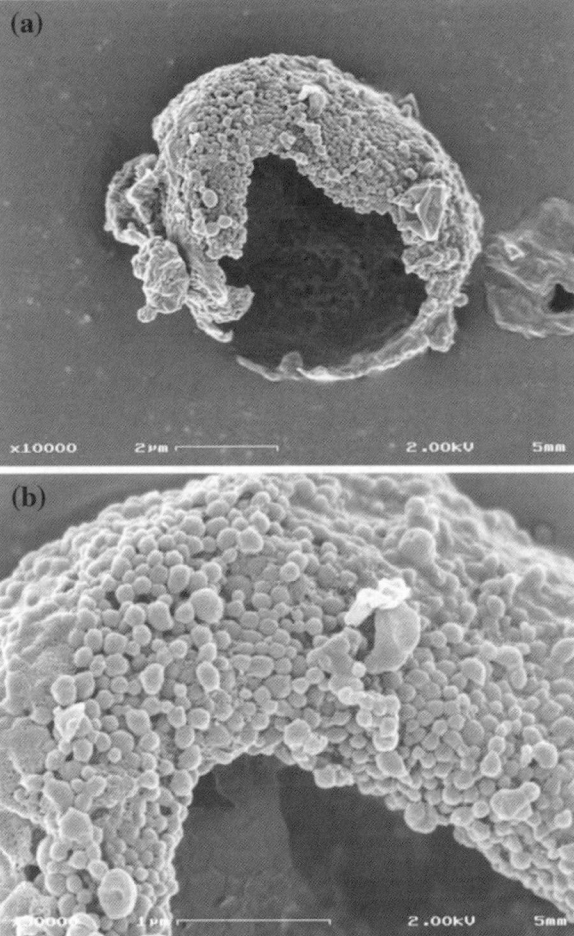

Figure 5.2.5 Scanning electron micrographs of spray dried rifampicin porous particle containing 80% nanoparticles by weight (PNAP80). A magnification of the surface of the PNAP indicates a shell of aggregated nanoparticles with structure intact (scale bars represent: (a) 2 μm, (b) 1 μm). Reproduced with permission from ref. 60.

This was observed both in bronchoalveolar lavage and lung tissue indicating evidence for delayed release from the nanoparticles. Conversely, after pulmonary delivery of an aerosol without nanoparticles, rifampicin could not be detected in the lungs at eight hours. This drug delivery system showed the potential to provide therapeutic advantages for diseases where both systemic and local treatment of lungs may improve efficacy, such as tuberculosis.

To reduce the mucociliary clearance, enzymatic degradation and phagocytosis by alveolar macrophages, Li and coauthors developed microparticles containing nanoparticles (size within the nanometre range, including solid lipid nanoparticles), produced by co-spray drying nanoparticles with bulking agents and dispersibility enhancers.[61]

Thymopentin (TP5) is a synthetic pentapeptide (Arg-Lys-Asp-Val-Tyr) which corresponds to the residues 32–36 of the 49 amino acid thymopoietin. As an immunomodulator, TP5 has been successfully used to treat a variety of diseases, including primary and secondary immune deficiency, autoimmune diseases, infections and cancers. However, because of the extensive metabolism in the gastrointestinal tract and the short half-life ($\leq$30 seconds) in plasma the pulmonary administration route was proposed as a non-invasive alternative for systemic drug delivery.

TP5-loaded SLNs (TP5-SLNs) were prepared by double emulsion technique. TP5 solution (40 mg mL^{-1}) containing sodium cholate was added to a chloroform and ether solution (1 : 1, v/v) containing glyceryl monostearate and soybean phosphatidylcholine. A poloxamer 188 solution was added to this primary water in oil emulsion (W/O) leading to the formation of a water-in-oil-in-water emulsion (W/O/W). The organic solvent was removed by evaporation under reduced pressure using a rotary evaporator.

Then, inhalable microparticles were prepared by spray drying of an aqueous suspension containing mannitol/leucine/TP5-SLNs at a ratio of 1/3/1 (w/w). TP5 was labelled with fluorescein isothiocyanate (FITC-TP5).

The *in vivo* study was performed by randomly dividing rats into four groups of 5 animals each. The rats in group 1 were given FITC-TP5 solution by tail vein injection. The rats in group 2 through group 4 received an intrapulmonary insufflation of FITC-TP5 solution, FITC-TP5-SLNs suspension and FITC-TP5 microparticles, respectively. Results showed that the C_{max}, AUC and MRT were significantly increased after the pulmonary delivery of the microparticles, as compared with the IV TP5 group. These results showed that the TP5 absorption was markedly enhanced after the entrapment in the SLNs, thus improving its bioavailability. Also, the long MRT demonstrated the sustained drug release of the SLNs. In a pharmacodynamic study, superoxide dismutase activity and lymphocyte subsets were chosen as the biomarkers to reflect the therapeutic efficacies of different TP5 formulations. The results demonstrated that the TP5 microparticles had a remarkable immunomodulating effect on immunodepressed rats compared with IV TP5 solution.

In conclusion, these results illustrated that the inhalable SLN possessed an ample potential for the pulmonary delivery of the TP5 peptide and that mannitol and leucine, used for their protective effect of nanoparticles, did not affect the release properties.

5.2.5 Conclusions

Nanoparticles represent a technology for drug property modification without chemistry: solubility, permeability, immunogenicity and stability can be optimized in order to adapt the drug properties to the intended delivery and activity. The application of this technology to inhalation is recent. It is still not widely performed, owing to the evident constraints created by the need for product delivery to the lung. This deposition is performed through a process that is made up of three steps, namely dosing, aerosolizing and deposition. In addition, the three steps require a device that constitutes a component of the product together with the formulation. Nanoparticles have to be transiently transformed in order to obtain their deposition in the lung, where an aerodynamic size between 1 and 5 microns is required.

The need to discover new formulations of respirable particles that can carry the nanosystems is the unavoidable requirement for nanoparticle use in inhalation. This nanoparticle engineering has to employ safe excipients and the preference in this sense goes to lipids compared to polymers.

Nanoparticles are very promising for making feasible the prolongation of drug release through inhalation. In conclusion, besides being respirable the formulation must first and foremost be safe.

References

1. J. S. Patton and P. R. Byron, *Nat. Rev. Drug Discov.*, 2007, **6**, 67.
2. D. C. Thompson in Pharmaceutical inhalation aerosol technology, ed. A. Hickey, Marcel Dekker, New York, 2004, p 29.
3. G. Scheuch and R. Siekmeier, *J. Physiol. Pharmacol.*, 2007, **58**, 615.
4. B. L. Laube, *Resp. Care*, 2005, **50**, 1161.
5. V. Alabraba, A. Farnsworth, R. Leigh, P. Dodson, S. C. L. Gough and T. Smyth, *Diabetes Technol. Ther.*, 2009, **11**, 427.
6. E. C. Walvoord, A. de la Peña, S. Park, B. Silverman, L. Cuttler, S. R. Rose, G. Cutler, S. Drop and J. J. Chipman, *J. Clin. Endocrinol. Metab.*, 2009, **94**, 2052.
7. Y. Roth, J. S. Chapnik and P. Cole, *Ann. Otol. Rhinol. Laryngol.*, 2003, **112**, 264.
8. U. Griesenbach, D. M. Geddes and E. W. Alton, *Gene Ther.*, 2006, **13**, 1061.
9. R. B. Moss RB, C. Milla, J. Colombo, F. Accurso, P. L. Zeitlin, J. P. Clancy, L. T. Spencer, J. Pilewski, D. A. Waltz, H. L. Dorkin, T. Ferkol, M. Pian, B. Ramsey, B. J. Carter, D. B. Martin and A. E. Heald, *Hum. Gene Ther.*, 2007, **18**, 726.
10. S. Azarmi, W. H. Roa and R. Lobenberg, *Adv. Drug Deliv. Rev.*, 2008, **60**, 863.

11. M. M. Bailey and C. J. Berkland, *Med. Res. Rev.*, 2008, **29**, 196.

12. T. C. Carvalho, J. I. Peters and R. O. Williams III, *Int. J. Pharm.*, 2011, **406**, 1.

13. C. O'Callaghan and P. W. Barry, *Thorax*, 1997, **52**(Suppl 2), S31.

14. O. Nerbrink, M. Dahlbäk and H. C. Hansson, *J. Aerosol Med.*, 1994, **7**, 259.

15. K. M. G. Taylor and O. N. M. McCallion, *Int. J. Pharm.*, 1997, **153**, 93.

16. S. L. Tiano and R. N. Dalby, *Pharm. Dev. Technol.*, 1996, **1**, 261.

17. K. Nikander, M. Turpeinen and P. Wollmer, *J. Aerosol Med.*, 1999, **12**, 47.

18. T. S. Wiedmann, L. DeCastro and R. W. Wood, *Pharm. Res.*, 1997, **14**, 112.

19. K. O. Ostrander, H. W. Bosch and D. M. Bondanza, *Eur. J. Pharm. Biopharm.*, 1999, **48**, 207.

20. N. Hernandez-Trejo, O. Kayser, H. Steckel and R. H. Müller, *J. Drug Targeting*, 2005, **13**, 499.

21. M. Keller, J. Jauernig, F. C. Lintz and M. Knoch in *Respiratory Drug Delivery VIII*, ed. R. N. Dalby, P. R. Byron, J. Peart and S. J. Farr, Davis Horwood International Publishing, Godalming, 2002, p. 197.

22. W. K. Kraft, B. Steiger, D. Beussink, J. N. Quiring, N. Fitzgerald, H. E. Greenberg and S. A. Waldman, *J. Clin. Pharmacol.*, 2004, **44**, 67–72.

23. L. A. Dailey, T. Schmehl, T. Gessler, M. Wittmar, F. Grimminger, W. Seeger and T. Kissel, *J. Controlled Release*, 2003, **86**, 131.

24. L. A. Dailey, E. Kleemann, M. Wittmar, T. Gessler, T. Schmehl, G. Roberts, W. Seeger and T. Kissel, *Pharm. Res.*, 2003, **20**, 2011.

25. C. Jacobs and R. H.Müller, *Pharm. Res.*, 2002, **19**, 189.

26. W. Abdelwahed, G. Degobert, S. Stainmesse and H. Fessi, *Adv. Drug Deliv. Rev.*, 2006, **58**, 1688.

27. C. B. Packhauser, K. Lahnstein, J. Sitterberg, T. Schmehl, T. Gessler, U. Bakowsky, W. Seeger and T. Kissel, *Pharm. Res.*, 2008, **26**, 129.

28. M. Nassimi, C. Schel, H. D. Lauenstein, R. Hussein, H.G. Hoymann, W. Koch, G. Pohlmann, N. Krug, K. Sewald, S. Ritinghausen, A. Braun and C. Müller-Goymann, *Eur. J. Pharm. Biopharm.*, 2010, **75**, 107–116.

29. Y. Kawashima, H. Yamamoto, H. Takeuchi, S. Fujioka and T. Hino, *J. Controlled Release*, 1999, **62**, 279.

30. H. Yamamoto, Y Kuno, S. Sugimoto, H, Takeuchi and Y. Kawashima, *J. Controlled Release*, 2005, **102**, 373.

31. F. Gagnadoux, J .Hureaux, L. Vecellio, T. Urban, A. Le Pape, I. Valo, J. Montharu, V. Leblond, M. Boisdron-Celle, S. Lerondel, C. Majoral, P. Diot, J. L. Racineux and E. Lemarie, *J. Aerosol. Med.*, 2008, **21**, 61.

32. R. R. Patolla, M. Chougule, A. R. Patel, T. Jackson, P. N. V. Tata and M. Singh, *J. Controlled Release*, 2010, **144**, 233.

33. M. A. Videira, M. F. Botelho, A. C. Santos, L. F. Gouveia, J. J. Pedroso de Lima and A. J. Aleida, *J. Drug. Targeting*, 2002, **10**, 607.

34. A. Zahoor, S. Sharma and G. K. Khuller, *Int. J. Antimicrob. Agents*, 2005, **26**, 298.

35. W. Yang, K. P. Johnston and R. O. Williams III, *Eur. J. Pharm. Biopharm.*, 2010, **75**, 33.

36. T. O'Riordan, *Resp. Care*, 2002, **47**, 1305.
37. R. Gupta, M. Hindle, P. R. Byron, K. Cox and D. D. McRae, *Aerosol Sci. Technol.*, 2003, **37**, 672.
38. P. Chattopadhyay, B. Y. Shekunov, D. Yim, D. Cipolla, B. Boyd and S. Farr, *Adv. Drug Deliv. Rev.*, 2007, **59**, 444.
39. S. P. Newman, *Resp. Care*, 2005, **50**, 1177.
40. M. Crampton, R. Kinnersley and J. Ayres, *J. Aerosol Med.*, 2004, **17**, 33.
41. P. A. Dickinson, S. W. Howells and I. W. Kellaway, *J. Drug. Targeting*, 2001, **9**, 295.
42. B. K. Nyambura, I. W. Kellaway and K. M. G. Taylor, *Int. J. Pharm.*, 2009, **375**, 114.
43. B. Bharatwaj, L. Wu, J. A. Whittum-Hudson and S. R. P. da Rocha, *Biomaterials*, 2010, **31**, 7376.
44. P. Selvam, I. M. El-Sherbiny and H. D. C Smyth, *J. Aerosol Med.*, 2011, **24**, 25.
45. J. D. Engstrom, J. M. Tam, M. A. Miller, R. O. Williams III and K. P. Johnston, *Pharm. Res.*, 2009, **26**, 101.
46. J. M. Tam, J. D. Engstrom, D. Ferrer, R. O. Williams III and K. P. Johnston, *J. Pharm. Sci.*, 2010, **99**, 3150.
47. M Chougule, B Padhi and A Misra, *Int. J. Nanomed.*, 2007, **2**, 675.
48. N El-Gendy, EM Gorman, EJ Munson and C Berkland, *J. Pharm. Sci.*, 2009, **98**, 2731.
49. L. Thorsson and D. Geller, *Respir. Med.*, 2005, **99**, 836.
50. T. L. Rogers, A. C. Nelsen, M. Sarkari, T. J. Young, K. P. Johnston and R. O. Williams III, *Pharm. Res.*, 2003, **20**, 485.
51. Z. Yu, T. L. Rogers, J. Hu, K. P. Johnston and R. O. Williams III, *Eur. J. Pharm. Biopharm.*, 2002, **54**, 221.
52. J. Hu, K. P. Johnston and R. O. Williams III, *Int. J. Pharm.*, 2004, **271**, 145.
53. M. El-Sherbiny and H. D. C. Smyth, *J. Microencapsulation*, 2010, **27**, 657.
54. M. El-Sherbiny, S. McGill and H. D C. Smyth, *J. Pharm. Sci.*, 2010, **99**, 2343.
55. M. El-Sherbiny and H. D. C. Smyth, *Int. J. Pharm.*, 2010, **395**, 132.
56. T. Lebhardt, S. Roesler, H. P. Uusitalo and T. Kissel, *Eur. J. Pharm. Biopharm.*, 2011, **78**, 90.
57. L. Ely, W. Roa, W. Finlay and R. Löbenberg, *Eur. J. Pharm. Biopharm.*, 2007, **65**, 346.
58. Grenha, C. Remuñán-López, E. L. Carvalho and B. Seijo, *Eur. J. Pharm. Biopharm.*, 2008, **69**, 83.
59. Grenha, B. Seijo and C. Remuñán-López, *Eur. J. Pharm. Sci.*, 2005, **25**, 427.
60. J. C. Sung, D. J. Padilla, L. Garcia-Contreras, J. L. Verberkmoes, D. Durbin, C. A. Peloquin, K. J. Elbert, A. J. Hickey and D. A. Edwards, *Pharm. Res.*, 2009, **26**, 1847.
61. Y. Z. Li, X. Sun, T. Gong, J. Liu, J. Zuo and Z. R. Zhang, *Pharm. Res.*, 2010, **27**, 1977.

Section 6
Nanostructures Overcoming the Skin Barrier

CHAPTER 6.1

Physiological and Mechanistic Issues of the Skin Barrier

M. J. ALVAREZ-FIGUEROA* AND PABLO M. GONZÁLEZ

Departamento de Farmacia, Facultad de Química, Pontificia Universidad
Católica de Chile, Vicuña Mackenna 4860, Santiago, Chile
*E-mail: mjalvare@uc.cl

6.1.1 The Skin

The skin is the largest and most accessible organ in the body. It is highly elastic, resistant to tensile stress with a remarkable regenerative capacity. Typically, its surface area is about 1.2–1.3 m^2, with less than 2mm thickness, accounting for 16% body weight in an adult person. It is considered as a highly perfused organ, receiving around a third of the blood coming from the systemic circulation. Normal skin has a pH around 5 showing circadian variations of no more than 0.5 units during the day.[1–3]

Skin can be divided into 3 layers: epidermis, the external one; hypodermis, the deepest one; and dermis in the middle of both. The rate limiting barrier for diffusion is the *stratum corneum*, the outer-most epidermal sublayer.

Stratum corneum is mainly composed of lipids with water content around 20%.[4] Common lipids include glucolipids, sterol esters, triglicerides, cholesterol sulfate and hidrocarbon compounds. While ceramides, free sterols, and free fatty acids can also be present, phospholipids are absent.[5,6] However, its precise composition and thickness is not homegeous throughout the body, being especially thicker at the palms of the hand and plants of the feet, where keratin filaments are more densely packed.[7,8] *Stratum corneum* cells, corneocytes, are metabolically inactive due to the absence of a nucleus or

RSC Drug Discovery Series No. 22
Nanostructured Biomaterials for Overcoming Biological Barriers
Edited by Maria Jose Alonso and Noemi S. Csaba
© The Royal Society of Chemistry 2012
Published by the Royal Society of Chemistry, www.rsc.org

organelles and their high-keratin content cytosol.[5,7] This stratum has an essential role on the transport of water and chemicals across the skin. Being predominantly lipophilic, it is particularly impermeable to hydrophilic drugs.[4]

The second skin layer, dermis, is highly vascularized and permeable to solutes, with a high expression of metabolic enzymes. Elastin and collagen fibers in this layer are responsible for the mechanical support of skin.[9,10] Appendages such as hair follicles, sebaceous and sweat glands can be found at this level.[2,8,9]

The deepest layer, hypodermis, is rich in adiposities, acting as a fat reservoir that helps maintaining body temperature.[8,11]

6.1.2 Transdermal Drug Delivery

The use of topical skin products has been practiced for centuries. However, using skin as a route to deliver drugs for a systemic effect is fairly recent. In fact, the first transdermal delivery system to be approved for systemic use dates from 1979, and consisted of a patch loaded with scopolamine for the treatment of nausea and vomiting associated to motion sickness.[12]

One of the main advantages of delivering drugs *via* the transdermal route is avoiding hepatic first-pass metabolism. Additionally, constant drug plasma levels are easier to maintain reducing the potential for side effects associated with fluctuations in plasma concentrations.[12,13] Besides, it offers a simple and comfortable therapeutic regime to patients allowing for once-a-day or even once-a-month dosing, reducing intra- and inter-patient variability, and promoting patient compliance. Finally, by simply removing the vehicle the clinical or toxic effect can be rapidly finished.[1,13–15] However, to take advantage of these benefits, sophisticated delivery systems must be developed most of which are not suitable when rapid onset of action or high drug plasma levels are sought.[1]

The considerations above suggest that potent drugs with short half-life, extensive first-pass metabolism, narrow therapeutic window, that are chemically unstable in gastrointestinal fluids are good candidates for transdermal delivery.[13,16,17] The actual delivery rate from a transdermal system is linearly related to the surface area, and there is general agreement about the maximum daily dose that can be absorbed is no more than 10 mg.[15,18] Currently, there are about 40 transdermal products in the market carrying around 20 drugs with diverse chemical structure (Table 6.1.1).[12]

6.1.3 Routes for Transdermal Absorption

For a drug to reach the systemic circulation it must first cross the *stratum corneum*. Then, it will diffuse through the epidermis, and into the dermis, to be ultimately absorbed into the body. Since dermis is rich in capillaries that offer no resistance to mass transport, sink condition is thought to prevail. For this

Table 6.1.1 Summary of drugs for transdermal delivery in the market.

Buprenorphine	Methylphenidate
Capsaicin	Nicotine
Clonidine	Nitroglycerin
Diclofenac	Oxybutynin
Estradiol	Piroxicam
Estradiol/Northindrone	Progestin/estrogen
Estrogen/Progesterone	Rivastigmine
Fentanyl	Rotigotine
Granisetron	Scopolamine
Ketoprofen	Selegiline
Lidocaine	Testosterone

reason, the rate-limiting barrier for transdermal absorption of drugs is the *stratum corneum*.[8]

The amount of drug that permeates through *stratum corneum* is dependent on its keratin content and hydration condition. In fact, while absorption is lowered by either high keratin or low water content, high hydration levels favor permeation. This is important to keep in mind if one considers that a totally hydrated *stratum corneum* can incorporate water up to 5–6 times its own weight.[5,8,19]

Mechanistically, the overall flux of a given drug through the *stratum corneum* is the result of two parallel permeation routes: transcellular and intercellular. In the transcellular route drugs have to move across the corneocytes and the intracellular protein matrix they contain. On the other hand, the intercellular route entails the diffusion of solutes around cells and through the lipid extracellular matrix.[9,20]

Alternatively, drugs can be absorbed into the dermis by using the conducts of appendages such as hair follicles and ecrine glands. However, since these routes represent only a small fraction of the total skin surface area (0.1–1%), their contribution to the total transdermal mass transport is very limited. Nevertheless, ions, high molecular weight, and polar polyfunctional drugs enter the body predominantly by this route. Iontophoresis, a technique that applies an electric current through the skin, forces molecules to move through these conducts so their contribution accounts for up to 50% of the total transport.[9,21]

The following sections will focus on drug properties and drug delivery strategies that promote permeation through *stratum corneum*, the rate-limiting step for transdermal drug absorption.[8]

6.1.4 Factors Determining Transdermal Permeability

To be able to predict the absorption of a molecule through the skin several elements have to be considered and can be grouped into three main factors: (1) physicochemical properties of drugs; (2) physicochemical properties of the delivery system; and (3) physiological and physiopathological conditions of the skin.

6.1.4.1   Physicochemical Properties of Drugs

Charge state: The ionization state of an acid-base drug is a function of its pK_a, the pH of the formulation, and the pH of skin (around 5 for a healthy adult). At this pH, the net negative charge of *stratum corneum* favors permeation of cationic compounds compared to neutral ones.[22]

Solubility in the vehicle: Solution-based systems allow for a more rapid permeation through skin while a suspension-based system permits a continuous flux of drug.[16]

Molecular weight: Transdermal absorption is inversely related to molecular weight, where large molecules (*i.e.* $M_w > 500$ Da) show very limited permeability.[12,23,24] However, this molecular weight cut-off has to be considered with caution since the three-dimensional structure of compounds can also impact transport.[12]

Lipophilicity: The ease of permeation through *stratum corneum* is determined by a precise balance between the hydrophilicity and lipophilicity of a compound (*i.e.* *n*-octanol partitioning, $K_{o/w}$). In fact, a log $K_{o/w}$ in the range from 1 to 3 has been shown to be optimal for transdermal absorption.[12,25]

A study on skin permeability (k_p) of more than 90 compounds, published in 1992, proposed a simple model to predict transdermal absorption of compounds based on both molecular weight (or volume, MV) and log $K_{o/w}$ (equation 6.1.1):[26]

$$\text{Log } k_p = -2.7 + 0.71 \log K_{o/w} - 0.0061 \times MV \tag{6.1.1}$$

6.1.4.2   Physicochemical Properties of the Delivery System

The physicochemical properties of the delivery system can be modified by incorporating substances aimed at modulating the rate and extension of drug absorption. Two classical strategies include increasing the water content of *stratum corneum* to enhance drug permeability and promoting the transdermal permeability *via* permeation enhancers (see below).

6.1.4.3   Physiological and Physiopathological Conditions of the Skin

Skin condition: Healthy normal skin is a very efficient barrier that limits the transepithelial loss of water and protects the body against a wide variety of injuries such as chemical substances, microbial colonization, and ultraviolet radiation. Any of these elements, or others such as a trauma that compromises the skin integrity, will diminish its barrier capacity.[23]

Hydration condition: Even though a higher hydration level promotes transdermal absorption, a dry skin condition can, paradoxically, also facilitate permeation of substances across the *stratum corneum* probably by altering the structure of the barrier.[27]

Metabolic capacity: Both epidermis and dermis are metabolically active layers and can play a role on drug biotransformation before reaching systemic circulation.[14,28] In fact, esterases are widely expressed at this level, suggesting that transdermal delivery of ester-prodrugs could be an interesting strategy to enhance bioavailability.[29,30] The metabolic capacity of skin has been estimated to be up to 5% of a topically administered drug.[23]

Age: There is a normal decline of bodily functions with age. However, the barrier capacity of *stratum corneum* stays fairly constant throughout the years. This arises from a status in which the decline in production of intercellular lipid production is compensated by the slower turn-over of epidermal cells which in turn leads to a thicker *stratum corneum*.[31]

Ethnicity: Skin characteristics such as size and cohesion of corneocytes, pH, lipid content, microflora, and water retention capacity show great diversity among ethnicity.[32] This is responsible for the idiosyncratic variability seen in the transdermal absorption of the same drug. For this reason, it is recommended to include subjects that represent some ethnic diversity when clinical studies of transdermal products are conducted. Similarly, one should be aware of the morphological, histological, and biochemical difference between animal species when their skin is used as an *in vitro* model. Even though skin from primates, pigs, mice, rats, guinea pigs, and snakes has been commonly used in studies, pig skin has shown better predictability over other animal models.[33] This can be due, at least in part, to the similarities between human and pig skin, such as lipid content of *stratum corneum*; thickness of epidermis and their sublayers; structure and density of hair follicles; and skin lipophilicity (*i.e.* π parameter).[33,34] On the other hand, pig skin shows fewer dermal elastic fibers than human skin;[34] although this difference may not impact predictability since the rate-limiting step is given by *stratum corneum* permeability.

Anatomical zone: There are important differences in thickness and lipid content of epidermis among different body regions that can have a direct impact in transdermal drug absorption. Thinner skin, with a lower lipid content, and higher blood perfusion facilitates drug permeation.[30]

6.1.5 Strategies for Transdermal Drug Delivery

The different technological approaches to develop transdermal drug delivery represent applications of three basic strategies that can be classified as: (1) formulations including substances or vehicles that promote absorption (*e.g.* permeation enhancers, microemulsions, and liposomes); (2) devices that promote the permeation of solutes by physical means without disrupting the epidermal layer (*e.g.* iontophoresis); (3) systems that facilitate the permeation of solutes across the epidermis by disrupting the barrier using physical or mechanical means (*e.g.* sonophoresis, electroporation, and microneedles).[15]

6.1.5.1   Permeation Enhancers

A considerable amount of effort has been directed towards developing specific chemical substances, and their mixtures, that can be used as promoters of transdermal drug absorption.[8,16,35] This group of compounds shows great chemical diversity and includes water-miscible organic solvents, surfactants, fatty acids, and natural oils such as terpenes.

Permeation enhancers may act by one or more of the following mechanisms:[8,13,16–18,20,36–38]

1.   Disrupting the structural organization of lipids in the *stratum corneum*; dimethylsulfoxide (DMSO), Azone®, terpenes, fatty acids, and alcohols.
2.   Interacting with intracellular proteins; DMSO, decylmethylsulfoxide, and ionic surfactants. These compounds are able to alter the helical conformation of keratin decreasing its resistance to mass transfer.
3.   Favoring drug partitioning into the *stratum corneum* by altering its lipophilicity; ethanol.
4.   Increasing the hydration level of *stratum corneum*; surfactants. These compounds favor skin wetting by decreasing the surface tension of the vehicle.
5.   Increasing drug solubility in *stratum corneum*; water-miscible organic solvents

One drawback of permeation enhancers as additives for drug delivery is their irritant effect on skin. It has been shown that oleic acid not only disrupts organization of the lipid barrier but also influences immune cells in the epidermis.[39] Additionally, promoters can increase permeability of other components in the formulation making toxic reactions more likely to appear.[13,17] Terpenes are considered to be less toxic and irritant compared to surfactants and other synthetic compounds. In fact, some of them are included in the GRAS (Generally Recognized As Safe) list of the FDA.[37,40]

6.1.5.2   Microemulsions

Microemulsions are dispersions of two immiscible components, water and oil, stabilized by a mixture of surfactant and co-surfactant.[20,41–44] These are thermodynamically stable systems that form spontaneously, have low viscosity and very small globule size.[20]

Microemulsions enhance transdermal drug penetration by different mechanisms:

1.   Solubilization of active ingredients. Having high solubilizing capacity, microemulsions act as a drug reservoir which can deliver the drug with

zero-order kinetics. This increases their efficiency as drug delivery systems compared to other common vehicles such as gels, lotions, and emulsions.[45,46]

2. Increase of drug thermodynamic activity. This property of microemulsions favors the partitioning of drugs into the *stratum corneum*.[47-49]

3. Reduction of the barrier competence of epidermis.[47-49]

The presence of volatile components together with differences in absorption rate between non-volatile ingredients can lead to changes in the composition of the microemulsion after topical administration. This so-called "effective formulation" can have increased irritant and toxic potential compared to the original one, which may cause permanent skin damage.[20]

Table 6.1.2 summarizes some model drugs that have been incorporated into transdermal microemulsions.[20,50]

Future challenges are using microemulsions to deliver hydrophilic biomolecules such as peptides, proteins, DNA, and RNA for dermal and transdermal delivery. Aqueous-based colloidal phase systems are promising technological approaches.[51]

6.1.5.3 Liposomes

Liposomes are the most extensively studied nanosized vectors. Even though they were originally described more than 80 years ago, studies on their use for transdermal delivery started in the early-80s.[3]

Liposomes are lipid bilayer vesicles typically containing cholesterol and phospholipids, with average diameter of 75 nm or more, that enclose an aqueous compartment where drugs can be loaded.[3] Depending on the manufacturing process uni- or multilamellar liposomes can be prepared.[51]

Table 6.1.2 Model drugs that have been incorporated into transdermal microemulsions.

5-Fluorouracil	Methotrexate
8-Methoxsalen	Nifedipin
Apomorphine hydrochloride	Niflumic acid
Ascorbic acid	Piroxicam
Ascorbyl palmitate	Prilocaine hydrochloride
Diclofenac	Propanolol
Diphenhydramine hydrochloride	Prostaglandin E1
Estradiol	Retinoic acid
Felopine	Sodium ascorbyl phospate
Glucose	Sodium fluorescein
Hematoporphyrin	Sodium salicylate
Indomethacin	Sucrose
Ketoprofen	Triptolide
Lidocaine	

Commonly, liposomes are used to achieve a dermal effect since their limited transdermal permeation leads them to accumulate in the *stratum corneum*.[18,51,52]

Some drugs that have been loaded into liposomes include lidocaine, tetracaine, ciclosporin A, insulin, diclofenac sodium, triamcinolone acetonide, levonorgestrel, ethinylestradiol, low-molecular-weight heparin, zidovudine, and iodide.[51,52]

Currently, a new generation of vesicles has emerged as promising carriers for transdermal delivery. Among them we can mention transfersomes, flexosomes, ethosomes, niosomes, vesosomes, invasomes, and polymerosomes. Vesicular delivery systems will be discussed in greater detail in the next chapter.

6.1.5.4   Iontophoresis

Iontophoresis is the process of increasing the penetration of ions and neutral substances through the skin by applying an external electric current.[12,53-55]

Mechanistically, iontophoresis facilitates the transdermal transport of positive species since the net negative charge of epidermal proteins that predominates a pH 7.4 preferentially enhances the permeability of cations. Therefore, skin can be considered as as permselective to positive ions, with sodium showing the most extensive transdermal flux. In fact, sodium flux accounts for 60% of the total transepithelial current (*i.e.* transport number of 0.6) representing more than twice that of chloride, the main skin anion.[56]

Basically, an iontophoretic system consists of a power source connected to an anodic and a cathodic compartment. The power source sets the current intensity at a given value and, when both compartments make contact with the *stratum corneum*, the skin ions begin to flow closing the circuit. Typically, cationic drugs must be placed in the anodic compartment since they are repelled when the electric current is applied, forcing them to migrate towards the cathode. Likewise, active anionic species must be included in the cathodic compartment so they can move into the anodic one. Optimal current intensity should be determined on a case-by-case basis considering both the surface area and the exposure time. Additionally, the amount and nature of ions in the donor compartment can be easily manipulated, to modify the ionic strength in order to modulate drug absorption.[57-60] Ultimately, the actual amount of drug transported into the skin will depend on its molecular weight and relative mobility.[57-61]

The iontophoretic current enhances transdermal drug absorption by three basic mechanisms:[59,62-65]

1.   Induction of skin depolarization increasing its permeability.
2.   Electrorepulsion of ions that forces them to move into the skin.
3.   Solvent convection through a charged membrane (*i.e.* skin) "drags" cationic and even neutral molecules along with it. This phenomenon is called electroosmosis.

A great number of studies have shown the iontophoretic enhancement of drug transdermal delivery of numerous chemically-diverse drugs belonging to different therapeutic classes. Examples include lidocaine, bupivacaine, fentanyl, methotrexate, cisplatin, vinblastine sulfate, haloperidol, chlorpromazine, piroxicam, diclofenac sodium, ropinirole, metochlopramide, acyclovir, and insulin.[50,66–69] Currently, lidocaine is commercially available as a transdermal iontophoretic patch (Vyteris, Inc., Fairlawn, NJ, USA).[64]

6.1.5.5 Sonophoresis

Sonophoresis employs ultrasound energy to promote the absorption of drugs through the *stratum corneum*. High-frequency sound waves in the range from 2 to 100 kHz have been successfully studied, with ultrasound frequency being proportional to the amount of drug that is able to transverse the skin. The detailed mechanisms by which ultrasound enhances permeation are currently not fully understood.[15,17,70]

Sonophoresis has been used as a means to enhance the transdermal delivery of a series of structurally diverse active compounds, either for systemic or local action. Some examples include: indomethacin, ketorolac, ibuprofen, ketoprofen, lidocaine, Cyclosporin A, insulin, dexamethasone, digoxin, calcein. To date, most of these studies have been conducted on animal models.[70]

6.1.5.6 Electroporation

Electroporation is a technique that increases epidermal permeability to drugs by applying a high voltage pulse to skin (25–2000 V). This electric potential causes a rapid and transitory increment in the number of water-filled pores in the *stratum corneum*. Electroporation has been used as a delivery strategy for small drug-like molecules with diverse physicochemical properties, such as fentanyl (low MW), metoprolol (hidrophilic), and timolol (lipophilic); together with high MW heparin (glycosaminoglycan), and calcitonin (peptide).[4,71–76] For any given drug, transport efficacy strongly depends on voltage intensity, pulse number and duration.[4,77]

6.1.5.7 Microneedles

Microneedles create a micropore through the *stratum corneum* allowing drugs to reach deeper layers of the skin immediately. They are very small structures designed to penetrate no deeper than 100 µm, making them a minimally invasive and pain-free drug delivery system. Microneedles can be designed with a variety of different shapes, sizes, and materials (silicon, glass, stainless steel, titanium, nickel-iron, and polymers), being either solid or hollowed.[12,15,78,79] While hollowed microneedles allow for easy delivery of drugs into the skin through an inner channel, solid ones are first embedded in a solution of active ingredient to coat them.[15,78] Typically, microneedles are incorporated into mini patch-like systems that can be either discarded or biodegraded by the body (rapid or sustained release, respectively) after application.[15,78] A micropore is left at the application site after removal of the delivery device. This pore can close after 12 to 18 hours depending on the size of the microneedle used.[79]

Based on their design, microneedles can be classified into 4 groups:[80]

1. Solid microneedles that puncture the skin increasing its permeability.
2. Solid microneedles coated with powdered drugs or vaccines that can be deposited and dissolved into the skin.
3. Microneedles made of biodegradable or water-soluble polymers such as polylactic co-glycolic acid (PGLA), carboxymethyl-cellulose (CMC), polyvinyl-pyrrolidine (PVP), or maltose. These systems can incorporate encapsulated vaccines for rapid or controlled release.
4. Hollowed microneedles to infuse drugs in a way that resembles a hypodermic injection.

Numerous publications have shown the potential of microneedles as a transdermal drug delivery system. The permeation of biomolecules such as desmopressin, plasmid DNA, insulin, human growth hormone, heparin, and oligonucleotides; vaccines including influenza, hepatitis B or C, diphtheria, anthrax, and human papillomavirus; and drugs like methotrexate, and nicardipine, has been successfully characterized using both *in vitro* and *in vivo* models.[12,81–85]

References

1. Y.W. Chien, *STP Pharma Sciences.*, 1991, **1**, 5.
2. L. C. Junqueira and J. Carneiro, *Histología Básica.* Marson, S.A., España. 4ta Edición, 1996.
3. G. Cevc and U. Vierl, *J. Controlled Release,* 2010, **141**, 277.
4. R. H. Guy, *Pharma Res.*, 1996, **13**, 1765.
5. P. M. Elias in *Topical Drug Delivery Formulations*, ed. D. W. Osborne and A. H. Amann, Marcel Dekker, Inc., New York, 1990, p 13–28.
6. P. M. Elias, *J. Controlled Release*, 1991, **15**, 199.
7. K. Menos and M. Duggan in *Skin Delivery Systems: Transdermals, dermatologicals and cosmetic actives*, ed. J. J. Wille, Blackwell Publishing, USA, 2006, p 25–42.
8. Y. W. Chien, *Novel Drug Delivery Systems*, Marcel Dekker, Inc., New York, 1982.
9. M. H. Abraham, H. Chadha and R. C. Mitchell, *J. Pharm. Pharmacol.*, 1995, **47**, 8.
10. R. H. Guy and J. Hadgraft, *J. Controlled Release*, 1987, **4**, 237.
11. M. A. González in *Topics in Pharmaceutical Sciences*, ed. D. J. A. Crommelin and K.K. Midha, Medpharm Scientific Publishers, Stuttgart, 1992, p 119.

12. R.K. Subedi, S.Y. Oh, M. Chun and H. Choi, *Arch. Pharm. Res.*, 2010, **33**, 339.
13. V. Kneep, J. Hadgraft, J. and R. H. Guy, *Crit. Rev. Ther. Drug Carrier Syst.*, 1987, **4**, 13.
14. R. H. Guy J. Hadgraft and D.A. Bucks, W. *Xenobiotica*, 1987, **17**, 325.
15. R. H. Guy, *Handb. Exp. Pharmacol.*, 2010, **197**, 399.
16. E. R. Cooper and D. C. Patel in *Topical Drug Delivery Formulations*, ed. D. W. Osborne, and A.H. Amann, Marcel Dekker, Inc., New York, 1990, p 1.
17. H. Guy, *Pharm. Res.*, 1996, **13**, 1765.
18. B. Barry, *DDT*, 2001, **6**, 967.
19. M. Walter, T. A. Hulme, M.G. Rippon, R. S. WalmsleyS. Gunnigle, M. Lewin, and S. Winnsey, *J. Pharm. Sci.*, 1997, **86**, 1381.
20. Kogan and N. Garti, *Adv. Colloid Interface Sci.*, 2006, **123–126**, 369.
21. M. S. Roberts, *Clin. Exp. Pharmacol. Physiol.*, 1997, **24**, 874.
22. Barry, *Eur. J. Pharmacol. Sci.*, 2001, **14**, 101.
23. Barry in *Farmacia, la ciencia del diseño de las formas farmacéuticas*, ed. M. E. Aulton, Elsevier, España S.A., 2004 second edition, p 499.
24. J. D. Bos and M. M. Meinardi, *Exp. Dermatol.*, 2000, **9**, 165.
25. Xin-Sheng, D. Cun-Zheng, X. Zhi-Dong and Y. Bao-An, *Biol. Pharm. Bull.*, 2008, **31**, 1045.
26. R.O. Potts and R.H. Guy, *Pharm. Res.*, 1992, **9**, 663.
27. P. Brisson, *Can. Med. Assoc. J.*, 1974, **110**, 1182.
28. Ramachandran and D. Fleisher, *Adv. Drug Deliv*ery, 2000, **42**, 197.
29. Gysler, B. Kleuser, W. Sippl, K. Lange, H. C. Korting, H-D. Höltje and M. Schäfer-Korting, *Pharm. Res.*, 1999, **16**, 1386.
30. S.A. Hotchkiss, *Dermal Metabolism*, Marcel Dekker, New York. 2008.
31. H. Tagami, *Arch. Dermatol. Res.*, 2008, **300**, 1.
32. Flagothier, C. Piérard-Franchimont and G.E. Piérard, *Rev. Med. Liege.*, 2005, **60**, 53.
33. Godin and E. Touitou, *Adv. Drug Delivery Rev.*, *2007*, **59**, 1152.
34. M. Walter, T.A. Hulme, M.G. Rippon, R.S. Walmsley, S. Gunnigle, M. Lewin and S. Winnsey, *J. Invest. Dermatol.*, 1967, **86**, 1381.
35. T. Akimoto, T. Aoyagi, J. Minoshima, and Y. Nagase, *J. Controlled Release*, 1997, **49**, 229.
36. K. Walters in *Transdermal Drug Delivery: Development Issues and Research Initiatives*, ed J. Hadgraft, and R. H. Guy, Marcel Dekker, Inc., New York. 1989.
37. B. Sapra, S. Jain and A. K. Tirway, *AAPS J.*, 2008, **10**, 120.
38. S. Gao and J. Singh, *J. Controlled Release*, 1998, **51**, 193.
39. Touitou, B. Godin, Y. Karl, S. Bujanover and Y. Becker, *J. Controlled Release,* 2002, **80**, 1.
40. M. Aqil, A. Ahad, Y. Sultana and A. Ali, *Drug Discovery Today*, 2007, **12**, 1061.
41. Attwood in *Colloidal Drug Delivery Systems,* ed J. Kreuter, Marcel Dekker, Inc., New York, 1994.

42. H. B. Lawrence in *Pharmaceutical Dosage Forms: Disperse Systems Volume 2.*, ed A. Lieberman, M. M. Rieger and G. S. Banker, Marcel Dekker, Inc., New York, 1989.
43. J. Shaw, *Colloid and Surface Chemistry*, Butterworth–Heinemann Ltd., Oxford. 4th Edition, 1992.
44. W. Warisnoicharoen, A. B. Lansley, and M. J. Lawrence, *Int. J. Pharm.*, 2000, **198**, 7.
45. M. Eccleston in *Encyclopedia of Pharmaceutical Technology*, ed. J. Swarbrick, and J.C. Boylan, Marcel Dekker, Inc., U.S.A., Vol. 9, 1994.
46. M. R. Gasco in *Industrial Applications of Microemulsions*, ed. C. Solans and H. Kunieta, Marcel Dekker, Inc., New York, 1997.
47. B. Baroli, A. López-Quintela, M. B. Delgado-Charro, A.M. Fadda, and J. Blanco-Méndez, *J. Controlled Release*, 2000, **69**, 209.
48. M. B. Delgado-Charro, G. Iglesias-Vilas, J. Blanco-Méndez, M. A. López-Quintela, J.P. Marty and R. H. Guy, *Eur. J. Pharm. Biopharm.*, 1997, **43**, 37.
49. D. W. Osborne, J. I. Ward, and K. J. O'Neill, *J. Pharm. Pharmacol.*, 1991, **43**, 451.
50. M. J. Alvarez-Figueroa and J. Blanco-Méndez, *Int. J. Pharm.*, 2001, **215**, 57.
51. R. H. H. Neubert, *Eur. J. Pharm. Biopharm.*, 2011, **77**, 1.
52. M. A. Elsayed, O. Y. Abdallah, V. F. Naggar and N. M. Khalafallah, *Int. J. Pharm.*, 2007, **332**, 1.
53. Moll, and P. Knoblauch, *Drug. Dev. Ind. Pharm.*, 1993, **19**, 1143.
54. P. Santi, P. L. Catellani, G. Massimo, G. Zanardi, G. and P. Colombo, *Int. J. Pharm.*, 1993, **92**, 23.
55. V. Srinivasam, S. M. Sims, W. I. Higuchi, C. R. Behl, A. W. Malick and S. Pons in *Pulsed and Self-regulated Drug Delivery,* ed J. Kost, CRS Press. 1990.
56. N. H. Yoshida and M. S. Roberts, *Adv. Drug Delivery Rev.*, 1992, **9**, 239.
57. R.R. Burnette in *Transdermal Drug Delivery*, ed J. Hadgraft and R. H. Guy, Marcel Dekker, Inc., New York. 1989.
58. L. P. Gargarosa, and J. M. Hill, *Int. J. Pharm.*, 1995, **123**, 159.
59. B. H. Sage and J. E. Riviere, *Adv. Drug Delivery Rev.*, 1992, **9**, 265.
60. P. Singh and H. I. Maibach, *Clin. Pharmacokinet.,* 1994, **26**, 327.
61. P. G. Green, *J. Controlled Release*, 1996, **41**, 33.
62. M. J. Pikal, *Adv. Drug Delivery Rev.*, 1992, **9**, 201.
63. P. Santi, P. L. Catellani, G. Massimo, G. Zanardi and P. Colombo, *Int. J. Pharm.*, 1993, **92**, 23.
64. Y. Wang, R. Thakur, Q. Fan and B. Michniak, *Eur. J. Pharm. Biopharm.*, 2005, **60**, 179.
65. R. H. Guy, Y. N. Kalia, M. B. Delgado-Charro, V. Merino, A. Y. López and D. Marro, *J. Controlled Release*, 2000, **64**, 129.
66. Y. N. Kalia, A. Naik, J. Garrison and R. H. Guy, *Adv. Drug Delivery Rev.*, 2004, **56**, 619.

67. M. J. Alvarez-Figueroa, M. B. Delgado-Charro and J. Blanco-Méndez, *Int. J. Pharm.*, 2001, **212**, 101.
68. M. J. Alvarez-Figueroa, I. Araya-Silva and C. Díaz-Escobar, *Pharm. Dev. Technol.*, 2006, **11**, 371.
69. M. J. Alvarez-Figueroa and J. V. González-Aramundiz, *Pharm. Dev. Technol.*, 2008, **13**, 271.
70. J. J. Escobar-Chávez, D. Bonilla-Martínez, M. A. Villegas-González, I. M. Rodríguez-Cruz and C. L. Domínguez-Delgado, *J. Pharm. Sci.*, 2009, **12**, 88.
71. D. A. Edwards, M. R. Prausnitz, R. Langer and J. C. Weaver, *J. Controlled Release*, 1995, **34**, 211.
72. R. Vanbever, M. Prausnitz and V. Prèat, *Pharm. Res.*, 1997, **14**, 638.
73. R. Denet, R. Vanbever and V. Préat, *Adv. Drug Delivery Rev.*, 2004, **56**, 659.
74. R. Denet and V. Préat, *J. Controlled Release*, 2003, **88**, 253.
75. M. R. Prausnitz, *Biol. Technol.*, 1995, **13**, 1205.
76. R. Vanbever, M. A. Leroy and V. Préat, *J. Controlled Release*, 1998, **50**, 225.
77. J. J. Escobar-Chávez, D. Bonilla-Martínez, M. A. Villegas-González and A. Revilla-Vásquez, *J. Clin. Pharmacol.*, 2009, **49**, 1262.
78. M. R. Prausnitz, *Adv. Drug Delivery Rev.*, 2004, **56**, 581.
79. Kalluri, C. S. Kolli and A. K. Banga, *AAPS J.*, 2011, **13**, 473.
80. J. Escobar-Chávez, D. Bonilla-Martínez, M. A. Villegas-González, E. Molina-Trinidad, N. Casas-Alancaster and A. L. Revilla-Vázquez, *J. Clin. Pharmacol.,*2011, **51**, 964.
81. Gupta, H. S. Gill, S. N. Andrews and M. R. Prausnitz, *J. Controlled Release*, 2011, **154**, 148.
82. N. Roxhed, B. Samel, L. Nordquist, P. Griss and G. Stemme, *IEEE Trans. Biomed. Eng.*, 2008, **55**, 1063.
83. S. S. S. Lanke, C. S. Kolli, J. G. Stromma and A. K. Banga, *Int. J. Pharm.*, 2009, **365**, 26.
84. V. Vemulapalli, Y. Yang, P. M. Friden and A. K. Banga, *J. Pharm. Pharmacol.*, 2008, **60**, 27.
85. L. Morefield, R. F. Tammariello, B. K. Purcell, P. L. Worsham, J. Chapman , L. A. Smith, J. B. Alarcon, J. A. Mikszta and R. G. Ulrich, *J. Immune Based Ther. Vaccines*, 2008, **6**, 1.

CHAPTER 6.2

Nanostructures Overcoming the Skin Barrier: Drug Delivery Strategies

NATHALIE SCHLEICH AND VÉRONIQUE PRÉAT*

Université catholique de Louvain, Louvain Drug Research Institute, Pharmaceutics and Drug Delivery, Avenue Mounier, 73 bte B1.73.12, 1200 Brussels, Belgium
*E-mail: veronique.preat@uclouvain.be

6.2.1 Topical and Transdermal Drug Delivery

The application of nanoparticles on the skin is an emerging field for topical delivery of active ingredients into the skin for dermatological and cosmetic applications as well as for transdermal drug delivery across the skin for systemic treatment.[1–8] The review will focus on the use of nanocarriers in medical applications for drug delivery through or into the skin.

Nanoparticles designate particles with all dimensions between 1 and 100 nm (American Standard Institute). Nanomedicines have been defined as nanometre size scale complex systems consisting of at least two components, one of which is the active ingredient (European science foundation). Most of the nanoparticles used in the pharmaceutical sciences are spherical drug carriers with a range size between 10 and 200 nm. According to their structure, we can divide them into two groups: the nanocapsules showing typical core-shell structures and the homogeneous nanospheres.[2] Nanoparticles are also distinguished according to their composition: lipidic (e.g. liposomes, deform-

RSC Drug Discovery Series No. 22
Nanostructured Biomaterials for Overcoming Biological Barriers
Edited by Maria Jose Alonso and Noemi S. Csaba
© The Royal Society of Chemistry 2012
Published by the Royal Society of Chemistry, www.rsc.org

able vesicles, solid lipid nanoparticles), polymeric (degradable or not), metallic (*e.g.* gold and silver nanoparticles), mineral (*e.g.* TiO$_2$, ZnO) and organic nanoparticles.[3] In this review we will focus on lipid- and polymer-based nanoparticles used as drug carriers. These nanocarriers have been shown to enhance drug permeation and expand the range of molecules that can be delivered by the transdermal route.[1-8]

The history of using vesicular/nanoparticular systems for drug delivery to and through the skin started three decades ago.[4] The field of nanoparticle delivery to the skin is fragmented in term of the drugs investigated. The lipid- and polymer-based nanoparticles have been mainly investigated and developed for small drug delivery which has been covered in a recent excellent review.[2,5,6] Besides the delivery of small drugs, typical applications of nanoparticles in the skin are (i) TiO$_2$ which have a size of 10–60 nm as sunscreen, (ii) silver nanoparticles to provide slow release of silver ions that have antimicrobial and wound healing properties, (iii) transdermal delivery of macromolecules, (iv) adjuvant (carrier and/or immunomodulator) to induce and enhance immune response for topical vaccine applications.

An overriding concern is the safety of any applied nanoparticles. Due to their nanosize, nanoparticles have completely different possibilities to interact with the body than microparticles, in particular uptake by cells or release of cytotoxic products. The most important factor for this potential nanotoxicity is the lack of biodegradability of nanomaterials which can be taken up and retained in the reticuloendothelial system.[6,7] This underlines the need for extensive toxicity monitoring.

6.2.2 Skin Barrier and Penetration Pathways of Drug Loaded Nanoparticles

The major role of the skin is to act as a defensive barrier to threats form external environment. It provides physical, immunological, metabolic and UV protective barriers including a natural barrier against particle penetration.[6] The difficulty for drug delivery into or through the skin lies in crossing the *stratum corneum* which provides an efficient barrier. Hence, many methods such as the use of chemical enhancers or physical methods have been developed to overcome this problem (see Chapter 6.1). Yet, these methods can damage the skin barrier. Therefore, drug-loaded nanoparticles have been tested and have shown promising results in terms of penetration into or across the skin.

There is ongoing debate on penetration of nanoparticles and nanomaterials in the skin.[8] Drugs can penetrate into the skin by three penetration pathways: the intercellular route through the lipid bilayers of the *stratum corneum*, the transcellular route and the follicular route into the hair follicles (Figure 6.2.1). Free drug penetration occurs mainly by the transcellular route: small molecules move freely within the intercellular spaces by passive diffusion. Their lipophilicity and their molecular weight/volume govern their diffusion rate.

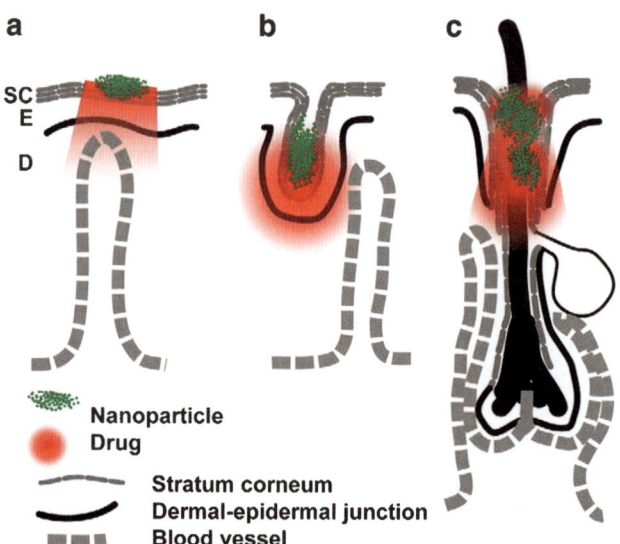

a

SC
E

D

b

c

Nanoparticle
Drug
Stratum corneum
Dermal-epidermal junction
Blood vessel

Figure 6.2.1 Sites in skin for nanoparticle delivery. Topical delivery of drug-loaded nanoparticles takes place in three major sites: *stratum corneum* (SC) surface (a), furrow (dermatoglyphs) (b), and openings of hair follicles (infundibulum) (c). The nanoparticles are shown in green and the drug in red. Other sites for delivery are the viable epidermis (E) and dermis (D).[6]

In contrast, free movement of particles is physically restricted in the horny layers.

The potential of nanoparticle skin penetration is negligible. Most environmental nanosized particles such as viruses, bacteria, allergens, dust or particles do not penetrate the skin unless the skin barrier is disrupted.[6] Drug delivery with nanocarriers targeting the deeper layer of the skin has to overcome this significant barrier: nanoparticle delivery into the viable epidermis and dermis without barrier modification has had limited success unless the skin barrier is compromised or diseased or unless deformable particles are used.[6] There is no indication that solid particles larger than 100 nm in size can pass the healthy skin barrier.[9] Other reports support the hypothesis that nanoparticles <10–20 nm in diameter are unlikely to penetrate through the *stratum corneum* into the viable skin but will accumulate in the hair follicle openings, especially after massage. These are opportunities to deliver therapeutic molecules with nanocarriers, and significant uptake does occur after skin damage and in certain diseased skin.[6,10]

The permeation route of encapsulated drugs can be affected by the nanocarriers. Nanoparticles can penetrate the hair follicle providing an efficient way for follicular drug delivery and/or a reservoir for sustained drug delivery.[1] Interest in pilosebaceous units is directed towards their utilization as reservoirs for localized therapy.[9] In contrast to the conventional transcellular pathway, the follicular pathway is especially favorable for highly hydrophilic and high molecular weight substances as well as for particle-based delivery.[9] Penetration of particles in hair follicles has been reported to be higher than that of free solutions. Their penetration is size-dependent. The optimal size is around 300–600 nm corresponding to the size of the cuticula of the hairs. As penetration is not homogenous, it has been hypothesized that "open" and "closed" hair follicles can be distinguished.[1]

6.2.3 Lipid-based Nanoparticles for Topical and Transdermal Drug Delivery

Lipid-based colloid systems possess a number of features advantageous for topical drug delivery. They are composed of natural physiological and completely biodegradable lipids exhibiting low toxicity and excellent tolerability. Hence, they are well suited for use on damaged or inflamed skin because they are based on nonirritative and nontoxic lipids. The small particle size facilitates contact of the encapsulated drug with the skin. Occlusion due to film formation contributes to enhance drug permeation into the skin by increased hydration (Figure 6.2.2).[11] Enhancement of chemical stability of active ingredients after incorporation in lipid nanocarriers has been reported.[7]

6.2.3.1 Liposomes

Conventional liposomes are vesicles composed of phospholipids which form single or multiple lipid bilayers. They may include cholesterol to modify liposome rigidity. They contain a hydrophilic core and a lipophilic compartment within the lipid bilayers. Their size varies depending on the method of preparation from 25–50 nm for small unilamellar vesicles to 50–500 nm for large unilamellar vesicles or even higher for multilamellar vesicles.

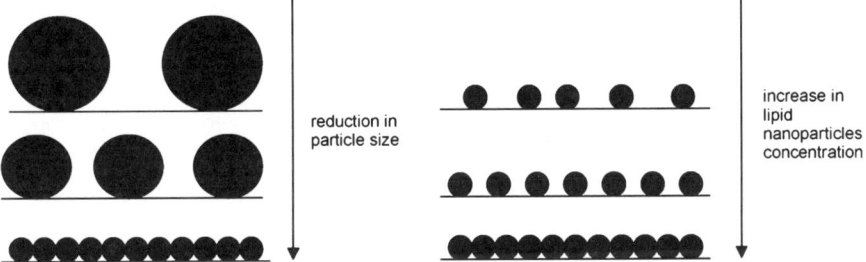

Figure 6.2.2 The occlusion factor of lipid nanoparticles depends on various factors: at identical lipid content, reducing the particle size leads to an increase in particle number, the film becomes denser (left) and therefore the occlusion factor increases. At a given particle size, increasing the lipid concentration increases particle number and density of the film (right) which also leads to a higher occlusion factor.[7]

Physicochemical properties influence the penetration of a drug encapsulated in liposomes into the skin. The penetration decreases with increasing diameter.[12] The thermodynamic state of the liposomes, such as the critical phase transition temperature which depends on the composition of the liposomes, in particular the percentage of cholesterol and the structure of the phospholipids, also plays a role. Liposomes in a liquid crystal phase are preferred to liposomes in a gel phase.[5,12,13]

The mechanism of liposomal skin delivery is the subject of debate. Several nonexclusive mechanisms have been described (Figure 6.2.3). (i) The liposomes can be adsorbed on the *stratum corneum*. They can fuse with the lipid domain of the *stratum corneum* inducing a lipid exchange between the phospholipid bilayers of the liposomes and the *stratum corneum*. (ii) They can disintegrate on the skin surface. The lipid molecule can act as permeation enhancer by disrupting the packing and fluidity of *stratum corneum* lipid bilayers. (iii) The intact liposomes could penetrate into the *stratum corneum*. (iv) The liposomes can penetrate into the pilosebaceous unit and can act as a drug reservoir.[4] For conventional liposomes, it is generally accepted that most of the liposomes do not penetrate the skin as vesicles.

Liposomes were initially used to improve skin deposition and reduce systemic effects.[4,14] Steroidal drugs in liposomes have been extensively studied

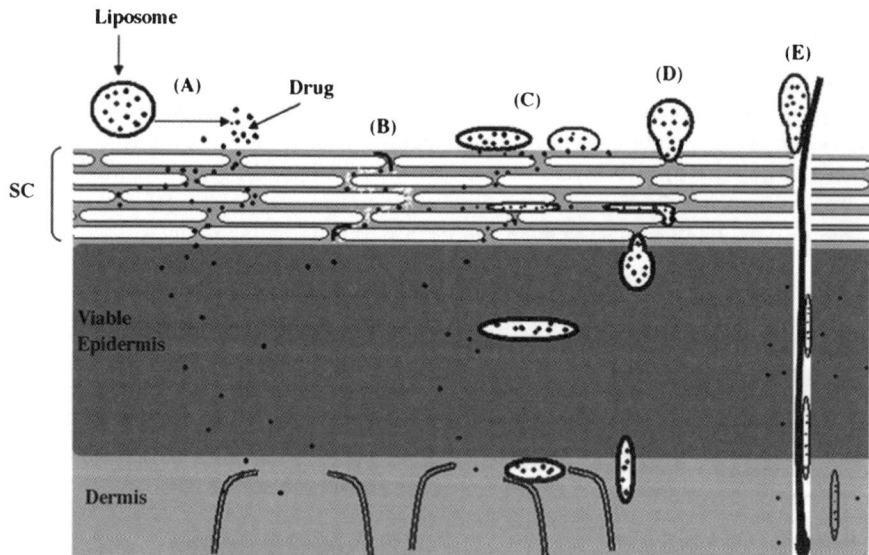

Figure 6.2.3   Possible mechanisms of action of liposomes as skin drug delivery systems. (A) is the free drug mechanism, (B) is the penetration enhancing process of liposome components, (C) indicates vesicle adsorption to and/or fusion with the *stratum corneum* (SC) and (D) illustrates intact vesicle penetration into or into and through the intact skin (not to scale).[4]

but other drugs are now commercially available (*e.g.* Econazole or diclofenac) or under investigation for acne, psoriasis, infections or other skin diseases.[2,12]

6.2.3.2 Niosomes

Niosomes are nonionic surfactant-based vesicles which have the same bilayer structure as liposomes.[15] Surfactants are known to enhance skin permeability by disrupting the *stratum corneum* lipids. The delivery mechanism by niosomes may be a combination of the permeation enhancement effect of the surfactant on the skin and increased flexibility conferred to the vesicle structure in the presence of surfactant.[5] Deformable nonionic surfactant vesicles can partition deep into the *stratum corneum* layers mainly intact and release the drug in the viable layers of the skin.[16]

6.2.3.3 Deformable Vesicles

The most recent vesicle design is elastic-liquid state vesicles with high membrane elasticity.

Ethanol is known to act as a permeation enhancer and has been widely used in lipid or surfactant-based vesicles, forming so called ethosomes.[17] Penetration to deep skin and systemic absorption have been reported for a wide range of drug encapsulated in ethosomes.[5] The enhanced absorption does not result from ethanol only. Inclusion of a high level of ethanol (30%) in the phospholipid structure creates bilayers that are flexible, allowing deposition in the *stratum corneum* bilayers which are further disrupted by ethanol and phospholipid.

Deformable liposomes, also called elastic vesicles or Transfersomes®, are composed of phospholipids (*e.g.* phosphatidylcholine) and contain a single chain surfactant with a high radius of curvature (*e.g.* sodium deoxycholate) that acts as an "edge activator" to destabilize the lipid bilayers and increases the deformability of the vesicles.[18] The basic principle was developed by Cevc and confirmed by other groups with many different vesicle compositions. The Transferomes® have been reported to penetrate the skin intact. They are claimed to squeeze through channels in the *stratum corneum* and penetrate into deeper skin because the surfactant acts as an edge activator. The transport of Transfersomes® seems to be driven by the osmotic gradient across the skin, as penetration is higher under occlusive conditions.[5,18]

The permeation of many drugs encapsulated in elastic vesicles has been extensively studied and has been recently reviewed.[5] Most of the studies were achieved *in vitro*, but *in vivo* data confirm that drug permeation is enhanced for a topical as well as a systemic effect. There is evidence that elastic vesicles can deliver an enhanced amount of both small and large therapeutic agents into and across the skin. Transferosome® products (Idea AG, DE) are now under advanced phase clinical trials, *e.g.* for ketoprofen delivery for osteoarthritis management.

6.2.3.4 Solid Lipid Particles

Solid lipid nanoparticles are produced by replacing the liquid lipids of an o/w emulsion by a blend of solid lipids that are solid at room and body temperatures. They are composed of (i) solid lipids (0.1 to 30% w/w) dispersed in an aqueous medium with stabilizing surfactants (0.5 to 5% w/w), referred to as solid lipid nanoparticles, or (ii) a mix of solid and liquid lipids, often referred to as nanostructured lipid carriers. Most of the solid lipid nanoparticles are produced by hot or cold high pressure homogenization.[7]

The mechanisms of increased drug permeation could be attributed to several factors. Film formation on the skin and subsequent occlusion effect and skin hydration have been reported (Figure 6.2.2).[7] In general, the lipids do not penetrate the horny layer but a follicular uptake has been reported. Increased contact surface of the active ingredient with corneocytes, release of surfactant and prolonged release might also contribute.

There are cosmetic products containing solid lipid nanoparticles currently on the market. Solid lipid nanoparticles are also being investigated to improve the treatment of skin diseases such as atopic eczema, acne, psoriasis or rheumatoid arthritis[7].

6.2.4 Polymeric Nanoparticles for Topical and Transdermal Drug Delivery

Nanoparticles based on natural and synthetic polymers have been developed for drug delivery into or across the skin. The most widely used nanoparticles in skin delivery are chitosan, a cationic polysaccharide, or synthetic polyesters, such as poly-ε-caprolactone, poly(lactic acid), poly(glycolic acid) and their co-polymers poly(lactic-co-glycolic acid). All these polymers are biocompatible and biodegradable.[3,19] There are different ways of distribution of the active compound within the nanostructure: drugs can be homogeneously distributed in the polymer matrix, incorporated in the core or in the shell of the capsule or adsorbed on the surface of the particle. Dendrimers, highly ordered and regularly branched globular macromolecules have been less studied.

The use of polymeric nanoparticles as drug carriers can offer several advantages. Indeed, nanoparticles can increase the stability of the encapsulated drug, permit a controlled release and show particular properties useful in cosmetology such as flowing easily, being transparent and pleasant to the touch.[3,20] There are several patterns of drug release from the particles depending on the way of distribution of the active compounds in the particle: (i) desorption of the surface-bound/adsorbed drug; (ii) diffusion through the nanoparticle matrix; (iii) diffusion through the polymer wall; (iv) nanoparticle matrix erosion; (v) a combined erosion/diffusion process.[21] These patterns are accompanied by different release rates: burst release (fast release of a certain drug amount) and prolonged release (controlled release).[2]

Polymeric nanoparticles do not usually permeate the horny layer. Nevertheless, they penetrate into the hair follicles. Consequently, nanoparticles can accumulate in hair follicles and a high local concentration is then observed. Thereafter, polymeric particles can act as a reservoir and release the incorporated drug. This transappendageal pathway is of great importance since the storage time of the carrier can last about 10 days.[3,6,22,23] Hence, it constitutes the major route for transdermal delivery of polymeric nanoparticles even if this pathway is sometimes neglected, given the only 0.1% of the total skin surface area that hair follicles occupy. Since the corneocyte layer in the lower follicular tract is incomplete, drugs can diffuse easier in the underlying tissue and induce a systemic effect.

Polymeric nanoparticles have been less studied than lipid-based particles. Stable nanoparticles allow an insight into the processes of skin permeation and are of great relevance for the understanding of nanoparticulate drug delivery in the skin.[2] The most common use of polymeric nanoparticles under investigation is still the topical drug delivery. Biodegradable polymeric nanoparticles are commonly investigated for carrying sunscreen agents. They stay on the *stratum corneum* enabling the formation of a more effective protective film on the skin surface.[24] They can improve sunscreen filter adherence to skin and, in this way, enhance their activity. Nanoparticles have also been investigated to improve skin penetration of several compounds used for transdermal delivery.

Some new strategies have been developed to improve transdermal drug delivery. For instance, some amphiphilic graft co-polymer nanoparticles seem to provide better permeation *via* the intercellular pathway,[25] or the use of microneedles with the particles.[26]

6.2.5 Lipid- and Polymer-based Nanocarriers in Dermatology

Many applications of lipid- and polymer-based nanocarriers have been investigated in preclinical and clinical studies, and some formulations are marketed. The first topical liposomal product introduced into the market in 1988 was an econazole formulation for the treatment of dermatomycosis (Pevaryl lipogel®).[27]

Suncreens scatter, reflect and absorb UV radiation. Physical filters such as TiO_2 and ZnO are widely commercialized as nanoparticles to improve their cosmetic acceptance and their performance. However, safety concerns have been raised due to their extensive use without knowledge about the extent of absorption and fate. Lipid-based nanocarriers are used in sunscreens (i) as a stabilizer of the active ingredient and (ii) for prolonged effect resulting in less frequent applications. Liposomes containing methoxycinnamate and other filters have been approved in Europe for the prophylaxis of actinic keratosis in high risk patients (Daylong Actinica®).[2,12,27]

Standard treatment of acne includes topical treatment with benzoyl peroxide, tretinoin, adapalene and antibiotics. Several studies indicate that tretinoin encapsulated in lipid-based nanoparticles maintains its activity while decreasing skin irritation. Liposomal benzoyl peroxide is more efficient than conventional formulation with a marked reduction in adverse effects. Similar data with liposomes or SLN containing antibiotics or antiandrogen have been reported.[2,7,27]

Topical corticosteroid therapy is the first choice for the treatment of dermatitis. Many studies have demonstrated higher drug levels in the skin and lower systemic availability following application of lipid-based nanoparticles compared with conventional formulations of corticosteroids. Consequently, better efficacy, lower side effects and lower frequency of applications have been observed. Moreover, the lipid particles can contribute to the restoration of the lipid barrier.[2,7,12,27]

Psoriasis might also benefit from the use of nanocarriers for topical drug delivery. Several preclinical and clinical studies have reported that encapsulation of antipsoriatic drugs, such as dithranol, and immunosuppressive drugs, such as cyclosporine A, enhances drug deposition in the skin.[12]

As nanoparticulate systems of an appropriate size can target the hair follicle, they are particularly adapted for the treatment of hair diseases such as alopecia.

6.2.6 Conclusions

The number of products using nanomaterials has increased because of the unique properties of nanosizing. Encapsulation using nanoparticulate systems is an increasingly implemented strategy in drug targeting and delivery which has been used for topical administration aiming at enhancing drug transport into and across the skin barrier.

Lipid-based nanoparticular systems have been more investigated than polymer-based particles and are marketed for topical delivery of active ingredients for cosmetic products and dermatological products. The conventional liposomes have been shown to enhance drug deposition in the skin layers while decreasing systemic exposure. Novel elastic vesicles can enhance the topical and transdermal delivery of a wide range of drugs.

There is a need to understand the interactions of nanoparticles with the skin and their transport mechanisms into the skin to better understand how the formulation can be optimized and how its composition can be modified to target a specific site in the skin or to achieve a transdermal rather than a localized delivery. There is convincing evidence that irrespective of nanomaterials used, most nanoparticles do not cross the *stratum corneum* and accumulate in hair follicles.[3]

References

1. J. Lademann, H. Richter, S. Schanzer, F. Knorr, M. Meinke, W. Sterry and A. Patzelt, *Eur. J. Pharm. Biopharm.*, 2011, **77**, 465.

2. H. C. Korting and M. Schafer-Korting, *Handb. Exp. Pharmacol.*, 2010, 435.
3. P. Desai, R. R. Patlolla and M. Singh, *Mol. Membr. Biol.*, 2010, **27**, 247.
4. G. M. El Maghraby and A. C. Williams, *Exp. Opin. Drug Delivery*, 2009, **6**, 149.
5. H. A. Benson, *Curr. Drug Delivery*, 2009, **6**, 217.
6. T. W. Prow, J. E. Grice, L. L. Lin, R. Faye, M. Butler, W. Becker, E. M. T. Wurm, C. Yoong, T. A. Robertson, H. P. Soyer and M. S. Roberts, *Adv. Drug Delivery Rev.*, 2011, **63**, 470.
7. J. Pardeike, A. Hommoss and R. H. Muller, *Int. J. Pharm.*, 2009, **366**, 170.
8. B. Baroli, *J. Pharm. Sci.*, 2010, **99**, 21.
9. H. Wosicka and K. Cal, *J. Dermatol. Sci.*, 2010, **57**, 83.
10. On regulatory aspects of nanomaterials, Scientific Committee on consumer products, European Commission, 2008.
11. E. Gonzalez-Mira, S. Nikolic, M. L. Garcia, M. A. Egea, E. B. Souto and A. C. Calpena, *J. Pharm. Sci.*, 2011, **100**, 242.
12. L. J. de Leeuw, H. C. de Vijlder, P. Bjerring and H. A. Neumann, *J. Eur. Acad. Dermatol. Venereol.*, 2009, **23**, 505.
13. D. Cosco, C. Celia, F. Cilurzo, E. Trapasso and D. Paolino, *Expert Opin. Drug Delivery*, 2008, **5**, 737.
14. G. M. El Maghraby, B. W. Barry and A. C. Williams, *Eur. J. Pharm. Sci.*, 2008, **34**, 203.
15. I. F. Uchegbu, A. Schatzlein, G. Vanlerberghe, N. Morgatini and A. T. Florence, *J. Pharm. Pharmacol.*, 1997, **49**, 606.
16. P. L. Honeywell-Nguyen, G. S. Gooris and J. A. Bouwstra, *J. Invest. Dermatol.*, 2004, **123**, 902.
17. E. Touitou, N. Dayan, L. Bergelson, B. Godin and M. Eliaz, *J. Controlled Release*, 2000, **65**, 403.
18. G. Cevc, *Adv. Drug Delivery Rev.*, 2004, **56**, 675.
19. M. L. Hans and A. M. Lowman, *Curr. Opin. Solid State Mater. Sci.*, 2002, **6**, 319.
20. A. Wiesenthal, L. Hunter, S. Wang, J. Wickliffe and M. Wilkerson, *Int J. Dermatol.*, 2011, **50**, 247.
21. K. S. Soppimath, T. M. Aminabhavi, A. R. Kulkarni and W. E. Rudzinski, *J. Controlled Release*, 2001, **70**, 1.
22. J. Lademann, F. Knorr, H. Richter, U. Blume-Peytavi, A. Vogt, C. Antoniou, W. Sterry and A. Patzelt, *Skin Pharmacol. Physiol.*, 2008, **21**, 150.
23. M. Schneider, F. Stracke, S. Hansen and U. F. Schaefer, *Dermatoendocrinol.*, 2009, **1**, 197.
24. R. Alvarez-Roman, G. Barre, R. H. Guy and H. Fessi, *Eur. J. Pharm. Biopharm.*, 2001, **52**, 191.
25. J. Xing, L. Deng, J. Li and A. Dong, *Int. J. Nanomed.*, 2009, **4**, 227.
26. W. Zhang, J. Gao, Q. Zhu, M. Zhang, X. Ding, X. Wang, X. Hou, W. Fan, B. Ding, X. Wu, X. Wang and S. Gao, *Int. J. Pharm.*, 2010, **402**, 205.
27. D. Papakostas, F. rancan, W. Sterry, U. Blume-Peytavi and A. Vogt, *Arch. Dermatol. Res.*, 2011, **303**, 533.

Section 7
Nanostructures Overcoming the Blood-Brain Barrier

CHAPTER 7.1

Nanostructures Overcoming the Blood-Brain Barrier: Physiological Considerations and Mechanistic Issues

AIKATERINI LALATSA, ANDREAS G. SCHÄTZLEIN AND IJEOMA F. UCHEGBU*

UCL School of Pharmacy, 29–39, Brunswick Square, London, WC1N 1AX, UK
*E-mail: ijeoma.uchegbu@ucl.ac.uk

7.1.1 The Blood-Brain Barrier

Neurological diseases, such as cancers, neurodegenerative conditions, infections, pain and psychiatric disorders, are a leading cause of disability, morbidity and mortality. At any time 1.5 billion people worldwide are suffering from some form of central nervous system disorder, with this number predicted to reach 1.9 billion by 2020, if curative treatments fail to emerge.[1] Delivery of drugs to the brain is difficult, as the vast majority of therapeutic agents are excluded from the brain by the blood-brain barrier (BBB).[2] Poor delivery to the brain has been described as a bottleneck in the development of CNS drugs[3] and is responsible for long drug development times and their high failure rate in clinical trials.

The presence of a barrier that segregates the brain from the circulatory blood was proposed one century ago, following Ehrlich's experimental

RSC Drug Discovery Series No. 22
Nanostructured Biomaterials for Overcoming Biological Barriers
Edited by Maria Jose Alonso and Noemi S. Csaba
© The Royal Society of Chemistry 2012
Published by the Royal Society of Chemistry, www.rsc.org

observation that most peripheral organs could be stained by the intravenous injection of a water soluble dye with the exception of the brain and spinal cord.[4] Edwin Goldman's follow-up experiments confirmed the presence of a barrier as injection of trypan blue directly into the cerebrospinal fluid (CSF) stained all cell types in the brain but failed to penetrate into the periphery.[5] Brain capillaries have evolved to limit the transport of macromolecules (including plasma proteins[6]) and cells between the blood and the brain to protect the brain against circulating toxins or infectious agents. The BBB is a unique membranous barrier that tightly segregates the brain from the circulating blood and thus by regulating the constancy of the internal environment of the brain within very precise limits, allows for the neuronal functions of the CNS to optimally take place.[7] However, the same mechanisms that protect it against exogenous compounds need to be overcome to provide useful treatments for neurological diseases.

The CNS consists of blood capillaries, which are structurally different from the blood capillaries in other tissues. The BBB is formed at the level of the endothelial cells of the cerebral capillaries (Figure 7.1.1) and is comprised of two plasma membranes in series, which are the luminal and the abluminal membranes of the brain capillary endothelium that are separated by about 0.3 μm of endothelial cytosol.[8] Anatomically, the endothelial cells of the BBB are distinguished from those in the periphery by increased mitochondrial content,[9] a lack of fenestrations,[10] minimal pinocytotic activity[11] and the presence of tight junctions.[12,13] These tight junctions between the capillary cells (zonula occludens)[14] are produced by the interaction of several transmembrane proteins that project into and seal the paracellular pathway resulting in the BBB's impermeability (Figure 7.1.2).[13,15] The interaction of these junctional

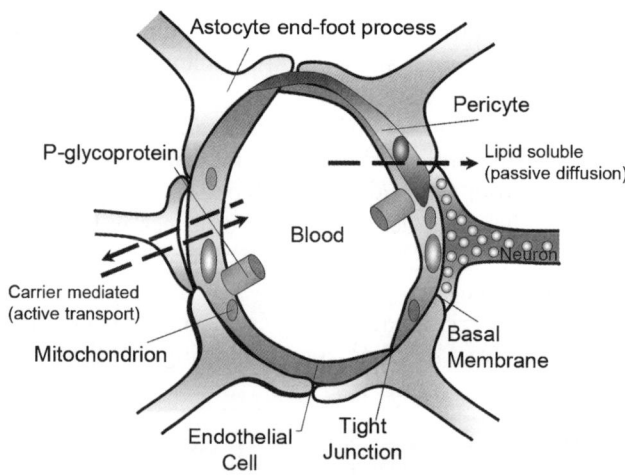

Figure 7.1.1   A schematic diagram of the neurovascular unit/cell association forming the BBB.

proteins, especially occludin and claudin, is complex and effectively blocks an aqueous route of free diffusion for polar solutes from blood along these potential paracellular pathways, denying free access to these solutes into the interstitial fluid. Occludin is a 60–65 kDa protein that has four transmembrane domains with the carboxyl and amino terminals oriented to the cytoplasm and two extracellular loops that span the intercellular cleft of cerebral endothelial cells.[16] It increases electrical resistance in tight junctions containing tissues, a property mediated by the second extracellular loop domain.[17] The junctional adhesion molecules (JAMs) and the endothelial selective adhesion molecule (ESAM) are members of the immunoglobulin superfamily and are believed to

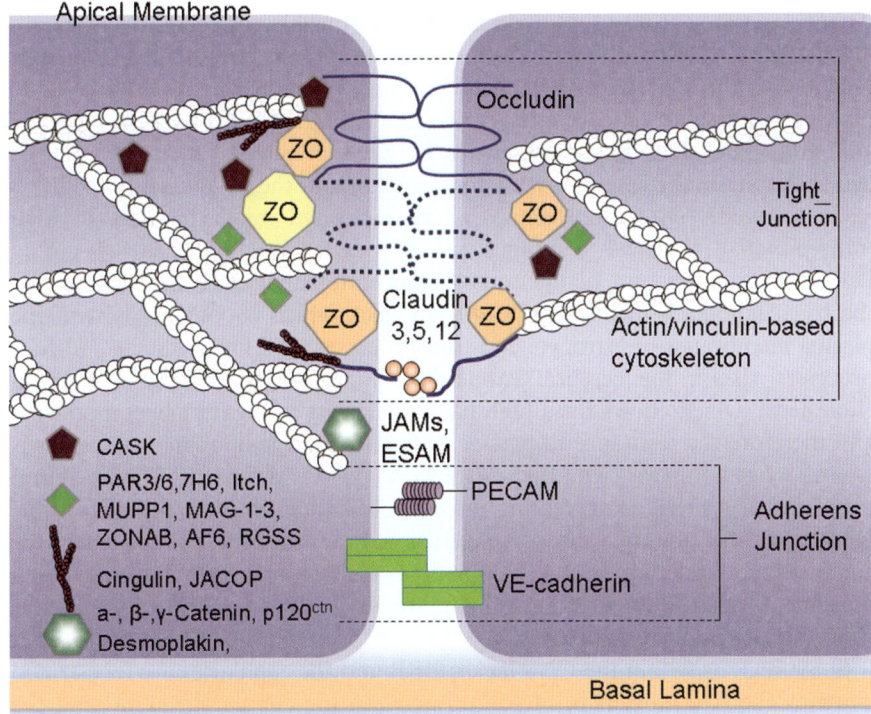

Figure 7.1.2 Molecular composition of endothelial tight junctions. Key; JAMs: junctional adhesion molecules, ESAM: endothelial selective adhesion molecule, ZO: zonula occludens 1, 2, and 3, CASK: Ca²⁺dependent serine protein kinase, JACOP: junction-associated coiled-coil protein, MUPP1: multi-PDZ-protein 1, PAR 3/6: Partitioning defensive proteins 3 and 6, MAG-1-3: membrane-associated guanylate kinase, ZONAB: ZO-1-associated nucleic acid-binding protein, AF6: afadin, RGS5: regulator of G-protein signalling 5 (RGS5), VE-cadherin: vascular endothelial cadherin, PECAM: platelet-endothelial cell adhesion molecule, p120ctn : p120 catenin, Itch: E3ubiquitin protein ligase. Modified from ref. 13 and 15.

mediate the early attachment of adjacent cell membranes *via* homophilic interactions. Within the cytoplasm are many first-order adaptor proteins, including zonula occludens 1, 2, and 3 (ZO) and Ca $^{2+}$dependent serine protein kinase (CASK), that bind to the intramembrane proteins. Among the second-order adaptor molecules, cingulin is important, and junction-associated coiled-coil protein (JACOP) may also be present. Signaling and regulatory proteins include multi-PDZ-protein 1 (MUPP1), the partitioning defensive proteins 3 and 6 (PAR 3/6), MAGI-1-3 (membrane-associated guanylate kinase with inverted orientation of protein–protein interaction domains), ZO-1-associated nucleic acid-binding protein (ZONAB), afadin (AF6), and regulator of G-protein signalling 5 (RGS5). All of these adaptor and regulatory/signalling proteins control the interaction of the membranous components with the actin/vinculin-based cytoskeleton. In epithelial cells, tight and adherens junctions are strictly separated from each other, but in endothelial cells these junctions are intermingled. The most important molecule of the endothelial adherens junctions is the vascular endothelial cadherin (VE-cadherin). In addition, the platelet-endothelial cell adhesion molecule (PECAM) mediates homophilic adhesion. The chief linker molecules between adherens junctions and the cytoskeleton are the catenins, with desmoplakin and the p120 catenin (p120^{ctn}) also involved. The endothelium is characterised by exhibiting a high transepithelial electrical resistance in the region of 1500–2000 Ω cm^2.[19] Pericytes (granular and filamentous[18]) and endothelial cells are sheathed by the basal lamina, a membrane 30–40 nm thick which is contiguous with the plasma membranes of astrocyte end-foot processes, and the astrocyte end-foot processes sheath the cerebral capillaries.[13] The plasma bilayer is high in cholesterol, which allows for a high packing density of membrane components and therefore the resistance to passive diffusion of large lipophilic molecules is greater when compared with the resistance to passive diffusion of non-cerebral capillaries.[20] As well as preventing the entry of solutes into the brain from the blood, the BBB does not allow free diffusional movement of solutes out of the CNS either, hindering free diffusion in a bidirectional way and restricting movement of solutes to transcellular transport (passive diffusion, carrier transport or endocytotic processes).

Some regions within the CNS lack a BBB and the capillaries are fenestrated allowing the free movement of solutes between the circulating blood and the surrounding interstitial tissue. These areas around the ventricles of the brain are collectively known as the circumventricular organs (CVO) and comprise of the choroid plexus, the median eminence, the neurohypophysis, the pineal gland, and the organum vasculosum of the lamina terminalis, the subfornical organ, the subcommisural organ and the area postrema. However, the relative surface area of the permeable fenestrated capillaries of the CVO compared to the tight BBB capillaries is 1 : 5000, making these high permeability areas unable to influence the bulk composition of the brain extracellular fluid and an unrealistic route for drug entry into the brain.[21]

The cells forming the avascular arachnoid membrane that envelopes the whole CNS, also possess tight junctions that seal the paracellular difussional pathway between these cells.[2]

Despite human brain capillaries having an estimated total length of 650 km and a total surface are of 12 m^2,[22] the BBB is very efficient and makes the brain practically inaccessible to polar molecules. Thus, the endothelial cells are required to maintain a high level of expression of transport proteins for essential polar metabolites such as glucose and amino acids to facilitate their entry into the brain.[23]

Apart from being a physical barrier, the BBB is also an enzymatic barrier making drug delivery even more challenging as active agents are exposed to cytosolic and membrane associated enzymes, such as γ-glutamyl transpeptidase, alkaline phosphatase, aromatic acid decarboxylase, dipeptidyl(amino)-peptidase IV, and aminopeptidase A and N which are directed at metabolizing neuroactive agents[24–27] and any agents associating with the barrier.

Additionally, there are a number of efflux transporters located on the luminal membranes of the brain capillary endothelial cells.[28–32] These ATP binding cassette (ABC) transporters transport lipophilic compounds out of the brain[29] against a concentration gradient and among these transporters the P-glycoprotein efflux pump (ABCB1) and the breast cancer resistance protein (ABCG2) are important for drug efflux out of the brain while ABCC1 is important for efflux from the cerebrospinal fluid to the blood.[33]

The BBB is effective at keeping most molecules out of the brain but there is evidence that the BBB is compromised in certain disease states such as cancer,[34] where blood vessels supplying brain tumour tissue are reported to be devoid of a BBB. The BBB is also reinforced in certain diseases such as epilepsy, where there is an over expression of the P-glycoprotein efflux pump at the BBB[35] and an alteration of the BBB, specifically the activity of the ABC transporters, has been implicated in the pathogenesis of neurodegenerative diseases such as Alzheimer's Disease and Parkinson's Disease.[36]

Despite the impenetrability of the BBB, there are a number of highly controlled routes across the BBB (see Figure 7.2.1 in Chapter 7.2). Polar solutes transport across the paracellular pathway are severely restricted by the presence of tight junctions. However, leukocytes and ions may cross the BBB adjacent to, or by modifying, the tight junctions. Lipophilic or amphiphilc solutes may passively diffuse through the large surface of the lipid cell membrane and cross the endothelium. Greater lipid solubility favors this process. Additionally, carrier-mediated influx, which may be passive or secondarily active, can transport many essential polar molecules such as glucose, amino acids, small peptides, amines, monocarboxylates, choline and nucleosides into the CNS. Active efflux carriers may intercept some of these passively penetrating solutes and pump them out of the endothelial cell (*e.g.* Cyclosporine A, Vinca alkaloids). On the other hand, receptor-mediated trancytosis (RMT) can transport macromolecules such as peptides and proteins across the cerebral endothelium. Examples are transferrin, insulin, leptin, cytokines, viruses (*e.g.* Herpes Simplex virus). Finally, adsorptive-mediated transcytosis (AMT) appears to be induced nonspecifically by positively charged macromolecules such as cationised albumin resulting in transport across the BBB. Drug delivery across the brain endothelium depends on making use of the latter four pathways, while most CNS active agents enter *via* passive diffusion. These aspects are reviewed in detail in the following sections.

7.1.2 Transcellular (Lipophilic) Diffusion

7.1.2.1 Lipophilicity

A wide range of low molecular weight (< 700 Da, but typically < 300 Da) lipophilic molecules readily transverse the BBB by passive transport or diffusion through the endothelial plasma membranes.[37–39] There is an established relationship between the rate at which a solute enters the CNS and its lipid solubility, which takes into account both the ionised and neutral species available in solution.[40,41] These lipophilic compounds only accumulate in the brain if they are not substrates for the ABC efflux transporters.[42]

Polar surface area (PSA) is also an important molecular descriptor for passive diffusion across the BBB.[43] Based on the inverse linear relationship between experimental data for penetration of the brain and the dynamic polar surface area of 45 drug molecules, it was found that most orally administered drugs that cross the BBB by passive diffusion have a PSA below 60–70 Å^2.[43,44]

Weak hydrogen-bonding potential (< 6 hydrogen bonds), lipophilicity and small size (ideally < 300 Da) along with the absence of free rotatable bonds are favourable to BBB penetration.[6,45] A tendency to form more than 6 hydrogen bonds increases the free energy requirements of moving from an aqueous into the lipid phase of the cell membrane[39,40] and thus hampers transcellular diffusion from the blood into the brain endothelial cells. Ionisation state plays an important role in the extent of CNS penetration with basic molecules being generally more CNS penetrant than neutral molecules, followed by zwitterions and finally acidic molecules which are the least CNS penetrant.[39] Thus bases, which are positively charged at physiological pH, have an advantage over acids, where BBB penetration is concerned, due to interactions with the negatively charged glycocalyx and phospholipid head groups of the cell membrane facilitating their entry.[39,46]

7.1.2.2 Plasma Protein Binding

Extensive plasma protein binding (largely binding to albumin) compromises transport across the BBB,[47,48] as the unbound fraction is the fraction that is available to cross the BBB. Molecular weight and lipophilicity both positively influence plasma protein binding. Generally, as molecular weight (MW) increases, plasma protein binding increases.[49] Molecules with a MW of < 300 Da are 72% bound, while those with a MW between 300–500 are 54% bound and those with MW between 500–700 are 98.2% bound.[39] Lipophilicity has a major influence on plasma protein binding, with an increase in protein binding seen with increasing lipophilicity.[39,50,51] Additionally ionisation state also affects protein binding, with acids being more plasma protein bound followed by neutral compounds, then zwitterions and finally bases.[39] For basic and zwitterionic compounds it can be noted that even though they are less protein bound than acids, they can achieve equivalent levels of binding with a sufficient increase in their lipophilic character.[39]

7.1.2.3   Brain Tissue Binding

Although it has received less attention than plasma protein binding, brain tissue binding has an important influence on the activity of CNS active compounds, as only the brain free unbound fraction is available to elicit a pharmacological response. Thus, both the extent of brain permeation and the extent of brain tissue binding need to be considered for successful CNS delivery of a molecule. Increasing molecular weight (< 700 Da) and increasing lipophilicity appears to increase brain tissue binding.[39]

7.1.2.4   Efflux Transporters – ATP Binding Cassette Transporters

Even if a molecule with sufficient lipophilicity and molecular weight characteristics does enter the brain, it is likely to be removed from the brain and the cerebral endothelium by active efflux transporters, *e.g.* the ABC transporters.[29] Both lipidisation[6] and an increase in molecular weight[39] make a compound more likely to be a substrate for P-glycoprotein. P-glycoprotein, present in high density in the luminal membrane of the brain microvessel endothelia,[32] has been implicated in preventing the brain accumulation of a number of lipophilic drugs including colchicine, vinblastine, paclitaxel, loperamide and cyclosporine.[52–54]

7.1.2.5   Exploiting Transcellular Pathways

In order to promote transcellular transport, drug molecules must be of low molecular weight (< 700 Da and ideally < 300 Da), lipophilic, not ABC efflux transporter substrates, not extensively plasma protein bound and maintain a high concentration gradient of the unbound fraction between blood and the brain. These requirements are achieved by: (a) controlled molecular lipidisation, (b) enhancing compound metabolic stability within the peripheral circulation and (c) diminishing drug affinity for the ABC efflux transporter. Strategies to increase the level of the unbound fraction within the plasma by reducing plasma protein binding have not received much attention.

7.1.2.5.1   Lipidisation

Lipidisation is achieved by either removal of a polar group or the addition of a nonpolar group. These structural changes significantly enhance the drug's lipid solubility and have been shown to increase the drug's ability to cross the BBB. For example, diamorphine, a diacetyl derivative of morphine, crosses the BBB about 100 times more easily than its parent drug just by being more lipophilic than its parent drug.[22]

A number of peptide drugs have been made more lipophilic and as a result more BBB permeable. Lipophilicity may be conferred by the acetylation of the N-terminus, amidation of the C-terminus, and methylation, all structural modifications which reduce hydrogen bonding potential, increase lipophilicity and in turn increase BBB transport.[55] Trimethylation of the phenylalanine group of DPDPE [(H_2N-Tyr^1-D-Pen^2-Gly^3-Phe^4-D-Pen^5-OH, (D-Pen^2,D-Pen^5)-enkephalin, d-penicillamine (2,5)-enkephalin] resulted in enhanced lipophilicity and BBB permeability while also a reduced affinity for the P-glycoprotein pump; although these changes were not accompanied by an increase in analgesia,[56] indicating that the site and type of modification chosen are vitally important for delivery enhancement and the maintenance of pharmacological activity. There is some evidence that confining the acylation or alkylation to the terminal amino acids results in an increase in lipophilicity and membrane permeability, while minimally interfering with receptor binding. Acyl derivatives of DADLE (H_2N-Tyr^5-D-Ala^4-Gly^3-Phe^2-Leu-OH, D-Ala,D-Leu-enkephalin),[57,58] DPDPE,[59] thyrotropin-releasing hormone (TRH),[59,60] and insulin[61] synthesised by creating an amide bond in the C- or N-terminal or the free amine of a lysine residue have thus shown improved absorption through artificial and biological membranes and enhanced enzymatic stability, while retaining pharmacological activity. Selective lipidisation has also been shown to increase receptor activity. Dimethylation of the tyrosine residue on the opioid peptide DPDPE resulted in a 10-fold increase in the potency over the non-methylated DPDPE at the delta opioid receptor and a 35-fold increase in potency at the μ-opioid receptor, while substantial delta receptor selectivity was still maintained.[62]

Modifications of a BBB-impermeable polypeptide, horseradish peroxidise (HRP), with lipophilic (stearoyl) or amphiphilic (Pluronic block copolymer) moieties considerably enhanced the transport of this polypeptide across the BBB, while maintaining its enzymatic activity.[63] Stearoyl modifications of HRP improved its penetration by about 60% but also increased clearance from blood.[63] Whereas modifications of HRP with amphiphilic block copolymer moieties [the Pluronics – poly(oxyethylene)-co-poly(oxypropylene) block copolymers] through degradable disulphide links resulted in compounds that were most effectively transported across *in vitro* brain microvessel endothelial cell monolayers and also resulted in efficient delivery of HRP to the brain.[63] Pluronic modifications increased penetration of the BBB and had no significant effect on clearance so that uptake by brain was almost doubled.[63]

The oral bioavailability of a dermorphin tetrapeptide analogue (N^α-1-iminoethyl-Tyr-D-Met-O^2-Phe-MeβAla-COOH) was improved by increasing its lipophilicity. Lipophilicity was increased by esterification of the C-terminal carboxyl group and/or acylation of the phenolic hydroxyl group on tyrosine.[64] Dermorphin is a natural opiate tetrapeptide that was isolated from the skin of the South American frog. It binds selectively to the μ opioid receptor, with a potency 30–40 times that of morphine.[65] Dermorphin tetrapeptide analogues are known to be highly stable against enzymatic degradation and their low oral bioavailability is due to poor intestinal absorption.[64,66–68] Derivatives with acylation of the phenolic hydroxyl group on tyrosine[1] were more active than derivatives with esterification of the C-terminal carboxyl group,[64] as measured by the tail pressure test.

Halogenation of peptides such as DPDPE,[69,70] DPLPE (D-Pen²,L-Pen⁵-enkephalin)[71] and biphalin[72] significantly enhances their lipophilicity and BBB permeability and this enhancement is dependent upon the type of conjugated halogen (Cl, Br, F, I). Addition of chlorine on the Phenylalanine[4] residue of DPDPE led to a significant increase in BBB permeability.[69,70]

Lipophilic modifications must be carried out with care as they are known to increase the ability of compounds to cross the BBB while decreasing the amount of compound accumulated by the brain[63] and have even been known to decrease a compound's intrinsic activity. To prevent this paradox, the site of modification or attachment of moieties to increase lipophilicity must be carefully selected, as receptor binding affinity may be diminished if alterations are within the pharmacophore region, thus reducing biological activity.[73] On the other hand, enhancement of lipophilicity alone may not necessarily improve BBB transport. Factors such as molecular size, stability, intracellular sequestration, non-target organ uptake, volume of distribution, P-glycoprotein efflux activity (increased with increased lipophilicity) must also be considered if one is to optimise any pharmacological advantages accruing from an increase in lipophilicity, as must the rate of bioconversion for lipidic prodrugs.

7.1.2.5.2 Prodrugs

A number of lipidic prodrugs have been synthesised in an effort to facilitate across BBB transport. These compounds are specifically designed to be cleaved *in vivo* prior to interaction of the resulting drug with the receptor. Ideally, a prodrug is enzymatically stable in the blood, but rapidly degraded to the active parent compound when it reaches the site of action.

There is evidence that esterification and amidation are useful strategies in prodrug design. Esterification or amidation of amino, hydroxyl, or carboxylic acid containing drugs may greatly enhance lipophilicity/brain permeability[74,75] and receptor binding. Hydrolysis of the ester or amide bond releases the active compound once the prodrug enters the brain.[74] Both aromatic benzoyl esters[74] and branched chain tertiary butyl esters[76] have shown stability in plasma, while being cleaved within the CNS.[73] The addition of a phenylalanine (lipophilic) amino acid as a cleavable unit to DPDPE at the amino terminal resulted in enhanced permeability to the BBB.[77] Chains of nonpolar amino acids could further enhance lipophilicity. However, the balance between molecular size and degree of lipophilic enhancement must be optimum for a successful approach and individualised for a particular molecule.

Another prodrug approach used the redox system,[78,79] in which a lipophilic attachment (*e.g.*, 1,4-dihydrotrigonelline) is converted to the hydrophilic quaternary ammonium form, effectively "locking" the drug beyond the BBB.[80] The drug is conjugated to a methyldihydropyridine carrier and subsequently oxidised by NADH-linked dehydrogenases in the brain resulting in a quaternary ammonium salt, which cannot cross back through the BBB endothelium.[81] This strategy has been applied to a wide variety of drugs, such as steroids, antivirals, neurotransmitters, anticonvulsants, thyrotropin releasing hormones and a leucine[5]-enkephalin analogue.[80,82] With this strategy, a careful choice of the cleavable moieties and their point of attachment to the active pharmacophores is a necessity for an optimised pharmacokinetic profile.

A further example of the approach being applied to enkephalins is the formation of a lipophilic prodrug of DADLE. This prodrug was created by esterification of the free C-terminal with cholesterol and by amidation of the free N-terminus with the 1,4-dihydrotrigonelline.[83] The resulting compound enters the brain more readily than DADLE, due to its increased lipophilicity. Subsequent to brain entry the compound undergoes enzyme-mediated oxidation to the hydrophilic, membrane-impermeable trigonellinate ion, trapping the prodrug within the brain.[83] Increased antinociception was demonstrated on intravenous administration of this prodrug, when the prodrug was administered in a formulation containing a mixture of dimethylsulfoxide, ethanol, and a 50% aqueous 2-hydroxypropyl-β-cyclodextrin solution.[83]

7.1.2.5.3 Lipidisation and Plasma Protein Binding

Lipidisation with these and other strategies still poses certain limitations. Since highly water insoluble drugs may be extensively plasma protein-bound, there is the potential for a reduction in the amount of free or exchangeable drug in the plasma; a situation which ultimately compromises brain uptake.[84] *In vivo* studies of BBB permeability can differ from what has been predicted from *in vitro* studies, due to clearance, serum protein binding, and tissue sequestration; all of which are able to adversely influence *in vivo* results.

7.1.2.5.4 Enhancing Metabolic Stability

Brain uptake by passive diffusion may also be increased by reducing drug clearance rates from the plasma; this can be achieved by inhibiting catalytic enzymes. Structural changes aimed at reducing enzymatic degradation *in vivo* can involve the simple addition of moieties that chemically protect the targeted bond from attack; replacement of enzyme sensitive bonds altogether or global changes that modify the peptide or drug conformation, such that it is no longer recognised by the enzyme concerned. These modifications require extensive analytical investigation to define the site of enzymatic cleavage as well as the enzymes that act upon a specific drug.

Conformation changes (peptide backbone constrains or cyclisation[85–89] and the use of D-amino acids or a residue containing an unnatural side chain[90,91]) are often carried out in an effort to reduce enzymatic degradation. These methods reduce hydrogen bonding, increase lipophilicity, reduce the molecule's hydrodynamic radius and often significantly increase peptide half-life, leading to an increase in peptide bioavailability. For example the use of disulphide bridge constrained peptide analogues has been shown to significantly reduce enzymatic degradation[70,92] and also enhance the specificity for receptor subtypes. In the case of the linear methionine[5]-enkephalin, the conversion into the cyclic analogue DPDPE resulted in a δ-opioid-specific peptide[93] with a saturable mode of transport across the BBB.[94–96] Due to the incorporation of 2 D-penicillamine (D-Pen) residues and conformational restriction by a disulphide bridge, DPDPE is enzymatically stable.

A further method that has been used to enhance metabolic stability is the co-administration of enzyme inhibitors. A well-known example is the use of DOPA-decarboxylase (L-amino acid decarboxylase) inhibitors to improve L-DOPA (L-3,4 dihydroxyphenylalanine) brain delivery, as L-DOPA uses the large-neutral amino acid transporter and is converted to dopamine in the brain. However, DOPA-decarboxylase is not only present in the brain, but is also present in the periphery, resulting in peripheral conversion of L-DOPA to dopamine; a conversion that reduces CNS delivery and efficacy, while also increasing its side-effects significantly.[3]

7.1.2.5.5 ABC Efflux Transporter Inhibition

To by-pass the ABC efflux transporters two strategies have been employed: (a) co-administration of an efflux transporter inhibitor or (b) design of a drug analogue that is not a substrate for the efflux transporters. Several inhibitors have been developed: for example the co-administration of the P-glycoprotein inhibitor, valspodar (SDZ PSC-833), increases the brain levels of paclitaxel and reduces tumour volume in mice after oral administration.[97] However, the effects of long term P-glycoprotein inhibition are not known and potentially toxic substrates may also gain entry into the various cell types with this approach.

Designing drug analogues that show reduced affinity for the transporter is a more popular approach. Polymer conjugation is one way of generating molecules with low affinity for the efflux transporters. Pegylation of the opioid peptide DPDPE with a 2 kDa PEG [poly(ethylene glycol)] moiety on the terminal tyrosine decreased clearance, reduced plasma protein binding, first-pass elimination and P-glycoprotein binding,[98] while increasing the analgesic effect following intravenous administration.[99] The main disadvantage of pegylation of peptides and drugs is that the drugs are made more hydrophilic and larger in size, all attributes that confound delivery to the CNS. Furthermore, there is a real possibility of a loss of activity with an improper choice of PEG (length, branching), choice of linker and/or choice of attachment site.

7.1.3   Carrier Mediated Transport

The endothelial cells rely upon transport proteins to facilitate the entry of essential polar nutrients, such as glucose and amino acids, into the brain.[6] The endothelial cells forming the BBB (as well as the CVO) exhibit a polarised expression of transport proteins in the luminal (part of the cell's membrane facing the bloodstream) and the abluminal (side exposed to the actual brain tissue) membranes of the endothelial cells, with some transporters expressed exclusively in one of these interfacial membranes and some in the other, whereas some are present on both membranes.[21,100,101] As some transporters are uni-directional and some bi-directional in their transport of solutes across the cell membrane, this polarization means that some solutes can be preferentially transported into the brain and some out of the brain. The formation of tight junctions between endothelial cells may also act as a barrier in the cell membrane, preventing both transport protein and lipid rafts in the membrane from exchanging between luminal and the abluminal membrane domains and therefore preserving the polarity of the BBB.[2,6] Several carrier systems have been described in brain capillaries such as those for small peptides, hexoses, monocarboxylic acids, amino acids, organic anions and cations, neurotransmitters and nucleosides (Table 7.1.1).[6,102] Utilisation of these carrier systems expressed at the BBB might be a useful strategy for the CNS delivery of drugs.

7.1.3.1   Exploiting Carrier Mediated Processes

Exploitation of the various carrier systems at the BBB is a strategy that can be used to deliver drugs to the brain.[21,138,139] Levodopa, a lipid insoluble precursor of dopamine, indicated for the treatment of Parkinson's disease, is an excellent example of a compound that exploits these endogenous transporters. L-DOPA contains the carboxyl and α-amino groups that allows it to be transported across the BBB by the large neutral amino acid carrier.[73,140] Additionally D,L-2-amino-7-bis[(2-chloroethyl)amino]-1,2,3,4-tetrahydro-2-naphthoic acid (D,L-NAM), an anticancer amino acid drug also possesses a high affinity for the neutral amino acid carrier[141] making brain uptake of this compound for the treatment of brain tumours a possibility. Chemical groups may be designed and attached to specific pharmacologically active compounds rendering them substrates for these endogenous carriers, as in the case of Biphalin, an enkephalin analogue (H_2N-Tyr[1]-D-Ala[2]-Gly[3]-Phe[4]-NH-NH-Phe[4]-Gly[3]-D-Ala[2]-Tyr[1]), that has been shown to use the neutral amino acid carrier.[142] The carrier is specific for large neutral amino acids and it recognizes a carboxylic acid group and an amino group covalently linked to the same carbon atom, which is a characteristic of an α amino group or a conformation that closely resembles this grouping (as in the case of baclophen and gabapentin). A bulky hydrophobic group on the molecule is required to interact with the cell membrane in order to align the amino and carboxylic groups to the active receptor site, thus excluding amino acids such as glycine and alanine.

Some carriers are very selective in their stereochemical substrate requirements such as the Glucose transporter 1 (GLUT-1) carrier. Only molecules that closely resemble D-glucose are transported *via* GLUT-1. For example the GLUT-1 transporter is thought to be responsible for the transport of L-serynyl-β-D-glucoside analogues of methionine enkephalin across the BBB. These compounds produce marked and long-lasting analgesia after intraperitoneal injection in mice.[143,144] As well as being a substrate for the GLUT-1 transporters, the analgesia enhancement seen on glycosylation of opioid peptides may be due to a number of other properties: increased bioavailability of glycopeptides due to their higher metabolic stability,[145] reduced clearance[146] as well as improved BBB transport.[144] The enhanced BBB transport of glycopeptides is not due to passive diffusion as octanol/saline distribution studies expectedly revealed that the addition of glucose significantly reduced lipophilicity, thereby reducing passive diffusion.[144] The enhanced transport of glycosylated peptides across the BBB may not be due solely to the activities of the GLUT-1 receptor however, as while glycosylated analogues of methionine[5]-enkephalin are transported *via* GLUT-1,[138,143] endocytotic mechanisms have also been shown to be involved in the transport of glycosylated peptides.[147] It has also been proposed that some glycopeptides promote a negative membrane curvature leading to an increased rate of endocytosis, which in turn results in enhanced BBB transport.[148]

The hexose and large neutral amino acid carriers have the highest carrying capacity and presently are the best candidates to exploit for delivery of substrates to the brain.[6,73] Additional examples are provided in the following chapters (Chapters 7.2. and 7.3.). Peptides, which generally require low concentrations to induce effects can utilise the low capacity carriers.[73] It must be stated that targeting drugs to a specific nutrient transporter will require a thorough knowledge of both the drug and the transporter for the strategy to yield favourable results.

Glutathione (GSH) has also been used as a vector.[149] Glutathione tagged PEGylated liposomes deliver antiviral drugs to the brain *via* the glutathione transporter[149,150] and glutathione coated poly-(lactide-co-glycolide) (PLGA) nanoparticles (NPs) of paclitaxel have been tested for the treatment of brain cancers.[151] Additionally glutathione tagged PLGA nanoparticles were able to deliver coumarin-6 across the BBB.[151]

Table 7.1.1 Endogenous transporters controlling the penetration of certain molecules across the BBB.

Carrier Systems	Abbreviation	BBB location	Orientation	Susbtrate	Reference
Hexose	GLUT1	L, A	Blood to brain	D-Glucose (facilitative, bi-directional)	103
	GLUT1	L, A	Blood to brain	DHA (dehydroascorbic acid) (facilitative)	104
Monocarboxylic acids	MCT1	L, A	Blood to brain	Lactic acid, pyruvic acid	105, 106
Thyroid hormone	MCT8	L, A	Blood to brain	T3 thyroid hormone (facilitative)	107–109
Organic anion	OAT1	A	Endothelium to brain	17β-estradiol-d-17-β-glucoronide	110
	OAT3	A, and possibly L	Endothelium to brain	para-aminohippuric acid, probenecid, benzylpenicililn, cimetidine	111, 112
Organic anion transporting polypeptide	OATP1	A	Endothelium to brain	Opioid agonists (N-tyrosinated peptides), pravastatin, glucoronide conjugates, aldosterone, thyroxine, tridothyronine	113, 114
	OATP2	L	Blood to endothelium	Opioid agonists (deltorphin, DPDPE) thyroxine, tridothyronine, digoxin	114–116
	OATP2B1	L	Blood to endothelium	Estrone-3-sulfate (organic anion/ bicarbonate exchangers)	117, 118
Novel organic cation transporter	OCTN2	L, A	Blood to endothelium, Endothelium to brain	Carnitine (organic cation / proton exchange)	119
Peptide transport systems 1–5	PTS1–5	L	Blood to brain	Methionine enkephalin, Tyr-MIF-1, TRH, DSIP, α-melanocyte-stimulating hormone, leucine enkephalin (partially)	120–123
Glutathione	GSH	L	Blood to endothelium	Glutathione	124

Table 7.1.1 (*Continued*)

Carrier Systems	Abbreviation	BBB location	Orientation	Substrate	Reference
Amino acid	Large neutral (LAT1)	L, A	Blood to brain	Large neutral amino acids: asparagine, glutamate, histidine, isoleucine, leucine, methionine, phenylalanine, threonine, tryptophan, tyrosine, valine, DOPA, cysteine, serine (facilitative, bi-directional)	125, 126
	Excitatory (EAAT1, 2, 3)	A	Brain to endothelium	Anionic amino acids, glutamate, aspartate (sodium dependent)	126
	Cationic (CAT1/y+) and (CAT3)	L	Blood to endothelium	Basic L-amino acids: arginine, lysine, ornithine (sodium independent)	127
	Neutral-α	A	Brain to endothelium	Small neutral amino acids: glycine, alanine, asparagine, proline, serine, glutamine (sodium dependent)	128
	Neutral-β	L, A	Brain to endothelium	Taurine, β-alanine (sodium dependent)	129, 130
	ASCT1, ASCT2	A	Brain to endothelium	L-alanine, serine, cysteine (sodium dependent)	131
Neurotransmitter	GAT	A	Brain to endothelium	GABA (sodium dependent)	132
	SERT	L, A	Brain to endothelium	Serotonin	133
	NET	A	Brain to endothelium	Norepinephrine	133
Nucleoside	ENT1	L	Blood to endothelium	Thymidine (facilitative, equilibrative)	134
	ENT2	L	Blood to endothelium	Adenosine, uridine	135
	CNT1	A	Endothelium to brain	Pyrimidines	136, 137
	CNT2	A	Endothelium to brain	Purines	136, 137
	CNT3	A	Endothelium to brain	Pyrimidines and purines	136, 137

7.1.4 Receptor Mediated Transcytosis

Endocytosis is the main route of cellular entry for large molecular weight compounds and several peptides and proteins are transported across the BBB (Table 7.1.2) *via* receptor mediated transcytosis (RMT). The transport mechanisms have been well characterised. Binding of the ligand to its specific membrane receptor on the cell surface induces a modification of the receptor protein and triggers the formation of invaginations that can be clathrin coated (coated pits visualised using electron microscopy); these caveoli trigger the formation of endocytotic vesicles.[147] These endocytotic vesicles may fuse with an endosome (pre-lysosomal compartment with an acidic pH) and dissociation of the ligand from the receptor takes place, allowing the free receptor to be recycled to the cell surface.

Transferrin receptors, which are diffusely distributed over the entire plasma membrane, migrate to coated pits only after binding to their ligand. The low-density lipoprotein (LDL) receptors are predominantly localised at the membrane surface where coated pits are found even when a ligand is not bound to them.[152] The ligand containing vesicles can be either exocytosed leading to transport across the BBB, fused with a lysosome leading to intracellular degradation,[147] or can bind to a second intracellular receptor as in the case of the transfer of iron from transferrin to intracellular ferritin.[169] Another intracellular pathway may involve trafficking of endosomes, containing intact receptor ligand, to the inner saccule of the Golgi complex, where the

Table 7.1.2 Transport of macromolecules across the BBB.

Transport System	Receptor	Ligand	Direction	Reference
Insulin	Insulin	Insulin	Blood to brain	153, 154
Insulin-like growth factors	Insulin	IGF I/II	Blood to brain	155, 156
Transferrin	TfR	Transferrin	Blood to brain	157–159
Melanotransferrin	MTfR	Melanotransferrin	Blood to brain	160
Leptin	Leptin	Leptin	Blood to brain	161
Tumour Necrosis Factor	TNFα	TNFα	Blood to brain	162
Epidermal Growth Factor	EGF	EGF	Blood to brain	163
Immunoglobulin G	IgG	IgG	Blood to brain	164
Interleukin	IL	IL1α, IL1β, IL6	Blood to brain	165, 166
Apolipoprotein E	ApoER2	Lipoproteins and ApoE bound molecules	Blood to brain	167
LDL-receptor-related protein 1 and 2	LRP1	Lipoproteins, Amyloid-β	Blood to brain, brain to blood	167
	LPR2	ApoE, Melanotransferrin	Blood to brain	168
Diptheria toxin receptor	DTR	Diptheria toxin, CRM 197	Blood to brain	168

enzymes can cause dissociation of the ligand from the receptor, and the separated ligand may then be exported in vesicles destined for lysosomal degradation.[102] Exocytosis and the avoidance of the lysosomal pathway may be a special feature of the BBB compared to other types of cells and tissues as transcytosis of a number of macromolecules is a homeostatic requirement.[6] Receptor-mediated endocytosis across the BBB *in vivo* has been shown for a few peptides and proteins, such as: insulin, transferrin, certain cytokines[165,166] and leptin.[166,170]

7.1.4.1 Exploiting Receptor Mediated Transcytosis

Conjugating biomacromolecules or nanoparticles to these transport ligands (Table 7.1.2), is a strategy that has been successfully used to deliver biomacromolecules across the BBB, as extensively reviewed in Chapters 7.2. and 7.3. Earlier research concentrated on receptors that transport large endogenous molecules to the brain, such as transferrin or human insulin. Recent research suggests that even a signalling receptor such as the nicotinic acetylcholine receptor may be used as a portal for BBB delivery (Table 7.1.3).

The vector used should have sufficient high affinity for the receptor, but be able to allow the cargo to be released in the brain parenchyma. The endogenous ligand should not compete with the delivery vector for receptor occupancy at the BBB, thus a careful consideration of the relative binding affinities and the physiological levels of the endogenous ligand need to be made. Transferrin is thought not to be a suitable vector as its plasma concentration is > 1000 fold higher than the K_d of 5.6 nM. Ideally the vector should not be pharmacologically active (*e.g.* insulin is not suitable). The vector conjugate should have a high receptor affinity and the type of linker or spacer used may influence this receptor affinity. The brain uptake of the vector conjugate should be high enough to allow for a therapeutic dose to be administered after correction for the amount of drug contained in the brain blood capillaries. Uptake of > 2% of the injected dose per gram of brain (mouse) has been suggested as a reasonable target.[171] The conjugate must retain the drug activity or release the drug in the brain parenchyma to elicit its pharmacological response. Finally, ideally the receptor should be specifically expressed in the brain capillary endothelium in order to target the conjugate exclusively to the brain, eliminating loss of dose in the periphery and any unwanted peripheral side-effects. It should be noted that brain receptor exclusivity is not usual however.

7.1.4.1.1 Transferrin and Insulin Receptors

Therapeutic compounds are able to cross the BBB after association or conjugation to specific transferrin and insulin receptor antibodies[172] or to the relevant endogenous ligand, forming what have been termed Molecular Trojan Horses.

Table 7.1.3 Technologies exploiting the receptor mediated uptake pathway.[a]

Technology	Drugs Delivered	Route	Pharmacokinetic data	Pharmacodynamic data	Stage of development	Ref.
Single-domain brain-targeting llama antibody fragments targeting the TMEM30A receptor	Single domain antibody (MW = 14 kDa)	IV	Detected in brain, liver, lung and kidney	No data	PC	214–217
Fusion proteins targeted to the HIR and EPO receptor by RMT	Iduronate-2-sulfatase, iduronidase, EPO	IV	2.1% of injected enzyme dose per gram of brain in Rhesus monkeys	73% reduction in intracellular lysosomal inclusion bodies	PC	190–192, 218–221
Fusion proteins with chimeric mAb against mouse TfR	TNF-α inhibitor, avidin-biotinylated Ab-amyloid peptide (1–40), GDNF	IV	2.1% injected dose per g of brain in mice	Decrease in Parkinson's Disease symptoms in a Parkinson's Disease mouse mode and 272% increase in striatal tyrosine hydroxylase enzyme activity in mice	PC	190, 220, 222–224
Bidirectional vectors comprising one part for entry by RMT and one part for eventual exit by RMT	Ab amyloid peptide (1–40), α-synuclein, huntington protein, PrP prion protein, West Nile envelope protein, tumour necrosis factor related apoptosis inducing ligand (TRAIL), Nogo A, HER 2, EGFR, HGF, oligodendrocyte surface antigen	IV	No data	No data	PC	195, 224

Table 7.1.3 (*Continued*)

Technology	Drugs Delivered	Route	Pharmacokinetic data	Pharmacodynamic data	Stage of development	Ref.
OX26 pegylated immunoliposomes against the TfR	Daunomycin	IV	No data	No data	PC	225, 226
OX26 sialic acid immunoliposomes against the TfR	Plasmid DNA	IV	No data	No data	PC	227
Angiopep 1–7 against the LRP-1 receptor	Leptin and leptin analogues	IP	Increased level in brain parenchyma	7-Fold reduction in food intake after 4 hours with diet induced obese mice	PC	228
Angiopep 1–6 against the LRP-1 receptor	Neurotensin and neurotensin analogues	IV	Increased level in the brain parenchyma	Hypothermia in mice and rats and analgesia.	PC	229
Angiopep-2 drug conjugates for delivery across the LRP1 receptor	Paclitaxel, IgG	IV	Increase in Ang1005 (Angiopep-2-paclitaxel) in a mouse intracranial tumour model. Increase in brain IgG levels	Increase in survival in a rat intracranial tumour model and tumour growth retardation in a mouse intracranial tumour model	P1, P2	209, 230–232
Surfactant coated particles for delivery across the LDL receptor	Loperamide, dalargin, doxorubicin, NGF	IV	Increased levels of drug in the brain parenchyma	Increase in antinociception as measured by the tail flick assay with dalargin and loperamide. Decrease in tumour blood vessel density with doxorubicin. Decrease in extrapyramidal symptoms and Parkinson's symptoms in a Parkinson's Disease mouse model	PC	233–239

Table 7.1.3 *(Continued)*

Technology	Drugs Delivered	Route	Pharmacokinetic data	Pharmacodynamic data	Stage of development	Ref.
Surfactant coated PLGA nanoparticles for delivery across LDL receptor	Doxorubicin	IV	N/A	Increased survival in a rat tumour model	PC	240
Nanoparticles with covalently coupled ApoE for delivery across the LDL receptor	Dalargin	IV	No Data	Increase in antinociception as measured by the tail flick test	PC	203
Drugs with adsorbed Apo A1, B100 and E3 for delivery across the LDL receptor	Loperamide	IV	No data	Increase in antinociception as measured by the tail flick test	PC	204
A chimeric CNS-targeting polypeptide with a BBB receptor binding domain (APOE, APOA, APOB) and a payload domain attached to a viral delivery system for delivery across the LDL receptor	Glucocerebrosidase	IV	Enzyme found in neuronal lysosomes	Increase in glucocerebrosidase activity	PC	241
Carriers linked to an RVG-derived peptide for delivery across the nicotinic aceylcholine receptor a7 subunit	siRNA or shRNA	IV	Brain gene silencing in mouse models	Increase in survival in NOD/ SCID mice challenged with JEV after administration of shRNA	PC	211, 242–244

Table 7.1.3 (*Continued*)

Technology	Drugs Delivered	Route	Pharmacokinetic data	Pharmacodynamic data	Stage of development	Ref.
Carriers linked with a CRM197 ligand or a CRM197 conjugate for delivery across the diphtheria-toxin receptor	Ribavirin	IV	Brain accumulation of ribavirin in hamsters	Reduced mortality in a hamster model of flavivirus infection (acute encephalitis, polyomyelitis-like syndrome)	PC	150
Bispecific humanised antibody with one arm comprising a low affinity anti-transferrin receptor antibody and the other arm comprising the high-affinity anti-BACE1 antibody (anti-TfR/BACE1)–transferrin receptor	Anti-BACE1 antibody	IP, IV	0.8% of injected dose per gram of anti-TfR-BACE1 antibody in the brain	Increase in cognitive memory in single transgenic hAPP mice	Ph2	179, 180, 245, 246

[a]Key: AChR: Acetylcholine receptor, APOE: apolipoprotein E, APOA: apolipoprotein A, APOB: apolipoprotein B, BACE1: beta-secretase 1 enzyme, BBB: blood-brain barrier, EGFR: epidermal growth factor receptor, EPO: erythropoetin, Fab: fragment, antigen binding region of an antibody, GDNF: glial-derived neurotrophic factor, GFP: green fluorescent protein, hAPP: human amyloid precursor protein, HER 2: human epidermal growth factor receptor 2, HGF: hepatocyte growth factor; HIR: human insulin receptor, IP: intraperitoneal, IV: intravenous, JEV: Japanese enkephalitis virus, LDL: low-density lipoprotein, LRP: lipoprotein receptor-related protein, MAb: monoclonal antibody, miRNA: micro RNA, NCAM: neural cell adhesion molecule, NGF: nerve growth factor, P1: Phase 1, P2: Phase 2, PC: Preclinical, RGV: rabies virus glycoprotein, PLGA: poly (lactic-co-glycolic acid), shRNA: short-hairpin RNA, siRNA: small interference RNA, SOD-1: superoxide dismutase gene

The TfR (transferrin receptor) mediates uptake of transferrin bound iron and is expressed in many organs (liver, spleen, lung and brain), which makes it a non-specific receptor for brain targeting. However, when gene delivery was the desired objective, a combination of the TfR targeting ligand with brain specific gene promoters provided sufficient specificity and the gene was only expressed in the brain.[173] Competition of the transferrin (Tf) conjugate with high levels of endogenous transferrin makes its use as a brain delivery vector problematic. However, pegylated albumin nanoparticles loaded with azidothymidine conjugated to Tf have been shown to increase the percentage of injected azidothymidine in the rat brain compared to pegylated nanoparticles without the TF vector. Loperamide adsorbed on albumin nanoparticles covalently coupled to Tf or to the anti-TfR monoclonal antibody (mAb) (OX26), *via* a PEG spacer, resulted in significant antinociceptive effects.[174] Furthermore, three weeks of intravenous therapy with OX26 immunoliposomes in which OX26 is attached to the end of a poly(ethylene glycol) spacer and loaded with a glial-derived neurotrophic factor, led to near complete recovery of rats with neurotoxin-induced Parkinson's disease.[175] These OX26 tagged liposomes have also been used to deliver low molecular weight P-glycoprotein efflux substrates such as digoxin.[176] Recently, the use of OX26 antibody conjugates has been shown to localise drug principally within the brain endothelial cells and not in the post-capillary compartment,[177] presumably because dissociation of this antibody from its specific receptor may be difficult due the high affinity of the antibody for the TfR. It is possible that antibodies with a lower binding affinity may prove more useful.

Improvements in antibody design have resulted in successful brain delivery and a protein comprising a single-chain therapeutic antibody for the treatment of Alzheimer's Disease fused with a TfR mAb was significantly taken up into the brain at levels that compare favourably (3.5% injected dose per g) with those obtained with the therapeutic antibody alone and a mAb with no TfR specificity (0.06% injected dose per g).[178]

Recently, a novel CNS delivery strategy was proposed for antibodies involving a bispecific mouse antibody with one arm comprising a low affinity anti-transferrrin receptor antibody and the other arm comprising the high affinity anti-BACE1 antibody.[179] The enzyme β-secretase (BACE1) processes the amyloid plaques present in Alzheimer's disease patients. Thus, blocking its activity would lead to a reduction in production of the aggregation prone Aβ peptides, thus decreasing amyloid plaque formation slowing Alzheimer's disease progression. In wild-type mice, the bispecific antibody penetrates the brain crossing the BBB more easily than did mono-specific anti-BACE1,[180] reaching up to 10-fold higher concentrations and significantly lowering endogenous Aβ-40 levels (Table 7.1.3). This strategy can potentially be used to deliver active therapeutic antibodies across the blood-brain barrier, and might be broadly applicable to numerous CNS disorders especially if one considers that antibodies currently in trials for Alzheimer's disease tend to enter the brain poorly, necessitating fairly high doses.

The HIR (human insulin receptor) has also been used as an individual target or in combination with the TfR for brain delivery even through there is a potential for such methods to interfere with insulin metabolism. HIR or TfR mediated uptake has been achieved using a variety of conjugates, notably: vasoactive intestinal peptide conjugated to TfR mAbs (VIP-TR mAb),[181] brain derived neurotrophic factor[182] or fibroblast growth factor-2[183] conjugated to the HIR mAbs (BDNF-HIR mAb or FGF2-HIR mAb, respectively), and epidermal growth factor (EGF)[184] or amyloid β1-40 peptide[185] or siRNA[186] conjugated to the TfR mAb. In addition, fusion proteins between β-galactosidase[187] and neurotrophin[188] and the HIR mAb have been produced and patents have been filed protecting a bifunctional fusion antibody construct comprising the HIR mAb and antibodies for other BBB specific receptors.[189–192]

When chimeric HIR-mAbs fused with a tumour necrosis factor decoy receptor were administered intravenously to Rhesus monkeys, 3% of the intravenous radiolabel dose (15 mg kg^{-1}) was found in the brain, approximately 30% higher than that of a nonspecific fusion protein.[193] However quite high levels (45%) of the dose were detected in the liver 2 hours after dosing.[193] The HIR mAbs have also been used to deliver particulate (liposome encapsulated) agents across the BBB. HIR mAbs conjugated to the end of 2000 Da PEG chains on pegylated liposomes have been shown to deliver siRNA specific for tyrosine hydrolase across the BBB,[194] leading to disease modification in a rat Parkinson's disease model. Ever more sophisticated delivery modalities are being employed to deliver proteins and other larger molecules to the brain and to this end, a novel trifunctional fusion antibody comprising HIR for brain entry, amyloid beta for disruption of amyloid plaques and the neonatal Fc receptor for exit out of the brain has also been developed.[195]

7.1.4.1.2 Leptin Receptor

Leptin is a 16 kDa protein produced in white peripheral adipocytes which binds to the leptin receptor in the choroid plexus and on the brain capillary endothelial cells, where it is taken up into the brain parenchyma.[196,197] The leptin receptor can be saturated in obese patients that have elevated levels of leptin (K_d of the receptor is similar to normal serum levels). Leptin$_{61-90}$ provided the highest uptake in rats similar to the endogenous protein and has been explored for gene therapy. Leptin$_{61-90}$ decorated pegylated poly-lysine dendrimer nanoparticles complexed with DNA showed higher uptake in the brain compared to control nanoparticles and the attendant increased gene expression in the brain.[197]

7.1.4.1.3 Low-Density Lipoprotein Receptor and Low-Density Lipoprotein Receptor Related Proteins 1 and 2

Low density lipoprotein receptor (LDLR) binds lipoprotein particles carrying apoliporotien E (ApoE) and apoliprotein B100 (ApoB100), internalising these particles in the process by endocytosis.[198,199] Lipoprotein Receptor Related Protein (LRP) receptor is a multifunctional lipoprotein receptor which interacts with a great variety of ligands such as ApoE, tissue plasminogen activator, amyloid precursor protein (APP), lactoferrin and others, mediating their endocytosis and is expressed in many tissues and the CNS.[200] LRP is expressed in the cerebellum, on neuronal cells, and in astrocytes and is over expressed in malignant astrocytomas, especially glioblastomas.[201] These receptors are able to facilitate the transport of compounds across the BBB. The covalent attachment of ApoE to particles facilitates their uptake by brain capillary endothelial cells and their transfer to the brain parenchyma.[202–204]

The LRP1 receptor has been implicated in the transport of melanotransferrin[205] across the BBB and it has a higher capacity than the TfR and may be exploited to achieve across the BBB delivery. This is exemplified by the transport of doxorubicin conjugates of melanotransferrin across the BBB and the resultant increase in survival of an intracranial tumour mouse model.[205] Lactoferrin is normally present at very low physiological levels and so endogenous lactoferrin is unlikely to compete with a vector conjugate making it a good targeting ligand. Lactoferrin conjugated *via* a PEG spacer to polyamidoamine dendrimers delivered a neurotrophic factor gene to a rotenone-induced rat model of Parkinson's disease with some neuroprotective effects.[206]

A number of workers have exploited ligand peptide domains instead of using the whole ligand and have achieved promising results. The administration of a dipalmitoylated ApoE-derived neuropeptide conjugated to pegylated immunoliposomes resulted in uptake into brain capillary endothelial cells.[207] A lentivirus encoding for a fusion protein comprising human glucocerebrosidase and a 38 amino acid LDLR binding domain of ApoB or the 17 amino acid binding domain of ApoE was administered intraperitoneally to mice, forming a depot in the liver secreting the fusion protein leading to glucocerebrosidase activity in the brain.[198]

One company, Angiochem, has exploited the use of ligand-derived peptides to create medicines for the treatment of brain tumours. BBB permeable peptides based on the structure of aprotinin called Angiopeps are a family of 19 amino acids peptides, derived from the kunitz domain of aprotinin, that have a high transcytosis rate *via* the LRP-1 and LRP-2 receptors.[208] The use of these peptides as BBB-shuttles is also reviewed in the next chapter (Chapter 7.2). In brief, chemical conjugation of the peptide vector (Angiopep-2) to 3 molecules of paclitaxel (ANG1005) resulted in a 100 fold greater delivery of the drug to the brain in an *in-situ* rat brain perfusion assay leading to increased survivial in a mouse intracranial tumour model.[209] Two phase I clinical trials have now been completed on ANG1005 with primary glioma or secondary brain metastases patients who had failed standard therapy. Similar conjugates of Angiopep-2 with doxorubicin (ANG1007) and etoposide (ANG01009) have been shown to deliver higher brain levels especially in the halves of the brain containing the tumour. More than 1% of the injected dose per gram of brain was delivered to the part of the brain with the tumour. Angiopep-2 has been also studied for gene delivery when covalently linked to a PEG spacer linked to a PAMAM dendrimer.[210] 0.25% of injected dose was delivered to the brain with the majority of the dose detected in the kidneys at 2 hours post administration. It is clear from the foregoing that the lipoprotein receptor is a promising target for across BBB delivery.

7.1.4.1.4 Nicotinic Acetylcholine Receptor (nAChR)

A 29 amino acid peptide derived from the rabies glycoprotein (RGV29) when linked *via* a short triglycine spacer to D-arginine-9-mer (9R) and complexed with siRNA resulted in specific gene silencing within the brain on intravenous injection to a mouse.[211] Uptake of RGV29-9R to the brain was blocked by bungarotoxin, a substrate for the nicotine acetylcholine receptor, suggesting that RVG29 entered the brain *via* this receptor and involved the a7 subunit of this receptor.[211] Brain cellular prion protein gene silencing has also been achieved with an RVG29 cationic liposome siRNA complex on intravenous administration.[212] It has been suggested that the $GABA_B$ receptor may be involved in RVG29 uptake as bungarotoxin is a substrate of $GABA_B$ and the brain accumulation of an RVG29 PAMAM dendrimer gene complex (RVG29 covalently linked to a PEG spacer) was not blocked by nicotinic agonists or antagonists.[213]

7.1.4.1.5 Diphtheria Toxin Receptor

The receptor-specific protein vector CRM 197 is a nontoxic mutant diphtheria toxin that has been used for human vaccination and is a proposed vector for brain delivery. CRM197 fused with horseradish peroxidise or coated on liposomes loaded with horseradish peroxidise was transcytosed across the BBB in a guinea pig model.[168] Potential immunogenicity problems make CRM197 unsuitable for chronic applications.

7.1.5 Adsorptive Mediated Transcytosis

While RMT involves specific plasma membrane receptors, cationic large molecular weight biopharmaceuticals can be taken up by the brain *via* adsorptive mediated transcytosis (AMT). AMT requires an excess positive charge on the molecule at physiological pH, which allows it to interact electrostratically with anionic sites on the cell surface, *i.e.* acidic glycoproteins (type IV collagen, laminin, fibronectin and heparin sulphate),[247] triggering endocytosis and subsequent transcytosis.[248] The processes following endocytosis are similar to the processes associated with RMT. However, AMT has a higher capacity for transport compared to RMT. Cationised albumin is known to utilise this pathway to gain entry to the brain,[249,250] along with avidin,[251] histone,[252] cationised polyclonal bovine immunoglobulin,[253] E-2078 (small dynorphin-like basic peptide)[254] and the cell penetrating peptides HIV transactivator of transcription (TAT) protein[255] and other arginine-rich peptides, such as SynB5 (**RGGRLAYLRRRWAVLGR**) and pAnt-(43–58) (**RQIKIWFQNRRMKWKK**).[256] The arginine content of these oligomers is a critical factor[257] and the important structural features of guadinium-rich cell-penetrating vectors are now better understood.[258] However, the concentration of cationic peptides in the brain may be limited by the fact that cationic agents are more readily taken up by the liver and kidney, so that the actual mass taken into the brain is minimal – less than 0.1% of intravenously injected dose.[259] To prevent peripheral organ uptake, researchers have tried masking the cell-penetrating vector with another oligopeptide, which is designed to be cleaved off at the target tissue by specific extracellular proteases, and expose the cationic vector to promote absorption.[260] Lipidisation of the cationic polypeptide has also been used as a strategy to enhance transcytosis of a myristoylated polyarginine vector.[261] Neurotrophic factors covalently linked to naturally occurring polyamines such as putrescine, spermidine and spermine have been also shown to have increased BBB permeability.[262]

7.1.5.1   Exploiting Absorptive Mediated Transcytosis

Cationisation of peptides is a method of increasing membrane permeability *via* absorptive mediated endocytosis (AME)[263–266] as the BBB endothelium possesses anionic sites that attract cationic substances to the membrane surface.[267] Ebiratide, an adrenocorticotrophic hormone (ACTH) analogue, and a dynorphin-like analgesic peptide, E-2078, are polycationic peptides at physiologic pH that are transferred across the BBB *via* AME.[268–270] Based on the applicability of peptide cationic charge for across BBB transport the μ-selective [D-Arg[2]]-demorphin tetrapeptide analogues H-Tyr-D-Arg-Phe-Sar-OH, and H-Tyr-D-Arg-Phe-β-Ala-OH (TAPA) have been developed.[271] These peptides show potent analgesic activity with low physical and psychological dependence.[272] TAPA was reported to cross the BBB *via* AME triggered by binding of the peptides to negatively charged sites on the surface of brain capillary endothelial cells.[273] However, two additional [D-Arg[2]]-demorphin analogues were designed (Nα-amidino-Tyr-D-Arg-Phe-β-Ala-OH (ADAB) and Nα-amidino-Tyr-D-Arg-Phe-Meβ-Ala-OH (ADAMB) that exhibited a slower degree of analgesic onset which could be linked to a parallel decrease in AME across the BBB.[265] AME is not very specific, but the higher capacity of AME, compared to receptor-mediated transcytosis, could be a favourable property for the delivery of peptides to the brain.

Cationised albumin displayed a longer serum half-life and a general selectivity for the brain.[152] Additionally, when cationised albumin was conjugated to β-endorphin, it yielded increased uptake into isolated brain endothelial cells, as compared to β-endorphin alone by AME.[249] However, cationised albumin has been shown to be significantly cleared by the kidney and liver, posing a potential toxicological threat as well.[152,252,274] This approach is also non-specific when compared to tissue uptake, unless additionally coupled to a selective vector.

Unfortunately, cationised proteins have been shown to induce immune complex formation with membranous nephropathy,[275,276] and general non-specific increased cerebral and peripheral vascular permeability,[263,264,277] limiting the therapeutic applications of this strategy.

7.1.6 Conclusions

The BBB is a highly specialized structure, which controls the entry of solutes into the brain; limiting brain distribution to less than a tenth of one percent of the intravenous dose in most cases. This limitation makes the treatment of CNS disorders very difficult and the factors which affect the distribution of drugs to the brain are listed in Table 7.1.4. However, a number of highly controlled transport systems exist at the boundary between the blood supply and the brain – the blood-brain barrier (BBB) – and these may be exploited to deliver therapeutic molecules and therapeutic nanoparticles across the BBB to the brain. These transport processes are *via*: passive diffusion, carrier mediated transport, receptor mediated endocytosis and absorptive endocytosis. Of these processes only the endocytotic transport mechanisms have been proven effective in delivering drugs to the brain that would not normally access the brain parenchyma – namely antibodies and other therapeutic macromolecules, while the other transport modalities are suitable for low molecular weight (ideally less than 300 Da) lipid soluble drugs in the case of passive diffusion and some low molecular weight drugs in the case of carrier mediated systems.

Table 7.1.4 Factors affecting the brain delivery of drugs.

Molecular properties	*BBB characteristics*	*ADME*
Molecular weight		
Ionisation state		
Lipophilicity		
Hydrogen bonding		
Molecular aggregation		Protein binding
Concentration gradient	Membrane charge	Brain tissue binding
Amino acid composition of peptides and proteins	Carrier transport systems	Volume of distribution
Conformation	Receptors	Metabolic stability
Flexibility	Lipid composition	Clearance rate
Molecular folding	Cerebral blood flow	Plasma concentration
Efflux protein substrate	Pathology	
Receptor/ carrier affinity		
Intracellular enzymatic stability		

References

1. W. M. Pardridge, *Brain drug targeting: the future of brain development*, 1st edn, Cambridge University Press, Cambridge, 2001.
2. D. J. Begley, *Pharmacol. Ther.*, 2004, **104**, 29.
3. W. M. Pardridge, *NeuroRx*, 2005, **2**, 3.
4. P. Ehrlich, *Das Sauerstoff-Bedurfnis des Organismus: Eine Farbenanalytische Studie*, Berlin, Hirschwald, 1885.
5. E. E. Goldmann, *Abhandl Konigl preuss Akad Wiss*, 1913, **1**, 1.
6. N. J. Abbott *et al.*, *Neurobiol. Dis.*, 2010, **37**, 13.
7. P. Calvo *et al.*, *Pharm. Res.*, 2001, **18**, 1157.
8. W. M. Pardridge, *Peptide Drug Delivery to the Brain*, Raven Press Ltd, New York, 1991.
9. W. H. Oldendorf, M. E. Cornford and W. J. Brown, *Ann. Neurol.*, 1977, **1**, 409.
10. J. Fenstermacher *et al.*, *Ann. N. Y. Acad. Sci.*, 1988, **529**, 21.
11. R. Sedlakova, R. R. Shivers and R. F. Del Maestro, *J. Submicrosc. Cytol. Pathol.*, 1999, **31**, 149.
12. U. Kniesel and H. Wolburg, *Cell Mol. Neurobiol.*, 2000, **20**, 57.
13. B. T. Hawkins and T. P. Davis, *Pharmacol. Rev.*, 2005, **57**, 173.
14. M. W. Brightman and T. S. Reese, *J. Cell Biol.*, 1969, **40**, 648.
15. H. Wolburg *et al.*, *Neurosci. Lett.*, 2001, **307**, 77.
16. M. Furuse *et al.*, *J. Cell Biol.*, 1993, **123**, 1777.
17. V. Wong and B. M. Gumbiner, *J. Cell Biol.*, 1997, **136**, 399.
18. M. Tagami *et al.*, *Stroke*, 1990, **21**, 1064.
19. A. M. Butt, H. C. Jones and N. J. Abbott, *J. Physiol.*, 1990, **429**, 47.
20. A .Schirmacher *et al.*, *Bioelectromagnetics*, 2000, **21**, 338.
21. D. J. Begley, *J. Pharm. Pharmacol.*, 1996, **48**, 136.
22. A. Misra *et al.*, *J. Pharm. Pharmacol. Sci.*, 2003, **6**, 252.
23. D. J. Begley and M. W. Brightman, in *Peptide transport and Delivery into the Central Nervous System. Progress in Drug Research.*, ed. L.P.-T. Prokai, Birkhauser Verlag, Basel, Switzerland, 2003, p. 39.
24. A. Minn *et al.*, in *The Blood-brain Barrier and Drug Delivery to the CNS.*, ed. D. J. Begley, M. W. B. Bradbury and J. Kreuter, Dekker, New York, 2000, p. 145.
25. N. J. Abbott and I. A. Romero, *Mol. Med. Today*, 1996, **2**, 106.
26. J. Brownless and C. H. Williams, *J. Neurochem.*, 1993, **60**, 793.
27. A .Minn *et al.*, *Brain Res. Brain Res. Rev.*, 1991, **16**, 65.
28. D. J. Begley, *Curr. Pharm. Design*, 2004, **10**, 1295.
29. T. Terasaki and K. Hosoya, *Adv. Drug Deliv. Rev.*, 1999, **36**, 195.
30. M. Demeule *et al.*, *Vascular Pharmacol.*, 2002, **38**, 339.
31. H. Kusuhara and Y. Sugiyama, *NeuroRx: J. Am. Soc. Exp. NeuroTherapeutics*, 2005, **2**, 73.

32. A. Tsuji, in *The Blood-Brain barrier and Drug Delivery to the CNS*, ed. D. J. Begley, M. W. B. Bradbury and J. Kreuter, Marcel Dekker NY, 2000, p. 121.
33. S. Shen and W. Zhang, *Rev. Neurosci.*, 2010, **21**, 29.
34. A. Tsuji, *Ther. Drug Monit.*, 1998. **20**, 588.
35. W. Loscher and H. Potschka, *NeuroRx: J. Am. Soc. Exp. NeuroTherapeutics*, 2005, **2**, 86.
36. A. ElAli and D. M. Hermann, *Neuroscientist*, 2011, **17**, 423.
37. V. A. Levin, *J. Med. Chem.*, 1980, **23**, 682.
38. X. Liu *et al.*, *Drug Metab. Dispos.*, 2004, **32**, 132.
39. M. P. Gleeson, *J. Med. Chem.*, 2008, **51**, 817.
40. D. E. Clark, *Drug Discovery Today*, 2003, **8**, 927.
41. S. D. Kramer, *Pharm. Sci. Technol. Today*, 1999, **2**, 373.
42. D. J. Begley, in *Blood-spinal Cord and Brain Barriers in Health and Disease*, ed. H. S. Sharma and J. Westman, Elsevier, San Diego, 2004, p. 83.
43. J. Kelder *et al.*, *Pharm. Res.*, 1999, **16**, 1514.
44. H. Lennernas and E. Lundgren, *Drug Discovery Today: Technol.*, 2004, **1**, 417.
45. X. Fu, Z. Song and W. Liang, A Predictive Model for Blood-Brain Barrier Penetration, in *Internet Electronic Conference of Molecular Design*, 2004.
46. G. Gerebtzoff and A. Seelig, *J. Chem. Inf. Model.*, 2006, **46**, 2638.
47. L. Di, E. H. Kerns and G. T. Carter, *Exp. Opin. Drug Discovery*, 2008, **3**, 677.
48. R. Nau, F. Soergel and H. Eiffert, *Clin. Microbiol. Rev.*, 2010, **23**, 858.
49. M. H. Abraham, *Chem. Soc. Rev.*, 1993, **22**, 73.
50. K. Valko *et al.*, *J. Pharm. Sci.*, 2003, **92**, 2236.
51. N. A. Kratochwil *et al.*, *Biochem. Pharmacol.*, 2002, **64**, 1355.
52. N. Drion *et al.*, *J. Neurochem.*, 1996, **67**, 1688.
53. E. M. Kemper *et al.*, *Clin. Cancer Res.*, 2003, **9**, 2849.
54. C. S. Hughes *et al.*, *J. Neurooncol.*, 1998, **37**, 45.
55. E. G. Chikhale *et al.*, *Pharm. Res.*, 1994, **11**, 412.
56. K. A. Witt *et al.*, *J. Neurochem.*, 2000, **75**, 424.
57. A. Bak, *et al.*, *Pharm. Res.*, 1999, **16**, 24.
58. T. Uchiyama, *et al.*, *Pharm. Res.*, 2000, **17**, 1461.
59. H. Bundgaard and J. Moss, *Pharm. Res.*, 1990, **7**, 885.
60. A. Yamamoto, *Nippon Rinsho*, 1998, **56**, 601.
61. H. Asada *et al.*, *J. Pharm. Sci.*, 1995, **84**, 682.
62. D. W. Hansen, Jr. *et al.*, *J. Med. Chem.*, 1992, **35**, 684.
63. E. V. Batrakova *et al.*, *Bioconjug. Chem.*, 2005, **16**, 793.
64. T. Ogawa *et al.*, *Chem. Pharm. Bull. (Tokyo)*, 2003, **51**, 759.
65. M. Broccardo *et al.*, *Br. J. Pharmacol.*, 1981, **73**, 625.
66. Y. Sasaki *et al.*, *Neuropeptides*, 1985, **5**, 391.
67. M. Marastoni *et al.*, *J. Med. Chem.*, 1987, **30**, 1538.
68. K. Chaki *et al.*, *Life Sci.*, 1990, **46**, 1671.

69. S. J. Weber *et al.*, *J. Pharmacol. Exp. Ther.*, 1993, **266**, 1649.
70. S. J. Weber *et al.*, *J. Pharmacol. Exp. Ther.*, 1991, **259**, 1109.
71. C. L. Gentry *et al.*, *Peptides*, 1999, **20**, 1229.
72. T. J. Abbruscato *et al.*, *J. Pharmacol. Exp. Ther.*, 1996, **276**, 1049.
73. K. A. Witt and T. P. Davis, *AAPS J.*, 2006, **8**, E76.
74. M. K. Ghosh and A. K. Mitra, *Pharm. Res.*, 1992, **9**, 1173.
75. D. M. Lambert *et al.*, *J. Pharm. Belg.*, 1995, **50**, 194.
76. N. H. Greig *et al.*, *J. Chromatogr.*, 1990, **534**, 279.
77. D. L. Greene *et al.*, *J. Pharmacol. Exp. Ther.*, 1996, **277**, 1366.
78. N. Bodor, E. Shek and T. Higuchi, *Science*, 1975, **190**, 155.
79. L. Prokai, K. Prokai-Tatrai and N. Bodor, *Med. Res. Rev.*, 2000, **20**, 367.
80. N. Bodor and P. Buchwald, *Adv. Drug Delivery Rev.*, 1999, **36**, 229.
81. M. E. Brewster, K. S. Estes and N. Bodor, *J. Med. Chem.*, 1988, **31**, 244.
82. L. Prokai *et al.*, *J. Am. Chem. Soc.*, 1994, **116**, 2643.
83. N. Bodor *et al.*, *Science*, 1992, **257**, 1698.
84. Audus, K.L., *et al. Adv. Drug Res.*, 1992. **23**, 3.
85. G. M. Pauletti *et al.*, *Adv. Drug Delivery Rev.*, 1997, **27**, 235.
86. G. T. Knipp *et al.*, *Pharm. Res.*, 1997, **14**, 1332.
87. B. Wang *et al.*, *J. Pept. Res.*, 1999, **53**, 370.
88. R. T. Borchardt, *J. Controlled Release*, 1999, **62**, 231.
89. C. Toniolo, *Int. J. Pept. Protein Res.*, 1990, **35**, 287.
90. T. Uchiyama *et al.*, *J. Pharm. Sci.*, 1998, **87**, 448.
91. D. Roemer and J. Pless, *Life Sci.*, 1979, **24**, 621.
92. S. J. Weber *et al.*, *J. Pharmacol. Exp. Ther.*, 1992, **263**, 1308.
93. R. J. Knapp *et al.*, *J. Pharmacol. Exp. Ther.*, 1991, **258**, 1077.
94. K. Kalyanasundaram and J. K. Thomas, *J. Am. Chem. Soc.*, 1977, **99**, 2039.
95. S. Liao *et al.*, *J. Med. Chem.*, 1998, **41**, 4767.
96. R. D. Egleton and T. P. Davis, *J. Pharm. Sci.*, 1999, **88**, 392.
97. S. Fellner *et al.*, *J. Clin. Invest.*, 2002, **110**, 1309.
98. C. Chen and G. M. Pollack, *J. Pharmacol. Exp. Ther.*, 1998, **287**, 545.
99. K. A. Witt *et al.*, *J. Pharmacol. Exp. Ther.*, 2001, **298**, 848.
100. A. L. Betz, J. A. Firth and G. W. Goldstein, *Brain Res.*, 1980, **192**, 17.
101. K. Mertsch and J. Maas, *Curr. Med. Chem: Cent. Nerv. Syst. Agents*, 2002, **2**, 187.
102. I. Brasnjevic *et al.*, *Prog. Neurobiol.*, 2009, **87**, 212.
103. A. K. Kumagai, K. J. Dwyer and W. M. Pardridge, *Biochim. Biophys. Acta*, 1994, **1193**, 24.
104. J. C. Vera *et al.*, *Nature*, 1993, **364**, 79.
105. W. H. Oldendorf, *Am. J. Physiol.*, 1973, **224**, 1450.
106. Y. Kido *et al.*, *Pharm. Res.*, 2000, **17**, 55.
107. L. M. Roberts *et al.*, *Endocrinology*, 2008, **149**, 6251.
108. H. Heuer *et al.*, *Endocrinology*, 2005, **146**, 1701.
109. E. C. Friesema *et al.*, *Mol. Endocrinol.*, 2006, **20**, 2761.
110. D. Sugiyama *et al.*, *J. Pharmacol. Exp. Ther.*, 2001, **298**, 316.

111. R. Kikuchi *et al.*, *J. Pharmacol. Exp. Ther.*, 2003, **306**, 51.
112. S. Ohtsuki *et al.*, *J. Neurochem.*, 2002, **83**, 57.
113. B. Gao *et al.*, *J. Pharmacol. Exp. Ther.*, 2000, **294**, 73.
114. B. Gao *et al.*, *J. Histochem. Cytochem.*, 1999, **47**, 1255.
115. C. Reichel *et al.*, *Gastroenterology*, 1999, **117**, 688.
116. H. Kusuhara *et al.*, *J. Biol. Chem.*, 1999, **274**, 13675.
117. T. Nozawa *et al.*, *J. Pharmacol. Exp. Ther.*, 2004, **308**, 438.
118. M. Grube *et al.*, *Mol. Pharmacol.*, 2006, **70**, 1735.
119. I. Tamai *et al.*, *J. Biol. Chem.*, 1998, **273**, 20378.
120. W. A. Banks and A. J. Kastin, *Prog. Brain Res.*, 1992, **91**, 139.
121. W. A. Banks and A. J. Kastin, *Alcohol*, 1997, **14**, 237.
122. V. Ganapathy and S. Miyauchi, *AAPS J.*, 2005, **7**, E852.
123. G. A. Maresh *et al.*, *Brain Res.*, 1999, **839**, 336.
124. R. Kannan *et al.*, *J. Neurochem.*, 1999, **73**, 390.
125. R. J. Boado *et al.*, *Proc. Natl. Acad. Sci. U. S. A.*, 1999, **96**, 12079.
126. W. H. Oldendorf and J. Szabo, *Am. J. Physiol.*, 1976, **230**, 94.
127. Q. R. Smith, *J. Nutr.*, 2000, **130**, 1016S.
128. M. M. Sanchez del Pino, R. A. Hawkins and D. R. Peterson, *J. Biol. Chem.*, 1992, **267**, 25951.
129. I. Tamai *et al.*, *Biochem. Pharmacol.*, 1995, **50**, 1783.
130. J. Komura *et al.*, *J. Neurochem.*, 1996, **67**, 330.
131. N. Zerangue and M. P. Kavanaugh, *J. Biol. Chem.*, 1996, **271**, 27991.
132. Q. R. Liu *et al.*, *J. Biol. Chem.*, 1993, **268**, 2106.
133. K. Wakayama *et al.*, *Neurosci. Res.*, 2002, **44**, 173.
134. C. M. Anderson *et al.*, *J. Neurochem.*, 1999, **73**, 867.
135. M. Griffiths *et al.*, *Biochem. J.*, 1997, **328 (Pt 3)**, 739.
136. H. Lu C. Chen and C. Klaassen, *Drug Metab. Dispos.*, 2004, **32**, 1455.
137. I. Aymerich *et al.*, *Biochem. Soc. Trans.*, 2005, **33**, 216.
138. A. Tsuji and I. Tamai, in *Introduction to the Blood-Brain Barrier: Methodology, biology, and pathology*, ed. W. M. Pardridge, Cambridge University Press, Cambridge, UK, 1998, p. 243.
139. A. Tsuji, *NeuroRx*, 2005, **2**, 54.
140. L. A. Wade and R. Katzman, *J. Neurochem.*, 1975, **25**, 837.
141. Y. Takada *et al.*, *Cancer Chemother. Pharmacol.*, 1991, **29**, 89.
142. S. A. Thomas *et al.*, *J. Pharmacol. Exp. Ther.*, 1997, **280**, 1235.
143. R. Polt *et al.*, *Proc. Natl. Acad. Sci. U. S. A.*, 1994, **91**, 7114.
144. R. D. Egleton *et al.*, *Brain Res.*, 2000, **881**, 37.
145. M. F. Powell *et al.*, *Pharm. Res.*, 1993, **10**, 1268.
146. J. F. Fisher *et al.*, *J. Med. Chem.*, 1991, **34**, 3140.
147. R. D. Broadwell, B. J. Balin and M. Salcman, *Proc. Natl. Acad. Sci. U. S. A.*, 1988, **85**, 632.
148. R. D. Egleton *et al.*, *Tetrahedron: Asymmetry*, 2005, **16**, 65.
149. P. J. Gaillard, Glutathione-based drug delivery system, WO/2010/095940, PCT/NL2010/050082, 2010, p. 52.

150. P.J. Gaillard, Targeted intracellular deliveery of antiviral agents, WO/ 2008/118013, PCT/EP, 2008, p. 65.
151. W. Geldenhuys *et al.*, *J. Drug Targeting*, 2011, **19**, 837.
152. U. Bickel, T. Yoshikawa and W. M. Pardridge, *Adv. Drug Delivery Rev.*, 2001, **46**, 247.
153. M. van Houten *et al.*, *Endocrinology*, 1979, **105**, 666.
154. H. J. Frank and W. M. Pardridge, *Diabetes*, 1981, **30**, 757.
155. H. J. Frank *et al.*, *Diabetes*, 1986, **35**, 654.
156. K. R. Duffy, W.M. Pardridge and R.G. Rosenfeld, *Metabolism*, 1988, **37**, 136.
157. C. C. Visser *et al.*, *J. Drug Targeting*, 2004, **12**, 145.
158. W. A. Jefferies *et al.*, *Nature*, 1984, **312**, 162.
159. J. B.Fishman *et al.*, *J. Neurosci. Res.*, 1987, **18**, 299.
160. M. Demeule *et al.*, *J. Neurochem.*, 2002, **83**, 924.
161. W. A. Banks and C. L. Farrell, *Am. J. Physiol. Endocrinol. Metab.*, 2003, **285**, E10.
162. W. Pan and A. J. Kastin, *Exp. Neurol.*, 2002, **174**, 193.
163. W. Pan and A. J. Kastin, *Peptides*, 1999, **20**, 1091.
164. B. V. Zlokovic *et al.*, *Exp. Neurol.*, 1990, **107**, 263.
165. W. A. Banks, *Nutrition*, 2001, **17**, 434.
166. W. A. Banks *et al.*, *J. Pharmacol. Exp. Ther.*, 2001, **299**, 536.
167. J. Herz and P. Marschang, *Cell*, 2003, **112**, 289.
168. P. J. Gaillard, A. Brink and A. G. De Boer, *Int. Congr. Series*, 2005, **1277**, 185.
169. M. C. Willingham *et al.*, *Proc. Natl. Acad. Sci. U. S. A.*, 1984, **81**, 175.
170. W. A. Banks, *Curr. Pharm. Des.*, 2001, **7**, 125.
171. W. M. Pardridge, *J. Drug Targeting*, 2010, **18**, 157.
172. W. M. Pardridge, *Pharm. Res.*, 2007, **24**, 1733.
173. N. Shi *et al.*, *Proc. Natl. Acad. Sci. U. S. A.*, 2001, **98**, 12754.
174. K. Ulbrich *et al.*, *Eur. J. Pharm. Biopharm.*, 2009, **71**, 251.
175. Y. Zhang and W.M. Pardridge, *Pharm. Res.*, 2009, **26**, 1059.
176. J. Huwyler *et al.*, *J. Drug Targeting*, 2002, **10**, 73.
177. T. Moos and E.H. Morgan, *J. Neurochem.*, 2001, **79**, 119.
178. R. J. Boado *et al.*, *Mol. Pharm.*, 2010, **7**, 237.
179. J. K. Atwal *et al.*, *Sci. Transl. Med.*, 2011, **3**, 84ra43.
180. Y. J. Yu *et al.*, *Sci. Transl. Med.*, 2011, **3**, 84ra44.
181. U. Bickel *et al.*, *Proc. Natl. Acad. Sci. U. S. A.*, 1993, **90**, 2618.
182. Y. Zhang and W. M. Pardridge, *Brain. Res.*, 2001, **889**, 49.
183. D. Wu *et al.*, *J. Drug Targeting*, 2002, **10**, 239.
184. Y. Zhang *et al.*, *Clin. Cancer Res.*, 2004, **10**, 3667.
185. A. Kurihara and W. M. Pardridge, *Bioconjug. Chem.*, 2000, **11**, 380.
186. C. F. Xia, R. J. Boado and W. M. Pardridge, *Mol. Pharm.*, 2009, **6**, 747.
187. Y. Zhang and W. M. Pardridge, *J. Pharmacol. Exp. Ther.*, 2005, **313**, 1075.
188. R. J. Boado and W. M. Pardridge, *Drug Metab. Dispos.*, 2009, **37**, 2299.

189. W. M. Pardridge and R. J. Boado, Fusion proteins for delivery of GDNF to the CNS, WO/2009/070597, PCT/US2008/084718, 2009, p. 164.

190. W. M. Pardridge and R. J. Boado, Compositions and methods for blood-brain barrier delivery of IGG-decoy receptor fusion proteins, WO/2010/108048, PCT/US2010/027882, 2010, p. 63.

191. W. M. Pardridge and R. J. Boado, Methods and compositions for increasing iduronate 2-sulfatase activity in the CNS, WO/2011/044542, PCT/US2010/052113, 2010, p. 66.

192. W. M. Pardridge and R. J. Boado, Fusion proteins for delivery of erythropoetin to the CNS, WO/2011/088409, PCT/US2011/021418, 2011, p. 129.

193. R. J. Boado *et al.*, *J. Biotechnol.*, 2010, **146**, 84.

194. W. M. Pardridge, *Adv. Drug Delivery Rev.*, 2007, **59**, 141.

195. R. J. Boado *et al.*, *Bioconjug. Chem.*, 2007, **18**, 447.

196. W. A. Banks *et al.*, *Peptides*, 1996, **17**, 305.

197. G. L. Barrett, J. Trieu and T. Naim, *Regul. Pept.*, 2009, **155**, 55.

198. B. J. Spencer and I. M. Verma, *Proc. Natl. Acad. Sci. U. S. A.*, 2007, **104**, 7594.

199. B. Dehouck *et al.*, *J. Cell Biol.*, 1997, **138**, 877.

200. G. W. Rebeck *et al.*, *Neuron*, 1993, **11**, 575.

201. M. Yamamoto *et al.*, *Cancer Res.*, 1997, **57**, 2799.

202. A. Zensi *et al.*, *J. Controlled Release*, 2009, **137**, 78.

203. J. Kreuter *et al.*, Nanoparticles made of protein with coupled apolipoprotein E for penetration of the BBB and methods for the production thereof, US/2004/0131692, 2004.

204. J. Kreuter *et al.*, Transport of drugs via the blood-brain barrier by means of apolipoproteins, WO/2008/095652, PCT/EP2008/000822, 2008, p. 25.

205. D. Karkan *et al.*, *PLoS One*, 2008, **3**, e2469.

206. R. Huang *et al.*, *Brain Res. Bull.*, 2010, **81**, 600.

207. I. Sauer *et al.*, *Biochim. Biophys. Acta*, 2006, **1758**, 552.

208. M. Demeule *et al.*, *J. Neurochem.*, 2008, **106**, 1534.

209. A. Regina *et al.*, *Br. J. Pharmacol.*, 2008, **155**, 185.

210. W. Ke *et al.*, *Biomaterials*, 2009, **30**, 6976.

211. P. Kumar *et al.*, *Nature*, 2007, **448**, 39.

212. B. Pulford *et al.*, *PLoS One*, 2010, **5**, e11085.

213. M. E. Wilkins, X. Li and T. G. Smart, *J. Biol. Chem.*, 2008, **283**, 34745.

214. A. Muruganandam *et al.*, *FASEB J.*, 2002, **16**, 240.

215. A. Muruganandam *et al.*, Single-Domain Brain-Targeting antibody fragments derived from Llama antibodies, WO/2002/057445, US 2011/0171720, 2001, p. 70.

216. A. Muruganandam *et al.*, Single-Domain Brain-Targeting antibody fragments derived from LLAMA antibodies, US 2009/0162422, 2009, p. 43.

217. A. Abulrob, D. Stanimirovic and A. Muruganandam, Blood-brain barrier epitopes and uses thereof, US 2009/0047300, 2009, p. 43.

218. J. Z. Lu *et al.*, *Biotechnol. Bioeng.*, 2011, **108**, 1954.
219. R. J. Boado *et al.*, *Mol. Pharm.*, 2011, **8**, 1342.
220. Q. H. Zhou *et al.*, *Bioconjug. Chem.*, 2011, **22**, 1611.
221. W. M. Pardridge and R. J. Boado, Methods and compositions for increasing alpha-iduronidase activity in the CNS, WO/2009/018122, PCT/US2008/07112, 2009, p. 60.
222. Q. Zhou *et al.*, *J. Pharmacol. Exp. Ther.*, 2011, **339**, 618.
223. A. Fu *et al.*, *Brain Res.*, 2011, **1352**, 208.
224. W. M. Pardridge and R. J. Boado, Agents for blood-brain barrier delivery, WO/2008/022349, PCT/US2007/076316, 2008, p. 154.
225. J. Huwyler, J. Yang and W. M. Pardridge, *J. Pharmacol. Exp. Ther.*, 1997, **282**, 1541.
226. W. M. Pardridge and J. Huwyler, Transport of liposomes across the blood-brain barrier, WO/1998/022092, PCT/US1997/021352, 1997, p. 38.
227. D. Thakker and M. E. Benz, Patent Application, 2007, WO/2008/033253.
228. J.-P. Castaigne *et al.*, Leptin and Leptin analog conjugates and uses thereof, WO/2010/063123, PCT/CA2009/001780, 2009, p. 70.
229. J.-P. Castaigne *et al.*, Conjugates of Neurotensin of Neurotensin analogs and uses thereof, WO/2010/063122, PCT/CA2009/001779, 2009, p. 79.
230. F. C. Thomas *et al.*, *Pharm. Res.*, 2009, **26**, 2486.
231. K. Beliveau *et al.*, Aprotinin polypeptides for transporting a compound across the blood-brain barrier, WO/2006/086870, PCT/CA2005/001158, 2005, p. 73.
232. K. Beliveau *et al.*, Use of aprotinin polypeptides as carriers in pharmaceutical conjugates, WO/2007/009229, PCT/CA2006/001165, 2007, p. 100.
233. J. Kreuter *et al.*, *Brain Res.*, 1995, **674**, 171.
234. R. N. Alyaudtin *et al.*, *J. Drug Targeting*, 2001, **9**, 209.
235. R. N. Alyautdin *et al.*, *Pharm. Res.*, 1997, **14**, 325.
236. S. Wohlfart *et al.*, *Int. J. Pharm.*, 2011, **415**, 244.
237. K. B. Kurakhmaeva *et al.*, *J. Drug Targeting*, 2009, **17**, 564.
238. J. Kreuter *et al.*, Drug targeting system, method for preparing same and its use, WO/1995/22963, PCT/EP1995/000724, 1995, p. 35.
239. J. Kreuter *et al.*, Drug targeting to the nervous system by nanoparticles, US/2006/6117454, 2000, p. 13.
240. J. Kreuter *et al.*, Polylactide nanoparticles, WO/2007/110152, PCT/EP2007/002198, 2007, p. 53.
241. I. M. Verma and B. J. Spencer, Compositions and methods for targeting a polypeptide to the central nervous system, US/2010/0015117, 2009, p. 30.
242. P. Kumar *et al.*, *PLoS Med.*, 2006, **3**, e96.
243. S. Kumar and V.A. Arankalle, *PLoS One*, 2011, **5**, e8615.
244. M. Narasimhaswamy, P. Shankar and P. Kumar, Method for delivery across the blood brain barrier, WO/2008/054544, PCT/US2007/012152, 2007, p. 152.

245. A. Pfeifer *et al.*, Humanized antibodies against amyloid beta, WO/2008/156622, PCT/US2008/007318, 2007, p. 151.
246. A. Pfeifer *et al.*, Monoclonal anti beta amyloid antibody, WO/2008/156621, PCT/US2008/007317, 2008, p. 153.
247. A. W. Vorbrodt, *J. Neurocytol.*, 1989, **18**, 359.
248. I. Sauer *et al.*, *Biochemistry*, 2005, **44**, 2021.
249. A. K. Kumagai, J. B. Eisenberg, and W. M. Pardridge, *J. Biol. Chem.*, 1987, **262**, 15214.
250. W. M. Pardridge *et al.*, *J. Pharmacol. Exp. Ther.*, 1990, **255**, 893.
251. W. M. Pardridge and R. J. Boado, *FEBS Lett.*, 1991, **288**, 30.
252. W. M. Pardridge, D. Triguero and J. Buciak, *J. Pharmacol. Exp. Ther.*, 1989, **251**, 821.
253. D. Triguero *et al.*, *Proc. Natl. Acad. Sci. U. S. A.*, 1989, **86**, 4761.
254. T. Terasaki *et al.*, *Pharm. Res.*, 1991, **8**, 815.
255. A. D. Frankel and C.O. Pabo, *Cell*, 1988, **55**, 1189.
256. G. Drin *et al.*, *J. Biol. Chem.*, 2003, **278**, 31192.
257. N. Schmidt *et al.*, *FEBS Lett.*, **584**, 1806.
258. P. A. Wender *et al.*, *Adv. Drug Delivery Rev.*, 2008, **60**, 452.
259. H. J. Lee and W.M. Pardridge, *Bioconjug. Chem.*, 2001, **12**, 995.
260. T. Jiang *et al.*, *Proc. Natl. Acad. Sci. U. S. A.*, 2004, **101**, 17867.
261. W. Pham *et al.*, *Neuroimage*, 2005, **28**, 287.
262. J. F. Poduslo, G. L. Curran and J. S. Gill, *J. Neurochem.*, 1998, **71**, 1651.
263. V. M. Vehaskari *et al.*, *J. Clin. Invest.*, 1984, **73**, 1053.
264. J. E. Hardebo and J. Kahrstrom, *Acta Physiol. Scand.*, 1985, **125**, 495.
265. Y. Deguchi *et al.*, *J. Pharmacol. Exp. Ther.*, 2004, **310**, 177.
266. W. M. Pardridge, R. J. Boado and Y. S. Kang, *Proc. Natl. Acad. Sci. U. S. A.*, 1995, **92**, 5592.
267. S. Chakrabarti and A. A. Sima, *Microvasc. Res.*, 1990, **39**, 123.
268. T. Terasaki *et al.*, *J. Pharmacol. Exp. Ther.*, 1989, **251**, 351.
269. T. Terasaki *et al.*, *Pharm. Res.*, 1992, **9**, 529.
270. J. Yu *et al.*, *J. Pharmacol. Exp. Ther.*, 1997, **282**, 633.
271. M. Ukai *et al.*, *Eur. J. Pharmacol.*, 1995, **287**, 245.
272. P. Paakkari *et al.*, *J. Pharmacol. Exp. Ther.*, 1993, **266**, 544.
273. Y. Deguchi *et al.*, *J. Neurochem.*, 2003. **84**, 1154.
274. P. Bergmann, R. Kacenelenbogen and A. Vizet, *Clin. Sci. (Lond.)*, 1984, **67**, 35.
275. S. G. Adler *et al.*, *J. Clin. Invest.*, 1983, **71**, 487.
276. S. S. Feng *et al.*, *Curr. Med. Chem.*, 2004, **11**, 413.
277. Z. Nagy, H. Peters and I. Huttner, *Lab. Invest.*, 1983, **49**, 662.

CHAPTER 7.2

Drug Delivery Strategies: BBB–Shuttles

R. PRADES[a], M. TEIXIDÓ*[a] AND E. GIRALT*[a,b]

[a] Institute for Research in Biomedicine (IRB Barcelona), Parc Científic de Barcelona, Baldiri Reixac 10, 08028 Barcelona, Spain; [b] Department of Organic Chemistry, University of Barcelona, Martí Franquès 1, 08028 Barcelona, Spain
*E-mail: meritxell.teixido@irbbarcelona.org; ernest.giralt@irbbarcelona.org

7.2.1 Introduction

Brain disorders account for more than one third of the total burden of diseases in Europe, where about a quarter of the population is affected.[1] More than 600 disorders are characterized by progressive nervous system dysfunction and these are often associated with the atrophy of affected central peripheral structures.[2] Despite the enormous potential market for the pharmaceutical industry, the targeting of drugs and diagnostic agents to the central nervous system (CNS) is challenged by the presence of the blood-brain barrier (BBB). More than 98% of all small molecules do not cross the BBB, while no large-molecule drugs, including peptides, recombinant proteins and monoclonal antibodies, cross this barrier.[3] In fact, of the 7000 molecules in the Comprehensive Medical Chemistry (CMC) database, less than 5% are active in the CNS. The activity of these drugs is generally limited to only four conditions, namely affective disorders, chronic pain, insomnia, and epilepsy.[4] If affective disorders are excluded from this list, less than 1% of the drugs are pharmacologically active in the brain.[5] Given this scenario, it is clear that there is an urgent need to develop efficient delivery vectors (BBB-shuttles) able to transport different types of cargo across the BBB.

RSC Drug Discovery Series No. 22
Nanostructured Biomaterials for Overcoming Biological Barriers
Edited by Maria Jose Alonso and Noemi S. Csaba
© The Royal Society of Chemistry 2012
Published by the Royal Society of Chemistry, www.rsc.org

7.2.2 Transport Through the BBB

The BBB is located in brain capillaries. In addition to the BBB, the choroid plexus epithelium presents a second barrier at the blood-cerebrospinal fluid (CSF) interface. The estimated surface area of the BBB is approximately 20 m^2, which makes it about 1000 time larger than the blood-CSF and the brain-CSF barrier.[6] Therefore, the BBB is considered the most important barrier to tackle in order to achieve drug delivery to the brain.

This barrier is composed of a continuous layer of highly specialized vascular endothelial cells whose primary function is to maintain brain homeostasis. The BBB is a highly efficient physical and enzymatic barrier, restricting and regulating the passage of substances from the blood to the brain, thus providing an optimal environment for the function of this organ. The barrier effect is achieved in part by a complex network of tight junctions between endothelial cells[7–9] that hinders paracellular transport. Furthermore, transcellular transport from blood to brain is also limited as a result of low vesicular transport, high metabolic activity, lack of fenestrae,[10] and the expression of a variety of enzymes. These include cytosolic forms, and enzymes on the extracellular membrane of endothelial cells also contribute to the restrictive nature of the BBB.[11] In addition to all these features, this barrier also expresses efflux pumps, such as the P-glycoprotein (P-gp) efflux pump, in the luminal plasma membrane of endothelial cells. This is an ATP-dependent pump that prevents the accumulation of an extensive variety of potential drugs and hydrophilic compounds in the CNS.[12] All these characteristics are induced and maintained by complex interactions between the brain endothelial cells, astrocytes, neuronal endings and pericytes.[7,13,14] As a result of the restrictive role of the BBB, the delivery of many potential therapeutic agents to the brain is limited (for complementary information see Chapter 7.1).

Despite the protective and isolation function of the BBB, the brain is not fully insulated. It requires essential nutrients and ions to perform its functions and to maintain the integrity of the BBB. This barrier holds several influx mechanisms that ensure basic nutrients for the CNS (Figure 7.2.1). These mechanisms can be divided into passive and active transport systems, and they are potential pathways for the delivery of neurotherapeutic drugs and diagnostic agents into the CNS.

Passive diffusion is a non-energy dependent and non-saturable mechanism. Gaseous molecules such as O_2 and CO_2 and small lipid-soluble compounds (generally <300–500 Da) can diffuse passively to cross the BBB. In contrast, active transport is an energy-dependent and saturable mechanism that can be divided into three types: carrier-mediated transport, absorptive-mediated transcytosis, and receptor-mediated transcytosis. Active transport is usually the mechanism by which highly specific compounds, which can be hydrophilic and large, enter the CNS.

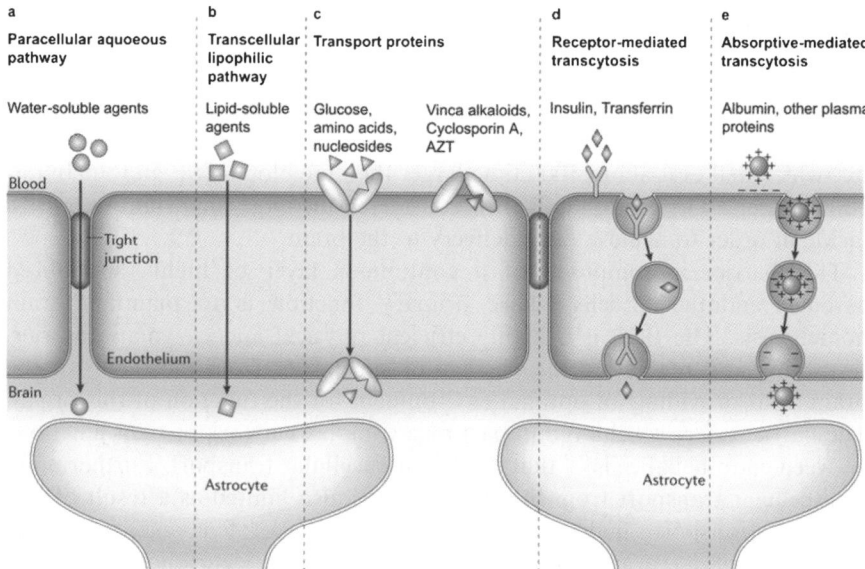

a
Paracellular aquoeous pathway

Water-soluble agents

Blood

Tight junction

Endothelium

Brain

b
Transcellular lipophilic pathway

Lipid-soluble agents

c
Transport proteins

Glucose, amino acids, nucleosides

Vinca alkaloids, Cyclosporin A, AZT

Astrocyte

d
Receptor-mediated transcytosis

Insulin, Transferrin

e
Absorptive-mediated transcytosis

Albumin, other plasma proteins

Astrocyte

Figure 7.2.1   Pathways across the blood-brain barrier. (A) The paracellular pathway is very limited because of the presence of tight junctions between endothelial cells, one of the main features of the BBB. (B) The transcellular pathway or passive diffusion, like the paracellular pathway, is limited as a result of the features of the BBB. This pathway normally is limited to small highly lipophilic molecules (< 300–500 Da). (C) Carrier-mediated transport is one of the main nutrient entrance routes to the brain. It is a highly sterospecific pathway that relies on the recognition of a substrate by a carrier protein. (D) Receptor-mediated transcytosis is a vesicle-based transport pathway known to be the mechanism of entrance of transferrin, insulin and leptin to the brain. (E) Absorptive-mediated transcytosis is the route by which highly cationic substances that interact with negative charged areas of the cell membrane enter the brain. This interaction could lead transcytosis; however, this type of vesicular transport is highly down-regulated in the BBB to protect the brain from non-specific exposure to polycationic compounds. (Reproduced with the permission from Nature Reviews Neuroscience).

7.2.3   Drug Delivery to the Brain

Drug delivery to the brain can be done by invasive, pseudo-invasive, or non-invasive strategies.

7.2.3.1   Invasive Strategies

Invasive strategies are neurosurgical-based and are the most aggressive way to deliver drugs. These procedures usually involve drilling a hole surgically in the head of the patient and delivering the drug by intracerebroventricular injection, intracerebral injection or convection-enhanced diffusion.

Intracerebroventricular injection: the drug is injected into the CSF compartment and it penetrates the brain parenchyma *via* diffusion. This strategy is not valid for the administration of all drugs since the CSF pool in the human brain is turned over every 4–5 hours. Regarding drugs, the kinetics of bulk is several orders of magnitude higher than the kinetics of diffusion. Consequently, most of the drugs injected following this procedure do not exert a pharmacological effect in the brain. However, there are some exceptions. This is the case of opioid peptides[15] and other pharmacological agents whose receptor is near the ependymal surface of the brain (area involved in the production of CSF).

Intracerebral injection: this approach, like intraventricular injection, relies on the diffusion of the drug after injection. Here the drug is injected directly into the brain. This technique is valid only for local delivery since diffusion in brain tissue takes place very slowly and it is not possible to achieve significant distribution of the drug in the entire brain.[16]

Convection-enhanced diffusion: a catheter implanted in the brain is connected to a pump, which drives fluid flow in the brain at a prescribed infusion rate and the drug administered is distributed *via* bulk flow. Intense astrogliotic reaction in the region of the fluid flow has been reported in brain following convection-enhanced diffusion treatment.[17]

7.2.3.2 Pseudo-invasive Strategies

These non-neurosurgical-based strategies involve the disruption of the nasal epithelial barrier or the BBB in order to achieve pharmacological effects after drug administration. This section includes trans-nasal drug delivery and the temporal disruption of the BBB.

Trans-nasal drug delivery: once the molecule crosses the nasal epithelial barrier and enters the subcutaneous space of the nose, it can diffuse across the arachnoid membrane and enter the CSF compartment of the olfactory region. Small lipophilic molecules not only cross the BBB *via* passive diffusion but also cross the nasal epithelial barrier and the arachnoid membrane. In situations involving the administration of a drug that is hydrophilic or that has a molecular weight higher than 400 Da, the disruption of the epithelial barrier may be required in order to achieve transport.[18] However drugs such as Davunetide (Allon Therapeutics Inc.), a neuroprotective peptide for the treatment of Alzheimer's disease, schizophrenia cognitive impairment, and frontotemporal dementia that is currently in clinical development, is successfully delivered to the CNS after trans-nasal administration.

Temporal disruption of the BBB: in some treatments, access of the therapeutic agent to the CNS is achieved by BBB disruption. This can be done by the coinjection of the drug with a solvent/adjuvant that promotes the temporal breakdown of this barrier. Compounds such as mannitol are known to induce disruption by osmotic shrinkage of endothelial cells.[19] Other compounds, such ethanol and dimethylsulfoxide, or detergents such as SDS and polysorbate-80,[20–23] also have the capacity to compromise the BBB integrity. Intracarotid acid pH injection, low temperatures, high-dose free fatty acid,[24–26] localized hyperthermia[27] and focused ultrasound (FUS) can also be used to locally and reversibly open the BBB.[28,29] However, several cases have been reported in which FUS has damaged brain tissue.[30]

Temporal breakdown of the BBB implies severe risks because the brain becomes exposed not only to the drug administered but also to potential neurotoxic, endogenous and exogenous compounds found in plasma. Plasma proteins such as albumin are toxic for brain cells, and astrogliolisis is induced when the brain comes into contact with blood.[31] Moreover, BBB disruption is associated with chronic neuropathological changes, cerebral vasculopathy, and seizures.[32–34] These complications are especially critical when dealing with chronic patients.

In summary, invasive and pseudo-invasive strategies, especially neurosurgical-based ones and temporal disruption of the BBB, imply risk of infection and toxicity. Furthermore, they require highly skilled personnel.

7.2.3.3 Non-invasive Drug Delivery to the Brain

Given the rich vascularity of the brain, intravenous (i.v.) injection is the ideal route for drug administration. The total capillary length in the human brain is estimated to be 650 km,[35] with capillaries encasing virtually every brain cell.[36] However, drug delivery to the brain after i.v. injection is the most challenging approach because of the presence of the BBB. Once the drug is in the blood stream after injection it can overcome the BBB by means of various endogenous transport mechanisms present in the barrier itself.

Although the BBB is highly restrictive, in some pathologies its integrity can be compromised. It is well known that in the BBB located near a tumor, the endothelial cells and other cells implied in the maintenance of the properties of the barrier show significant abnormalities. For instance, human brain astrocytes fail to express or express a non-functional form of occludin (one of the proteins involved in the formation of tight junctions). An increase in the number and size of pinocytic vacuoles has been also reported.[37–40] In spite of the local disruption induced by the tumor, treatment of this type of cancer remains a challenge because of the nature of the disease.[41] Moreover, it is difficult to reach the invasive front of the tumor, given that this part of it does not induce BBB breakdown.[42] BBB disruption has also been observed during brain infections. An increase in BBB permeability occurs during an inflammatory process. The immune response is characterized by a rapid production and release of inflammatory mediators. These molecules alter the structure and function of the BBB. Cytokines, predominantly TNF-α and IL-1, and metalloproteinases contribute to opening the BBB.[43]

However, in standard conditions, and in most brain pathologies, the BBB is not disrupted. In this scenario, several drug delivery strategies can be applied to achieve transport through the barrier. These range from chemical modifications of the drug to achieve greater lipophilicity to structural modifications to resemble a natural ligand of a transport system expressed on the BBB, as disclosed in Chapter 7.1. An alternative to chemical modifications on the drug is the linkage of the drug/cargo to a BBB-shuttle. The linkage between the cargo and the BBB-shuttle can be performed either by covalent[44–53] or non-covalent interactions.[54–56] These will depend on the nature of the cargo and BBB-shuttle used; the crucial issue is that both entities retain their functionality after linkage.

7.2.3.3.1 Passive Diffusion

This is a non-saturable and spontaneous transport process. It has been reported that some small lipophilic molecules (< 300–500 Da) cross the BBB by this mechanism, but most molecules do not. The classical approach of medicinal chemistry to circumvent this limitation is the modification of the molecules in order to increase their lipid solubility by blocking hydrogen bond-forming groups on the parent drug molecule (for review see Chapter 7.1). A gain in lipid solubility increases permeation across all biological membranes of the body, including the BBB, thus potentially leading to side effects. Moreover, lipidization may enhance binding to plasma proteins, which could offset the increase in membrane permeation caused by greater lipid solubility. However, many plasma protein-bound drugs are available for transport across the BBB *in vivo via* a mechanism of enhanced dissociation at the brain capillary endothelial surface.[6]

Recent years have witnessed several key advances in the development of peptides as therapeutic agents and drug carriers for brain disorders. Although from a classical point of view passive diffusion has been considered probable only for small molecules, peptides are no longer considered too large to cross the BBB. Many large peptides cross this barrier by simple diffusion.[57–62] Therefore, molecular weight does not appear to be the only determinant for peptide penetration into the brain. Indeed, some peptides have already been used to shuttle drugs into the CNS (Table 7.2.1). Diketopiperazines (DKPs) transport baicalin (Figure 7.2.2A) and dopamine across a very restrictive *in vitro* BBB model (BBB-PAMPA).[63] Highly *N*-methylated phenylalanine-rich peptides have been described to transport small cargos such as L-dopa, 4-aminobutanoic acid (GABA), nipecotic acid (Nip) and 5-aminolevulinic acid (Ala) in the same model.[64,65] In general, shuttles with the capacity to cross the BBB by passive diffusion do so because of its unique physico-chemical

Table 7.2.1 Examples of BBB-shuttling by passive diffusion.

BBB-Shuttle	*Cargo*	*Reference*
DKPs	Baicalin and dopamine	63
NMePhe-rich peptides	L-Dopa, 4-aminobutanoic acid, nipecotic acid and 5-aminolevulinic acid	64, 65

A)

B)

C)

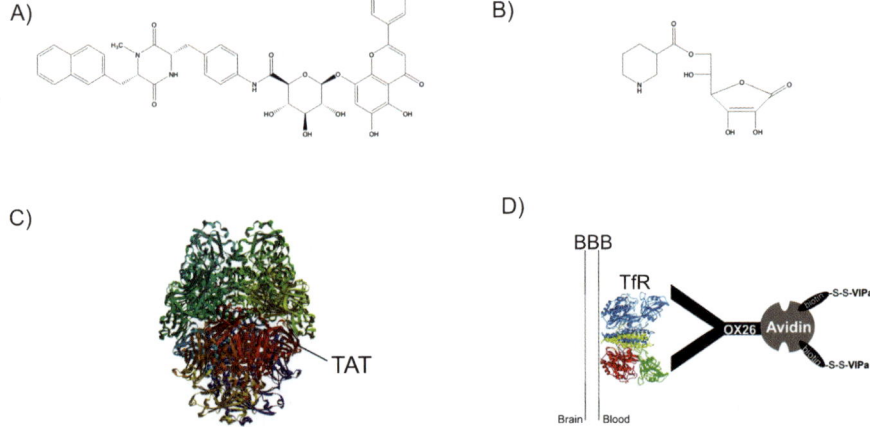

D)

Figure 7.2.2   A range of BBB-shuttles. (A) From ref. 63, diketopiperazynes (cyclic dipeptides) are potential BBB-shuttles that have the capacity to cross the BBB due to their unique physico-chemical properties, small molecules such as baicalin or dopamine can be transported to the brain by means of this type of BBB-shuttles. (B) From ref. 69, ascorbic acid can be use as a BBB-shuttle for molecules such as nipecotinic acid, the fact that this transport is highly sterospecific limits the transport to small molecules. (C) From ref. 84, cell-penetrating peptides can be use as a BBB-shuttle to transport different cargos across the BBB, as it is a vesicle-based mechanism the cargo is not limited to small molecules, different molecules ranging from small molecules, enzymes to nanoparticles have been successfully delivered to the brain using this approach. (D) From ref. 112, monoclonal antibodies or peptides can be use to deliver a wide range of cargos to the brain. In this example OX26 mAb carrying VIPa.

properties. If the cargo linked to the BBB-shuttle is a small molecule that does not affect the properties of the shuttle, then the latter will be able to transport the cargo into the CNS. In contrast, if the cargo affects the physico-chemical properties of the shuttle, which is likely to happen when a cargo such as a nanoparticle, a protein or a peptide is attached, the cargo will not penetrate the BBB.

7.2.3.3.2   Carried-mediated Transport (CMT)

This is a saturable and energy-dependent mechanism. Endothelial membranes contain several specific transport (carrier) proteins that provide essential nutrients and endogenous compounds to the brain. These transport systems include the glucose transporter, large neutral and cationic amino acid transporters, and nucleoside transporters, to mention a few.[66] For more information regarding the endogenous transporters that control the penetration of molecules across the BBB see Table 7.1.1 in Chapter 7.1. CMT systems are potential points of drug entry to the brain, but these are highly stereospecific pore-based transporters, and there are significant structural requirements for transport affinity. The medicinal chemistry approach to achieve drug delivery to the brain by means of CMT has been the modification of the structure of drugs to resemble the native substrate of the transporter. The challenge is to preserve the activity of the drug after these modifications.[3]

An interesting approach is to link a cargo to a substrate of a given endogenous transporter system. Since CMT is highly substrate structure-specific, it is difficult to make a drug that is normally not transported across the BBB transportable by the simple coupling to a substrate of a given transporter. However, there are some examples of drug delivery into the CNS that include linking the drug to a substrate of a CMT system (Table 7.2.2). Bonina et al.[67] showed that the conjugation of Nip to tyrosine, which enters the brain through the large neutral amino acid transporter (LAT1) system, had significant dose-dependent anticonvulsant activity when it was systematically administrated to mice. Unconjugated Nip did not show this effect, proving that tyrosine can be used as a BBB-shuttle to ferry Nip to the brain. In another study, Gynter et al.[68] demonstrated that tyrosine can also shuttle ketoprofen to the brain by means of LAT1. Manfredini et al.[69] reported that the Nip-ascorbic acid conjugation, which was proposed to enter the brain by sodium-dependent transporters (SVCT2), is better absorbed compared to the non-conjugated parent drug and shows an anticonvulsant effect in a mouse model.

Glycosylation of the linear opioid peptide Tyr-D-Thr-Gly-Phe-Leu-Ser-NH$_2$ on Ser[6] or the attachment of various sugar moieties to the cyclic opiod peptide enkephalin increases their penetration and allows the resulting glycopeptide analogs to function as analgesic drugs. This finding could be attributed, at least in part, to the transport of the analogs by the glucose transporter GLUT-1, and to the organic anion-transporting polypeptide transporter (OATP).[70,71] Glutathione has been also exploited for the delivery of L-dopa, dopamine and adamantamine.[72,73]

CMT requires high affinity. Consequently, the cargo conjugated to a BBB-shuttle that crosses the barrier by this mechanism should preserve the substrate capacity to be recognized by the receptor. Even when recognition is preserved,

Table 7.2.2 Examples of BBB-shuttling by carried-mediated transcytosis.

BBB-Shuttle	Cargo	Carrier system	Reference
L-Tyrosine	Nipecotic acid and ketoprofen	LAT1	67, 68
Ascorbic acid	Nipecotic acid	SVCT2	69 (Figure 7.2.2.B)
Glucose	Ketoprofen and indomethacin	GLUT-1	74
Glycosilation	Linear opiod peptide (Ser[6])	OATP	71
Glycosilation	Cyclic opiod peptide (D-Cys[2,5], Ser[6], Gly[7])	μ-opioid receptor	70
Glutathione	L-Dopa, dopamine and adamantamine	glutathione receptor	72, 73

the drug-BBB-shuttle construct might have insufficient affinity for a BBB transporter to achieve significant uptake and/or transport across the BBB. In addition, it might be limited by competition with endogenous substrates. Since CMT is highly sterospecific, it is not particularly amenable for the transport of large-molecule therapeutics.

7.2.3.3.3   Absorptive-mediated Transcytosis (AMT)

This is also an energy-dependent mechanism that is used by cell-penetrating peptides (CPPs) and other cationic compounds to overcome the BBB. This process does not involve any plasma membrane receptor, and although there is some controversy regarding the exact mechanism of transcytosis, it is known that it begins with the interaction of the cationic residues of a compound (a CPP for example) with the negative charges on the plasma membrane surface area (*e.g.* proteoglycans).[75] This interaction triggers endocytosis, which can lead to transcytosis. As detailed in Chapter 7.1, several drugs have been described to enter the brain *via* this mechanism.[76,77] It is also the mode of entrance of large compounds, such as cationized albumin and histone.[78,79]

Several CPPs are now known, such as TAT, penetratin, pegelin and oligoarginine. These four overcome cellular membranes in a very efficient manner.[80–83] Moreover, some also have the capacity to cross the BBB.[84–86] This property has been used for the delivery of diverse cargos to the brain (Table 7.2.3), mainly by using TAT peptide. Schwartz *et al.*[84] fused this peptide to the 116 kDa β-galactosidase protein (β-Gal; Figure 7.2.2C) and assessed tissue distribution after intraperitoneal injection in mice. There was strong β-Gal activity in all the tissues analyzed, including liver, kidney, heart muscle, lung and brain. In contrast, mice injected with control β-Gal showed no activity of this protein in the brain 4 hours after injection.

Ritanovir is one of the pharmaceutical agents approved by the FDA for the treatment of HIV infection. However, this drug is recognized by the P-gp, thereby limiting its transport across the BBB. Encapsulated ritonavir in poly(L-lactide) nanoparticles (NPs) conjugated to TAT peptide bypasses the efflux action of P-gp and enhances its permeability in MDCK-MDR1 and MDCK-wt cell lines. Furthermore, *in vivo* experiments in mice revealed an 800-fold higher brain ritonavir level with TAT-conjugated NPs than with the free ritonavir and

Table 7.2.3   Examples of BBB-shuttling by absorptive-mediated transcytosis.

CPP	Cargo	Reference
Penetratin	Doxorubicin	86, 88, 89
TAT	Bcl-xL fusion protein, β-galactosidase, chaperon heat shock protein 70, poly(L-lactide) nanoparticles carrying ritonavir	84, 87, 90–92
Pegelin (SynB1)	Doxorubicin, benzylpenicillin, dalargin	86, 93, 94
Pegelin (SynB3)	Morphine-6-β-D-glucuronide, paclitaxel	51, 95

about 7-fold higher accumulation than the non-conjugated NPs. TAT-conjugated NPs not only overcome the BBB, but also enhance the uptake and sustain the retention of the drug in the CNS, critical aspects for therapeutic efficacy of antiretroviral drug therapy.[87] More information regarding the use of surface-modified nanoparticles with ligands is comprised in the next chapter (Chapter 7.3).

Since AMT is a non-carrier dependent process, the use of CPPs as BBB shuttles is not limited to small cargos and there is no competition with endogenous compounds present in the blood stream. However, this type of vesicular transport is actively down-regulated in the BBB to protect the brain from non-specific exposure to polycationic compounds. Specific targeting by AMT to the CNS cannot be expected because of the lack of specificity of shuttle targeting, which may lead to widespread absorption in many tissues.

7.2.3.3.4   *Receptor-mediated Transcytosis (RMT)*

According to numerous authors, this is one of the most promising mechanisms to transport cargos across the BBB. Although CMT is a highly selective route to overcome the BBB it is limited to small cargos. The main advantage of RMT is that it is not limited by the size of the cargos. This system enables large molecules such as peptides, proteins and nucleic acids to specifically enter the brain, since this approach uses vesicle-based transport rather than a stereoselective carrier. Three types of receptors are present on brain endothelial cells: bi-directional receptors, reverse receptors, and receptor-mediated endocytosis systems.[3] In terms of the delivery of compounds to the brain, the most interesting are the bi-directional receptors. Some of these receptors are well known and have been used for the delivery of various types of cargo to the CNS including nanostructures (Table 7.2.4). In this chapter the most important receptors will be covered, namely transferrin receptor, insulin receptor, low-density lipoprotein receptor, membrane-bound precursor heparin binding epidermal growth factor, nicotinic acetylcholine receptor, lactotransferrin receptor and leptin receptor.

In general, RMT occurs in three steps: receptor-mediated endocytosis of the compound at the luminal (blood) side, movement through the endothelial cytoplasm, and exocytosis of the compound at the abluminal (brain) side of the brain capillary endothelium.[96]

Transferrin Receptor (TfR)
This is perhaps the most widely studied receptor for drug targeting. TfR is a type II transmembrane glycoprotein consisting of two 90 kDa subunits linked covalently by two disulfide bridges. Each subunit can bind one transferrin (Tf) molecule,[97] the native ligand of TfR, an 80 kDa iron-binding blood plasma glycoprotein that controls the level of free iron in biological fluids. TfR is expressed in practically all the cells of the body, including brain capillaries and it mediates the delivery of iron to the brain.[98] It is also expressed in choroid

Table 7.2.4   Examples of BBB-shuttling by receptor-mediated transcytosis.

BBB-Shuttle	Cargo	Receptor	Reference
Tf	Albumin nanoparticles carrying azidothymidine or loperamide	TfR	172, 173
Tf	Liposome carrying *nido*-carborane	TfR	174
Tf	Quantum dots	TfR	175
OX26	Liposome carrying human cytomegalovirus promoter virus	TfR	176
OX26	Lipid nanocapsules labeled with ^{188}Re lipophilic complex	TfR	177
OX26	Poly(β-L-malic acid) carrying antisense oligonucleotides	TfR	178
OX26	Chitosan-PEG nanoparticles loaded with the peptide Z-DEVD-FMK (inhibitor caspase-3 enzyme)	TfR	179
OX26	Liposome carrying daunomycin	TfR	114, 180
OX26	Peptide nucleic acids	TfR	181
OX26	Radio-labeled liposome	TfR	182
OX26	Liposome carrying plasmid DNA	TfR	183
OX26	Liposome carrying tyrosine hydroxylase	TfR	184
OX26	Fibroblast growth factor	TfR	55, 185
OX26	Liposome carrying plasmid DNA encoding β-galactosidase	TfR	186
OX26	Brain-derived neurotrophic factor	TfR	113, 187, 188
OX26	Liposome carrying plasmid encoding luciferase or β-galactosidase	TfR	189
OX26	Vasoactive intestinal peptide analog (VIPa)	TfR	54, 112
OX26	Epidermal growth factor	TfR	190
OX26	Polyethylen glycol-poly(ε-caprolactone) carrying NC-1900 (peptide)	TfR	54
OX26	Radiolabeled Aβ$^{[1-40]}$	TfR	191
OX26	Chitosan nanoparticles	TfR	179
8D3	β-galactosidase	TfR	115
8D3	Liposome carrying β-galactosidase or luciferase plasmid	TfR	192, 193
8D3	Liposome carrying short hairpin RNA	TfR	194
8D3	Peptide nucleic acid	TfR	195
8D3	Radiolabeled Aβ$^{[1-40]}$	TfR	196
83-14	Short hairpin RNA	IR	194
83-14	Radiolabeled Aβ$^{[1-40]}$	IR	125
83-14	Liposome carrying β-galactosidase or luciferase plasmid	IR	197
83-14	α-L-iduronidase	IR	124
Chimeric 83-14	Brain-derived neurotrophic factor	IR	123
Melanotransferrin	Paclitaxel, doxorubicin	LRP	137, 138
Angiopep	Micelles carrying amphotericin B	LRP	198

Table 7.2.4 (*Continued*)

BBB-Shuttle	Cargo	Receptor	Reference
Angiopep	PAMAM dendrimers complexed with DNA	LRP	142
Angiopep	Paclitaxel, doxorubicin, etoposide	LRP	140, 141
RVG	Small interfering RNA	AChR	56
RVG	PAMAM dendrimers loaded with DNA	AChR	162
Leptin (61–90)	poly-L-lysine dendrigraft	LR	168
Lf	poly(ethylenglycol)-poly(lactide) nanoparticles	LfR	50
Lf	Procationic liposomes	LfR	171

plexus epithelial cells and neurons.[97] However, its expression differs depending on the cell type.[99] In general, highly proliferating cells, such as cancer cells, overexpress TfR because of their increased metabolism.[100,101] This makes TfR very attractive for drug targeting, and explains why it has received so much attention for drug delivery.

Targeting the brain by means of TfR can be achieved in several ways. The simplest one is using Tf as BBB-shuttle. Visser *et al.*[102] showed that Tf conjugated to horseradish peroxidase (HRP), a 40 kDa enzyme, is efficiently taken up by BBB endothelial cells *in vitro*. The same author also demonstrated that liposomes loaded with HRP and tagged with Tf are suitable for drug delivery in these cells.[103] However, the potential use of Tf as a BBB-shuttle is limited to *in vitro* applications because TfR is almost saturated under physiological conditions as a result of high endogenous plasma concentrations of Tf.[104] The use of a large amount of cargos tagged with Tf as a BBB-shuttle to shift endogenous Tf could alter iron homeostasis and thereby lead to serious physiological problems.

Vectors that target TfR but that do not overlap with the native binding site of Tf are promising tools for the delivery of cargos into the CNS because they could allow delivery without altering iron levels. Examples of this type of vector are the murine monoclonal antibody (mAb) OX26[105] and the rat mAb 8D3.[106] The mechanism of action of these mAbs is not fully elucidated, and some controversy remains about this issue.[97,105,107–110] Although the mechanism of transcytosis of OX26 is not yet fully elucidated, drug delivery to the brain *via* TfR is efficient. Delivery of various types of cargo to the brain by means of OX26 and 8D3 has been reported.

The vasoactive intestinal polypeptide (VIP) is a potent cerebral vasodilator when applied topically to pial vessels.[111] However, an *in vivo* study showed no neuropharmacological effect. This finding is attributed to the presence of the BBB, which limits peptide transport to the brain. Bickel *et al.*[112] reported how a VIP analog (VIPa) was coupled to OX26 by means of the avidin-biotin system in order to increase the brain uptake of the peptide (Figure 7.2.2D). Internal carotid artery perfusion of the biotinylated VIPa coupled to avidin-OX26 conjugate in rats resulted in a pharmacological effect on the CNS *in vivo*. There was a 65% increase in blood flow, whereas biotinylated VIPa without the brain transport vector was ineffective.

Brain-derived neurotrophic factor (BDNF) is a potential neuropharmaceutical with potential for the treatment of a variety of neurodegenerative disorders. However, like other neurotrophins and protein-based therapeutics, BDNF does not undergo significant transport through the brain capillary endothelial wall. A conjugate of OX26 and streptavidin (SA) to BDNF increased brain uptake of this drug.[113]

Human basic fibroblast growth factor (bFGF) is an endogenous polypeptide with potent neuroprotective effects in a variety of preclinical studies. The dose of intravenous bFGF required to achieve neuroprotection in the brain is relatively large, and the peripheral side effects of bFGF have limited clinical trials of this peptide as a neuroprotective agent. Conjugation of mono-biotinylated bFGF to OX26 modified with SA was shown to maintain high binding affinity for the bFGF receptor in cultures of BHK-21 cells. bFGF conjugated to OX26 showed decreased peripheral organ distribution, and a 5-fold increase in brain uptake with respect to unconjugated bFGF following intravenous injection in rats.[55]

OX26 has also been used as a targeting moiety anchored to the surface of nanostructures (for more information see Chapter 7.3.) Huwyer *et al.*[114] conjugated liposomes carrying the antineoplastic agent daunomycin to OX26 through a polyethylene glycol (PEG) linker. Brain delivery of free daunomycin or conventional liposomes is poor because of their rapid clearance from the circulation. The plasma clearance of liposomes was reduced 66-fold by PEG conjugation; however, there was no brain uptake of PEG-liposomes. Coupling of OX26 to PEG-liposomes partially reversed the effect of PEG conjugation and resulted in a 5-fold increase in plasma clearance. However, brain delivery of immunoliposomes was greater than that of free daunomycin, conventional liposomes and PEG-conjugated liposomes. Since a single liposome can carry >10 000 daunomycin molecules, the use of PEG-conjugated immunoliposomes increases the drug-carrying capacity of the monoclonal antibody by up to 4 logarithmic orders of magnitude.

Zang *et al.*[115] have used the 8D3 mAb for the delivery of β-Gal *in vivo*. The enzyme was conjugated to the rat mAb *via* a streptavidin-biotin linkage, and was injected intravenously in adult mice. Enzymatic activity was measured in the brain after 1 hour. A 10-fold increase in brain uptake of the enzyme with respect to unconjugated β-Gal was detected.

The main drawback of OX26 and 8D3 is that they are antibodies against murine and rat TfR, respectively, and do not bind to the human TfR. Consequently, this technology cannot be transferred to clinical practice. Rat antibodies cause immunogenic reactions in humans unless they are humanized. The preparation of humanized or chimeric antibodies is complex and in some cases may lead to a loss of affinity for the target receptor.[116] An alternative to Tf or mAb for targeting human TfR is the use of peptides, which have relatively small molecular weights and relatively low cytotoxicity and immunogenicity.[117] Lee *et al.*[118] used the phage display technique to find small peptidic ligands for human TfR. The authors obtained a 7- and 12-mer peptide that bound to TfR in a different binding site to Tf and that triggered internalization. Several groups have been working with these peptides, mainly for cancer therapy, and their capacity to target TfR have been validated.[119,120] Moreover, it has been observed that in the presence of Tf, the uptake of the peptides by TfR is increased. Using one of the sequences for the delivery of gold NPs into the brain, we have obtained positive results (Prades *et al.*, submitted). Other groups have also used phage display to develop new peptides against human TfR.[121,122]

Insulin Receptor (IR)

Together with TfR, IR is among the receptors most studied and it has been widely characterized for drug targeting to the brain. IR is a transmembrane receptor that belongs to the large class of tyrosine kinase receptors. IR comprises two α subunits and two β subunits. These subunits are joined by disulfide bridges to form a cylindrical structure. The binding of insulin to IR leads to a rearrangement of the structure of the receptor and induces a complex cellular response by phosphorylating proteins on their tyrosine residues. Although endocytosis is not required for insulin action, it is important for removing insulin from the cell surface so that the target cell responds to the hormone in a time-limited fashion. The endocytotic mechanism of IR has been exploited for the targeting of drugs to the brain.

The *in vivo* application of insulin as a BBB-shuttle, as occurs with Tf, is limited because of the amount of endogenous insulin present in the blood stream. To achieve an effect using insulin as a shuttle, large amounts of shuttle will be required to shift endogenous insulin. This requirement is potentially dangerous for human applications, since changes in glucose homeostasis can lead to a lethal overdose or coma.

W. Pardridge's group has wide experience in the delivery of a range of cargoes to the CNS by means of IR. A murine mAb (known as 83-14) against human IR (mAb-HIR) was found to bind with very high affinity to IR of the human BBB, and to undergo rapid transport across the barrier *in vivo*. Fusion of BDNF to the mAb-HIR allows neurotrophin to undergo receptor-mediated transport across the BBB *via* HIR in Rhesus monkey following intravenous injection. The concentration of the BDNF-mAb-HIR in brain was >10-fold higher than the endogenous concentration of BDNF in this organ. This increase in brain BDNF, following intravenous administration, is not achievable with native BDNF because this form does not cross the BBB.[123]

Mucopolysaccharidosis Type I, also known as Hurler's Syndrome, is a lysosomal storage disorder that affects the brain. The mainstay of treatment is enzyme replacement therapy with recombinant enzymes. However, the missing enzyme, α-L-iduronidase (IDUA), does not cross the BBB. The fusion of IDUA to the mAb-HIR allows BBB transfer of the enzyme following the intravenous injection in Rhesus monkey.[124]

Radiolabeled $A\beta^{[1-40]}$ (first 40 amino acids of β-amyloid peptide) is a potential radiopharmaceutical peptide for neurodiagnostic quantitation of the Aβ amyloid burden in the brains of living subjects with Alzheimer's disease. However, ^{125}I-$A\beta^{[1-40]}$ does not cross the BBB. ^{125}I-$A\beta^{[1-40]}$ was conjugated to the mAb-HIR by means of streptavidin-biotin technology. Its biological activity was retained despite the conjugation to mAb-HIR. The complex still binds the neuritic plaques in sections of brains affected by Alzheimer's disease to a degree comparable to that of the unconjugated biotinylated $A\beta^{[1-40]}$. Intravenous injection in monkeys showed that the brain uptake of the radiopharmaceutical peptide without conjugation to the BBB delivery system was negligible. However, a marked increase in brain uptake of the radiopharmaceutical peptide conjugated to the streptavidin-mAb-HIR BBB system was observed.[125]

Although the BBB-shuttle used by Pardridge's group in these examples was a mAb rather than insulin, it is not exempt of risk for human applications since it could alter glucose homeostasis. Moreover, as occurs with TfR, IR is widely expressed in body tissues. The use of a BBB-shuttle that targets IR will deliver its cargo not only to the brain but also to many other peripheral tissues. In addition to these drawbacks, the 83-14 mAb cannot be used in humans owing to immunogenic reactions. However, genetically engineered effective forms of the mAb have been produced, which may allow for drug and gene delivery to the human brain. In this regard, Pardridge's group produced a humanized form of 83-14 mAb.[126] After humanization, the affinity of the mAb decreased 27% with respect to the murine mAb-HIR. The capacity of the humanized mAb to interact with HIR was tested using an *in vitro* model system of the human BBB. The antibody was proved to show affinity for human IR. The humanized mAb was also labeled with ^{125}I and intravenously administered to Rhesus monkeys. Brain scanning showed that this antibody was rapidly transported into all parts of the primate brain.[127]

Low-density Lipoprotein Receptor-related Protein-1 and -2 (LRP1 and LRP2) Receptors

Low-density lipoprotein receptor-related protein (LRP) has been reported to mediate the transport of substrates across endothelial cells of the BBB.[128–130] In fact, LRP1 and LRP2 have been explored as a gate of entry for various compounds into the CNS, in a similar way as for TfR and IR.

LRP1 and LRP2 are found in the membrane of several types of cells, and are ubiquitously expressed in the CNS.[131–133] LRPs are involved in a number of functions and are multiligand binding receptors.[134] Many substrates are shared between the two receptors, although each receptor also has specific ligands. A unique feature of LRP is its rapid endocytosis rate, with half of the receptors on the cell surface internalizing within 30 s.[135]

Melanotransferrin (or P97), a cell-surface glycoprotein highly expressed in melanoma cells, is a specific substrate for LRP1. P97 shares sequence similarity and iron-binding properties with members of the Tf superfamily. Béliveau et al.[136] found that radiolabeled P97 was highly accumulated in mouse brain following intravenous injection and *in situ* brain perfusion. This observation opens up the possibility to use P97 as a BBB-shuttle. In fact, P97 has been successfully used for the delivery of paclitaxel and doxorubicin to rats and mice.[137,138]

Aprotinin (Trasylol®, Bayer), a monomeric globular polypeptide derived from bovine lung tissue that inhibits trypsin and related proteolytic enzymes, is a ligand for LRP1 and LRP2. Trasylol was administered by injection to reduce bleeding during complex surgery. Its main effect is the slowing down of fibrinolysis. Trasylol was permanently withdrawn from the market in May 2008; studies suggested that its use increases the risk of complications or death. However, this drug is still in use for highly specific research purposes. It is known that aprotinin crosses the BBB.[139] This capacity opens up the possibility to deliver cargos to the brain using this protein as a BBB-shuttle. Demeule *et al.*[139] attempted to identify the minimal sequence of aprotinin required for transport across this barrier. This was addressed by searching for sequence homologies with other proteins known to interact with the same receptor. Alignment of protein sequences allowed the authors to design more than 96 human aprotinin-derived peptides (called Angiopeps). These peptides were tested using a BBB *in vitro* model and *in situ* brain perfusion. Angiopep-2 (19-mer peptide) transcytosis was found to be 7-fold higher than that of aprotinin. The same results were obtained also *in vivo*. The use of Angiopep-2 as a BBB-shuttle for the delivery of cargos across the BBB has been evaluated by conjugating it with paclitaxel. In mice, the conjugate inhibited the growth of orthotopic human glioblastoma (U87 MG) brain tumors more potently than paclitaxel alone and significantly increased animal survival rates.[140] In another study, Angiopep-2 was conjugated to doxorubicin and etoposide,[141] two potent anticancer molecules. Both agents killed cancer cell lines *in vitro* with similar IC_{50} values and with an apparently similar cytotoxicity mechanism as unconjugated doxorubicin and etoposide. The two conjugates exhibited dramatically higher BBB influx rate constants than the unconjugated drugs and pooled within brain parenchymal and tumor tissue. Passage through the BBB for these two agents was not increased in mice lacking the ABCB1 efflux pump, thereby implying that their transport bypasses this pump system. This feature distinguishes the new agents from unconjugated doxorubicin and etoposide.

Ke *et al.*[142] have also used Angiopep-2 as a BBB-shuttle for the delivery of cargos to the brain. In this study the authors performed nucleic acid delivery to the brain using a PEG-modified polyamidoamine (PAMAM) dendrimer decorated with Angiopep-2 (PAMAM-PEG-Angiopep-2). Several experiments were performed. After administration of PAMAM-PEG-Angiopep-2 by a tail injection, accumulation of the dendrimer was found in the mouse brain 2 hours after administration.

Receptor-associated protein (RAP) is a 39 kDa protein that functions as a specialized endoplasmatic reticulum chaperone, assisting in the folding and trafficking of members of the LRP receptor family. RAP has a high transport capacity through the BBB. Its permeability through this barrier is higher than the permeability of melanotransferrin and Tf.[143] Exogenously applied RAP can be endocytosed efficiently by all members of the LRP receptor family and could therefore serve as a vehicle for RMT delivery of other proteins into the brain *in vivo*.[144]

Regarding TfR and IR, the application of vectors that interact with the LRP should further studied to unveil possible side effects and to predict the safety of the use of the LRP1/LRP2 receptors for the targeting of drugs to the brain in humans. Complex interactions of LRP1 and LPR2 with Alzheimer's disease have been described. LRP1 is believed to be associated with the processing and metabolism of the β-amyloid precursor peptide and the β-amyloid peptide uptake by neurons.[145,146] LRP2 has also been reported to mediate the uptake of β-amyloid complexed to apoJ and apoE.[147,148]

Membrane-bound Precursor Heparin-binding Epidermal Growth Factor (Diphtheria Toxin) Receptor

The membrane-bound precursor of heparin-binding epidermal growth factor (HB-EGF), also known as diphtheria toxin receptor, has also been used to deliver compounds to the CNS. HB-EGF is constitutively expressed on the BBB, neurons and glial cells.[149] In ischemic strokes and gliomas, it is up-regulated on cerebral blood vessels,[150,151] which may lead to a site-selective improvement of the therapeutic efficacy of the targeted ligand into the brain. CRM197, a non-toxic mutant of diphtheria toxin, has been used for many years as an immunological adjuvant or a carrier protein in human vaccinations against bacterial infection.[152] This implies that this drug has a long record of safe use in human treatments. Moreover, it shows a certain degree of antitumor activity.[153,154] It is well known that CRM197 interacts with the HB-EGF and for this reason has been proposed as a BBB-shuttle for the delivery of cargos into the CNS.

Gaillard *et al.*[155,156] conjugated CRM197 to HRP and to pegylated liposomes containing HRP. These systems were evaluated in an *in vitro* model of the BBB where transcytosis was observed, although at a slow rate (in the order of hours). This transcytosis process is believed to occur by a caveolae-mediated pathway.[157] *In vivo* studies were also performed in guinea pigs injected with the HRP-CRM197 conjugate. The HRP reaction product was observed mainly in brain blood vessels, thereby corroborating the *in vitro* data that indicated a slow rate of transcytosis.

CRM197 has been safely applied to humans[158] and therefore could be useful for the transport of drugs to the human brain.

Nicotinic Acetylcholine Receptor (AChR)

The rabies virus shows a high degree of neurotropism *in vivo*. A 29-amino acid peptide derived from rabies virus glycoprotein (RVG) is known to inhibit the binding of the snake-venom toxin α-bungarotoxin (BTX) to the nicotinic acetylcholine receptor (AChR) in solution,[159,160] a receptor widely expressed in the brain, including endothelial cells of brain capillaries.[161] The use of this peptide as a potential vector for the delivery of cargos to the brain was successfully explored by Kumar *et al.*[56] for the delivery of small interfering siRNA. The siRNA-RVG system was injected in mice and after 10 hours specific delivery of the cargo was observed. The authors proposed an RMT process by means of the α7 subunit of the AChR. In another study, Liu *et al.*[162] also used the same peptide for the delivery of PAMAM dendrimers loaded with DNA. These authors achieved delivery of labeled DNA into the mouse brain. In this study, the authors claimed that transcytosis did not occur exclusively by means of the a α7 subunit of the AChR, as it was pointed out by Kumar *et al.*, but proposed a relevant contribution of the GABA$_B$ receptor RVG uptake.

Leptin Receptor (LR)
Leptin is a 16 kDa protein hormone secreted into the blood stream by adipocytes. This hormone plays a key role in regulating intake and energy expenditure, including appetite and metabolism. It is reported that when administered peripherally, leptin stimulates brain cells[163,164] and that CSF leptin levels are strongly correlated to plasma levels. LR is believed, at least in part, to mediate the uptake of leptin to the brain.[165,166]

With the leptin sequence in hand, Barret *et al.*[167] designed an overlapping set of peptides spanning the 146 amino acids of the mature human leptin protein. These authors found that the peptide corresponding to the position 61–90 possesses a high brain plasma ratio, equivalent to that of leptin. This peptidic sequence (61–90) was used by Liu *et al.*[168] as a BBB-shuttle for pegylated poly-L-lysine dendrigraft for brain-targeted gene delivery. Accumulation of the polymer in the mouse brain was detected 80 minutes after injection.

Lactoferrin Receptor (LfR)
Lactoferrin (Lf) is a multifunctional protein that belongs to the Tf family. It is a globular glycoprotein folded into two globular lobes, each containing one iron-binding site. It is known that Lf crosses the BBB *via* LfR.[169,170] Lf has been used as a BBB-shuttle for the delivery of cargos to the brain. Hu *et al.*[50] conjugated polyethylene glycol-poly(lactide) (PEG-PLA) NPs to Lf. Intravenous injection to mice gave a 3-fold accumulation of the conjugated NPs in the mouse brain with respect to the unconjugated PEG-PLA NPs. Chen *et al.*[171] also used Lf as a BBB-shuttle for procationic liposomes (PCL). The injection of the PCL-Lf conjugate in mice produced a 2.3-fold accumulation in brain mice with respect to unconjugated PCL. Given the nature of PCL, it is plausible that the PCL-Lf conjugate crosses the BBB by RMT and AMT.

7.2.4 Conclusions and Perspectives

There is an urgent need to develop BBB-shuttles to deliver cargos for a range of therapeutic and diagnostic applications to the brain. However, the presence of the BBB implies that there is no straightforward solution. The sections of this chapter show that although drugs can be delivered using diverse strategies, in general these achieve a low efficiency with respect to the dose injected. In addition, while some shuttles are useful for a given cargo, they are not suitable for others. Moreover, it is difficult to achieve selective delivery of the cargo to the brain and simultaneously avoid possible side effects.

It will be most interesting to develop a universal BBB-shuttle with the capacity to transport any type of cargo into the CNS, and moreover in a highly selective and efficient way. As an ideal requirement, this universal shuttle should be safe for human applications and easy to produce. The most convenient mechanism that fulfills these requirements is RMT. The targeting of cargos to a receptor that is unambiguously expressed and/or overexpressed in the BBB, and that in addition does not interfere in the binding of any endogenous ligand of the selected receptor would be the perfect scenario.

The use of BBB-shuttles that interact with a given receptor but do not overlap with the native binding site of the endogenous ligand is of great interest from a therapeutic point of view, since body homeostasis will not be altered.

The use of combinatorial antibody libraries can be useful for the discovery of antibodies that target specific receptors of the BBB endothelium in order to trigger transcytosis.[199] The screening of these antibody libraries can be done in the presence of endogenous ligands present in the blood stream, which could allow identification of hits that do not overlap with the binding site of the endogenous ligands. Given that the antibodies described in this chapter were either of rodent origin or partially humanized, undesired immunogenic reactions in human patients may arise. Thus for human therapeutic application, fully human compatible antibodies are preferable.[200]

Peptides offer an interesting alternative to the use of antibodies or proteins as BBB-shuttles. The phage display technique is, for example, an excellent way to identify potential peptides against a specific receptor.[118,121,122] The exploration of fragments of proteins known to interact with a BBB receptor and that trigger transcytosis is also a suitable approach to find peptidic BBB-shuttles.[139,167] Peptides as BBB-shuttles offer the following advantages: (i) they have relatively small molecular weights; (ii) they can be synthesized easily and relatively inexpensively; (iii) they show relatively low cytotoxicity and immunogenicity; and (iv) they are degraded *in vivo* to naturally occurring compounds.[117] From the receptors that have been explored to date, TfR is the most convenient because is has been widely characterized, it is expressed on the BBB (unfortunately not in a exclusive way) and antibodies (not humanized) and peptides that interact with the receptor in a site that does not overlap with the native binding site of Tf are already available.[105,118]

RMT also fulfills the requirement of universal BBB-shuttle since its mechanism implies a vesicular transport system. We have seen that different types of cargos targeted by a BBB-shuttle to one of the receptors expressed on the BBB have been successfully ferried to the brain. These cargos range from small molecules to proteins, NPs, liposomes, dendrimers, siRNA and DNA.

The design of a highly efficient BBB-shuttle is perhaps the most challenging issue in BBB-shuttle development, not only because of the BBB and its characteristics but also because of the presence of the reticule endothelial system (RES) and the non-specific binding of the cargo-BBB-shuttle system to plasma proteins. Secondary targeting of the cargo inside the brain must also be taken into account. Nanotechnology could play a key role in overcoming these limitations. In terms of drug delivery, the use of platforms such as NPs, liposomes or hydrogels can allow the loading of large amounts of drugs. Moreover, these platforms can be functionalized with several copies of the BBB-shuttle and also other vectors for secondary targeting into a specific area of the brain. Fine-tuning can be also performed on the platform surface, for example with the incorporation of several molecules of PEG, which have been proven to increase the circulation time of molecules and compounds in blood by reducing the unspecific binding to plasma proteins and avoiding the RES.

References

1. B. Jonssen, *Eur. J. Neurol.*, 2005, **12**, 8–9.
2. B. V. Zlokovic, *Neuron*, 2008, **57**, 178–201.
3. W. M. Pardridge, *NeuroRx*, 2005, **2**, 3–14.
4. A. K. Ghose, V. N. Viswanadhan and J. J. Wendoloski, *J. Comb. Chem.*, 1999, **1**, 55–68.
5. C. A. Lipinski, *J. Pharmacol. Toxicol. Methods*, 2000, **44**, 235–249.
6. W. M. Pardridge, in *Brain Drug Targeting: The Future of Brain Drug Development*, Cambridge University Press, Cambridge, U.K., 2001, 1–12.
7. L. L. Rubin and J. M. Staddon, *Annu. Rev. Neurosci.*, 1999, **22**, 11–28.
8. M. W. Brightman and T. S. Reese, *J. Cell Biol.*, 1969, **40**, 648–677.
9. T. S. Reese and M. J. Karnovsky, *J. Cell Biol.*, 1967, **34**, 207–217.
10. W. M. Pardridge, in *Peptide Drug Delivery to the Brain*, ed. W. M. Pardridge, Raven Press, New York, 1991, pp. 52–98.
11. N. Bodor and P. Buchwald, *Adv. Drug Delivery Rev.*, 1999, **36**, 229–254.
12. T. Terasaki and K. Hosoya, *Adv. Drug Delivery Rev.*, 1999, **36**, 195–209.
13. L. E. DeBault and P. A. Cancilla, *Science*, 1980, **207**, 653–655.
14. S. H. Hendry, E. G. Jones and M. C. Beinfeld, *Proc. Natl. Acad. Sci. U. S. A.*, 1983, **80**, 2400–2404.
15. W. M. Pardridge, *Pharm. Res.*, 2007, **24**, 1733–1744.
16. C. E. Krewson, M. L. Klarman and W. M. Saltzman, *Brain Res.*, 1995, **680**, 196–206.
17. Y. Ai, W. Markesbery, Z. Zhang, R. Grondin, D. Elseberry, G. A. Gerhardt and D. M. Gash, *J. Comp. Neurol.*, 2003, **461**, 250–261.
18. H. Yamazumi, *Nippon Jibiinkoka Gakkai Kaiho*, 1989, **92**, 608–616.
19. B. Zunkeler, R. E. Carson, J. Olson, R. G. Blasberg, H. DeVroom, R. J. Lutz, S. C. Saris, D. C. Wright, W. Kammerer, N. J. Patronas, R. L. Dedrick, P. Herscovitch and E. H. Oldfield, *J. Neurosurg.*, 1996, **85**, 1056–1065.
20. J. P. Hanig, J. M. Morrison, Jr. and S. Krop, *Eur. J. Pharmacol.*, 1972, **18**, 79–82.

21. R. D. Broadwell, M. Salcman and R. S. Kaplan, *Science*, 1982, **217**, 164–166.
22. A. Saija, P. Princi, D. Trombetta, M. Lanza and A. De Pasquale, *Exp. Brain Res.*, 1997, **115**, 546–551.
23. M. N. Azmin, J. F. Stuart and A. T. Florence, *Cancer Chemother. Pharmacol.*, 1985, **14**, 238–242.
24. W. H. Oldendorf, B. E. Stoller, T. A. Tishler, J. L. Williams and S. Z. Oldendorf, *Am. J. Physiol.*, 1994, **267**, H2229–2236.
25. B. Oztas and M. Kucuk, *Neurosci. Lett.*, 1995, **190**, 203–206.
26. L. Sztriha and A. L. Betz, *Brain Res.*, 1991, **550**, 257–262.
27. K. Kakinuma, R. Tanaka, H. Takahashi, M. Watanabe, T. Nakagawa and M. Kuroki, *J. Neurosurg.*, 1996, **84**, 180–184.
28. F. Wang, Y. Cheng, J. Mei, Y. Song, Y. Q. Yang, Y. Liu and Z. Wang, *J. Ultrasound Med.*, 2009, **28**, 1501–1509.
29. H. L. Liu, M. Y. Hua, P. Y. Chen, P. C. Chu, C. H. Pan, H. W. Yang, C. Y. Huang, J. J. Wang, T. C. Yen and K. C. Wei, *Radiology*, **255**, 415–425.
30. K. Hynynen, N. McDannold, N. A. Sheikov, F. A. Jolesz and N. Vykhodtseva, *Neuroimage*, 2005, **24**, 12–20.
31. A. Nadal, E. Fuentes, J. Pastor and P. A. McNaughton, *Proc. Natl. Acad. Sci. U. S. A.*, 1995, **92**, 1426–1430.
32. T. S. Salahuddin, B. B. Johansson, H. Kalimo and Y. Olsson, *Acta Neuropathol.*, 1988, **77**, 5–13.
33. A. S. Lossinsky, A. W. Vorbrodt and H. M. Wisniewski, *J. Neurocytol.*, 1995, **24**, 795–806.
34. E. A. Neuwelt and S. I. Rapoport, *Fed. Proc.*, 1984, **43**, 214–219.
35. R. Cecchelli, V. Berezowski, S. Lundquist, M. Culot, M. Renftel, M. P. Dehouck and L. Fenart, *Nat. Rev. Drug Discovery*, 2007, **6**, 650–661.
36. F. Schlachetzki, Y. Zhang, R. J. Boado and W. M. Pardridge, *Neurology*, 2004, **62**, 1275–1281.
37. K. E. Schlageter, P. Molnar, G. D. Lapin and D. R. Groothuis, *Microvasc. Res.*, 1999, **58**, 312–328.
38. T. A. Martin and W. G. Jiang, *Histol. Histopathol.*, 2001, **16**, 1183–1195.
39. M. C. Papadopoulos, S. Saadoun, D. C. Davies and B. A. Bell, *Br. J. Neurosurg.*, 2001, **15**, 101–108.
40. M. C. Papadopoulos, S. Saadoun, C. J. Woodrow, D. C. Davies, P. Costa-Martins, R. F. Moss, S. Krishna and B. A. Bell, *Neuropathol. Appl. Neurobiol.*, 2001, **27**, 384–395.
41. I. Brigger, C. Dubernet and P. Couvreur, *Adv. Drug Delivery Rev.*, 2002, **54**, 631–651.
42. J. J. Vredenburgh, A. Desjardins, J. E. Herndon, 2nd, J. Marcello, D. A. Reardon, J. A. Quinn, J. N. Rich, S. Sathornsumetee, S. Gururangan, J. Sampson, M. Wagner, L. Bailey, D. D. Bigner, A. H. Friedman and H. S. Friedman, *J. Clin. Oncol.*, 2007, **25**, 4722–4729.
43. A. M. Wolka, J. D. Huber and T. P. Davis, *Adv. Drug Delivery Rev.*, 2003, **55**, 987–1006.

44. F. Bi, J. Zhang, Y. Su, Y. C. Tang and J. N. Liu, *Biomaterials*, 2009, **30**, 5125–5130.
45. L. Josephson, C. H. Tung, A. Moore and R. Weissleder, *Bioconjug. Chem.*, 1999, **10**, 186–191.
46. W. Cai and X. Chen, *Nat. Protoc.*, 2008, **3**, 89–96.
47. J. M. de la Fuente and C. C. Berry, *Bioconjug. Chem.*, 2005, **16**, 1176–1180.
48. E. Fernandez-Megia, J. Correa, I. Rodriguez-Meizoso and R. Riguera, *Macromolecules*, 2006, **39**, 2113–2120.
49. L. Polito, D. Monti, E. Caneva, E. Delnevo, G. Russo and D. Prosperi, *Chem. Commun.*, 2008, 621–623.
50. K. Hu, J. Li, Y. Shen, W. Lu, X. Gao, Q. Zhang and X. Jiang, *J. Controlled Release*, 2009, **134**, 55–61.
51. J. Temsamani, C. Bonnafous, C. Rousselle, Y. Fraisse, P. Clair, L. A. Granier, A. R. Rees, M. Kaczorek and J. M. Scherrmann, *J. Pharmacol. Exp. Ther.*, 2005, **313**, 712–719.
52. J. H. Lee, K. Lee, S. H. Moon, Y. Lee, T. G. Park and J. Cheon, *Angew. Chem., Int. Ed. Engl.*, 2009, **48**, 4174–4179.
53. S. Pujals, N. G. Bastus, E. Pereiro, C. Lopez-Iglesias, V. F. Puntes, M. J. Kogan and E. Giralt, *ChemBioChem*, 2009, **10**, 1025–1031.
54. D. Wu and W. M. Pardridge, *J. Pharmacol. Exp. Ther.*, 1996, **279**, 77–83.
55. D. Wu, B. W. Song, H. V. Vinters and W. M. Pardridge, *J. Drug Targeting*, 2002, **10**, 239–245.
56. P. Kumar, H. Wu, J. L. McBride, K. E. Jung, M. H. Kim, B. L. Davidson, S. K. Lee, P. Shankar and N. Manjunath, *Nature*, 2007, **448**, 39–43.
57. R. A. Gray, D. G. Vander Velde, C. J. Burke, M. C. Manning, C. R. Middaugh and R. T. Borchardt, *Biochemistry*, 1994, **33**, 1323–1331.
58. T. J. Abbruscato, S. A. Thomas, V. J. Hruby and T. P. Davis, *J. Pharmacol. Exp. Ther.*, 1997, **280**, 402–409.
59. W. A. Banks, A. V. Schally, C. M. Barrera, M. B. Fasold, D. A. Durham, V. J. Csernus, K. Groot and A. J. Kastin, *Proc. Natl. Acad. Sci. U. S. A.*, 1990, **87**, 6762–6766.
60. A. J. Kastin, V. Akerstrom, L. Hackler and J. E. Zadina, *J. Neurochem.*, 2000, **74**, 385–391.
61. D. Dogrukol-Ak, W. A. Banks, N. Tuncel and M. Tuncel, *Peptides*, 2003, **24**, 437–444.
62. A. J. Kastin and V. Akerstrom, *J. Pharmacol. Exp. Ther.*, 1999, **289**, 219–223.
63. M. Teixido, E. Zurita, M. Malakoutikhah, T. Tarrago and E. Giralt, *J. Am. Chem. Soc.*, 2007, **129**, 11802–11813.
64. M. Malakoutikhah, M. Teixido and E. Giralt, *J. Med. Chem.*, 2008, **51**, 4881–4889.
65. M. Malakoutikhah, R. Prades, M. Teixido and E. Giralt, *J. Med. Chem.*, **53**, 2354–2363.

66. A. Tsuji and I. I. Tamai, *Adv. Drug Delivery Rev.*, 1999, **36**, 277–290.
67. F. P. Bonina, L. Arenare, F. Palagiano, A. Saija, F. Nava, D. Trombetta and P. de Caprariis, *J Pharm. Sci.*, 1999, **88**, 561–567.
68. M. Gynther, K. Laine, J. Ropponen, J. Leppanen, A. Mannila, T. Nevalainen, J. Savolainen, T. Jarvinen and J. Rautio, *J. Med. Chem.*, 2008, **51**, 932–936.
69. S. Manfredini, B. Pavan, S. Vertuani, M. Scaglianti, D. Compagnone, C. Biondi, A. Scatturin, S. Tanganelli, L. Ferraro, P. Prasad and A. Dalpiaz, *J. Med. Chem.*, 2002, **45**, 559–562.
70. R. D. Egleton, S. A. Mitchell, J. D. Huber, J. Janders, D. Stropova, R. Polt, H. I. Yamamura, V. J. Hruby and T. P. Davis, *Brain Res.*, 2000, **881**, 37–46.
71. R. D. Egleton, S. A. Mitchell, J. D. Huber, M. M. Palian, R. Polt and T. P. Davis, *J. Pharmacol. Exp. Ther.*, 2001, **299**, 967–972.
72. F. Pinnen, I. Cacciatore, C. Cornacchia, P. Sozio, A. Iannitelli, M. Costa, L. Pecci, C. Nasuti, F. Cantalamessa and A. Di Stefano, *J. Med. Chem.*, 2007, **50**, 2506–2515.
73. S. S. More and R. Vince, *J. Med. Chem.*, 2008, **51**, 4581–4588.
74. M. Gynther, J. Ropponen, K. Laine, J. Leppanen, P. Haapakoski, L. Peura, T. Jarvinen and J. Rautio, *J. Med. Chem.*, 2009, **52**, 3348–3353.
75. A. W. Vorbrodt, *J. Neurocytol.*, 1989, **18**, 359–368.
76. Y. Deguchi, T. Naito, T. Yuge, A. Furukawa, S. Yamada, W. M. Pardridge and R. Kimura, *Pharm. Res.*, 2000, **17**, 63–69.
77. Y. Deguchi, Y. Miyakawa, S. Sakurada, Y. Naito, K. Morimoto, S. Ohtsuki, K. Hosoya and T. Terasaki, *J. Neurochem.*, 2003, **84**, 1154–1161.
78. A. K. Kumagai, J. B. Eisenberg and W. M. Pardridge, *J. Biol. Chem.*, 1987, **262**, 15214–15219.
79. W. M. Pardridge, D. Triguero and J. Buciak, *J. Pharmacol. Exp. Ther.*, 1989, **251**, 821–826.
80. M. Green and P. M. Loewenstein, *Cell*, 1988, **55**, 1179–1188.
81. A. D. Frankel and C. O. Pabo, *Cell*, 1988, **55**, 1189–1193.
82. D. Derossi, G. Chassaing and A. Prochiantz, *Trends Cell Biol.*, 1998, **8**, 84–87.
83. S. Futaki, T. Suzuki, W. Ohashi, T. Yagami, S. Tanaka, K. Ueda and Y. Sugiura, *J. Biol. Chem.*, 2001, **276**, 5836–5840.
84. S. R. Schwarze, A. Ho, A. Vocero-Akbani and S. F. Dowdy, *Science*, 1999, **285**, 1569–1572.
85. S. J. Bolton, D. N. Jones, J. G. Darker, D. S. Eggleston, A. J. Hunter and F. S. Walsh, *Eur. J. Neurosci.*, 2000, **12**, 2847–2855.
86. C. Rousselle, P. Clair, J. M. Lefauconnier, M. Kaczorek, J. M. Scherrmann and J. Temsamani, *Mol. Pharmacol.*, 2000, **57**, 679–686.
87. K. S. Rao, M. K. Reddy, J. L. Horning and V. Labhasetwar, *Biomaterials*, 2008, **29**, 4429–4438.

88. C. Rousselle, M. Smirnova, P. Clair, J. M. Lefauconnier, A. Chavanieu, B. Calas, J. M. Scherrmann and J. Temsamani, *J. Pharmacol. Exp. Ther.*, 2001, **296**, 124–131.

89. M. Mazel, P. Clair, C. Rousselle, P. Vidal, J. M. Scherrmann, D. Mathieu and J. Temsamani, *Anticancer Drugs*, 2001, **12**, 107–116.

90. G. Cao, W. Pei, H. Ge, Q. Liang, Y. Luo, F. R. Sharp, A. Lu, R. Ran, S. H. Graham and J. Chen, *J. Neurosci.*, 2002, **22**, 5423–5431.

91. E. Kilic, G. P. Dietz, D. M. Hermann and M. Bahr, *Ann. Neurol.*, 2002, **52**, 617–622.

92. T. R. Doeppner, F. Nagel, G. P. Dietz, J. Weise, L. Tonges, S. Schwarting and M. Bahr, *J Cereb. Blood Flow Metab.*, 2009, **29**, 1187–1196.

93. C. Rousselle, P. Clair, J. Temsamani and J. M. Scherrmann, *J. Drug Targeting*, 2002, **10**, 309–315.

94. C. Rousselle, P. Clair, M. Smirnova, Y. Kolesnikov, G. W. Pasternak, S. Gac-Breton, A. R. Rees, J. M. Scherrmann and J. Temsamani, *J. Pharmacol. Exp. Ther.*, 2003, **306**, 371–376.

95. E. Blanc, C. Bonnafous, P. Merida, S. Cisternino, P. Clair, J. M. Scherrmann and J. Temsamani, *Anticancer Drugs*, 2004, **15**, 947–954.

96. W. M. Pardridge, *J. Neurovirol.*, 1999, **5**, 556–569.

97. T. Moos and E. H. Morgan, *Cell. Mol. Neurobiol.*, 2000, **20**, 77–95.

98. S. Skarlatos, T. Yoshikawa and W. M. Pardridge, *Brain Res.*, 1995, **683**, 164–171.

99. Z. M. Qian, H. Li, H. Sun and K. Ho, *Pharmacol Rev.*, 2002, **54**, 561–587.

100. T. Miyamoto, N. Tanaka, Y. Eishi and T. Amagasa, *Int. J. Oral Maxillofac. Surg.*, 1994, **23**, 430–433.

101. H. N. Keer, J. M. Kozlowski, Y. C. Tsai, C. Lee, R. N. McEwan and J. T. Grayhack, *J. Urol.*, 1990, **143**, 381–385.

102. C. C. Visser, S. Stevanovic, L. Heleen Voorwinden, P. J. Gaillard, D. J. Crommelin, M. Danhof and A. G. De Boer, *J. Drug Targeting*, 2004, **12**, 145–150.

103. C. C. Visser, S. Stevanovic, L. H. Voorwinden, L. van Bloois, P. J. Gaillard, M. Danhof, D. J. Crommelin and A. G. de Boer, *Eur. J. Pharm. Sci.*, 2005, **25**, 299–305.

104. P. A. Seligman, *Prog. Hematol.*, 1983, **13**, 131–147.

105. W. M. Pardridge, J. L. Buciak and P. M. Friden, *J. Pharmacol. Exp. Ther.*, 1991, **259**, 66–70.

106. H. J. Lee, B. Engelhardt, J. Lesley, U. Bickel and W. M. Pardridge, *J. Pharmacol. Exp. Ther.*, 2000, **292**, 1048–1052.

107. W. M. Pardridge, J. Eisenberg and J. Yang, *Metabolism*, 1987, **36**, 892–895.

108. Y. Zhang and W. M. Pardridge, *J. Neurochem.*, 2001, **76**, 1597–1600.

109. T. Moos and E. H. Morgan, *J. Neurochem.*, 2004, **88**, 233–245.

110. R. Deane, W. Zheng and B. V. Zlokovic, *J. Neurochem.*, 2004, **88**, 813–820.

111. L. I. Larsson, J. Fahrenkrug, O. Schaffalitzky De Muckadell, F. Sundler, R. Hakanson and J. R. Rehfeld, *Proc. Natl. Acad. Sci. U. S. A.*, 1976, **73**, 3197–3200.

112. U. Bickel, T. Yoshikawa, E. M. Landaw, K. F. Faull and W. M. Pardridge, *Proc. Natl. Acad. Sci. U. S. A.*, 1993, **90**, 2618–2622.
113. W. M. Pardridge, D. Wu and T. Sakane, *Pharm. Res.*, 1998, **15**, 576–582.
114. J. Huwyler, D. Wu and W. M. Pardridge, *Proc. Natl. Acad. Sci. U. S. A.*, 1996, **93**, 14164–14169.
115. Y. Zhang and W. M. Pardridge, *J. Pharmacol. Exp. Ther.*, 2005, **313**, 1075–1081.
116. W. M. Pardridge, *Brain Drug Targeting: The Future of Brain Drug Development*, Cambridge University Press, Cambridge, 2001.
117. S. L. Lo and S. Wang, *Biomaterials*, 2008, **29**, 2408–2414.
118. J. H. Lee, J. A. Engler, J. F. Collawn and B. A. Moore, *Eur. J. Biochem.*, 2001, **268**, 2004–2012.
119. S. Oh, B. J. Kim, N. P. Singh, H. Lai and T. Sasaki, *Cancer Lett.*, 2009, **274**, 33–39.
120. L. Han, R. Huang, S. Liu, S. Huang and C. Jiang, *Mol. Pharm.*, **7**, 2156–2165.
121. H. Xia, B. Anderson, Q. Mao and B. L. Davidson, *J. Virol.*, 2000, **74**, 11359–11366.
122. F. I. Staquicini, M. G. Ozawa, C. A. Moya, W. H. Driessen, E. M. Barbu, H. Nishimori, S. Soghomonyan, L. G. Flores, 2nd, X. Liang, V. Paolillo, M. M. Alauddin, J. P. Basilion, F. B. Furnari, O. Bogler, F. F. Lang, K. D. Aldape, G. N. Fuller, M. Hook, J. G. Gelovani, R. L. Sidman, W. K. Cavenee, R. Pasqualini and W. Arap, *J. Clin. Invest.*, 2011, **121**, 161–173.
123. R. J. Boado, Y. Zhang and W. M. Pardridge, *Biotechnol. Bioeng.*, 2007, **97**, 1376–1386.
124. R. J. Boado, Y. Zhang, C. F. Xia, Y. Wang and W. M. Pardridge, *Biotechnol. Bioeng.*, 2008, **99**, 475–484.
125. D. Wu, J. Yang and W. M. Pardridge, *J. Clin. Invest.*, 1997, **100**, 1804–1812.
126. M. J. Coloma, H. J. Lee, A. Kurihara, E. M. Landaw, R. J. Boado, S. L. Morrison and W. M. Pardridge, *Pharm. Res.*, 2000, **17**, 266–274.
127. R. J. Boado, Y. Zhang and W. M. Pardridge, *Biotechnol. Bioeng.*, 2007, **96**, 381–391.
128. M. Shibata, S. Yamada, S. R. Kumar, M. Calero, J. Bading, B. Frangione, D. M. Holtzman, C. A. Miller, D. K. Strickland, J. Ghiso and B. V. Zlokovic, *J. Clin. Invest.*, 2000, **106**, 1489–1499.
129. S. Ito, S. Ohtsuki and T. Terasaki, *Neurosci. Res.*, 2006, **56**, 246–252.
130. R. D. Bell, A. P. Sagare, A. E. Friedman, G. S. Bedi, D. M. Holtzman, R. Deane and B. V. Zlokovic, *J. Cereb. Blood Flow Metab.*, 2007, **27**, 909–918.
131. S. K. Moestrup, J. Gliemann and G. Pallesen, *Cell Tissue Res.*, 1992, **269**, 375–382.
132. G. Bu, E. A. Maksymovitch, J. M. Nerbonne and A. L. Schwartz, *J. Biol. Chem.*, 1994, **269**, 18521–18528.
133. M. Ishiguro, Y. Imai and S. Kohsaka, *Brain Res. Mol. Brain Res.*, 1995, **33**, 37–46.

134. P. May, J. Herz and H. H. Bock, *Cell Mol. Life Sci.*, 2005, **62**, 2325–2338.
135. Y. Li, M. P. Marzolo, P. van Kerkhof, G. J. Strous and G. Bu, *J. Biol. Chem.*, 2000, **275**, 17187–17194.
136. M. Demeule, J. Poirier, J. Jodoin, Y. Bertrand, R. R. Desrosiers, C. Dagenais, T. Nguyen, J. Lanthier, R. Gabathuler, M. Kennard, W. A. Jefferies, D. Karkan, S. Tsai, L. Fenart, R. Cecchelli and R. Beliveau, *J. Neurochem.*, 2002, **83**, 924–933.
137. R. Gabathuler, G. Arthur, M. Kennard, Q. Chen, S. Tsai, J. Yang, W. Schoorl, T. Z. Vitalis and W. A. Jefferies, *Int. Congr. Ser.*, 2005, **1277**, 171–184.
138. D. Karkan, C. Pfeifer, T. Z. Vitalis, G. Arthur, M. Ujiie, Q. Chen, S. Tsai, G. Koliatis, R. Gabathuler and W. A. Jefferies, *PLoS One*, 2008, **3**, e2469.
139. M. Demeule, A. Regina, C. Che, J. Poirier, T. Nguyen, R. Gabathuler, J. P. Castaigne and R. Beliveau, *J. Pharmacol. Exp. Ther.*, 2008, **324**, 1064–1072.
140. A. Regina, M. Demeule, C. Che, I. Lavallee, J. Poirier, R. Gabathuler, R. Beliveau and J. P. Castaigne, *Br. J. Pharmacol.*, 2008, **155**, 185–197.
141. C. Che, G. Yang, C. Thiot, M. C. Lacoste, J. C. Currie, M. Demeule, A. Regina, R. Beliveau and J. P. Castaigne, *J. Med. Chem.*, **53**, 2814–2824.
142. W. Ke, K. Shao, R. Huang, L. Han, Y. Liu, J. Li, Y. Kuang, L. Ye, J. Lou and C. Jiang, *Biomaterials*, 2009, **30**, 6976–6985.
143. W. Pan, A. J. Kastin, T. C. Zankel, P. van Kerkhof, T. Terasaki and G. Bu, *J. Cell Sci.*, 2004, **117**, 5071–5078.
144. W. S. Prince, L. M. McCormick, D. J. Wendt, P. A. Fitzpatrick, K. L. Schwartz, A. I. Aguilera, V. Koppaka, T. M. Christianson, M. C. Vellard, N. Pavloff, J. F. Lemontt, M. Qin, C. M. Starr, G. Bu and T. C. Zankel, *J. Biol. Chem.*, 2004, **279**, 35037–35046.
145. R. Deane, S. Du Yan, R. K. Submamaryan, B. LaRue, S. Jovanovic, E. Hogg, D. Welch, L. Manness, C. Lin, J. Yu, H. Zhu, J. Ghiso, B. Frangione, A. Stern, A. M. Schmidt, D. L. Armstrong, B. Arnold, B. Liliensiek, P. Nawroth, F. Hofman, M. Kindy, D. Stern and B. Zlokovic, *Nat. Med.*, 2003, **9**, 907–913.
146. R. Deane, Z. Wu and B. V. Zlokovic, *Stroke*, 2004, **35**, 2628–2631.
147. B. V. Zlokovic, *Life Sci.*, 1996, **59**, 1483–1497.
148. B. V. Zlokovic, C. L. Martel, E. Matsubara, J. G. McComb, G. Zheng, R. T. McCluskey, B. Frangione and J. Ghiso, *Proc. Natl. Acad. Sci. U. S. A.*, 1996, **93**, 4229–4234.
149. K. Mishima, S. Higashiyama, Y. Nagashima, Y. Miyagi, A. Tamura, N. Kawahara, N. Taniguchi, A. Asai, Y. Kuchino and T. Kirino, *Neurosci. Lett.*, 1996, **213**, 153–156.
150. K. Mishima, S. Higashiyama, A. Asai, K. Yamaoka, Y. Nagashima, N. Taniguchi, C. Kitanaka, T. Kirino and Y. Kuchino, *Acta Neuropathol.*, 1998, **96**, 322–328.
151. N. Tanaka, M. Sasahara, M. Ohno, S. Higashiyama, Y. Hayase and M. Shimada, *Brain Res.*, 1999, **827**, 130–138.

152. S. T. Sigurdardottir, K. Davidsdottir, V. A. Arason, O. Jonsdottir, F. Laudat, W. C. Gruber and I. Jonsdottir, *Vaccine*, 2008, **26**, 4178–4186.
153. S. Buzzi, D. Rubboli, G. Buzzi, A. M. Buzzi, C. Morisi and F. Pironi, *Cancer Immunol. Immunother.*, 2004, **53**, 1041–1048.
154. T. Kageyama, M. Ohishi, S. Miyamoto, H. Mizushima, R. Iwamoto and E. Mekada, *J. Biochem.*, 2007, **142**, 95–104.
155. P. J. Gaillard, A. Brink and A. G. De Boer, *Int. Congr. Ser.*, 2005, 185–198.
156. P. J. Gaillard and A. G. de Boer, *J. Controlled Release*, 2006, **116**, e60–62.
157. P. Wang, Y. Xue, X. Shang and Y. Liu, *Cell Mol. Neurobiol.*, **30**, 717–725.
158. P. J. Gaillard, C. C. Visser and A. G. de Boer, *Expert Opin. Drug Delivery*, 2005, **2**, 299–309.
159. S. Leonard and D. Bertrand, *Nicotine Tob. Res.*, 2001, **3**, 203–223.
160. T. L. Lentz, *J. Mol. Recognit.*, 1990, **3**, 82–88.
161. C. Gotti and F. Clementi, *Prog. Neurobiol.*, 2004, **74**, 363–396.
162. Y. Liu, R. Huang, L. Han, W. Ke, K. Shao, L. Ye, J. Lou and C. Jiang, *Biomaterials*, 2009, **30**, 4195–4202.
163. C. F. Elias, J. F. Kelly, C. E. Lee, R. S. Ahima, D. J. Drucker, C. B. Saper and J. K. Elmquist, *J. Comp. Neurol.*, 2000, **423**, 261–281.
164. T. Hosoi, T. Kawagishi, Y. Okuma, J. Tanaka and Y. Nomura, *Endocrinology*, 2002, **143**, 3498–3504.
165. A. J. Kastin and W. Pan, *Regul. Pept.*, 2000, **92**, 37–43.
166. B. Burguera and M. E. Couce, *Physiol. Behav.*, 2001, **74**, 717–720.
167. G. L. Barrett, J. Trieu and T. Naim, *Regul. Pept.*, 2009, **155**, 55–61.
168. Y. Liu, J. Li, K. Shao, R. Huang, L. Ye, J. Lou and C. Jiang, *Biomaterials*, **31**, 5246–5257.
169. C. Fillebeen, L. Descamps, M. P. Dehouck, L. Fenart, M. Benaissa, G. Spik, R. Cecchelli and A. Pierce, *J. Biol. Chem.*, 1999, **274**, 7011–7017.
170. R. Q. Huang, W. L. Ke, Y. H. Qu, J. H. Zhu, Y. Y. Pei and C. Jiang, *J. Biomed. Sci.*, 2007, **14**, 121–128.
171. H. Chen, L. Tang, Y. Qin, Y. Yin, J. Tang, W. Tang, X. Sun, Z. Zhang, J. Liu and Q. He, *Eur. J. Pharm. Sci.*, **40**, 94–102.
172. V. Mishra, S. Mahor, A. Rawat, P. N. Gupta, P. Dubey, K. Khatri and S. P. Vyas, *J. Drug Targeting*, 2006, **14**, 45–53.
173. K. Ulbrich, T. Hekmatara, E. Herbert and J. Kreuter, *Eur. J. Pharm. Biopharm.*, 2009, **71**, 251–256.
174. Y. Miyajima, H. Nakamura, Y. Kuwata, J. D. Lee, S. Masunaga, K. Ono and K. Maruyama, *Bioconjug. Chem.*, 2006, **17**, 1314–1320.
175. G. Xu, K. T. Yong, I. Roy, S. D. Mahajan, H. Ding, S. A. Schwartz and P. N. Prasad, *Bioconjug. Chem.*, 2008, **19**, 1179–1185.
176. H. Zhao, G. L. Li, R. Z. Wang, S. F. Li, J. J. Wei, M. Feng, Y. J. Zhao, W. B. Ma, Y. Yang, Y. N. Li and Y. G. Kong, *J. Int. Med. Res.*, **38**, 957–966.
177. A. Beduneau, F. Hindre, A. Clavreul, J. C. Leroux, P. Saulnier and J. P. Benoit, *J. Controlled Release*, 2008, **126**, 44–49.

178. B. S. Lee, M. Fujita, N. M. Khazenzon, K. A. Wawrowsky, S. Wachsmann-Hogiu, D. L. Farkas, K. L. Black, J. Y. Ljubimova and E. Holler, *Bioconjug. Chem.*, 2006, **17**, 317–326.
179. Y. Aktas, M. Yemisci, K. Andrieux, R. N. Gursoy, M. J. Alonso, E. Fernandez-Megia, R. Novoa-Carballal, E. Quinoa, R. Riguera, M. F. Sargon, H. H. Celik, A. S. Demir, A. A. Hincal, T. Dalkara, Y. Capan and P. Couvreur, *Bioconjug. Chem.*, 2005, **16**, 1503–1511.
180. A. Schnyder, S. Krahenbuhl, J. Drewe and J. Huwyler, *J. Drug Targeting*, 2005, **13**, 325–335.
181. T. Suzuki, D. Wu, F. Schlachetzki, J. Y. Li, R. J. Boado and W. M. Pardridge, *J. Nucl. Med.*, 2004, **45**, 1766–1775.
182. S. Gosk, C. Vermehren, G. Storm and T. Moos, *J. Cereb. Blood Flow Metab.*, 2004, **24**, 1193–1204.
183. Y. F. Zhang, R. J. Boado and W. M. Pardridge, *Pharm. Res.*, 2003, **20**, 1779–1785.
184. Y. Zhang, F. Calon, C. Zhu, R. J. Boado and W. M. Pardridge, *Hum. Gene Ther.*, 2003, **14**, 1–12.
185. B. W. Song, H. V. Vinters, D. Wu and W. M. Pardridge, *J. Pharmacol. Exp. Ther.*, 2002, **301**, 605–610.
186. N. Shi, R. J. Boado and W. M. Pardridge, *Pharm. Res.*, 2001, **18**, 1091–1095.
187. Y. Zhang and W. M. Pardridge, *Stroke*, 2001, **32**, 1378–1384.
188. Y. Zhang and W. M. Pardridge, *Brain Res.*, 2001, **889**, 49–56.
189. N. Shi and W. M. Pardridge, *Proc. Natl. Acad. Sci. U. S. A.*, 2000, **97**, 7567–7572.
190. A. Kurihara, Y. Deguchi and W. M. Pardridge, *Bioconjug. Chem.*, 1999, **10**, 502–511.
191. Y. Saito, J. Buciak, J. Yang and W. M. Pardridge, *Proc. Natl. Acad. Sci. U. S. A.*, 1995, **92**, 10227–10231.
192. C. Zhu, Y. Zhang, Y. F. Zhang, J. Yi Li, R. J. Boado and W. M. Pardridge, *J. Gene Med.*, 2004, **6**, 906–912.
193. N. Shi, Y. Zhang, C. Zhu, R. J. Boado and W. M. Pardridge, *Proc. Natl. Acad. Sci. U. S. A.*, 2001, **98**, 12754–12759.
194. Y. Zhang, Y. F. Zhang, J. Bryant, A. Charles, R. J. Boado and W. M. Pardridge, *Clin. Cancer Res.*, 2004, **10**, 3667–3677.
195. H. J. Lee, R. J. Boado, D. A. Braasch, D. R. Corey and W. M. Pardridge, *J. Nucl. Med.*, 2002, **43**, 948–956.
196. H. J. Lee, Y. Zhang, C. Zhu, K. Duff and W. M. Pardridge, *J. Cereb. Blood Flow Metab.*, 2002, **22**, 223–231.
197. Y. Zhang, F. Schlachetzki and W. M. Pardridge, *Mol. Ther.*, 2003, **7**, 11–18.
198. K. Shao, R. Huang, J. Li, L. Han, L. Ye, J. Lou and C. Jiang, *J. Controlled Release*, **147**, 118–126.
199. A. Muruganandam, J. Tanha, S. Narang and D. Stanimirovic, *FASEB J.*, 2002, **16**, 240–242.
200. P. Holliger and P. J. Hudson, *Nat. Biotechnol.*, 2005, **23**, 1126–1136.

CHAPTER 7.3

Drug Delivery Strategies: Nanostructures for Improved Brain Delivery

MARIA DE LA FUENTE[a,c], MARIA V. LOZANO[a], IJEOMA F. UCHEGBU[b] AND ANDREAS G. SCHÄTZLEIN*[a]

[a] Department of Pharmaceutical and Biological Chemistry, School of Pharmacy, University of London, UK; [b] Department of Pharmaceutics, School of Pharmacy, University of London, UK; [c] Department of Pharmaceutical Technology, University of Santiago de Compostela, Spain
*E-mail: andreas.schatzlein@pharmacy.ac.uk

7.3.1 Engineering of Nanostructures for Brain Delivery

The effective nature of the blood-brain-barrier (BBB) frequently precludes pharmacotherapy or requires the use of invasive procedures to bypass the BBB and deliver the drugs directly into the cerebrospinal fluid or brain parenchyma. While this method may be acceptable for the treatment of acute, life-threatening conditions, it does not allow routine use, *e.g.*, for chronic drug therapy. Nanostructures as delivery 'devices' offer several conceptual advantages that could help to overcome some of these challenges:

• Packaging in containers

Depending on their specific composition and structure, drugs and macromolecules with different physicochemical properties (*e.g.*, charge, solubility, molecular weight) can be accommodated in these nanocontainers,

RSC Drug Discovery Series No. 22
Nanostructured Biomaterials for Overcoming Biological Barriers
Edited by Maria Jose Alonso and Noemi S. Csaba
© The Royal Society of Chemistry 2012
Published by the Royal Society of Chemistry, www.rsc.org

sometimes even simultaneously as drug combinations. This packaging, encapsulation and release of the cargo is controlled by physical chemical properties and interactions within the biomaterials, which in turn are engineered using the individual molecules' chemistry.[1,2]

- Control of interactions

    Due to their minuscule size, nanostructures have a very large surface area relative to their volume; this makes their surface and its properties one of the key factors determining biological behaviour. Nanoparticles formation is typically governed by the chemistry and physical chemistry that drive covalent and non-covalent association of individual molecules into nanostructures. These allow fine-tuning of surfaces to control, for example, the adsorption of proteins in biological fluids or the nanoparticle binding to cell surfaces: reduced binding can lead to increased nanoparticle plasma half-life,[3] while increased adsorption of proteins such as ApoE protein target particles to the BBB.[4] Modulation of such interactions is used to control pharmacokinetics and biodistribution of nanocontainers and, consequently, the encapsulated drug.

- Targeting and specificity

    In addition to the non-specific interactions, nanostructures can also facilitate selective binding and active targeting to specific biological structures and motifs. This process depends on the presence of a suitable ligand that can recognise and bind to a cell surface receptor. Ligands targeted to receptors on the surface of the BBB endothelial cells[5] (*e.g.*, ApoE binding to the low-density lipoprotein receptor-related protein) can then mediate uptake across the BBB *via* endocytosis.[6]

    Some of the nanosystems more commonly used in an effort to enhance BBB transport of therapeutics are highlighted below, and information regarding their ability to improve brain delivery in animal models compiled in Table 7.3.1.

- **Polymeric nanoparticles** are typically based on a non-toxic polymeric matrix in which the drug payload is dissolved or dispersed and/or adsorbed onto the nanoparticle surface.[1,7] The polymers should be degraded in the body or rapidly excreted to avoid accumulation. For brain targeting, nanoparticles made from PBCA (polybutylcyanoacrylate), PLA/PLGA (polylactic / polylactic-co-glycolic acids) or CS (chitosan) are used widely to deliver active pharmaceutical ingredients (API) such as low MW hydrophobic drugs, peptides, proteins, or even genes.[4,8,9]
- **Solid lipid nanoparticles (SLN)** are typically formed from lipids and stabilizers. These tend to be non-toxic as they are efficiently broken down by lipid metabolism.[10] SLN are for example useful for the delivery of peptides and small molecules after intranasal and intravenous administration.[11–14]
- **Nanoemulsions** are typically formed when an oily phase is dispersed into a polar/aqueous solvent (o/w nanoemulsions) with the help of emulsifying

agents. Coating of these oil droplets with a polymer creates nanocapsules. This can improve stability and add functionality to the nanoparticle such as mucoadhesion mediated by CS,[15] which can increase central nervous system (CNS) bioavailability of the encapsulated drugs.[16]

- **Amphiphilic polymers** of a block-type or comb-type architecture can self-assemble into nanostructures such as micelles, *e.g.* palmitoylated quaternised glycolchitosan forms micelles that enhance transport of some drugs across the BBB by a factor of ten.[17] Polymeric micelles made from 1,2-distearoyl-sn-glycero-3-phosphoethanolamine-*N*-[methoxy(polyethylene glycol)-2000] (PE-PEG) and surface decorated with Angiopep2 to target the lipoprotein receptor LRP at the BBB, also enhance brain bioavailability.[18]

- **Liposomes** and similar aggregates formed by self-assembly of amphiphilic lipids allow drug encapsulation in the bilayer membrane or within the aqueous core. Composed of one (unilamellar) or several (multilamellar) lipid bilayers they are frequently used for brain delivery purposes.[19,20] Targeted liposomes decorated with integrins can even bind to monocytes to piggyback across the BBB.[21]

- **Dendrimers** are non-stochastic, monodisperse globular polymers possessing a highly branched architecture. This results in a high number of groups being available near the molecule 'surface' to facilitate multivalent attachment of functional ligands, tracers or drugs.[22] In particular, covalent modification of PAMAM dendrimers with *e.g.*, folate, transferrin, lactoferrin, Angiopep2, or the virus glycoprotein (RVG29), has been used in an effort to facilitate BBB transport.[23–27]

- Other promising nanostructures include **nanotubes** and **nanofibres**. These low-dimensional nanostructures have a large axial ratio, reminiscent of some natural organisms and structures, such as rod shaped bacteria or viruses, microtubules, or axons and dendrites.[28,29] It is thought that such shapes may have advantages for CNS delivery after systemic but also local delivery.[30]

7.3.2 BBB Mechanisms and Methods

While other barriers such as the blood-cerebrospinal fluid barrier (BCSFB) or the CSF-to-brain barrier also contribute to brain homeostasis, the BBB is considered the most important route of transport into the brain because of its several thousand times larger surface area. As described in Chapter 7.1, the BBB is a dynamic structure expressing multiple types of transporters, receptors and enzymes, aimed at minimising free diffusion of substances into the CNS while at the same time facilitating transport of essential nutrients, ions and metabolites.[31–34]

Development of methods for improved delivery to the CNS depends on accurate mechanistic understanding of the barrier as well as methods that allow testing of novel systems and approaches. However, assessing transport to the brain is a challenging undertaking; *in vivo* measurements are technically difficult, resources intensive, and sometimes fraught with ethical problems. For

Table 7.3.1

Type of nanostructure	Targeting	Cargo	Model	Key observations	Ref.
Invasive Administration					
Implants					
Acetate phthalate nanoparticles	—	Dopamine	Rat	Sustained release of drug over a period of 30 days.	57
PLGA nanofibres	—	Paclitaxel	Intracranial mouse glioma (U87MG)	Significant decrease in the tumour proliferation index after 41 days of treatment and in comparison to sham and placebo controls	30
Microinjections					
PEGylated lysine polypeptide rod-like nanoparticles	—	pGDNF	Rat	Stable expression of GDNF up to 3 weeks post-injection	60
Liposomes	—	Angiotensin-1-7/ angiotensin II	Rat	Increased duration of pressor effects compared to free peptide ("day" vs. "minute")	58, 59
Liposomes	Transferrin/ receptor	pNGF	Rat model of brain injury	Enhanced transgene expression from targeted lipoplexes, neuroprotective and repairing effect in brain injury model	172
PAMAM dendrimers	Folate/receptor	anti-EGFR ODN	Intracranial rat glioma (C6 cells)	Increased survival time of targeted PAMAM vector (20 days) compared to untargeted (17 days) or oligofectamine (16 days)	23
Convection enhanced delivery					
PEGylated lipidic nanocapsules	—	Ferrociphenol	Intracranial rat glioma (9L cells)	In combination with radiotherapy ferrociphenol-loaded lipidic nanosystem increases survival time by 60%	78
Poly(ethylene glycol)-β-poly(aspartic acid) block copolymer micelles	—	Doxorubicin	Intracranial rat glioma (9L cells)	Micelles improve survival time compared to PEGylated liposomes (36 vs. 16.6 days)	76

Table 7.3.1 (Continued)

Type of nanostructure	Targeting	Cargo	Model	Key observations	Ref.
Liposomes	—	Irinotecan/ Doxorubicin	Intracranial rat glioma (U87MG, U251)	CED combination therapy with liposomal doxorubicin and irinotecan improves survival compared to either formulation alone: U251MG tumours >100 vs. 85/81 days; U87MG 62 vs. 24/30 days)	72
Liposomes	—	Irinotecan	Rat bearing an intracranial glioma (U87MG)	Liposomal formulation allows 26 × increase of the dose resulting in improved efficacy	73
Liposomes	—	Topotecan	Intracranial rat glioma (U87MG, U251)	Improved survival times in U87MG or U251MG tumours but less effective in larger U87MG tumours	74
Liposomes	—	Doxorubicin	Intracranial rat glioma (U87MG, U251 cells)	Liposomal doxorubicin increases survival time compared to liposomes alone, IV administered liposomal dox or free drug (84 vs. 47, 45, 50 days)	75
PAMAM dendrimers	EGF/receptor	Methotrexate	Intracranial rat glioma (F98)	Cetuximab-conjugated PAMAM dendrimers show targeting to EGFR positive tumours but no improvement in surival times	79
Intrathecal					
Liposomes	—	Morphine	Mouse/rat	Increased drug levels in brain lead to prolonged antinociceptive effect	86
Liposomes	—	Carmustin	Meningeal gliomatosis rat (C6)	Increased survival from drug-loaded liposomes compared to free drug	87

Table 7.3.1 (*Continued*)

Type of nanostructure	Targeting	Cargo	Model	Key observations	Ref.
Liposomes	—	Fasudil	Aneurysmal subarachnoid haemorrhage (rat)	Drug-loaded liposomes significantly reduce vasoconstriction in the rat basilar artery compared to blank liposomes (87.7% and 66.3%, respectively)	88
Non-invasive Administration					
Intravenous					
Human serum albumin nanoparticles	Apolipoprotein E / Low-density lipoprotein receptor	Loperamide	ICR Mouse	Enhanced antinociceptive effect compared to free drug	6
Human serum albumin nanoparticles	Monoclonal antibodies (OX26, R17217) & transferrin/receptor	Loperamide	ICR Mouse	Both transferrin and anti-transferrin antibodies showed similar antinociceptive effect	171
Human serum albumin nanoparticles	Monoclonal antibody (29B4) & insulin/receptor	Loperamide	ICR Mouse	Both insulin and anti-insulin antibodies showed similar antinociceptive effect, targeting blocked by free antibody	195
Solid lipid nanoparticles	—	Camptothecin	Mouse	Improved absolute bioavailability, with (AUC 10.4-fold greater when the drug was encapsulated and with respect to the solution	14
Immunonanocapsules	Antibodies (OX26) & fragments & transferrin/receptor	Radioactivity	Wistar rat	Significant accumulation of the targeted systems, more efficient the whole antibody	196
Liposomes	RMP-7/B2 receptors	Nerve growth factor	SD rat	Three times higher %ID when the system was targeted	197

Table 7.3.1 (*Continued*)

Type of nanostructure	Targeting	Cargo	Model	Key observations	Ref.
PEG-PAMAM dendrimers	Angiopep-2/low-density lipoprotein receptor	Radioactivity, pEGFP	Balb/c mouse	Higher brain accumulation and transfection efficiency	26
PAMAM dendrimers	Lactoferrin /receptor	GFP	Balb/c mouse	Lactoferrin promotes the nanoparticle uptake, part of the dose in the parenchyma, the rest remains at the BBB	161
Liposomes	Transferrin/receptor, RI7217/, COG133, angiopep-2, CRM197/low-density lipoprotein receptor, heparin-binding epidermal growth factor	Radioactivity	Balb/c mouse	Only RI7217 could efficiently target liposomes to the brain with %ID values of 0.18%	194
TMCS-conjugated PLGA nanoparticles	—	Coenzyme Q10	Mouse-APP/PS1 transgenic mouse	Neuroprotection and memory improvement compared to uncoated formulation or free drug; 6-coumarin loaded nanoparticles observed in brain section/fluorescence microscope	132
Poly(D,L-lactide-co-glycolide) nanoparticles	Similopioid peptide/ sialic acid receptors	Fluorescent/ loperamide	Albino rat	Long-term antinociception; >14% of the injected dose in brain; nanoparticles in brain parenchyma/confocal microscopy	198

Table 7.3.1 (*Continued*)

Type of nanostructure	Targeting	Cargo	Model	Key observations	Ref.
TMCS nanoparticles	—	anti-neuroexcitation peptide (ANEP)	Mouse	Enhanced brain biodistribution and retention time for the encapsulated neuropeptide; fluorescence in brain sections with FITC-labelled ANEP/TMCS nanoparticles	137
Human serum albumin nanoparticles	Apolipoprotein A-I/scavenger receptor	Unloaded	SV 129 mouse/Wistar rat	Nanoparticles are taken up by brain capillary endothelial cells and accumulated in brain tissue; HSA nanoparticles observed in different brain parenchyma areas/EM	199
GCPQ micelles	—	Propofol	Mouse	Anasthesia 10 × increased for same dose compared to commercial formulations	17, 124
Immunoliposomes	Transferrin receptor/antibodies (OX26)	Unloaded	P18 Rat	*In situ* perfusion/intravenous tail vein; most of the dose (84–86%) remains in the brain capillary endothelial cells; no transport towards brain parenchyma	186
Oral					
Polysorbate-coated-PBCA nanoparticles	—	Dalargin	Mouse	Polysorbate coating (but not dextran or poloxamer) coating of nanoparticles allows anesthesia after oral administration	46
Polysorbated/PEG-double coated PBCA nanoparticles	—	Dalargin	Mouse	Improved analgesia by combination of polysorbate and PEG nanoparticle coating	47

Table 7.3.1 (*Continued*)

Type of nanostructure	Targeting	Cargo	Model	Key observations	Ref.
PVA-PLG nanoparticles	—	Antituberculosis drugs (rifampicin + isoniazid + pyrazinamide + ethambutol)	Mouse – cerebral *M. tuberculosis* infection	Enhanced duration of effect related to sustained drug exposure achieved by drug encapsulation into nanoparticles	48
Intranasal					
CS nanoparticles	—	Estradiol	Rat	Increased drug concentrations in CSF compared to systemic administration of the nanoparticles	107
MePEG-PLA nanoparticles	—	Nimodiopine	Rat	Improved nose to CNS transport for the encapsulated *vs.* free drug (1.6–3.3-fold)	108
CS nanoparticles	—	Didanosine	Rat	Improved nose to CNS transport for the encapsulated drug *vs.* free drug (1.2–1.6-fold)	109
PLA nanoparticles	—	Thyrotropin-releasing hormone	Rat – kindling model of temporal lobe epilepsy	Anti-seizure activity comparable to intraprenchymal implants (microdisks)	111
PLA nanoparticles	—	Neurotoxin-I	Rat	CNS concentration of the neurpeptide improved compared to systemic administration (1.5-fold)	112
PEG-PLA nanoparticles	WGA lectin /*N*-acetyl-D-glucosamine and sialic acid residues	Vasoactive intestinal peptide	Rat – AF64A-induced cholinergic inhibition model	Improved learning function from VIP-loaded WGA-decorated nanoparticle compared to unmodified nanoparticles	115

in vitro experiments on the other hand, the challenge is to create models that accurately predict transport across this complex multi-component barrier.[35–37]

For small molecules, *in silico* methods and cell-free *in vitro* models (*e.g.*, Immobilized Artificial Membrane Chromatography (IAM), Parallel Artificial Permeability Assay (PAMPA)) can provide initial screening tools;[36,38] their applicability for studying the BBB transport of drugs with nanocarriers remains to be validated. However, *ex vivo* models, including isolated brain capillaries,[39] primary cell based cell culture models,[40,41] or immortalised cells[42] provide more established tools in this context. Nevertheless, no single model currently exists which reproduces the BBB faithfully, and important aspects of the transport process, such as the interplay between pharmacokinetics, plasma protein binding, and CNS efflux, are difficult to reproduce using this approach.

Therefore, *in vivo* experiments still provide the most reliable means of assessing CNS bioavailability.[36,43] A selection of more specialist techniques ranging from highly invasive procedures such as isolated perfusion and microanalysis to non-invasive imaging techniques, such as MRI and SPECT, have been developed. Taken together, these can give a reasonably accurate prediction of CNS bioavailability granted that the underlying mechanisms are accurately represented and studies are supported by pharmacodynamic readouts.[43]

We have divided the strategies for CNS delivery with the help of nanostructures into those which simply improve pharmacokinetics and increase transport by achieving higher blood levels and those which actually increase relative transport by either minimising the barrier, *e.g.* by bypassing it, or by increasing cross barrier transport.

7.3.3 Enhanced CNS Activity Through Improved Systemic Bioavailability

Although the BBB is a highly efficient barrier, some drugs, in particular low MW lipophilic compounds, show a limited intrinsic permeability. The concentration of such drugs found in the brain therefore corresponds to the level of drug found in the blood: an increase in blood levels will lead to increased amounts crossing the BBB (but may also risk increasing peripheral side effects). However, depending on the drug it can be challenging to try to achieve the desired drug pharmacokinetics because of *e.g.*, poor bioavailability or limited drug solubility. Nanostructures have been engineered with the specific aim to overcome such problems and improve delivery of drugs to the systemic circulation. While intravenous injections and infusions with nanoparticulate systems may allow higher drug concentrations to be reached, we will focus on the more convenient and practically relevant mucosal/oral administration routes.

Conceptually, absorption of drugs *via* the respiratory mucosa has the advantage of avoiding degradation in the gut and first pass metabolism in the liver, and nanostructures have indeed been shown to significantly enhance bioavailability of encapsulated drugs and macromolecules when administered to the nasal mucosa.[15,44] In contrast to the olfactory epithelium, which provides a potential direct nose-to-brain route (see section 7.3.4.2.), the respiratory (nasal) mucosa does not possess any direct connection to the brain. However, transmucosal systemic delivery of drugs such as morphine, can enhance CNS drug activity due to the improved pharmacokinetics,[45] a fact that can complicate the interpretation of data from direct nose-to-brain studies.

The oral route remains the most important and attractive mode of administration of drugs, attractive because of good patient compliance and relatively low costs. These considerations are particularly important for those CNS conditions that are likely to require chronic treatment, such as the degenerative or psychiatric disorders like Alzheimers, schizophrenia or Parkinsons. However, oral delivery of drugs to the brain is particularly challenging, as delivery across the BBB in this case also requires efficient uptake in the gut. In particular, for many of the biopharmaceutical compounds such as peptides both barriers are difficult to overcome. Nevertheless, there are some reports showing CNS analgesic effects after oral delivery of the peptide dalargin (MW 725 Da) encapsulated in poly(butylcyanoacrylate) (PBCA) nanoparticles coated with polysorbate 85,[46] or as an improved version with both polysorbate 80 and PEG 2000.[47] Similarly, the oral delivery of a combination of four anti-tuberculosis drugs (rifampicin + isoniazid + pyrazinamide + ethambutol) also leads to a significant increase in brain bioavailability and residence time of the encapsulated drugs (9 days *vs.* one day for free drug), a result confirmed in therapeutic experiments performed in *M. tuberculosis* infected mice.[48]

The mechanism of action for these carriers is not entirely clear but likely to involve increased systemic drug levels rather than BBB activity of the orally administered nanoparticles. It is equally unclear whether nanoparticles are able to cross the intact (nasal) mucosa to reach the systemic circulation.[49,50] It seems currently likely that most of the observed effects are related to reduced degradation as well as enhanced solubilisation, dissolution, adhesion, and penetration.[15,49,51]

7.3.4   Bypassing the Blood-Brain Barrier

An alternative strategy to enhancing drug concentrations in the brain is to bypass it in an invasive or non-invasive fashion *i.e.*, by gaining access to the brain through surgery and by injection, or by delivering to specialised sites exempted from the BBB control, such as the olfactory nerve/bulb.

7.3.4.1   Invasive Administration

Given the hermetically sealed nature of the brain, bypassing the BBB often involves invasive methods,[52] already described in Chapter 7.2. The challenges for these approaches are twofold: sustained controlled release of the drug is critical to minimise the need for continued intervention and the associated damage. In addition, it can be important to ensure distribution of the drug throughout the brain: while local delivery of therapeutics circumvents the need for high systemic doses the distribution of the drug from a local source is highly restricted, making it difficult to achieve sustained therapeutic drug levels throughout the brain.

A wide range of non-degradable and biodegradable polymers have been developed as *intracranial implantable depots*, typically in the form of drug loaded wafers, disks or microparticles. Such depots can release drugs over a period of days to months.[52–54] One or the first systems, the Gliadel® wafers (formed from polifeprisan 20 and carmustine), are used to suppress cancer growth after intracranial surgery in patients with newly diagnosed high-grade malignant glioma. As an adjunct to surgery and radiation the approach prolongs the mean survival time from 11.6 months (placebo group) to 13.9 months (treated patients), with up to 11% of the treated patients surviving for more than 3 years (2% in the placebo group).[55] While sustained release systems and implants such as wafers are not nanosized their material properties are sometimes governed and designed based on supramolecular interactions on the nanoscale. These systems intended for sustained release include paclitaxel-loaded PLGA nanofibres fabricated as disks suitable for implantation; the system maintains therapeutic drug concentrations for 42 days even in distant brain regions[30] and is as least as efficacious against orthotopic tumour models as a microparticulate PLGA formulation entrapped in an alginate hydrogel.[56] Alginate entrapment of nanoparticles was also used to achieve 30 days release of dopamine from acetate phthalate nanoparticles implanted into the frontal lobe in rats to treat Parkinson disease.[57]

A potentially somewhat less invasive means of local delivery involves the use of site-specific *microinjections*. PEGylated liposomes encapsulating short-lived peptides (angiotensin-1-7 and angiotensin II) were injected into the rostral ventrolateral medulla of rats resulting in therapeutic effects being extended from minutes (free peptides) to days (peptide-loaded PEGylated liposomes).[58] The improvement for the PEGylated liposomes was ascribed to their 'stealth' character, which is thought to result in reduced uptake by endocytic cells and improved distribution within the brain parenchyma. Another important property to consider in this context involves the fluidity of the bilayer, which will affect release of the drug.[59]

Microinjection has also been applied for gene delivery purposes. A PEGylated lysine polypeptide was used for the compaction of pDNA, forming rod-like nanoparticles with a very small diameter (8–11 nm). After injection into the rat striatum, cells were successfully transfected with a plasmid encoding for glial cell line–derived neurotrophic factor (pGDNF). Overexpression of GDNF was recorded for up to 3 weeks.[60] Nerve growth factor (NGF) transfection was also achieved using transferrin coated cationic liposomes which had a superior therapeutic effect compared to 'naked' cationic liposomes.[172] In addition to the targeting of the transferrin receptor in neurons it may also be important that the coated particles would probably have a tendency to distribute further as they would be less charged and therefore show less non-specific interactions. In an attempt to increase interaction with cancer cells for targeted gene delivery nanocarriers were developed using folate-linked PAMAM dendrimers (G5).[23] Injection of these nanoparticles complexed with antisense oligonucleotides (ODN) against EGFR resulted in an increased survival (20 days) of glioma bearing rats compared to non-targeted complexes (17.75 days).[23]

Convection-enhanced delivery (CED)

The histology and morphology of the brain parenchyma severely limit organ-wide distribution of locally administered drugs by convection and diffusion. This makes pharmacotherapy of conditions that affect the majority of the brain challenging. Similarly, conditions where there is a potential for a local spread such as *e.g.*, invasive gliomas, present a challenge even after the BBB has been bypassed by implants or injections.

In convection-enhanced delivery (CED) drugs are pumped into the brain parenchyma through an intracranial catheter under positive pressure to create a flow of fluid (convection) that can carry material to distant areas.[61,62] Convection, in contrast to diffusion, is independent of molecular weight and is therefore also applicable for the delivery of larger therapeutics such as monoclonal antibodies, targeted toxins, proteins, viruses, and even nanocarriers.[63] Low drug solubility and non-specific binding/uptake near the catheter are main limitations of CED that nanostructures can potentially help to overcome. Consequently, when engineering nanostructures for CED administration, the encapsulation strategy, size and surface properties are critical, as they will directly affect the ability to carry and distribute associated cargo through the brain.

The brain's extracellular matrix (ECM) is the conduit through which drugs and nanostructures must distribute after entering the brain.[63] For nanoparticle sizes approaching the conduits intrinsic size, exclusion limits and a mechanical retention effect can confine nanoparticles to the injection site. For example, polystyrene nanoparticles of 20 nm have a 9-fold higher distribution volume than those with a size of 200nm.[64] A threshold of around 200nm was also identified when liposomes were injected to rats, with the ideal size being smaller than 100 nm.[65]

An interesting strategy to increase CED is based on the use of hyperosmotic solutions (*e.g.*, sugars or PEG), which are thought to induce an osmolarity-mediated dilation of the extracellular space and/or reduced back flow by virtue of their increased viscosity.[66–69] Similarly, an increase in the administered dose of nanoparticles is thought to increase penetration due to additional osmotic effects.[65]

In terms of surface charge, neutral particles distribute better, while cationic particles show non-specific binding and thus restricted mobility which can be counteracted by charge shielding with *e.g.*, PEG [65,70]

Studies of the distribution of PEG coated liposomes administered by CED using magnetic resonance image (MRI), fluorescent techniques, and radio-activity show that liposomes efficiently distribute throughout normal rodent brains and, most importantly, in intracranial xenografts.[65,66,70–72] While liposomes diffused through the tumour and even reach the surrounding brain tissue that contained invasive tumour cells[71] this distribution was found to be xenograft-dependent.[72] For illustration see Figure 7.3.1.

The use of 'stealth' formulations prolongs liposome half-life[65,66] and, consequently, drug half-life resulting in an improved therapeutic effect.[73,74] Similarly, the reduced tendency for interactions appears to give CED PEG formulations a clear therapeutic advantage over CED infusion of the free drug, or intravenous administration of the same formulation, *e.g.* for the treatment of gliomas.[72–75] PEGylated polymeric micelles conjugated with doxorubicin have similarly been demonstrated to be highly effective in prolonging survival of rats bearing intracranial 9L syngeneic tumours.[76] The therapeutic advantage of a prolonged half-life after encapsulation is also illustrated by a study where [188]Re perrhenate encapsulated in PEGylated lipidic nanocapsules was retained up to 99% in the brain 12 hours after administration, in contrast to the 20% of the drug solution.[77] When used to deliver the poorly soluble ferrociphenol (Fc-diOH) this strategy also improves survival of glioma-bearing rats.[78]

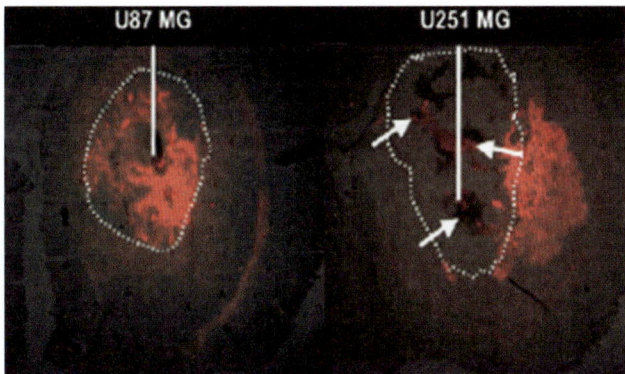

Figure 7.3.1   Brain sections of animals bearing intracranial gliomas (U87MG and U251MG, respectively). After CED of 20µL, fluorescent liposomes (red fluorescence) distributed extensively throughout the U87MG brain tumour xenografts, but the same amount distributed mainly outside U251MG brain tumour xenografts and within necrotic areas (white line represents infusion catheter placement; dotted line delineates tumour margin; white arrows show necrotic areas in U251 brain tumours). The observed tumour-specific distribution highlights one contribution to variability of therapeutic efficacy between tumour models. Despite the better distribution of liposomes in U87MG tumours, the fast growth of U87MG, the lower sensitivity to the therapy and the slow release of therapeutics from liposomes, all contributed to a reduced efficacy in U87MG *versus* U251MG tumours. Adapted from ref. 72 with permission.

CED delivered nanosystems can additionally be targeted to a particular cell population using specific ligands: a PAMAM dendrimer methotrexate conjugate targeted to the epidermal growth factor receptor (EGFR and EGFRvIII isoforms) using antibody conjugation (cetuximab) shows a 5.5-fold improved in retention in EGFR-positive *versus* EGFR-negative tumours (F98 glioma-bearing rats) 24 hours after administration.[79] The lack of improved survival times in this case was attributed to potentially inefficient release of the conjugated drug.

In addition to having shown some benefit, CED delivery of drugs incorporated into nanostructures appears in general to be well tolerated. Furthermore, the CED methodology seems to transfer well between species. In the dog, CED administration of liposomes is by and large well tolerated so that the technique is being considered for treatment of spontaneous gliomas in canines.[80] The linear correlation between the infusion volume and distribution volume of CED delivered liposomes in rodents has been confirmed in non-human primates.[70,81]

Overall, results to date suggest that CED in combination with nanotechnology offers a promising strategy for bypassing the BBB that can provide therapeutic benefits compared to other local delivery strategies. Naturally, the invasive nature of the technology suggests that its application will be predominantly in life threatening situations such as for the treatment of brain tumours.

Intrathecal Injection

The subarachnoid space between the arachnoid membrane and the *pia mater* surrounds the brain and the spinal cord and is occupied by cerebrospinal fluid (CSF). As the CSF-brain barrier is far more permeable than BBB/BCSFB, intrathecal administration by direct injection into the subarachnoid space offers an alternative approach to accessing the brain. The subarachnoid space can be accessed through injection between the lumbar vertebrae (below L2) or through surgically implanted indwelling catheters that reach the brain's ventricles and can be refilled through a subcutaneous reservoir port.[82,83] One of the limitations of this particular approach is the rapid washout of the free drug due to the rapid exchange of CSF fluid (4–5 times per day).[84] In this case, nanostructures can potentially (i) increase the local drug concentration at the CSF and thus improve diffusion into the brain parenchyma, (ii) provide an improved interaction with the brain parenchyma or (iii) minimise drug washout from the CSF.

The intrathecal route is particularly interesting for the administration of opioids for pain management: morphine encapsulation in liposomes leads to a side-effect free increase of drug concentration in the brain as well as an extended duration of action.[85,86] As the intrathecal route is also suited for the treatment of spinal diseases and disseminated meningeal diseases, it is not surprising that liposomes have also been explored for the treatment of meningeal gliomatosis with carmustine, leading to an improved effect in a rat model of the disease. Further pharmacokinetic data in dogs suggests this could be linked to the increased half-life of the nanomedicine.[87] Interesting data were also reported for the treatment of cerebral vasospasm following subarachnoid haemorrhage using microinjection of the protein kinase inhibitor Fasudil encapsulated in liposomes.[88] This approach has also shown some promise for the application of gene therapy as the intrathecal administration of DNA lipoplexes produced significant levels of expression of reporter and therapeutic genes in non-parenchymal cells lining the brain and the spinal cord.[89] Finally, in addition to potentially allowing bypass of the BBB, intrathecal injections also present advantages for the local treatment of spinal cord diseases, as has been shown recently with PLGA nanoparticles dispersed into a biopolymer matrix composed by hyaluronan and methylcellulose for local delivery of neuroprotective and neuroregenerative factors for treatment of major spinal cord injuries.[90]

Given the highly invasive nature of this route and the associated risk for complications,[91] it seems likely that this nanomedicines administration route will remain relevant for the most serious medical conditions only. If successful, targeting of systemically administered drugs to the choroid plexus for access to the CSF could provide a less invasive alternative.[92]

7.3.4.2 Nose-to-brain Transport

The nose offers a potentially interesting route for nanomedicine-mediated delivery of drugs to the brain. Apart from providing important advantages for increasing systemic bioavailability (section 7.3.3),[15] there is also evidence that the nasal route may provide a direct way to the brain, bypassing the BBB.[93] This approach is based on the initial observation that particulate materials can gain direct access to the brain from the olfactory mucosa in the nose: for example, long-term exposure to particles such as coal dust can lead to accumulation of material in the brain of mine workers. Furthermore, some neurotropic viruses have been shown to invade the brain through olfactory nerves, also suggesting a direct route from the nose to the brain exists that bypasses the BBB.[94,95]

Filaments of the olfactory nerve form a network in specialised region of the nose called olfactory mucosa, which is responsible for our sense of smell. These neurons of the olfactory nerve belong to the peripheral nervous system; however, their axons terminate in the olfactory bulb, which is an extension of the forebrain and part of the CNS. Thus, olfactory neurons connect the nasal cavity directly to the brain. Substances could therefore in principle be carried *e.g.*, by axonal transport along these structures from the nose to the olfactory bulb using the trigeminal neural pathway, or a combination of several pathways, depending on the drug/formulation.[94,96–98]

Given the anatomical and histological differences of the nose between species, it is reassuring that intranasal administration of solutions of morphine, neurotrophic peptides, or growth factors, leads to transport to the olfactory bulb/brain parenchyma and to therapeutic effects in experimental animal models and humans.[97,99–103]

However, to gain access to the neuronal network of the olfactory the barrier of the nasal mucosal needs to be overcome. Bioavailability of the administered drugs in the brain, therefore, is typically lower than 1%, depending on drug molecular weight, dissociation, and lipophilicity. Specific challenges for this route of administration include the limited volume that can be administered to the nasal cavity, the small size of the olfactory mucosa *vs.* nasal mucosa, and the limitations in drug absorption due to low dose and short retention times.[44,93,94,104]

Extensive work has been carried out during the last decades regarding the use of nanostructures for transmucosal delivery of drug to the systemic circulation.[15] Consequently, similar nanoparticle compositions to those reported for the transmucosal delivery of drugs tend to also be employed for nose-to-brain delivery. The properties considered to be important for those nanostructures include mucoadhesion and biodegradability. Chitosan (CS), as well as lactic polymers, poly-lactic acid (PLA) and poly-lactic-glycolic acid (PLGA), or derivatives, are frequently the polymers of choice.

There is evidence that nanosystems can increase the bioavailability of low MW hydrophobic molecules after intranasal administration: transport of the fluorescent dye pyrene (MW 202, log$P$ 4.88) to the olfactory mucosa and olfactory bulb was shown after administration in methoxypoly (ethylene glycol)–poly(lactic acid) (MePEG-PLA) nanoparticles.[105,106] Similarly, administration of estradiol (MW 376, log$P$ 5.5) in electrostatically cross-linked CS nanoparticles in rats leads to higher CSF concentrations than intravenous administration.[107] The transport of nimodipine (MW 418, log$P$ 2.36) was also improved by a factor of 1.6–3.3-fold due to its encapsulation into MePEG-PLA nanoparticles, compared to a nasal solution.[108]

However, as low molecular weight hydrophobic compounds tend to have some intrinsic, if limited, BBB permeability, the enhancement seen could simply be due to enhanced mucosal transport and improved systemic bioavailability (as mentioned in section 7.3.3). Nevertheless, there is also some evidence to suggest that nasal delivery with nanoparticles may also increase transport of more hydrophilic molecules: encapsulation of didanosine (MW 236, log$P$ −0.84) in CS nanoparticles leads to a 20–60% increase in the CNS concentration after intranasal administration in rats.[109] Furthermore, a PLA nanoparticle formulation of thyrotropin-releasing hormone (TRH, MW 362, log$P$ −2.53) demonstrated anti-epileptic effects comparable to those achieved with a locally implanted TRH microdisk carrier.[110,111] PLA nanoparticles were also used to deliver the polypeptide Neurotoxin-I (NT-I, MW 6.9kDa) resulting in a 1.5 fold increase after nasal administration *vs.* intravenous injection.[112]

The interaction of the nanoparticles with the mucus layer and their ability to cross this barrier and interact with the underlying epithelia is thought to be a critical step for the efficacy of the nanoparticle/drug transport.[15,50] This notion is supported by the results of Gao *et al;*[113] these authors reported the use of wheat germ agglutinin (WGA) lectin to promote binding of conjugated nanoparticles to *N*-acetyl-D-glucosamine and sialic acid, which are both highly abundant in the nasal cavity. The amount of coumarin (MW 146, log*P* 1.8) in the brain doubled when delivered in WGA nanoparticles compared to the unmodified nanostructures. The presence of the WGA lectin is thought to facilitate the active transport of nanoparticles through the nasal/respiratory mucosa not only to the brain but also into the blood.[113] Interestingly, when the WGA lectin was replaced by the *ulex europeus agglutinin I* (UEA-I), which specifically recognises L-fucose residues on the apical surface of olfactory axons, the enhanced uptake of encapsulated coumarin in rats was limited to the brain,[114] suggesting that nanostructures could be tailored for the specific site of the nasal interaction and cell type involved. Even for larger molecules, such as peptides, a sustained high local drug concentration appears to be important, as suggested by the 3.5–5.6 fold increase in the peak concentration of vasoactive intestinal peptide VIP (MW 3325) obtained from adhesive WGA-decorated compared to blank nanoparticles.[115] The specificity of the distribution of the decorated nanoparticles in the olfactory mucosa of rats is illustrated in Figure 7.3.2.

Overall, while the exact mechanisms underlying intranasal drug delivery to the CNS are not entirely understood, an accumulating body of evidence demonstrates the capacity of nanostructures to increase the direct nose-to-brain transport of substances with a variety of molecular weights and different solubility properties. However, it is unclear whether the nanosystems actually reach the brain or rather facilitate transport of the payload *e.g.* by maintaining high local drug concentrations. It can also be difficult to separate the systemic pharmacokinetic effects (section 7.3.3 nose-to-blood-to-brain) from the direct nose-to-brain effects. Sometimes properties likely to increase mucosal transport of drugs could actually hinder transport of the nanoparticles. For example, CS is thought to open tight junctions[116] but its mucoadhesive properties mean that coating of fluorescent polystyrene nanoparticles with chitosan reduces transport into olfactory epithelial cells as the particles are retained in the mucus.[105] The authors conclude that axoplasmatic transport would probably require small (< 100 nm) particles that are not retained by the mucus.

While there is clear evidence that nanoparticles can facilitate drug transport to the brain after intranasal administration, more studies are needed to help to distinguish between *direct* CNS transport along the neuronal structures and *indirect* effects mediated *e.g.* through enhanced nasal adsorption.

7.3.5   Crossing the Blood-Brain Barrier

Strategies for overcoming the BBB itself can be divided into a number of conceptually different approaches that either aim to improve drug bioavailability by *reducing the effectiveness of the barrier* or seek to *increase the*

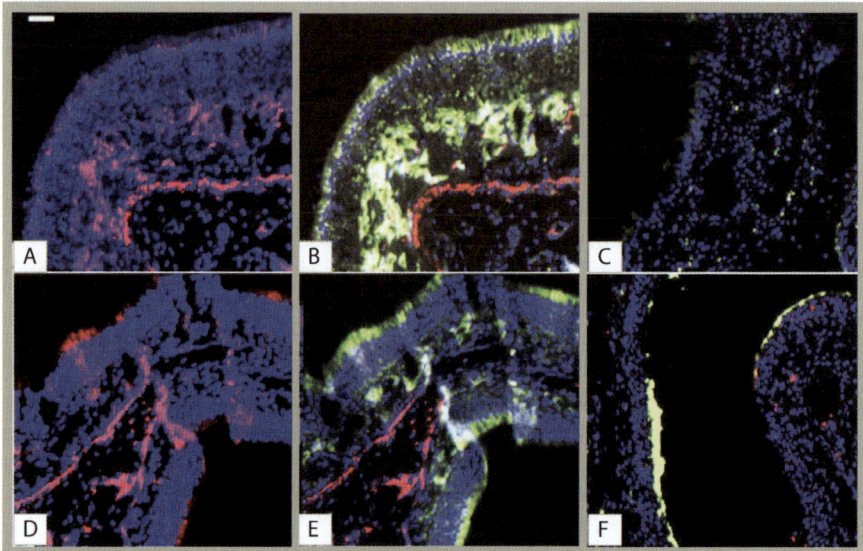

Figure 7.3.2   Distribution of 6-coumarin-loaded WGA-modified particles (WGA-NP) and unmodified nanoparticles (NP) in the nasal mucosa. The olfactory mucosa was labelled with anti-nerve specific enolase antibody (A, D). Two hours following intranasal administration of 6-coumarin-loaded WGA-NP, green fluorescence was observed both on the surface of olfactory mucosa and in the connective tissue surrounding the olfactory nerve bundles (B) while in the same visual field, significantly less green fluorescence was detected on the respiratory mucosa (C). But in the case of intranasal administration of NP, comparative amounts of green fluorescence were observed on both the olfactory (E) and respiratory mucosa (F). These results suggested that WGA increased the affinity of nanoparticles to the olfactory mucosa. Although evidence of the access of intact WGA-NP to the CNS is still needed, an improvement in brain drug delivery, following intranasal application of vasoactive intestinal peptide-loaded WGA-NP, was reported in the same work. Adapted from ref. 115 with permission.

transport of the carrier or the active pharmaceutic ingredient (API) across the intact barrier.

One *strategy to reduce the barriers functionality* is to effect a general but temporary disruption of the barrier *e.g.* by using hyperosmolar solutions, localised ultrasound, or vasoactive compounds such as bradykinin.[117] Nanostructures currently do not play an important role in those approaches and we will not discuss this strategy in any detail.

Another '*barrier reduction*' strategy relies on interfering with the function of the barrier by targeting the active-efflux transporters (AETs, P-glycoprotein (P-gp), Multidrug Resistance Proteins (MRP) or Breast Cancer Resistance Proteins (BCRP and ABCG2)) residing in the brain endothelial cells, which eliminate compounds that have crossed the BBB.[118] Nanostructures are not thought to be substrates for these transporters and drugs encapsulated or associated with the nanostructures are effectively 'hidden' from the AETs. When nanostructures are carried across the BBB this effect can consequently contribute to an overall increase in the CNS concentration of drugs. Once released, the free drug again is a substrate for the AETs and will undergo rapid elimination. As this effect is reliant on the effectiveness of nanoparticle transport across the BBB and difficult to quantify we will not consider it separately in our discussion. However, while nanostructures themselves may not interact with AETs there is some evidence that the constituent molecules of some nanoparticles such as Pluronic micelles can modulate activity of AETs; in this case AET inhibition may become an important element of the overall effect.[119,120] For the transport of digoxin with such micelles it was shown that the P-gp inhibition by monomeric block-copolymer molecules was probably central to its activity.[121] It is important to bear in mind that the AET system is saturable and its transporters have relatively low substrate specificity; consequently, competition in the presence of co-substrates or high concentrations of substrate molecules can temporarily overwhelm the transporters. These factors may contribute to the observation that an increase in dose/concentration can result in a non-linear increase of drug concentration in the brain.

Depending on the properties of the API, a number of *strategies for increasing the CNS bioavailability* of drugs can be considered for APIs that have an intrinsic BBB permeability. Nanostructures may be used to maximise this intrinsic transport, while for those compounds that have an extremely low permeability the aim may be to achieve transport of the nanostructure in order to increase bioavailability of the encapsulated drug.

Another view of these systems is more focused on the fundamental *physiological transport mechanisms* operational at the BBB that the nanostructures aim to exploit to enhance transport. These have been covered in detail in Chapter 7.1 but in principle include:[122]

- **Passive diffusion.** This is practically limited to compounds which show some intrinsic permeability *via* transcellular diffusion, as the effectiveness of tight junctions precludes paracellular diffusion.
- **Solute carriers.** These are responsible for carrying specific water soluble nutrients such as glucose, amino acids, or nucleosides across the BBB.
- **Transcytosis.** Vesicle mediated transport occurs after receptor mediated or adsorption mediated endocytotic uptake.
- **Cell migration.** Cells, *i.e.* of mononuclear cells, can extravasate at the BBB to gain access to the brain parenchyma.

Cell penetrating peptides, also known as protein transduction domains (PTDs), membrane translocating sequences (MTSs), or Trojan peptides, are also promising moieties to increase brain uptake of molecules/nanostructures which do not rely on these physiological transport mechanisms acting as BBB-shuttles and are covered in detail in Chapter 7.2.

Finally, from the point of view of *engineering nanostructures* to exploit these underlying mechanisms one can also divide the approaches into those that involve the use of specific ligands for which a reciprocal recognition occurs at the BBB and those, which operate without the use of such ligands.

Although the discussion will be following the latter division, the systems can in reality be categorised according to all three paradigms. Furthermore, even if a full mechanistic understanding of all process involved existed it is quite likely that more than one mechanism/aspect will be relevant to the overall effects observed.

7.3.5.1   Non-specific Blood-Brain Barrier Interaction

The systems covered in this section do not possess specific targeting moieties and can increase BBB bioavailability of drugs and biopharmaceuticals based on a number of mechanisms. As discussed in section 7.3.3, drugs that have an intrinsic permeability can cross the BBB in small amounts, typically *via* passive (transcellular) diffusion. Fick's law $\left(J = -D\dfrac{\partial\phi}{\partial\chi} \right)$ makes it clear that one way to increase the flux (*J*) of the compound is to increase its concentration ($\partial\varphi^{\partial\phi}$) and to minimise the diffusion distance ($\partial\chi$). Another concept that is relevant here is that of solubility, *e.g.* the transport of a lipophilic compound that would diffuse along cellular membranes could be limited by its low water solubility because only low concentrations can be reached in the blood. Solubility can also limit cross-barrier transport when the barrier consists of different milieus, *i.e.* lipophilic membrane and aqueous environment, through which the drug needs to diffuse.

Consequently, maximising the *local* drug concentration at the BBB and minimising the diffusion distance and number of environments a drug has to overcome to allow *passive transcellular diffusion* can facilitate increased transport of intrinsically permeable compounds.

Using small polymeric micelles made from amphiphilic glycolchitosan (GCPQ), this strategy has been demonstrated recently by our group.[17] The lipophilic intravenous anaesthetic propofol (MW 178, log*P* 4) is encapsulated in highly dynamic polymeric micelles,[123] which can adhere to brain capillary endothelial cells but do not cross into the brain.[124] In this fashion the particles containing a high local concentration of the propofol adhere directly to the cell surface of the endothelial cells to allow direct passive diffusion of the drug. This strategy can increase the pharmacodynamic response (sleep time) to propofol by a factor of ten for formulations using the same dose and concentration as the marketed Diprivan product,[123] as depicted in Figure 7.3.3. Formulations containing the drug in a more concentrated form give rise to an even longer pharmacodynamic response.

While this system demonstrates that significant improvements can be achieved using relatively simple nanoparticulate systems it is not always easy to unravel the underlying mechanisms (see section 7.3.2). Pluronic block copolymers micelles increase the delivery of agents such as digoxin,[125] and

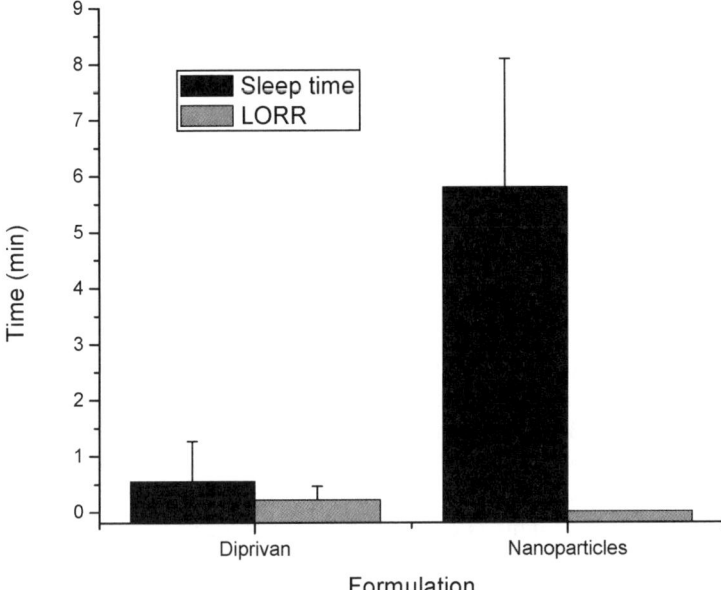

Figure 7.3.3   Pharmacodynamic activity recorded after tail vein injection to mice of propofol-loaded GCPQ nanoparticles or the commercial formulation Diprivan (0.4mg of propofol per mouse). The sleep times obtained with the GCPQ-propofol formulation were up to 10 times those obtained when using Diprivan (black columns). Moreover, a loss of righting reflex time (LORR, grey columns) could not be recorded for the GCPQ-propofol formulation, as animals were asleep by the end of the injection administration period. These results evidence that delivery of the centrally acting drug propofol across the blood-brain barrier is rapid and efficient when loaded into GCPQ micelles, and significantly superior to the marketed available formulations. Adapted from ref. 17 with permission.

leptin[126] at least partly based on the mechanism described above. However, studies in primary cultured bovine brain microvessel endothelial cells (BBMEC) and a comparison in P-gp knock-out and wild type suggest that at least for the Pluronic copolymers, in contrast to the GCPQ micelles,[17] P-gp inhibition plays a major role (see above).[121] A recent review illustrates clearly that nanostructures and/or their constituent molecules are not necessarily inert but can in fact have a wide range of biological effects, including efflux pump, signal transduction, and mitochondrial metabolism.[120]

Such studies demonstrate that optimised passive diffusion can yield remarkable increases in BBB transport without the need for nanostructures to actually cross the BBB. However, in the case of APIs that do not show any appreciable BBB permeability, *e.g.*, biopharmaceuticals like nucleic acids or antibodies, it seems likely that carrier transport would be an important factor.

Carrier mediated transport does not necessarily rely on any *specific* interaction with a corresponding receptor at the BBB: **adsorption mediated**

transcytosis (AMT) utilises *non-specific* adsorption mechanisms and subsequent endocytosis for BBB transport, as also discussed in chapters 7.1.and 7.2. One of the more commonly observed interactions relies on electrostatic interaction of cationic nanostructures with the negatively charged endothelial glycocalyx. Binding to sialo-glycoconjugates and heparin sulfate proteoglycans present on the luminal side of the endothelial cells thus triggers endocytosis of the complex.[127] The particles then follow the transcytotic pathways of the BBB endothelial cells resulting in transport, followed by exocytosis at the apical side of the barrier. This transport was firstly observed for protamine, a polycationic protein, which overcomes the BBB but also carries other cargo proteins with it.[128]

The effect has also been demonstrated after cationic modification of albumin. Cationic albumin, (CBSA) unlike native albumin (BSA), can overcome the BBB[129] and can even serve as a targeting moiety for PEGylated liposomes, as the enhanced accumulation of liposomes in porcine brain capillary endothelial cells (BCECs) and freshly isolated porcine brain capillaries suggests.[130] This is confirmed by the observation of CBSA linked PLA-PEG nanoparticles which mediate uptake by rat BCECs and increased transport of an encapsulated fluorescent marker to the brain.[131] Another cationic polymer, tri-methylated chitosan (TMCS), when used to coat PLGA nanoparticles allows transport of the lipophilic ubiquinone (MW 863) at a therapeutic level (neuroprotective effect/memory function) in a transgenic model.[132] TMCS nanoparticles also enhance transport of encapsulated anti-neuroexcitation peptide (ANEP) in the brain of mice.[133] Cationic GCPQ micelles, as described before, significantly increased the pharmacodynamics response of the encapsulated drug propofol.[17]

Solid lipid nanoparticles are generally considered highly biocompatible[11] and have been shown to enhance transport of *e.g.*, campothecin.[14] Negatively charged SLN nanoparticles (200 nm, ζ-potential -69 mV) increased the blood-to-brain AUC ratio 10.4 fold and also increased transport of a lipophilic fluorodeoxyuridine derivative[134] and the amyotrophic lateral sclerosis (ALS) drug riluzole (MW 234, log$P$ 2.3).[135] Apart from the mechanisms discussed above relying on electrostatic interaction of cationic nanostructures with the negatively charged endothelial glycocalyx, it has been suggested that the presence of lipids in the carrier improves transport: for example, the coating of positively charged nanoparticles in a phospholipid bilayer led to a 3- or 4-fold improved transport in *in vitro* model of the BBB.[136]

One important concept to bear in mind when discussing the underlying mechanisms for the enhanced transport observed with nanostructures is that they are not inert. Not only can the particles or their constituent parts have an effect on cells (as seen for Pluronic) but the reverse also holds true: interaction with the body can also alter the nanostructures. For example, the adsorption of proteins from the blood stream can alter the particle surface or can induce release of the cargo or aggregation of the particles. Such effects can have a dramatic effect on nanoparticle pharmacokinetics and biodistribution and can transform an apparently 'non-specific' uptake mechanism into a targeted one, as is the case when adsorption of apolipoproteins retargets particles to the brain (see below).

7.3.5.2 Ligand/Receptor-mediated BBB Interaction

In this section, nanostructures that facilitate transport by exploiting specific interactions between a ligand (typically covalently linked to a drug, biopharmaceutical, or nanostructure) and a corresponding 'receptor' at the BBB are discussed in more detail. The specific physiological transport mechanisms at the BBB that utilise ligand recognition are the solute carriers and endocytotic receptors described earlier (see Chapters 7.1 and 7.2).

The ligand-receptor interaction conceptually fulfils a number of functions including targeting to the BBB, binding/localisation to endothelial cells, internalisation and transcytosis, and, finally, release. Frequently, the ligands used will target endothelial cell surface receptors, which, after binding, internalises the ligand and its associated cargo and helps to shuttle it across the cell to the parenchymal side of the barrier. However, not all ligand/receptor combination actually combine *all* of these functions, *i.e.* some receptors may be ideally suited to internalising material at the BBB but will not actually allow BBB *specific* targeting because they are also present and active in other parts of the body. Other receptor/ligand combinations may be effective at internalising the ligand and associated material but may fail to release it after transcytosis or fail to recycle to the luminal site.

In the previous section, nanostructures that enhance transport of drugs possessing an intrinsic permeability by increasing their local concentration at the BBB, *e.g.*, by using non-specific binding at to the barrier, have been discussed. Ligand receptor interactions can be used to achieve the same ends, *i.e.* to simply localise the material at the BBB and so *indirectly* increase exposure. Alternatively, they could aim to exploit specific transport mechanisms and thus *directly* enhance uptake. Depending on the carrier and the properties of the drug it is likely that some systems will actually utilise both mechanisms to a varying degree,[198] which could be difficult to quantify (see section 7.3.5.1). The remainder of this section will focus on the latter approach, *i.e.* the use of ligand receptor interactions for targeting and to exploit specific transport mechanisms.

The high metabolic demands of the brain mean that effective transport systems need to be active at the luminal and/or abluminal membranes of the brain endothelia. Thus, one of the key concepts behind the use of nanostructures to enhance BBB bioavailability of drugs is based on the idea that they can serve as carriers for therapeutics. If one can exploit the mechanisms to allow transport of the nanostructures they could provide a generic way to carry the associated drugs into the brain.

A figure frequently used to assess performance of a system is the percentage of the injected dose found in the brain. The brain in rodents weighs around 1–2% of the body weight; in the absence of any barrier and for an evenly distributed drug one would expect around 1% of the dose in the brain with higher values suggesting accumulation. In the case of morphine, only around 0.02% of the administered dose, equivalent to 1 molecule out of 5000, reaches the brain.[117] For comparison, actively targeted nanosystems are reported to deliver in the order of 0.01% to 0.5% of the dose to the brain.[43] These observations suggest that nanosystems can be quite efficient at mediating uptake. However, the specificity of this targeting is limited, as the corresponding relative dose can reach as high as 75% for other tissues like the liver. In addition to off-target accumulation by the nanosystems due to the ubiquitous presence of the target, the recognition and sequestration of particles by the reticulo-endothelial system can skew biodistribution away from the brain. Nevertheless, if the specificity of the drug is good and the pharmacodynamic response selective it is possible to achieve significant enhancement in pharmacological activity by targeting such non-selective receptors at the BBB.

In terms of transporter classes we will consider specifically the solute carrier-mediated transport (SCT), and receptor-mediated transcytosis. AETs (*e.g.* P-glycoprotein (Pgp), Multidrug Resistance Proteins (MRP)) are less relevant here because of their primarily protective function of pumping drugs and xenobiotics out.

Solute Carrier-mediated Transport

These systems mediate the transport of *small hydrophilic nutrients* across the BBB (see Chapters 7.1 and 7.2). The respective transporters are expressed at the luminal (blood) and basolateral (brain) side of the endothelial cells. The transporters of the SCT system are substrate-selective and include the GLUT1 glucose transporter, the LAT1 large neutral aminoacid transporter, the CAT1 cationic aminoacid transporter, the MCT1 monocarboxylic acid transporter or the CNT2 adenosine transporter.[137,138] In order to successfully enhance the transport of drugs to the brain by taking advantage of the SCTs, the structure of the combined drug has to mimic that of the substrate/ligand.[139,140]

L-DOPA provides an excellent example for the use of these types of transporters. Dopamine, the precursor, cannot cross the BBB, but after chemical modification the alpha-amino acid, L-DOPA, is able to utilise the LAT1 transporter. Although only 5% of the dose reaches the brain, L-DOPA is successfully used to treat Parkinson's disease. The structural requirements for substrates of this specific SCT have been defined and are being applied to other drugs[141,142] including melphalan,[143] gabapentin,[142] or alkylating agents.[144] Due to the mechanism of transport and the limitations on substrate molecular weight (only low MW), the modification of the drug molecule itself requires medicinal chemistry to balance receptor affinity and drug activity (this strategy is described in detail in Chapter 7.1). In addition, as each drug molecule will be carried across individually, the overall capacity depends on drug binding affinity, as well as transporter capacity, with LAT1, GLUT1, and MCT1 having the highest capacity.

The GLUT1 receptor has also been used to target vesicles to the BBB: glucose targeted niosomes facilitate BBB transport of vasoactive intestinal peptide (VIP)[145] and mannose modified liposomes that of the neuroprotective drug cytidine 5'-diphosphocholine.[146] Given the intrinsic substrate size limitation of these receptors, it is not clear to what extent the nanoparticles were actually taken up by these receptors or whether the receptor binding simply served to localise the nanoparticles at the BBB to allow transport of drug or carrier by other means.

Receptor-mediated Transcytosis (RMT)

In order to function, the brain also needs access to nutrients and factors, which do not fit into the SCT paradigm of a water-soluble small substrate. This is the case, for example, for some of the larger signalling molecules (*e.g.* insulin), as well as molecules that require transport proteins or carriers such as iron or lipids. These types of molecules/aggregates typically are taken up utilising receptor-mediated endocytosis. The most frequently studied receptors in this context include the transferrin receptors (TfR), insulin receptors (INSR), leptin receptors, insulin-like growth factor receptors, the LRP family (including LDL, very LDL, LRP1, megalin, apoE-receptor 2 (LRP8) and the mosaic LDLR-related protein LR11) (see Chapter 7.1).[147,148]

The transport capacity of these receptors is determined by a number of factors, such as the receptor number, activity and turnover, as well as ligand affinity, and the concentration of potentially competing endogenous ligand in the blood.[149,150]

The receptor numbers and activity are not fixed but change in response to physiological and pathological conditions. For example, TfR expression is usually maximised in tissues that are rapidly dividing[151] or have a high demand of iron,[152] its expression in capillary endothelia is upregulated under ischemia conditions.[153] Similarly, the heparin-binding epidermal growth factor-like growth factor receptor (HB-EGF) or diphtheria toxin receptor (DTR) are strongly upregulated by ischemia or in gliomas.[154,155]

Many of the most studied receptor/ligand combinations come from the family of iron-binding glycoproteins, like transferrin, melanotransferrin and lactoferrin, which are involved in the supply and homeostasis of iron to the brain.[156] As the uptake functionality of these receptors is also essential for other tissues there is a tendency for the receptors to be distributed widely throughout the body. For example, the transferrin receptor is not only expressed in brain capillary endothelial cells, choroid plexus epithelial cells, or brain parenchyma cells but also in red blood cells, monocytes, hepatocytes, intestinal cells, erythroid cells; a fact which would make selective brain targeting using these ligand/receptor combinations a challenge.

Furthermore, in the case of physiologically important nutrients, the ligand/nutrients would potentially be present in the blood stream at relatively high concentrations. Plasma concentrations of the endogenous ligands are regulated and respond to physiological and pathological changes. Depending on the specific receptor/ligand combination and affinity, the endogenous substrate or ligand will be in competition with the targeted nanosystems. A high level of endogenous ligand occupancy of the receptors would then reduce overall efficiency or could lead to variable uptake depending on the concentration of the endogenous ligand. A receptor thought to be particularly affected by this is the transferrin receptor (TfR) for which the plasma concentration of endogenous ligand is relatively high.[138,156]

In contrast, the other iron binding proteins, lactoferrin and melanotransferrin (also known as human melanoma-associated antigen p97), have low plasma concentrations of ≤5 nM and ≤80 pM, respectively[147,157] and competition with the ligands of targeting moieties are much less likely for these molecules.[158] Melanotransferrin and lactoferrin have also been used as ligands for very small nanostructures, such as drug conjugates (the use of ligands as BBB-shuttles is described in Chapter 7.2) with adriamycin (10-fold higher concentration)[159,160] as well as for the transport of DNA in PAMAM:DNA nanoparticles for gene delivery, respectively.[25] In the latter study, lactoferrin was about twice as efficient as transferrin but it is unclear what role the lack of competing endogenous ligand played, as both specific and non-specific mechanisms due to the positive charge contributed to the improvement.[161]

While nutrient based ligands potentially have the problem of competing endogenous ligands, targeting of receptors for signalling molecules can be hampered by the pharmacodynamic activity of the exogenous ligands. For example, the insulin receptor (INSR) is highly expressed on BBB capillaries but the potential hypoglycaemic effect of a insulin targeting ligand limits its practical use.[5] In such cases, the use of alternative endogenous ligands, *e.g.*, insulin-like growth factor (IGF), could be used for targeting, as it also interacts with receptors IGF1R and IGF2R at the BBB level but has milder hypoglycaemic effects.[5]

Despite the functional overlap between the iron-binding proteins there tends to be little cross reactivity, *e.g.* melanotransferrin does not bind to the transferrin receptor, but to a member of the LRP family, while lactoferrin interacts with a member of the LRP family and with the megalin receptor.[138,160]

The LRP receptor interacts with a range of secreted and cell surface bound ligands which, apart from lactoferrin and melanotransferrin, also include the apolipoproteins (apoE). Apolipoproteins are involved in the solubilisation and transport of cholesterol and other fats in the form of lipoproteins. While the lipids form the core, the apolipoproteins cover the surface of these nanometre-sized assemblies (≤10nm–≥500 nm). After binding to receptors of the LRP family (low-density lipoprotein (LDL) receptor-related proteins), the cells take up the whole assembly by endocytosis.[199] Thus, uptake of nanomeric structures is part of the physiological mechanism of action for these receptors, making them a natural choice for the delivery of nanoparticles. Their amphiphilic nature means that apolipoproteins can be adsorbed onto the nanoparticles to facilitate BBB delivery.[162]

Interestingly, the adsorption process can also occur in the blood stream after injection.[163] The coating appears sufficient to mediate uptake to allow a pharmacodynamic response, which is reduced upon injection in ApoE deficient ApoEtm1Unc mice.[162] The coating process is dependent on the particular nanostructures and its surface properties. For example, poly(alkyl cyanoacrylate) or solid lipid nanoparticles coated with surfactants, such as polysorbate, in particular polysorbate 80, or some poloxamers, like 188, appear to be highly susceptible to ApoE coating and subsequent transport across the BBB.[164,165] The picture is complicated further by the fact that different apolipoprotein isoforms differ in their binding behaviour: for example, ApoA1 appears to bind preferentially to cyanoacrylate nanoparticles coated with polysorbate-80 and poloxamer 188, resulting in enhanced delivery of doxorubicin.[166] In addition, ApoB and ApoE appear to be more efficient than ApoA, ApoJ, or ApoC.[162,163] These observations highlight the difficulties of clearly defining mechanisms of action of particles *in vivo*, as particles can undergo dynamic changes after injection, which can alter their properties and biodistribution and may lead to some targeting effects even in non-targeted nanostructures. For a more comprehensive overview regarding protein adsorption and its role in drug delivery see ref. 3 and 167.

Surface modification with hydrophilic polymers such as PEG are frequently utilised as a way to increase plasma half-life of particles; in those situations targeting ligands are typically conjugated to the terminal ends of the PEG chains. This approach has been used successfully to target loperamide encapsulating nanoparticles using ApoE,[6] but also other ligands such as lactoferrin to target for example poly(lactic acid) nanoparticles,[168] poly(ε caprolactone) polymersomes,[169] PAMAM dendrimers,[170] or human serum albumin nanoparticles.[171]

De Novo Ligands for RMT

While endogenous ligands have been frequently used as targeting moiety of choice,[6,160,172] potential advantages of non-endogenous, *de novo* ligands are emerging. Approaches for the identification of such ligands include peptide fragments of endogenous ligands,[173] phage display derived peptides,[174,175] or engineered antibodies.[176]

Such novel ligands potentially can target epitopes which do not coincide with those occupied by endogenous ligands. This could thus minimise problems such as competition from endogenous ligand or agonist activity at receptors with signalling functionality.

For example, monoclonal antibodies that are internalised but do not activate the insulin receptor, such as HIRMAb, a humanised recombinant form of a murine anti-human IR anti-body, have been used to target immunoliposomes for non-viral gene delivery to the brain in mice[177] and primates,[178] as well as for the brain delivery of non-covalently linked siRNA (biotin/avidin),[179] or of larger peptides as recombinant fusion proteins.[180,181]

Antibodies, in addition to targeting epitopes independent of the endogenous ligand, can be selected to have higher affinity for the receptor than the natural ligand itself, as is the case for the transferrin binding antibody OX26.[182] These antibodies have been used to target macromolecules (neurotrophic factors or β-galactosidase) to the brain using the avidin-biotin technology[108,155] and also for the targeting of nanostructures such as liposomes, *e.g.*, genetic therapy in a Parkinson's disease model in rats.[183] Other nanosystems targeted with these antibodies include poly(ethylene glycol) chitosan nanoparticles[184] and polymeric vesicles formed by self-assembling of amphiphilic block copolymers.[185]

The use of antibodies also allows a dual-targeting approach for more than one target, as illustrated for PEGylated liposomes with anti-EGFR siRNA which were targeted to the transferrin receptor (rat 8D3 mAb) and simultaneously to the insulin receptor (murine 83-14 mAb) leading to improved survival and a reduction in vascular density.[177]

Recently, Ulbrich *et al.* have reported the covalent modification of albumin nanoparticles with transferrin monoclonal antibodies (OX26 and R17217) to facilitate the BBB transport of loperamide, showing higher antinociceptive effects than the non-modified particles.[171] However, for small molecules it is sometimes difficult to distinguish between effects due to the localisation of a high concentration of the drug at the endothelial barrier and those related to actual receptor mediated endocytosis of the HSA nanoparticles. Furthermore, *in situ* perfusion studies suggest that OX26 targeted immunoliposomes bind to the barrier much better than particles targeted using a non-specific anti-body (16× higher), but do not actually cross the BBB.[186] Efficient RMT involves recognition/binding and endocytosis/transcytosis, but also dissociation of the protein–receptor conjugate and recycling. High affinity binding of the antibody could potentially limit release and recycling/transcytosis. Yu *et al.* were recently able to demonstrate that BBB transport, and particular transcytosis *via* the transferrin receptor, could be enhanced by utilising antibodies with a reduced affinity.[176]

Angiopeps are a family of peptides derived from endogenous LRP ligands like secreted β-amyloid protein (APP) or aprotinin, with which they target the 'Kunitz protease inhibitor domain', the likely site for LRP binding. Angiopep-2 has emerged as one of the most promising new ligands due to its high transcytosis capacity, which is partly mediated by the LRP1 receptor.[173] As described in Chapter 7.2 (BBB-shuttles), ANG1005, a conjugate of Angiopep-2 linked to three paclitaxel molecules, has been shown to have 100-fold greater uptake into the brain compared to the free drug, resulting in prolonged survival in orthotopic brain tumour models.[187,188] The drug is currently in the early stages of clinical development, with encouraging Phase I/II data having been reported.[189] Other formulations that are currently in the pipeline are Angiopep-2 conjugates of doxorubicin (ANG1007) and etoposide (ANG1009).[190] The Angiopep-2 ligand has also been applied to nanostructures, such as dendrimers, micelles and nanoparticles with encouraging results in animal models of diseases such as cancer or infections of the CNS.[18,26,191]

The ligand CRM197 is a non-toxic derivative of the diphtheria toxin that mediates specific transport in cell culture and enhanced transport to the brain after *in vivo* administration,[192] *e.g.*, after conjugation to a model protein (horseradish peroxidase) or bound to liposomes. Anti-diphtheria antibodies are quite common due to vaccination and may pose a challenge to the wider therapeutic use of this ligand; more recent observations of pharmacodynamic effects and a potential weak toxicity of CRM197 may also be a concern.[193] Interestingly, in a recent comparison of ligands for immunoliposomes targeting, including transferrin, the ApoE fragment COG133, angiopep-2, CRM197, and the murine transferrin receptor antibody RI7217, the latter ligand was the only one to significantly enhance brain uptake,[194] as illustrated in Figure 7.3.4.

7.3.6 Conclusions

The need for therapies for neurodegenerative and psychiatric disorders is increasing in developed and developing countries; a number of biopharmaceuticals are emerging as new classes of agents that may be suitable for tackling these complex conditions. Taken together, these observations highlight an ever-increasing need for tools to overcome the BBB and deliver these agents to the CNS.

Nanostructures have demonstrated the ability to overcome the BBB using a number of strategies, including reducing its effectiveness by interfering with AETs, by bypassing it using invasive and non-invasive means, and by enhancing transport of a drug, or a drug and its carriers, using non-specific and specific strategies. Nanostructures have been shown quite clearly to confer advantages for the transport of a range of drugs and biopharmaceuticals but it has proven experimentally challenging to quantitatively understand the role of specific mechanisms or nanoparticle properties.

Our understanding of the mechanisms underlying BBB transport is increasingly providing the rationale for a systematic engineering of nanosystems to address this medical need. However, the BBB is a highly complex biological barrier that cannot be understood only by analysis of its constituent parts. This makes the development of improved models and methods that allow studying of BBB pathophysiology and of nanoparticle transport at the BBB one of the key requirements for this field. If the potential advantages of nanomedicines for BBB delivery are to be translated into the clinical predictive models that accurately reflect mechanisms and interactions, relevant nanostructure need to be validated and correlated with work in experimental animals and disease models. Finally, the correlation between these models and the transport across the human BBB needs to be established.

Furthermore, the absence of standardised assays and benchmarking tools that would allow a comparison between the various systems is increasingly

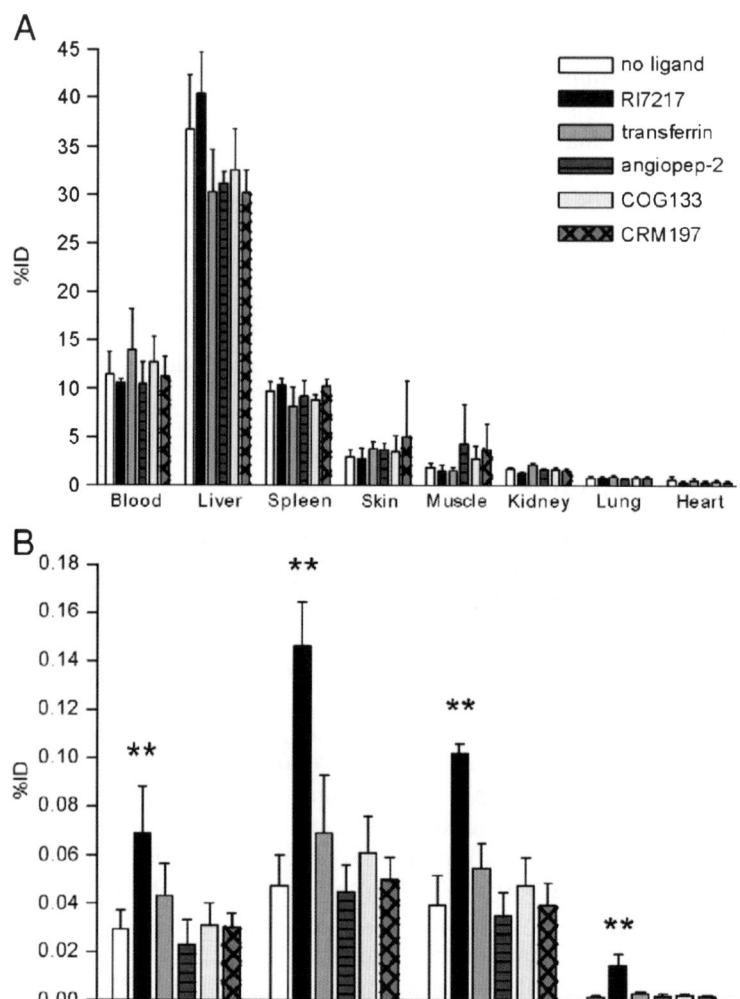

Figure 7.3.4   Comparative biodistribution study of liposomes modified with targeting moieties towards brain delivery. Levels were obtained 12 hours after injection and expressed as percentage of injected dose (% ID). (A) Distribution in blood and major organs, in which it can be observed that the targeting modification does not modify the accumulation pattern of the systems in the different tissues. (B) Distribution in the brain: total cerebellum, total cerebrum, cerebral parenchyma and cerebral capillaries. The only ligand that significantly accumulates in the brain is the RI7217 antibody, showing 4.5 times increase over the untargeted liposomes. The importance of this paper is the comparison performed within the same delivery system of different targeting moieties, which helps to understand their performance. Adapted from ref. 194 with permission.

holding back this field. Publications tend to demonstrate an enhancement from the use of the nanosystems as compared to the free drug, but very rarely are studies comparing across a number of systems to allow a ranking. Variations in experimental methods/protocols between groups, as well as the lack of common model cargos, make it impossible to simply rank systems according to the percentage of drug in the brain.

Having the research tools described above in place would allow rational development of optimised systems in which distinct functional elements have been tuned to maximise transport. These systems would, for example, make it much easier to understand the role of affinity for the various nanoparticle strategies; does it need to be tuned to allow recycling of some receptors, should it be as high as possible to simply bind to the barrier *etc.*? From such studies a more complete picture will emerge of what the requirements for nanostructures actually are.

It is also probably not too early to consider the long-term toxicological consequences of the use of such strategies. The fact that a polymer is broken down easily on systemic injection does not actually say much about its behaviour when taken up into the brain. It may not be desirable to overwhelm the brain with nanoparticles that are efficiently taken up at the BBB. It also seems likely that the easily accessible rodent models will not be able to fully elucidate the long term risks to *e.g.* cognitive function that may be associated with enhanced transport into the BBB.

The medical need for improved pharmacotherapies of CNS conditions will drive the development of nanostructures as delivery tools. However, in order for nanomedicines to be able to deliver on the initial promise shown by systems such ANG1005, a concerted effort of researchers across a number of fields will be required.

References

1. Y. E. Koo, G. R. Reddy, M. Bhojani, R. Schneider, M. A. Philbert, A. Rehemtulla, B. D. Ross and R. Kopelman, *Adv. Drug DeliveryDelivery Rev.*, 2006, **58**, 14, 1556.
2. J. W. Yoo, N. Doshi and S. Mitragotri, *Adv. Drug DeliveryDelivery Rev.*, 2011, **63**(14-15), 1247.
3. P. Aggarwal, J. B. Hall, C. B. McLeland, M. A. Dobrovolskaia and S. E. McNeil, *Adv. Drug DeliveryDelivery Rev.*, 2009, **61**, 6, 428.
4. J. Kreuter, *Adv. Drug DeliveryDelivery Rev.*, 2001, **47**, 1, 65.
5. A. Béduneau, P. Saulnier and J. P. Benoit, *Biomaterials*, 2007, **28**, 33, 4947.
6. K. Michaelis, M. M. Hoffmann, S. Dreis, E. Herbert, R. N. Alyautdin, M. Michaelis, J. Kreuter and K. Langer, *J. Pharmacol. Exp. Ther.*, 2006, **317**, 3, 1246.
7. L. Brannon-Peppas and J. O. Blanchette, *Adv. Drug Delivery Rev.*, 2004, **56**, 11, 1649.
8. J. C. Olivier, *NeuroRx*, 2005, **2**, 1, 108.

9. A. Mistry, S. Stolnik and L. Illum, *Int. J. Pharm.*, 2009, **379**, 1, 146.
10. A. J. Almeida and E. Souto, *Adv. Drug Delivery Rev.*, 2007, **59**, 6, 478.
11. P. Blasi, S. Giovagnoli, A. Schoubben, M. Ricci and C. Rossi, *Adv. Drug Delivery Rev.*, 2007, **59**, 6, 454.
12. I. P. Kaur, R. Bhandari, S. Bhandari and V. Kakkar, *J. Controlled Release*, 2008, **127**, 2, 97.
13. S. Patel, S. Chavhan, H. Soni, A. K. Babbar, R. Mathur, A. K. Mishra and K. Sawant, *J. Drug TargetingTargeting*, 2011, **19**, 6, 468.
14. S. C. Yang, L. F. Lu, Y. Cai, J. B. Zhu, B. W. Liang and C. Z. Yang, *J. Controlled Release*, 1999, **59**, 3, 299.
15. N. Csaba, M. Garcia-Fuentes and M. J. Alonso, *Expert Opin. Drug Delivery*, 2006, **3**, 4, 463.
16. M. Kumar, A. Misra, A. K. Babbar, A. K. Mishra, P. Mishra and K. Pathak, *Int. J. Pharm.*, 2008, **358**, 1–2, 285.
17. X. Qu, V. V. Khutoryanskiy, A. Stewart, S. Rahman, B. Papahadjopoulos-Sternberg, C. Dufes, D. McCarthy, C. G. Wilson, R. Lyons, K. C. Carter, A. Schätzlein and I. F. Uchegbu, *Biomacromolecules*, 2006, **7**, 12, 3452.
18. K. Shao, R. Huang, J. Li, L. Han, L. Ye, J. Lou and C. Jiang, *J. Controlled Release*, 2010, **147**, 1, 118.
19. Z. A. Tokes, A. K. St Peteri and J. A. Todd, *Brain Res.*, 1980, **188**, 1, 282.
20. A. Schnyder and J. Huwyler, *NeuroRx*, 2005, **2**, 1, 99.
21. E. Afergan, H. Epstein, R. Dahan, N. Koroukhov, K. Rohekar, H. D. Danenberg and G. Golomb, *J. Controlled Release*, 2008, **132**, 2, 84.
22. M. J. Cloninger, *Curr. Opin. Chem. Biol.*, 2002, **6**, 6, 742.
23. C. Kang, X. Yuan, F. Li, P. Pu, S. Yu, C. Shen, Z. Zhang and Y. Zhang, *J. Biomed. Mater. Res. A*, 2009, **93**, 2, 585.
24. H. He, Y. Li, X. R. Jia, J. Du, X. Ying, W. L. Lu, J. N. Lou and Y. Wei, *Biomaterials*, 2011, **32**, 2, 478.
25. R. Huang, W. Ke, Y. Liu, C. Jiang and Y. Pei, *Biomaterials*, 2008, **29**, 2, 238.
26. W. Ke, K. Shao, R. Huang, L. Han, Y. Liu, J. Li, Y. Kuang, L. Ye, J. Lou and C. Jiang, *Biomaterials*, 2009, **30**, 36, 6976.
27. Y. Liu, R. Huang, L. Han, W. Ke, K. Shao, L. Ye, J. Lou and C. Jiang, *Biomaterials*, 2009, **30**, 25, 4195.
28. J. L. Gilmore, X. Yi, L. Quan and A. V. Kabanov, *J. Neuroimmune Pharmacol.*, 2008, **3**, 2, 83.
29. H. S. Yoo, T. G. Kim and T. G. Park, *Adv. Drug Delivery Rev.*, 2009, **61**, 12, 1033.
30. S. H. Ranganath, Y. Fu, D. Y. Arifin, I. Kee, L. Zheng, H. S. Lee, P. K. Chow and C. H. Wang, *Biomaterials*, 2010, **31**, 19, 5199.
31. Z. Redzic, *Fluids Barriers CNS*, 2011, **8**, 1, 3.
32. B. Engelhardt and L. Sorokin, *Semin. Immunopathol.*, 2009, **31**, 4, 497.
33. P. Ballabh, A. Braun and M. Nedergaard, *Neurobiol. Dis.*, 2004, **16**, 1, 1.
34. N. J. Abbott, L. Ronnback and E. Hansson, *Nat. Rev.*, 2006, **7**, 1, 41.

35. M. Lemaire and S. Desrayaud, *Int. Congress Series*, 2005, **1277**, 32.
36. J. Mensch, J. Oyarzabal, C. Mackie and P. Augustijns, *J. Pharm. Sci.*, 2009, **98**, 12, 4429.
37. U. Bickel, *NeuroRx*, 2005, **2**, 1, 15.
38. A. R. Mehdipour and M. Hamidi, *Drug Discovery Today*, 2009, **14**, 21–22, 1030.
39. G. W. Goldstein, A. L. Betz, P. D. Bowman and K. Dorovini-Zis, *Ann. N. Y. Acad. Sci.*, 1986, **481**, 202.
40. M. Culot, S. Lundquist, D. Vanuxeem, S. Nion, C. Landry, Y. Delplace, M. P. Dehouck, V. Berezowski, L. Fenart and R. Cecchelli, *Toxicol In Vitro*, 2008, **22**, 3, 799.
41. T. Terasaki, S. Ohtsuki, S. Hori, H. Takanaga, E. Nakashima and K. Hosoya, *Drug Discovery Today*, 2003, **8**, 20, 944.
42. Y. Omidi, L. Campbell, J. Barar, D. Connell, S. Akhtar and M. Gumbleton, *Brain Res.*, 2003, **990**, 1–2, 95.
43. I. Van Rooy, S. Cakir-Tascioglu, W. E. Hennink, G. Storm, R. M. Schiffelers and E. Mastrobattista, *Pharm. Res.*, 2011, **28**, 3, 456.
44. H. R. Costantino, L. Illum, G. Brandt, P. H. Johnson and S. C. Quay, *Int. J. Pharm.*, 2007, **337**, 1–2, 1.
45. L. Illum, P. Watts, A. N. Fisher, M. Hinchcliffe, H. Norbury, I. Jabbal-Gill, R. Nankervis and S. S. Davis, *J. Pharmacol. Exp. Ther.*, 2002, **301**, 1, 391.
46. U. Schroeder, P. Sommerfeld and B. A. Sabel, *Peptides*, 1998, **19**, 4, 777.
47. D. Das and S. Lin, *J. Pharm. Sci.*, 2005, **94**, 6, 1343.
48. R. Pandey and G. K. Khuller, *J. Antimicrob. Chemother.*, 2006, **57**, 6, 1146.
49. A. des Rieux, V. Fievez, M. Garinot, Y. J. Schneider and V. Preat, *J. Controlled Release*, 2006, **116**, 1, 1.
50. S. K. Lai, Y. Y. Wang and J. Hanes, *Adv. Drug Delivery Rev.*, 2009, **61**, 2, 158.
51. C. Prego, M. Garcia, D. Torres and M. J. Alonso, *J. Controlled Release*, 2005, **101**, 1–3, 151.
52. P. P. Wang, J. Frazier and H. Brem, *Adv. Drug Delivery Rev.*, 2002, **54**, 7, 987.
53. A. J. Sawyer, J. M. Piepmeier and W. M. Saltzman, *Yale J. Biol. Med.*, 2006, **79**, 3–4, 141.
54. E. Fournier, C. Passirani, C. N. Montero-Menei and J. P. Benoit, *Biomaterials*, 2003, **24**, 19, 3311.
55. M. Westphal, Z. Ram, V. Riddle, D. Hilt and E. Bortey, *Acta Neurochir. (Wien)*, 2006, **148**, 3, 269.
56. S. H. Ranganath, I. Kee, W. B. Krantz, P. K. Chow and C. H. Wang, *Pharm. Res.*, 2009, **26**, 9, 2101.
57. S. Pillay, V. Pillay, Y. E. Choonara, D. Naidoo, R. A. Khan, L. C. du Toit, V. M. Ndesendo, G. Modi, M. P. Danckwerts and S. E. Iyuke, *Int. J. Pharm.*, 2009, **382**, 1–2, 277.
58. F. Frezard, N. M. Silva-Barcellos and R. A. Dos Santos, *Regul. Pept.*, 2007, **138**, 2–3, 59.

59. N. M. Silva-Barcellos, S. Caligiorne, R. A. dos Santos and F. Frezard, *J. Controlled Release*, 2004, **95**, 2, 301.

60. D. M. Yurek, A. M. Fletcher, G. M. Smith, K. B. Seroogy, A. G. Ziady, J. Molter, T. H. Kowalczyk, L. Padegimas and M. J. Cooper, *Mol. Ther.*, 2009, **17**, 4, 641.

61. M. A. Vogelbaum, *J. Neurooncol.*, 2005, **73**, 1, 57.

62. D. S. Bidros, J. K. Liu and M. A. Vogelbaum, *Future Oncol.*, 2010, **6**, 1, 117.

63. E. Allard, C. Passirani and J. P. Benoit, *Biomaterials*, 2009, **30**, 2302.

64. M. Y. Chen, A. Hoffer, P. F. Morrison, J. F. Hamilton, J. Hughes, K. S. Schlageter, J. Lee, B. R. Kelly and E. H. Oldfield, *J. Neurosurg.*, 2005, **103**, 2, 311.

65. J. A. MacKay, D. F. Deen and F. C. Szoka, Jr., *Brain Res.*, 2005, **1035**, 2, 139.

66. C. Mamot, J. B. Nguyen, M. Pourdehnad, P. Hadaczek, R. Saito, J. R. Bringas, D. C. Drummond, K. Hong, D. B. Kirpotin, T. McKnight, M. S. Berger, J. W. Park and K. S. Bankiewicz, *J. Neurooncol.*, 2004, **68**, 1, 1.

67. R. Saito, M. T. Krauze, C. O. Noble, M. Tamas, D. C. Drummond, D. B. Kirpotin, M. S. Berger, J. W. Park and K. S. Bankiewicz, *J. Neurosci. Methods*, 2006, **154**, 1–2, 225.

68. B. Perlstein, Z. Ram, D. Daniels, A. Ocherashvilli, Y. Roth, S. Margel and Y. Mardor, *Neuro. Oncol.*, 2008, **10**, 2, 153.

69. K. B. Neeves, A. J. Sawyer, C. P. Foley, W. M. Saltzman and W. L. Olbricht, *Brain Res.*, 2007, **1180**, 121.

70. R. Saito, M. T. Krauze, J. R. Bringas, C. Noble, T. R. McKnight, P. Jackson, M. F. Wendland, C. Mamot, D. C. Drummond, D. B. Kirpotin, K. Hong, M. S. Berger, J. W. Park and K. S. Bankiewicz, *Exp. Neurol.*, 2005, **196**, 2, 381.

71. R. Saito, J. R. Bringas, T. R. McKnight, M. F. Wendland, C. Mamot, D. C. Drummond, D. B. Kirpotin, J. W. Park, M. S. Berger, K. S. Bankiewicz, *Cancer Res.*, 2004, **64**, 7, 2572.

72. M. T. Krauze, C. O. Noble, T. Kawaguchi, D. Drummond, D. B. Kirpotin, Y. Yamashita, E. Kullberg, J. Forsayeth, J. W. Park and K. S. Bankiewicz, *Neuro. Oncol.*, 2007, **9**, 4, 393.

73. C. O. Noble, M. T. Krauze, D. C. Drummond, Y. Yamashita, R. Saito, M. S. Berger, D. B. Kirpotin, K. S. Bankiewicz and J. W. Park, *Cancer Res.*, 2006, **66**, 5, 2801.

74. R. Saito, M. T. Krauze, C. O. Noble, D. C. Drummond, D. B. Kirpotin, M. S. Berger, J. W. Park and K. S. Bankiewicz, *Neuro. Oncol.*, 2006, **8**, 3, 205.

75. Y. Yamashita, R. Saito, M. T. Krauze, T. Kawaguchi, C. Noble, D. C. Drummond, D. B. Kirpotin, J. W. Park, M. S. Berger and K. S. Bankiewicz, *Targeted Oncology*, 2006, **1**, 2, 79.

76. T. Inoue, Y. Yamashita, M. Nishihara, S. Sugiyama, Y. Sonoda, T. Kumabe, M. Yokoyama and T. Tominaga, *Neuro. Oncol.*, 2009, **11**, 2, 151.

77. E. Allard, F. Hindre, C. Passirani, L. Lemaire, N. Lepareur, N. Noiret, P. Menei and J. P. Benoit, *Eur. J. Nucl. Med. Mol. Imaging*, 2008, **35**, 10, 1838.
78. E. Allard, D. Jarnet, A. Vessieres, S. Vinchon-Petit, G. Jaouen, J. P. Benoit and C. Passirani, *Pharm. Res.*, 2010, **27**, 1, 56.
79. G. Wu, R. F. Barth, W. Yang, S. Kawabata, L. Zhang and K. Green-Church, *Mol. Cancer Ther.*, 2006, **5**, 1, 52.
80. P. J. Dickinson, R. A. LeCouteur, R. J. Higgins, J. R. Bringas, B. Roberts, R. F. Larson, Y. Yamashita, M. Krauze, C. O. Noble, D. Drummond, D. B. Kirpotin, J. W. Park, M. S. Berger and K. S. Bankiewicz, *J. Neurosurg.*, 2008, **108**, 5, 989.
81. M. T. Krauze, S. R. Vandenberg, Y. Yamashita, R. Saito, J. Forsayeth, C. Noble, J. Park and K. S. Bankiewicz, *Exp. Neurol.*, 2008, **210**, 2, 638.
82. J. Z. Kerr, S. Berg and S. M. Blaney, *Crit. Rev. Oncol. Hematol.*, 2001, **37**, 3, 227.
83. K. H. Simpson and I. Jones, *J. Opioid Manag.*, 2008, **4**, 5, 293.
84. W. M. Pardridge, *Fluids Barriers CNS*, 2011, **8**, 1, 7.
85. T. Nishiyama, R. J. Ho, D. D. Shen, T. L. Yaksh, *Anesth. Analg.*, 2000, **91**, 2, 423.
86. G. J. Grant, M. Cascio, M. I. Zakowski, L. Langerman and H. Turndorf, *Anesth. Analg.*, 1995, **81**, 3, 514.
87. I. Kitamura, M. Kochi, Y. Matsumoto, R. Ueoka, J. Kuratsu and Y. Ushio, *Cancer Res.*, 1996, **56**, 17, 3986.
88. Y. Takanashi, T. Ishida, T. Meguro, M. J. Kirchmeier, T. M. Allen and J. H. Zhang, *J. Clin. Neurosci.*, 2001, **8**, 6, 557.
89. C. Meuli-Simmen, Y. Liu, T. T. Yeo, D. Liggitt, G. Tu, T. Yang, M. Meuli, S. Knauer, T. D. Heath, F. M. Longo and R. J. Debs, *Hum. Gene Ther.*, 1999, **10**, 16, 2689.
90. M. D. Baumann, C. E. Kang, C. H. Tator and M. S. Shoichet, *Biomaterials*, 2010, **31**, 30, 7631.
91. K. A. Follett, R. L. Boortz-Marx, J. M. Drake, S. DuPen, S. J. Schneider, M. S. Turner and R. J. Coffey, *Anesthesiology*, 2004, **100**, 6, 1582.
92. A. M. Gonzalez, W. E. Leadbeater, M. Burg, K. Sims, T. Terasaki, C. E. Johanson, E. G. Stopa, B. P. Eliceiri and A. Baird, *BMC Neurosci*, 2011, **12**, 4.
93. L. Illum, *Eur. J. Pharm. Sci.*, 2000, **11**, 1, 1.
94. S. Mathison, R. Nagilla, U. B. Kompella, *J. Drug Targeting*, 1998, **5**, 6, 415.
95. L. T. Webster and A. D. Clow, *J. Exp. Med.*, 1936, **63**, 3, 433.
96. D. D. Hunter and R. D. Dey, Neuroscience, 1998, **83**, 2, 591.
97. R. G. Thorne, G. J. Pronk, V. Padmanabhan and W. H. Frey, 2nd, *Neuroscience*, 2004, 127, 2, 481.
98. S. V. Dhuria, L. R. Hanson and W. H. Frey, 2nd, *J. Pharm. Sci.*, 2010, **99**, 4, 1654.
99. G. K. Reznik, *Environ. Health Perspect.*, 1990, **85**, 171.

100. R. Pietrowsky, C. Struben, M. Molle, H. L. Fehm and J. Born, *Biol. Psychiatry*, 1996, **39**, 5, 332.
101. J. Born, T. Lange, W. Kern, G. P. McGregor, U. Bickel and H. L. Fehm, *Nat. Neurosci.*, 2002, **5**, 6, 514.
102. W. H. Frey, J. Liu, X. Chen, R. G. Thorne, J. R. Fawcett, T. Ala and Y.-E. Rahman, *Drug Delivery*, 1997, **4**, 2, 87.
103. U. E. Westin, E. Bostrom, J. Grasjo, M. Hammarlund-Udenaes and E. Bjork, *Pharm. Res.*, 2006, **23**, 3, 565.
104. C. Bitter, K. Suter-Zimmermann and C. Surber, *Curr. Probl. Dermatol.*, 2011, **40**, 20.
105. A. Mistry, S. Z. Glud, J. Kjems, J. Randel, K. A. Howard, S. Stolnik and L. Illum, *J. Drug Targeting*, 2009, **17**, 7, 543.
106. Q.-Z. Zhang, L.-S. Zha, Y. Zhang, W.-M. Jiang, W. Lu, Z.-Q. Shi, X.-G. Jiang and S. K. Fu, *J. Drug Targeting*, 2006, **14**, 5, 281.
107. X. Wang, N. Chi and X. Tang, *Eur. J. Pharm. Biopharm.*, 2008, **70**, 3, 735.
108. Y. Zhang and W. M. Pardridge, *Brain Res.*, 2006, **1111**, 1, 227.
109. A. M. Al-Ghananeem, H. Saeed, R. Florence, R. A. Yokel and A. H. Malkawi, *J. Drug Targeting*, 2010, **18**, 5, 381.
110. M. J. Kubek, D. Liang, K. E. Byrd and A. J. Domb, *Brain Res.*, 1998, **809**, 2, 189.
111. M. C. Veronesi, Y. Aldouby, A. J. Domb and M. J. Kubek, *Brain Res.*, 2009, **1303**, 151.
112. Q. Cheng, J. Feng, J. Chen, X. Zhu and F. Li, *Biopharm. Drug Dispos.*, 2008, **29**, 8, 431.
113. X. Gao, W. Tao, W. Lu, Q. Zhang, Y. Zhang, X. Jiang and S. Fu, *Biomaterials*, 2006, **27**, 18, 3482.
114. X. Gao, J. Chen, W. Tao, J. Zhu, Q. Zhang, H. Chen and X. Jiang, *Int. J. Pharm.*, 2007, **340**, 1–2, 207.
115. X. Gao, B. Wu, Q. Zhang, J. Chen, J. Zhu, W. Zhang, Z. Rong, H. Chen and X. Jiang, *J. Controlled Release*, 2007, **121**, 3, 156.
116. O. Felt, P. Buri and R. Gurny, *Drug Dev. Ind. Pharm.*, 1998, **24**, 11, 979.
117. M. A. Bellavance, M. Blanchette and D. Fortin, *AAPS J.*, 2008, **10**, 1, 166.
118. S. Eyal, P. Hsiao and J. D. Unadkat, *Pharmacol. Ther.*, 2009, **123**, 1, 80.
119. D. W. Miller, E. V. Batrakova, T. O. Waltner, V. Alakhov and A. V. Kabanov, *Bioconjug. Chem.*, 1997, **8**, 5, 649.
120. E. V. Batrakova and A. V. Kabanov, *J. Controlled Release*, 2008, **130**, 2, 98.
121. E. V. Batrakova, D. W. Miller, S. Li, V. Y. Alakhov, A. V. Kabanov and W. F. Elmquist, *J. Pharmacol. Exp. Ther.*, 2001, **296**, 2, 551.
122. N. J. Abbott, A. A. Patabendige, D. E. Dolman, S. R. Yusof and D. J. Begley, *Neurobiol. Dis.*, 2010, **37**, 1, 13.
123. S. Ahmad, B. F. Johnston, S. P. Mackay, A. G. Schatzlein, P. Gellert, D. Sengupta and I. F. Uchegbu, *J. R. Soc. Interface*, 2010, **7** Suppl 4, S423.

124. J. Moger, N. Garret, A. Lallatsa, M. de la Fuente, A. Schatzlein and I. F. Uchegbu, unpublished data, 2011.

125. E. V. Batrakova, S. Li, V. Y. Alakhov, D. W. Miller and A. V. Kabanov, *J. Pharmacol. Exp. Ther.*, 2003, **304**, 2, 845.

126. T. O. Price, S. A. Farr, X. Yi, S. Vinogradov, E. Batrakova, W. A. Banks and A. V. Kabanov, *J. Pharmacol. Exp. Ther.*, 2010, **333**, 1, 253.

127. F. Herve, N. Ghinea and J. M. Scherrmann, *AAPS J.*, 2008, **10**, 3, 455.

128. W. M. Pardridge, J. L. Buciak, Y. S. Kang and R. J. Boado, *J. Clin. Invest.*, 1993, **92**, 5, 2224.

129. A. K. Kumagai, J. B. Eisenberg and W. M. Pardridge, *J. Biol. Chem.*, 1987, **262**, 31, 15214.

130. M. Thole, S. Nobmann, J. Huwyler, A. Bartmann and G. Fricker, *J. Drug Targeting*, 2002, 10, 4, 337.

131. W. Lu, Y. Zhang, Y. Z. Tan, K. L. Hu, X. G. Jiang, S. K. Fu, J. Controlled Release, 2005, **107**, 3, 428.

132. Z. H. Wang, Z. Y. Wang, C. S. Sun, C. Y. Wang, T. Y. Jiang and S. L. Wang, *Biomaterials*, 2010, **31**, 5, 908.

133. S. Wang, T. Jiang, M. Ma, Y. Hu and J. Zhang, *Int. J. Pharm.*, 2010, **386**, 1–2, 249.

134. J. X. Wang, X. Sun and Z. R. Zhang, *Eur. J. Pharm. Biopharm.*, 2002, **54**, 3, 285.

135. M. L. Bondi, E. F. Craparo, G. Giammona and F. Drago, *Nanomedicine*, 2010, **5**, 1, 25.

136. L. Fenart, A. Casanova, B. Dehouck, C. Duhem, S. Slupek, R. Cecchelli and D. Betbeder, *J. Pharmacol. Exp. Ther.*, 1999, **291**, 3, 1017.

137. M. M. Patel, B. R. Goyal, S. V. Bhadada, J. S. Bhatt and A. F. Amin, *CNS Drugs*, 2009, **23**, 1, 35.

138. J. Lichota, T. Skjorringe, L. B. Thomsen and T. Moos, *J. Neurochem.*, 2010, **113**, 1, 1.

139. D. D. Allen, P. R. Lockman, K. E. Roder, L. P. Dwoskin and P. A. Crooks, *J. Pharmacol. Exp. Ther.*, 2003, **304**, 3, 1268.

140. W. M. Pardridge, *Mol. Interventions*, 2003, **3**, 2, 90.

141. Q. R. Smith, S. Momma, M. Aoyagi and S. I. Rapoport, *J. Neurochem.*, 1987, **49**, 5, 1651.

142. H. Uchino, Y. Kanai, D. K. Kim, M. F. Wempe, A. Chairoungdua, E. Morimoto, M. W. Anders and H. Endou, *Mol. Pharmacol.*, 2002, **61**, 4, 729.

143. E. M. Cornford, D. Young, J. W. Paxton, G. J. Finlay, W. R. Wilson and W. M. Pardridge, *Cancer Res.*, 1992, **52**, 1, 138.

144. Q. R. Smith, ed. A. G. Boer, *Int. Congr. Series*, 2005, p. 63.

145. C. Dufes, F. Gaillard, I. F. Uchegbu, A. G. Schätzlein, J. C. Olivier and J. M. Muller, *Int. J. Pharm.*, 2004, **285**, 1–2, 77.

146. S. Ghosh, N. Das, A. K. Mandal, S. R. Dungdung and S. Sarkar, *Neuroscience*, 2010, **171**, 4, 1287.

147. S. Rothenberger, M. R. Food, R. Gabathuler, M. L. Kennard, T. Yamada, O. Yasuhara, P. L. McGeer and W. A. Jefferies, *Brain Res.*, 1996, **712**, 1, 117.
148. W. Stockinger, E. Hengstschlager-Ottnad, S. Novak, A. Matus, M. Huttinger, J. Bauer, H. Lassmann, W. J. Schneider and J. Nimpf, *J. Biol. Chem.*, 1998, **273**, 48, 32213.
149. W. S. Prince, L. M. McCormick, D. J. Wendt, P. A. Fitzpatrick, K. L. Schwartz, A. I. Aguilera, V. Koppaka, T. M. Christianson, M. C. Vellard, N. Pavloff, J. F. Lemontt, M. Qin, C. M. Starr, G. Bu and T. C. Zankel, *J. Biol. Chem.*, 2004, **279**, 33, 35037.
150. A. G. de Boer, P. J. Gaillard, *Annu. Rev. Pharmacol. Toxicol.*, 2007, **47**, 323.
151. L. Kuhn, H. Schulman and P. Ponka, in *Iron Transport and Storage*, ed. P. Ponka, H. Schulman and R. C. Woodworth, CRC Press, Boca Raton, 1990, p. 149.
152. J. Harford, T. A. Rouault and R. D. Klausner, in *Iron Metabolism in Health and Disease*, ed. J. H. Brock, J. W. Halliday, M. J. Pippard and L. W. Powell, Saunders, London, 1994, p. 123.
153. N. Omori, K. Maruyama, G. Jin, F. Li, S. J. Wang, Y. Hamakawa, K. Sato, I. Nagano, M. Shoji and K. Abe, *Neurol. Res.*, 2003, **25**, 3, 275.
154. K. Mishima, S. Higashiyama, A. Asai, K. Yamaoka, Y. Nagashima, N. Taniguchi, C. Kitanaka, T. Kirino and Y. Kuchino, *Acta Neuropathol.*, 1998, **96**, 4, 322.
155. Y. Zhang and W. M. Pardridge, *Brain Res.*, 2001, **889**, 1–2, 49.
156. Z. M. Qian, H. Li, H. Sun and K. Ho, *Pharmacol. Rev.*, 2002, **54**, 4, 561.
157. M. R. Food, S. Rothenberger, R. Gabathuler, I. D. Haidl, G. Reid and W. A. Jefferies, *J. Biol. Chem.*, 1994, **269**, 4, 3034.
158. M. Demeule, J. Poirier, J. Jodoin, Y. Bertrand, R. R. Desrosiers, C. Dagenais, T. Nguyen, J. Lanthier, R. Gabathuler, M. Kennard, W. A. Jefferies, D. Karkan, S. Tsai, L. Fenart, R. Cecchelli and R. Beliveau, *J. Neurochem.*, 2002, **83**, 4, 924.
159. D. Karkan, C. Pfeifer, T. Z. Vitalis, G. Arthur, M. Ujiie, Q. Chen, S. Tsai, G. Koliatis, R. Gabathuler and W. A. Jefferies, *PLoS One*, 2008, **3**, 6, e2469.
160. R. Gabathuler, G. Arthur, M. Kennard, Q. Chen, S. Tsai, J. Yang, W. Schoorl, T. Z. Vitalis and W. A. Jefferies, *Int. Congress Series*, 2005, **1277**, 171.
161. R. Huang, W. Ke, L. Han, Y. Liu, K. Shao, L. Ye, J. Lou, C. Jiang and Y. Pei, *J. Cereb. Blood Flow Metab.*, 2009, **29**, 12, 1914.
162. J. Kreuter, D. Shamenkov, V. Petrov, P. Ramge, K. Cychutek, C. Koch-Brandt and R. Alyautdin, *J. Drug Targeting*, 2002, **10**, 4, 317.
163. H. R. Kim, K. Andrieux, S. Gil, M. Taverna, H. Chacun, D. Desmaële, F. Taran, D. Georgin and P. Couvreur, *Biomacromolecules*, 2007, **8**, 3, 793.
164. J. Kreuter, *J. Nanosci. Nanotechnol.*, 2004, **4**, 5, 484.

165. T. M. Göppert and R. H. Müller, *J. Drug Targeting*, 2005, **13**, 3, 179.
166. B. Petri, A. Bootz, A. Khalansky, T. Hekmatara, R. Müller, R. Uhl, J. Kreuter and S. Gelperina, *J. Controlled Release*, 2007, **117**, 1, 51.
167. W. Norde, *Adv. Colloid Interface Sci.*, 1986, **25**, C, 267.
168. K. Hu, J. Li, Y. Shen, W. Lu, X. Gao, Q. Zhang and X. Jiang, *J. Controlled Release*, 2009, **134**, 1, 55.
169. Z. Pang, L. Feng, R. Hua, J. Chen, H. Gao, S. Pan, X. Jiang and P. Zhang, *Mol. Pharm.*, 2010, **7**, 6, 1995.
170. R. Q. Huang, Y. H. Qu, W. L. Ke, J. H. Zhu, Y. Y. Pei and C. Jiang, *FASEB J.*, 2007, **21**, 4, 1117.
171. K. Ulbrich, T. Hekmatara, E. Herbert and J. Kreuter, *Eur. J. Pharm. Biopharm.*, 2009, **71**, 2, 251.
172. M. T. Girão da Cruz, A. L. C. Cardoso, L. P. de Almeida, S. Simões and M. C. Pedroso de Lima, *Gene Ther.*, 2005, **12**, 16, 1242.
173. M. Demeule, A. Regina, C. Ché, J. Poirier, T. Nguyen, R. Gabathuler, J. P. Castaigne and R. Béliveau, *J. Pharmacol. Exp. Ther.*, 2008, **324**, 3, 1064.
174. R. Pasqualini and E. Ruoslahti, *Nature*, 1996, **380**, 6572, 364.
175. I. van Rooy, S. Cakir-Tascioglu, P. O. Couraud, I. A. Romero, B. Weksler, G. Storm, W. E. Hennink, R. M. Schiffelers and E. Mastrobattista, *Pharm. Res.*, 2010, **27**, 4, 673.
176. Y. J. Yu, Y. Zhang, M. Kenrick, K. Hoyte, W. Luk, Y. Lu, J. Atwal, J. M. Elliott, S. Prabhu, R. J. Watts and M. S. Dennis, *Sci. Transl. Med.*, 2011, **3**, 84, 84ra44.
177. Y. Zhang, Y. F. Zhang, J. Bryant, A. Charles, R. J. Boado and W. M. Pardridge, *Clin. Cancer Res.*, 2004, **10**, 11, 3667.
178. C. Chu, Y. Zhang, R. J. Boado and W. M. Pardridge, *Pharm. Res.*, 2006, **23**, 7, 1586.
179. C. F. Xia, R. J. Boado and W. M. Pardridge, *Mol. Pharm.*, 2009, **6**, 3, 747.
180. R. J. Boado, E. K. W. Hui, J. Zhiqiang Lu and W. M. Pardridge, *J. Pharmacol. Exp. Ther.*, 2010, **333**, 3, 961.
181. W. M. Pardridge and R. J. Boado, *Pharm. Res.*, 2009, **26**, 10, 2227.
182. W. M. Pardridge, *J. Controlled Release*, 2007, **122**, 3, 345.
183. Y. Zhang and W. M. Pardridge, *Pharm. Res.*, 2009, **26**, 5, 1059.
184. Y. Aktaş, M. Yemisci, K. Andrieux, R. N. Gürsoy, M. J. Alonso, E. Fernandez-Megia, R. Novoa-Carballal, E. Quiñoá, R. Riguera, M. F. Sargon, H. H. Çelik, A. S. Demir, A. A. Hincal, T. Dalkara and Y. Çapan, P. Couvreur, *Bioconjug. Chem.*, 2005, **16**, 6, 1503.
185. Z. Pang, W. Lu, H. Gao, K. Hu, J. Chen, C. Zhang, X. Gao, X. Jiang and C. Zhu, *J. Controlled Release*, 2008, **128**, 2, 120.
186. S. Gosk, C. Vermehren, G. Storm and T. Moos, *J. Cereb. Blood Flow Metabol.*, 2004, **24**, 11, 1193.
187. A. Régina, M. Demeule, C. Ché, I. Lavallée, J. Poirier, R. Gabathuler, R. Béliveau and J. P. Castaigne, *Br. J. Pharmacol.*, 2008, **155**, 2, 185.

188. F. C. Thomas, K. Taskar, V. Rudraraju, S. Goda, H. R. Thorsheim, J. A. Gaasch, R. K. Mittapalli, D. Palmieri, P. S. Steeg, P. R. Lockman and Q. R. Smith, *Pharm. Res.*, 2009, **26**, 11, 2486.

189. M. Demeule, R. Gabathuler, G. Yang, C. Che, A. Regina, A. Abulrob, H. Sartelet, D. Stanimirovic, R. Beliveau, J.-P. Castaigne, Abstract #3784: Development of a new peptide vector technology, Angiopep, for the transport of therapeutics to the brain, in *AACR Annual Meeting*, 2009. Denver.

190. C. Ché, G. Yang, C. Thiot, M. C. Lacoste, J. C. Currie, M. Demeule, A. Régina, R. Béliveau and J. P. Castaigne, *J. Med. Chem.*, 2010, 53, 7, 2814.

191. H. Xin, X. Jiang, J. Gu, X. Sha, L. Chen, K. Law, Y. Chen, X. Wang, Y. Jiang and X. Fang, *Biomaterials*, 2011, **32**, 4293.

192. P. J. Gaillard, A. Brink and A. G. de Boer, *Int. Congress Series*, 2005, **1277**, 185.

193. P. J. Gaillard and A. G. de Boer, *J. Controlled Release*, 2006, **116**, 2, e60.

194. I. Van Rooy, E. Mastrobattista, G. Storm, W. E. Hennink and R. M. Schiffelers, *J. Controlled Release*, 2011, **150**, 1, 30.

195. K. T. Ulbrich, T. Knobloch and J. Kreuter, *J. Drug Targeting*, 2011, **19**, 2, 125.

196. A. Béduneau, F. Hindré, A. Clavreul, J. C. Leroux, P. Saulnier and J. P. Benoit, *J. Controlled Release*, 2008, **126**, 1, 44.

197. Y. Xie, L. Ye, X. Zhang, W. Cui, J. Lou, T. Nagai and X. Hou, *J. Controlled Release*, 2005, **105**, 1–2, 106.

198. G. Tosi, A. V. Vergoni, B. Ruozi, L. Bondioli, L. Badiali, F. Rivasi, L. Costantino, F. Forni and M. A. Vandelli, *J. Controlled Release*, 2010, **145**, 1, 49.

199. A. Zensi, D. Begley, C. Pontikis, C. Legros, L. Mihoreanu, C. Buchel and J. Kreuter, *J. Drug. Targeting*, 2009, **18**, 10, 842.

Section 8
Overcoming Biological Barriers with Parenteral Nanomedicines

CHAPTER 8.1

Overcoming Biological Barriers with Parenteral Nanomedicines: Physiological and Mechanistic Issues

LIN ZHU, SARA MOVASSAGHIAN AND
VLADIMIR P. TORCHILIN*

Mugar Life Science Building, Department of Pharmaceutical Sciences and
Center for Pharmaceutical Biotechnology and Nanomedicine, Northeastern
University, 360 Huntington Ave, Boston, MA, 02115, USA
*E-mail: v.torchilin@neu.edu

8.1.1 Introduction

Nanotechnology has brought deep impacts and new possibilities to our lives, particularly, in the area of drug development and its application. Like other materials, a medicine's physical and chemical properties as well as its biological properties can be altered dramatically at the nanoscopic scale. To construct nano-structured medicines or nanomedicines, nano-sized pharmaceutical carriers are indispensable and include liposomes, micelles, nanoparticles, dendrimers, inorganic carriers and bioconjugates.[1-3] These drug carriers or delivery systems can be separated into two categories based on their structure: particulate and macromolecular delivery systems. Small or macromolecular drug molecules can be loaded into nano-sized drug delivery systems by adsorption, electrostatic interaction, or covalent binding. Commonly, the

RSC Drug Discovery Series No. 22
Nanostructured Biomaterials for Overcoming Biological Barriers
Edited by Maria Jose Alonso and Noemi S. Csaba
© The Royal Society of Chemistry 2012
Published by the Royal Society of Chemistry, www.rsc.org

applications and therapeutic outcomes of a nanomedicine's use are determined by the properties of the drug carrier, while the properties of drug molecules are negligible since they are small in quantity and can be isolated or surrounded by carrier matrices.[4] In addition to their small size, nanocarriers may impart superior properties to therapeutic agents compared to the drug molecules themselves, since the architecture and components of nanocarriers provide extra opportunities for engineering and modification, especially of their surface.[5] Surface modification of nanocarriers is usually made to control their biological properties in a desirable fashion and make them perform various therapeutic or diagnostic functions simultaneously, resulting in the increased stability, blood circulation half-life, bioavailability, targetability, as well as the minimization of non-specific protein binding, immunogenicity, undesirable distribution, and other side-effects.[5] Because of these many combined advantages, the design and application of nanomedicines has grown dramatically during the last few decades. Among these many studies, parenteral administration has been the major delivery route for pharmaceutical, therapeutic and diagnostic agents using nanocarriers. However, to achieve efficacious clinical outcomes, the barriers on the pathway to the target sites, such as endothelium, cell membrane and digesting enzymes have to be overcome. In this chapter, we will discuss the physiological barriers, the delivery strategies and mechanisms and the applications of the most commonly used parenteral nanomedicines.

8.1.2 Biological Barriers for Parenteral Nanomedicines

Successful delivery of exogenous molecules into biological systems that have developed natural barriers and adopted a defense system during evolution can be extremely difficult to achieve with the unmodified drug molecules themselves. Although the barriers for parenteral medications, skin and its underlying associated tissues, are readily breached by needles, there are several obstacles to be overcome before a nanomedicine reaches its desired site and has an action. These include inactivation during blood's circulation, capture by the reticuloendothelial system (RES, also known as mononuclear phagocyte system (MPS)), glomerular filtration and excretion, and cellular and subcellular barriers (Figure 8.1.1).

8.1.2.1 Biological Membranes and Matrices

First of all, nanomedicines must cross the biological membranes and matrices from administration sites to intracellular targets, including extracellular, cellular and intracellular biological membranes and matrices. Following systemic administration, medicines go sequentially across the endothelium into the vascular system, intercellular tissue junctions, and plasma membrane of the target cells, followed by escape from the endosome and entry of the subcellular organelles.[3] For other parenteral administration routes of

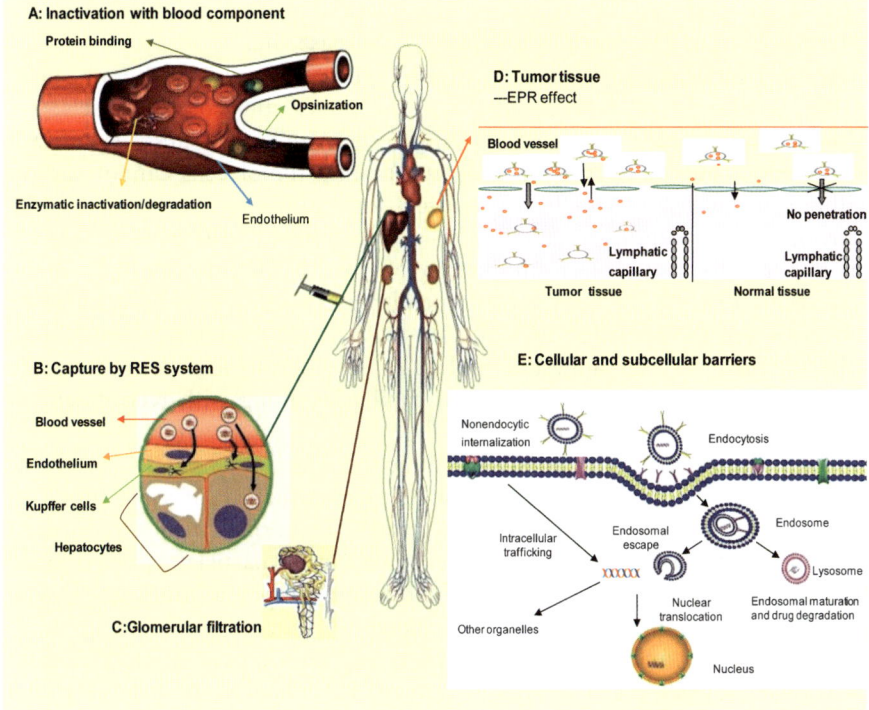

Figure 8.1.1 Biological barriers to parenteral nanomedicines.

injections, such as subcutaneous, intratumoral and intramuscular, except for local therapeutic effects, nanomedicines may travel beyond the extracellular matrices to the vascular system and their efficacy may be lost. However, vaccination is different. Nano-sized vaccines can induce a strong systemic immune response and less toxicity when administered intramuscularly or subcutaneously.[6] If target sites are located in the central nervous system (CNS), the tight junctions between the endothelial cells of CNS blood vessels, the so called blood brain barrier (BBB) must be breached (see Section 7).

On the way to the target, the first barrier of a systemically administered nanomedicine is the endothelium of the blood vessels. It plays an important role in the transport of molecules between the blood and the surrounding tissues, in the recruitment of neutrophils, lymphocytes and monocytes, and in participation in neovascularization (Figure 8.1.1A). Under normal healthy conditions, the endothelium is semipermeable only for small water soluble molecules and is typically impermeable to nanomedicines *via* simple diffusion. However, in certain pathological conditions, for example, with inflammatory processes or in tumors, endothelial cells (ECs) activated by proinflammatory cytokines produce cell adhesion molecules, chemokines, and cytokines, resulting in an increased local permeability and recruitment of leukocytes.

Such high permeability enhances the possibility for nanomedicines to extravasate from the vascular system to the disease site.

Another physiological change in chronic inflammatory diseases and cancer is the generation of new capillaries from pre-existing blood vessels, *i.e.* neovascularization or angiogenesis.[7] Angiogenesis in tumor sites is significant since uncontrolled overgrowing and proliferating tumor cells cannot survive without a sufficient supply of oxygen and nutrients provided by the neovasculature. Commonly, angiogenesis is prompted by the hypoxic stressor on tumor cells that trigger production of growth factors, resulting in the activation of ECs.[8] The basal membrane of pre-existing vessels is degraded by proteases and the activated ECs migrate towards the angiogenic stimuli and, with the help of cell adhesion molecules, proliferate and form tube-like structures that mature into functional capillaries along with recruitment of additional necessary proteins and cells.[8]

It has been reported that most solid tumors contain a rich highly permeable vascular system. Such an altered vascular system ensures a sufficient supply of nutrients and oxygen for rapid tumor growth. In normal tissues, the leaked cells, blood components, and exogenous molecules or particles can be drained and cleared by the lymphatic system. However, lymphatic drainage is weak and not sufficient to remove the leakage in tumor tissues, leading to accumulation of macromolecules or particles. This unique anatomical and pathophysiological phenomenon is termed an enhanced permeability and retention (EPR) effect (Figure 8.1.1D), which has been extensively explored in diagnostic and therapeutic applications to tumors[9] as well as in inflammatory and infarcted tissues.[9,10] For drug delivery, macromolecules ($> 40\ 000$ Da) or particles including liposomes, micelles and other nanoparticles excluded from loss by glomerular filtration with a prolonged systemic circulation time, can extravasate through the blood vessels and accumulate in tumor tissues.[11] However, the architecture of large tumors is heterogeneous. The highly vascularized tissue that shows an EPR effect is commonly located at the outer part of a tumor, while the central area is typically hypovascular and contains more necrotic tissues, resulting in a diminished EPR effect. To effectively deliver drugs into larger tumors, macromolecular drugs can be administered by hydrodynamic injection under hypertensive conditions to improve their intratumoral distribution.[12] Since NO and NO-releasing compounds such as nitroglycerin were reported to increase the extravasation of nanoparticles, these compounds have also been used to improve the homogeneity of drug distribution inside large tumors, resulting in an efficient EPR effect.[12]

After escape from vessels, nanomedicines travel through various extracellular matrix (ECM) containing cells and connective tissues, before reaching target cells. The ECM, composed mainly of proteins and glycosaminoglycans, supports and segregates the surrounded resident cells and provides them a favorable microenvironment. Collagens are the most abundant ECM proteins in most animals. In normal conditions, the ECM may not be a serious barrier for drug delivery. The collagen content in the ECM is low enough so that nanomedicines can diffuse through ECM at a considerable rate. However, in fibrotic diseases, for example, liver fibrosis, abnormal synthesis overwhelms the degradation of collagen, leading to the accumulation of fibrillar collagens type I and III in the ECM, and ECM remodeling.[13,14] In this condition, excessive tight connective ECM may be a considerable barrier for nanomedicines because of the increased distance from capillaries to target cells and the loss of sinusoidal fenestrae or interstitial space.[15] These pathological changes suggest that large particulate nanomedicines may not be suitable for drug delivery to target cells in fibrotic tissues, although such large particles accumulate predominantly in the liver.

The third set of barriers includes the plasma membrane and intracellular barriers of the target cells (Figure 8.1.1E). The cell membrane is a semipermeable phospholipid bilayer embedded with functional proteins and that allows only specific types of molecular transport. These molecules commute between the cells and extracellular fluid *via* simple diffusion, pore transport, membrane fusion, or endocytic processes (pinocytosis, phagocytosis, or endocytosis). For nanomedicines, however, simple diffusion and pore transport are limited since nanomedicines are relatively large. The internalization mechanisms for nanomedicines vary and depend on their surface properties and the types of cells involved. Commonly, viruses or virus-like particles can be taken up *via* membrane fusion since they have a similar lipid bilayer membrane. Soluble polymeric delivery systems may be taken up *via* pinocytosis while the large molecules and particulate nanomedicines may undergo pinocytosis, endocytosis or phagocytosis. In fact, the cellular uptake process is sophisticated enough to involve multiple internalization pathways. In general, the endocytic process starts at a cell membrane indentation triggered by external stimuli followed by the formation of endocytic vesicles with coating proteins, such as caveolae. The endocytic vesicles engulfing external substances fuse with some organelles to form endosomes that release some of the membrane components and receptor proteins for recycling. The endosome is an acidic organelle with a pH of about 5–6. Following recruitment of digesting enzymes, the lysosome forms and the contained foreign substances undergo digestion. Effective nanomedicines must escape this processing and enter the cytoplasm before digestion. Methods aimed at enhanced endosomal/lysosomal escape have been designed to improve the intracellular drug delivery and efficacy.[16–18]

Upon endosomal escape, some medicines have their biological action in cytoplasmic organelles, while others, especially genetic materials such as plasmid DNAs or antisense oligonucleotides (ODNs), are translocated into the nucleus to initiate the transcription process. For eukaryotic cells, most of the genetic materials of the host are restricted to the nucleus enclosed by its nuclear envelope. This nuclear membrane is the final barrier for transgene expression. The mass exchange between the cytoplasm and nucleus is regulated by nuclear pore complexes (NPCs) presented in the nuclear membrane.[19] Molecules less than 9 nm in diameter can diffuse passively through the membrane *via* inner pores of NPCs. However, many nanomedicines, *e.g.* plasmids or their condensates (lipoplexes or polyplexes) can not diffuse freely through the pores and require an active transport process.[20]

8.1.2.2   The Immune System

Besides biological membranes and matrices, the body's defense system, consisting of digesting enzymes, immune molecules, and immune cells, efficiently removes exogenous substances from biological fluids (Figure 8.1.1A and B). Protein/peptide and nucleic acid-based nanomedicines are easily degraded by proteinases and nucleases. To improve their stability, strategies which create a surrounding shield to prevent an enzyme's access to these molecules are commonly used.[3] Except for the stability issue, the efficacy of nanomedicines is significantly influenced by recognition and capture by the immune system, especially the RES (Figure 8.1.1B). Phagocytic cells are the major engulfing cells of the RES. They occur mainly in the liver and spleen, and include monocytes in blood, Kupffer cells in liver, and macrophages in spleen and other tissues. The "plain" or unshielded nanomedicines with a weakly biocompatible surface are likely to be considered as foreign substances and opsonized by binding proteins and complement molecules, leading to enhanced binding with scavenger receptors of phagocytic cells. As a result, such medicines are engulfed and rapidly removed from the blood circulation, lose their opportunity to access the target cells and so do not complete their intended effects. It is well known that the foreign aspects of macromolecules, such as a conserved sequence of nucleic acids or the "non-self" structures of proteins, can be recognized as invasive aliens by the innate immune system. These macromolecules are engulfed by the immune cells and activate the complement system, resulting in negligible or null therapeutic effects especially after multiple intravenous injections.[21,22] Other aspects, such as surface charge, also influence the RES's uptake of nanomedicines. Positively charged macromolecules or nanoparticles are likely to attach binding proteins, immune cells, and other negatively charged biological surfaces *via* ionic interaction, resulting in a low bioavailability and shorter plasma half-life. In contrast, the negatively charged and neutral nanomedicines typically have a relatively longer circulation time although some negatively charged particles have been reported to be taken up rapidly by the RES.[23] To mask the charged surface and prevent phagocytic clearance by the RES, hydrophilic polymers with long flexible chain(s), such as PEG, have been used to modify nanomedicines by forming a hydrated polymer-coated surface.

8.1.2.3   Pharmacokinetic (PK) Barriers

Like conventional medicines, the *in vivo* fate of nanomedicines is strongly influenced by their pharmacokinetic properties. As we mentioned earlier, "plain" nanomedicines are readily taken up by the RES, resulting in a short half-life in the blood circulation. The composition and physicochemical properties of nanomedicines have a significant influence on their *in vivo* distribution. Lipophilic molecules or lipid-based formulations accumulate preferentially in the liver *via* uptake by adipocytes. For example, conjugation with cholesterol increases a drug's lipophilicity resulting in increased liver targeting.[24] For water soluble molecules including nucleic acids, proteins, peptides, and some polymers, the kidney plays an important role in their disposition after systemic administration. In general, macromolecules with a molecular weight of less than 40 000 Da are susceptible to clearance by glomerular filtration (Figure 8.1.1C).[11] The Cy3 labeled siRNAs were found to accumulate predominantly in the kidney 20 min post i.v. injection.[25] The macromolecules that are too large to go through the glomerular pore will more likely accumulate in the liver because of its loose endothelial structure and high blood perfusion rate. However, the administration technique also affects a nanomedicine's distribution. The hydrodynamic injection of DNA or RNA, regardless of its molecular size, causes high accumulation in the well-perfused internal organs, *e.g.* the liver.[26,27]

Particle size and surface charge of nanomedicines also has an influence on their *in vivo* behavior. Decreasing a medicine's size to the nanoscale leads to an exponential increase in its surface area, resulting in a higher likelihood of interaction between a nanomedicine and the biological environment[28] that alters the absorption, distribution, metabolism and excretion (ADME) of the nanomedicine. At the cellular level, nanosizing may change the endocytic pathway, efficiency of cellular uptake, and subcellular events following treatement.[29] Generally, particulate delivery systems target the liver *via* the engulfment by Kupffer cells after systemic administration. Whereas, to deliver a drug to hepatocytes requires drug carriers to have a size of less than 100 nm for effective extravasation through the sinusoidal endothelium and the space of Disse, particularly in fibrotic livers.[30] To target tumors, particles smaller than the cutoff size of pores in tumor vessels were found to be especially effective.[31] Positively charged molecules/particles easily target the endothelium and RES.[12,32] Lipoplexes formed by plasmid DNA and cationic lipids undergo aggregation to some extent in the presence of the ions and negatively charged proteins in blood. Their larger size and positive charge result in the rapid clearance and high expression of DNA in first-pass organs, especially in the lung.[33]

However, in pathological conditions, the alteration of PK parameters must be considered for drug delivery. For example, hepatic and/or renal failure may result in abnormally long circulation of the molecules which undergo hepatic and/or renal elimination. In general, to improve their PK properties, "plain" nanomedicines have to be surface-functionalized to shield various moieties and increase their biocompatibility.

8.1.3 Delivery Strategies and Applications of Parenteral Nanomedicines

With these barriers in mind, effective carriers for parenteral nanomedicines are expected to minimize detrimental interactions and enhance the beneficial interactions with these biological barriers. The strategic aim is to protect drug molecules from degradation/inactivation upon administration, prevent undesirable adverse effects, and increase drug bioavailability and targetability. The commonly used nanocarriers, which include liposomes, micelles, nanoparticles, dendrimers, carbon nanotubes, quantum dots, gold nanoparticles, and bioconjugates, are constructed largely of biocompatible or biodegradable and easily made materials such as lipids, polymers, peptides or proteins, and of inorganic materials, *e.g.* carbon and metal elements. However, such "plain" nanomedicines will be taken up mainly by macrophages of the RES but will not target other desired cell types. For example, conventional liposomes have been reported to deliver dichloromethylene diphosphonate, an apoptotic agent, to the liver and spleen *via* the ingestion by macrophages leading to macrophage apoptosis.[34,35] To bypass the phagocytosis and improve their performance, these "plain" nanocarriers can be modified with a variety of moieties to endow them with different surface properties and functionalities (Figure 8.1.2).[5]

8.1.3.1   Blood Circulation Time and Passive Targeting

As is well known, the intrinsic properties of "plain" medicines often induce the innate and adaptive defenses, resulting in a short blood circulation time. A basic and most important strategy to boost effectiveness is to prolong the blood circulation time of nanomedicines, and thus provide more time and opportunity for their cellular interaction. Long circulation of nanomedicines can be achieved *via* surface modification with biocompatible, soluble and hydrophilic polymers with highly flexible chains, such as PEG,[36] polyvinyl alcohol (PVA),[37] poly(vinyl pyrrolidone) (PVP)[38] and *N*-(2-hydroxypropyl) methacrylamide (HPMA).[39] PEGylation is a well known technique that can efficiently protect nanomedicines from enzymatic degradation as well as prolong their blood circulation time *via* so called "steric stabilization". It was shown that covalent linkage of PEG to siRNA significantly increased its stability for up to 16 h compared to 4 h for "naked" siRNA in the presence of 50% FBS.[40] Incorporation of dioleoyl-*N*-(monomethoxy polyethyleneglycol succinyl)-phosphatidylethanolamine (PEG5000-PE) into large unilamellar liposomes resulted in more than a 10-fold increase of their blood circulation half-life in a mouse model.[41] To increase the solubility and biocompatibility of single wall carbon nanotube (SWNT), 1,2-distearoyl-*sn*-glycero-3-phosphoethanolamine-*N*-[amino(polyethylene glycol)-2000] (DSPE-PEG2000) was incorporated *via* noncovalent adsorption using a sonication and centrifugation procedure.[42] These PEGylated SWNTs exhibited relatively long blood circulation times and low uptake by the RES after i.v. injection into mice.[26]

The so-called "stealth" property of PEG is probably due to its ability to increase surface hydrophilicity, shield the surface charge and other active groups (or ligands), and enhance the repulsive forces between nanomedicines and biological surfaces, which prevent nanocarriers from non-specific binding

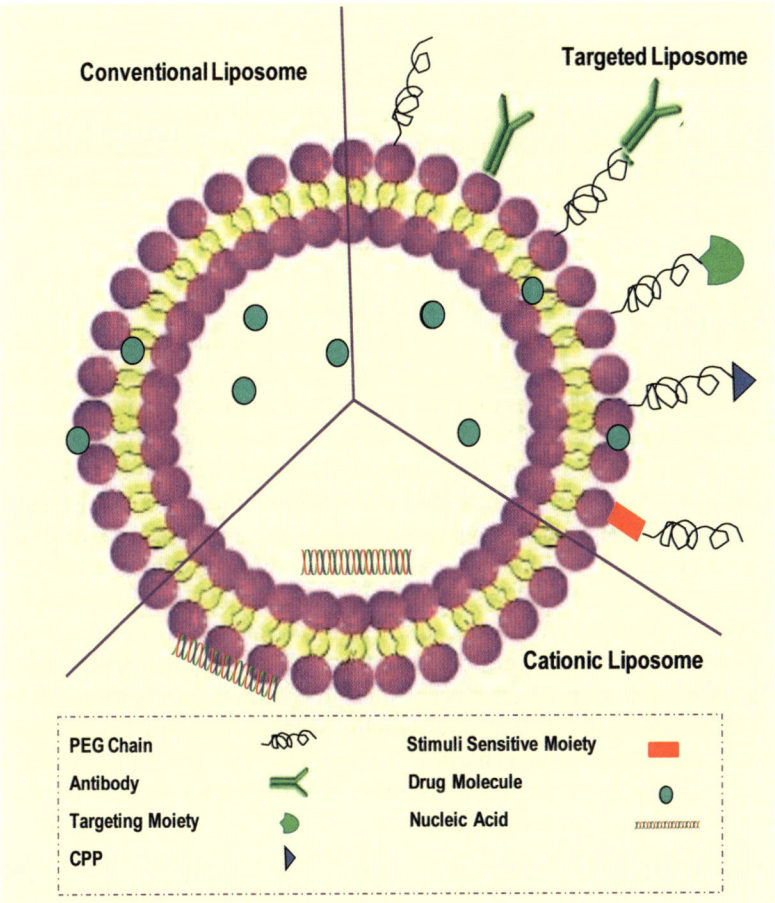

Figure 8.1.2 Delivery strategies for parenteral nanomedicines (liposomes as an example).

with plasma proteins and other blood components such as opsonins and slow their capture by the RES, resulting in prolonged circulation.[43] However, circulation time is determined by not only the surface protection but by the particle size of nanomedicines. In a mouse model, the long circulation time was observed only in small PEG5000-PE modified liposomes with a size of less than 200 nm, while larger liposomes benefited little from PEGylation, with about 35% of injected dose accumulation in the spleen.[44] The RES uptake increased progressively with increase in size of the PEGylated liposomes. PEGylated liposomes of approximately 160–220 nm had the longest circulation time with a minimum liver uptake and moderate spleen uptake in rabbits.[45] The circulation time and the degree of steric hindrance of the liposomes were directly proportional to the chain length of the incorporated PEG-PE (PEG5000-PE > PEG2000-PE > PEG750-PE). However, PEG5000-

PE reduced the activity of other surface functional groups probably due to its long chain and strong shielding effects.[44,46,47]

The rapid renal clearance (by glomerular filtration) of nanomedicines is a cause of a short blood half-life (Figure 8.1.1C). Macromolecules with a size below 40 000 Da readily undergo glomerular filtration and are removed from blood (*e.g.* oligonucleotide and siRNA). PEGylation of oligonucleotides significantly prolonged oligonucleotides' elimination half-life ($t_{1/2\beta}$) from about 35 min to 119 min post i.v. injection in rats due to its steric hindrance, although the molecular weights of the PEGylated nucleic acids were still less than the molecular weight cut-off of the glomerulus.[48,49]

PEGylated nanomedicines, such as long-circulating liposomes (LCL), slowly accumulate in certain tissues and organs, such as tumors or infarcts, due to their optimized particle sizes and the EPR effect in these target tissues (*i.e.* by passive targeting). Doxil® (doxorubicin in long-circulating PEGylated liposomes) has shown less RES uptake, longer circulation time, higher intratumoral drug concentrations, and better therapeutic effects than equivalent doses of non-PEGylated liposome encapsulated doxorubicin or free doxorubicin in a number of animal and human tumor studies.[50] To extravasate from the blood vessel to tumor tissue, the size of nanomedicines should be less than the effective pore size of the "leaky" endothelium. It ranges from 200–600 nm in most peripheral tumors.[31] The liposomes with a diameter of less than 400 nm readily penetrated the tumor interstitium, whereas liposomes of more than 600 nm were excluded from the extravascular space in a tumor xenograft mouse model.[31] However, for macromolecules with molecular weight from 25 000 to 160 000 Da, the tumor vessels are demonstrably less selective than normal vessels, due to the large pores in the vessel wall.[31]

To target cells in fibrotic tissues, a small particle size is preferred since the high level of "scar"-like collagen in the ECM narrows the interstitial space and elongates the drug delivery path to target cells. Compared to particulate delivery systems, the macromolecular delivery systems, *i.e.* drug-polymer conjugates, are smaller and more attractive for drug delivery to the resident cells except macrophages in the fibrotic liver. Small molecules, doxorubicin[51] and PAP19 (an imatinib derivative)[52] have been conjugated to mannose-6-phosphate-modified human serum albumin (M6P-HSA) leading to hepatic stellate cell (HSC) targeting and inhibition of the liver fibrosis induced by bile duct ligation (BDL) post-systemic administration. Galactose- or M6P-modified PEGylated ODN and siRNA, have also been designed to target hepatocytes or HSCs in the fibrotic liver.[49,53]

8.1.3.2 Surface Modification with Targeting Moieties

Although the PEGylation can significantly reduce the irrelevant interactions with the immune system and enhance passive targeting efficiency, the performance of PEGylated medicines can be further improved. To achieve a more specific targeting effect with PEGylated nanocarriers, it is better to further modify them with specific ligands for receptors on the surface of target cells. In the presence of long circulation, such ligands have greater opportunities to specifically recognize and bind to these receptors. So far, a variety of different receptors have been characterized. Various monoclonal antibodies (mAb) bind specifically with antigens on tumor cells, to serve as the tumor targeting ligands. MAb 2C5, a nucleosome-specific monoclonal antibody, recognizing various tumor cells but not normal cells *via* their surface-bound nucleosomes improved drug-loaded pharmaceutical nanocarriers targeting to these various tumor cells.[54,55] Modification with transferrin or the antibodies against transferrin receptor which is overexpressed in many tumor cells, significantly improved the tumor targeting of PEGylated liposomes.[56] Anti-human epidermal growth factor receptor 2 (HER-2) antibodies can bind specifically with HER-2 receptor overexpressed mainly in breast and ovarian cancer, resulting in tumor targeting and significant anticancer activity.[57,58] The proteins or peptides containing the Arg-Gly-Asp (RGD) sequence can be used as the ligands to target integrins in vascular systems.[59]

Except for proteins and peptides, some small molecules are capable of binding with surface receptors of various types of cells, leading to specific cell targeting. Folate-modified liposomes delivered doxorubicin[60] as well as antisense oligonucleotides[61] to various tumor cells *via* binding to the overexpressed folate receptor. Direct conjugation of α-tocopherol (vitamin E)[62] or cholesterol[24] with siRNA significantly increased its lipophilicity and led to high hepatic accumulation. Nanocarriers including particulate and macromolecular systems modified with sugar moieties lead to hepatic cell targeting. Galactosylated liposomes[63] and oligonulceotide conjugates[49] targeted hepatocytes *via* asialoglycoprotein receptors, while mannosylated liposomes delivered drug to non-parenchymal cells, such as macrophages,[64] *via* mannose receptors. *P*-Aminophenyl-6-phospho-α-D-mannopyranoside modified bovine serum albumin (M6P-BSA)[65] or PEGylated siRNA (M6P-PEG-siRNA)[53] deliver their loads specifically to HSCs *via* overexpressed mannose 6-phosphate/insulin like growth factor II (M6P/IGF II) receptor in the fibrotic liver.

To fully expose ligands to cellular receptors, the ligand molecules are commonly attached to nanocarriers *via* a PEG spacer, although these ligands have also been reported to directly attach to the surface of "plain" carriers.[46,58,65] Attachment of ligands to the distal end of a PEG chain may preclude the influence of "steric hindrance" on the ligand–receptor interaction.[47] To perform the surface modification, a variety of terminal functionalized PEG or PEG-PE polymers have been introduced and some have been commercialized. For example, the amphiphilic polymer, *p*-nitrophenylcarbonyl-PEG-PE (pNP-PEG-PE), was designed to react with the amino group of proteins, peptides or other ligands and form a stable, nontoxic carbamate bond.[21,66]

8.1.3.3   Stimuli-sensitive Targeting

It should be kept in mind that the stable PEGylation of nanocarriers may not always enhance drug delivery. At the cellular level, besides possible interference with internalization, PEGylation has been found to interfere with intracellular events, such as endosomal escape,[67] which impairs the expected clinical outcomes associated with both passive and active targeting. Therefore, the protective effect of PEG chains should be avoided by the target cells during internalization and the endocytic processes. It is well known that solid tumors have a lower pH and a higher temperature as well as altered enzyme levels in their microenvironment. Thus, stimulus sensitive nanocarriers have been designed, which respond to these microenvironmental conditions including both internal (*e.g.* pH,[68,69] enzymatic[63,70]) and external (*e.g.* magnetic field,[71,72] temperature[71,73] and light[74]) stimuli resulting in either release of their loaded drugs or enhancement of their cellular uptake.

A variety of ester and hydrazone moieties have been covalently incorporated between the PEG and the nanocarriers to endow pH sensitivity to the carriers (Figure 8.1.3) since these bonds are fairly stable at a pH of about 7.3–7.5 in blood, but easily hydrolyzed at a pH of 6 or below of a tumor mass or endocytic vacuole.[68,69] Therefore, these carriers will remain intact in the blood circulation while the PEG chains are finally detached from the nanocarrier by cleavage of the pH-sensitive bonds within the acidic targets. The pH-sensitive PEG-hydrazone-phosphatidylethanolamine (PEG-Hz-PE)-based micelles and liposomes designed have rapid and significant detachment of the PEG at a pH of 5–6 compared to pH 8.0 (Figure 8.1.4), resulting in a cellular uptake similar to PEG-free liposomes *in vitro* (Figure 8.1.5).[68] Similarly, the macromolecular prodrug with an ester bond between PEG and its load, a galactose-PEG-oligonucleotide conjugate, was fairly stable when incubated with blood serum, but the oligonucleotide was released after incubation with a pH 5.5 buffer.[49] To increase the efficiency of intracellular drug delivery, the disulfide bond has also been widely used as the cleavable linkage in nanocarriers, especially for gene delivery, including disulfide bond-containing lipoplexes[75] and PEGylated siRNA,[40,53] which can be cleaved in the intracellular reductive microenvironment, leading to release of intact nucleic acids.

The alteration in the composition and expression level of local tissue enzymes, such as matrix metalloproteinases (MMPs) during liver fibrosis, provide an opportunity to deliver drug molecules to the pathological sites *via* an enzyme-triggered mechanism.[76] Within the MMP family, overexpressed MMP2 and MMP9 have been widely recognized as involved in the invasion, progression and metastasis of most human tumors.[70] The MMP-sensitive peptide sequence, the octapeptide (Gly-Pro-Leu-Gly-Ile-Ala-Gly-Gln) has been used to synthesize an MMP2-cleavable PEG-PE to endow PEGylated liposomes enzyme sensitivity for the delivery of N^4-octadecyl-1-β-D-arabino-furanosylcytosine (NOAC) to tumor cells.[63] The same sequence was also used to construct the MMP2-sensitive albumin doxorubicin conjugate[70] to deliver doxorubicin to liver cells of the fibrotic liver.

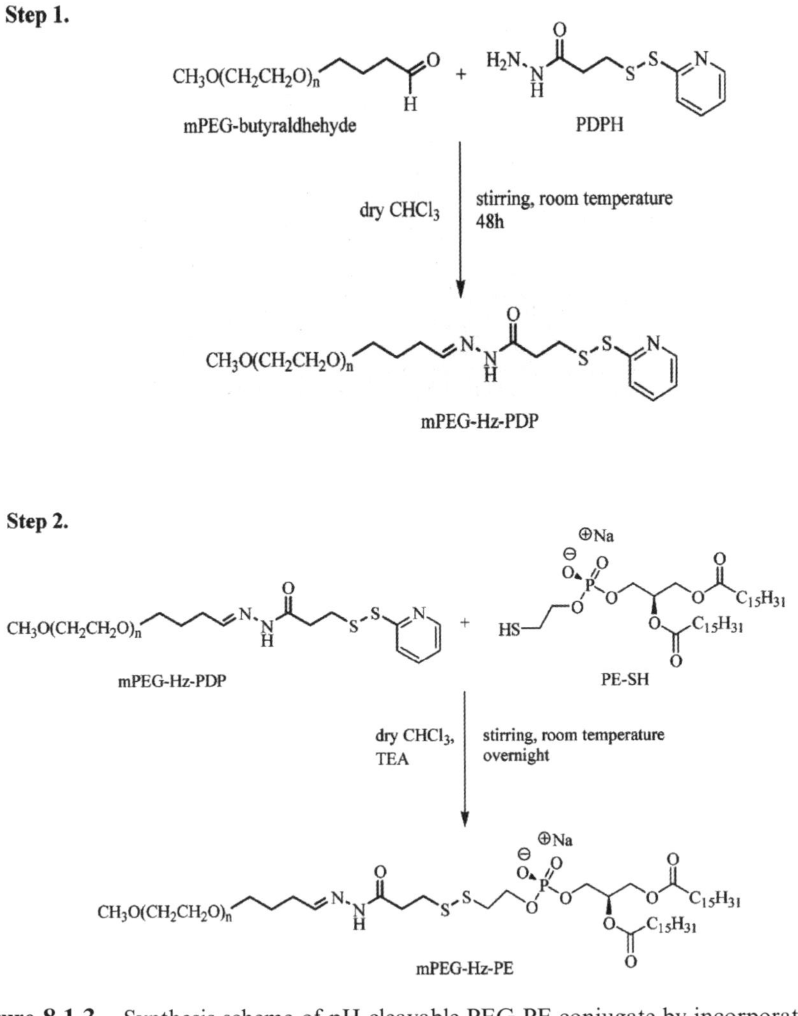

Step 1.

CH₃O(CH₂CH₂O)ₙ — mPEG-butyraldehyde

+ PDPH

dry CHCl₃ | stirring, room temperature 48h

mPEG-Hz-PDP

Step 2.

CH₃O(CH₂CH₂O)ₙ — mPEG-Hz-PDP

+ PE-SH

dry CHCl₃, TEA | stirring, room temperature overnight

mPEG-Hz-PE

Figure 8.1.3 Synthesis scheme of pH-cleavable PEG-PE conjugate by incorporating a pH-sensitive hydrazone bond (from ref. 68).

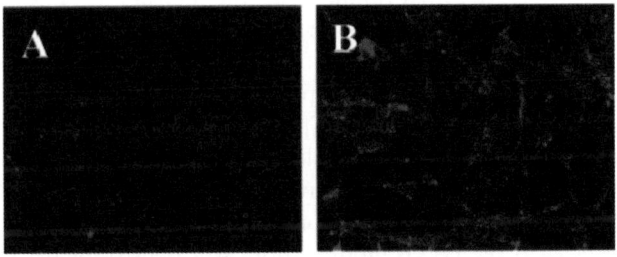

Figure 8.1.4 Internalization of TATp-modified pH-sensitive rhodamine-labeled micelles at pH 8.0 (A) and pH 5.0 (B) for 30 min in NIH 3T3 cells (from ref.68).

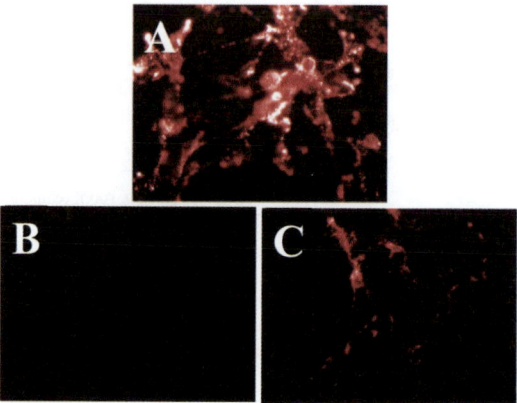

Figure 8.1.5   Internalization of TATp-modified pH-sensitive rhodamine-labeled liposomes in U-87 MG astrocytoma cells. (A) PEG-free liposomes, (B) 18 mol% pH-noncleavable liposomes at pH 7.4, and (C) 18 mol% pH-cleavable lipsomes pretreated with pH 5.0 buffer for 20 min (from ref. 68).

8.1.3.4   Intracellular Delivery and Endosomal Escape

Although the targeting ligands significantly increase the specificity of drug disposition, the efficiency of intracellular delivery of nanomedicines can be further improved to overcome the cellular barriers. Cell-penetrating proteins and peptides (CPPs), such as transactivating transcriptional activator (TAT) proteins and peptides, have been used for intracellular delivery of a broad variety of cargos ranging from small molecules to various particulate nanocarriers.[5] TATp has proven to be efficient for intracellular transduction either *via* electrostatistic interaction and hydrogen bonding when coupling with small molecules or *via* energy-dependent macropinocytosis when coupling with macromolecules or nanoparticles (Figure 8.1.6).[47] However, direct contact between the positively charged protein transduction domain (PTD) of TATp and the negative residues of the cell-surface proteoglycans or glycosaminoglycans is required for internalization.[47,77,78] TATp conjugated with HPMA or HPMA-doxorubincin at the density of one peptide per polymer rapidly and efficiently transported its cargo to both the cytoplasm and nucleus *via* a nonendocytotic, concentration independent process in A2780 human ovarian carcinoma cells.[16] For particulate delivery systems, simply due to their large size, the density of TATp per particle is expected to be higher than that of the macromolecular delivery system to mediate an efficient transduction. Dextran-coated superparamagnetic iron oxide particles with a mean size of about 40 nm were transported into the cytoplasm and nucleus of lymphocytes *in vitro* by TATp modification at a density of 6.7 peptides per particle.[79] To deliver doxorubicin containing unilamellar liposomes with a size of about 100 nm, at least 100 PTD had to be attached on a liposome's surface to trigger efficient cellular uptake.[80] The transduction ability of CPPs provides

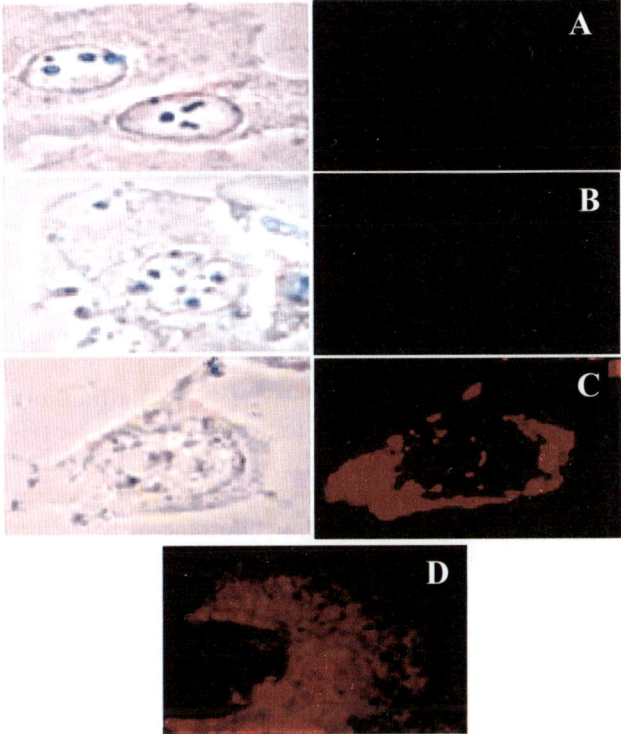

Figure 8.1.6   Bright field (left) and fluorescent (right) microscopy of cell cultures treated with different preparations of rhodamine-PE labeled liposomes ($\times 600$). (A) BT20 cells and LIP-pNP-PEG(3000), (B) BT20 cells and LIP-NGPE-TAT, (C) BT20 cells and LIP-pNP-PEG(3000)-TAT, and (D) mid-point view of confocal microscopy of BT20 cells treated with LIP-pNP-PEG(3000)-TAT. Abbreviation: LIP, liposomes; pNP, *p*-nitrophenylcarbonyl; NGPE, *N*-glutaryl-PE (from ref. 47).

the opportunity for cytoplasmic delivery of therapeutic molecules *via* nonendocytic processes that avoids the major cellular and subcellular obstacles.[77]

Since the cellular uptake of negatively charged macromolecules, especially nucleic acids, is more difficult than that of small molecules, cationic carriers have been developed and widely used to condense and particulate nucleic acids in lipoplexes by cationic liposomes or in polyplexes by cationic polymers *via* electrostatic interaction, to enhance internalization by target cells. Cationic lipids or liposomes are the most commonly used transfection vectors for the delivery of plasmid DNA, ODN and siRNA. Most cationic lipids contain three basic components, including (i) a positively charged hydrophilic head group, such as a mono- or polyamine, which interacts electrostatically with nucleic acids, leading to their condensation, (ii) a hydrophobic domain, such as a fatty acid or cholesterol, which helps in formation of liposomes and exchange of lipids with cell membrane, and (iii) a linker, such as an ester, ether, amide, or carbamate, which binds the head group to the hydrophobic moiety and determines the stability and biodegradability of the cationic lipids.[3] Besides their structures, formulation factors such as the charge ratio between cationic lipids and nucleic acids and the types and content of co-lipids in cationic liposomes significantly influence the transfection efficiency of nucleic acids. The luciferase plasmid could be completely condensed into positively charged lipoplexes with a particle size of about 120nm by pyridinium cationic lipids at a charge ratio (N/P) of 3 : 1 and above, leading to more than a 10-fold increase of gene expression in Chinese Hamster Ovary (CHO) cells compared to the lipoplexes formed at a charge ratio of 1 : 1.[81] L-α-Dioleoylphosphatidylethanolamine (DOPE) and cholesterol are the most commonly used co-lipids in these liposome preparations. The incorporation of fusogenic lipids, such as DOPE, enhances the endosomal escape by its fusion with endosomal lipid bilayers[82,83] *via* the transformation of the lamellar (L_α) DOPE-containing complex to the hexagonal phase (H_{II}).[84] The use of cholesterol as a co-lipid enhances *in vivo* gene expression, probably due to its efficient interaction with the cell membrane in the presence of serum.[85]

In the endocytic pathway, escape from endosome or lysosome plays a key role by its influence on the efficiency of intracellular delivery. Fusogenic peptides including TATp, either conjugated with macromolecules[16] or incorporated into nanocarriers,[17] enhance the endosomal release of internalized nanomedicines, due to pH-dependent destabilization of endosomal membranes.[86] The fusogenic peptide from the herpes simplex virus mediated 100% of membrane fusion at pH 4.5 at a peptide/lipid molar ratio of 0.005 : 1 compared to only 80% of membrane fusion at pH 7.4 at the molar ratio of 0.05 : 1 in the lipid mixing assay. Incorporation of this fusogenic peptide into lipoplexes increased transgene expression up to 30-fold in human cell lines.[17] The nuclear localization capability of some fusogenic peptides has also been observed in gene delivery studies.[17,87] In acidic conditions, protonated polyethyleneimine (PEI) dramatically increased the influx of protons, chloride ions and water into the endosome leading to the endosome collapse and release of its content to cytoplasm, a so-called "proton sponge" effect, when it was incorporated in the gene delivery carriers.[18] Besides the individual use of PEI for a gene delivery carrier, conjugation of PEI with other polymers, such as cholesterol, also showed significant endosomal release and transgene expression of loaded plasmids.[88,89]

It has been estimated that less than 1/1000 of a naked plasmid DNA dose can translocate to the nucleus. With PEI this number can increase to 1/100 (which is still pretty low).[90] More enhanced nuclear transport of gene medicines can be achieved by the effective interaction between NPCs and the specific protein sequence of so-called nuclear localization signals (NLS). NLS peptides are commonly short basic amino acid sequences. For instance, mixing plasmid with a peptide containing the NLS sequence, PKKKRKV, resulted in nuclear uptake of plasmid up to 100-fold faster than that of a reversed NLS sequence upon injection of the plasmid/NLS complexes into the cytoplasm of zebrafish embryos.[91] Conjugation of a single NLS, PKKKRKVEDPYC, to a luciferase plasmid increased luciferase expression by 10 to 1000-fold irrespective of cationic carrier and cell type.[92] However, multi-NLS peptides per plasmid inhibited nuclear transport if the nucleic acid was longer than the distance separating adjacent pores.[92]

8.1.3.5 Multifunctional Nanomedicines

Although the use of individual functionalities has already benefited drug delivery, a better clinical outcome should be expected if these functionalities are combined together reasonably and coordinatedly in an optimal fashion for maximum targeted drug delivery.[5] PEGylation of immunoliposomes improved tissue targeting compared to either PEGylated liposomes or immunoliposomes.[54,55,93] Thermosensitive magnetoliposomes (TMs) have shown enhanced magnetic targeting and temperature-sensitive drug release, leading to significant improvement of the drug's PK profile.[71,94] Recently, a drug delivery platform, which possesses multifunctionalities, including longevity, stimuli sensitivity, targetability, intracellular penetration, and visualization, *i.e.* a multifunctional pharmaceutical nanocarrier, has been proposed and shown potential in many therapeutic and diagnostic applications.[5] Liposomal and micellar drug delivery systems containing PEG chains, a targeting ligand (mAb 2G4), TATp and the pH-cleavable PEG-PE have been designed and tested.[68] After systemic administration of this "smart" multifunctional nanocarrier, several steps become involved in the delivery process, including: (i) escape of the PEGylated nanocarrier from the RES and accumulation in the tumor sites by the EPR effect; (ii) binding of mAb 2G4 on the distal end of a long PEG chain specifically to the nucleosomes on the surface of tumor cells, resulting in the accumulation of nanocarriers around tumor cells; (iii) at acidic extracellular pH in tumor sites, cleavage of the linker between long chain PEG and the nanocarriers, resulting in the exposure of TATp on the distal end of the short PEG chain; and (iv) interaction of TATp with the tumor cell membrane and triggering of the internalization of the nanocarriers. A similar strategy has been introduced into macromolecular delivery systems. PEGylated nucleic acid conjugates containing the targeting ligand, such as galactose and M6P, and a stimulus-sensitive bond, such as a pH-cleavable ester and reducible disulfide bonds, showed increased enzymatic stability, prolonged circulation time, cell-specific accumulation, and stimulus triggered drug release.[49,53]

8.1.4 Conclusion

To design a successful parenteral nanomedicine, the biological barriers, including physical barriers, immune defense systems, and pathological changes, on the path to the target have to be well understood. Unique physicochemical and biological properties make nanomedicines superior to traditional medicines at either penetrating or circumventing biological barrier, and has resulted in enhanced therapeutic effects and minimized side-effects. Surface modification with functional moieties is currently the mostly widely used strategy aimed at further improvement of the performance of nanomedicines in terms of circulation time, targeting effect, specific stimuli response, cellular and intracellular penetration, and visualization. For precise site-specific delivery of nanomedicine, the combined use of these functional moieties, *i.e.* multifunctional nanocarriers, in a coordinated manner is highly recommended.

References

1. V. P. Torchilin, *Nat. Rev. Drug Discov.*, 2005, **4**, 145–160.
2. V. P. Torchilin, *Pharm. Res.*, 2007, **24**, 1–16.
3. L. Zhu and R. I. Mahato, *Expert Opin. Drug Delivery*, 7, 1209–1226.
4. T. M. Allen and P. R. Cullis, *Science*, 2004, **303**, 1818–1822.
5. V. P. Torchilin, *Adv. Drug Delivery Rev.*, 2006, **58**, 1532–1555.
6. A. Brave, K. Ljungberg, B. Wahren and M. A. Liu, *Mol. Pharm.*, 2007, **4**, 18–32.
7. P. Carmeliet, *Nat. Med.*, 2003, **9**, 653–660.
8. R. K. Jain, *Nat. Med.*, 2003, **9**, 685–693.
9. H. Maeda, J. Wu, T. Sawa, Y. Matsumura and K. Hori, *J. Controlled Release*, 2000, **65**, 271–284.
10. A. N. Lukyanov, W. C. Hartner and V. P. Torchilin, *J. Controlled Release*, 2004, **94**, 187–193.
11. Y. Takakura, R. I. Mahato and M. Hashida, *Adv. Drug Delivery Rev.*, 1998, **34**, 93–108.
12. J. Fang, H. Nakamura and H. Maeda, *Adv. Drug Delivery Rev.*, 2011, **63**, 136–151.
13. R. C. Benyon and J. P. Iredale, *Gut*, 2000, **46**, 443–446.
14. M. J. Arthur, *Am. J. Physiol. Gastrointest. Liver Physiol.*, 2000, **279**, G245–249.
15. R. Bataller and D. A. Brenner, *J. Clin. Invest.*, 2005, **115**, 209–218.
16. A. Nori, K. D. Jensen, M. Tijerina, P. Kopeckova and J. Kopecek, *Bioconjug. Chem.*, 2003, **14**, 44–50.
17. Y. Tu and J. S. Kim, *J. Gene Med.*, 2008, **10**, 646–654.
18. O. Boussif, F. Lezoualc'h, M. A. Zanta, M. D. Mergny, D. Scherman, B. Demeneix and J. P. Behr, *Proc. Natl. Acad. Sci. U. S. A.*, 1995, **92**, 7297–7301.
19. E. A. Nigg, *Nature*, 1997, **386**, 779–787.
20. D. S. Goldfarb, J. Gariepy, G. Schoolnik and R. D. Kornberg, *Nature*, 1986, **322**, 641–644.
21. V. P. Torchilin, A. N. Lukyanov, Z. Gao and B. Papahadjopoulos-Sternberg, *Proc. Natl. Acad. Sci. U. S. A.*, 2003, **100**, 6039–6044.
22. A. D. Judge, V. Sood, J. R. Shaw, D. Fang, K. McClintock and I. MacLachlan, *Nat. Biotechnol.*, 2005, **23**, 457–462.

23. S. D. Li and L. Huang, *Mol. Pharm.*, 2008, **5**, 496–504.
24. J. Soutschek, A. Akinc, B. Bramlage, K. Charisse, R. Constien, M. Donoghue, S. Elbashir, A. Geick, P. Hadwiger, J. Harborth, M. John, V. Kesavan, G. Lavine, R. K. Pandey, T. Racie, K. G. Rajeev, I. Rohl, I. Toudjarska, G. Wang, S. Wuschko, D. Bumcrot, V. Koteliansky, S. Limmer, M. Manoharan and H. P. Vornlocher, *Nature*, 2004, **432**, 173–178.
25. A. Santel, M. Aleku, O. Keil, J. Endruschat, V. Esche, G. Fisch, S. Dames, K. Loffler, M. Fechtner, W. Arnold, K. Giese, A. Klippel and J. Kaufmann, *Gene Ther.*, 2006, **13**, 1222–1234.
26. Z. Liu, W. Cai, L. He, N. Nakayama, K. Chen, X. Sun, X. Chen and H. Dai, *Nat. Nanotechnol.*, 2007, **2**, 47–52.
27. T. Suda and D. Liu, *Mol. Ther.*, 2007, **15**, 2063–2069.
28. K. L. Aillon, Y. Xie, N. El-Gendy, C. J. Berkland and M. L. Forrest, *Adv. Drug Delivery Rev.*, 2009, **61**, 457–466.
29. J. Rejman, V. Oberle, I. S. Zuhorn and D. Hoekstra, *Biochem. J.*, 2004, **377**, 159–169.
30. R. I. Mahato, A. Rolland and E. Tomlinson, *Pharm. Res.*, 1997, **14**, 853–859.
31. F. Yuan, M. Dellian, D. Fukumura, M. Leunig, D. A. Berk, V. P. Torchilin and R. K. Jain, *Cancer Res.*, 1995, **55**, 3752–3756.
32. R. B. Campbell, D. Fukumura, E. B. Brown, L. M. Mazzola, Y. Izumi, R. K. Jain, V. P. Torchilin and L. L. Munn, *Cancer Res.*, 2002, **62**, 6831–6836.
33. N. S. Templeton, D. D. Lasic, P. M. Frederik, H. H. Strey, D. D. Roberts and G. N. Pavlakis, *Nat. Biotechnol.*, 1997, **15**, 647–652.
34. M. Naito, H. Nagai, S. Kawano, H. Umezu, H. Zhu, H. Moriyama, T. Yamamoto, H. Takatsuka and Y. Takei, *J. Leukoc. Biol.*, 1996, **60**, 337–344.
35. N. Van Rooijen, *J. Immunol. Methods*, 1989, **124**, 1–6.
36. V. S. Trubetskoy and V. P. Torchilin, *Adv. Drug Delivery Rev.*, 1995, **16**, 311–320.
37. H. Takeuchi, H. Kojima, H. Yamamoto and Y. Kawashima, *J. Controlled Release*, 2000, **68**, 195–205.
38. V. P. Torchilin, T. S. Levchenko, K. R. Whiteman, A. A. Yaroslavov, A. M. Tsatsakis, A. K. Rizos, E. V. Michailova and M. I. Shtilman, *Biomaterials*, 2001, **22**, 3035–3044.
39. R. Satchi-Fainaro, M. Puder, J. W. Davies, H. T. Tran, D. A. Sampson, A. K. Greene, G. Corfas and J. Folkman, *Nat. Med.*, 2004, **10**, 255–261.
40. S. H. Kim, J. H. Jeong, S. H. Lee, S. W. Kim and T. G. Park, *J. Controlled Release*, 2006, **116**, 123–129.
41. A. L. Klibanov, K. Maruyama, V. P. Torchilin and L. Huang, *FEBS Lett.*, 1990, **268**, 235–237.
42. N. W. Kam, M. O'Connell, J. A. Wisdom and H. Dai, *Proc. Natl. Acad. Sci. U. S. A.*, 2005, **102**, 11600–11605.
43. V. P. Torchilin, *AAPS J.*, 2007, **9**, E128–147.
44. A. L. Klibanov, K. Maruyama, A. M. Beckerleg, V. P. Torchilin and L. Huang, *Biochim. Biophys. Acta*, 1991, **1062**, 142–148.

45. V. D. Awasthi, D. Garcia, B. A. Goins and W. T. Phillips, *Int. J. Pharm.*, 2003, **253**, 121–132.
46. A. Mori, A. L. Klibanov, V. P. Torchilin and L. Huang, *FEBS Lett.*, 1991, **284**, 263–266.
47. V. P. Torchilin, R. Rammohan, V. Weissig and T. S. Levchenko, *Proc. Natl. Acad. Sci. U. S. A.*, 2001, **98**, 8786–8791.
48. K. Cheng, Z. Ye, R. V. Guntaka and R. I. Mahato, *Mol. Pharm.*, 2005, **2**, 206–217.
49. L. Zhu, Z. Ye, K. Cheng, D. D. Miller and R. I. Mahato, *Bioconjug. Chem.*, 2008, **19**, 290–298.
50. A. Gabizon and F. Martin, *Drugs*, 1997, **54 Suppl 4**, 15–21.
51. R. Greupink, H. I. Bakker, W. Bouma, C. Reker-Smit, D. K. Meijer, L. Beljaars and K. Poelstra, *J. Pharmacol. Exp. Ther.*, 2006, **317**, 514–521.
52. T. Gonzalo, L. Beljaars, M. van de Bovenkamp, K. Temming, A. M. van Loenen, C. Reker-Smit, D. K. Meijer, M. Lacombe, F. Opdam, G. Keri, L. Orfi, K. Poelstra and R. J. Kok, *J. Pharmacol. Exp. Ther.*, 2007, **321**, 856–865.
53. L. Zhu and R. I. Mahato, *Bioconjug. Chem.*, **21**, 2119–2127.
54. A. N. Lukyanov, T. A. Elbayoumi, A. R. Chakilam and V. P. Torchilin, *J. Controlled Release*, 2004, **100**, 135–144.
55. T. A. ElBayoumi and V. P. Torchilin, *Clin. Cancer Res.*, 2009, **15**, 1973–1980.
56. O. Ishida, K. Maruyama, H. Tanahashi, M. Iwatsuru, K. Sasaki, M. Eriguchi and H. Yanagie, *Pharm. Res.*, 2001, **18**, 1042–1048.
57. J. W. Park, D. B. Kirpotin, K. Hong, R. Shalaby, Y. Shao, U. B. Nielsen, J. D. Marks, D. Papahadjopoulos and C. C. Benz, *J. Controlled Release*, 2001, **74**, 95–113.
58. J. W. Park, K. Hong, D. B. Kirpotin, G. Colbern, R. Shalaby, J. Baselga, Y. Shao, U. B. Nielsen, J. D. Marks, D. Moore, D. Papahadjopoulos and C. C. Benz, *Clin. Cancer Res.*, 2002, **8**, 1172–1181.
59. E. Ruoslahti, *Annu. Rev. Cell Dev. Biol.*, 1996, **12**, 697–715.
60. A. Gabizon, A. T. Horowitz, D. Goren, D. Tzemach, H. Shmeeda and S. Zalipsky, *Clin. Cancer Res.*, 2003, **9**, 6551–6559.
61. C. P. Leamon, S. R. Cooper and G. E. Hardee, *Bioconjug. Chem.*, 2003, **14**, 738–747.
62. K. Nishina, T. Unno, Y. Uno, T. Kubodera, T. Kanouchi, H. Mizusawa and T. Yokota, *Mol. Ther.*, 2008, **16**, 734–740.
63. T. Terada, M. Iwai, S. Kawakami, F. Yamashita and M. Hashida, *J. Controlled Release*, 2006, **111**, 333–342.
64. C. D. Muller and F. Schuber, *Biochim. Biophys. Acta*, 1989, **986**, 97–105.
65. Z. Ye, K. Cheng, R. V. Guntaka and R. I. Mahato, *Biochemistry*, 2005, **44**, 4466–4476.
66. V. P. Torchilin, T. S. Levchenko, A. N. Lukyanov, B. A. Khaw, A. L. Klibanov, R. Rammohan, G. P. Samokhin and K. R. Whiteman, *Biochim. Biophys. Acta*, 2001, **1511**, 397–411.
67. S. Mishra, P. Webster and M. E. Davis, *Eur. J. Cell Biol.*, 2004, **83**, 97–111.

68. R. M. Sawant, J. P. Hurley, S. Salmaso, A. Kale, E. Tolcheva, T. S. Levchenko and V. P. Torchilin, *Bioconjug. Chem.*, 2006, **17**, 943–949.
69. A. A. Kale and V. P. Torchilin, *Bioconjug. Chem.*, 2007, **18**, 363–370.
70. A. M. Mansour, J. Drevs, N. Esser, F. M. Hamada, O. A. Badary, C. Unger, I. Fichtner and F. Kratz, *Cancer Res.*, 2003, **63**, 4062–4066.
71. L. Zhu, Z. Huo, L. Wang, X. Tong, Y. Xiao and K. Ni, *Int. J. Pharm.*, 2009, **370**, 136–143.
72. H. Nobuto, T. Sugita, T. Kubo, S. Shimose, Y. Yasunaga, T. Murakami and M. Ochi, *Int. J. Cancer*, 2004, **109**, 627–635.
73. D. Needham, G. Anyarambhatla, G. Kong and M. W. Dewhirst, *Cancer Res.*, 2000, **60**, 1197–1201.
74. Y. Wan, J. K. Angleson and A. G. Kutateladze, *J. Am. Chem. Soc.*, 2002, **124**, 5610–5611.
75. F. Tang and J. A. Hughes, *Bioconjug. Chem.*, 1999, **10**, 791–796.
76. P. Meers, *Adv. Drug Delivery Rev.*, 2001, **53**, 265–272.
77. V. P. Torchilin, *Adv. Drug Delivery Rev.*, 2008, **60**, 548–558.
78. J. S. Wadia, R. V. Stan and S. F. Dowdy, *Nat. Med.*, 2004, **10**, 310–315.
79. C. L. Kaufman, M. Williams, L. M. Ryle, T. L. Smith, M. Tanner and C. Ho, *Transplantation*, 2003, **76**, 1043–1046.
80. Y. L. Tseng, J. J. Liu and R. L. Hong, *Mol. Pharmacol.*, 2002, **62**, 864–872.
81. L. Zhu, Y. Lu, D. D. Miller and R. I. Mahato, *Bioconjug. Chem.*, 2008, **19**, 2499–2512.
82. D. C. Litzinger and L. Huang, *Biochim. Biophys. Acta*, 1992, **1113**, 201–227.
83. H. Farhood, N. Serbina and L. Huang, *Biochim. Biophys. Acta*, 1995, **1235**, 289–295.
84. J. Smisterova, A. Wagenaar, M. C. Stuart, E. Polushkin, G. ten Brinke, R. Hulst, J. B. Engberts and D. Hoekstra, *J. Biol. Chem.*, 2001, **276**, 47615–47622.
85. K. Crook, B. J. Stevenson, M. Dubouchet and D. J. Porteous, *Gene Ther.*, 1998, **5**, 137–143.
86. E. Mastrobattista, G. A. Koning, L. van Bloois, A. C. Filipe, W. Jiskoot and G. Storm, *J. Biol. Chem.*, 2002, **277**, 27135–27143.
87. N. J. Caron, S. P. Quenneville and J. P. Tremblay, *Biochem. Biophys. Res. Commun.*, 2004, **319**, 12–20.
88. D. A. Wang, A. S. Narang, M. Kotb, A. O. Gaber, D. D. Miller, S. W. Kim and R. I. Mahato, *Biomacromolecules*, 2002, **3**, 1197–1207.
89. S. Han, R. I. Mahato and S. W. Kim, *Bioconjug. Chem.*, 2001, **12**, 337–345.
90. H. Pollard, J. S. Remy, G. Loussouarn, S. Demolombe, J. P. Behr and D. Escande, *J. Biol. Chem.*, 1998, **273**, 7507–7511.
91. D. Gorlich and I. W. Mattaj, *Science*, 1996, **271**, 1513–1518.
92. M. A. Zanta, P. Belguise-Valladier and J. P. Behr, *Proc. Natl. Acad. Sci. U. S. A.*, 1999, **96**, 91–96.
93. V. P. Torchilin, A. L. Klibanov, L. Huang, S. O'Donnell, N. D. Nossiff and B. A. Khaw, *FASEB J.*, 1992, **6**, 2716–2719.
94. E. Viroonchatapan, M. Ueno, H. Sato, I. Adachi, H. Nagae, K. Tazawa and I. Horikoshi, *Pharm. Res.*, 1995, **12**, 1176–1183.

CHAPTER 8.2

Drug Delivery Strategies: Polymer Therapeutics

RICHARD M. ENGLAND, INMACULADA
CONEJOS–SÁNCHEZ AND MARÍA J. VICENT*

Polymer Therapeutics Lab, Centro de Investigación Príncipe, Felipe. Av.
Autopista del Saler 16, E–46012 Valencia, Spain
*E-mail: mjvicent@cipf.es

8.2.1 Introduction

Drug pharmacology, and therefore therapeutic efficacy, in the treatment of disease is heavily influenced by its route of administration into the body. In fact, the same drug can produce different results depending on the way it is administered. As a consequence, several factors must be considered before medication, namely drug properties (*i.e.* solubility, concentration, potency), the disease to be treated and residence time required for a desired therapeutic effect.[1] Whilst oral administration of therapeutics is currently the leading method for administration in chronic diseases mainly due to patient compliance, parenteral delivery is still a major requirement for many in order to achieve the desired pharmacological effect since absorption in the digestive tract can be low or unpredictable. One of its major advantages lies in that with intravenous (IV), subcutaneous (SC) or intramuscular (IM) injection, there is rapid drug distribution in the blood stream. However, for the same reason, some therapeutic agents could be either toxic (due to an acute overdose as a result of a fast response), and/or quickly excreted leading to diminishing patient compliance as repeated injections over prolonged periods of time could

RSC Drug Discovery Series No. 22
Nanostructured Biomaterials for Overcoming Biological Barriers
Edited by Maria Jose Alonso and Noemi S. Csaba
© The Royal Society of Chemistry 2012
Published by the Royal Society of Chemistry, www.rsc.org

then be required. Whilst rapid excretion of therapeutics is a major biological barrier to overcome, there are many other barriers that are associated with parenteral administration, in particular the limited transport to the desired diseased tissue or biological compartments (considering also organelle compartmentalisation at cellular level), a limited extravasation in specific sites (particularly blood brain barrier (BBB)) or the poor stability and even immunogenicity of some kinds of bioactives under certain biological environments. One method to negate some of these issues is to conjugate the therapeutic through a rationally designed covalent bond to a biologically compatible polymer, thus utilising the beneficial properties of polymers as carriers. These nano–sized macromolecular systems are known as 'Polymer Therapeutics'.

8.2.1.1 Polymer Therapeutics. Concept and Current Status

'Polymer Therapeutics' was coined by Prof. Ruth Duncan to define a family of new chemical entities (NCEs) considered the first polymeric nanomedicines.[2] Polymer therapeutics comprise a variety of complex macromolecular systems, their common feature being the presence of a rationally designed covalent chemical bond between a water–soluble polymeric carrier (with or without inherent activity) and the bioactive molecule(s). Drug conjugation to a polymer not only enhances its aqueous solubility but also changes its pharmacokinetics at the whole organism and even subcellular level, with the possibility to clearly enhance drug therapeutic value.[3–5] With a careful selection of the carrier, conjugation can yield long–circulating macromolecular systems that display passive targeting to diseased tissues, such as tumours or inflamed areas, due to the leakiness of angiogenic tumour blood vessels by the enhanced permeability and retention (EPR) effect.[6] It also limits uptake to the endocytic route, favouring lysosomotropic delivery with subsequent transfer of drug out of the endosomal/lysosomal compartment providing the opportunity to bypass mechanisms of drug resistance associated with membrane efflux pumps.[7]

'Polymer Therapeutics' can be subdivided into five general categories: polymeric drugs, polymer–protein conjugates, polyplexes, polymeric micelles and polymer–drug conjugates.[2] These nanosystems hold features such as; a tailored drug loading, an incorporation of drug combinations or intelligent linker design tailored for specific biological conditions within the target organ or cell. Furthermore, the versatility of synthetic chemistry and inclusion of bioresponsive elements provides scope for polymer therapeutics when in comparison with other nanopharmaceutics.[2]

With several products in the market (PEGylated proteins and polymeric drugs) and more than 16 polymer–drug conjugates already in clinical development (mainly as anticancer agents *via* parenteral administration, Table 8.2.1),[5] a second generation of conjugates is now focused on improved polymer structures,[8] polymer–based combination therapy,[9] and novel molecular targets,[10] which will further progress this platform technology.

In parallel, there is a growing effort to improve synthesis and characterisation methodology to better understand the structure—activity relationships for defining safety and efficacy of these new therapeutics to meet with regulatory demand.[11] First examples with polymer conjugates show that the panacea of enabling specific and individualised therapy through nanomedicine is becoming a feasible approach.[12] An exhaustive review on polymer therapeutics is out of the scope of this book chapter; however, considering the exponentially growing interest in the field there is a vast quantity of literature available and highly recommended to read.[5,9,13,14]

This chapter describes some of the various biological barriers that limit the efficacy of therapeutics after parenteral administration and how through competent rational design this platform technology can help to overcome these barriers for selected bioactives in the treatment of various diseases.

8.2.2   Parenteral Delivery of Polymer Therapeutics

The parenteral methods for polymer therapeutics covered in this chapter includes the use of intravenous (IV), subcutaneous (SC), intramuscular (IM), intratumoural (IT) and intraperitoneal (IP), each of which has a direct advantage over the other. For example, SC administration has advantages such that the patient may self–administer the therapeutic (as in the case of insulin for diabetes) and that large portions of the body are available for injection preventing injury in one specific site. One major disadvantage to SC is the irregular or slow absorption of therapeutic. IV is the most recognised in this field and has the advantage of the most rapid uptake of therapeutic and drug bioavailability in comparison to other methods. Downsides to IV include trauma to veins and possible infections at the site of administration. An intermediate between SC and IV is IM, which has a slower onset of action than IV but faster than SC due to the high vascularisation of muscle tissue, but also has the associated downside that it causes pain at the site of injection. IT provides the most direct route when treating solely a solid tumour and can help with diffusion of the therapeutic within the interstitial space and chaotic lymph vessels.

Parenteral administration of polymer therapeutics takes advantage of the many inherent properties of these systems, which confers them with several advantages for treating life threatening diseases, such as cancer or neurodegenerative diseases.

8.2.2.1   Intravenous

Intravenous (IV) is the preferred route of administration when using polymer therapeutics as it profits from the fastest drug distribution in comparison to other parenteral routes explored (*i.e.* SC, IM or IP), and therefore could enhance the benefits of intrinsic targeting mechanisms such as the EPR effect.[15]

Table 8.2.1 Representative examples of polymer therapeutics on the market and in clinical development by parenteral administration route.[a,5,9,14]

Product	Company	Trade Name	Indication	Admin. route	Status
Polymeric drugs/sequestrants					
Copolymer of Glu Ala Tyr	Teva	Copaxone®	Multiple sclerosis	SC	Market
Polymeric micelles					
Pluronics ®-based formulation of doxorubicin	Supratek Pharma Inc.	SP1049C	Cancer–upper GI, NSCLC, colorectal	IV	Phase III
Paclitaxel block copolymer micelle	NanoCarrier Co.–Nippon Kayaku Co.	NK 105	Cancer–stomach (Nanocarrier® technology)	IV	Phase II
Cisplatin block copolymer micelle	NanoCarrier Co.–TOUDAI TLO Ltd	NK 6004 Nanoplatin	Cancer (Nanocarrier® technology)	IV	Phase I/II
Oxaliplatin block copolymer micelle	NanoCarrier Co.–TOUDAI TLO Ltd	NC 4010	Cancer(Nanocarrier® technology)	IV	Phase I
Polymer–protein conjugates					
Styrene maleic anhydride–neocarzinostatin (SMANCS)	Yamanouchi	Zinostatin Stimaler®	Hepatocellular carcinoma	Intrahepatic artery	Market 1990 (Japan)
PEG-adenosine deaminase	Enzon	Adagen®	SCID	IM	Market 1990
PEG-asparaginase	Enzon	Oncospar®	Acute lymphocytic leukaemia	IV / IM	Market 1990

Table 8.2.1 (*Continued*)

Product	Company	Trade Name	Indication	Admin. route	Status
PEG–interferon alfa–2b	Schering	PEGINTRON®	Chronic hepatitis C	SC	Market 2001
PEG–interferon alfa–2a	Roche	PEGASYS®	Chronic hepatitis C	SC	Market 2002
PEG–human G–CSF	Amgem	Neulasta®	Febrile neutropenia	SC	Market 2002
PEG–HGH antagonist	Pfizer	Somavert®	Acromegaly	SC	Market 2003
PEG–antiTNF Fab	UCB Pharma	Cimzia®	Chron's disease Arthritis	SC	Market 2008
PEG–uric acid enzyme	Savient Pharmaceuticals	Kristexxa™ (pegloticase)	Chronic gout	IV	Market 2009 Market 2010
PEG–antiVEGFR2 Fab fragments	UCB Pharma	CDP 791	NSCLC	SC	Phase II
PEG–haemoglobin	Sangart	Hemospan®/MP4OX	Delivery of CO/O_2 in trauma patients	IV	Phase II
Polymer–aptamer					
PEG–aptamer (aptanib)	OSI–Eyetech	Macugen®	AMD	Intravitreous	Market 2004
PEG–anti-platelet–binding function of von Willebrand Factor	Archemix	ARC1779	Thrombotic microangiopathies	IV	Phase II
PEG–anti–PDGF aptamer combination with Lucentis®	Ophthotech	E10030	AMD	Intravitreous	Phase II
Polymer–drug conjugate					
PGA–paclitaxel	Cell Therapeutics Inc.	Xyotax™, OPAXIO™	NSCLC, ovarian and esophageal cancer	IV	Filed EMEA Phase III

Table 8.2.1 (*Continued*)

Product	Company	Trade Name	Indication	Admin. route	Status
HPMA copolymer–Dox	Pfizer	FCE 28068	Breast, lung & colon cancer	IV	Phase II
HPMA copolymer–Dox–galactosamine	Pfizer	FCE 28069	Hepatocellular carcinoma	IV	Phase I/II
Polymer–cyclodextrin nanoparticle–siRNA	Calando Pharmaceutical	CALAA–01	Solid tumours	IV	Phase I
PEG–docetaxel	Nektar	NKTR–105	Solid tumours	IV	Phase I
PHF/Fleximer® – camptothecin	Mersana	XMT–1001	Solid tumours	IV	Phase I
Polymer conjugated–cyclodextrin–camptothecin (Cyclosert® technology assembled into a nanoparticle)	Calando Pharmaceutical	IT–101	Solid tumours	IV	Phase I/II
PEG–irinotecan	Nektar	NKTR–102	Colorectal, breast, ovarian and cervical cancers	IV	Phase II
HPMA copolymer platinate	Access Pharmaceuticals	ProLindac®	Ovarian cancer	IV	Phase II

*a*PGA: poly-L-glutamic acid, PEG: poly(ethyleneglycol), HPMA; *N*-(2-hydroxypropyl)methacrylamide, PHF or Fleximer®: poly(1-hydroxymethylethylene hydroxymethylformal). SCID: severe combined immune deficiency syndrome. AMD: age–related macular degeneration. NSCLC: non–small cell lung cancer. G–CSF: colony stimulating factor. HGH: human growth hormone

Major obstacles to overcome after IV injection include tissue–specific issues, particularly BBB, tumour microenvironment, cell compartmentalisation, hepatic clearance or renal filtration.

8.2.2.1.1   The Blood Brain Barrier (BBB)

Drug delivery to the central nervous system (CNS) by means of a systemic approach is a highly challenging prospect in drug development. As has been mentioned in this book, physiological and enzymatic features of BBB together with its efflux activity play key roles when designing drug delivery platforms to cross the brain endothelium by means of IV route. Polymer therapeutics possess all the requirements necessary for diagnosis and treatment of CNS disorders and brain tumours after IV administration and therefore, constitute a promising field as BBB targeted drug delivery systems. Characteristics such as (i) specific strategies for active targeting by covalent binding of ligands or monoclonal antibodies of brain receptors, (ii) availability of different regions in the polymer backbone which serve a specific purpose, and (iii) sustained or controlled release of conjugated drugs through cleavable linkers in specific environmental situations, can be tailored to achieve this goal. It is noteworthy to emphasise that the polymer therapeutics studies described here are still at an early stage of development but the results are encouraging for future possibilities in achieving successful delivery of therapeutics to the CNS. Some of the main strategies adopted for crossing the BBB include innovative polymer–based therapeutics with covalently bound targeting ligands to mimic endogenous molecules (commonly named "Trojan Horses").[21] These are able to cross the BBB through receptors that are highly expressed at the endothelium, such as; the Low–Density Lipoprotein (LDL) receptor family, transferrin or insulin. Polymer therapeutics adopting this strategy include: *polymeric drugs*, *polymeric micelles*, *polyplexes* (non–viral vectors for DNA delivery) and *polymer conjugates* with linear or branched structures. Regarding polymer structures adopted by these compounds, few linear examples manage to cross the BBB; most of them have a vesicular conformation which favours the passage. Examples of the polymers exploited for polymer therapeutics construction are shown in Figure 8.2.1.

Nanosized *polymeric micelles* are created when individual polymer chains (unimers) aggregate above a threshold concentration and temperature when dissolved in aqueous solution (known as the critical micelle concentration (CMC) and critical micelle temperature (CMT), respectively). Hydrophobic drugs can be solubilised within its core,[22,23] or covalently conjugated.[24] Studies involving polymeric micelles of Pluronic® block copolymers consisting of hydrophilic poly(ethylene oxide) (PEO) and hydrophobic poly(propylene oxide) (PPO) blocks (PEO–*b*–PPO–*b*–PEO) have been established as promising carriers for CNS drug delivery. The inherent amphiphilic character of this polymer gives surfactant–like properties and therefore interacts with hydrophobic surfaces and biological membranes.[25] In addition to the

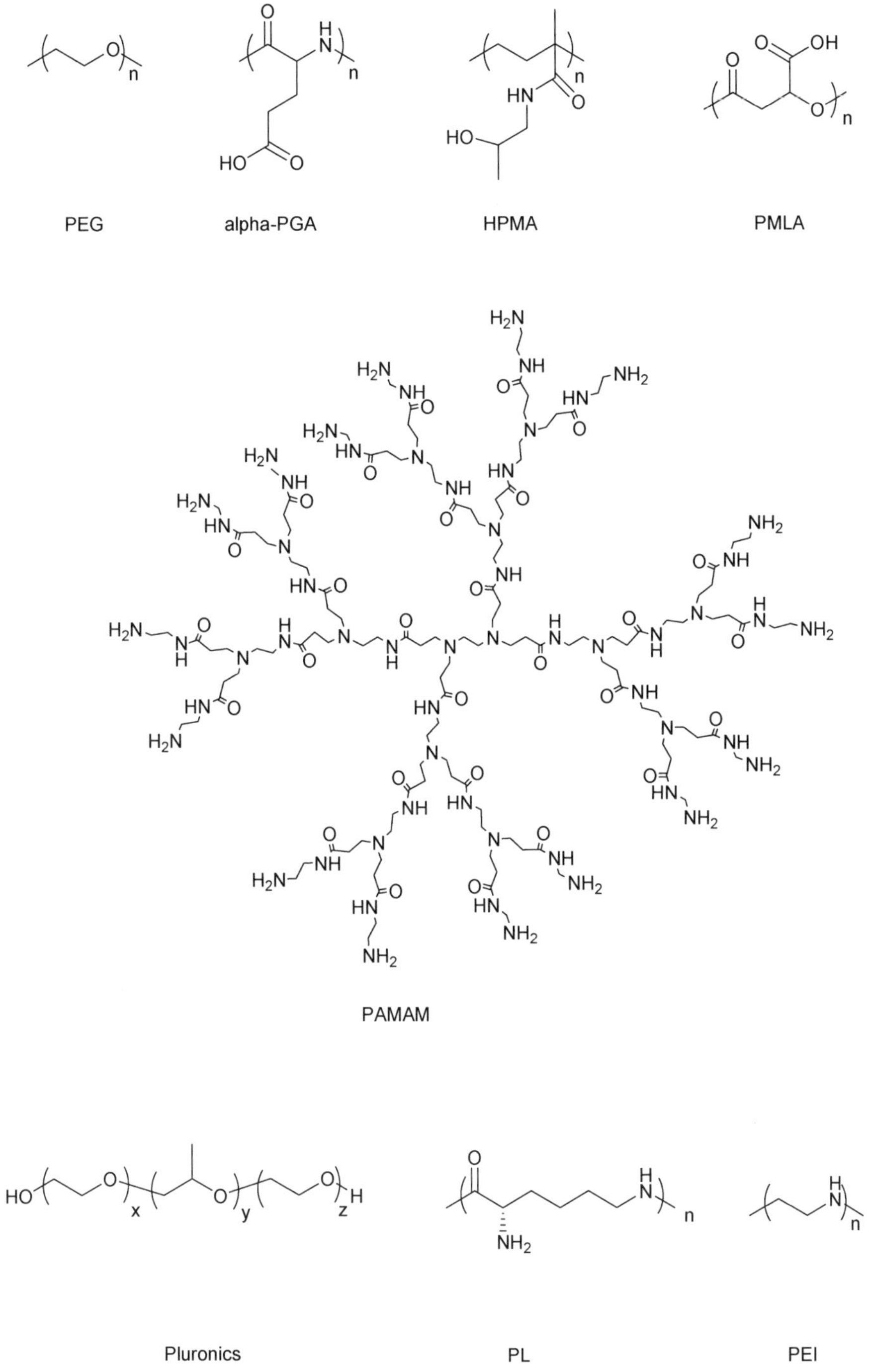

PEG

alpha-PGA

HPMA

PMLA

PAMAM

Pluronics

PL

PEI

Figure 8.2.1 Representative polymers used in the design of polymer therapeutics. PGA: poly–L–glutamic acid, PEG: poly(ethyleneglycol), PEI: poly-ethyleneimine, PAMAM: polyaminoamines, PMLA: poly(maleic acid), HPMA: *N*–(2–hydroxypropyl)methacrylamide, PL: poly–L–lysine.

advantages conferred to the drug by micellisation, it has been discovered that Pluronic® unimers exhibit biological response modifying activities in drug formulations, therefore acting as a *polymeric drug*.[26] Depending on the aggregation state (unimer or micelle), this copolymer was shown to utilise multiple pathways to enter cells.[27] Pluronic® unimers are able to incorporate into membranes by changing their microviscosity and subsequently the translocation into cells is possible through caveolae–mediated endocytosis.[27] Once there, they can alter multiple cellular functions which includes; inhibition of mitochondrial respiration, ATP depletion, inhibition of drug–efflux transporters (*e.g.* P–glycoprotein, multidrug resistance proteins and breast cancer resistance proteins), apoptotic signal transduction and gene expression.[25] Individual results from this make it possible to elucidate their activity within the brain.

Successful studies have shown brain accumulation through conjugation of targeting moieties *i.e.* polyclonal antibodies against brain $\alpha 2$–glycoprotein or insulin, which allowed BBB crossing through receptor mediated endocytosis of Pluronics®, either carrying a drug or a fluorescent probe.[22] They have also managed to increase analgesic effects with several opioid peptides[28] or polypeptide delivery to the brain *via* a degradable disulfide linker (horseradish peroxidise, HRP as a model protein[29] and more recently leptin[30]). In addition, Pluronics® are able to stimulate transcriptional activation of gene expression *in vitro* and *in vivo*.[25] Finally, to confirm that this system has future promise in CNS drug delivery it must be remarked that a Pluronics®–based formulation of doxorubicin (SP1049C, Supratek Pharma) has advanced to Phase III in clinical trials to treat highly resistant tumours and could be the first FDA–approved polymeric micelle (see Table 8.2.1).[31]

Ljubimova *et al.* have designed a novel biodegradable nanoplatform based on poly(β–L–malic acid) (PLMA) for IV treatment of gliomas.[32,33] This *polymer conjugate* uses a tandem combination of monoclonal antibodies (mAbs) to enhance drug targeting across: (i) the BBB, thanks to an anti–mouse transferrin antibody and (ii) the blood–tumour barrier (BTB) where antibody 2C5 targets the tumour cell surface antigen. Through hierarchical addition, different molecules can be covalently attached at the same time to the polymer, *e.g.* with endosomal disrupting units, fluorescent dyes, drugs and PEG chains (as spacers and also as protection modules). Recent studies have shown that antisense oligonucleotides for tumour–specific protein inhibition (laminin–411) and the introduction of a trileucine endosome escape moiety have achieved specific accumulation in brain tumours and suppression of intracranial glioma growth.[34] PMLA–based delivery systems constitute an auspicious tool to treat brain tumours and for tumour imaging. However, this has drawbacks regarding the difficulties of chemical modification and the resulting technical limitations for high scale production.[35]

Regarding dendrimers, most research has been focused on glioma treatment. These macromolecules contain symmetrically arranged branches arising from a multifunctional core where repeated reaction sequences add a precise number of terminal groups at each step or generation.[36] Polyamidoamines (PAMAM) dendrimers have been evaluated for brain drug delivery. Sarin *et al.* have studied how PAMAM size influences its accumulation within malignant glioma cells by magnetic resonance and fluorescence imaging after IV administration, concluding that sizes of 11.1 to 11.9 nm in diameter were able to transverse BTB of malignant gliomas.[37]

These polymeric systems have also recently been tested *in vitro* for neuroinflammation treatment by its conjugation through a disulfide bond to *N*–acetyl–cystein (NAC), an anti–inflammatory agent.[38] Glutathione present in cell cytoplasm triggers drug release. However, future *in vivo* studies must be completed to prove its validity in brain disease, thus showing successful BBB crossing, safety and drug efficacy.

An increasingly popular field in the treatment of brain disorders is gene therapy using *polyplexes* as non–viral vectors. It should be remarked that cellular trafficking of these nanomaterials strongly depends on their composition, size, shape and surface characteristics as well as the cell line used for *in vitro* studies.[27] The surfaces of polyplexes are typically modified with hydrophilic and biocompatible polymers such as PEG or poly(hydroxy-propylmethacrylate) (PHPMA) for *in vivo* applications, with the aim of increasing colloidal stability and solubility after its systemic delivery by preventing aggregation and interaction with plasma proteins due to steric repulsions.[39] To improve the activity, acid–labile linkers between PEG and a polycationic block have been tested to expose the positively charged polyplex (after endosome hydrolysis) to enhance membrane disruption and cytosolic escape.[40] The design of polyplexes is often complemented by addition of brain–specific ligands for targeting neurodegenerative disorders. It has been shown that targeting moieties should be linked in distal ends of PEG to avoid masking receptor–ligand interaction.[41] Introduction of Nuclear Localisation Signal peptides (NLS) into polyplexes has been also attempted to achieve DNA delivery.[42] For example, recent studies pre–mixing NLS–PEG–trisacridine conjugate with DNA and biodegradable polyamines demonstrated significantly enhanced transfection into brain capillary endothelial cells.[43]

Due to the cationic character of PAMAM, its use as a non–viral vector has been highly exploited. However, PAMAM safety has always compromised its clinical development for systemic applications. Numerous targeting ligands have been conjugated to them to examine their potential for BBB passage. Angiopep–2 is an amino acid sequence endocytosed by the low–density lipoprotein receptor–related protein (LRP–1) and to date is the most advanced technology in brain drug delivery. Its conjugation to Paclitaxel (ANG1005, Angiochem Inc., Canada) is in two Phase I clinical trials for recurrent malignant gliomas and brain metastases of advanced cancer.[44,45] Angiopep–2 conjugated PEG–PAMAM dendrimer nanoparticles were observed to be internalised through clathrin– and caveolae–mediated endocytosis and partially by macropinocytosis. *In vivo* assays confirmed its passage through BBB and subsequent accumulation, showing higher efficiency in gene expression than non–modified dendrimers.[46] Transferrin–conjugated PEG–PAMAM dendrimer promoted gene expression in the brains of mice after IV administration.[47]

Cell–based therapy transport involves polymer therapeutics, but not as the main actors. Batrakova *et. al.* have developed a PEI–PEG complexed nanozyme loaded into bone marrow derived monocytes for Parkinson's disease treatment.[48] This system was able to attenuate oxidative stress due to sustained release of the antioxidant enzyme.

Future possibilities for polymer therapeutics for brain delivery include the need of smart vector development and critical evaluation of currently existing ones. Also required is active targeting through novel ligand development in order to make the most of surface–mediated transcytosis as it is believed this barrier constitutes a promising non–invasive strategy.

Numerous and fruitful studies have been carried out in the area of parenteral drug delivery for treating CNS disorders, leading to the development of sophisticated systems that allow drug targeting and sustained or controlled release of parenteral medicines in animal models. Currently, most of the systems described are finding their way into the clinical arena but in diseases other than those associated with CNS delivery, which undoubtedly will help to open doors for treatment of brain disorders.[12]

8.2.2.1.2   Solid Tumours

Barriers to Drug Delivery Across Solid Tumours

The use of polymer therapeutics to treat solid tumours is the most well known application for this type of nanopharmaceutics, with examples already on the market since the 90s and many in clinical trials. Solid tumours could be considered as organ–like entities, and are divisible into three main subcompartments: vascular; interstitial and cellular.[49,50] Each of these subcompartments accounts for several biological barriers that have to be overcome by therapeutics in order to successfully treat this severe disease. Transport of a therapeutic agent from the systemic circulation following IV administration to cancer cells is a three–step process. First, macromolecules flow to different regions of tumours *via* blood vessels. They must then cross the vessel wall, and finally, penetrate through the interstitial space to reach the target cells (Figure 8.2.2).[51] Tumours show a highly chaotic and leaky vessel organisation exhibiting heterogeneous hyperpermeability and defective lymphatic drainage compared to normal tissue.[52] Polymer therapeutics as well as other nanopharmaceutics have been widely exploiting these pathophysiological characteristics to achieve targeted drug delivery to tumours.[6] There are many other factors to consider when designing anticancer polymer therapeutics. Tumour barriers such as hypoxia, interstitial hypertension or low extracellular pH could stand as obstacles to achieve an adequate drug delivery; however, with a good rationale of design it is possible to benefit from these intrinsic differences and achieve a targeted delivery.[52]

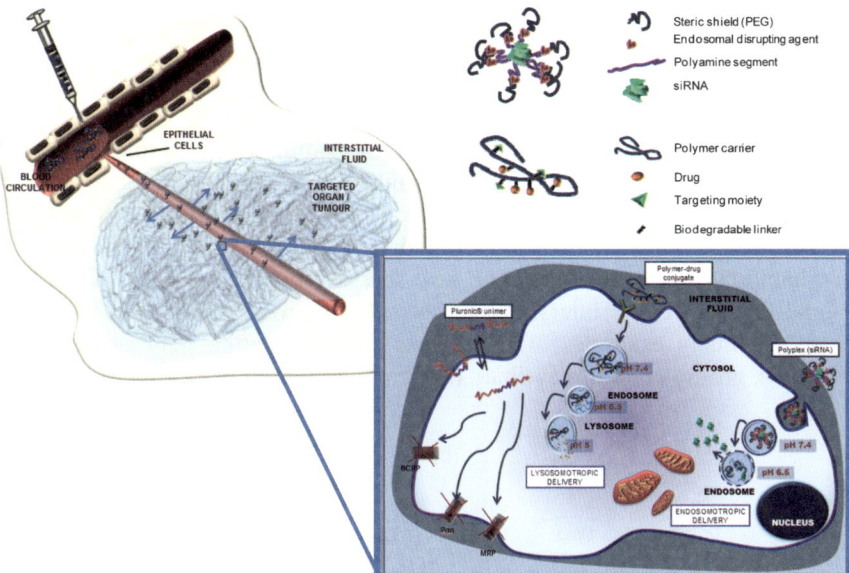

Figure 8.2.2 Scheme of biological barriers to overcome after parenteral administration of model polymer therapeutics; from blood vessels to cellular organelles.

Tumour Vasculature

The vasculature that supplies solid tumours often has several abnormalities in structure when compared with normal physiological vessels. This can lead to tortuous and/or hyperpermeable vessels and in some areas there exists a pore cut–off size ranging from 380–780 nm (within different tumour models).[53,54] This irregular permeability with large cut–off sizes in comparison to normal vasculature, permits for specificity when delivering polymeric nanopharmaceutics with hydrodynamic diameters within that cut–off range, which cannot pass through normal blood vessel walls. All polymer therapeutics except one (FCE 28069, known as PK2) that have already been transferred to clinics rely on the EPR effect (Table 8.2.1), and now the newer and more advanced systems are still mostly designed to fit this description.[3,4,10] Unfortunately, as mentioned, the vasculature of tumours is heterogeneous and complicated with regions of necrosis where the area is poorly vascularised. Simply utilising the hyperpermeability of this vasculature alone may not be possible to treat the entire tumour.

Hypoxia

Poor vasculature results in reduced oxygen availability and thus to intratumoural hypoxia, where cell survival and growth is mediated through a hypoxia–inducible factor (HIF) dependent transcriptional program.[55] The lack of vasculature is a challenging barrier to overcome in the treatment of solid tumours. However, acidic conditions arise from anaerobic processes associated with HIF activity and can be used advantageously as a release mechanism for polymer therapeutics that contain acid labile linkers (Figure 8.2.3).

Polymer–conjugates designed following this rationale, are stable at neutral pHs but can release the cargo under mild acidic conditions. One method described by Sirova *et al.*[56] includes the conjugation of doxorubicin (Dox) to a copolymer of *N*–(2–hydroxypropyl)methacrylamide (HPMA) and 6–methacrylamidohexanohydrazide *via* an acid sensitive hydrazone bond, where the intended target is the interstitial space in the tumour. Within the literature there are many examples of hydrazone bond formation between the C13 carbonyl group of anthracyclines (*i.e.* conjugation of Dox or daunomycin

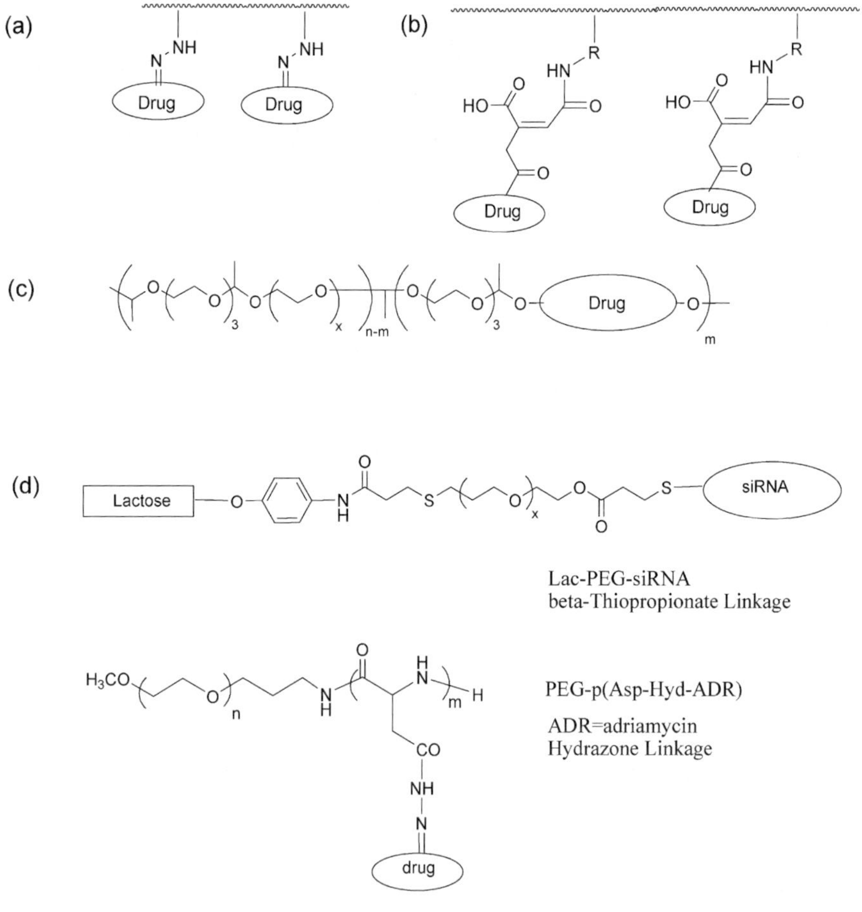

Figure 8.2.3 Bioresponsive pH–labile bonds that allow for specific therapeutic release from the polymer showing: (a) hydrazone bond, (b) *cis*–aconitic bond (c) diethylstilbestrol (DES)–polyacetal (redrawn from ref. 67); (d) polymeric micelle examples (redrawn from ref. 70, 128).

(Dau) with polymers containing hydrazide functionalities).[57–61] Amide bonds, either directly to the polymer or with a *cis*–aconityl residue containing spacers have also been used for preparation of polymer–drug conjugates with an enzymatic or acidic pH release mechanism, respectively.[62–65] Additionally, a further method for acid sensitive release is through the use of acetal bonds, such as the used in the design of PEG–polyacetal conjugates that consist of tripartite structures; a water–soluble polymer, an anticancer agent and a pendant linker.[66,67] These polyacetal structures have the added benefit of allowing the bioactive agent with the right functionalities (*i.e.* diol bearing drugs) within the polymer mainchain resulting in a large enough polymer therapeutic to utilise the EPR effect, whilst in presence of an acidic environment degrading into the smaller subunits on release of the drug without the need for a biodegradable linker.[64] More examples of all these pH–responsive strategies are found in the design of polymeric micelles (Figure 8.2.3).[68–70]

Importantly, the release of drugs in the methods mentioned may not occur solely within the interstitial space but most likely following endocytosis where the pH within endosomes and lysosomes is also acidic. (This process will be mentioned in more detail in section 2.1.3.)

Interstitial Pressure

The remaining major biological barrier to treating a solid tumour is to overcome the interstitial pressure, which is typically higher than in normal tissue and thus limits the passive diffusion of therapeutics into the tumour. In fact, experiments have shown that the concentration of Dox decreases exponentially with distance from tumour blood vessels, decreasing to half its perivascular concentration at a distance of about 40 to 50 μm.[71] Given that the mean distance from blood vessels to regions of hypoxia is in excess of this then polymer therapeutics have to be capable of crossing this space. Combination therapies have been developed to help improve the accumulation and penetration of both therapeutics and polymer–drug conjugates.

Delivery of therapeutics to the interstitial space can be enhanced in a number of ways. One of the most interesting approaches has been recently described by Frost and colleagues from Halozyme Therapeutics Inc.[72] They focus on the therapeutic potential of enzymatic remodeling of the tumour microenvironment through Hyaluronan (HA) depletion using PEG–recombinant human HA–degrading enzyme (rHuPH20) conjugate (PEGPH20) as therapeutic agent. PEGPH20 enhanced both docetaxel and Doxil activity in PC3 tumours ($p < 0.05$) but did not significantly improve the activity of docetaxel in low–HA prostate DU145 tumours. These results support enzymatic remodeling of the tumour stroma with PEGPH20 to treat tumour characterized by the accumulation of HA.[72]

The presence of vascular endothelial growth factor (VEGF), identified as a vascular permeability factor facilitates the extravasation of blood plasma components into the interstitial space of tumour tissue.[73,74] Other methods include work by Maeda's group using nitroglycerin to enhance the hyperpermeability of vasculature in solid tumours,[75,76] and also the use of angiotensin II,[77] which elevates systemic blood pressure and improves the quantity of therapeutic that reaches the target. These combinational therapies have yielded promising results, and when in combination with polymer–drug conjugations may lead the way for future treatment of solid tumours.

It is worth mentioning in this section that inducing hyperthermia in solid tumours in combination with therapeutics is a method that takes advantage of the abnormal vasculature in a solid tumour. Where in normal tissue the vasculature is able to dilate to increase blood perfusion and thus regulate the temperature, the vasculature in a tumour is unable to efficiently vasodilate and the temperature continues to rise.[78–80] Applicators for inducing local hyperthermia in a tumour have been developed for clinical use and include using microwave arrays[81] and ultrasound transducers.[82] Although there are no known examples of polymer therapeutics combined with these techniques inducing hyperthermia, this combination could be a positive step forward in the field for improved tumour drug delivery.

8.2.2.1.3 Barriers at the Cellular Level

Once the macromolecule is accumulated in the damaged tissue (thanks to EPR effect and/or active targeting), the next step is to reach the cellular inside. For many polymer therapeutics the target goal is delivery of the drug to the nucleus for interaction with the DNA (although not true of every therapeutic), but before reaching this organelle there are a number of cellular barriers to overcome. The first barrier to a therapeutic is the cell membrane, the fluid phospholipid bilayer sheet that excludes structures such as proteins from the cell interior, whilst retaining nanostructures such as ribosomes within. Normal uptake of nanoconjugates is achieved by endocytosis, governed by interactions with proteins and receptors on the cell membrane surface. However, there are exceptions to this with some polymer therapeutics that use electrostatic interactions to disrupt the cell membrane and gain entrance, for example in the case of polypropylenimine.[83]

The pH of the endosomal compartment drops during maturation to values of ≤5.5 because of the action of ATP–dependent proton pumps located in the membrane of the compartment.[84] As discussed, this drop in pH can be utilised for the breakdown of acid cleavable linkers for the release of a therapeutic within a cell. Other pathways for release are possible within lysosomes as they contain a host of enzymes that are present for the breakdown of complex molecules such as lipids and proteins. For example, the clinical performance of Opaxio[TM], a PGA–PTX conjugate in Phase III trials as anticancer agent, highly depends on the levels of the lysosomal enzyme cathepsin B. In fact, after finding a relation between estrogen levels and cathepsin B expression in cancer patients, the latest clinical trials with Opaxio[TM] have been designed only to enrol women with estrogen levels above 25pg mL^{-1} to ensure PTX release and therefore antitumour efficacy.[85,86]

Summarising, the performance of a polymer therapeutic is usually dependent on: efficient endosomal escape, retrograde trafficking and/or late endocytic/lysosomal activation for pharmacological activity.[87]

On the other hand, it is important to recognise that the combination of hydrolytic enzymes and an acidic pH makes the intravesicle environment a hostile place for the therapeutic and can result in the inactivation of low molecular weight (MW) drugs as well as degradation of proteins, peptides and oligonucleotides that arrive there unless they are able to quickly escape to the cytosol. Escape from endosomal compartments can be achieved by disrupting the membrane bilayer. The use of pH–responsive endosomolytic systems is one of the preferred approaches to enhance cytosolic delivery for bioactives such as proteins, peptides or siRNA sequences. The most successful example of this approach is CALAA–01, a transferrin targeted siRNA conjugate that consists of a self–assembling cyclodextrin–containing polycation and anti–EWS–FLI1 siRNA–based nanoparticles (30–40 nm) pioneered by Davis *et al.* and developed by Calando Pharmaceuticals in Phase I clinical trials (Table 8.2.1).[36] Calando Pharmaceuticals, Inc. is developing this system into a proprietary delivery technology for clinical application that is called RONDEL™ for RNAi/Oligo nanoparticle delivery (http://www.calandopharma.com/). Similar strategy has been also followed by Kataoka *et al.* by means of polymeric micelles in order to enhace cytosolic delivery of siRNA sequences[69] (Figure 8.2.3d).

Recently, a diblock polymer consisting of HPMA with dimethylaminoethyl methacrylate, propylacrylic acid and butyl methacrylate was synthesised using a controlled radical polymerisation technique, RAFT (Reversible Addition Fragmentation chain–Transfer), to give a carrier polymer with vesicle membrane disruptive properties.[88] This polymer was conjugated with a proapoptotic peptide and tests *in vitro* with HeLa cervical carcinoma cells showed markedly improved results over the peptide alone. Some peptides are also known to disrupt the endosomal membrane, where upon protonation in the acidic environment they adopt an α–helical structure from a random coil structure.[89]

Beyond escaping the endosomal membrane barrier, the ultimate barrier is to deliver a therapeutic into the nucleus of the cell, particularly in gene therapy. It was reported in 1982 that anthracyclines can pass into the nucleus by passive diffusion from the cytoplasm through the nuclear pore complex (NPC).[90] Polycations such as PEI,[91–93] chitosan,[94–97] and PL[98,99] have shown initial success in interacting and facilitating the entry of genetic material into the nucleus. Recently, Ma *et al.*[100] investigated the modification of PEI of varying MW with conjugated triamcinolone acetonide (a nuclear localisation signal) as a potential gene carrying agent. They showed that PEI 1800 combined with the signalling agent had apparently improved transgenic activity in comparison to PEI without the signalling agent. With gene therapy in mind, this kind of method has sparked a lot of interest in the field for using polymers as gene–carriers. This is particularly true because of the associated risks of using viral vectors, such as a reactivation of the virus, toxicity, gene control and targeting, which are either not present with polymers or can be controlled through the modification of the polymer properties such as MW and architecture and also through conjugation of site specific markers or receptors.

In cases where endocytosis is diminished in the cell, polymer–directed enzyme prodrug therapy (PDEPT) is a useful tool for drug delivery.[5] This system encompasses the combination of a polymer–drug conjugate carrying a single bioactive agent with a polymer–enzyme conjugate. The aim of the combination relies on appropriate and selective drug release from the polymer–drug conjugate at the desired site and is independent of endocytosis and the intracellular trafficking mechanisms.

Finally, another approach when these conditions are present is the use of plasma membrane disruptors such as melittin[101] or cell penetrating peptides also known as 'Trojan Horses'.[102]

8.2.2.1.4   Hepatic Barriers

For all nanopharmaceuticals, the avoidance of the liver can be extremely important for the treatment of diseases in other parts of the body and therefore is a major barrier to overcome. To avoid degradation of carefully designed materials that may contain enzyme cleavable linkers or a readily degradable drug or protein it is important that they can go "unnoticed" by the liver to increase their circulation time in the blood to reach their site of action. This is achieved with so called "stealth" systems, which are typically polymer conjugates consisting of, or modified with, hydrophilic neutral polymers such as PEG. In particular, within the polymer therapeutics family, polymeric drugs and polymer conjugates are the preferred forms to avoid liver accumulation mainly due to their size and shape (smaller than 30 nm). If the carrier used is not a neutral polymer this fact could change as polyanions are highly recognised by macrophages, therefore the surface charge of the conjugates is an important parameter to determine in order to predict a correct body distribution. To improve biodistribution, the polymer could also be modified to enhance their stealth properties, one such example is with the use of PAMAM dendrimers modified with PEG, which provides a stealth effect and much higher biocompatibility.[103]

Unlike solid tumours, where the majority of polymeric materials are designed to utilise the hyperpermeability of the vasculature in order for specific targeting, major organs like the liver are targeted by using either the physiological conditions, or by organ–specific enzymes to release the active pharmaceutical agent. The liver has a high blood perfusion rate due to its dual blood supply and the result is that a therapeutic will arrive within minutes after administration. On the other hand, for the same reason the liver is a primary location for metastases to occur and thus a greater challenge to treat successfully. Consequently, improving the specific delivery of polymeric prodrugs by adding binding motifs for targeting the liver has been a key goal for novel therapeutics. In fact, the unique active–targeted polymer–drug conjugate in clinical trials used this strategy, HPMA copolymer–Dox–Galactosamine (PK2, FCE28069) in Phase I/II.[104] The galactosamine mediates efficient liver targeting *via* the asialoglycoprotein receptor (ASGPR) of hepatocytes.[105] Following the same strategy, Huang *et al.*[106] describe a linear dendritic block copolymers (PAMAM–PEG) coupled with Dox with or without a liver specific galactose ligand. Other recent examples of hepatic targeting polymeric prodrugs include; polyethylene sebacate–Dox with pull-alan as an ASGPR,[107] and a HPMA oligonucleotide bioconjugate with a mannose–6–phophate motif for hepatic stellate cell targeting.[108]

8.2.2.1.5 Renal Barriers

The physical properties and particle size of therapeutics that enter the body through parenteral delivery routes are critical aspects in specific targeting, as has been explained. The majority of polymer therapeutics are renally excreted including biodegradable as well as non–biodegradable materials with hydro-dynamic volumes lower than the cut off size for renal clearance (a key parameter to consider for an adequate rational design). Therefore, a major barrier for therapeutics is to ensure damaged tissue targeting at the same time as an adequate renal clearance. Most molecular–sized drugs administered are cleared by the renal system through filtration and then excretion into urine as they have masses less than 25 to 50 kDa, which loosely equates to diameters of approximately 5 nm or less.[109] This excretion of low MW therapeutics can be reduced through the conjugation of the therapeutic to a polymer with a suitable MW or configuration that results in a hydrodynamic radius in excess of the kidney exclusion limit. In the case of proteins, the well–known PEGylation approach is one of the most common methods for avoiding renal excretion, with demonstrated clinical benefits regarding protein pharmacokinetic profile. PEG has been well established as an excellent carrying agent for therapeutics and more can be read in exhaustive reviews on this topic.[110,111] In addition to the size of the conjugates, the architecture of the polymer has an effect on the retention and biodistribution in the body. Sadekar *et al.*[112] have conducted a comparison of iodine–125 labelled HPMA copolymers and branched PAMAM dendrimers of varying MW for biodistribution. The tumour accumulation and excretion of the polymers was affected by the size, but also notable is that a HPMA linear copolymer and a generation 6.0 dendrimer with similar MW were not excreted equally, with the dendrimer showing retention in the kidney for more than 1 week and the linear polymer not exhibiting any kidney accumulation/entrapment.

It is important to note that whilst overcoming the barrier of rapid renal excretion of therapeutics by increasing the MW of polymers with a non–biodegradable backbone one could induce toxic side effects, such as lysosomal storage disease or macromolecular syndrome, as in the case of high MW PEG in the liver. Therefore, the design of polymeric prodrugs to have long circulation time but which are able to be excreted by the body is of upmost importance.

8.2.2.2 Subcutaneous and Intramuscular Administration

Subcutaneous (SC) delivery involves drug introduction by injection into a layer of fatty tissue from where macromolecules can be absorbed. SC injections have a lower ratio of absorption and slower onset of action than IM or IV injections. Drug delivery through this path is influenced by molecule size (where larger molecules results in slower penetration), viscosity (high viscosity impedes body fluids diffusion), vascularity of the tissue at the site of injection, and quantity of fatty tissue (which determines the amount of drug adsorbed). SC may entail local complications such as area irritation at the injection site and so this must be varied to avoid unabsorbed drug accumulation. Onset of action of IM is faster than SC but slower than IV, because muscles (often deltoids of gluteus) are better vascularised. In SC as well as in the IM route, drugs are injected into an interstitial environment and later absorbed into the circumvent vasculature or drain into the lymphatic system. Compounds with higher MW (> 5 kDa) enter into lymphatic capillaries and subsequently systemic circulation.[113,114] Accumulation into the lymphatic system is an interesting route for targeting in metastasis treatment because the lymphatic system is a common route for tumour cell migration. Both paths provide a prolonged and sustained effect, but IM is preferred when higher amounts of drug are needed.

Copaxone® is the first polymeric SC administered drug to reach the market. This random copolymer based on L–alanine, L–lysine, L–glutamic acid and L–tyrosine with MW within the range of 5 to 10 kDa is used for multiple sclerosis treatment.[115] Several PEGylated proteins for treatment of diseases other than cancer are also in clinical routines and are administered SC or IM to allow prolonged therapeutic efficiency.[116] It has been also proposed that PEG possibly avoids interaction with components of the subcutaneous interstitium allowing skin permeation.[117]

Other recent examples of SC or IM administered conjugates include novel PEGylated proteins, HA anti–tumour conjugates and targeted HPMA copolymer combination conjugates.[118–121]

8.2.2.3 Intratumoural

The intratumoural (IT) route to administration entails direct drug injection inside the tumour site. This technique has not been established yet in clinical practice due to its invasive characteristic, relatively rapid clearance of topically applied drugs from tumours and dose–limiting toxicities in surrounding areas. In spite of all these disadvantages, IT has been evaluated with standard anticancer agents. By means of a drug delivery system, there is an improvement in the therapeutic index, which ultimately leads to a higher concentration within the tumour and higher specificity for targeting. As a general example of this route, an *in vivo* evaluation of HPMA copolymer–Dox conjugates was reported showing antitumour activity.[122] Further to this, Chun *et al.*[123] developed a thermosensitive poly(organophosphazene)–PTX conjugate hydrogel for IT injection. They reported an increased therapeutic efficacy *via* a controlled and sustained drug release at the tumour site. Finally, a dextran–Dox conjugate bearing acid–labile linkages was IT administered *in vivo* and showed tumour regression.[124]

It is important to remark the importance of the IT route in brain delivery. In fact, IT is a technique widely employed in the treatment of gliomas as it entails avoidance of drug passage through the restrictive BBB. Covalent drug binding to micellar nanosystems has been explored for this purpose. Inoue and co-workers conjugated Dox to the aspartic acid residue of PEG–block–poly(aspartic acid) copolymer. The *polymeric micelles* formed by the resultant Dox–conjugated poly(aspartic acid) block were delivered IT to the brain through convention–enhanced delivery (CED). According to their report, micellar Dox infused by CED resulted in prolonged median survival compared with free Dox.[125] PAMAM dendrimers have also been evaluated *via* IT injection in order to target the epidermal growth factor receptor (EGFR), a cell surface receptor that is frequently overexpressed in brain tumours.[126] Attachment of monoclonal antibody Cetuximab gave higher accumulation in the brain tumour.

8.2.2.4 Intraperitoneal

Intraperitoneal (IP) drug administration conveys injection into the peritoneum, the serous membrane which covers the abdominal cavity. It is widely used for chemotherapeutic agent administration in patients with intra–abdominal malignancies, *i.e.* ovarian and gastrointestinal cancers. The rate and amount of drug transferred in the perintoneum depends mainly on peritoneal inflammation, surface areas, and peritoneal blood flow.[127] However, the drug is confined to tissues and organs in peritoneal surroundings unlike IV where the drug is dispersed throughout the blood circulation. Polymer therapeutics can play an important role when injected into the peritoneal cavity as the rate and extent of drug absorption from the cavity to the site of disease is improved and provides advantages for antitumour activity.

IP administration could be advantageous for specific local application such as peritoneal dialysis in order to avoid systemic distribution of potential damaging substances (in order to avoid haemolytic character or anticoagulant effect). In this context, and among the first polymer therapeutics to exploit this route, is found dextrin–2–sulfate (MW = 25 kDa).[11]

8.2.3 Conclusions

Current state of the art polymer therapeutics relies on strong foundations coming from 30 years of interdisciplinary research from the bench to the bedside; they can be considered amongst the most successful polymeric nanomedicines with great capability to surpass specific biological barriers. There are a growing number of polymer therapeutics that are products and also entering clinical development as both novel treatments and imaging agents. They are used as Nano–sized Medicines in the form of individual agents or conjugates or as components of complex, self–assembling nanoparticles and micelles. Consequently, an exponentially growing industrial pipeline is currently available in big pharmaceutical companies as well as in small biotechnologies devoted to specific nanoconjugates.

Looking at the number of biological barriers to overcome in order to achieve a successful targeted drug delivery after parenteral administration, it is widely accepted that four main strategies will drive the future research in the field: (1) use of new molecular targets in cancer as well as other diseases, (2) design of polymer–based combination therapy, (3) control of polymeric platforms and their conformational behavior in solution and (4) exhaustive physico–chemical characterisation essential to transform a promising conjugate into a candidate for clinical evaluation following regulatory indications. Therefore, there is a need for continued development of validated analytical techniques for characterisation of these complex nano–sized medicines and validated methods for establishing preclinical monitoring and safety. Increased understanding of polymer therapeutic features that govern clinical risk–benefit is leading to an appreciation of clinical biomarkers that will open new possibilities for personalised therapy and optimised clinical trial design.

Acknowledgements

We acknowledge support by Centro de Investigación Príncipe Felipe and MICINN (EUI2008–03904, CTQ2007–60601, CTQ2010–18195, FPU grant AP2007–01665). MJV is a Ramon y Cajal researcher.

References

1. J. F. Coelho, P. C. Ferreira, P. Alves, R. Cordeiro, A. C. Fonseca, J. R. Gois and M. H. Gill, *The EPMA Journal*, 2010, **1**, 164–209.
2. R. Duncan, *Nat. Rev. Drug Discov.*, 2003, **2**, 347.
3. R. Duncan, *Nat. Rev. Cancer*, 2006, **6**, 688–701.
4. M.J. Vicent and R. Duncan, *Trends Biotechnol.*, 2006, **24**, 39–47.
5. M. J. Vicent, H. Ringsdorf and R. Duncan, *Adv. Drug Deliv. Rev.*, 2009, **61**, 1117–1120.
6. A. K. Iyer, G. Khaled, J. Fang and H. Maeda, *Drug. Discov. Today*, 2006, **11**, 812–818.
7. R. Duncan, *Biochem. Soc. Trans.*, 2007, **35**, 56–60.
8. M. Barz, R. Luxenhofer, R. Zentel and María J. Vicent, *Polym. Chem.*, 2011, **2**, 1900–1918.
9. F. Greco and M. J. Vicent, *Adv. Drug Deliv. Rev.*, 2009, **61**, 1203–1213.

10. J. Sanchis, F. Canal, R. Lucas and M. Jesus, *Nanomedicine*, 2010, **5**, 915–935.
11. R. Gaspar and R. Duncan, *Adv. Drug Deliv. Rev.*, 2009, **61**, 1220–1231.
12. J. H. Sakamoto, A. L. van de Ven, B. Godin, E. Blanco, R. E. Serda, A. Grattoni, A. Ziemys, A. Bouamrani, T. Hu, S. I. Ranganathan, E. De Rosa, J. O. Martinez, C. A. Smid, R. M. Buchanan, S. Y. Lee, S. Srinivasan, M. Landry, A. Meyn, E. Tasciotti, X. Liu, P. Decuzzi and M. Ferrari, *Pharmacol. Res.*, 2010, **62**, 57–89.
13. E. Segal and R. Satchi–Fainaro, *Adv. Drug Deliv. Rev.*, 2009, **61**, 1159–1176.
14. R. Duncan, *Adv. Drug Deliv. Rev.*, 2009, **61**, 1131–1148.
15. N. J. Abbott, L. Ronnback and E. Hansson, *Nat. Rev. Neurosci.*, 2006, **7**, 41–53.
16. W. M. Pardridge, *Mol. Interv.*, 2003, **3**, 90–105, 151.
17. M. M. Patel, B. R. Goyal, S. V. Bhadada, J. S. Bhatt and A. F. Amin, *CNS Drugs*, 2009, **23**, 35–58.
18. A. Beduneau, P. Saulnier and J. P. Benoit, *Biomaterials*, 2007, **28**, 4947–4967.
19. T. Bansal, N. Akhtar, M. Jaggi, R. K. Khar and S. Talegaonkar, *Drug. Discov. Today*, 2009, **14**, 1067–1074.
20. P. Debbage, *Curr. Pharm. Des.*, 2009, **15**, 153–172.
21. W. M. Pardridge, *Discov. Med.*, 2006, **6**, 139–143.
22. A. V. Kabanov, E. V. Batrakova, N. S. Melik–Nubarov, N. A. Fedoseev, T. Y. Dorodnich, V. Y. Alakhov, V. P. Chekhonin, I. R. Nazarova and V. A. Kabanov, *J. Control. Release*, 1992, **22**, 141–157.
23. E. V. Batrakova, T. Y. Dorodnych, E. Y. Klinskii, E. N. Kliushnenkova, O. B. Shemchukova, O. N. Goncharova, S. A. Arjakov, V. Y. Alakhov and A. V. Kabanov, *Br. J. Cancer*, 1996, **74**, 1545–1552.
24. M. Yokoyama, M. Miyauchi, N. Yamada, T. Okano, Y. Sakurai, K.Kataoka and S.Inoue, *Cancer Res.*, 1990, **50**, 1693–1700.
25. E. V. Batrakova and A. V. Kabanov, *J. Control. Release*, 2008, **130**, 98–106.
26. A. V. Kabanov and V. Y. Alakhov, *Crit. Rev. Ther. Drug Carrier Syst.*, 2002, **19**, 1–72.
27. G. Sahay, D. Y. Alakhova and A. V. Kabanov, *J. Control. Release*, 2010, **145**, 182–195.
28. K. A. Witt, J. D. Huber, R. D. Egleton and T. P. Davis, *J. Pharmacol. Exp. Ther.*, 2002, **303**, 760–767.
29. E. V. Batrakova, S. V. Vinogradov, S. M. Robinson, M. L. Niehoff, W. A. Banks and A. V. Kabanov, *Bioconjug. Chem.*, 2005, **16**, 793–802.
30. T. O. Price, S. A. Farr, X. Yi, S. Vinogradov, E. Batrakova, W. A. Banks and A. V. Kabanov, *J. Pharmacol. Exp. Ther.*, 2010, **333**, 253–263.
31. A. K. Sharma, L. Zhang, S. Li, D. L. Kelly, V. Y. Alakhov, E. V. Batrakova and A. V. Kabanov, *J. Control. Release*, 2008, **131**, 220–227.

32. M. Fujita, B. S. Lee, N. M. Khazenzon, M. L. Penichet, K. A. Wawrowsky, R. Patil, H. Ding, E.Holler, K. L. Black and J. Y. Ljubimova, *J. Control. Release*, 2007, **122**, 356–363.

33. J. Y. Ljubimova, M. Fujita, A. V. Ljubimov, V. P. Torchilin, K. L. Black and E. Holler, *Nanomedicine*, 2008, **3**, 247–265.

34. H. Ding, S. Inoue, A. V. Ljubimov, R. Patil, J. Portilla–Arias, J. Hu, B. Konda, K. A. Wawrowsky, M. Fujita, N. Karabalin, T. Sasaki, K. L. Black, E. Holler and J. Y. Ljubimova, *Proc. Natl. Acad. Sci. USA*, 2010, **107**, 18143–18148.

35. J. A. Portilla–Arias, M. Garcia–Alvarez, J. A. Galbis and S. Munoz–Guerra, *Macromol. Biosci.*, 2008, **8**, 551–559.

36. R. Duncan, *Targeting and Intracellular Delivery of Drugs*, Wiley–VCH Verlag GmbH & Co. KGaA, Weinheim, 2006.

37. H. Sarin, A. S. Kanevsky, H. Wu, K. R. Brimacombe, S. H. Fung, A. A. Sousa, S. Auh, C. M. Wilson, K. Sharma, M. A. Aronova, R. D. Leapman, G. L. Griffiths and M. D. Hall, *J. Transl. Med.*, 2008, **6**, 80.

38. Y. E. Kurtoglu, R. S. Navath, B. Wang, S. Kannan, R. Romero and R. M. Kannan, *Biomaterials*, 2009, **30**, 2112–2121.

39. R. J. Christie, N. Nishiyama and K. Kataoka, *Endocrinology*, 2009, **151**, 466–473.

40. H. Yu and E. Wagner, *Curr. Opin. Mol. Ther.*, 2009, **11**, 165–178.

41. A. Kichler, *J. Gene. Med.*, 2004, **6 Suppl 1**, S3–10.

42. R. J. Christie, N. Nishiyama and K. Kataoka, *Endocrinology*, 2009, **151**, 466–473.

43. H. Zhang and S. V. Vinogradov, *J. Control. Release*, 2010, **143**, 359–366.

44. A. G. de Boer and P. J. Gaillard, *Clin. Pharmacokinet.*, 2007, **46**, 553–576.

45. A. Regina, M. Demeule, C. Che, I. Lavallee, J. Poirier, R. Gabathuler, R. Beliveau and J. P. Castaigne, *Br. J. Pharmacol.*, 2008, **155**, 185–197.

46. W. Ke, K. Shao, R. Huang, L. Han, Y. Liu, J. Li, Y. Kuang, L. Ye, J. Lou and C. Jiang, *Biomaterials*, 2009, **30**, 6976–6985.

47. R. Q. Huang, Y. H. Qu, W. L. Ke, J. H. Zhu, Y. Y. Pei and C. Jiang, *FASEB J.*, 2007, **21**, 1117–1125.

48. E. V. Batrakova, S. Li, A. D. Reynolds, R. L. Mosley, T. K. Bronich, A. V. Kabanov and H. E. Gendelman, *Bioconjug. Chem.*, 2007, **18**, 1498–1506.

49. R. K. Jain, *Cancer Research*, 1987, **47**, 3039–3051.

50. V. Vogel, *Nanotechnology: Nanomedicine Volume 5*, Wiley VCH Verlag GmbH & Co. KGaA, Weinheim, Germany, 2009.

51. R. K. Jain and T. Stylianopoulos, *T. Nat. Rev. Clin. Oncon.*, 2010, **7**, 653–664.

52. M. K. Danquah, X. A. Zhang and R. I. Mahato, *Adv. Drug Deliv. Rev.*, DOI: 10.1016/j.addr.2010.11.005.

53. S. K. Hobbs, W. L. Monsky, F. Yuan, W. G. Roberts, L. Griffith, V. P. Torchilin and R. K. Jain, Proceedings of the National Academy of Sciences of the United States of America, 1998.

54. F. Yuan, M. Dellian, D. Fukumura, M. Leuning, D. D. Berk, P. Yorchilin and R. K. Jain, *Cancer Research*, 1995, **55**, 3752.

55. A. Ao, H. Wang, S. Kamarajugadda and J. Lu, *Proc. Natl. Acad. Sci. USA*, 2008, **105**, 7821–7826.

56. M. Sirova, T. Mrkvan, T. Etrych, P. Chytil, P. Rossmann, M. Ibrahimova, L. Kovar, K. Ulbrich and B. Rihova, *Pharm. Res.*, 2010, **27**, 200–208.

57. P. Chytil, T. Etrych, J. Kriz, V. Subr and K. Ulbrich, *Eur. J. Pharm. Sci.*, 2010, **41**, 473–482.

58. P. C. A. Rodrigues, T. Roth, H. H. Fiebig, C. Unger, R. Mʋlhaupt and F. Kratz, *Bioorganic & Medicinal Chemistry*, 2006, **14**, 4110–4117.

59. L. Kovar, T. Etrych, M. Kabesova, V. Subr, D. Vetvicka, O. Hovorka, J. Strohalm, SJ. klenar, P. Chytil, K. Ulbrich and B. Rihova, *Tumor Biol.*, 2010, **31**, 233–242.

60. T. Etrych, J. Strohalm, L. Kovar, M. Kabesova, B. Rihova and K. Ulbrich, *J. Control. Release*, 2009, **140**, 18–26.

61. H. Yuan, K. Luo, Y. S. Lai, Y. J. Pu, B. He, G. Wang, Y. Wu and Z. W. Gu, *Mol. Pharm.*, 2010, **7**, 953–962.

62. N. Lavignac, J. L. Nicholls, P. Ferruti and R. Duncan, *Macromol. Biosci.*, 2009, **9**, 480–487.

63. J. H. Park, Y. W. Cho, Y. J. Son, K. Kim, H. Chung, S. Y. Jeong, K. Choi, C. R. Park, R. W. Park, I. S. Kim and I. C. Kwon, *Colloid Polym. Sci.*, 2006, **284**, 763–770.

64. D. Gaal and F. Hudecz, *Eur. J. Cancer*, 1998, **34**, 155–161.

65. H. Kamada, Y. Tsutsumi, Y. Yoshioka, Y. Yamamoto, H. Kodaira, S. Tsunoda, T. Okamoto, Y. Mukai, H. Shibata, S. Nakagawa and T. Mayumi, *Clinical Cancer Research*, 2004, **10**, 2545–2550.

66. R. Tomlinson, J. Heller, S. Brocchini and R. Duncan, *Bioconjug. Chem.*, 2003, **14**, 1096–1106.

67. M. J. Vicent, R. Tomlinson, S. Brocchini and R. Duncan, *J. Drug Target.*, 2004, **12**, 491–501.

68. H. Q. Yin and Y. H. Bae, *Eur. J. Pharm. Biopharm.*, 2009, **71**, 223–230.

69. V. A. Sethuraman, M. C. Lee and Y. H. Bae, *Pharm. Res.*, 2008, **25**, 657–666.

70. Y. Bae, N. Nishiyama, S. Fukushima, H. Koyama, M. Yasuhiro and K. Kataoka, *Bioconjug. Chem.*, 2005, **16**, 122–130.

71. A. J. Primeau, A. Rendon, D. Hedley, L. Lilge and I. F. Tannock, *Clinical Cancer Research*, 2005, **11**, 8782–8788.

72. C. B. Thompson, H. M. Shepard, P. M. O'Connor, S. Kadhim, P. Jiang, R. J. Osgood, L. H. Bookbinder, X. Li, B. J. Sugarman, R. J. Connor, S. Nadjsombati and G. I. Frost, *Mol. Cancer Ther.*, 2010, **9**, 3052–3064.

73. D. R. Senger, S. J. Galli, A. M. Dvorak, C. A. Perruzzi, V. S. Harvey and H. F. Dvorak, *Science*, 1983, **219**, 983–985.

74. H. Maeda, T. Sawa and T. Konno, *J. Control. Release*, 2001, **74**, 47–61.

75. H. Maeda, *J. Control. Release*, 2010, **142**, 296–298.

76. H. Maeda, T. Seki and J. Fang, *Br. J. Cancer*, 2008, **67**, 975–980.
77. A. Nagamitsu, K. Greish and H. Maeda, *Jpn. J. Clin. Oncol.*, 2009, **39**, 756–766.
78. H. S. Reinhold and B. Endrich, *Int. J. Hyperthermia*, 1986, **2**, 111–137.
79. B. Emami and C. W. Song, *Int. J. Radiat. Oncol. Biol. Phys.*, 1984, **10**, 289–298.
80. C. W. Song, H. J. Park, C. K. Lee and R. Griffin, *Int. J. Hyperthermia*, 2005, **21**, 761–767.
81. P. F. Turner, *Microwave Theory Tech.*, 1986, **34**, 508–513.
82. C. J. Diederich and K. Hynynen, *Ultrasound Med. Biol.*, 1999, **25**, 871–887.
83. N. A. Stasko, C. B. Johnson, M. H. Schoenfisch, T. A. Johnson and E. L. Holmuhamedov, *Biomacromolecules*, 2007, **8**, 3853–3859.
84. M. E. H. El-Sayed, A. S. Hoffman and P. S. Stayton, *Expert Opin. Biol. Ther.*, 2005, **5**, 23–32.
85. S. D. Chipman, F. B. Oldham, G. Pezzoni and J. W. Singer, *Int. J. Nanomed.*, 2006, **1**, 375–383.
86. P. Sabbatini, M. W. Sill, D. O'Malley, L. Adler and A. A. Secord, *Gynecol. Oncol.*, 2008, **111**, 455–460.
87. F. P. Seib, A. T. Jones and R. Duncan, *J. Control. Release*, 2007, **117**, 291–300.
88. C. L. Duvall, A. J. Convertine, D. S. W. Benoit, A. S. Hoffman and P. S. Stayton, *Mol. Pharm.*, 2010, **7**, 468–476.
89. N. M. Moore, C. L. Sheppard and S. E. Sakiyama–Elbert, *Acta Biomaterialia*, 2009, **5**, 854–864.
90. T. Skovsgaard and N. I. Nissen, *Pharmacol. Therapeut.*, 1982, **18**, 293–311.
91. S. Choi and K. D. Lee, *J. Control. Release*, 2008, **131**, 70–76.
92. S. Patnaik, M. Arif, A. Pathak, R. Kurupati, Y. Singh and K. C. Gupta, *Nanomed.–Nanotechnol. Biol. Med.*, 2010, **6**, 344–354.
93. H. L. Jiang, R. Arote, D. Jere, Y. K. Kim, M. H. Cho and C. S. Cho, *Mater. Sci. Technol.*, 2008, **24**, 1118–1126.
94. N. Duceppe and M. Tabrizian, *Expert Opin. Drug Deliv.*, 2010, **7**, 1191–1207.
95. S. F. Peng, M. T. Tseng, Y. C. Ho, M. C. Wei, Z. X. Liao and H. W. Sung, *Biomaterials*, 2011, **32**, 239–248.
96. H. Bordelon, A. S. Biris, C. M. Sabliov and W. T. Monroe, *J. Nanomater.*, 2011, 9.
97. L. X. Liu, Y. Y. Bai, D. W. Zhu, L. P. Song, H. Wang, X. Dong, H. L. Zhang and X. G. Leng, *J. Biomed. Mater. Res. Part A*, 2011, **96A**, 170–176.
98. W. C. Sun and P. B. Davis, *J. Control. Release*, 2010, **146**, 118–127.
99. J. D. Ramsey, H. N. Vu and D. W. Pack, *J. Control. Release*, 2010, **144**, 39–45.
100. K. Ma, M. X. Hu, M. Xie, H. J. Shen, L. Y. Qiu, W. M. Fan, H. Y. Sun, S. Q. Chen and Y. Jin, *J. Gene. Med.*, 2010, **12**, 669–680.

101. N. Lavignac, M. Lazenby, J. Franchini, P. Ferruti and R. Duncan, *Int. J. Pharm.*, 2005, **300**, 102–112.

102. R. L. Juliano, R. Alam, V. Dixit and H. M. Kang, *Wiley Interdiscip. Rev.–Nanomed. Nanobiotechnol.*, 2009, **1**, 324–335.

103. H. Yang, S. T. Lopina, L. P. DiPersio and S. P. Schmidt, *J. Mater. Sci.–Mater. Med.*, 2008, **19**, 1991–1997.

104. L. W. Seymour, D. R. Ferry, D. Anderson, S. Hesslewood, P. J. Julyan, R. Poyner, J. Doran, A. M. Young, S. Burtles and D. J. Kerr, *J. Clin. Oncol.*, 2002, **20**, 1668–1676.

105. P. J. Julyan, L. W. Seymour, D. R. Ferry, S. Daryani, C. M. Boivin, J. Doran, M. David, D. Anderson, C. Christodoulou, A. M. Young, S. Hesslewood and D. J. Kerr, *J. Control. Release*, 1999, **57**, 281–290.

106. J. Huang, F. Gao, X. Tang, J. Yu, D. Wang, S. Liu and Y. Li, *Polym. Int.*, 2010, **59**, 1390–1396.

107. S. A. Guhagarkar, R. V. Gaikwad, A. Samad, V. C. Malshe and P. V. Devarajan, *Int. J. Pharm.*, 2010, **401**, 113–122.

108. N. N. Yang, Z. Y. Ye, F. Li and R. I. Mahato, *Bioconjug. Chem.*, 2009, **20**, 213–221.

109. B. Bhushan, Springer Handbook of Nanotechnology. Second Edition, Springer, Heidelberg, 2006.

110. F. M. Veronese and G. Pasut, *Drug. Discov. Today*, 2005, **10**, 1451–1458.

111. G. Pasut and F. M. Veronese, *Isr. J. Chem.*, 2010, **50**, 151–159.

112. S. Sadekar, A. Ray, M. Jant–Amsbury, C. M. Peterson and H. Gandehari, *Biomacromolecules*, 2010.

113. M. H. Y. Takakura, H. Sezaki, Lymphatic transport after parenteral drug administration, Boca Raton, 1992.

114. W.B. Liechty, D.R. Kryscio, B.V. Slaughter and N. Peppas, *Annu. Rev. Chem. Biomol. Eng.*, 2010, **1**, 149.

115. K. P. Johnson, B. R. Brooks, C. C. Ford, A. Goodman, J. Guarnaccia, R. P. Lisak, L. W. Myers, H. S. Panitch, A. Pruitt, J. W. Rose, N. Kachuck and J. S. Wolinsky, *Mult. Scler.*, 2000, **6**, 255–266.

116. F. M. Veronese, *PEGylated Protein Drugs: Basic Science and Clinical Applications*, Birkhauser Publishing, Basel, 2009.

117. J. das Neves, M. M. Amiji, M. F. Bahia and B. Sarmento, *Adv. Drug Deliv. Rev.*, 2010, **62**, 458–477.

118. D. E. Mager, B. Neuteboom and W. J. Jusko, *Pharm. Res.*, 2005, **22**, 58–61.

119. G. Pasut, A. Mero, F. Caboi, S. Scaramuzza, L. Sollai and F. M. Veronese, *Bioconjug. Chem.*, 2008, **19**, 2427–2431.

120. S. Cai, Y. Xie, N. M. Davies, M. S. Cohen and M. L. Forrest, *J. Pharm. Sci.*, 2010, **99**, 2664–2671.

121. E. Segal, H. Pan, P. Ofek, T. Udagawa, P. Kopeckova, J. Kopecek and R. Satchi–Fainaro, *PLoS One*, 2009, **4**, 5233.

122. T. Lammers, P. Peschke, R. Kuhnlein, V. Subr, K. Ulbrich, P. Huber, W. Hennink and G. Storm, *Neoplasia*, 2006, **8**, 788–795.

123. C. Chun, S. M. Lee, S. Y. Kim, H. K. Yang and S. C. Song, *Biomaterials*, 2009, **30**, 2349–2360.
124. T. M. J. Khandare, *Prog. Polym. Sci.*, 2006, **31**.
125. T. Inoue, Y. Yamashita, M. Nishihara, S. Sugiyama, Y. Sonoda, T. Kumabe, M. Yokoyama and T. Tominaga, *Neuro. Oncol.*, 2009, **11**, 151–157.
126. E. R. Gillies and J. M. Frechet, *Drug. Discov. Today*, 2005, **10**, 35–43.
127. K. Chaudhary, S. Haddadin, R. Nistala and C. Papageorgio, *Curr. Clin. Pharmacol.*, 2010, **5**, 82–88.
128. M. Oishi, Y. Nagasaki, K. Itaka, N. Nishiyama and K. Kataoka, *J. Am. Chem. Soc.*, 2005, **127**, 1624–1625.

CHAPTER 8.3

Drug Delivery Strategies: Lipid Nanocapsules

G. BASTIAT[a,b], S. HIRSJÄRVI[a,b] AND J. P. BENOIT*[a,b,c]

[a] LUNAM Université, MINT – Micro et nanomédecines biomimétiques, Angers, France; [b] UMR_S 1066, Institut de biologie en santé - IRIS, 4 rue Larrey CHU, 49933 Angers Cedex 9, France; [c] CHU, Angers, France
*E-mail: jean-pierre.benoit@univ-angers.fr

8.3.1 Introduction

Over the past few decades, the development of nanomedicines has been significant with the discovery of new drugs, new targets and also new drug delivery systems. The link between nanotechnological development and medicine can provide medical and pharmaceutical benefits, particularly in the field of cancer research. Nanosized drug delivery systems based on liposomes, micelles, polymer or lipidic nanoparticles, polymer-drug conjugates, and dendrimers[1] (also known as nanovectors or nanocarriers) can modify poor biopharmaceutical characteristics of drug molecules (solubility, biological half-life, and specific targeted delivery) and improve existing treatments. However, the benefits of these systems following parenteral administration appear limited. For example, the rapid elimination of nanocarriers from blood circulation, due to opsonisation and non-specific uptake by the mononuclear phagocyte system (MPS), is one of the barriers to overcome. In order to develop a long-term circulating system, poly(ethylene) glycol (PEG) can be grafted onto the nanoparticle surface to decrease MPS recognition.[2]

RSC Drug Discovery Series No. 22
Nanostructured Biomaterials for Overcoming Biological Barriers
Edited by Maria Jose Alonso and Noemi S. Csaba
© The Royal Society of Chemistry 2012
Published by the Royal Society of Chemistry, www.rsc.org

The concept of lipid nanocapsules (LNCs) has been recently patented by our group (UMR_S 1066, University of Angers, France). These nanocarriers, with a diameter ranging from 20 to 100 nm, can mimic lipoproteins. LNCs are generally composed of an oily core (medium-chain triglycerides) surrounded by a mixed layer of lecithins and a pegylated surfactant. LNCs are produced by a solvent-free, soft-energy method and they possess a good degree of stability in an aqueous environment of up to 18 months (liposomes are often produced by a process involving organic solvents and they are unstable in biological fluids).

Due to their PEGylated shell and also due to further developments, LNCs can provide an interesting platform for carrying different, active pharmaceutical ingredients: hydrophobic as well as hydrophilic moieties, DNA, and other biological macromolecules. Surface modification can be carried out in order to improve the mean circulation time (by being coated with longer PEG chains), or to perform active targeting (by grafting ligands). These modifications are described in this chapter. Moreover, physicochemical and biological characteristics such as blood profile and the cellular uptake of LNCs, as well as applications in the field of passive and active targeting, local cancer therapy, and local treatment of inner ear diseases, are illustrated.

8.3.2 Easy and Safe Method for the Formulation of Lipid Nanocapsules

The formulation principle of LNCs is based on the phase-inversion temperature (PIT) method first described by Shinoda *et al.* and developed by Hertault *et al.*[3–5] The three principal components for making the LNCs are the oily phase (Labrafac®, Labrafil®,...), the aqueous phase and non-ionic surfactants (Mainly Lipoid® and Solutol HS-15,...). All these components are approved by the Food and Drug Administration (FDA) for parenteral administration. Depending on the composition of the medium, the LNC hydrodynamic diameter can vary from 20 to 100 nm, with a very narrow range of dispersity (polydispersity indices (PdI) less than 0.1–0.3). To summarise, the LNCs are constituted of an oily core and a membrane made up of lecithin molecules (Lipoid®) and PEGylated surfactants (Solutol HS-15), with PEG chains towards the aqueous phase.[6]

The monolayer surfactant structure and the oily core of LNCs allow lipophilic or amphiphilic drugs and dyes to be easily incorporated inside the nanocarriers. During the last decade, a large variety of drugs have been encapsulated in LNCs: amiodarone,[7] ibuprofen,[8] indinavir,[9] etoposide,[10] paclitaxel,[11–15] tripentone,[16] and derivatives of 4-hydroxy tamoxifen combined with ferrocen.[17–19] These drugs have been incorporated with an encapsulation rate above 90%, demonstrating the efficacy of encapsulation.

With the recent developments of *in vivo* fluorescence imaging, standard lipophilic fluorescent probes have been encapsulated in LNCs: Nile Red,[13,20,21] DiO,[19] cholesteryl BODIPY®,[22] and coumarin-6.[23]

A challenge for LNCs is the encapsulation of hydrophilic drugs. This approach has been partially achieved with the encapsulation of two kinds of molecules of interest: radionuclides (^{99m}Tc, ^{188}Re, ^{125}I and ^{111}In) for imaging and radiotherapy,[13,24–27] and the encapsulation of plasmids for gene therapy.[22,29,30] The strategy has consisted of preformulating the molecules with a lipophilic moiety to solubilise them in the oily phase. Recently, to establish a universal protocol for the encapsulation of hydrophilic drugs or dyes, new formulation processes have been developed. The first was to design LNCs with a bidimensional network of oil, surfactant and polyurea around the aqueous core.[30] Various hydrophilic species with different molecular weights (*e.g.* methylene blue and isothiocyanate fluorescein-labelled BSA protein) can be encapsulated with a great efficacy. Moreover, the simultaneous encapsulation of a hydrophilic dye (methylene blue in the core) and a lipophilic dye (Sudan Red III in the shell) can be achieved. The second system was based on the formulation of reverse micelle-loaded LNCs.[31] Using this last method, fluorescein sodium salt has been successfully encapsulated in LNCs without any release of the hydrophilic dye being observed for 24 h in the suspending medium.

Finally, a way to improve nanoparticle targeting is to cover the surface with ligands that are specific to certain receptors implied in the specific pathology. Until now, a variety of ligands have been tested with different nanoparticles (as reviewed comprehensively *e.g.* by Huynh *et al.*[32]). Such ligands include vitamins (folic acid), sugars (galactose), peptides (arginine-glycine-aspartate, so-called RGD), and proteins (epidermal growth factor, transferrin) among which are antibodies and their fragments, nucleic acids (aptamers), and extracellular matrix receptors (heparin sulfate, chondroitin sulfate, hyaluronan).

Nanoparticle surface modification may be necessary in order to obtain better targeting. Perrier *et al.* reported some methods for the functionalisation of nanoparticles.[33] The first one is the post-insertion process. Amphiphilic molecules can be transferred from their micellar assembly to the nanoparticle surface by their incorporation into the external layer of the particle. Amphiphilic molecules based on 1,2-distearoyl-sn-glycero-3-phosphoethanol-amine-*N*-(polyethylene glycol) (DSPE-PEG) with various PEG lengths have been post-inserted into the external layer of LNCs.[34] The presence of functions at the terminal position of the PEG chain is of interest because it enables the grafting of other molecules after the completion of post-insertion. This procedure allows hydrophilic fluorescent probes (fluorescein-5-isothiocyanate and Rhodamine B) to be fixed to the LNC surface *via* a terminal amino function.[35] This function has also been used to attach galactose to the surface in order to target hepatocytes.[28] Finally, DSPE-PEG with maleimide function (DSPE-PEG-maleimide) has been post-inserted onto LNCs, and OX26 antibody (with a thiolation pretreatment) or OX26 antibody Fab' fragments have subsequently been attached by the thiol-maleimide reaction.[36] Examples of targeted LNCs are presented later in this chapter. The second physical process applied to LNCs is the layer-by-layer (LBL) approach. The aims of this approach are: (i) to coat the nanoparticles and to modify their colloidal stability and drug release kinetics; (ii) to entrap molecules within the layers. Hirsjärvi *et al.* used this approach to adsorb sequentially fondaparinux sodium or dextran sulphate and chitosan on the LNC surface.[37] This study showed that hydrophilic therapeutic molecules can be incorporated into the LNC shell by electrostatic interaction. At the same time, the molecules can be protected from the external environment. In addition to physical methods, chemical approaches can be used to modify the particle surface. Some reactions have been reported by Perrier *et al.*: (i) traditional coupling methods (carbodiimine chemistry, transacylation), (ii) olefin metathesis (acyclic diene metathesis polymerisation, cross-metathesis, enyne metathesis, ring-opening cross-metathesis, ring-opening metathesis polymerisation), (iii) copper(I)-catalysed azide-alkyne cycloaddition and click chemistry, and (iv) the Diels-Alder reaction.[33]

8.3.3   Lipidic Nanocapsules and Blood Circulation

After the intravenous administration of foreign objects such as nanoparticles, various biological processes are activated to ensure the protection of the body. Foreign objects are rapidly removed from the blood circulation system due to their recognition by the mononuclear phagocyte system (MPS), *e.g.* Kupffer cells in the liver, macrophages in the spleen, and bone-marrow.[2] Moreover, the adsorption of certain plasma proteins (opsonisation) enhances this recognition.[38] Nevertheless, some physicochemical and structural properties of particles (the nature of the components, their size, the apparent electrical charge, hydrophilicity *etc.*) can diminish this elimination and increase their stealth characteristics. Preferably, stealth nanoparticles should be small (less than 150 nm) and present a neutral and hydrophilic surface.

8.3.3.1   Unmodified LNCs and Blood Circulation

LNCs can be regarded as satisfactory nanocarriers owing to the nature of the compounds used in their formulation, to their optimal nanoscale size range, and their high-density, surface PEG layer. Using ^{188}Re/^{99m}Tc-labelled LNCs, Ballot *et al.* measured the systemic circulation time of LNCs.[25] Free radionuclide complexes were quickly accumulated in the liver (90% of the injected dose (ID) at 5 min). On the contrary, the half-life time ($t_{1/2}$) for the LNCs was determined to be about 21 ± 1 minutes, and after 24h, about 10% of the ID was still detected in the circulation. Uptake of the LNCs was predominant in the liver and the spleen 30 minutes after injection. To evaluate the integrity of LNCs in systemic circulation, Lacoeuille *et al.* studied the pharmacokinetics and biodistribution of LNCs containing a ^{14}C radio-labelled oil (marker of the LNC core) or ^{14}C radio-labelled surfactant (marker of the LNC shell).[11] After intravenous injection, both radio-labelled LNCs exhibited the same type of clearance profile, with a longer residence time in the circulation ($t_{1/2}$ about 2.5h) compared to the $^{188}Re/^{99m}Tc$-labelled LNCs. Analysing the organs, the LNCs were mainly found in the liver.

Vonarbourg *et al.* investigated complement system activation and macrophage uptake of LNCs.[22] Due to the high density of PEG on the surface, very low complement activation was observed whatever the LNC size. However, by increasing the LNC size from 25 nm to 100 nm, the stealth properties of the particles decreased slightly. The size increase seemed to have more of an impact on macrophage uptake (increased uptake with increased size), which was explained by the changing conformation of the PEG chains on the LNC surface.

LNC toxicity was evaluated on a murine model by the intravenous administration of LNCs and paclitaxel-loaded LNC formulations.[15] The maximum tolerated dose (MTD) and the lethal dose (LD50) of the drug-loaded LNCs were 8-fold and 11-fold, respectively, higher than those displayed by paclitaxel alone under the commercialised form Taxol®. After autopsy, no abnormalities were observed in the mean weights of the organs and histological studies revealed that the livers, spleens, lungs and kidneys were normal. Moreover, no lipid accumulation was observed in the liver, spleen or kidneys. Complete blood count was in agreement with normal values for healthy animals. Thus, the LNCs were well-tolerated with no sign of toxicity, whatever the intravenous injection plan.

8.3.3.2 Modified LNCs and Blood Circulation

Hoarau *et al.* studied the pharmacokinetics of post-inserted LNCs, particularly with DSPE-PEG2000 and DSPE-PEG5000 (2000 and 5000 Da PEG chain, respectively). After intravenous injection to rats, the post-inserted LNCs exhibited longer circulation properties: $t_{1/2}$ was about 8 h for the LNC containing 6mol% PEG5000 or 10mol% PEG2000.[39]

The encapsulation of DNA inside nanoparticles prolongs its blood circulation time by protecting it against the outer environment. Vonarbourg *et al.* encapsulated DNA lipoplexes in standard LNCs and showed their long-circulating capacity in comparison to free lipoplexes of which only 1% was detected in the blood 5 minutes after injection.[21] About 40% of the DNA nanocapsules (DNA-LNC) were still circulating in the blood after 1h. To improve the blood circulation of the DNA-LNCs, Morille *et al.* performed a post-insertion of DSPE-PEG2000 on the carrier surface.[29] Weaker complement activation was obtained with the post-inserted DNA-LNCs when compared to the DNA-LNCs. The $t_{1/2}$ was evaluated to be about 7.1 h when 10 mmol of DSPE-PEG2000 was post-inserted on the DNA-LNCs, which represented a significant increase in blood circulation time together with an increase in the area under the curve (AUC) of about 194% (compared to the DNA-LNCs without modification). Thus, an increase in the length of the PEGylated chains on the LNC shell improves circulation time, a prerequisite for achieving passive targeting.

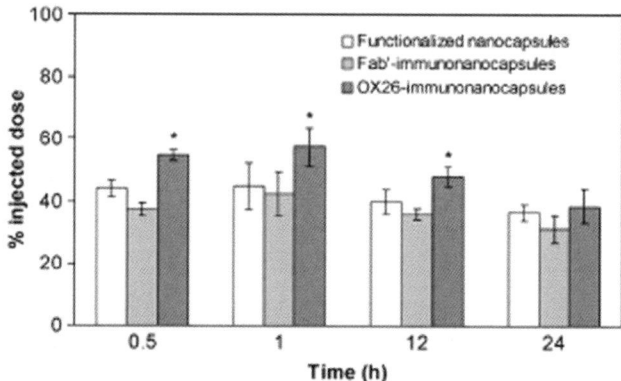

Figure 8.3.1   Liver uptake (ID% of radioactivity in the liver) of functionalised [188]Re-
loaded LNCs and [188]Re-loaded immuno-nanocapsules (adapted from
ref. 27). Values represent means $\pm$ SEM ($n = 5$); *p < 0.05.

In order to assess active targeting, ligands have been attached to the carrier
surface. Beduneau *et al.* grafted antibodies (OX26 antibody) and antibody
fragments (Fab') on the surface of LNCs which were previously post-inserted
by DSPE-PEG2000-maleimide.[27] 24 h after intravenous injection to healthy
rats, approximately 10% of ID remained in circulation. The elimination of
these immuno-nanocapsules was faster than that of LNCs without any ligands
attached. The pharmacokinetic parameters obtained with the Fab'-immuno-
nanocapsules were intermediate between the functionalised nanocapsules
(DSPE-PEG200-maleimide) and the OX26-immuno-nanocapsules, due to the
absence of Fc portions on the grafted moieties. The high concentration (60% of
ID) of OX26-immuno-nanocapsules found in the liver was in line with their
fast clearance (Figure 8.3.1). The targeting ligands on LNCs induce rapid
recognition by the MPS. In future formulation development, in order to
improve vectorisation, it will be necessary to find a balance between long
circulation time and active-targeting properties.

8.3.4   Lipid Nanocapsules and Cellular Uptake

In the development of nanomedicines, it is important to understand the
interactions between the carriers and the targeted cells. Several studies have
reported that LNCs can cross cell biological barriers and thus improve the
therapeutic effect of encapsulated molecules.

8.3.4.1   P-glycoprotein Inhibition

In tumour treatment, one major challenge is multidrug resistance due to P-
glycoprotein (P-gp). P-gp, located in the cell membrane, is an ATP-dependent
efflux pump eliminating anticancer drugs from the targeted tumour cells to the

extracellular matrix. This phenomenon prevents the accumulation of anticancer drugs within the cells. Lamprecht *et al.* tested etoposide-loaded LNCs and blank LNCs (hydrodynamic diameter from 25 to 100 nm) in different glioma cell lines (C6, F98 and 9L).[10] Etoposide-loaded LNCs inhibited P-gp. In parallel, sustained drug release from the LNCs led to a high level of drug accumulation inside the cells, resulting in the increase of *in vitro* cytotoxicity. When the cells were incubated with blank LNCs, P-gp inhibition also occurred and it was size and concentration-dependent. The LNC formulation with the smallest diameter was found to be the best inhibitor. Once internalised inside the cells, Solutol HS-15 was partially released from the LNC shell, preventing the action of P-gp. This property was investigated *in vivo* on a subcutaneous, F98 glioma model. With paclitaxel-loaded LNCs, the treatment by a single intratumoural injection statistically reduced the tumour volume and tumour mass compared to the free drug (Taxol®),[11] showing the importance of P-gp inhibition by the LNCs themselves.

8.3.4.2 Cellular Trafficking

The efficacy of a treatment is linked to the entry of the drug into the cell but also to the site of action of the transported molecule inside the cell (cytoplasm, nucleus, specialised organelles …). Therefore, knowledge of the fate of the nanocarriers inside cells is necessary. Paillard *et al.*, using radio and fluorescent-labelled LNCs, have highlighted three major points about glioma cells and LNC interaction.[13] Firstly, the internalisation of the LNCs in cells was rapid and based on an active process (2 minutes of exposure), mainly mediated through a cholesterol-dependent pathway, indicative of endocytosis. Secondly, the intracellular residence of the LNCs was long enough to ensure proper bioavailability of the drug inside the cell (independent on the cell nature). Finally, only a minor part of the LNCs was found in lysosomes (about 10% of the internalised LNC fraction after 2 h of exposure), which revealed a specific capacity of LNCs to escape lysosomes. For drugs that are sensitive to lysosomal degradation or that should reach extra-lysosomal organelles as targets, endo-lysosomal escape is fundamental. For drugs that require a certain cell cycle time, the long time residence of the carrier within cells is important.

In vivo transfection activity was tested on mice grafted with subcutaneous kidney tumours (HEK293β3 cells).[40] DNA-LNCs were administered intravenously, and luciferase expression was determined after dissection of the main organs responsible for elimination (liver, lung, spleen, kidney) and the tumour (Figure 8.3.2). Regardless the LNC formulation type, low luciferase expression was measured in liver, lung, spleen and kidney. However, strong luciferase expression in the tumour after injection of DSPE-PEG2000-coated DNA-LNCs was measured while no expression was observed with non-coated DNA-LNCs. Among the tested formulations, the DSPE-PEG2000-coated LNCs exhibited the longest blood circulation time together with the highest $t_{1/2}$. These properties, coupled with the enhanced permeability and retention (EPR)

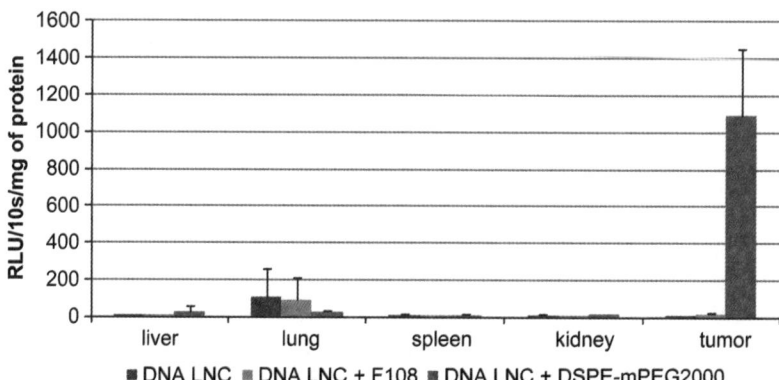

Figure 8.3.2   *In vivo* transfection efficiency 48 h after intravenous injection. DNA-
LNCs, F108-coated DNA-LNCs, or DSPE-PEG2000 DNA-LNCs.
Luciferase expression was tested in the liver, lungs, spleen, kidneys, and
the tumour (adapted from ref. 40).

effect (see below), could explain the better accumulation in the tumour tissue.
Finally, once internalised, the DSPE-PEG200-coated DNA-LNCs escaped
from endo-lysosomes without degradation and, thus, transfection was able to
occur. These DNA-LNCs appeared to be efficient vectors for the systemic
delivery of DNA, and the transfection was shown to be specific toward tumour
cells, with only minor transfection observed in other organs.

8.3.5   Lipid Nanocapsules and Passive Targeting *vs.* Active Targeting

8.3.5.1   Passive Targeting and the EPR Effect

The passive targeting concept is closely linked to the mean residence time of
the nanocarriers in systemic circulation. Long blood circulation times enable
the carriers to be more efficiently accumulated at the sites of action, such as
solid tumours. During the rapid growth of solid tumours, abnormal
development of blood vessels is observed: vasculature might be excessive but
it is irregular, leaky, dilated, and the endothelial cells are poorly aligned, and
the presence of fenestrations can be observed. In parallel, no or very little
lymphatic drainage is present. These properties are at the origin of the
enhanced permeation and retention (EPR) effect that helps the nanoparticles
to accumulate in tumours. Kahlid *et al.* showed, after systemic administration,
higher tumoural accumulation of docetaxel-loaded LNCs compared to the free
drug.[41] In this study, colon adenocarcinoma cells (C26) were subcutaneously
implanted in three different locations in mice. With longer PEG chains added
to the LNC surface, the drug accumulation in tumours was increased due to
longer blood circulation time. Morille *et al.* compared accumulation of DNA-
LNCs (with or without longer PEG chains on the surface) using another

tumour animal model (subcutaneously-implanted U87MG glioma cells).[29] Fluorescence imaging was used to evaluate the tissue distribution of the LNCs. Fluorescence intensity of the non-coated DNA-LNCs increased in the liver from 3 h after intravenous injection up to 24 h, whereas less accumulation in the liver was observed with 10 mmol DSPE-PEG2000-coated DNA-LNCs (Figure 8.3.3). At 24 h and 48 h after injection, the DSPE-PEG2000-coated DNA-LNCs displayed higher fluorescence intensity in the tumour compared to the non-coated DNA-LNCs.

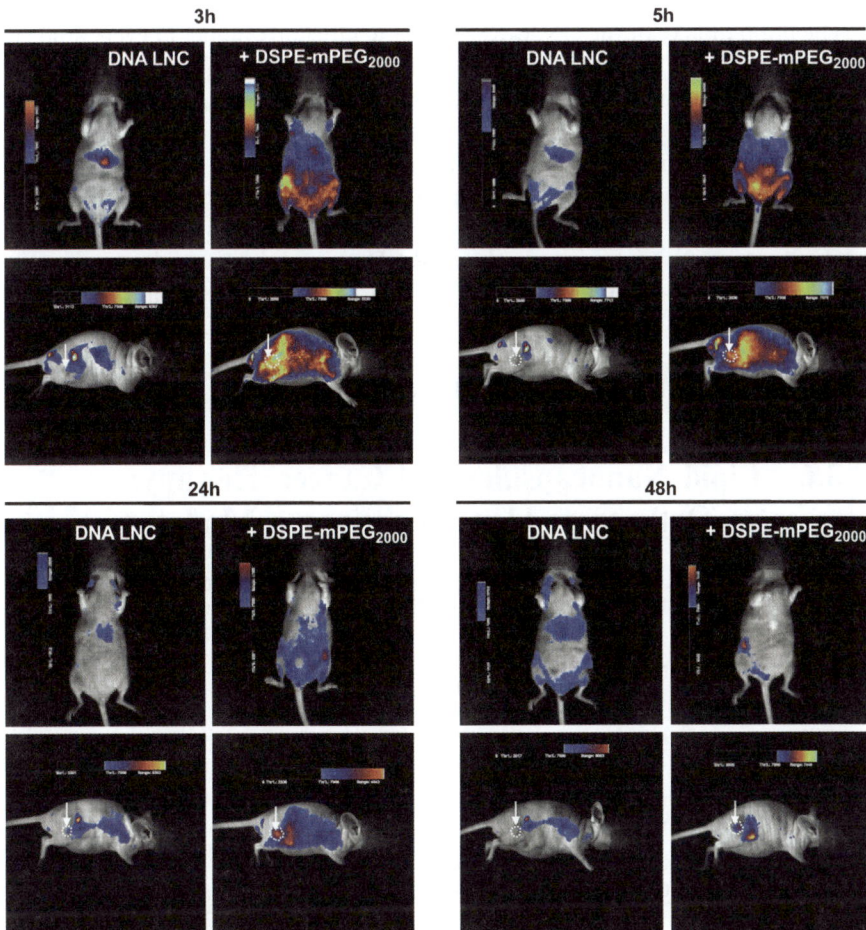

Figure 8.3.3 *In vivo* fluorescence imaging and optical images of athymic nude mice, bearing U87MG tumorus, after intravenous injection of DNA-LNCs or DSPE-PEG2000-coated DNA-LNCs. The coloured bar on the picture indicates the signal magnitude of the fluorescence emission from the animal. The tumour location is specified with a white arrow (adapted from ref. 29).

8.3.5.2   Active Targeting and Crossing the Blood-Brain Barrier

Gliomas are tumours located in the brain. To reach a glioma by systemic administration, the blood-brain barrier must be crossed. This necessitates active-targeting approaches for the treatment. As described previously, LNCs with monoclonal antibodies (OX26 MAb) or Fab' antibody fragments grafted on the surface have been evaluated for this purpose.[27] These immuno-nanocapsules bearing between 30–40 ligands/LNC were evaluated in brain targeting using a radionuclide ([188]Re). The retained antibody binds to the transferrin receptors (TfR), these being over-expressed on both endothelial and glioma brain cells.

The brain uptake of the OX26-LNCs was significantly higher at every studied time point when compared to the other formulations (Figure 8.3.4). For example, at 24 h, and compared to the non-targeted LNCs, the brain uptake of the Fab'-LNCs and the OX26-LNCs was 1.5 and 2-times higher, respectively. This enhanced brain accumulation was in the same range as that observed for OX26-immunoliposomes. This correlation demonstrates the interest of OX26 MAb in brain targeting, and the results also clearly demonstrate the ability of LNCs to target the brain *via* the TfR. Despite longer residence time in circulation, brain targeting using the Fab'-LNCs was less efficient in comparison to OX26-LNCs. A higher number of the Fab' fragments per LNC could increase the number of recognition sites and, consequently, improve brain targeting without significantly decreasing blood circulation time.

8.3.6   Lipid Nanocapsules and Cancer Therapy: an Orthotopic Hepatocarcinoma Model

Cancer therapy is one of the most important applications of LNCs. At the moment, several studies with *e.g.* bronchic, pancreatic, breast, colon cancer

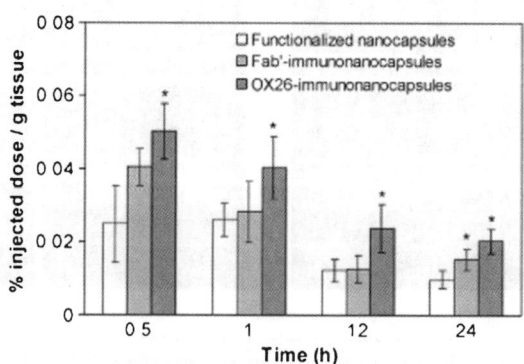

Figure 8.3.4   Brain targeting (ID% per g tissue) of functionalised [188]Re-loaded LNCs and [188]Re-loaded imuno-nanocapsules (adapted from ref. 27). Values represent means ± SEM ($n = 5$), $p < 0.05$.

and glioblastoma are under way. These studies use various ways to graft cancer cells in animals (principally subcutaneous as described in previous parts). Among the cancer models, orthotopic models with cancer cell implant in the organ are considered to mimic the best the real pathology.

Hepatocellular carcinoma (HCC) is considered to be one of the most common cancers in the world, and the most important malignant liver tumour. So far, no efficient therapy exists except surgical treatment (ablative treatment or hepatic transplantation) for some feasible cases. HCC is also known to be chemo-resistant to anticancer drugs by over-expressing P-gp. The therapeutic efficacy of paclitaxel-loaded LNC (Px-LNC) on an HCC orthotopic model *versus* a standard formulation of paclitaxel has been studied.[11]

To mimic clinical practices, the therapeutic scheme in the *in vivo* study consisted of four doses of paclitaxel (70mg m^{-2}) injected intravenously. First of all, similar results were found in the two control groups (*i.e.* saline control and blank-LNCs). On the contrary, the survival profiles of paclitaxel treated groups showed statistically significant differences with the saline control survival profile (Figure 8.3.5). Animals treated with the Px-LNCs showed the most significant increase in mean survival times compared to the controls, and the results were close to those obtained with the animals treated with

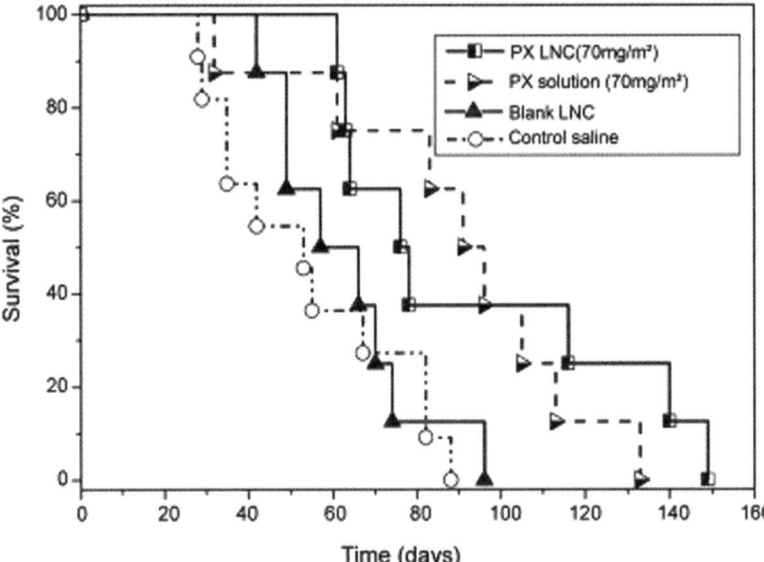

Figure 8.3.5 Survival curves of HCC tumour-bearing Wistar rats after various treatments: free paclitaxel solution (*n* = 8), blank LNCs (*n* = 8), Px-LNCs (*n* = 8) and saline control (*n* = 11). Each animal received one intravenous injection per week for 4 weeks. The starting point of the survival studies was fixed at the last day of treatment and the end point was determined by the death of the animal or weight loss for 3 consecutive weeks (adapted from ref. 11).

Px-solution. No animal from the control groups survived after 106 days (twice the median of the control saline group) whereas 25% of the animals treated with the free Px-solution and 37.5% of those treated with the Px-LNC survived for a longer time.

Using the same animal model, a recent study presented the effect of local radiotherapy by using radio-labelled LNCs (hepatic artery injection).[42] [188]Re-loaded LNCs were essentially confined to the liver, with an increasing uptake up to 24 h. On the contrary, [188]Re as a solution experienced a rapid stomach uptake. After the treatment with [188]Re-loaded LNCs, therapeutic efficiency was demonstrated by an increase in the median survival from 54 to 107%, as compared to the control groups. These results correlated well with magnetic resonance images (MRI) and histopathological analysis, demonstrating a slowdown in tumour progression. Moreover, HCCs are known to express angiogenic factors such as VEGF. Lower VEGF levels found in the [188]Re-LNC group, in comparison with the control groups, possibly reflect altered angiogenesis and, therefore, reduced tumour progression in the rat HCC model.

8.3.7   Lipid Nanocapsules and Inner Ear Diseases

Hearing loss is a major public health problem and it is ranked as the 8th commonest disease in the European Union. Damage of the sensory hair cells, nerve fibres and/or the supporting cells of the inner ear, are among the leading causes of hearing loss. Traditional treatment strategies for sensorineural hearing loss are unsatisfactory. Cochlear implants restore the auditory pathway in the context of severe hearing loss. Nevertheless, after implantation, hearing capacity varies. The challenge lies in the use of nanotechnology with local parenteral injection for the treatment of inner ear diseases.

Fluorescence-labelled LNCs with an FITC graft on the surface were tested *in vivo* in guinea pigs.[43] After local inner ear administration, the LNCs were detectable in various inner ear cell types: inner and outer hair cells, pillar cells, and cells located in the *stria vascularis*. Neither toxic effects nor structural and functional damage to the hair cells were observed. The hearing of the LNC-treated animals was completely preserved. Cytocochleograms did not reveal any *in vivo* pathological alterations at lower LNC concentrations, thus showing good biocompatibility and suitability of the LNC for inner ear drug delivery.

Another approach for LNC inner-ear administration was developed by Zou *et al.*[20] A gelfoam was saturated with fluorescence-labelled LNC (DIO or Nile Red (core) and FITC or RhB (shell)) and placed on the round window membrane of the animal after surgery (retro-auricular incision and a hole to attain the round window). Migration of the LNCs through the round window membrane was observed followed by the penetration of the LNCs into various cell types within the cochlea. After administration, within 30 minutes the LNCs reached the spiral ganglion, including the neurons and nerve fibres, Corti organ, and the lateral wall. This rapid LNC diffusion was explained by a paracellular pathway. Finally, faster and more extensive uptake of the LNCs in the inner hair cells, compared to the outer hair cells, was observed. This finding suggested that the main pathway for the LNCs to reach the hair cells was the "nerve pathway" (passage from the inner hair cells and pillar cells to the outer hair cells).

Concerning biocompatibility, Zhang *et al.* demonstrated that, when applied to the middle ear cavity, LNCs did not provoke hearing impairment, cell death, or morphological changes in the inner ear.[44] Cochlear neural elements such as synaptophysin, ribbon synapses, and neurofilament-200 were not affected by the administration of the LNCs. This suggested that LNCs are a potential vector for the delivery of drugs and proteins to inner ear targets.

8.3.8 Conclusions

Lipid nanocapsule technology presents many advantages in the pharmaceutical area for therapeutic and imaging applications. The formulation process is easy to perform and no toxic solvents are needed during preparation. Lipophilic drugs can be encapsulated in the oily core of the LNCs with high encapsulation yields, whereas the encapsulation of hydrophilic drugs remains feasible *via* double-emulsion or reverse-micelle formulations. Surface modifications of LNCs by physical approaches are easy to carry out in mild conditions. These modifications allow adding functionality to the LNCs without modifying the encapsulation rate.

Various *in vivo* studies have proved the potential of LNCs as drug delivery and imaging systems after parenteral administration. With LNCs, it is possible to formulate long-circulating nanocarriers for passive targeting to solid tumours. In tumour targeting, LNCs are able to enter cells and inhibit the action of P-gp. Moreover, LNCs can escape from endo-lysosomes and remain for a sufficiently long time in the cytoplasm to ensure drug action. This has been observed *in vivo* with DNA-LNCs with an increased transfection after systemic injection. The blood-brain barrier can be bypassed with actively-targeted LNCs, mediated by antibody or antibody fragments. The treatment of orthotopic HCC with LNCs containing anticancer drugs or radionuclides *via* intravenous or intra-arterial administration have been found to be encouraging. Finally, LNCs have been applied for local treatment in the inner ear. The LNCs were able to reach the round window from the middle ear to the inner ear, and they were found to be distributed in the cells related to hearing loss pathologies.

The LNCs presented in this chapter were suitable nanocarriers for local or systemic administration, whatever the pathology and the organs involved in the disease. At the moment, several modified LNC formulations are being tested *in vitro* and *in vivo* for different diseases. These formulations include LNCs that are actively targeted with the help of surface-attached peptides, or LNCs carrying siRNA.

References

1. K. Letchford and H. Burt, *Eur. J. Pharm. Biopharm.*, 2007, **65**, 259.
2. S. M. Moghimi, A. C. Hunter and J. C. Murray, *Pharmacol. Rev.*, 2001, **53**, 283.
3. K. Shinoda and H. Saito, *Colloids Interface Sci.*, 1969, **30**, 258.
4. A. Vonarbourg, P. Saulnier, C. Passirani and J.-P. Benoit, *Electrophoresis*, 2005, **26**, 2066.
5. B. Heurtault, P. Saulnier, B. Pech, J.-E. Proust, J. Richard and J.-P. Benoit, *Patent No. WO02688000*, 2000.
6. B. Heurtault, P. Saulnier, B. Pech, J.-P. Benoit and J.-E. Proust, *Coll. Surf. B: Biointerfaces*, 2003, **30**, 225.
7. A. Lamprecht, Y. Bouligand and J.-P. Benoit, *J. Controlled Release*, 2002, **84**, 59.
8. A. Lamprecht, J. L. Saumet, J. Roux and J.-P. Benoit, *Int. J. Pharm.*, 2004, **278**, 407.
9. M. Pereira de Oliveira, E. Garcion, N. Venisse, J.-P. Benoit, W. Couet and J. C. Olivier, *Pharm. Res.*, 2005, **22**, 1898.
10. A. Lamprecht and J.-P. Benoit, *J. Controlled Release*, 2006, **112**, 208.
11. F. Lacoeuille, F. Hindre, F. Moal, J. Roux, C. Passirani, O. Couturier, P. Cales, J. J. Le Jeune, A. Lamprecht and J.-P. Benoit, *Int. J. Pharm.*, 2007, **344**, 143.
12. E. Garcion, A. Lamprecht, B. Heurtault, A. Paillard, A. Aubert-Pouessel, B. Denizot, P. Menei and J.-P. Benoit, *Mol. Cancer Ther.*, 2006, **5**, 1710.
13. A. Paillard, F. Hindré, C. Vignes-Colombeix, J.-P. Benoit and E. Garcion, *Biomaterials*, 2010, **31**, 7542.
14. E. Roger, F. Lagarce, E. Garcion and J.-P. Benoit, *Eur. J. Pharm. Sci.*, 2010, **40**, 422.
15. J. Hureaux, F. Lagarce, F. Gagnadoux, M.-C. Rousselet, V. Moal, T. Urban and J.-P. Benoit, *Pharm. Res.*, 2010, **27**, 421.
16. A. Malzert-Freon, S. Vrignaud, P. Saulnier, V. Lisowski, J.-P. Benoit and S. Rault, *Int. J. Pharm.*, 2006, **320**, 157.
17. E. Allard, C. Passirani, E. Garcion, P. Pigeon, A. Vessieres, G. Jaouen and J.-P. Benoit, *J. Controlled Release*, 2008, **130**, 146.
18. E. Allard, N. T. Huynh, A. Vessières, P. Pigeon, G. Jaouen, J.-P. Benoit and C. Passirani, *Int. J. Pharm.*, 2009, **379**, 317.
19. E. Allard, D. Jarnet, A. Vessieres, S. Vinchon-Petit, G. Jaouen, J.-P. Benoit and C. Passirani, *Pharm. Res.*, 2010, **27**, 56.
20. J. Zou, P. Saulnier, T. Perrier, Y. Zhang, T. Manninen, E. Toppila and I. Pyykkö, *J. Biomed. Mater. Res. B Appl. Biomater.*, 2008, **87**, 10.
21. A. Vonarbourg, C. Passirani, L. Desigaux, E. Allard, P. Saulnier, O. Lambert, J.-P. Benoit and B. Pitard, *Biomaterials*, 2009, **30**, 3197.
22. A. Vonarbourg, C. Passirani, P. Saulnier, P. Simard, J.-C. Leroux and J.-P. Benoit, *J. Biomed. Mater. Res. A*, 2006, **78**, 620.

23. M. Roger, A. Chevreul, M.-C. Venier-Julienne, C. Pasirani, L. Sindji, P. Schiller, C. Montero-Menei and P. Menei, *Biomaterials*, 2010, **31**, 8393.
24. E. Jestin, M. Mougin-Degraef, A. Faivre-Chauvet, P. Remaud-Le Saec, F. Hindre, J.-P. Benoit, J. F. Chatal, J. Barbet and J. F. Gestin, *Q. J. Nucl. Med. Mol. Imaging*, 2007, **51**, 51.
25. S. Ballot, N. Noiret, F. Hindre, B. Denizot, E. Garin, H. Rajerison and J.-P. Benoit, *Eur. J. Nucl. Med. Mol. Imaging*, 2006, **33**, 602.
26. E. Allard, F. Hindre, C. Passirani, L. Lemaire, N. Lepareur, N. Noiret, P. Menei and J.-P. Benoit, *Eur. J. Nucl. Med. Mol. Imaging*, 2008, **35**, 1838.
27. A. Beduneau, F. Hindre, A. Clavreul, J.-C. Leroux, P. Saulnier and J.-P. Benoit, *J. Controlled Release*, 2008, **126**, 44.
28. M. Morille, C. Passirani, E. Letrou-Bonneval, J.-P. Benoit and B. Pitard, *Int. J. Pharm.*, 2009, **379**, 293.
29. M. Morille, T. Montier, P. Legras, N. Carmoy, P. Brodin, B. Pitard, J.-P. Benoit and C. Passirani, *Biomaterials*, 2010, **31**, 321.
30. N. Anton, P. Saulnier, C. Gaillard, E. Porcher, S. Vrignaud and J.-P. Benoit, *Langmuir*, 2009, **25**, 11413.
31. N. Anton, H. Mojzisova, E. Porcher, J.-P. Benoit and P. Saulnier, *Int. J. Pharm.*, 2010, **398**, 204.
32. N. T. Huynh, E. Roger, N. Lautram, J.-P. Benoît and C. Passirani, *Nanomedicine*, 2010, **5**, 1415.
33. T. Perrier, P. Saulnier and J.-P. Benoit, *Chem. Eur. J.*, 2010, **16**, 11516.
34. T. Perrier, P. Saulnier, F. Fouchet, N. Lautram and J.-P. Benoit, *Int. J. Pharm.*, 2010, **396**, 204.
35. J. Zou, P. Saulnier, T. Perrier, Y. Zhang, T. Manninen, E. Toppila and I. Pyykko, *J. Biomed. Mater. Res. B: Applied Biomat.*, 2008, **87**, 10.
36. A. Beduneau, P. Saulnier, F. Hindre, A. Clavreul, J.-C. Leroux and J.-P. Benoit, *Biomaterials*, 2007, **28**, 4978.
37. S. Hirsjarvi, Y. Qiao, A. Royere, J. Bibette and J.-P. Benoit, *Eur. J. Pharm. Biopharm.*, 2010, **76**, 2010.
38. C. Passirani and J.-P. Benoit, in *Biomaterials for delivery and targeting of proteins and nucleic acids*, CRC Press, Inc., Boca Raton, 2005, **76**, p 187.
39. D. Hoarau, P. Delmas, S. David, E. Roux and J.-C. Leroux, *Pharm. Res.*, 2004, **21**, 1783.
40. M. Morille, C. Passirani, S. Dufort, G. Bastiat, B. Pitard, J. L. Coll and J.-P. Benoit, *Biomaterials*, 2011, **32**, 2327.
41. M.N. Khalid, P. Simard, D. Hoarau, A. Dragomir and J.-C. Leroux, *Pharm. Res.*, 2006, **23**, 752.
42. C. Vanpouille-Box, F. Lacoeuille, J. Roux, C. Aube, E. Garcion, N. Leparuer, F. Oberti, F. Bouchet, N. Noiret, E. Garin, J.-P. Benoit, O. Couturier and F. Hindre, *Plos One*, 2011, **6**, e16926.
43. V. Scheper, M. Wolf, M. Scholl, Z. Kadlekova, T. Perrier, H.-A. Klok, P. Saulnier, T. Lenarz and T. Stover, *Nanomed.*, 2009, **4**, 623.
44. Y. Zhang, W. Zhang, M. Löbler, K. P. Schmitz, P. Saulnier, T. Perrier, I. Pyykkö and J. Zou, *Int. J. Pharm.*, 2011, **404**, 211.

Section 9
Drug Delivery in Tissue Engineering

CHAPTER 9.1

Drug Delivery in Tissue Engineering: General Concepts

T. SIMÓN-YARZA[a], E. GARBAYO[a], E. TAMAYO[a], F. PRÓSPER[b] AND M. J. BLANCO-PRIETO*[a]

[a] Pharmacy and Pharmaceutical Technology Department, School of Pharmacy, University of Navarra, Pamplona, Spain; [b] Hematology Service and Area of Cell Therapy, Clínica Universidad de Navarra, Foundation for Applied Medical Research, University of Navarra, Pamplona, Spain
*E-mail: mjblanco@unav.es

9.1.1 Tissue Engineering: Definition and Objectives

Tissue and organ failure is a serious problem that will only increase as the population grows and ages. Treatment options include transplantation, surgical repair, artificial prostheses, mechanical devices and, in a few cases, drug therapy.[1] Even if organ transplantation has achieved significant advances in recent years, many limitations and unsolved issues still remain. For instance, transplants are limited by the critical donor shortage and the need for lifelong immunosuppression. Tissue engineering (TE) offers potential to ameliorate the limitations of current therapy and represents the future of transplantation in medicine. TE seeks to create, repair and/or replace tissues and organs by using cells, biomaterials and biologically active molecules administered alone or in combination.[2] Depending on the tissue or organ to be repaired, one of these components will dominate, but typically a combination of two or more is required. The main objective of TE could therefore be defined as to restore, maintain or enhance tissue and organ function.[2] This multidisciplinary field

RSC Drug Discovery Series No. 22
Nanostructured Biomaterials for Overcoming Biological Barriers
Edited by Maria Jose Alonso and Noemi S. Csaba
© The Royal Society of Chemistry 2012
Published by the Royal Society of Chemistry, www.rsc.org

integrates aspects of engineering with biology and medicine. The term "TE" arose in the late 1980s.[3] Researchers have been examining the potential of cell-based therapies to reconstruct functional tissue for almost 30 years. Unfortunately, little success has been reported in the literature using cell-based therapies, mainly due to the poor survival of grafted cells during the first few days after transplant. Cells were next administered in combination with growth factors (GFs) that improved their survival and differentiation in order to overcome poor cell survival. But limited evidence of success was again described. These findings led researchers to consider that the ideal type of TE product would be one which involved a biodegradable polymer support into which cells of the appropriate phenotype are introduced, along with some suitable drugs such as GFs, which improve their survival and differentiation. These three fundamental components (cells, engineered materials, and signaling molecules) are also known as the TE triad (Figure 9.1.1). The parallel development of artificial biomaterials obtained from biologically derived and synthetic materials contributed decisively to the early development of TE.

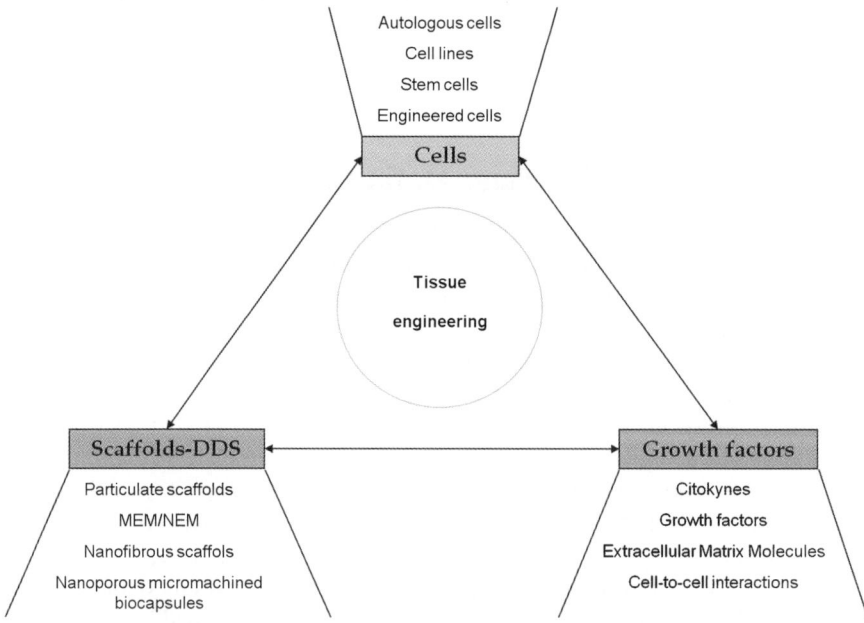

Figure 9.1.1   The TE triad includes scaffold, cells and soluble regulators (*e.g.* GFs). Depending on the tissue to be repaired, one of these 3 components will dominate, but typically a combination of two or more will be required. The ideal TE product would combine a biodegradable support into which cells of the appropriate phenotype are introduced, along with some suitable drugs such as GFs, which improve their survival and differentiation. DDS: Drug delivery systems; MEM/NEM: micro/nanoelectromechanical.

9.1.1.1 Cell Sources in Tissue Engineering

Cell therapies currently being investigated include autologous primary cells, cell lines and various stem cells, those involving cells from a donor (allogeneic cells) and transgenic cells from other species (xenogenic cells). Unlike other cell types, stem cells, characterized by their capacity for self-renewal and for their ability to differentiate into specialized cell types, can be expanded relatively easily *in vitro*. This makes it possible to obtain a large quantity of cells in a relatively short period of time and to have a potentially limitless supply of cells, which makes this a very attractive option for TE applications. They can be classified according to their origin into:[4,5]

–Embryonic stem cells (ESCs): these are pluripotent cells with the capacity to differentiate into almost any kind of cell type. However, harvesting of ESCs requires destruction of human embryos, raising significant ethical concerns.
–Adult stem cells: two types of stem cells obtained from adults are being used in TE therapy.

- Mesenchymal stromal cells (MSCs), which can be easily obtained from bone marrow, adipose tissue or umbilical cord blood, are multipotent stem cells that can differentiate into a variety of cell types. Interestingly, MSCs secrete large quantities of cytokines and GFs that are both immunomodulatory and trophic and that are responsible to a great extent for the beneficial effect of MSCs, as they also mediate angiogenesis stimulation besides decreasing inflammation. This feature makes MSCs extremely attractive for use in TE.
- Tissue-derived specific stem cells with a more restricted differentiation capacity contribute to sustain cell turnover in their respective tissues. They include epithelial stem cells, hepatic and pancreatic stem cells, endothelial progenitor cells (EPCs), cardiac progenitor cells (CPCs), neural stem cells (NSCs), *etc.*

An emerging potential source of adult stem cells is to be found in induced pluripotent stem cells (iPSCs), which are artificially generated stem cells.[6] iPSCs are generated from terminally differentiated somatic cells *via* nuclear reprogramming. These cells may overcome the ethical problems of ESCs and can potentially be derived from any adult patient, with a predicted similar pluripotency of ESCs. Many researchers believe iPSCs hold promise for routine research and future medical use. However, the scientific community should be cautious because some iPSCs limitations have already been described. Indeed, there are some important questions that must be addressed with respect to alterations of cell-specific regulatory pathways, differences in gene expression and epigenetic control, which may result in tissue chimerism or cell malfunction.[7,8]

Moreover, when combining cells with scaffolds in TE, it is important to assess the genotypic instability, tumorigenicity and phenotypic profile of the cells in the critical manufacturing process to ensure the safety profile of the product.

9.1.1.2   Bioactive Molecules in Tissue Engineering

GFs are naturally occurring regulatory proteins that affect cell and tissue function by influencing the differentiation of cells, altering their biochemical activity, and regulating their rate of proliferation. Nowadays, over 100 GFs have already been identified. Some of them have been produced by recombinant technology but only a few have been approved for clinical use in humans due to the difficulties in obtaining large amounts of purified GF together with the higher costs associated with their production. Recently, in different pathologies it has been reported that multiple GFs delivery turns out to be more effective than the administration of a single GF since it mimics better the natural microenvironments of tissue formation and repair.[9,10] The difficulty of this approach principally concerns finding the optimized GF ratio, each factor at a physiological dose and in a specific spatiotemporal pattern. Moreover, the direct administration of GFs has clear limitations. Infusion of GFs into the systemic circulation or direct injection in the injured tissue has failed on numerous occasions. To overcome these limitations, several technologies have been explored. Encapsulation of GFs in drug delivery systems (DDS), micro and nanoparticles for instance, may allow controlled, localized release of the GF to yield a desirable concentration over a certain period of time. DDS may also protect proteins from degradation, preserving their bioactivity during release.

GFs have recently attracted more interest since the discovery that the beneficial effect of MSCs may be mediated by their paracrine secretion.[11–13] The GFs secreted by the MSCs identified so far are vascular endothelial growth factor (VEGF), fibroblast growth factor-2 (FGF-2), hepatocyte growth factor (HGF), transforming growth factor beta (TGF-β), insulin-like growth factor 1 (IGF-1) and interleukins (as IL-10 and IL-23), among others.[12,13] The potential for administering the GFs responsible for the beneficial effect of MSCs, instead of stem cells, is currently being investigated. Importantly, advances in understanding the mechanism by which adult stem cells produce GFs may optimize their beneficial paracrine and autocrine effects. Further investigations are needed to address the potency of intrinsic paracrine factors released by stem cells as well as the optimal localization, timing, delivery, and stem cell type to be transplanted, allowing earlier and more effective clinical therapies.

The fabrication of large tissues and whole organs *de novo* remains a major challenge. Moreover, to successfully progress in TE, a methodology that combines empirical data with mathematical and statistical techniques and models is required. Once again, it is clear that TE is a multidisciplinary field that should be supported by many disciplines like engineering, materials science, informatics, biology and medicine (as will be discussed in section 9.1.5).

9.1.1.3 Scaffolds in Tissue Engineering

Scaffolds are artificial three-dimensional structures capable of supporting physiological activities of cells and tissue formation. They play an essential role in TE. The first materials used for scaffold fabrication were non-biodegradable polymers. However, the use of biodegradable polymers is preferred because they do not require surgical removal when the treatment is finished. Both natural (collagen, alginate, fibrin and chitosan among others) and synthetic materials [poly lactic-co-glycolic acid (PLGA), poly-ε-caprolactone (PCL), ethylene-vinyl acetate (EVA), poly(2-hydroxyethyl methacrylate) (pHEMA)] have been tested in clinical practice.[14,15] The introduction of synthetic scaffolds improved the application spectra of TE, since difficulties associated with the use of natural scaffolds, like complexities with purification, immunogenicity and pathogen transmission, were diminished. The next generation of biomaterials should be resorbable, bioactive and, once implanted into the body, would help the patient heal itself.[16] One of the first considerations when designing a scaffold is the choice of the material. Some of the most important aspects of the material that should be carefully taken into account are: (1) the maintenance of an appropriate shape after implantation, (2) the possibility of sterilization prior to implantation, (3) tissue compatibility and lack of immune response, (4) the appropriate degradation rate for the desired application and (5) the ability to provide a controlled release of the drug while supporting cell functions in the case of materials that encapsulate drugs. Interestingly, the function of biomaterials in the field of TE has evolved from the early ideas of an inert scaffolding device that served just as a physical support for cell adhesion and growth to one in which biomatrices are actively contributing to tissue regeneration. In this sense, the use of biomaterials has progressed from materials which may repair or replace diseased or damaged tissue, to the use of controlled 3D scaffolds in which cells can be seeded before implantation and which release GFs that can exert biological action over the cells or over the tissue. These living tissue constructs should be functionally, structurally and mechanically equal to (if not better than) the tissue they must replace.[17]

Materials or scaffolds can be used both as substrates for cells to improve their survival and differentiation because the majority of cells are anchorage-dependent, meaning they will die if an adhesion substrate is not provided, or as encapsulating materials that enclose cytokines, GFs or cells with the purpose of mimicking cell paracrine secretions. When the objective is to use the materials as substrates for cells, the main idea is to include novel biomaterials designed to direct cell organization, growth and differentiation in the process of forming functional tissue by providing physical, mechanical and chemical cues. In microencapsulation-based cell therapy the aim is to use the cells as natural biological mini-pumps that continuously produce and deliver therapeutic molecules. The microcapsule should protect encapsulated cells against the immune system and allow the diffusion of nutrients and oxygen towards the cells as well as the therapeutic agent in the opposite direction. The most complex challenges of this approach include controlling the GF release rate, the cell survival/replication rate within the capsule and the successful prevention of immune rejection, which hampers its reproducibility.[18-21]

9.1.2 Drug Delivery Systems: Definition and Objectives

The delivery of drugs is not a new field. The basic goal of controlled DDS is to deliver a biologically active molecule at a desired rate for a desired duration, so as to maintain the drug level in the body within the therapeutic window. DDS offer many advantages over conventional dosage forms. These systems, can, in general, provide targeted (cellular/tissue) delivery of drugs, improve bioavailability, solubilize drugs for intravascular delivery, and improve therapeutic agent stability against enzymatic degradation.[22] Progress made in the field of drug delivery has accelerated enormously over the past two decades in parallel with major discoveries of new low molecular weight active molecules, and, primarily, the introduction of the therapeutic products from biotechnology, such as proteins and peptides, which pose particular physicochemical and biopharmaceutical challenges for drug administration.

Polymer implants belong to the first DDS generations.[22] Drugs were either coated on the implant surface or incorporated into the polymer implant. These initially developed DDS were based on non-degradable polymers that needed implant removal. Thus, novel devices were proposed.

Since GFs delivery has become an area of growing interest in the TE field, the next section will review DDS and polymers used for protein delivery. Among DDS, the most common systems used for protein delivery are particulate systems including micro and nanoparticles[23] (nanoparticles will be reviewed later on this section). DDS protect protein from degradation and allow more than one protein delivery according to the need for a specific application. DDS can be designed in various shapes and sizes and made with different biodegradable and non-biodegradable materials offering multiple protein release profiles (see Figure 9.1.2).[24] Two possible delivery rates are feasible: continuous and pulsatile. For continuous protein delivery, several polymers have been investigated. First, non-biodegradable polymers, ethylene-vinyl acetate copolymers (EVAc) and silicones were tested.[24] Drug delivery from these polymers is based on diffusion (Figure 9.1.2 A and D). At present, the use of EVAc for protein controlled release is rare due to the non-biodegradable nature of the polymer. A novel cellular delivery silicon-based platform forming "nanoporous micromachined biocapsules" for cell encapsulation and immunoprotection is now under investigation, and represents a more recent approach using non-biodegradable polymers.[25] In the near future, the membrane will be a multifunctional system that incorporates a biosensing technology that release immunosuppressive agents in a controlled way, with self-cleaning capabilities and with the ability to modulate angiogenesis and tissue response *via* surface microarchitecture or immobilized GFs.[25] To avoid non-biodegradable material inconvenience, like polymer surgical removal once the drug is supplied, biodegradable polymers are preferred. These polymers

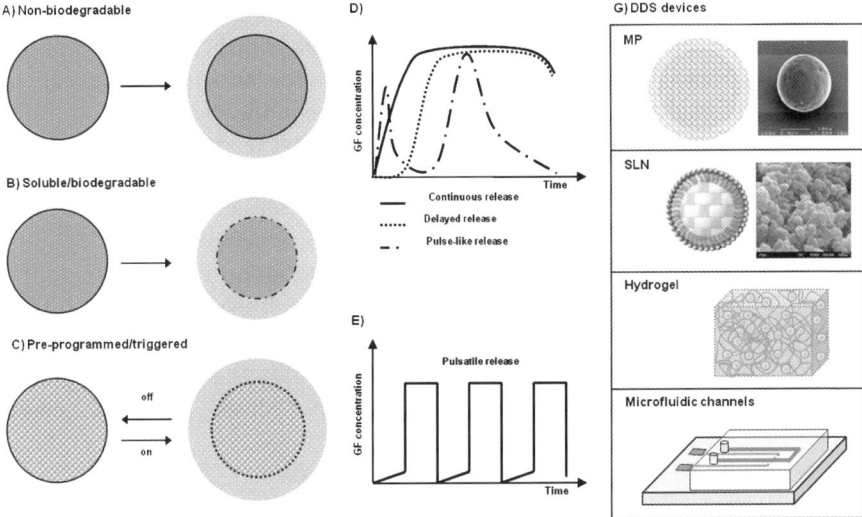

Figure 9.1.2   Types of DDS used in TE and their possible delivery profiles. GF release from non-biodegradable (A) and soluble/biodegradable (B) materials could be continuous, delayed or pulsatile-like (D). Programmed/triggered devices (C) can achieve pulsatile release after stimulus application (E). Schematic representation and scanning electron microscopy (SEM) images of different DDS used in TE (G): polymeric microparticle (MP), solid lipid nanoparticle (SLN), GF-loaded hydrogel and microfluidic channels.

release the drug by a combination of polymer degradation and drug diffusion (Figure 9.1.2 B and D). Depending on their nature, biodegradable polymers can be also classified into two categories: natural and synthetic. The lack of purity and homogeneity of natural ones and the disease transmission risk have decreased their use. Among synthetic biodegradable polymers, aliphatic polyesters like poly lactic acid (PLA) and PLGA attracted great interest due to their biodegradable properties, their satisfactory biocompatibility and the absence of significant toxicity.[26,27] They have suitable mechanical properties for use in micro and nano-encapsulation. These polymers are approved for use in humans by the US Food and Drug Administration (FDA) and have been used extensively in medicine in a variety of applications.[28] PLGA is totally biodegradable by bulk erosion. The biodegradation rate may vary from < 1 month to a period of a few years. The drug release rate can be modulated by the copolymer ratio and molecular weight variation. PLGA systems can be easily sterilized by γ-irradiation allowing their surgical use.[29,30] In total, more than 100 different molecules have been incorporated in PLGA microspheres since their first application[31,32] (see polymer microparticles in Figure 9.1.2 G) and different PLGA formulations for protein release are already on the market.[24] Other polymers investigated for protein delivery are polyanhydrides. They undergo surface erosion providing a better and easier control over the protein release kinetics through the material formulation.[24] Recently there has been growing interest in biodegradable polymers which interact with the external biological environment to stimulate tissue regeneration. This could be achieved by loading the scaffold with bioactive molecules or by scaffold surface modification.

As previously mentioned, proteins could be released in a pulsatile mode which is usually characterized by a rapid, transient release within a short time period followed by a pre-determined off-release interval (Figure 9.1.2 C and E). Depending on the principles that trigger the release, these devices can be classified into "programmed" and "triggered" delivery systems. In programmed DDS, the release is governed by the device inner mechanism and in triggered DDS by changes in the device physiological environment or by external stimuli (temperature changes, electric or magnetic field, ultrasound and irradiation).[24]

Apart from particulate systems, other DDS used for protein delivery in TE are hydrogels (Figure 9.1.2 G). They consist of a network of polymer chains that can absorb large amounts of water without dissolving. Natural polymers such as collagen, gelatin, fibrin, and glysocaminoglycans are the most commonly used hydrogels for tissue regeneration and drug delivery. They can be used in cartilage and vascular TE since they exhibit high diffusivity to oxygen and other water-soluble metabolites. In order to facilitate nutrient and soluble factor exchange, microfluidic channels are being used within synthetic cell laden hydrogels (Figure 9.1.2 G).[33] The channels are also potentially suitable to promote vascularisation.[34]

Bio-nanotechnology, which is a branch of nanotechnology and one of the most promising fields of nanomedicine, offers promising perspectives for the development of DDS with great potential in TE. This technology is based on the possibility of integrating bioactive compounds into nanoscale devices, generally characterized by sizes ranging from 1 to 1000 nm. Nanomedicines include therapeutic polymers, polymeric micelles, nanoemulsions, liposomes, nanosuspensions, and nanoparticles. Polymeric and lipid based nanosystems as carriers for drug molecules will be reviewed next in this section. Polymeric nanoparticles consist of a polymer matrix, usually formed by a biodegradable and biocompatible polymer and a bioactive molecule that can be entrapped within or immobilized into the polymeric matrix. On the other hand, lipid nanoparticles, either solid lipid nanoparticles (SLN; see Figure 9.1.2 G), nanostructured lipid carriers (NLC), lipid drug conjugates (LDC) or surface modified lipid based nanosystems, are composed of physiological lipids (for details of the manufacturing methods see ref. 35). These are capable of encapsulating hydrophobic and hydrophilic drugs, and they also provide protection against chemical, photochemical or oxidative degradation of drugs, as well as the possibility of sustained release of the incorporated drugs. Along with these last issues, the feasibility of scaling up for large scale production and the low cost of lipids as compared to biodegradable polymers or phospholipids have favoured their use as potential DDS.[35] One of the most exciting nanoparticle applications is the delivery of GFs in TE. Moreover, particulate systems developed for single drug delivery could be mixed and simultaneously administered to the target sites to facilitate multiple drug delivery and to achieve controlled release when necessary.

In parallel to the development of nanoparticulate delivery systems, significant improvements have also been made in the field of micro/nanoelectromechanical (MEM/NEM) device-based drug delivery.[36] For instance, implantable microchips containing nanosized reservoirs have been developed to deliver drugs for long time periods in a precisely controlled manner. Nanofibrous scaffolds made of different materials are further examples of nanosystems that are actually under investigation in TE as they exhibit a similar structure to protein nanofiber in the extracellular matrix (ECM). Within tissues, cells are surrounded by ECM that provides support to the cells and directs cell behavior *via* cell-ECM interactions. Thus, the ability to engineer biomaterials that emulate the complexity and functionality of ECM is essential for successful regeneration of tissues. There are three dominant nanofabrication methods: electrospinning, self-assembly and phase separation. In addition, biological factors can be incorporated in nanofibrous scaffolds by either modified electrospinning or self-assembly technology to use them as DDS. Adjustments to control the thickness and porosity of the polymer can be made in order to improve the release kinetics of the GFs. Another exciting advancement made with nanotechnology is the application of nanocoating or nanostructured surfaces that can improve the biocompatibility and adhesion properties of the existing biomaterials.[36] Nanotechnology can enable the design and fabrication of biocompatible scaffolds at the nanoscale and control the spatiotemporal release of biological factors, resembling native ECM, to direct cell behaviors, and eventually lead to the creation of implantable tissues.[36] In the next section the essential aspects to bear in mind when designing and developing DDS for TE applications will be discussed.

9.1.3 Drug Delivery as a Tool for Tissue Engineering

GFs delivery plays an important role in TE. Firstly, single GFs were administered as recombinant proteins by several methods: bolus injection; adsorbed on implant surfaces, in collagen sponges or in porous coatings; constant delivery *via* osmotic pumps or particulate systems; controlled release from encapsulating systems. However, these approaches mainly failed in obtaining successful regeneration because of the loss of bioactivity, location and need for multiple factors. Nowadays it is known that new tissue growth, both *in vivo* during development and *in vitro* in TE applications, is a complex multi-step process that requires exact spatial and temporal interaction between three essential elements: cells, multiple instructive molecules and a 3D ECM.[37] Therefore, the challenge is how to administer these elements to repair a specific organ or tissue successfully. At this point, the development of suitable delivery systems is critical and necessary. The drug delivery approach is the tool for the local administration of all the information useful to guide the functional tissue regeneration *in situ*, without harmful effects on the rest of the organism. So, in this way the repairing cells obtain data about the space (where they are), their identity, the functions to carry out, the duration of their actions and the state of the tissue (neighborhood) (see Figure 9.1.3). New strategies are being developed to modify the physico-chemical properties of the materials in order to achieve desirable features in each delivered element for tissue repair.[38]

–Scaffold: this acts as a multifunctional 3D support. It is the element which must mimic the ECM, allowing cell penetration, survival and activity, while controlling the volume, shape and extent of the newly growth tissue. Moreover the scaffold may act as a biochemical source, delivering biological signals with a predictable tuneable spatio-temporal pattern. Ultimately the scaffold must be compatible with the tissue to avoid rejection.

–Cells: they act as precursor cells which stimulate new tissue growth either directly (by *in situ* proliferation and differentiation) or indirectly (by secretion of paracrine factors to attract and stimulate local endogenous cells). The cells loaded into therapeutic devices could be stem cells or adult differentiated cells, which also could harbour therapeutic genes. Cells must be delivered into the scaffold along with the necessary signals to help survival and to modulate cell activity in the desirable direction. Moreover, in tissues bearing resident progenitor cells, a suitable therapeutic strategy would be to stimulate the mobilization and differentiation of these cells to the damage site by delivering the appropriate signals (specific GFs and cytokines).

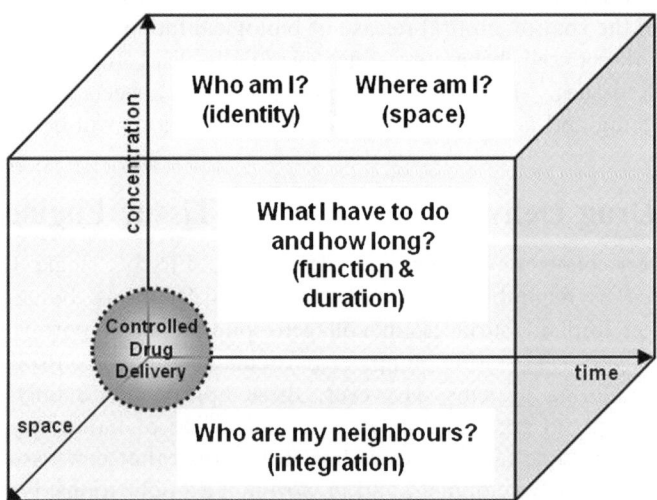

Figure 9.1.3   Drug delivery as a tool for TE. The ideal device is a multifunctional scaffold that mimics tissue architecture and releases biological orders by setting up spatio-temporal patterns. In this way, this scaffold supports regenerative cells to perform localized and functional tissue regeneration.

–Biological signals: the molecules that provide biological information can be GFs, cytokines, enzymes, antibodies, drugs, *etc.* They are delivered to modulate the cellular activity by stimulating tissue regeneration at different levels (survival, proliferation, migration, adhesion, secretion of factors, differentiation, *etc.*). These factors have to be supplied in a controlled manner to allow preservation of their bioactivity (since the injury tissue has high protease activity), control over the dose, localized delivery and a predetermined spatio-temporal release profile to the injury area for each factor. This means designing suitable carriers to achieve the most efficient loading in addition to the best release mechanism, depending on the therapeutic application (in the next section some examples of carriers developed to repair a specific organ or tissue are shown). To achieve this aim the biological agents may also undergo chemical modifications (such as derivation, dimerisation, oligomerisation, enzymatic cleavage, *etc.*) in order to modulate intrinsic molecular properties or molecule-carrier interactions. Moreover, as previously described, the release can be engineered to obtain continuous or pulsatile modes, wherein a specific concentration can be maintained for an extended duration or the release of a bioactive molecule can be triggered at the desired time.[24]

In summary, the drug delivery approach represents a highly versatile TE tool which allows the rational design of new medical devices for functional tissue regeneration *in situ* better than *ex vivo*. The construction of biologically inspired scaffolds involves substantial challenges, because drug delivery must evolve to assume increasing information about the molecular and cellular mechanisms which govern tissue regeneration. It is now clear that multiple GFs and cytokines acting in a spatio-temporal regulated sequence are required. However, what signals, what order and what duration may be specific to each tissue, and therefore controlled release methodologies must be fitted. Thus, new biomaterials besides new loading and release methods are being developed and assayed to answer crucial issues in TE: what, where, how and when to deliver in order to achieve efficient and safe tissue regeneration in each particular case. A great variety of carriers and scaffolds are being suited to the specific application: nanoparticles to pass vessels, microparticles for tissue residence, membranes for epithelial organs, scaffolds to rebuild ECM, *etc.* Moreover, once those concerns are solved and the device has been designed and tested, the drug delivery has to be scaled up for its use in clinical applications (this question is addressed below). Finally, several new challenges will appear when these "intelligent biosystems" are implanted in a pathological scenario in patients.

9.1.4 State of the Art of Drug Delivery in Tissue Engineering and Possible Applications

At the beginning of TE, the first steps were focused on cell-based therapy. In fact, the rise in the area is described as a result of the progress made in understanding cell biology and manipulation, including *in vitro* techniques.[39] However, as steps forward are being made in our knowledge of the natural repair processes, the need to provide the optimal environment to the cells that simulates the natural repair scene in a better way is now crucial. It is quite clear that the 3D conformation and the orchestrated delivery of a variety of factors are of great importance to achieve effective tissue repair.[40] But it is also necessary to supply these elements (matrix, stem cells and signalling molecules) in a localized, bioactive, harmless and controlled way. The pharmaceutical technology offers a great variety of DDS that can be customized to achieve the most effective delivery for each tissue.[41–44] The first step when treating a pathology using a novel TE approach would be to identify the cells and factors involved in the recovery process. The place, moment and necessary dose to act effectively should also be considered. The next step would be to design and manufacture the most appropriate DDS to achieve the desired environment for the cells. There are no fixed rules for this due to the complexity of the strategy. Thus, new DDS appear almost every month to treat the same diseases. Another thing to keep in mind is the complexity of the pathologies intended to be treated with TE. At first sight it is possible to differentiate between two main groups of disorders. The first one, apparently the simplest to treat, includes alterations for which the solution is the replacement of the original tissue architecture. Bone fractures, traumatic wounds, or cartilage diseases are included in this group. The second group needs the intervention of an unlimited number of GFs, chemokines and transcriptional factors to recover tissue functionality. More complex diseases, such as neurodegenerative disorders, cardiovascular disease or Diabetes Mellitus, are included in this group. However, when revising the up-to-date literature, the difference is not so clear. Nowadays researchers and clinicians consider that to repair any tissue efficiently and for life, it is necessary to include multiple factors and cells in the treatment. In this sense, the controlled delivery of factors involved in the regeneration of all the tissues is ordinarily being combined with cell therapy, so both the implanted cells and the damaged tissue receive signals co-acting in the regenerative process.[44,45] Interestingly, the delivery system architecture is now conceived not only as a delivery platform, but also as a well designed structure directly influencing tissue repair.[37]

9.1.4.1   Cartilage and Bone

The diseases or defects affecting cartilage or bone usually result from traumatic insult, congenital deformities or progressive degenerative diseases such as osteoarthritis, osteonecrosis and periodontitis, among others.[46–48] The first TE approaches to restore cartilage and bone consisted of tissue replacements (autografts, allografts, xenografts, and bone graft substitutes), GFs and synthetic devices, provided either as single constituents or in combination. Unfortunately, these reconstructive strategies showed numerous drawbacks such as immune rejection, complicated surgical procedures, chronic inflammation or lack of clinical predictability.[49,50] In an attempt to overcome the shortcomings, tissue-engineered constructs using preferably cultured auto-logous cells seeded onto a synthetic biodegradable polymer scaffold were developed as new therapeutic alternatives.

Regarding the cells, the ideal treatment should be the autologous graft of resident cells, chondrocytes for cartilage and osteocytes for bone regeneration. Nevertheless, the limited number and the difficulty of expanding these primary cells without de-differentiating[51] have forced researchers to use other ones. Since adult MSCs can be easily obtained from several sources (including synovium, SMSCs) and they can proliferate and differentiate into osteoblastic and chondrocytic precursors, TE has studied the possibility of MSCs combined with GFs administration to guide cellular differentiation *in vivo*, or *in vitro* pre-differentiation. For the *in vitro* cell pre-differentiation, some cell culture variables must be considered. For example, for chondrogenic induction, a high cell confluence, serum absence and the presence of TGF-β3, BMP-6 (bone morphogenetic protein-6), dexamethasone, ascorbic acid-2 phosphate and bovine serum albumine seem to be important. Other GFs like FGF, IGF, EGF (epidermal growth factor) and PDGF (platelet derived growth factor) can enhance cartilage growth.[52] For MSCs, osteoinduction, dexamethasone, ascorbic acid-2 phosphate and β-glycerol phosphate are added to the medium. It has also been described that BMP-2 and FGF-2 increase osteogenesis both *in vitro* and *in vivo*.[53]

The scaffold must fulfil some criteria for its use in bone or cartilage tissues such as controlled biodegradability, suitable mechanical strength and surface chemistry, ability to be processed in different shapes and sizes, and the ability to regulate cellular activities (delivering GFs and cytokines). While the cartilage is an avascular tissue, bone tissue regeneration requires vascularization, so the scaffold must be a vascularizable structure that provides pro-angiogenic factors to the tissue. Both natural (collagen, fibrinogen, starch, etc.) and synthetic polymers (polycarbonates, polyanhydrides, polyhydroxyacids, etc.) have been used to prepare scaffolds in a variety of forms, including fibrous structures, porous sponges, woven or non-woven meshes, and hydrogels.[52] More recent approaches are developing composites, combining several polymers to obtain a scaffold with optimal physiological and mechanical properties.[54] Some approaches assayed in cartilage and bone TE are reviewed below (bone regeneration by TE is described in more detail in Chapter 9.2).

Cartilage

TE represents a challenging approach for cartilage regeneration and repair due to the poor healing capacity of the cartilage.[55] Chemical and physical stimulations are very important for the induction and maintenance of chondrogenesis and for resilience (avoiding ossification and fibrosis). Moreover, if cells are embedded in a scaffold, the appropriate materials and architecture must be able to withstand shear forces due to mechanical load allowing effective movement.[56] The microfiber-based scaffolds most used since the 1990s are made of poly(α-hydroxy esters), especially PLA and PLGA. They allow the maintenance of chondrocyte phenotype and specific ECM production. Cell attachment can be improved by grafting bioactive proteins or peptides onto the polymers or modifying their surface chemistries.[57] Modern processing techniques, such as electrospinning, make it possible to construct nanofiber scaffolds, developed as a 3D scaffold, which simulate better ECM morphological properties.[58] Other systems proposed for cartilage engineering are hydrogels, where encapsulated chondrocytes can maintain their spherical conformation. Alginate, hyaluronan, chondroitin sulphate or polyethylene glycol (PEG) are some of the materials used to prepare these hydrogels. Hydrogels are also employed as DDS in combination with scaffolds, to co-administer cells and GFs obtaining the desired release profile of these. For instance, ligament-bone interface repair has been achieved by local delivery of fibrochondrocytes and BMP-2 into a poly(L-lactide-co-ε-caprolactone) scaffold combined with a heparin-based hydrogel.[59] Chondrogenic differentiation of human adipose tissue-derived stem cells has been observed using a variety of scaffolds: composite hydrogel encapsulating TGF-β,[60] a TGF-β1 loaded fibrin–poly(lactide-caprolactone) nanoparticulate complex,[61] gelatin-chitosan microspheres encapsulating TGF-β.[62] Differentiation into chondrocytes has also been obtained with injectable biodegradable hydrogel composites encapsulating bone marrow mesenchymal stem cell combined with gelatin microparticles containing TGF-β.[63] Another approach is the use of Pharmacological Active Microcarriers (PAMs), which are PLGA microparticles engineered to release TGF-β3 and with a biomimetic surface of fibronectin which improves MSCs adhesion and facilitates their *in vivo* differentiation.[64] Nowadays it is known that multiple factors are needed to stimulate cartilage repair efficiently. The regulation of their spatio-temporal distribution is also a crucial issue. Thus systems for sequential controlled release of multiple chondrogenic factors are currently being developed.[60,65]

Bone

Bone is a tissue that provides structural support for the body, supports muscular contraction, withstands load bearing and protects internal organs. It is a highly vascularized and mineralized tissue with a GF-dependent structure. Tissue regeneration does not occur normally in bone defects because of the lack of healthy cells and of the GFs secreted by them.[66] When a bone defect is present, the injury size must be taken into account because it will determine the complexity of the treatment. In minor defects, implanting a bioceramic scaffold could be sufficient, while major defects would need a more complex repair therapy.[67] As noted previously, for bone TE the most complete strategy would be to supply a 3D porous structure (able to withstand load and mechanical strain) with osteoconductive and osteoinductive properties, multiple factors controlled delivery systems (angiogenics, osteogenics and ECM constituents) and osteogenic precursor cells. However, multiple GFs delivery for bone repair must be still carefully examined, since it has been described that certain GFs combinations could lead to bone formation inhibition.[68] A delivery system widely used for bone TE studies and approved by the FDA is a PLGA scaffold incorporating BMP-2 and BMP-7-loaded microspheres.[69,70] Other studies have shown that BMP-2-loaded nanoparticles improved osteoinductive activity of human MSCs within a fibrin hydrogel scaffold.[63] Additionally, it has recently been shown that MSCs can also differentiate into osteoblasts when stimulated with nitrogenous bisphosphonate released from a complex controlled delivery system.[71] Some strategies for bone repair stimulating both osteogenesis and angiogenesis are being developed. An enhanced bone formation has been obtained using BMP-2 and PDGF released from combined brushite-chitosan system,[72] while the combination of VEGF and BMP-2 in different delivery systems has not yet produced conclusive results, probably due to the different spatio-temporal needs for angiogenic and osteogenic factors.[73,74]

Although more than 40 BMPs have already been identified and a large number of BMP carriers designed, very few have been further developed for clinical use, with only recombinant human BMP-2 and BMP-7 (also named Osteogenic protein-1 or OP-1) approved for use in patients.[75]

9.1.4.2 Cardiac Tissue

Cardiovascular diseases as a whole are the main cause of death in the world. Heart failure is the final consequence of detrimental progressive myocardial changes in the attempt at wound healing in the heart after ischemia. Current therapies driven to improve the myocardial function include pharmacological treatment, percutaneous intervention and surgery. Most of them help to mitigate the symptoms, but are not able to regenerate the tissue, to restore the heart function in a maintained form. Therefore, clinicians and researchers are now exploring new therapeutic approaches based on novel pharmaceutical and TE technologies that combine cells, genes or/and soluble factors. These strategies are able to induce the main processes involved in cardiac regeneration: angiogenesis, cardiomyogenesis and healthy healing. They represent a promising treatment for cardiovascular diseases and, at the same time, a great challenge since the reconstruction of different types of tissue vessels, cardiac muscle and ECM is needed. Furthermore, since the heart is a vital organ, new tissue formation is not sufficient and it is also necessary to integrate the different tissues restoring the synchronized beat for the rest of the life. The potential of stem cell therapy for cardiac regeneration has been assayed over the last decade. An array of cell types, including mononuclear bone marrow cells (BM-MNCs), MSCs from bone marrow, adipose tissue, skeletal muscle among others, EPCs, CPCs, ESCs, and lately iPSCs, has been proposed to achieve myocardial regeneration. Most of the pre-clinical and clinical studies have found improvement in the cardiac function, but the beneficial effects were mainly correlated to paracrine factor release from stem cells rather than proliferation or differentiation events after implantation, given their poor engraftment.[76] Therefore, efforts were directed to search for soluble mediators of cardiac repair. The most widely assayed molecules are pro-angiogenic factors to stimulate endogenous therapeutic angiogenesis, such as FGF-2, VEGF, PDGF or angiopoietins. However, the clinical trials carried out delivering these factors intravenously or intracoronary have shown limited beneficial effects, and have brought to light detrimental effects in some patients. These results could be attributed to low bioavailability/poor retention in the heart, loss of bioactivity (because of the short half-life of the angiogenic factors), need for multiple factors and the uncontrolled nature of the studies.[77] Stem cell therapy and paracrine factor therapy therefore need to be improved for their clinical translation. Although stem cells are being engineered (by genetic manipulation, preconditioning or pharmacological pre-treatment) to enhance their regenerative potential, several handicaps still remain to be solved. About 90% of cells delivered to the heart through a needle are lost to the circulation or leak back, and only about 10% of the remainder at the injection site will survive but they will probably not proliferate or repopulate the lost tissue. On the other hand, although the therapeutic proteins are being chemically modified to increase stability and activity, they require large-scale production with high cost, immunological concerns, potential adverse effects of therapeutic doses, route of delivery, repeated administration, need of multiple proteins, *etc.* At this point, the controlled delivery approach is a must. To date, several carriers based on biocompatible polymers, such as particulated systems (nano[78] or microparticles[79]) and hydrogels,[80] have been assayed for their angiogenic potential in animal models of ischemic heart disease with promising results. Currently biocompatible controlled release systems which would be able to deliver multiple bioactive factors creating pre-determined spatio-temporal gradients to mimic endogenous signals involved in cardiac repair are being developed. Moreover, in the attempt to mimic the cardiac environment and to guide revascularization, cardiomyogenesis and ECM remodeling, appropriate scaffolds are being explored. These matrices are made of biocompatible natural (as collagen, fibrin or gelatin) or artificial (as PLGA) polymers. They are nanostructured and designed with physico-mechanical properties suitable for the heart: porosity which allows growth of vessels and migration of precursor cells, high hydration, functionalized surface for the attachment of cells, possibility of establishing cell to cell communication (important cue to achieve synchronized beat). Approaches to myocardial TE using seeded scaffolds have shown improved cell survival, well-formed electromechanical junctions, synchronized contraction and vascularization in *in vitro* studies and after implantation in rodents.[81–84] In view of these promising results, pharmaceutical technology is currently researching materials and methods to construct intelligent scaffolds which host regenerative cells and release necessary signals to support their activities (survival, migration, proliferation, differentiation, *etc.*). Therefore TE is a newcomer for heart disease treatment that is full of therapeutic possibilities, and it appears it will be with us for a long time, since most probably the heart will be repaired in small steps before complete restoration.

9.1.4.3 Neurodegenerative Disorders

Neurodegenerative disorders are a varied group of conditions characterized by gradual, progressive cell loss from the brain and the spinal cord. The number of people affected by neurodegenerative disorders is rapidly increasing as the population ages. In general, brain damage has a profound effect on the quality of life of those affected. Current treatments are primarily focused on limiting damage and slowing degeneration, and most of them are only symptomatic, and function restoration is rarely achieved. This underlies the need for alternative therapies such as cell therapy that can correct the mechanism of the disease. Parkinson's disease (PD) is the most widely studied neurodegenerative disorder. Although the cause of PD remains unknown, the pathological manifestation involves the nigrostriatal dopaminergic pathway degeneration that causes progressive dopamine loss in the basal ganglia. Even if other regions may be affected, the localized dopaminergic neurodegeneration makes PD an ideal target for cell therapy. Huntington's disease (HD) is caused by polyglutamate expansion in the huntington protein and is therefore a heritable disorder. HD results in neuronal dysfunction and degeneration, mainly of medium spiny striatal GABAergic neurons. Alzheimer's disease is not an ideal target for cell therapy due to the multiple brain sites affected in this disease. Brain cell therapy may be one strategic approach for neurodegenerative disorders that would allow functional replacement of missing or damaged neurons by transplanting cells that may differentiate into the desired phenotype and integrate the host parenchyma, or alternatively rescue the affected neuronal population. Several teams have investigated the potential of neuronal precursors and stem cells (neural stem cells (NSC), MSCs, iPSCs) to prevent the loss of neurons or to replace them (for review see ref. 85–88). However, this approach has failed due to problems related to cell survival, control of cell fate, maintenance of a defined differentiated phenotype and proper cell engraftment after transplantation. TE may improve these issues.

Bioactive scaffolds are likely to reinforce the success of cell replacement therapies by providing a microenvironment that facilitates the survival, proliferation, differentiation and connectivity of transplanted and/or endogenous cells. Several proofs of concept have already been reported including particulate systems, hydrogels, self-assembling peptides, and electrospun nanofibres with even some clinical trials, as in the case of Spheramine for PD (for review see ref. 87, 89, 90). We should not forget, however, that the brain is a functionally complex organ, and in order to achieve its functional recovery, only cell growth and differentiation would not be enough. The correct reestablishment of the axonal connections and neuronal circuits would be also necessary. Tissue repair is also limited since, in relation to blood, cerebral spinal fluid is low in cellular nutrients, so it would need external supply. At this point, TE strategies could be a promising therapy for functional brain tissue regeneration. To this end it would be necessary to develop a scaffold that combines four main components: (1) a permissive growth substrate (hydrogel or micro/nanofiber, with microstructures such as pores, grooves or polymer fibres), (2) a neurostimulatory ECM (protein or peptide),

(3) the provision of neurotrophic factors, and (4) glial or support cells (Schwann cells, NSCs).[91] The scaffold should be biocompatible, immunulogically inert, conductive, biodegradable, and made of infection-resistant biomaterial to support neurite outgrowth. The GFs most widely used to stimulate neural regeneration are nerve growth factor (NGF),[92] brain cell-line derived neurotrophic factor (BDNF),[93] neurotrophin-3 (NT-3),[94] glial cell-line derived neurotrophic factor (GDNF)[95] and FGFs.[96] Each of them has multiple individual and shared functions. In brain TE, the structural support for axonal regeneration of a specific nerve population needs integrated polymeric and cellular delivery systems able to release therapeutic drugs and neurotrophic molecules. Advanced scaffold modifications, such as the biomimetic approach, would allow better control of cell survival, proliferation, migration, differentiation and *in vivo* engraftment. PAMs conveying a homologous subpopulation of MSCs, releasing NT3 and with a biomimetic coating of laminin, are a good example of an effective cell delivery vehicle for neuroprotection and repair after PD that integrates matrix, cells and soluble regulators in the same system. These PAMs improve cell survival, differentiation and repair capacities, therefore inducing functional recovery in animal models of PD.[97] Another approach recently developed is the use of scaffolds that incorporate electrically active nanomaterials which could provide an electrical stimulus (*i.e.* composites of zinc oxide particles embedded into a polymer matrix) and conductive polymers (as polypyrrole or Ppy, polyaniline or PANI, *etc.*).[98]

9.1.5   Translating Bench Side into a Bedside Reality

In this section, the steps that need to be followed to translate a bench side product into a bedside reality are reviewed. To introduce a TE product into the market, first of all it is necessary to comply with the regulatory aspects. The first TE product to reach the market was Carticel® (cartilage) in the USA in 1996.[99] Numerous products were then commercialized, most of them within the field of skin, cartilage and bone repair (see Table 9.1.1). However, there was no specific regulation for this kind of product in those years. In the following years, various regulations were developed for biological products and for medical devices. However, when dealing with TE products, which combine both biological products such as cells and cytokines and medical devices, the regulation was not clear enough. In addition, different committees have been created during the past 10 years in the European Union (EU)[100] and in USA,[101] and specific regulations have been prepared. In the USA, the Office of Combination Products (part of the FDA) is responsible for the regulatory oversight of combination products. On the other hand, in the EU, the Committee for advanced therapy plays a similar role. Briefly, what the normative indicates is that a combined product is going to be considered a biological product or a medical device, in terms of regulatory aspects, depending on its main action. In EU, when the engineering product contains

Table 9.1.1 Examples of commercialized TE products.

TISSUE	PRODUCT, COMPANY	SCAFFOLD	CELLS	GFs	USE
Cartilage	Carticel, Gencyme (1996)	Autologous periosteal patch	Autologous chondrocytes	–	Knee cartilage injury
	Bioseed-C, Biotissue Technologies (2001)	3D PLGA and Polydioxane	Autologous chondrocytes	–	Articular cartilage injury
	Hyalograft C autogaft, Ficia Advanced biopolymers (2008)	3D Hyaluran based	Autologous chondrocytes	–	Articular cartilage injury
Bone	OP-Putty, Stryker Biotech (2004)	Bovine collagen	–	OP-1	Posterolateral spinal fusion
	GEM 21S, Biomimetics Pharmaceuticals Incorporated (2005)	β-Tricalcium phosphate	–	rhPDGF	Dental/bone deffects
	INFUSE Bone graft, Wyeth Pharmaceuticals (2004)	Bovine collagen sponges	–	rhBMP-2	Spinal fusion in degenerative disc disease
Blood vessels	Vascugel, Pervasis (Phase III)	Porcine gelatin foam sponges	ECs	–	Vessel reconstruction
Nervous system	NeuraGen, Integra (2001)	Bovine type I collagen nerve conduits	–	–	Nerve injury

viable cells or tissues, it is automatically considered a medicinal product.[100] In the USA, to define the type of product the primary mode of action is studied.[101] To clarify the assignment process a document has been approved, the "Definition of Primary Mode of Action of a Combination Product: Proposed Rule". When starting the research and development of a product, a company should take these defining aspects into account, as they would probably determine the complexity of the requirements to target the market.[102] The agencies are also aware of the need of stimulate research in this area and to promote small and medium-sized companies. That is why the EU regulation also includes some incentives, such as scientific advice from the Committee for advanced therapy and a reduction of the fee for marketing authorization by 50% if the applicant is a hospital or a small or medium-sized enterprise.

But to enter the market, not only legal issues are involved. Cost-effectiveness analyses are needed too. In the technical report prepared by the European Science and Technology Observatory in 2003,[103] various already commercialized products were compared with conventional treatment in terms of cost and effectiveness. The conclusion was that, to that date, most of the TE products were more complex and expensive. It is therefore necessary to introduce more sophisticated products into the market with a clear benefit over current treatment, and moreover, which are able to treat disorders with no other known treatment. To achieve this, some scientific and technical problems need to be overcome to make it possible to scale up production of combined products to manufacture them on an industrial scale. The design of large-scale bioreactors to replicate cells and ensure 3D tissue growth, as well as the creation of optimal storage and preservation techniques, among other advances, will permit a significant cost reduction and the incorporation of new products in the market place.

It is also necessary to be aware of the nature of the TE sector in relation to the business model and strategy. The pharmaceutical sector is characterized by a major investment in Research and Development (R&D) and long development times balanced by large markets and exclusivity because of patent protection. On the other hand, the medical device sectors have lower investment in R&D and short development times, but the market is focused and patent protection is lower. The TE industry has high R&D costs and long developing times. That is why nowadays the market is still small. The scientific and technical advances leading to the inclusion of a wider range of products could completely alter this situation. TE is a very young field with a promising future. Therefore, it is still too early to know how this market will evolve.

9.1.6   Concluding Remarks and Future Perspectives

The TE approach has evolved from a static discipline to a dynamic field able to integrate the necessities of a particular scenario in which cells, scaffolds and GFs interact to mimic the physiological condition as far as possible. Effective TE requires appropriate component selection for cell growth and functionality. Release technology that takes into account both GF-vehicle and cell-scaffold functionalities is still in its infancy and is currently growing fast. The two major challenges that lie ahead are the selection of proper GFs cocktails and rigorous control of the relationship between concentration gradient and timing. The possible tools to continue developing in the near future are those that combine: (1) multiple factor delivery with different release patterns, (2) smart delivery systems, (3) drug and carrier functionalization. Concerted efforts have been and are still being made to achieve this ambitious challenge. However, considering current strategies for delivery of two or more GFs in TE, many more approaches can be attempted and a great deal of work remains to be done. Continual progress in understanding the basic biology of tissue formation in development, physiological and pathological conditions provides a vast amount of important information for the design of future GFs delivery systems. In order to improve delivery platform design, thorough knowledge of smart biomaterials, polymer surface modification techniques and an understanding of physical and genetic components within a cell are required. Systems engineering and systems biology will help in the developmental phase of new TE devices, modelling for instance the GFs delivery from the scaffolds, helping to find the best polymer combination to mimic a specific environment and the most appropriate GFs delivery for a specific pathology, including which GFs cocktail will be better. Scientists, funding agencies, and the regenerative profession need to join efforts to hasten the development of clinically available devices for regenerative therapies, bridging the gap between the lab bench and the bedside.

References

1. A. Persidis, *Nat. Biotechnol.,* 1999, **17**(5), 508.
2. R. Langer and J.P. Vacanti, *Science,* 1993, **260**(5110), 920.
3. J. Vacanti, *J. Pediatr. Surg.,* 2010, **45**(2), 291.
4. A.M. Riazi, S.Y. Kwon and W.L. Stanford, *Methods Mol. Biol.,* 2009, **482**, 55.
5. C. Rios, E. Garbayo, L.A. Gomez, K. Curtis, G. D'Ippolito and P.C. Schiller, *Stem Cells and their contribution to tissue repair* in *Stem Cell and Regenerative Medicine,* ed. H. Cheung, Bentham Science Publishers, USA, 2010,Vol. 1, p 9.
6. H. Inoue and S. Yamanaka, *Clin. Pharmacol. Ther.,* 2011, **89**(5), 655.
7. EMA/CAT, *Reflection paper on stem cell-based medicinal products.* 2011, European Medicines Agency Science Medicines Health.
8. S. Boue, I. Paramonov, M.J. Barrero and J.C. Izpisua Belmonte, *PLoS One,* 2010, **5**(9), 1.
9. S. Barrientos, O. Stojadinovic, M.S. Golinko, H. Brem and M. Tomic-Canic, *Wound Repair Regen.,* 2008, **16**(5), 585.
10. T.P. Richardson, M.C. Peters, A.B. Ennett and D.J. Mooney, *Nat. Biotechnol.,* 2001, **19**(11), 1029.
11. D.G. Phinney and D.J. Prockop, *Stem Cells,* 2007, **25**(11), 2896.

12. P.R. Crisostomo, T.A. Markel, Y. Wang and D.R. Meldrum, *Surgery,* 2008, **143**(5), 577.

13. D.J. Prockop, *Clin. Pharmacol. Ther.,* 2007, **82**(3), 241.

14. J.A. Hubbell, *Biotechnology (N. Y.),* 1995, **13**(6), 565.

15. D.S. Kohane and R. Langer, *Pediatr. Res.,* 2008, **63**(5), 487.

16. L.L. Hench and J.M. Polak, *Science,* 2002, **295**(5557), 1014.

17. R. Cortesini, *Transpl. Immunol.,* 2005, **15**(2), 81.

18. G.J. Lim, S. Zare, M. Van Dyke and A. Atala, *Adv. Exp. Med. Biol.,* 2010, **670**, 126.

19. Z.C. Liu and T.M. Chang, *Adv. Exp. Med. Biol.,* 2010, **670**, 68.

20. A. Murua, A. Portero, G. Orive, R.M. Hernandez, M. de Castro and J.L. Pedraz, *J. Controlled Release,* 2008, **132**(2), 76.

21. E. Santos, J. Zarate, G. Orive, R.M. Hernandez and J.L. Pedraz, *Adv. Exp. Med. Biol.,* 2010, **670**, 5.

22. E.R. Balmayor, H.S. Azevedo and R.L. Reis, *Pharm. Res.,* 2011, **28**(6), 1241.

23. C. Dai, B. Wang and H. Zhao, *Colloids Surf. B Biointerfaces,* 2005, **41**(2–3), 117.

24. M. Biondi, F. Ungaro, F. Quaglia and P.A. Netti, *Adv. Drug. Delivery Rev.,* 2008, **60**(2), 229.

25. L. Leoni and T.A. Desai, *Adv. Drug Delivery Rev.,* 2004, **56**(2), 21.

26. K.A. Athanasiou, G.G. Niederauer and C.M. Agrawal, *Biomaterials,* 1996, **17**(2), 93.

27. M.S. Shive and J.M. Anderson, *Adv. Drug Delivery Rev.* 1997, **28**(1), 5.

28. R.A. Jain, *Biomaterials,* 2000, **21**(23), 2475.

29. N. Faisant, J. Siepmann, P. Oury, V. Laffineur, E. Bruna, J. Haffner and J. Benoit, *Int. J. Pharm.,* 2002, **242**(1–2), 281.

30. N. Faisant, J. Siepmann, J. Richard and J.P. Benoit, *Eur. J. Pharm. Biopharm.,* 2003, **56**(2), 271.

31. J. Heller, *Biomaterials,* 1980, **1**(1), 51.

32. R.C. Mundargi, V.R. Babu, V. Rangaswamy, P. Patel and T.M. Aminabhavi, *J. Controlled Release,* 2008, **125**(3), 193.

33. M.P. Cuchiara, A.C. Allen, T.M. Chen, J.S. Miller and J.L. West, *Biomaterials,* 2010, **31**(21), 5491.

34. R.W. Barber and D.R. Emerson, *Altern. Lab. Anim.,* 2010, **38** Suppl. 1, 67.

35. A. Estella-Hermoso de Mendoza, M.A. Campanero, F. Mollinedo and M.J. Blanco-Prieto, *J. Biomed. Nanotechnol.,* 2009, **5**(4), 323.

36. J. Shi, A.R. Votruba, O.C. Farokhzad and R. Langer, *Nano Lett.,* 2010, **10**(9), 3223.

37. D.E. Discher, D.J. Mooney and P.W. Zandstra, *Science,* 2009, **324**(5935), 1673.

38. F. Quaglia, *Int. J. Pharm.,* 2008, **364**(2), 281.

39. M. Chen, M. Przyborowski and F. Berthiaume, *Crit. Rev. Biomed. Eng.,* 2009, **37**(4–5), 399.

40. D.F. Williams, *Trends Biotechnol.,* 2006, **24**(1), 4.

41. T. Yasukawa, Y. Tabata, H. Kimura and Y. Ogura, *Recent Pat. Drug Delivery Formul.,* 2011, **5**(1), 1.

42. S. Bhaskar, F. Tian, T. Stoeger, W. Kreyling, J.M. de la Fuente, V. Grazu, P. Borm, G. Estrada, V. Ntziachristos and D. Razansky, *Part. Fibre Toxicol.,* 2010, **7**, 3.

43. S. Kubinova and E. Sykova, *Nanomedicine (Lond.),* 2010, **5**(1), 99.

44. H.E. Davis and J.K. Leach, *Ann. Biomed. Eng.,* 2011, **39**(1), 1.

45. Y. Tabata, *J. R. Soc. Interface,* 2009, **6 Suppl 3**, S311.

46. D. Umlauf, S. Frank, T. Pap and J. Bertrand, *Cell. Mol. Life Sci.,* 2010, **67**(24), 4197.

47. T.D. Rachner, S. Khosla and L.C. Hofbauer, *Lancet,* 2011, **377**(9773), 1276.

48. N.L. Fazzalari, *Osteoporos. Int.,* 2011, **22**(6), 2003.

49. I.M. Khan, S.J. Gilbert, S.K. Singhrao, V.C. Duance and C.W. Archer, *Eur. Cell. Mater.,* 2008, **16**, 26.

50. M. Varkey, S.A. Gittens and H. Uludag, *Expert. Opin. Drug. Delivery,* 2004, **1**(1), 19.

51. E.E. Coates and J.P. Fisher, *Ann. Biomed. Eng.,* 2010, **38**(11), 3371.

52. C.K. Kuo, W.J. Li, R.L. Mauck and R.S. Tuan, *Curr. Opin. Rheumatol.,* 2006, **18**(1), 64.

53. M.B. Eslaminejad and P.E. Yazdi, *Yakhteh Medical Journal,* 2007, **9**(3), 158.

54. A.R. Boccaccini and J.J. Blaker, *Expert Rev. Med. Devices,* 2005, **2**(3), 303.

55. A.F. Steinert, S.C. Ghivizzani, A. Rethwilm, R.S. Tuan, C.H. Evans and U. Noth, *Arthritis Res. Ther.,* 2007, **9**(3), 213.

56. S. Miot, T. Woodfield, A.U. Daniels, R. Suetterlin, I. Peterschmitt, M. Heberer, C.A. van Blitterswijk, J. Riesle and I. Martin, *Biomaterials,* 2005, **26**(15), 2479.

57. Z. Ma, C. Gao, Y. Gong and J. Shen, *Biomaterials,* 2005, **26**(11), 1253.

58. W.J. Li, R. Tuli, C. Okafor, A. Derfoul, K.G. Danielson, D.J. Hall and R.S. Tuan, *Biomaterials,* 2005, **26**(6), 599.

59. J. Lee, W.I. Choi, G. Tae, Y.H. Kim, S.S. Kang, S.E. Kim, S.H. Kim, Y. Jung and S.H. Kim, *Acta Biomater.,* 2011, **7**(1), 244.

60. T.A. Holland, E.W. Bodde, V.M. Cuijpers, L.S. Baggett, Y. Tabata, A.G. Mikos and J.A. Jansen, *Osteoarthritis Cartilage,* 2007, **15**(2), 187.

61. Y. Jung, Y.I. Chung, S.H. Kim, G. Tae, Y.H. Kim, J.W. Rhie, S.H. Kim and S.H. Kim, *Biomaterials,* 2009, **30**(27), 4657.

62. J.E. Lee, S.E. Kim, I.C. Kwon, H.J. Ahn, H. Cho, S.H. Lee, H.J. Kim, S.C. Seong and M.C. Lee, *Artif. Organs,* 2004, **28**(9), 829.

63. K.H. Park, H. Kim, S. Moon and K. Na, *J. Biosci. Bioeng.,* 2009, **108**(6), 530.

64. C. Bouffi, O. Thomas, C. Bony, A. Giteau, M.C. Venier-Julienne, C. Jorgensen, C. Montero-Menei and D. Noel, *Biomaterials,* 2010, **31**(25), 6485.

65. T.A. Holland, Y. Tabata and A.G. Mikos, *J. Controlled Release,* 2005, **101**(1–3), 111.

66. J.R. Lieberman, A. Daluiski and T.A. Einhorn, *J. Bone Joint Surg. Am.,* 2002, **84-A**(6), 1032.

67. A. El-Ghannam, *Expert Rev. Med. Devices,* 2005, **2**(1), 87.

68. A.T. Raiche and D.A. Puleo, *Biomaterials,* 2004, **25**(4), 677.

69. F.B. Basmanav, G.T. Kose and V. Hasirci, *Biomaterials,* 2008, **29**(31), 4195.

70. P. Yilgor, N. Hasirci and V. Hasirci, *J. Biomed. Mater. Res. A,* 2010, **93**(2), 528.

71. X. Shi, Y. Wang, R.R. Varshney, L. Ren, F. Zhang and D.A. Wang, *Biomaterials,* 2009, **30**(23–24), 3996.

72. B. De la Riva, E. Sanchez, A. Hernandez, R. Reyes, F. Tamimi, E. Lopez-Cabarcos, A. Delgado and C. Evora, *J. Controlled Release,* 2010, **143**(1), 45.

73. D.H. Kempen, L. Lu, A. Heijink, T.E. Hefferan, L.B. Creemers, A. Maran, M.J. Yaszemski and W.J. Dhert, *Biomaterials,* 2009, **30**(14), 2816.

74. S. Young, Z.S. Patel, J.D. Kretlow, M.B. Murphy, P.M. Mountziaris, L.S. Baggett, H. Ueda, Y. Tabata, J.A. Jansen, M. Wong and A.G. Mikos, *Tissue Eng. Part A,* 2009, **15**(9), 2347.

75. B. Nussenbaum and P.H. Krebsbach, *Adv .Drug Delivery Rev.,* 2006, **58**(4), 577.

76. M. Mirotsou, T.M. Jayawardena, J. Schmeckpeper, M. Gnecchi and V.J. Dzau, *J. Mol. Cell. Cardiol.,* 2011, **50**(2), 280.

77. I. Zachary and R.D. Morgan, *Heart,* 2011, **97**(3), 181.

78. K.S. Oh, J.Y. Song, S.J. Yoon, Y. Park, D. Kim and S.H. Yuk, *J. Controlled Release,* 2010, **146**(2), 207.

79. F.R. Formiga, B. Pelacho, E. Garbayo, G. Abizanda, J.J. Gavira, T. Simon-Yarza, M. Mazo, E. Tamayo, C. Jauquicoa, C. Ortiz-de-Solorzano, F. Prosper and M.J. Blanco-Prieto, *J. Controlled Release,* 2010, **147**(1), 30.

80. H. Kobayashi, S. Minatoguchi, S. Yasuda, N. Bao, I. Kawamura, M. Iwasa, T. Yamaki, S. Sumi, Y. Misao, H. Ushikoshi, K. Nishigaki, G. Takemura, T. Fujiwara, Y. Tabata and H. Fujiwara, *Cardiovasc. Res.,* 2008, **79**(4), 611.

81. T. Shimizu, M. Yamato, Y. Isoi, T. Akutsu, T. Setomaru, K. Abe, A. Kikuchi, M. Umezu and T. Okano, *Circ. Res.,* 2002, **90**(3), e40.

82. P. Zammaretti and M. Jaconi, *Curr. Opin. Biotechnol.,* 2004, **15**(5), 430.

83. W.H. Zimmermann, I. Melnychenko and T. Eschenhagen, *Biomaterials,* 2004, **25**(9), 1639.

84. Y. Sapir, O. Kryukov and S. Cohen, *Biomaterials,* 2011, **32**(7), 1838.

85. C. Chen and S.F. Xiao, *Neurosci. Bull.,* 2011, **27**(2), 107.

86. L.W. Chen, *CNS Neurol. Disord. Drug Targets,* 2011, **10**(4), 449.

87. G.J. Delcroix, P.C. Schiller, J.P. Benoit and C.N. Montero-Menei, *Biomaterials,* 2011, **31**(8), 2105.

88. N. Joyce, G. Annett, L. Wirthlin, S. Olson, G. Bauer and J.A. Nolta, *Regen. Med.,* 2010, **5**(6), 933.

89. P.A. Walker, K.R. Aroom, F. Jimenez, S.K. Shah, M.T. Harting, B.S. Gill and C.S. Cox, Jr., *Stem Cell Rev.,* 2009, **5**(3), 283.

90. S.M. Willerth and S.E. Sakiyama-Elbert, *Adv. Drug Delivery Rev.,* 2007, **59**(4–5), 325.

91. R.V. Bellamkonda, *Biomaterials,* 2006, **27**(19), 3515.

92. P. Menei, J.M. Pean, V. Nerriere-Daguin, C. Jollivet, P. Brachet and J.P. Benoit, *Exp. Neurol.,* 2000, **161**(1), 259.

93. S. Koennings, A. Sapin, T. Blunk, P. Menei and A. Goepferich, *J. Controlled Release,* 2007, **119**(2), 163.

94. P.J. Johnson, S.R. Parker and S.E. Sakiyama-Elbert, *Biotechnol. Bioeng.,* 2009, **104**(6), 1207.

95. E. Garbayo, C.N. Montero-Menei, E. Ansorena, J.L. Lanciego, M.S. Aymerich and M.J. Blanco-Prieto, *J. Controlled Release,* 2009, **135**(2), 119.

96. D. Matsuse, M. Kitada, F. Ogura, S. Wakao, M. Kohama, J.I. Kira, Y. Tabata and M. Dezawa, *Tissue Eng. Part A,* 2011, **17**(15–16), 1993.

97. G.J. Delcroix, E. Garbayo, L. Sindji, O. Thomas, C. Vanpouille-Box, P.C. Schiller and C.N. Montero-Menei, *Biomaterials,* 2011, **32**(6), 1560.

98. J.T. Seil and T.J. Webster, *Wiley Interdiscip. Rev. Nanomed. Nanobiotechnol.,* 2010, **2**(6), 635.

99. C. De Bie, *Regen. Med.,* 2007, **2**(1), 95.

100. Regulation (EC) No 1394/2007 of the European Parliament and of the Council of 13 November 2007 on advanced therapy medicinal products and amending Directive 2001/83/EC and Regulation (EC) No 726/2004. 2007, p 121.

101. K. Hellman, *Tissue Engineerign: translating science to product*, in *Topics in tissue engineering*, ed. N. Ashammakhi, R. Reis and F. Chiellini, 2008, p 1.

102. M.H. Lee, J.A. Arcidiacono, A.M. Bilek, J.J. Wille, C.A. Hamill, K.M. Wonnacott, M.A. Wells and S.S. Oh, *Tissue Eng. Part B Rev.,* 2010, **16**(1), 41.

103. A. Bock, D. Ibarreta and E. Rodríguez-Cerezo, *Human tissue-engineered products*. 2003, European Science and Technology Observatory, p 58.

CHAPTER 9.2

Drug Delivery Strategies for Bone Regeneration

KYLE E. HAMMERICK, ANTONIOS G. MIKOS AND
F. KURTIS KASPER*

Department of Bioengineering, Rice University, Houston, TX 77251-1892,
USA
*E-mail: kasper@rice.edu

9.2.1 Introduction

The classical tissue engineering paradigm comprises three elements: a support or scaffold material, bioactive factors, and cells. Engineers are presented with the challenge of how to leverage nanoscale phenomena in this tissue engineering model to gain better control over the inclusion of biofactors for improved performance in the clinic. The nanometre regime is the fundamental unit of length over which cells and molecules interact with scaffold materials and their biological environments. Therefore, understanding the biology of these tissue-engineered systems at the nanoscale is of utmost importance. Nanostructures can enhance cellular synthesis of extracellular matrix proteins and growth factors as well as enable the retention and subsequent delivery of these extracellular matrix components as a bioactive drug that can accentuate osteogenic responses to implanted constructs.

This chapter will first review some novel properties of nanostructured materials. It will review the structure of bone and how nanoscale structures play a role in bone formation. Nanoproperties will be explored specifically as they relate to bone tissue engineering. Ultimately, the native proteins,

RSC Drug Discovery Series No. 22
Nanostructured Biomaterials for Overcoming Biological Barriers
Edited by Maria Jose Alonso and Noemi S. Csaba
© The Royal Society of Chemistry 2012
Published by the Royal Society of Chemistry, www.rsc.org

cytokines, and molecules pertinent to bone regeneration will be presented as a natural drug, and nanoscale phenomena will be explored for their utility in enhancing and retaining the drug that is the extracellular matrix.

9.2.2 Nanostructured Materials and Their Novel Properties

Nanostructures consist of materials fabricated with size scales ranging from the roughly 0.1 and 0.3 nm of atomic radii and bond lengths of individual molecules up to bulk materials of several hundred nanometres. At these extremely small length scales, the structural, electronic, optical and thermal behavior of materials transitions from being best described by classical mechanics to adhering closer to quantum behavior. Zero, one, and two dimensional structures such as fullerenes C_{60},[1] nanotubes,[2–4] and graphene[5,6] are classical representations of nanostructures, each with its own advantageous properties for tissue engineering.

The nanophysics of these small structures differ significantly from bulk properties. For example, melting temperatures in semiconductor nanocrystals are depressed as a function of decreasing particle size, since a larger fraction of the total number of atoms resides on the surface.[7] Not only are the properties altered at this length scale, but they are also highly tunable. The size of the nanostructure confines the electron wave functions and thus sets the energy scale.[8] Therefore, the band gap of particular semiconductor nanocrystals can be tuned by simply controlling the size of the crystal. The electronic properties of carbon nanotubes can also be changed from metallic to semiconducting depending on the diameter and conformation (armchair, zig-zag, or chiral). The conductivity of nanotubes is also vastly influenced by impurities binding to the nanotube and disturbing the wavefunction, therefore enabling its use as a resistive sensor.[9] The novel electronic and optical properties of nanostructures are vast and well suited to medical diagnostics and imaging modalities but have yet to be fully exploited in the area of tissue engineering.

Alternatively, the mechanical and chemical properties of nanostructures are likely to have a significant, immediate impact on technologies for drug release and tissue engineering. The reduced dimensionality can have a profound impact on the chemical properties of nanoscale structures. Due to the four-fold coordination of each carbon atom within diamond, the bond hybridization is sp^3. In graphene, the coordination is such that the sp^2 orbitals extend normal to the surface. This allows for pi bond stacking that enables chemical functionalization of carbon nanotubes with proteins or other molecules.[10,11] The mechanical strength of materials is also affected by the reduction in size scales. The Young's modulus of carbon nanotubes is on the order of 1050 GPa and the tensile strength is 150 GPa.[12–14] This compares very favorably with steel that has a Young's modulus of around 200 GPa and tensile strength of 0.4 GPa.[15] The extreme strength of these nanotubes makes them promising additives to biologic materials for dispersion strengthening of degradable polymers like chitosan.[16–18] In the physiologic environment, biomaterials must withstand physiologic loads while existing almost exclusively in a liquid environment. This makes the fluidic properties of biomaterials essential to their function. Nanostructures possess unique surface properties that influence their fluidic properties. Carbon nanotubes have very low surface friction, and fluid flow through carbon nanotube membranes occurs at rates several orders of magnitude greater than with traditional membranes.[19] These diverse and tunable properties ranging from the potential for surface modifications to the extreme strength of nanostructured materials make them an attractive tool for tissue engineering and drug delivery for tissue engineering applications.

9.2.3 Nanoscale Approaches to Drug Delivery for Tissue Engineering Bone

9.2.3.1 Nanostructure of Bone

A thorough understanding of the structure of bone is paramount for engineering systems that will improve bone regeneration. Hierarchical structures from the nanoscale to macroscale are found almost everywhere in nature, making nanostructures a critical component of form and function in biology.[20] By mimicking the composition, nanostructure, and properties of natural bone, one can create materials that have the potential to recapitulate the natural function of bone in the body. Biomineralization, the formation and patterning of inorganic mineral phases by living cells or organisms, is just one of the many self-assembly processes utilized by nature for nanomaterials synthesis. A fundamental understanding of the hierarchical structure of bone constituents and the process of skeletal development in vertebrates could be exploited for directed tissue formation.

Bone is composed of a number of materials spanning length scales from the microscale to nanoscale (Figure 9.2.1). Bone forms developmentally by two primary mechanisms, endochondral and intramembranous ossification. Briefly defined, endochondral bone formation is a process in fetal vertebrate development where the primary structure of the developing bone is formed in cartilaginous matrix. Chondrocytes responsible for the cartilage scaffold hypertrophy and begin to calcify the matrix. Osteoblasts and osteoblast precursors invade the structure and use the calcified matrix as a basis to further secrete osteoid and assemble mineralized matrix. Intramembranous ossification proceeds in the absence of an initial cartilaginous matrix whereby mesenchymal stem cells replace pre-existing embryonic connective tissue with mineralized osteoid. The resulting long bones that develop from the endochondral process are composed of a hard, dense outer layer of compact or cortical bone along the diaphysis that constitutes 80% of normal adult bone and a high porosity interior of trabecular bone.[21,22] The cortical bone is hierarchical in nature, with the basic functional unit being the osteon. The osteon is a cylindrical unit several millimetres long and approximately 200 μm

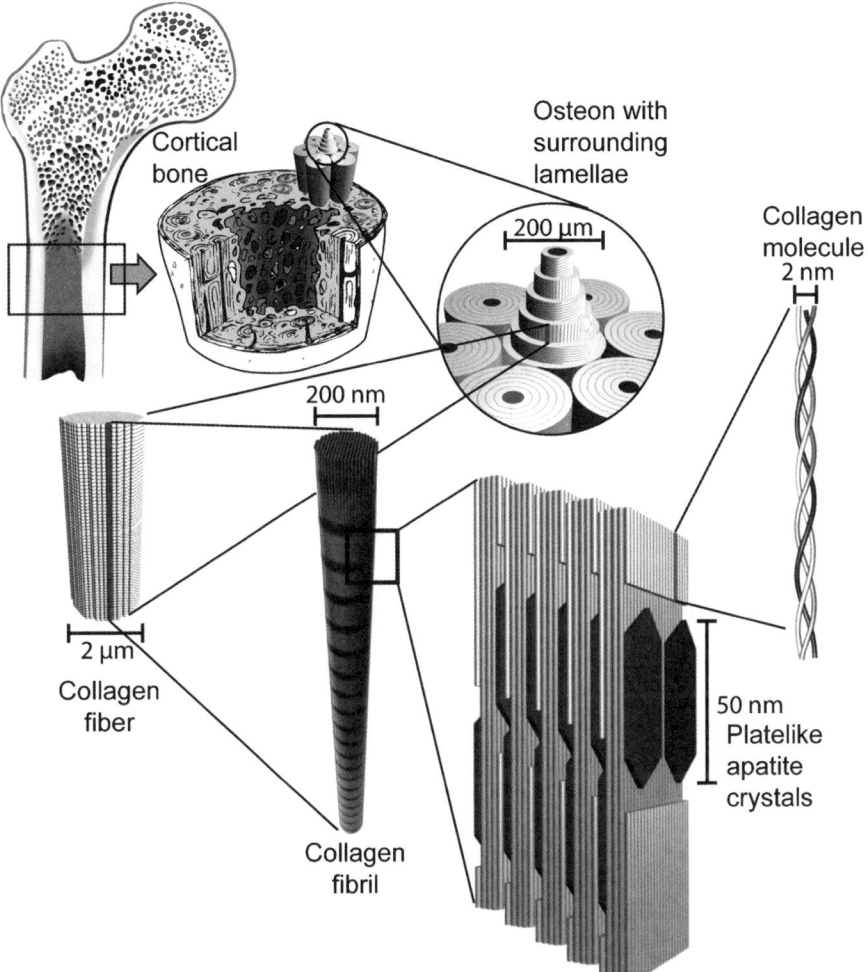

Figure 9.2.1 The hierarchical structure of bone. Cortical bone is composed of osteons with the Haversian canal in the center and lamellae of collagen fibers surrounding it. Collagen fibers, 2 μm, are composed of individual collagen fibrils, 200 nm, which are in turn made up of collagen molecules, 2 nm. Hydroxyapatite crystals that are approximately 2 nm thick by 25–50 nm wide are deposited in the holes within the collagen molecule array structures.

in diameter with the long axis parallel to the long axis of the long bone. Each osteon consists of concentric lamellae that are formed by the parallel arrangement of collagen fibrils ranging in size from 20–200 nm surrounding an inner Haversian canal that contains blood capillaries and nerves.[23,24] The collagen fibrils are further composed of individual collagen molecules, tropocollagen, that are typically 1.5 nm in diameter and 300 nm long in a triple helical configuration. The fibrils serve as a template for the organization of calcium-based organic salts and form the fundamental building block of vertebrate skeletons. Calcium occurs in the extracellular fluids of animals and humans where it is synthesized into the calcium phosphate phases (calcium phosphate dihydrate $CaHPO_4 \cdot 2\ H_2O$, octacalcium phosphate $Ca_8H_2(PO_4)_6 \cdot 5\ H_2O$, tricalcium phosphate $\beta\text{-}Ca_3(PO_4)_2$, and hydroxyapatite $Ca_5(PO_4)_3$) found in bone, which are responsible for the structural strength of the skeleton. Ossification, the development of the hard mineral component of bone, takes place as a product of osteoblasts that secrete a soft phase, osteoid, the previously described collagens and noncollagenous proteins that later mineralize and form the template for bone.[25] Eventually bone is a composite material composed primarily of non-stoichiometric hydroxyapatite (HAP) and collagens. The HAP takes the form of crystals oriented and aligned within the self-assembled collagen fibrils. The HAP crystallites in the bone are usually plate-like or cylindrical with characteristic dimensions of 1–4 nm thick by 50 nm long and 25 nm wide.[23,26] The organization of bone spans three or more orders of magnitude from large ~ 200 µm osteons with subunits of ~ 200 nm collagen fibrils augmented with 20 nm crystalized platelets of HAP.

9.2.3.2   Systems for Engineering Bone

The structure of bone and the interstitial perfusion within the Haversian canals suggests that regular vascular stimulation is an essential part of bone homeostasis. As a general rule, cells within mammalian tissues *in vivo* are not found more than ~ 200 µm from capillaries.[27] Cells much further from a nutrient source than this are typically metabolically inactive or become necrotic due to excessive metabolite concentrations and deficiencies in oxygen and other nutrients. Similarly, the maximum oxygen diffusion distance through cells *in vitro* is 100–200 µm.[28] To overcome these limitations of static culture conditions, researchers have turned to alternative dynamic culture conditions, such as spinner flasks and rotating wall vessels, to enable continuous nutrient replenishment and cellular stimulation. These types of culture conditions have proven somewhat effective at increasing nutrient diffusion into scaffolds, as demonstrated by increased cell proliferation and improved mineralization and increases in bone marker expression, however cultures of differentiating osteoblasts in spinner flasks have still shown a marked increase in cells and matrix at the periphery of the scaffold.[29] This demonstrates that mass transfer is still limited as the distance within the construct increases.[30] Many of these limitations can be overcome by perfusion bioreactors, which typically use a closed circuit media loop driven by a peristaltic pump to circulate culture media directly through porous scaffolding material laden with cells (Figure 9.2.2).[31–35]

9.2.3.3   Nanofluidics and Mechanical Effects

The perfusion culture of cells within bioreactors overcomes many of the mass transport limitations related to static cell culture toward the generation of

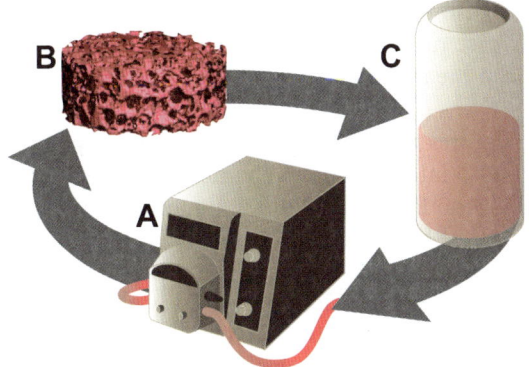

Figure 9.2.2　Schematic of a flow perfusion bioreactor. A closed circuit system consisting of (A) a perfusion pump, (B) a porous scaffold containing cells (usually in a perfusion chamber, shown here without any flow chamber for simplicity) connected to (C) a media reservoir to maintain proper pH via gas exchange and for changing media throughout the culture period.

three dimensional tissue engineered constructs and also has the added benefit of exposing the cells to mechanical forces. Shear stress by means of fluid flow has been demonstrated to be one of the primary mechanisms mediating the differentiation of mesenchymal stem cells and the synthetic response of osteoblasts.[36] Typical values of shear forces that have demonstrated stimulatory activity on osteoblastic cells *in vitro* are between 2–20 dynes cm^{-2}.[37] These values also agree well with the 8–30 dyne cm^{-2} estimated values that osteocytes experience in bone tissue affected by interstitial flow.[38] To study the influence of mechanical stresses on cell biology and metabolism it is important to characterize the influence of scaffold architecture on imposed shear stresses. However, the complicated architectures of the scaffolds make it difficult to analytically solve for the shear stresses experienced by cells within the scaffold. A first order approximation involves reducing the complexity of the system by approximating the tortuous passages within a porous scaffold as parallel cylindrical conduits of a fixed diameter. With this assumption, the shear forces acting on the walls of the scaffold can be described as a function of pore diameter and fluid flow rate according to the Hagen–Poiseuille formula.[31,39,40]

$$\tau_w = \frac{8 \cdot \mu \cdot u}{d_c}$$

$$(9.2.1)$$

This equation relates u, the flow rate through the pore, μ the dynamic viscosity of the perfused media (typical value at 37 °C, 0.77 cP),[41] and d_c the pore diameter to the wall shear stress, τ_w. Using this equation to examine values of shear for flow rates used in previous work demonstrates that the shear stresses increase rapidly as pore diameters approach the nanoscale, as is to be expected from the inverse relationship of shear stress with pore diameter (Figure 9.2.3).

As nanoscale features are incorporated into typically microscale scaffolds, further investigation into the flow characteristics in the nano regime is warranted. Steady-state Poiseuille flows maintain a laminar parabolic velocity profile down to several hundred nanometers and can still be predicted by the Hagen–Poiseuille formula.[42] However, at channels with dimensions below 200 nm, departures from Poiseuille flow were observed for certain solvents with slip lengths between 9 and 30 nm, but similar departures were not observed for water.[43] Fluid flow through nanochannels has been documented as very rapid.[44] Fluid flow through carbon nanotubes is theoretically predicted like flow through frictionless pipes and experimentally is faster than flow through other nanoscale materials.[45,46] Water fluid flow through carbon nanotubes with pore diameters of 1.3 to 2.0 nm showed dramatic enhancements in fluid flow rates, 560 to 8400 times greater than those calculated according to the Hagen–Poiseuille equation.[19] Accurately predicting fluid flow through

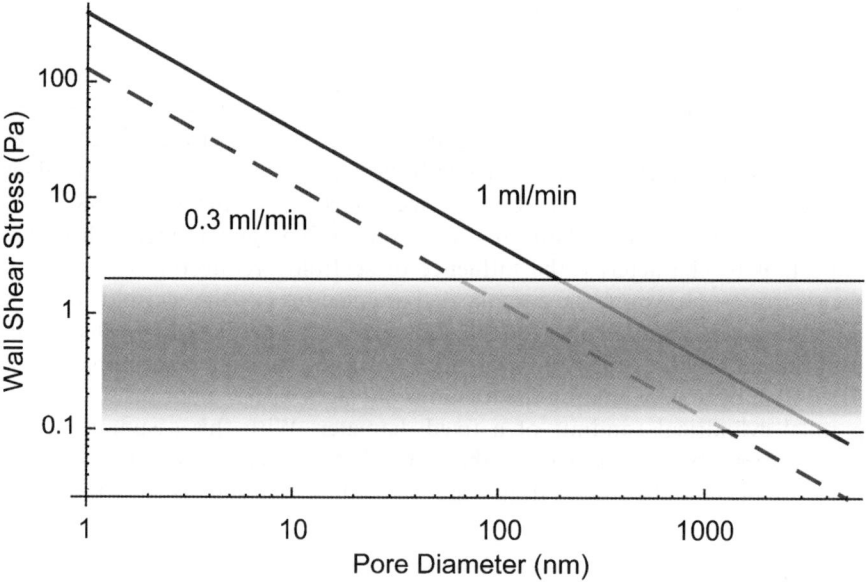

Figure 9.2.3   Inverse relationship of wall shear stress with pore diameter. The wall shear stress increases as the pore diameter within the scaffold constructs decreases as shown here for two physiologic flow rates. The shaded region illustrates the range of shear stresses 0.1–2 Pa that have typically elicited osteogenic responses from osteoblastic cells. The wall shear stress quickly surpasses the physiologic regime, however these extreme shear stresses may still be useful for artificially enhancing cellular osteogenic responses.

nanopores, therefore, requires the use of a hybrid of a continuum solution and a molecular dynamic solution.[47] While nanofluidic structures demonstrate improved transport properties relative to microfluidic channels, scaffolds for implantable constructs composed exclusively of nanoscale conduits are unrealistic microenvironments for cell-based tissue engineering, since they provide no porosity sufficient for the infiltration and residence of cells. In addition, one obvious problem with nanofluidic pores is the presence of large solute molecules and debris in bioreactor cell culture that can impede flow through the pore. Therefore, researchers are actively investigating ways to increase porosity in nanostructured materials by including secondary phases of sacrificial microparticles to generate additional microscale void spaces.[48] Nanostructures are also being implemented in bioreactor assemblies. Nanostructured gratings of ~ 300 nm have been used in a microfluidic bioreactor demonstrating that both the topography and the fluid flow influenced cell adhesion, migration, and morphology.[49] Constructs composed of micro and nanoscale features together make characterization of the shear stresses resulting from these hierarchical structures with the Hagen–Poiseuille equation untenable due to nonuniformity of pore sizes. Fortunately, the widespread use of computational fluid dynamcis (CFD) simulation software has made accurate analysis of the distribution of forces present in complex scaffold geometries such as these more accessible.[50-52]

9.2.3.4 Scaffold Architecture

Regardless of feature set (nano or micro) that imposes shear stresses on osteoblasts or mesenchymal stem cells, the end result is typically an increase in osteoblastic gene expression accompanied by increases in protein expression of key components of the bone extracellular matrix and ultimately increased mineralization.[53] Key parameters for tailoring the mechanical influence of perfused culture are the architecture and dimensions of the cell-containing scaffold material. The dimensions and structure of nanoscale polymeric biomaterials approximate the collagen phase of natural extracellular matrix (ECM) and therefore, nonwoven nanofiber matrices pose excellent ECM analogs.[54,55] There are numerous fabrication schemes for creating tissue engineering scaffolds, such as fiber drawing,[56] template synthesis,[57,58] phase separation,[59] self assembly,[60] and electrospinning.[61] Drawing is a process of extruding a viscoelastic polymer to a very long nanoscale fiber. Template synthesis is a process that uses a nanoporous membrane as a template to cast nanoscale fibers and then dissolve the template away. Polymers are often cast or other materials are applied by chemical vapor depositon or grown electrolytically in porous alumina that has pores of 60–300 nm. The template, in this case alumina, is then removed by dissolution in sodium hydroxide (NaOH) leaving behind the nanofibers. Phase separation consists of the extraction of a secondary phase from an emulsion after freezing and drying a polymer, creating a porous nano-foam. Self-assembly uses the energetically favorable ordering of molecules typically on surfaces to create patterns. Electrospinning is the continuous deposition of a polymer fiber from an extruded polymer solvent mixture under the forces generated by a strong electric field (Figure 9.2.4). Electrospinning is a versatile technology suitable for large scale production that can be applied to many polymer systems making it an attractive fabrication technology for tissue engineering scaffolds containing features ranging from the nanoscale up to the microscale.[62]

As the scale of the scaffold features approaches the size of the largest proteins, adsorption phenomena become very important.[63] A study conducted on poly(L-lactic acid) (PLLA) suggests that scaffolds with nanofibrous pore walls may mimic the ECM environment by adsorbing more serum proteins than similarly fabricated scaffolds with solid pore walls. Specifically, they adsorbed more fibronectin than the controls, and the adsorbed protein supported 1.7 times more osteoblast cell attachment (Figure 9.2.5).[64] A simple analysis can help elucidate the power of reducing dimensions with respect to increases in surface area. Electrospinning is essentially extruding a polymer to a long cylinder. If 1.0 mg of PCL is extruded to a 400 nm diameter fiber the theoretical length is $\sim 700\,000$ cm. Whereas extruding to a fiber size of 10 μm only results in a fiber ~ 1000 cm in length. During electrospinning a continuous fiber will randomly deposit on the collector causing some exposed surfaces of the fibers to become blocked. Ignoring those occluded surfaces and assuming all areas are exposed, results in a 25-fold increase in theoretical surface area of the nanostructured scaffold compared to microstructured scaffold. Assuming a constant protein adsorption rate per square cm of exposed polymer surface should yield a theoretical 25-fold greater amount of protein on nanostructured scaffolds than on microstructured scaffolds. This supports the use of nanostructures in engineering bone by increasing the retention of cell-synthesized proteins.

In addition to increasing the surface area of scaffolds, nanophase materials can be incorporated in bulk polymers to make them adequate candidates for the load bearing requirements of bone. For example, nanotubes can be used as reinforcement to improve the tensile modulus of microfiber scaffolds.[65] In bulk polymers, poly(lactide-caprolactone) (PLC) was strengthened by incorporation of 2 weight percent carbon nanotubes. Interestingly the lower percentage of nanotubes incorporated were better dispersed than the 5% carbon nanotubes. The nanotubes resulted in an increase in modulus by 100% and increased tensile strength by 160%. In addition, the presence of carbon nanotubes on the surface resulted in an increase in osteoblast viability and upregulated expression of *Runx2*, a transcription factor for bone.[66] In similar work PLC was reinforced with boron nitride nanotubes and the result was a 1370% increase in modulus and a 100% increase in tensile strength, accompanied by increased osteoblast viability compared to PLC alone.[67] Carbon nanofiber and polycarbonate composites also promoted osteoblast adhesion and viability as a function of fiber density promoting more osteoblast adhesion as the fiber diameter decreased.[68] These results corroborate earlier studies that show

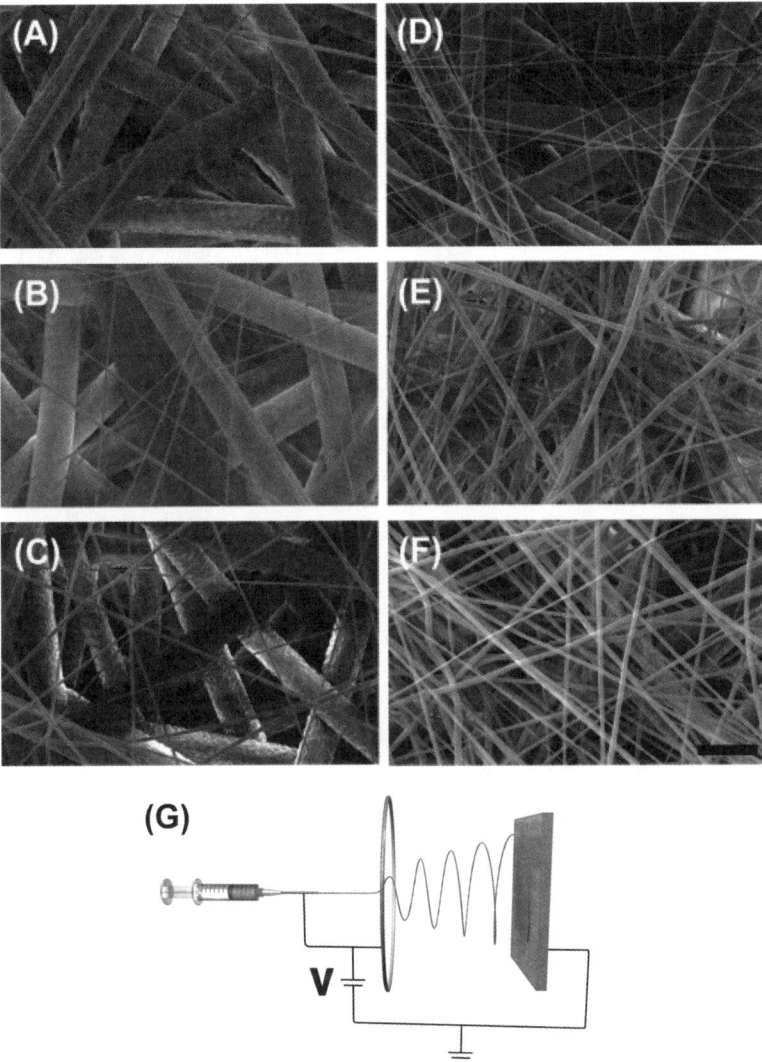

Figure 9.2.4 Electrospinning nanostructured polymer scaffolds. Scanning electron micrographs depicting sequential electrospinning of PCL nanofibers on an existing layer of 5 μm microfibers. The spinning time defines the fractional coverage of nanofibers: (A) 15 s, (B) 30 s, (C) 60 s, (D) 90 s, (E) 120 s, and (F) 300 s. (G) Schematic of an electrospinning apparatus where a polymer solvent solution is extruded under the force of a strong electric field through the positively charged spinnerette toward a negatively charged collector plate. Properties such as the spinnerette lumen diameter, distance between the spinnerette and collector, polymer solution fraction and solvent composition, feed rate of the polymer solution, and the electric field strength can be altered to achieve fibers that span a wide scale of sizes. Reproduced with permission from ref. 62.

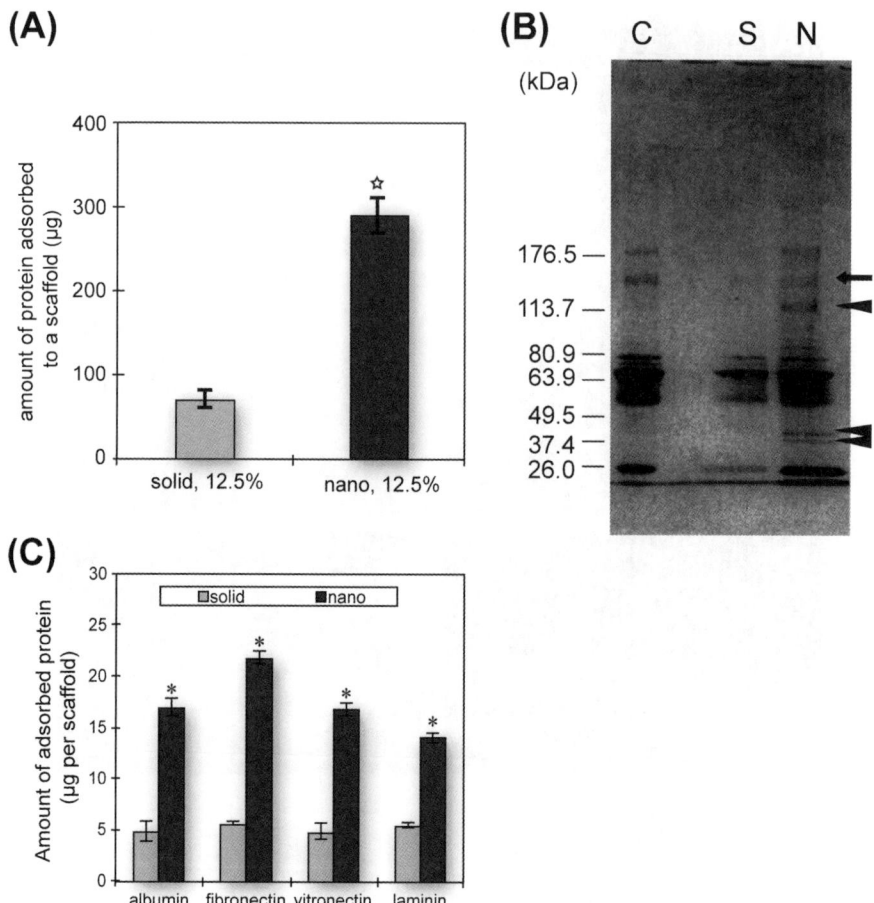

Figure 9.2.5   Protein adsorption by nanostructures. Nano-fibrous and solid-walled
PLLA matrices were exposed to serum and defined protein solutions of
fibronectin and vitronectin. (A) Nano-fibrous matrices adsorbed
significantly more protein than solid-walled matrices, $p < 0.05$, $n =$
4. (B) Polyacrylamide gels comparing lane C bovine serum proteins,
lane S the same proteins adsorbed on solid walled matrices, lane N
serum proteins adsorbed to nano-fibrous matrices. (C) Nano-fibrous
PLLA adsorbs greater amounts of proteins than solid-walled matrices
and adsorbs specifically fibronectin at a greater rate than laminin or
vitronectin, $p < 0.05$, $n = 4$. Reproduced with permission from ref. 64.

increased alkaline phosphatase activity and calcium deposition on 100 nm and
smaller carbon fibers compared to fibers larger than 100 nm.[69]

9.2.3.5 Nanotopography

Beyond indirectly interacting with bone cells through adsorbed proteins, nanostructures have shown efficacy in directing osteoblast cell morphology, proliferation, and migration. Nanostructured topography is typically more durable than chemical modifications of surfaces. Topography directly regulates cell morphology, and cell shape has long been recognized as a strong mediator of cell function.[70] The effects of topography to influence cell adhesion,[71] motility,[72] morphology,[73] cytoskeletal organization,[74] apoptosis,[75] differentiation,[76] and gene expression[77,78] have all been well documented. Cell-nanoscale material interactions can be exploited to induce osteoblastic differentiation of bone marrow derived mesenchymal stromal cells.[79] Titania surfaces fabricated by anodization resulted in nanotubular structures that supported greater proliferation, adhesion, and alkaline phosphatase expression in differentiating marrow stromal cells. Additionally, subcutaneous implantation of nanostructured titania did not result in significant inflammation or fibrosis.[80] Single- and multi-walled carbon nanotubes were modified to engineer the charge density on their surface. Nanotubes with neutral charges promoted the highest cell growth and production of hydroxyapatite.[81] The mechanisms for the various influences on cell behavior are most likely linked through mechanotransduction pathways and regulation of focal adhesion sites. Focal adhesions are the mechanical link between the substrate and the attached cell and therefore the mediator of cellular reactions to nanotopography.[82] Some evidence suggests that nanoscale features modulate integrin clustering and signaling downstream of focal adhesions.[83,84] The size of the nanofeatures appears to limit the size of the focal adhesion complex.[85] Expression of focal adhesion kinase (FAK) by osteoblasts on nano-pitted surfaces was higher on 14 and 29 nm feature sets than the FAK expression on flat and 45 nm PLLA. The cells also expressed higher levels of paxillin, vinculing and actin stress fibers than cells cultured on 45 nm features.[86] Increased focal adhesion size is typically associated with increased cytoskeletal tension and the subsequent recruitment of focal adhesion-associated signaling molecules. FAK is the most notable signaling molecule associated with mature adhesion sites.[87] Increases in the presence of nanostructures that induce spreading and increase focal adhesion signaling seem to enhance osteogenic differentiation.[88] Nanostructures induce signaling directly related to the size and structure of focal adhesion sites. The formation and evolution of focal adhesion sites also directly impacts the cytoskeleton of the cell. Changes in cytoskeletal tension initiate signaling cascades that often signal through RhoA dependent pathways.[89] More explicit patterning of a surface with adhesive islands rather than just nanoscale features reveals that there may be an optimal adhesive ligand spacing. Arnold *et al.* spaced gold nanodots with RGD, cell adhesive motifs, 28–85 nm apart. The 58–73 nm spaced RGD molecules were optimal for osteoblast adhesion and activation. Larger spacing reduced osteoblast adhesion, spreading, and focal adhesion formation.[90]

It is evident that there is a profound effect of topography directly on cell function and differentiation. It has been known for some time that cells try to minimize their surface energy by orienting to lines of minimal curvature.[91,92] It is becoming increasingly apparent, however, that the cellular response to substrates involves additional complexity and requires precise control of the structural composition down to the nanoscale. Nanostructures, and more specifically a defined set of nanostructures, may be used to pattern materials and scaffolds to elicit greater osteoblastic differentiation and responses from biomaterials.

9.2.4   Extracellular Matrix as a Drug

9.2.4.1   Drug Delivery for Bone Tissue Engineering

Typical drug delivery approaches at the nanoscale make use of the high surface area to volume ratios and easily exploited surface chemistries of tissue engineering scaffolds to attach or encapsulate bioactive molecules including growth factors and small molecules (Table 9.2.1). Nanoscale strategies aim to facilitate administration of drugs by overcoming biological barriers to transport, compatibility, immunogenicity, and bioavailability. As an example of the utility of nanoscale drug carriers, nanoparticles in the range of 70–200 nm have demonstrated the longest circulation times compared to very small, 10 nm particles, and particles larger than 200 nm.[93] Particles less than 10 nm were quickly cleared by the kidneys[94] and particles larger than 200 nm were sequestered in the spleen and removed by phagocytosis.[95] Using specific sizes of nanoparticle drug carriers therefore may result in prolonged administration of the contained therapeutic compound. Bone morphogenetic protein-2 (BMP-2) or osteogenic factors such as dexamethasone are the therapeutic

Table 9.2.1   Different approaches to drug delivery for bone tissue engineering at the nanoscale indicating the type of scaffold or carrier matrix and the active drug agent used to enhance bone formation.

Nanoscale System	Agent Released	Reference
Nanocrystalline hydroxyapatite	rhBMP-2, antibiotics	97, 98, 121, 122
Peptide amphiphile hydrogels	BMP-2	103
Calcium phosphate nanoparticles	BMP-2 plasmid, shRNA	99, 101
Electrospun nanofiber scaffolds	BMP-2, dexamethasone	102, 123
Hydrogel nanoparticles	BMP-2	104, 124
Bisphosphonate conjugated nanoparticles	Bone targeting	106, 125
Nanoparticle dendrimers	Dexamethasone	100, 126
PEG coated nanoparticles	BMP-2	127
Carbon nanotubes	Photoacoustic	128
Lipid nanoparticles	siRNA, BMP-2 plasmid	129–131

compounds most often used for orthopaedic tissue engineering. The incorporation of these factors in nanoparticles and coatings have shown efficacy in sustained delivery of these osteogenic factors. These nanoparticles integrated in orthopaedic implants have demonstrated improved union of bone defects and many technologies are now aimed at more effective and sustained release of biologics.[96] Nanocrystalline hydroxyapatite coatings increased adsorption of BMP-2, and tuning the HAP coating may also serve as a control scheme for the release of the growth factor from the scaffold surfaces.[97,98] Another calcium based nanoparticle containing a BMP-2 plasmid enhanced local nonviral gene delivery from alginate hydrogels and demonstrated enhanced mineralization *in vitro* in MC3T3-E1 preosteoblastic cell cultures.[99] Carboxymethylchitosan nanoparticles with amine terminated-poly(amidoamine) dendrimers were absorbed by rat bone marrow stromal cells and resulted in intracellular delivery of dexamethasone, enhanced mineralization, and were nontoxic.[100] A multilayer nanofilm of calcium phosphate was able to package and slowly release shRNA (small hairpin RNAs). The shRNA diffused from the nanolayers into the colonizing cells and generated intracellular siRNA against osteopontin and osteocalcin, inhibiting osteocalcin gene expression in human osteoblasts *in vitro*.[101] While this strategy is directly counter to osteogenesis, it may serve as a model system for ways to silence gene expression of osteogenic suppressors. Electrospun silk fibroin nanofibrous scaffolds, as mentioned earlier, have extremely high surface areas allowing for sustained BMP-2 release, which causes human mesenchymal stromal cells (hMSCs) to differentiate toward osteogenic lineages.[102] A different type of nanofibrous network, a peptide-amphiphile infused with a BMP-2 suspension, resulted in ectopic bone formation in a rat model.[103] A dispersed nanophase hydrogel complex of heparin-functionalized nanoparticles and fibrin gel was able to sustain BMP-2 release in a critical size rat calvarial defect model and improve the BMP-2 release over BMP-2 loaded fibrin gels without nanoparticles.[104] While most of these strategies are aimed at local delivery with either drug release or plasmid release directly in contact with the target cells, some strategies are aimed at systemic delivery with targeting sequences that can localize to bone and thus enhance healing. Bisphosphonates are a class of synthetic compounds with a high affinity for calcium crystals.[105] Gonzalez *et al.* demonstrated a C_{60} nanoparticle modified with a bisphosphonate group that exploits this affinity to target bone tissue for the delivery of fullerenes specifically to bone.[106] The incorporation of nanostructures in microstructured scaffolds offers a compromise between microscale structures for cell adhesion with nanoscale features for retention of extracellular matrix and improved cell contact area.[62] One engineering paradigm is to utilize nanostructures and flow perfusion bioreactor culture of MSCs to synthesize a bioactive ECM coating on normally biologically inert scaffold materials (Figure 9.2.6).

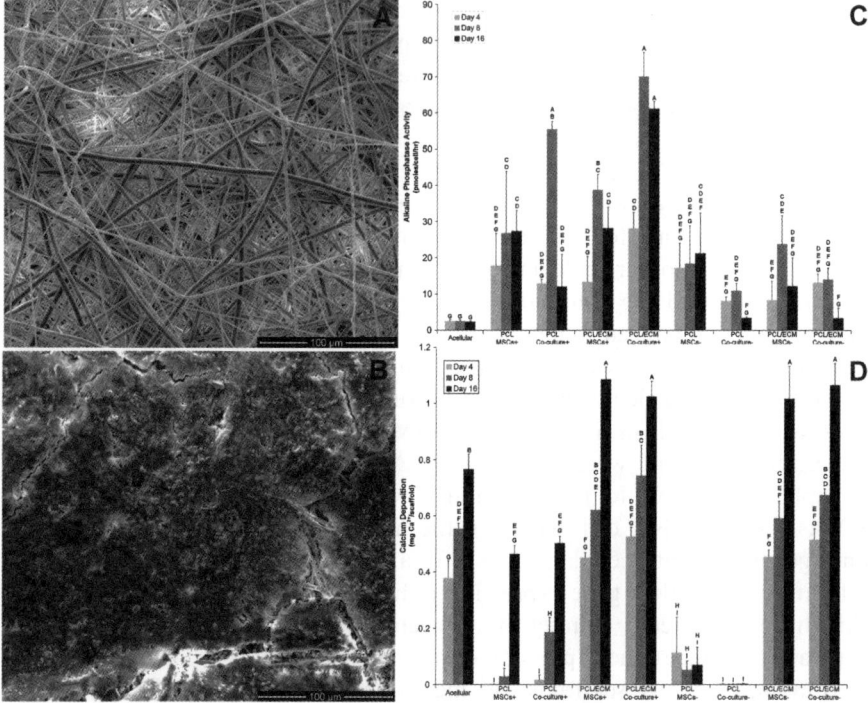

Figure 9.2.6   Extracellular matrix synthesized by rat mesenchymal stem cells (MSCs) on electrospun polymer matrices. (A) Scanning electron micrograph of PCL scaffolds and (B) bioactive PCL/ECM constructs after 12 days in perfusion culture. Scale bar represents 100 μm. The matrices that were rendered bioactive by *in vitro*-generated extracellular matrix supported the osteogenic differentiation of MSCs even in the absence of soluble osteogenic factors, as shown by enhanced (C) calcium accumulation and (D) increased alkaline phosphatase activity of PCL/ECM constructs relative to scaffolds lacking ECM. Reproduced with permission from ref. 132.

9.2.4.2   Mechanical Cues for Matrix Deposition

Earlier the mimicry of extracellular matrix scaled features was touted as an advantageous development in tissue engineering. However, if one can effectively deliver extracellular matrix as a constituent drug within a tissue engineered construct, the gains are likely even greater. This novel approach to nanoscale drug delivery comes from the recognition that native molecules ranging from growth factors to structural collagens are essentially biofactors or drugs that affect the function and development of the engineered system. This is a shift in philosophy from incorporating exogenous factors that can influence cell biology, to incorporating cells that synthesize their own relevant bioactive factors. The critical elements of this strategy consist of how to optimize scaffold properties that most importantly stimulate the appropriate biofactors and also retain them and effectively present them for greatest effect. One of the factors that appears to be essential for producing osteogenic matrix as discussed earlier is mechanical stresses induced by perfusion culture. Cyclic mechanical loading is essential for bone homeostasis.[107] Bone loss is observed in individuals subject to extended immobilization or weightlessness.[108–110] One of the primary hypotheses is that cyclic loading induces oscillatory interstitial fluid flow that acts upon mesenchymal stromal cell and osteoblast cell membranes, causing cell signaling events that result in bone maintenance.[37,111,112] Mechanotransduction as a process is well documented, but the specific sensor that transduces the mechanical shear stress is less well understood.[113] It has been shown that mesenchymal stem cells, when cultured on three-dimensional scaffolds and subjected to shear stresses resulting from flow perfusion, differentiated toward osteoblastic lineages and increased synthesis of osteogenic extracellular matrix. Titanium fiber mesh scaffolds with rat bone mesenchymal stromal cells showed enhanced proliferation of cells subject to 16 days of perfusion culture. The cells cultured on perfused matrices evolved much more mineralized extracellular matrix than those statically cultured on matrices.[34] In a similar study, decreasing the mesh size of the scaffold increased later differentiation of MSCs and increased matrix deposition resulting from fluid shear differences within the scaffolds.[114] Increasing fluid flow rates, which also increases shear stresses, increased calcium content of perfused constructs in a dose dependent manner.[115] Increasing fluid shear stress independent of flow rates was able to divorce the effect of enhanced fluidic transport from the effects of mechanical regulation. The constructs with higher viscosity perfusion media, and thus exposed to higher shear stresses, developed greater mineral content.[53] This demonstrates several important concepts: mesenchymal stem cells are mechanosensitive and respond to increasing shear stresses, and scaffold architecture is critical in engineering a cellular response.

Engineering scaffold designs that optimize the osteogenic response of cells is one part of the design challenge. These naturally produced extracellular matrix components may be used as a drug to render inert materials bioactive. To test this hypothesis, bone marrow derived mesenchymal stromal cells were cultured in perfusion bioreactors for 12 days generating extracellular matrix. The constructs were decellularized and reseeded with mesenchymal stromal cells. Compared to titanium fiber mesh scaffolds without pregenerated matrix, the bioactive constructs induced a 75-fold greater accumulation of mineralized matrix.[116] When similar constructs were decellularized with heat treatment to denature any active proteins or growth factors, the constructs induced significantly lower calcium accumulation, suggesting that in addition to the mineral phase there is an active biomolecular constituent responsible for accentuating the response of cells to the fiber mesh scaffolds with pregenerated matrix. Immunohistochemistry was conducted to determine the influence of culture parameters on the presence and distribution of bone-related growth factors in the pregenerated matrix. Transforming growth factor-β1, fibroblast growth factor-2, vascular endothelial growth factor, and bone morphogenetic protein-2 were all found in the constructs and their expression increased with increasing flow rate and culture duration.[117] After *in vitro* pregeneration of extracellular matrix, constructs were implanted in an ectopic site *in vivo* without any MSCs. Rat mesenchymal stromal cells were used to generate extracellular matrix on titanium mesh scaffolds for a duration of 16 days. The constructs were decellularized and implanted for up to 56 days intramuscularly in a rat model. Tissue infiltration was observed with improved blood vessel formation in the pregenerated ECM constructs as compared to the plain titanium scaffolds.[118] The influence of culture parameters such as scaffold geometry and perfusion rates have been extensively mapped to resultant cellular synthetic profiles. These results suggest that degradable matrices could also be activated by generating extracellular matrix within degradable poly(ε-caprolactone) scaffolds. Highly porous scaffolds were generated with varying amounts of nanofibers upon a matrix of microfibers *via* electrospinning. These scaffolds promoted the spreading but not necessarily attachment of rat MSCs. The nature of the bilayer scaffold reduced cell infiltration as the thickness of the nanofibrous layer increased. However flow perfusion was able to overcome some of the limitations of this configuration.[62] Nanostructures within microstrutured PCL were able to support the adhesion and growth of rat MSCs in perfusion culture.

Nanostructures incorporated into engineered bone grafts may act directly on the gene expression of osteoblasts to enhance bone formation or they may indirectly interact with the secreted bone forming growth factors from the osteoblasts, thus influencing their presentation, retention, and overall activity within the system. Most nano-phase materials used for tissue engineering applications in orthopaedics have focused on modulating cellular response for growth, differentiation, or adhesion to improve implant osteointegration.[119] As discussed above, nanometre scale features can interact with the cellular environment through protein adsorption and therefore influence adhesion of cells and molecular signal transduction that can control cell fate.[120] Further work is warranted to optimize bone graft engineering through the combined use of nanostructures to enhance the synthesis and retention of the drug that is the extracellular matrix.

9.2.5   Conclusion

Nanoscaled materials possess unique electromagnetic, optical, mechanical and chemical properties. Understanding these properties and how they interact with biological systems is proving an essential tool for engineering tissue constructs. The most pervasive and impactful nano-property for tissue engineering is the vast increase in surface area of scaffolds and implants constructed from nano-features. The increased surface area relative to traditional microscale scaffolds allows for increased conjugation of specific surface chemistries to enhance biocompatibility or delivery of growth factors and proteins to act as mitogens or direct cellular phenotype. The unique scale of nanoscale structures also approximates the scale of native collagens and large proteins resulting in greater cellular integration of engineered substrates with osteoblastic cell types and potentially increasing the rate and quality of extracellular matrix deposition. The augmentation of synthetic scaffold biocompatibility with bioactive matrix by the natural synthesis of cellular constituents on the nanostructured materials is a promising new approach to exploiting native matrix as a drug and the nanostructured scaffold as its carrier.

Acknowledgements

The authors thank Marily Mallison for her contributions to the figures. The authors also thank the Alliance for NanoHealth Postdoctoral Fellowship for their support of K. E. Hammerick. This work was further supported by the NIH R01 AR057083 grant.

References

1. H. W. Kroto, J. R. Heath, S.C. O'Brien, R. F. Curl and R. E. Smalley, *Nature*, 1985, **318**, 162.
2. S. Iijima, *Nature*, 1991, **354**, 56.
3. A. Rubio, J. L. Corkill and M. L. Cohen, *Phys. Rev. B*, 1994, **49**, 5081.
4. N. G. Chopra *et al.*, *Science*, 1995, **269**, 966.
5. K. S. Novoselov *et al.*, *Science*, 2004, **306**, 666.
6. K. S. Novoselov *et al.*, *Proc. Natl. Acad. Sci. U. S. A.*, 2005, **102**, 10451.
7. A. N. Goldstein, C. M. Echer CM and A. P. Alivisatos, *Science*, 1992, **256**, 1425.
8. L.E. Brus, *J. Chem. Phys.*, 1984, **80**, 4403.
9. S.-H. Jhi, S. G. Louie and M. L. Cohen, *Phys. Rev. Lett.*, 2000, **85**, 1710.
10. R. J. Chen, Y. Zhang, D. Wang and H. Dai, *J. Am. Chem. Soc.*, 2001, **123**, 3838.
11. R. J. Chen RJ *et al.*, *Proc. Natl. Acad. Sci. U. S. A.*, 2003, **100**, 4984.
12. A. Krishnan, E. Dujardin, T. W. Ebbesen, P. N. Yianilos andand M. M. J. Treacy, *Phys. Rev. B*, 1998, **58**, 14013.
13. M. M. J. Treacy, T. W. Ebbesen and J. M. Gibson, *Nature*, 1996, **381**, 678.
14. B. G. Demczyk *et al.*, *Mater. Sci. Eng. A*, 2002, **334**, 173.
15. Anonymous, *Engineering properties of steel*, ed. P. D. Harveyand, American Society for Metals, Metals Park, Ohio, 1982, pp v.
16. B. S. Harrison and A. Atala, *Biomaterials*, 2007, **28**, 344.
17. S. F. Wang, L. Shen, W. D. Zhang and Y. J. Tong, *Biomacromolecules*, 2005, **6**, 3067.
18. X. Shi *et al.*, *Biomaterials*, 2007, **28**, 4078.

19. J. K. Holt *et al.*, *Science*, 2006, **312**, 1034.
20. S. Mann, *Angew. Chem., Int. Ed. Engl.*, 2008, **47**, 5306.
21. T. J. Webster, Nanophase ceramics: The future orthopedic and dental implant material, in *Advances in Chemical Engineering*, Academic Press, Vol. 27, 2001, pp. 125.
22. R. H. Christenson, *Clin. Biochem.*, 1997, **30**, 573.
23. S. Weiner, W. Traub and H. D. Wagner, *J. Struct. Biol.*, 1999, **126**, 241.
24. J. Y. Rho, L. Kuhn-Spearing, and P. Zioupos, *Med. Eng. Phys.*, 1998, **20**, 92.
25. S. P. Bruder and A. I. Caplan, *Connect. Tissue Res.*, 1989, **20**, 65.
26. L. Wang, G. H. Nancollas, Z. J. Henneman, E. Klein E, and S. Weiner, *Biointerphases*, 2006, **1**, 106.
27. C. K. Colton, *Cell Transplant*, 1995, **4**, 415.
28. D. Fassnacht and R. Portner, *J. Biotechnol.*, 1999, **72**, 169.
29. V. I. Sikavitsas, G. N. Bancroft, and A. G. Mikos, *J. Biomed. Mater. Res.*, 2002, **62**, 136.
30. R. Portner, S. Nagel-Heyer, C. Goepfert, P. Adamietz and N. M. Meenen, *J. Biosci. Bioeng.*, 2005, **100**, 235.
31. G. N. Bancroft, V. I. Sikavitsas, and A. G. Mikos, *Tissue Eng.*, 2003, **9**, 549.
32. F. Zhao and T. Ma, *Biotechnol. Bioeng.*, 2005, **91**, 482.
33. M. T. Raimondi *et al.*, *Biomech. Model. Mechanobiol.*, 2002, **1**, 69.
34. J. van den Dolder *et al.*, *J. Biomed. Mater. Res., Part A*, 2003, **64**, 235.
35. R. L. Carrier *et al.*, *Tissue Eng.*, 2002, **8**, 175.
36. A. B. Castillo and C. R. Jacobs, *Curr. Osteoporos. Rep.*, 2010, **8**, 98.
37. M. V. Hillsley and J. A. Frangos, *Biotechnol. Bioeng.*, 1994, **43**, 573.
38. S. Weinbaum, S. C. Cowin and Y. Zeng, *J. Biomech.*, 1994, **27**, 339.
39. A. S. Goldstein, T. M. Juarez, C. D. Helmke, M. C. Gustin and A. G. Mikos, *Biomaterials*, 2001, **22**, 1279.
40. D. Katritsis *et al.*, *Prog. Cardiovasc. Dis.*, 2007, **49**, 307.
41. J. L. Moreira *et al.*, *Biotechnol. Prog.*, 1995, **11**, 575.
42. K. P. Travis, B. D. Todd BD and D. J. Evans, *Phys. Rev. E*, 1997, **55**, 4288.
43. J. T. Cheng and N. Giordano, *Phys. Rev. E*, 2002, **65**, 031206.
44. J. Pfahler, J. Harley, H. Bau and J. Zemel, *Sens. Actuators, A*, 1990, **22**, 431.
45. V. P. Sokhan, D. Nicholson and N. Quirke, *J. Chem. Phys.*, 2001, **115**, 3878.
46. V. P. Sokhan, D. Nicholson, and N. Quirke, *J. Chem. Phys.*, 2002, **117**, 8531.
47. Y. Xiaofan and Z. C. Zheng, *J. Fluids Eng.*, 2010, **132**, 061201.
48. Y. Y. Huang, D. Y. Wang, L. L. Chang LL, and Y. C. Yang, *J. Biomater. Sci. Polym. Ed.*, 2010, **21**, 1503.
49. Y. Yang, K. Kulangara, J. Sia, L. Wang and K. W. Leong, *Lab Chip*, 2011, **11**, 1638.

50. B. Porter, R. Zauel, H. Stockman, R. Guldberg and D. Fyhrie, *J. Biomech.*, 2005, **38**, 543.
51. F. Maes, P. Van Ransbeeck, II. Van Oosterwyck and P. Verdonck, *Biotechnol. Bioeng.*, 2009, **103**, 621.
52. R. Voronov, S. Vangordon, V. I. Sikavitsas and D. V. Papavassiliou, *J. Biomech.*, 2010, **43**, 1279.
53. V. I. Sikavitsas, G. N. Bancroft, H. L. Holtorf, J. A. Jansen and A. G. Mikos, *Proc. Natl. Acad. Sci. U. S. A.*, 2003, **100**, 14683.
54. L. S. Nair, S. Bhattacharyya and C. T, Laurencin, *Expert Opin. Biol. Ther.*, 2004, **4**, 659.
55. Z. Ma, M. Kotaki, R. Inai and S. Ramakrishna, *Tissue Eng.*, 2005, **11**, 101.
56. T. Ondarcuhu and C. Joachim, *Europhys. Lett.*, 1998, **42**, 215.
57. S. L. Tao and T. A. Desai, *Nano Lett.*, 2007, **7**, 1463.
58. C. R. Martin, *Science*, 1994, **266**, 1961.
59. P. X. Ma and R. Zhang, *J. Biomed. Mater. Res.*, 1999, **46**, 60.
60. G. M. Whitesides and B. Grzybowski, *Science*, 2002, **295**, 2418.
61. Q. P. Pham, U. Sharma and A. G. Mikos, *Tissue Eng.*, 2006, **12**, 1197.
62. Q. P. Pham, U. Sharma and A. G. Mikos, *Biomacromolecules*, 2006, **7**, 2796.
63. T. A. Horbett, Proteins: Structure, properties and adsorption to surfaces. *Biomaterials Science: An Introduction to Materials in Medicine*, ed. B. D.D. Ratner, A. S. Hoffman, F. J. Schoen and J. E. Lemons JE, Academic Press, New York, 1996, pp 133.
64. K. M. Woo, V. J. Chen and P. X. Ma, *J. Biomed. Mater. Res., Part A*, 2003, **67**, 531.
65. A. Martins *et al.*, *Tissue Eng. Part A*, 2010, **16**, 3599.
66. D. Lahiri *et al.*, *Appl. Mater. Interfaces*, 2009, **1**, 2470.
67. D. Lahiri *et al.*, *Acta Biomater.*, 2010, **6**, 3524.
68. R. L. Price, M. C. Waid, K. M. Haberstroh and T. J. Webster, *Biomaterials*, 2003, **24**, 1877.
69. K. L. Elias, R. L. Price, and T. J. Webster, *Biomaterials*, 2002, **23**, 3279.
70. J. Folkman and A. Moscona, *Nature*, 1978, **273**, 345.
71. J. O. Gallagher, K. F. McGhee, C. D. Wilkinson, and M. O. Riehle, *IEEE Trans. Nanobiosci.*, 2002, **1**, 24.
72. C. C. Berry, G. Campbell, A. Spadiccino, M. Robertson and A. S. Curtis, *Biomaterials*, 2004, **25**, 5781.
73. M. J. Dalby, M. O. Riehle, H. Johnstone, S. Affrossman, and A. S. Curtis *Biomaterials*, 2002, **23**, 2945.
74. P. Clark, P. Connolly, A. S. Curtis, J. A. Dow, and C. D. Wilkinson, *Development*, 1987, **99**, 439.
75. C. S. Chen, M. Mrksich, S. Huang, G. M. Whitesides and D. E. Ingber, *Science*, 1997, **276**, 1425.
76. M. Kalbacova, B. Rezek, V. Baresova, C. Wolf-Brandstetter and A. Kromka, *Acta Biomater.*, 2009, **5**, 3076.

77. M. J. Dalby *et al.*, *Exp. Cell Res.*, 2002, **276**, 1.
78. S. Oh S *et al.*, *Proc. Natl. Acad. Sci. U. S. A.*, 2009, **106**, 2130.
79. M. J. Dalby *et al.*, *Nat. Mater.*, 2007, **6**, 997.
80. K. C. Popat, L. Leoni, C. A. Grimes and T. A. Desai, *Biomaterials*, 2007, **28**, 3188.
81. L. P. Zanello, B. Zhao, H. Hu and R. C. Haddon, *Nano Lett.*, 2006, **6**, 562.
82. L. E. McNamara *et al.*, *J. Tissue Eng.*, 2010, **2010**, 120623.
83. M. J. Biggs, R. G. Richards, N. Gadegaard, C. D. Wilkinson and M. J. Dalby, *J. Mater. Sci. Mater. Med.*, 2007, **18**, 399.
84. M. J. Dalby *et al.*, *J. Cell. Biochem.*, 2007, **100**, 326.
85. M. J. Biggs *et al.*, *J. R. Soc. Interface*, 2008, **5**, 1231.
86. J. Y. Lim *et al.*, *Biomaterials*, 2007, **28**, 1787.
87. M. D. Schaller *et al.*, *Proc. Natl. Acad. Sci. U. S. Am.*, 1992, **89**, 5192.
88. M. J. Biggs *et al.*, *J. Biomed. Mater. Res., Part A*, 2009, **91**, 195.
89. R. McBeath, D. M. Pirone, C. M. Nelson, K. Bhadriraju and C. S. Chen, *Dev. Cell*, 2004, **6**, 483.
90. M. Arnold *et al.*, *ChemPhysChem*, 2004, **5**, 383.
91. G. A. Dunn and J. P. Heath, *Exp. Cell Res.*, 1976, **101**, 1.
92. G. A. Dunn and A. F. Brown, *J. Cell Sci.*, 1986, **83**, 313.
93. O. Ishida, K. Maruyama, K. Sasaki and M. Iwatsuru, *Int. J. Pharm.*, 1999, **190**, 49.
94. S. V. Vinogradov, T. K. Bronich and A. V. Kabanov, *Adv. Drug Deliv. Rev.*, 2002, **54**, 135.
95. S. Stolnik, L. Illum and S. S. Davis, *Adv. Drug Deliv. Rev.*, 1995, **16**, 195.
96. A. W. Yasko *et al.*, *J. Bone Joint Surg. Am.*, 1992, **74**, 659.
97. G. Xie, J. Sun, G. Zhong, C. Liu and J. Wei, *J. Mater. Sci. Mater. Med.*, 2010, **21**, 1875.
98. H. Autefage *et al.*, *J. Biomed. Mater. Res. B: Appl. Biomater.*, 2009, **91**, 706.
99. M. D. Krebs, E. Salter, E. Chen, K. A. Sutter and E. Alsberg, *J. Biomed. Mater. Res. Part A*, 2010, **92**, 1131.
100. J. M. Oliveira *et al.*, *Nanomedicine*, 2011, **7**, 914.
101. X. Zhang *et al.*, *Biomaterials*, 2010, **31**, 6013.
102. C. Li, C. Vepari, H. J. Jin, H. J. Kim and D. L. Kaplan, *Biomaterials*, 2006, **27**, 3115.
103. H. Hosseinkhani, M. Hosseinkhani, A. Khademhosseini and H. Kobayashi, *J. Controlled Release*, 2007, **117**, 380.
104. Y. I. Chung *et al.*, *J. Controlled Release*, 2007, **121**, 91.
105. H. Hirabayashi and J. Fujisaki, *Clin. Pharmacokinet.*, 2003, **42**, 1319.
106. K. A. Gonzalez, L. J. Wilson, W. Wu and G. H. Nancollas, *Bioorg. Med. Chem.*, 2002, **10**, 1991.
107. D. Carter and G. Beaupré, *Skeletal function and form : mechanobiology of skeletal development, aging, and regeneration*, Cambridge University Press, Cambridge, 2001.

108. H. Ohshima, *Clin. Calcium*, 2006, **16**, 81.
109. G. D. Whedon, *Calcif. Tissue Int.*, 1984, **36 Suppl 1**, S146.
110. R. D. Roer and R. M. Dillaman, *J. Appl. Physiol.*, 1990, **68**, 13.
111. K. M. Reich, C. V. Gay and J. A. Frangos, *J. Cell. Physiol.*, 1990, **143**, 100.
112. C. R. Jacobs *et al.*, *J. Biomech.*, 1998, **31**, 969.
113. C. R. Jacobs, S. Temiyasathit and A. B. Castillo, *Annu. Rev. Biomed. Eng.*, 2010, **12**, 369.
114. H. L. Holtorf, N. Datta, J. A. Jansen and A. G. Mikos, *J. Biomed. Mater. Res. A*, 2005, **74**, 171.
115. G. N. Bancroft *et al.*, *Proc. Natl. Acad. Sci. U. S. A.*, 2002, **99**, 12600.
116. N. Datta, H. L. Holtorf, V. I. Sikavitsas, J. A. Jansen and A. G. Mikos, *Biomaterials*, 2005, **26**, 971.
117. M. E. Gomes, C. M. Bossano, C. M. Johnston, R. L. Reis and A. G. Mikos, *Tissue Eng.*, 2006, **12**, 177.
118. Q. P. Pham *et al.*, *J. Biomed. Mater. Res. Part A*, 2009, **88**, 295.
119. K. Anselme, *Biomaterials*, 2000, **21**, 667.
120. J. El-Ali, P. K. Sorger and K. F. Jensen, *Nature*, 2006, **442**, 403.
121. T. Miyai *et al.*, *Biomaterials*, 2008, **29**, 350.
122. V. C. Martins, G. Goissis, A. C. Ribeiro, E. Marcantonio Jr. and M. R. Bet, *Artif. Organs*, 1998, **22**, 215.
123. A. Martins *et al.*, *Biomaterials*, 2010, **31**, 5875.
124. A. C. Docherty-Skogh *et al.*, *Plast. Reconstr. Surg.*, 2010, **125**, 1383.
125. G. Wang, C. Kucharski, X. Lin and H. Uludag, *J. Drug Targeting*, 2010, **18**, 611.
126. J. S. Choi *et al.*, *Int. J. Pharm.*, 2006, **320**, 171.
127. S. Zhang, C. Kucharski, M. R. Doschak, W. Sebald and H. Uludag, *Biomaterials*, 2010, **31**, 952.
128. B. Sitharaman, P. K. Avti, K. Schaefer, Y. Talukdar and J. P. Longtin, *Tissue Eng. Part A*, 2011.
129. M. O. Andersen *et al.*, *Mol. Ther.*, 2010, **18**, 2018.
130. I. Ono *et al.*, *Biomaterials*, 2004, **25**, 4709.
131. J. Park *et al.*, *Gene Therapy*, 2003, **10**, 1089.
132. R. A. Thibault, L. Scott Baggett, A. G. Mikos and F. K. Kasper, *Tissue Eng. Part A*, 2010, **16**, 431.

Section 10
Nanomedicine and Nanotoxicology

CHAPTER 10
Nanomedicine and Nanotoxicology

NOUR KARRA[a] AND JUERGEN BORLAK*[b]

[a] The Institute for Drug Research, The School of Pharmacy, Faculty of Medicine, The Hebrew University of Jerusalem, Jerusalem, Israel; [b] Center for Pharmacology and Toxicology, Hannover Medical School, Hannover, Germany
*E-mail: borlak.juergen@mh-hannover.de

10.1 Introduction

10.1.1 Nanomedicine and Nanoparticles (NPs) – the Good, the Bad and the Unknown

Nanotechnology and its application to medicine have evolved in recent years as multidisciplinary fields of research aimed at promoting advanced and efficient therapeutic and diagnostic strategies to various diseases. Notably, nanoscaled drug delivery systems (DDS) are becoming a widely investigated approach for their potential ability to target diseased tissues where physiologic or pathologic biological barriers prevent the efficient delivery of therapeutic drug concentrations. As such, these DDS have gained much interest particularly in the diagnosis and treatment of cancer tumors, where indiscriminate drug distribution and low drug concentration at the tumor tissue often limit treatment outcomes.

In the year 2011 and for cancer therapy and diagnosis only, there are more than 50 ongoing and recruiting clinical trials using nanoparticles (NPs).[1,2] The majority of NPs being tested for the treatment of various cancer types in clinical trials are "nab" type (nanoparticle albumin bound).[2] Nonetheless,

RSC Drug Discovery Series No. 22
Nanostructured Biomaterials for Overcoming Biological Barriers
Edited by Maria Jose Alonso and Noemi S. Csaba
© The Royal Society of Chemistry 2012
Published by the Royal Society of Chemistry, www.rsc.org

clinical trials are also performed with various other nanocarriers: nanoliposomal irrinotecan, superparamagnetic iron oxide NPs (SPIO) for the diagnosis of pre-operative stage of pancreatic cancer, transferrin targeted cyclodextrin based NPs for siRNA delivery (CALAA-01), as well as cholesterol based liposomes containing a Bik gene product (BikDD NPs) for patients with advanced pancreatic cancer. Additionally, clinical studies of radiofrequency-activated tumor ablation by lyso-thermosensitive liposomal doxorubicin (Thermodox®) have been performed.[2] The number of general nanotechnologies tested in patients is even bigger when taking into account other therapeutic indications. Examples of such technologies include plasmonic nanophotothermic therapy using transplantable ferro-magnetic NPs for atherosclerosis patients and insoluble antibacterial NPs for dental uses.[2]

With the rapid advances in nanotechnology and the translation of academic knowledge and findings from the laboratory bench to consumer products and the clinical practice of nanomedicine,[3] concerns about the potential hazards and toxicity of nanomaterials have surfaced, and are threatening to thwart their use. There is also a tendency to relate mainly to the reduction of non specific exposure and toxicity of drugs achieved by their nano-encapsulation in "clever" DDS. However, the potential "hidden" toxicities of the nanocarrier *per se*, which are sometimes sub-acute, are often disregarded. Thus, the unique size and surface properties of NPs may be perceived as a "double-edged sword" rendering NPs as both highly promising drug delivery systems able to effectively deliver high therapeutic drug doses at the target tissue, but also as potentially hazardous entities with tissue penetration properties possibly deleterious to cell organelles and functions. As a result, considerable attention is being drawn to toxicity screening of these systems. Hence, **nanotoxicology** is a rather new discipline of nanotechnology that has evolved in the past years for the study of the safety and potential toxicity of nanotechnology based developments.

Nanoparticles is a general term embracing versatile sub-micron colloidal carriers, with dimensions ranging from 1–1000 nm.[4] These carriers may be of diverse composition, shape and physicochemical characteristics dictating their *in vivo* behavior. Such nanoscaled constructs include mainly liposomes, nanoemulsions, polymeric micelles, solid lipid NPs, dendrimers and other stealth organic and inorganic NPs.

It is beyond the scope of this chapter to review the toxicology literature of all the nanoparticles currently investigated in nanomedicine, but rather the aim is to point out the potential safety issues that might result from the use of nanocarriers. The present chapter aims to provide a basic understanding of the main challenges currently faced by nanotoxicology assessment, as well as to present the key players dictating nanoparticle biological behavior, tolerance and biocompatibility. We will address the physicochemical characteristics and possible contributors to NPs toxicity, while highlighting the most investigated delivery routes of NPs. We will mostly review nanotoxicology of organic polymeric NPs, which are widely used in biomedical research, emphasizing recent works of special interest. Environmental nanotoxicology of combustion derived nanoparticles, fullerenes (carbon nanotubes) as well as inorganic NPs investigated for biomedical applications (quantum dots, gold NPs, silver NPs, iron oxide NPs and silica NPs) will not be addressed in the present chapter. For further information on the nanotoxicology of these nanomaterials, the reader is referred to other references.[5–14]

10.1.2 Nanotoxicology Regulation – Where Do We Stand?

Various questions and dilemmas arise when addressing regulatory aspects of nanomaterials and nanoparticles. To be able to set clear specifications as to the criteria that nanomaterials should fulfill when intended for human use, it is essential to ask, what are the existing regulatory policies regarding safety of nanomaterials?

Despite the fact that the U.S. invested around $1.5 billion in nanoscience in 2009, no actual Federal regulation yet exists. Notably, various efforts have been made to address the regulatory status in an attempt to formulate guidelines for the handling of nanomaterials. The Environmental Protection Agency issued a **White Paper** and a voluntary **Nanoscale Materials Stewardship Program**. California and Massachusetts both have State and Municipal regulations of nano health and safety. Additionally, in August 2006, the formation of an internal **FDA Nanotechnology Task Force** was announced. The Task Force was given the responsibility to determine the regulatory approaches that would enable the continued development of innovative, safe, and effective FDA-regulated products that use nanoscale materials.[15] However, no specific regulations were issued by the task force, and it is believed that existing regulations can govern nanomaterials. The European Union, on the other hand, is more inclined to precaution: The Scientific Committee on Emerging and Newly Identified Health Risks (SCENIHR) issued an opinion **Precautionary Principle** (PP) paper that was heavily commented on by the industry, stating that, *"Where the full extent of a risk is unknown, but concerns are so high that risk management measures are considered necessary,... measures must be based on the Precautionary Principle... The regulatory challenge is therefore to ensure that society can benefit from novel applications of nanotechnology, whilst a high level of protection of health, safety and the environment is maintained"*.

In view of these vague, somewhat contradictory positions adopted by the US FDA and by the EU, fundamental questions remain unanswered:

1. First, are the current safety regulations robust enough to handle risks of nanomaterials?
2. Are nanoscale materials equivalent to their micro or macroscale counterparts or should they be considered as "new substances"? Due to their small size, nanoparticulate materials acquire unique features that differ from their bulk source: higher surface to volume ratio, enhanced contact area with the biological surrounding, and enhanced internalization and

retention in cells. The nano-structure-dependent biological activity and toxicology are often not predicted by the bulk properties of the source materials in their macroscopic forms. Thus, it is becoming widely agreed within the field of nanotechnology that it is only logical to consider these materials as new substances, with different biological behaviors.[16,17]

3. Do polymeric materials that are "generally regarded as safe" (GRAS compounds) preserve their "inert" nature when combined with other compounds or re-formulated into nanoparticulate carriers? In his review from 2006, Alexander V. Kabanov addressed this issue and introduced examples where "simple and safe" polymeric materials that are in wide clinical use could independently alter genomic expression and even biological responses to drugs.[18] All the more so, when polymers, drugs and excipients are united into a more complex nanostructure; the new surface properties and diverse biodegradation products of the nanocarrier could lead to different interactions with biological membranes and barriers and to unpredictable behavior to that expected for the single constituents or their "simple" mixture. The author claimed that the wide misconception of considering polymer based nano scaled drug delivery systems as innocuous "biologically inert excipients" is undergoing major revision with the growing evidence of their interference in cellular responses at the gene level.[18]

4. Could multilayered, multifunctional nanomedical systems be considered by the FDA as a ***Combination product***? According to the FDA's definition in 21 CFR 3.2(e), a combination product might be considered as "*A product comprised of two or more regulated components, i.e., drug/device, biologic/device, drug/biologic, or drug/device/biologic, that are physically, chemically, or otherwise combined or mixed and produced as a single entity*".

This issue has become even more relevant with the development of multifunctional theragnostic NPs, comprising drug molecules, diagnostic probes and possibly targeting ligands on the NP's surface. The decision as to what would be the most appropriate approach to regulate such products is therefore becoming rather complex.

10.1.3   State of the Science – Challenges in Toxicity Assessment of NPs

1. Nanotoxicity data are available mainly from unintentional exposure to incidental nanomaterials, such as environmental or occupational exposures (for example combustion derived nanomaterials).[5,19,20] These data do not reflect potential toxicity as a result of controlled exposure to engineered nanoparticles that are intended for therapeutic applications and targeted nanomedicine. Despite the rapid development of sophisticated theragnostic nanoparticles, their safety evaluation has lagged far

behind, with only scarce comprehensive studies of their safety and toxicities.

2. Lack of appropriate methodologies for toxicity assessment: as previously emphasized by Dhawan and Sharma,[10] most of the methods currently used for toxicity assessment have been designed and standardized to fit chemical toxicity. However, NPs exhibit additional unique features that can challenge the use of classical toxicity assays (such as their size, surface area, charge, solubility, agglomeration, elemental purity, adsorption capacity, catalytic activities, magnetic and optical properties *etc.*), and sometimes interfere with *in vitro* assays, leading to false positive or false negative results.[10,21]

3. The latter problem makes it difficult to compare safety and toxicity findings obtained by different research groups. Even though a plethora of *in vitro, in vivo* and *ex vivo* methodologies is available for toxicity assessment (*e.g.* cytotoxicity, genotoxicity and, hemotoxicity), very few, if any, harmonized protocols exist for the testing of NPs biocompatibility.

4. There is an immense versatility of investigated NPs, constituted from various elements, both organic and inorganic, where the combinational possibilities of the composing elements are practically infinite. This infinite diversity makes it practically impossible to extrapolate findings and draw generalized conclusions and understandings as to the biological behavior of NPs without referring to their specific physicochemical characteristics and their correlations to observed biological effects.

10.1.4 Toxicity Screening Methodologies: Available Tools, Challenges, Limitations and Desirable Methodologies

Currently, many methodologies and models are being investigated and employed in nanotoxicology research. Conventional "classical" tools include:

1. *In vitro* **cell culture models:** cell cultures are widely used for toxicity assessment as they are time and cost effective, with the potential of providing valuable data on mechanisms of toxicity.[10] Co-cultures of various cell types are often used to study their mutual effects on the biological response to NPs, for example hepatocytes and kupffer cells,[22] or alveolar macrophages and lung cancer cells.[23] High content *in vitro* assays evaluating toxicity parameters and effects of various NPs employing several doses and multiple relevant cell lines have already been reported.[16,24] However, data obtained from these models may be of limited value as cultured cellular systems are devoid of complex cell–matrix interactions and hormonal effects present in *in vivo* settings, as well as intercellular mutual effects participating in the coordination of responses, such as chemotaxis and cell recruitment (*e.g.* macrophages).[10,25] Additionally, the combined results from multiple studies of various cells *in vitro* cannot be assumed to capture the behavior of the same cells arranged

in situ in a whole organ. Thus, at best, data from these studies may be considered only as extrapolatory, while extrapolation is not straightforward, and the complete lack of correlation between in *vitro* and in *vivo* findings may also be expected.[26] Furthermore, the doses usually used *in vitro* do not necessarily reflect the actual exposure doses in clinical realistic *in vivo* settings, thus compelling great caution in the interpretation and extrapolation of the results.[25] Additionally, many *in vitro* assays are highly influenced by the unique optical, fluorescent and magnetic properties of NPs,[10,27,28] which may lead to misinterpretation of results and false positive or negative results. Another limitation that has been pointed out by Fernandez-Urrusuno *et al.* is the short lifetime of *in vitro* cultures which could lead to underestimation of toxicological effects,[22] especially those associated with long term or chronic exposure.

2. **Three dimensional *in vitro* models:** other *in vitro* models attempt to replace costly *in vivo* models by expanding the capabilities of conventional cell culture models and mimicking biological natural interfaces between various cell types and tissues in specific organs.[16,29,30] These micro-engineered impressive models may indeed imitate several valuable aspects and functions of interest of the investigated organ, but they do not completely recapitulate the entire biological composition of a living tissue.[16]

3. ***In vivo/Ex vivo* models:** other authors have reported the use of combined *in vivo–ex vivo* models isolating specific cells or organs for the study of NPs effects and toxicities, either in non treated animals or post treatment with NPs.[22,31] To overcome the lack of sensitivity of blood aminotransferases in the investigation of liver toxicity, Fernandez-Urrusuno *et al.* have isolated rat hepatocytes following intravenous treatments with PACA NPs, and measured specific functions such as fructose metabolism and protein synthesis.[22]

4. ***In vivo* testing:** although time consuming, costly and ethically controversial, *in vivo* studies are perceived as the most accurate method for the investigation of nanotoxicity, particularly in light of the shortfall and limitations of *in vitro* models mentioned earlier. As pointed out by Fischer and Chan,[26] at this time, one cannot predict systematically neither the movement and locations of NPs after *in vivo* exposure, nor the final metabolic intracellular fate. Thus, the authors highlighted the knowledge that would be gained from *in vivo* studies, including the elucidation of biodistribution and residence time of NPs in specific organs and cells, eventually leading to more focused *in vitro* and *in vivo* studies, with the aim of deciphering the molecular basis of toxicity.[26]

5. **Proteomic-based tools and molecular biomarkers:** one of the limitations in toxicity assessment and risk prediction upon exposure to nanomaterials is the lack of sensitive yet non invasive toxicity biomarkers and the inability of conventional blood testing to recognize subjects at risk of nanotoxicity. Thus, additional reliable, accurate and predictive *in vivo* proteomic-based

approaches for the identification of molecular biomarkers of nanotoxocity are currently being sought. Most recently, Higashisaka *et al.,* suggested the measurement of the plasma acute phase proteins haptoglobin, C-reactive protein and serum amyloid A as highly sensitive biomarkers for predicting the exposure and toxicity of nanomaterials, specifically of the silica NPs evaluated in their study.[32] Despite the invaluable power potentially provided by proteomic-based strategies, it should be kept in mind that in real clinical settings, such biomarkers may also be influenced by other concomitant toxic exposures or co-morbidities, thus emphasizing the need to evaluate the specificity of these markers to nanotoxic exposure and effects and their clinical relevance.

6. **Toxicogenomics:** toxicogenomics is the use of transcriptomix and micro-assay based gene expression profiling in both investigative and predictive toxicology. This research field has been providing accumulative information on the nature of the interactions between nanomaterials and living cells at the genomic level aiming to provide an understanding of the origins of nanomaterials genotoxicity.[17,33] Nanotoxicogenomics may potentially provide a means to predict nanotoxicity prior the performance of classical toxicological endpoints such as histopathology or clinical chemistry,[33] thus saving time and money in the development of safe NPs. The toxicogenomic approach has been frequently applied in recent years for the evaluation of biological consequences of nano-sized DDS. Akhtar recently reviewed the toxicogenomics of cationic nanosystems intended for the delivery of siRNA.[33] The author deduced that indeed, various cationic nanosystems, either lipid or polymer based, not only possess delivery enhancing properties, but may also alter gene expression and influence the gene silencing magnitude elicited by the siRNA therapy. This, according to the author, compels the selection of a DDS also to be based on its toxicogenomic profile, among other characteristics.[33] For a detailed review on toxicogenomics of nanomaterials and nanoparticles the reader is referred to ref. 17, 18, 33 and 34.

Clearly, the toxicogenomic evaluation of NPs together with conventional assessment batteries of toxicity in animal models has become essential for the development of safe pharmaceutical formulations. A comprehensive risk assessment should be based on a broad spectrum of toxicity assays, characterizing the general and specific toxicities of NPs as well as their ecotoxicity, for example as performed by Robbens *et al.*[35]

The sheer diversity of sizes, shapes, structures and elemental composition of engineered NPs renders their broad toxicologic evaluation tedious and highly challenging. Furthermore, *in vivo* exposure studies for comprehensive nanotoxicity assessment are costly, time and labor consuming, all the more so when chronic exposure is evaluated. Therefore, the desirable state would infer a shift from an observational toxicology science to a predictive science, aiming at organizing the infinite space of accumulating data and translating it into useful knowledge, if possible, with the aid of theoretical paradigms supporting it. Such organization will be in the shape of a "matrix of relationships" correlating physicochemical properties and resultant bioactivity.[21] There is a growing aspiration to develop low cost, time sparing *in vitro* high throughput screening (HTS) assays, able to predict nanotoxocity, in the absence of complete toxicity testing. This need has been already outspoken by some authors, using terms such as the "control banding concept", originally proposed by Schulte *et al.*, as one example of a "matrix of relationship" system.[16,21]

However, despite the aspiration to achieve predictive methodologies and paradigms that can and will contribute to fast and rather easy toxicity screenings, this vision is far from being fully accomplished, and suitable *in vivo* models are considered the most reliable tool for toxicity assessment; we are still obliged to use a case by case testing approach of NP toxicology.

10.2 Physicochemical Characteristics of NPs and Their Correlations with Toxicity

The main characteristics of nanoparticles that are usually investigated include size, surface area, surface chemistry, solubility, porosity, shape, charge, elemental composition, and aggregation or agglomeration in relevant aqueous dispersions.

10.2.1 Size and Surface Area – Size Matters!

There are various size classifications when relating to particles in ambient air (Table 10.1). However, as previously mentioned, the general term "nanoparticles" comprises versatile sub-micron colloidal carriers, with dimensions ranging from 1–1000 nm.[4] The small size of NPs is the most important parameter that affects their interactions and uptake by biological systems, their enhanced tissue penetration properties and their circulation, biodistribution and elimination profiles.[10,36–38]

With regard to intravenously-administered NPs, small particles below 10 nm are cleared by renal excretion, while larger particles are cleared by the

Table 10.1 Size definitions used to describe particles in ambientô air.[5]

Particle definition	Description
PM 10	Particle mass fraction in ambient air with a mean diameter of 10 μm
PM 2.5	Particle mass fraction in ambient air with a mean diameter of 2.5 μm
Coarse particles	The mass fraction of PM10 bigger than 2.5μm
Ultrafine particles (UFPs) (PM 0.1)	The fraction of PM10 with a size cut off at 0.1 μm, and contains primary particles and agglomerates smaller than 100 nm

mononuclear phagocyte system (MPS) and by hepatic filtration.[38] The likelihood of opsonization of blood components to the surface of NPs and resultant liver uptake augments with the increase in particle size. NPs around 100 nm exhibit an improved circulation half life[38] compared to larger NPs. With regard to pulmonary delivery and lung deposition of NPs, it has been established that the deposition site and deposition mechanism, as well as the rate of particle phagocytosis and other clearance mechanisms are highly dependent on particle size.[39] This will be addressed further in the following section on pulmonary delivery of NPs.

On the cellular level, not only does size affect the extent of nanoparticle internalization into the cell, but also the specific pathway of internalization, the exocytosis of the internalized NPs and their resulting extent of intracellular retention. These issues have been reviewed by Harush-Frenkel *et al.*[40]

Size also dictates the number and most importantly the *surface area* of the NPs at a given volume or mass of the material. A decrease in NPs size leads to a dramatically increased surface area which becomes available to interact with biological components such as proteins and cell membranes. As NPs surfaces are involved in many catalytic and oxidative reactions, the increase in the surface area could result in increased toxicities.[41] There are various available technologies for the evaluation of NP size and surface area.[10] These include Dynamic Light Scattering (DLS), Transmission Electron Microscopy (TEM), Scanning Electron Microscopy (SEM), Cryo-TEM, Cryo-SEM, Atomic Force Microscopy (AFM), Nanoparticle Tracking and Analysis (NTA) and Brunauer–Emmett–Teller (BET). Some of the methods provide size measurements in aqueous media whilst others require sample drying under vaccum. Depending on the method used one can obtain information not only about size and size distribution, but also about agglomeration conditions in various media, the shape, morphology, surface area, sample topography, surface texture and roughness.[10]

10.2.2 Charge

A nanoparticle's surface can be neutral, positively or negatively charged. The surface charge could highly affect the *in vivo* biodistribution, accumulation in organs and toxicity profile of NPs. The positive surface charge of nanoparticulate systems is often exploited to elicit mucoadhesive properties, to promote gene delivery, and enhance drug–cell internalization, as a result of the electrostatic attraction of the carrier to the negatively charged cell membrane moieties.[42] Notably, increased cell internalization of a highly negatively charged hydrophilic macromolecule such as DNA was successfully achieved by condensing DNA with a cationic branched polymer, polyethyleneimine (PEI), resulting in the formation of positively charged particles at physiological pH.[43] Unfortunately, as a rule, it is widely accepted that polycations are substantially more toxic than non ionic and anionic polymers. PEI based polycations, often used for gene therapy or RNA interference *in vitro* as well as *in vivo*, have been shown to cause well-known adverse side effects, especially high cytotoxicity. It has already been reported that positively charged dendrimers and cationic macromolecules interact with blood components upon their uptake into the systemic circulation, destabilize cell membranes and eventually lead to cell lysis.[9,44-46]

Recently, charge effects of NPs on biological behavior and toxicity were observed in various *in vivo* studies. For example, Xu *et al.* demonstrated that the calculated half life of hemoglobin loaded polymeric nanoparticles (HbPNPs) was eight fold longer for cationized NPs as compared to non cationized NPs.[47] Cationized NPs also accumulated mainly in the liver, lungs and spleen 48 h after injection.[47] Surprisingly, in another study, the vascular thrombosis observed in the pulmonary circulation following intravenous injection of QDs was enhanced by negatively charged carboxyl coated QDs as compared to amine coated QDs.[48] The authors concluded that the higher negative charge of the carboxyl coated NPs triggered coagulation *via* contact activation, fibrin formation and platelet activation by thrombin.[48] Even though positively charged NPs are generally perceived as more hazardous,[5,9,18] anionic NPs could also present non benign side effects. The latter study only exemplifies the need to adopt a "case by case" approach for nanotoxicology assessment, and to avoid categorization of NPs and generalization of specific findings to other kinds of NPs with similar but not identical properties.

10.2.3   Surface Chemistry, Functionalization and Surface Reactivity

Polyethylene glycol (PEG) is a water soluble amphiphilic polymer clinically approved and widely used in nanomedicine, with a low potential for toxicity.[49,50] PEGylation is the covalent coating of PEG chains to another molecule (drug, protein) or carrier (NPs, liposomes). The idea of NPs grafting with the hydrophilic PEG chain moiety was proposed by Gref *et al.* as a means to "mask" the carrier from the host's immune system and circumvent the rapid opsonization and phagocytic clearance of NPs, hence prolonging their circulation in the blood.[51] Today, it is a well-documented fact that pegylated NPs, especially where the PEG chain is chemically attached, exhibit reduced phagocytosis and have longer residence time in the blood than non pegylated NPs. In fact, PEG-coated liposomes loaded with chemotherapeutic drugs (Doxil) have already been approved for clinical use.[52,53] The stealth properties provided by these PEG chains arise from a combination of mechanisms, including steric hindrance, flexibility and hydrophilicity of the particle's cell surface.[54] Additionally, a protective effect of pegylation against poly(amido-amine) (PAMAM) dendrimer-induced hemolysis was recently investigated by Wang *et al.*[55] The authors demonstrated that modification of PAMAM dendrimers with PEG chains of 5 kDa and 20 kDa molecular weights significantly enhanced the hemocompatibility of the nanocarrier.[55]

Additionally, the focus of recent research has been directed towards the active targeting of nanosized delivery systems for cancer treatment, by their

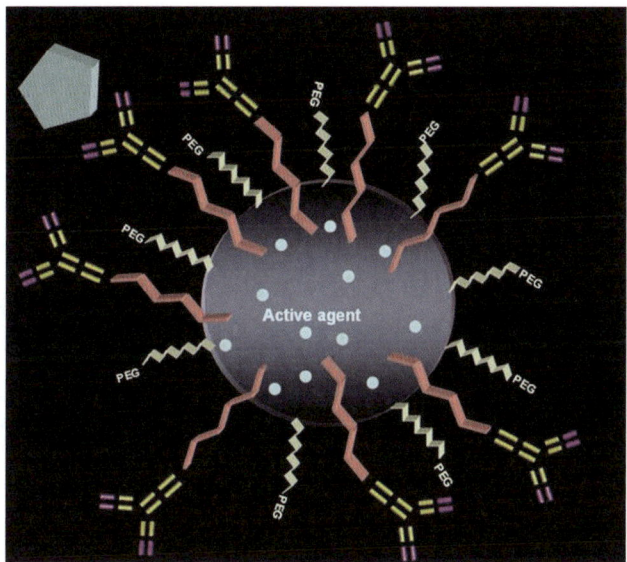

Figure 10.1 Illustration of nanoparticles encapsulating molecules of an active agent where the surface is functionalized with PEG chains and with monoclonal antibodies for cell specific drug delivery.

surface functionalization with various targeting ligands such as monoclonal antibodies, affibodies, peptides and aptamers.[3,56,57] These vector molecules are aimed to promote the specific binding and internalization of the carrier cargo into cancer cells, while avoiding collateral damage to adjacent healthy tissues (Figures 10.1–10.3). However, the inclusion of such targeting components often in high density on the surface of NPs might also increase their immunogenic potential. Additionally, even though the targeting of NPs is based on the recognition of a specific surface antigen which is over expressed on cancer cells, most targeted epitopes also exert a basal level of expression in normal cells, such as tyrosine kinase receptors (EGFR, HER2)[58] and EpCAM (epithelial cell adhesion molecule).[59] Furthermore, the surface expression of some epitopes could also be influenced by other co-morbidities, such as inflammatory processes.[59,60] These factors might lead to a change in the selectivity basis of the targeting approach.

10.2.4 Geometric Shape, Morphology and Surface Architecture

The *shape* of NPs has been found to also affect biological responses to NPs.[61,62] Rod or worm like NPs were reported to be less internalized by macrophages than spherical NPs, as complete phagocytosis seems to rely on the point of macrophage contact.[63] The efficient clearance of NPs may therefore be influenced by their shape.

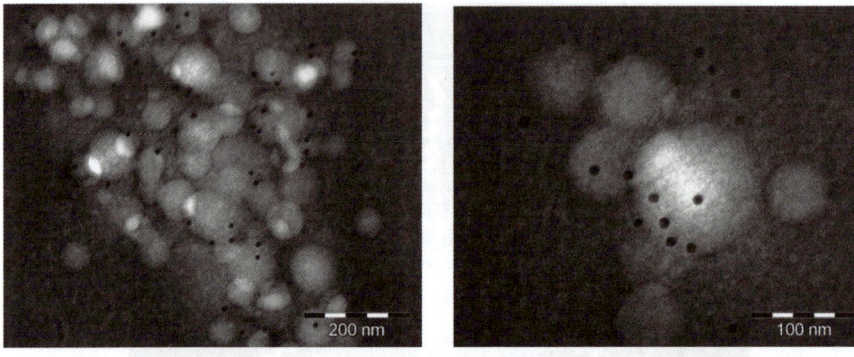

Figure 10.2   Transmission electron microscopy microphotography of trastuzumab
immunoNPs (at a density of 20 MAb molecules per particle) following
incubation over 1 h, using goat anti-human IgG secondary antibody
conjugated to 12 nm gold particle at different magnifications of 70 000
(left) and 140 000 (right). (Reprinted from ref. 170. Reproduced with
permission from Elsevier Limited.)

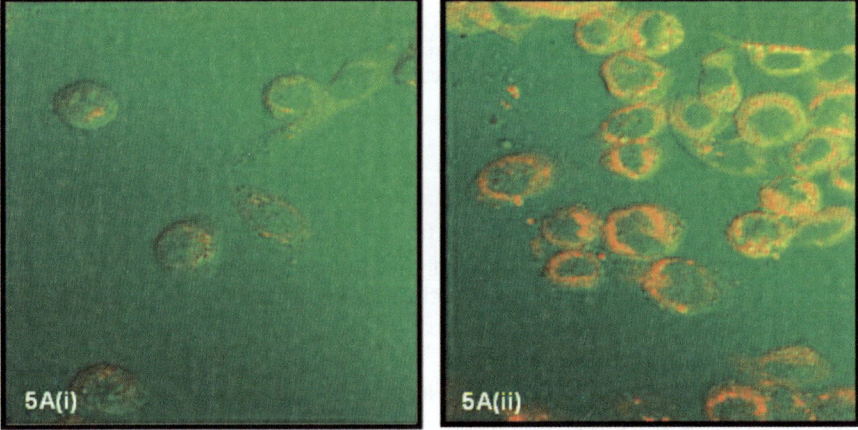

Figure 10.3   Cellular uptake to PC-3 cells of (i) coumarin-6 labeled NPs, (ii) pcpl
loaded trastuzumab immunoNPs as determined by CLSM. The figure
illustrates the enhanced internalization of targeted NPs functionalized
with trastuzumab monoclonal antibody to cancer cells over expressing
the HER2 receptor. (Reprinted from ref. 170. Reproduced with
permission from Elsevier Limited.)

Another *in vitro* study clearly exemplified the possible effect of not only the
chemical moiety used to coat the surface of NPs, but of the ***modular surface
design*** in terms of chain conformation, on the tolerance of NPs.[64]
Polysaccharide chain grafting on PACA NPs by a hair-like or end-on specific
conformation gave NPs superior non complement activating properties and
rendered NPs less cytotoxic as opposed to the same NPs with a side-on
conformation of the same coating chain.[64,65]

Similarly, Hamad *et al.* recently reported that **conformational states** of surface projected polyoxyethyleneoxide (PEO) chains on polystyrene NPs triggered complement activation differently. Alteration from "mushroom" to "brush" configuration of the poloxamine 908 PEO chains switched the complement activation pathway and reduced the level of various generated complement activation products.[66] It is also known that mushroom configurations of PEO chains on NPs are much more prone to sequestration by Kupffer cells, while brush configuration provides longer blood circulation to NPs.[67]

10.2.5 Chemical and Elemental Composition and Biodegradability

As mentioned earlier, various NPs are used in nanomedicine, including organic polymeric scaffolds, natural and synthetic, as well as inorganic elements. The latter include: gold NPs,[7] some of which are in clinical trials, silica NPs,[8] quantum dots (QDs) (Cd/Zn-selenides), metal oxide NPs such as iron oxide (some in clinical trials), and silver NPs. Literature data about the *in vitro* and *in vivo* toxicologic effects of various inorganic NPs were previously reviewed.[9,10] The nanotoxicity of engineered organic biodegradable polymer based NPs used in nanomedical research are given a distinct focus in this review, as these biodegradable materials are usually preferred for drug delivery purposes. These include natural and synthetic polymers, with neutral, positive or negative surface charge. Natural polymers include proteins, such as gelatin or albumin, and polysaccharides, such as dextran, cyclodextrin, alginates (anionic) and chitosan. For example, chitosan is a positively charged polysaccharide, used as an excipient in drug formulations. Due to its excellent mucoadhesive properties, it is widely employed as a paracellular permeability enhancer to increase absorption of hydrophilic drugs.[68–70] However, a novel derivative of chitosan was also synthesized as a new nanocarrier of hydrophobic intravenous drugs such as paclitaxel.[71]

The biodegradability of the NPs scaffolds is a critical contributor to their safety and biocompatibility. The degradation of NPs is needed to ensure the removal of polymeric material from the body, eventually through renal clearance. The persistence of non biodegradable NPs in tissues increases the chance of chronic inflammatory responses and long term toxicities. Thus, an optimal DDS with an ideal degradation profile should be compatible with the mechanism of defense of the organism while providing the desired drug release kinetics. Biodegradable polymers degrade *in vivo* to produce biocompatible or non toxic by-products, allowing the progressive release of the entrapped drug. Degradation products may be further metabolized or excreted *via* normal physiological pathways.[72] The extent and rate of biodegradation of polymeric NPs depend both on the polymer characteristics such as hydrophilicity, crystallinity, cross linking degree, amenability to enzymes or acid-base catalysis and the molecular weight of the employed polymer, as well as on NPs characteristics including size, surface area, surface hydrophobicity and surface porosity of the matrix. For example, a smaller particle size increases the surface area available for water penetration into the NPs, possibly resulting in faster degradation.

We will avoid an exhaustive description of the biodegradation, tolerance and toxicity data of too many to list polymers, but rather we will mention a few examples of the most currently investigated synthetic biodegradable polymers. These include poly lactide (PLA) and poly(D,L-lactide co-glycolide) (PLGA), polyalkycyanoacrylates (PACA), polycaprolactone, and polyanhydrides.

Poly lactide (PLA) and poly(D,L-lactide co-glycolide) (PLGA) are of great interest for drug and gene delivery.[73,74] In PLGA NPs, polymer degradation is biphasic, with an initial rapid degradation during the first 20–30 days, followed by a much slower phase.[40] Consistent with the hydrophilic nature of PLGA compared to lactide-only polymers, glycolic acid-rich PLGA copolymers (up to 70%) are generally amorphous in nature, exhibiting faster degradation. The hydrolytic cleavage of ester linkages in the PLA/PLGA chains results in the release of monomeric lactic and glycolic acids, which are formed at a very slow rate and easily metabolized through the Kreb's cycle as carbon dioxide and water. The role of enzymatic involvement in biodegradation is not clear. As the molecular weight of the polymer decreases, the degradation rate is higher because of the higher content of carboxylic groups at the end of polymer chains that accelerate the acid-catalyzed degradation.[40] **Polycaprolactone** NPs, on the other hand, may be more suitable for long-term delivery over a period of more than one year[75] as they are hydrolyzed at a slower rate than PLA and PLGA due to the crystalline hydrophobic nature of the polymer. Contrarily, **polyanhydrides** have been investigated for short-term controlled delivery as they are characterized by limited mechanical properties and rapid degradation *in vivo*.[75] This class of polymers also degrades into monomeric acid components and was reported to show minimal inflammatory reactions *in vivo*.[75,76]

The design of biodegradable NPs with **polyalkycyanoacrylate (PACA)** for *in vivo* drug delivery was first introduced by Couvreur *et al.*[77] PACA are bio-erodible polymers for which complete excretion will only occur if NPs are composed of low-molecular-weight polymers.[78] *In vivo*, the major degradation pathway of PACA involves the hydrolysis of the ester bond of the polymer's alkyl side chain catalyzed by esterases from serum, lysosomes, and pancreatic juice and the formation of alkylalcohol and poly(cyanoacrylic acid) components. Based on this mechanism, NPs are usually degraded within a couple of hours, depending on the polymer's alkyl side-chain length.[78] Most recently, doxorubicin NPs prepared using poly-iso-hexyl-cyanoacrylate polymer[79] were found to increase median survival rates by 17 months in hepatocellular carcinoma patients compared to patients receiving the current best of care.[80] On a safety note, this drug delivery system (Livatag®) was granted an orphan drug status by the FDA in 2005 but the phase II/III clinical trials were suspended in 2008 because of observed severe pulmonary adverse events in a frequency higher than expected. While the product is still on clinical hold, BioAlliance Pharma has recently announced the successful development of proprietary new intravenous administration of Livatag® validated in animal models, with reduced acute pulmonary adverse events[80] (see Chapter 11). This seemingly promising nano-sized delivery system exemplifies the urgent need for the careful monitoring of toxic effects in clinical trial phases and even in post marketing phases.

10.2.6 Agglomeration State

Nanoparticles may agglomerate into micrometre size particles when suspended in biological media, such as phosphate buffered saline, cell culture media during *in vitro* studies or even *in vivo*, upon contact with physiologic fluids such as the pulmonary surfactant. The degree of NPs dispersion state and agglomeration was found to affect bioactivity in various studies.[21]

10.3 Different Administration Routes of NPs – Pharmacokinetics and Biodistribution of NPs and Resulting Consequences for Toxicities

The various routes of exposure to NPs will dictate their interaction with biological barriers and components and will highly affect the specific fate of the NPs and their resulting toxicity. Thus, primary "macroscopic" mechanisms of toxicity will be determined by the initial and immediate interactions of NPs with their surrounding environment. NPs have been studied for a variety of delivery routes, including oral, [81] intravenous, pulmonary,[39] ocular[82] and transdermal.[83] In the present section we will review some *in vivo* studies assessing various aspects of NPs' toxicity following different administration routes.

10.3.1 Intravenous Delivery of NPs

Currently, a significant portion of the investigated nanocarriers is intended for intravenous delivery. In medical use, NPs are primarily designed to treat pathologies that present a disseminated nature, often involving more than one disease focus. This is particularly true for cancerous diseases. Therefore, intravenous injection of NPs for the delivery of therapeutic compounds is still one of the preferred routes of drug administration. As a result, prolonged circulation and moderate degradation rate that will provide enough time for the nanocarrier to reach the target tissue have become a prerequisite for the design of a successful nanoparticulate delivery system. As mentioned earlier, the most prominent factors affecting NPs' circulation time and biodegradation rate are related to their properties as a whole construct and to the nature of the constituent polymer. Upon intravenous injection, most of the NPs larger than 10 nm are recognized as foreign bodies and cleared by phagocytic cells, mainly cells of the mononuclear phagocyte system (MPS) and the polymorphonuclear leukocytes (PMN). Macrophages located in the reticuloendothelial system (RES), of which the Kupffer cells in the liver comprise 85–95% of the total intravascular phagocytic capacity, play a crucial role in the phagocytosis of injected particles. This rapid phagocytosis is generally attributed to an opsonization process of certain blood components (IgG, IgA and complement cascade components) to the surface of NPs, apparently by hydrophobic interactions. The likelihood of opsonization and liver uptake augments with the increase in particle size. NPs around 100 nm exhibit an improved circulation half life[38] compared to larger NPs. Notably, pegylation does not completely prevent or eliminate opsonization and phagocytosis, but may significantly delay and reduce these inevitable processes if appropriately designed.

In this regard, an early study by Fernandez-Urrusuno *et al.* recognized the risk of toxicity of intravenously delivered NPs as a result of their potential accumulation in RES organs as the liver.[22] The subacute toxicological effects of polyalkylcyanoacrylate NPs were studied in an *in vivo/ex vivo* model following 14 days of intravenous treatment. Modifications in hepatic function were detected after chronic administration of PACA NPs. However, these effects were reported by the authors to be reversible with the attenuation of treatment.[22]

Other possible toxicological consequences of systemically delivered NPs include hematological effects. Upon intravenous delivery, the earliest interactions of NPs will be with endothelial cells and blood components: RBCs, platelets, serum proteins, immunoglobulins, *etc.* It is therefore essential to investigate the hemocompatibility and hematological side effects of NPs (for example, blood clotting and hemolysis) as well as their immunogenicity. *Immunogenicity* is the ability of foreign antigens to induce humoral or cell mediated immune responses (complement activation, macrophage recruitment, and cytokines release). Various assays may be employed for NPs and other nanocarriers immunogenicity evaluation: immunopercipitation assays, IgG titres detection in blood by ELISA and complement activation.

Determination of hemolytic properties is actually one of the most common tests in studies of nanoparticle interactions with blood components.[27] Hemolytic properties can be determined *in vitro/ex vivo* by incubation of the nanocarrier with blood and determination of the percent of hemolysis induced by different incubation concentrations as well as morphologic changes in red blood cells (RBCs).[27] Alternatively, *in vivo* hemolysis assessment may be performed by complete blood count investigation (CBCs), including RBCs count, hemoglobin (Hb), hematocrit (HCT) and mean corpuscular hemoglobin (MCH).

For example, the effects of surface charge and size on nanoparticles hemocompatibility were investigated in an *ex vivo* study in human whole blood by Mayer *et al.*[84] Various parameters were assessed by the authors in response to non biodegradable polystyrene NPs of various sizes and charges. It was concluded that positive surface charge induced complement activation, but the specific investigated positive charge was not high enough to cause thrombocyte activation. Smaller size in anionic NPs on the other hand was found to induce thrombocyte and granulocyte activation and hemolysis. Anionic NPs with a hydrodynamic diameter larger than 60 nm appeared to be considerably less hematotoxic than smaller ones.[84]

A comprehensive biological evaluation of a chitosan derivative based nanocarrier, *N*-octyl-sulfate chitosan, was performed by Zhang *et al.* for the assessment of its biocompatibility for intravenous use.[71] Evaluation included monitoring of tissue distribution of the carrier and pharmacokinetic studies following intravenous delivery to rats. Furthermore, a series of safety studies was performed consisting of acute toxicity study and pathological examination in mice, intravenous stimulation study in rabbits, injection anaphylaxis study in guinea pigs and *ex vivo* hemolysis study.[71] The authors claimed that since no negative results were observed in these tests, the safety was persuasive for the potential intravenous use of this novel nanocarrier.[71]

10.3.2 Oral Delivery of NPs

As a relatively convenient and patient friendly method, oral delivery remains one of the preferred drug administration routes that will most likely dominate the drug delivery market for long.[85] However, the low and variable bioavailability of many of the small molecule drugs as well as proteins and peptides places a challenge on their effective delivery. Oral delivery of NPs could circumvent protein enzymatic degradation, Pgp efflux pumps and CYP3A4 that line along the gut wall. Thus, much effort has been invested in recent years to design nano scaled delivery systems for oral administration.[86–89] From a toxicological point of view, orally delivered NPs could induce toxic effects both locally in the gut wall, but also in disseminated organs, following systemic distribution. For example, gold NPs can cross the small intestine by persorption and be further distributed into the blood, brain, lung, heart, kidney, spleen, liver, intestine and stomach.[90] The pathway and extent of uptake of insoluble NPs are known to be size dependent.[91] Semete *et al.* have recently explored the tissue distribution, localization and effects of PLGA NPs and ZnO following oral administration to Balb/c mice.[92] Histopathology of the liver, lung, heart, spleen, kidneys, brain, intestines, stomach, pancreas and thymus revealed no pathological lesions suggestive of toxicity following the oral administration of PLGA NPs for 1, 5 and 10 consecutive days. ZnO NPs with a similar surface area, however, caused substantial deaths even at 24 h. Tissue distribution studies demonstrated a significant localization of PLGA NPs in the liver (40%) but also in the kidneys (25%) and to a lesser extent in the heart and brain ($\sim 12\%$). The authors concluded that even though NPs were shown to migrate from the GI tract, pass biological barriers such as the BBB and accumulate in various tissues, no pathological changes as lesions or inflammations could be detected even at high doses of 60 mg NPs per mouse. However, the authors did point out the need for pegylation of the PLGA NPs to reduce their accumulation in RES organs.

Finally, it was concluded that orally administered PLGA NPs, in the specific time frame and doses used in the study could be regarded as safe drug delivery systems.[92] In another work published by the same group, the *in vivo* uptake by macrophages and the immunogenicity of PEG and chitosan coated PLGA NPs were studied post oral administration to healthy mice.[93] It was demonstrated that the studied PLGA NPs were taken up by peritoneal macrophages after 5 daily consecutive oral administrations. However, the activation of monocytes to macrophages and NPs uptake did not elicit a substantial acute immune response, as the expression levels of inflammatory cytokines in plasma and in peritoneal lavage were similar to those observed in the saline control up to 24 h following a single dose.[93] Various works have also investigated the safety of nanoparticles loaded with different therapeutic entities. For example, Kalaria *et al.* have investigated the pharmacokinetic and toxicity behaviour of doxorubicin loaded PLGA NPs in rats following oral administration.[94] Doxorubicin NPs exhibited lower cardiotoxicity than the free drug solution, as observed by lower levels of MDA (malondialdehyde) and higher levels of the antioxidant enzymes catalase and SOD (superoxide dismutase). The blank PLGA NPs did not alter lipid peroxidation or antioxidant enzymes levels when compared to the control group, suggesting no oxidative damage caused by the nanocarrier *per se* under the studied conditions.[94] Similarly, blank NPs did not reduce heart weight and did not affect body weight when compared to the control group.[94] The reduced toxicity of doxorubicin NPs was attributed by the authors to the gradual release of the incorporated drug, but also to possible accumulation of NPs in tissues. The different plasma/tissue exposure ratio would also explain the lower toxicity observed with oral delivery of doxorubicin NPs as opposed to their intravenous administration.[94] This study emphasizes the importance of performing comprehensive pharmacokinetic and biodistribution studies. Furthermore, the *in vivo* evaluation of safety and efficacy of self-assembled insulin loaded chitosan NPs was recently reported by Sonaje *et al.*[95] To evaluate possible toxic effects of orally administered NPs, mice were treated with a daily dose of empty NPs for 14 days. No significant differences in body weight, in clinical signs or in hematological and biochemical parameters were observed between the NPs treated mice and the untreated controls. Furthermore, no inflammatory reactions or pathological changes in the liver, kidneys and intestinal segments were observed.[95] As to the paracellular transport mechanism of the released insulin, it was hypothesized by the authors that the NPs could effectively adhere to the mucosal surface while their constituted components were able to infiltrate into the mucosal cell membrane.[95] In another study by the same group, heparin loaded chitosan NPs were prepared using a similar simple ionic-gelation method.[96] Here, the biodistribution of the radiolabeled chitosan carrier was studied in rats following oral administration. As no significant radioactivity was found in internal organs, it was concluded that there exists a minimal absorption of this kind of carrier to the systemic circulation.[96] Similarly, further encouraging toxicity results were recently reported for the oral administration of repaglinide loaded poly(methyl methacrylate) (PMMA) NPs.[97] Healthy wistar albino rats treated with either 1 mg kg^{-1}, 2mg kg^{-1} or 5mg kg^{-1} NPs exhibited no pathological behavioral, hematological or biochemical changes compared to control rats. Moreover, no histopathological findings indicating nanotoxicity were observed in the brain, kidneys, liver or spleen tissues.[97] In two different works, Dandekar *et al.* reported the *in vivo* safety of orally delivered curcumin loaded polymeric nanoparticles of Eudragit S100 and HPMC with PVP hydrogel NPs, after acute and sub acute (prolonged duration) exposures. The safety of the oral formulations was also confirmed by various *in vivo* genotoxicity studies.[98,99]

10.3.3 Pulmonary Delivery of NPs

Nanoparticle delivery to the lungs is an attractive concept for the non invasive treatment of pulmonary diseases because it can cause retention of the particles in the lungs accompanied with a prolonged drug release.[100,101] Pulmonary delivery of NPs has gained much interest in research in the past years.[102–104] However, most of the toxicity data in the literature about inhaled NP is mainly available from epidemiologic toxicology studies of non therapeutic, airborne or combustion derived NPs. There is limited value to this data in the evaluation of possible toxicity and biofate of inhaled therapeutic NPs, as the former are mostly inorganic (metals or metals oxide), water insoluble and with different exposure doses.[105]

Most published work on pulmonary delivery of drug NPs focuses on polymeric NPs, liposomes and solid lipid NPs.[105,106,104,107] Indeed, various classes of polymers and ploymeric nanocarriers have been considered for pulmonary drug delivery; these include poly (ether-anhydrides), gelatin, PLGA, chitosan, modified branched polyesters and others.[102,108–112] NPs have been also employed to deliver siRNA to the lung using biodegradable cationic NPs based mainly on PEGylated poly(ethyleneimine) (PEI) and polyester derivatives.[113,114] Thus, inhalation of NPs is currently the subject of intense research and hence that of safety evaluations.[102,108,115,116 103,117]

The efficiency of inhaled nanoparticles depends on their deposition, as well as their subsequent fate in the lungs. The mechanism of deposition of nanoparticles in the small airways and alveoli is diffusional, referred to as Brownian diffusion.[39] The fate and clearance of nanoparticles depends on their solubility and on their depositing site.[39] First, NPs encounter the surfactant on top of the airway lining fluid of the epithelial cells. The surfactant layer composed of phospholipids and proteins is critical for the reduction of surface tension and prevention of alveoli collapse during breathing. Furthermore, proteins in the surfactant play a role in opsonization and clearance of particles by macrophages. Hence, it is important to investigate the interactions of NPs with the surfactant layer and their influence on its integrity and function. For example, gold and TiO$_2$ NPs may alter these parameters.[118,119] Recognizing the importance of the lung surfactant layer's function, Beck-Broichsitter *et al.* have studied *in vitro* the changes in the biophysical surfactant surface properties upon contact with NPs with different surface charge and hydrophobicity: negatively charged polystyrene (PS) NPs, PLGA NPs and positively charged Eudragit E100 (EU) NPs.[120] Their study revealed a dose dependent influence of negatively charged NPs on the biophysics of the surfactant, mainly on its ability to reduce surface tension within a short period of time, and particularly when hydrophobic PS NPs were employed. Interestingly, these effects were negligible for positively charged EU NPs. Overall, the negative surface charge and hydrophobicity induced higher adsorption rates of NPs to the positively charged proteins constituting the surfactant. Note, the NP doses used were unlikely to be reached *in vivo*. Therefore, the effects were not substantial and it is unlikely that these polymeric NPs will induce adverse effects on the surfactant function in healthy individuals.[120] In strong contrast, Al-Hallak *et al.* reported significant reduction of the stability of surfactant monolayer film *in vitro* by polysorbate 80-coated NPs as opposed to non coated NPs, which correlated with severe *in vivo* pulmonary toxicity and morbidity following pulmonary delivery of coated NPs to balb/c mice, as observed by shallow and slow breathing patterns.[121] This was also confirmed by lung histopathology which demonstrated deflated alveolar sacs, microhemorrhages from alveolar capillaries, neutrophils recruitment and edema in perivascular sacs.[121] It should be noted that polysorbate alone did not exhibit significant effect on the film collapse pressure. However, the authors declared that these results should not be generalized to all surfactant coatings or copolymers.[121] Thus, it is exemplified by these two works how various NPs with different surface properties (charge, hydrophobicity and surfactant coatings) might interact differently with lung components and induce extremely different clinical behavior. Again, the need for a case by case investigation is highlighted.

Biofate and Clearance Mechanisms of NPs in the Lungs

For relatively insoluble NPs, clearance is governed mainly by the mechanical removal of particles by phagocytizing alveolar macrophages (AM) and mucociliary transport and clearance (MCC). When NPs encounter the surfactant layer, they undergo enhanced wetting. More soluble NPs may partition and sink deeper into the bottom layer to become less susceptible to MCC and more likely adsorbed through the epithelium.[105] The most prevalent mechanism for solid particle clearance in the alveolar region where MCC is absent is mediated by alveolar macrophages, through phagocytosis of deposited particles mediated by opsonization and chemotaxis. Within 6–12 h after deposition in the alveoli, virtually all of the particles are phagocytosed, but it appears that the effectiveness of this process is highly dependent on particle size.[39] Once phagocytized, particles may be disintegrated or drained and accumulated in the lymphatic system. Smaller particles result in a slower rate of phagocytosis. As a result of their small size, defense against NPs would be less efficient as their recognition by macrophages is suggested to be impaired or less effective, possibly because of few opsonin molecules available per particle.[105,122,123] Particles that are not removed by phagocytosis, such as NPs in the deep lung where there are no macrophages, may readily gain access to epithelial and interstitial sites, blood circulation and even the lymphatic nodes.[39]

Thus, pulmonary delivered NPs could induce both local and systemic effects, depending on their lung retention and blood translocation. One of the possible causes of lung injury as a result of NP inhalation may be oxidative stress leading to activation of different pro-inflammatory factors.[76,124] ROS generation by NPs could be the result of active redox cycling (particularly with metal NPs) or functional oxidative groups on their surface.[125,126] Notably, the lung is a rich pool of ROS producer cells such as inflammatory phagocytes, neutrophils and macrophages. Thus, underlying cellular mechanisms of NPs induced pulmonary toxicity in the lung that could be the result of alveolar macrophages activation and phagocytosis, lead to ROS production and release of proteolytic enzymes, pro-inflammatory mediators and growth regulating proteins, which may lead to both acute and chronic inflammation.[6,127] As a result, stimulated epithelial and endothelial cells may further promote leukocyte recruitment into the lungs, and cytokine cascades are activated with interleukins and tumor necrosis factor pro-inflammatory signaling processes being upregulated.[6] Pulmonary inflammation as a direct toxicity parameter of pulmonary delivered NPs could result in pulmonary diseases (fibrosis) or exacerbation of pre-existing lung pathologies (asthma). Additionally, it may induce changes in membrane permeability, resulting in particle distribution outside the lung, and possibly affecting cardiovascular performance.

In this regard, Dailey et al.[116] investigated the pro-inflammatory potential of biodegradable NPs in the lungs. Their study demonstrated that pulmonary delivered NPs composed of diethylaminopropylamine poly(vinyl alcohol)-grafted poly(lactic-co-glycolic acid) (DEAPA-PVAL-g-PLGA), a derivative of the biocompatible biodegradable PLGA polymer, may possess a significantly decreased pro-inflammatory potential in the lungs, especially in polymorpho-nuclear (PMN) recruitment, as opposed to non-biodegradable polystyrene NPs of the same size and surface area.[116]

As previously mentioned, the surface charge of nanocarriers may affect their local tissue distribution and their safety profile and numerous studies have emphasized specifically the NP surface charge as an important parameter in nanotoxicity and *in vivo* NP performance.[47,48,113,114,128] With regard to pulmonary delivery, despite the fact that cationic charge may cause higher retention of NPs in cells,[40] Merkel et al. reported positive and encouraging results on the *in vivo* lung performance of cationic NPs in a murine model with only moderate pro-inflammatory effects.[113] Similarly, positively charged hydrophobically modified glycol chitosan (HGC) NPs were found to induce a transient neutrophilic pulmonary inflammation with subsequent recovery following intratracheal delivery to healthy mice.[129] The authors thus concluded that such a delivery system can be a successful candidate for pulmonary drug applications.[129]

Another important issue that needs to be addressed is the effect of repeated pulmonary exposures of the animals to NPs. Oberdorster et al.[130] noted that among the *in vivo* assays proposed for evaluation of pulmonary exposure, evaluation of the effects of multiple exposures is important. To address the two latter mentioned factors, Harush-Frenkel *et al.* have investigated the surface charge effects on the safety and tolerability of NPs following five consecutive days of local endotracheal delivery in a murine model of balb/c mice.[108] While Merkel *et al.* reported only moderate pro-inflammatory effects and absence of histological abnormalities in response to pulmonary delivery of cationic pegylated PEI nanocarriers,[113] positively charged stearylamine coated PEG-PLA NPs were found by Harush-Frenskel *et al.* to exhibit higher toxicity compared to anionic PEG-PLA NPs, as demonstrated by mortality, reduced weight gain and increased BAL macrophages and lymphocytes two weeks after the last dose.[108] Negatively charged NPs were reported by the authors to be better tolerated and safe, and may therefore be considered as drug carriers for pulmonary applications.[108] The authors however emphasized the need to avoid extrapolation of their findings to other cationic delivery systems.[108]

Nanoparticles may also form agglomerates due to adhesion to each other by weak forces (Van der Waals, electrostatic forces). It is possible that NPs agglomerates will reside on top of the airway fluid, and depending on the their dispersability and solubility in the lung fluid, they may remain unchanged at the fluid surface and behave like microparticles, or they may be dispersed into NPs and be absorbed into deeper airways.[105] Alternatively, NPs might also agglomerate once in contact with biological fluids.[131] The influence of the agglomeration state of gold NPs on *in vivo* pulmonary inflammation was recently studied by Gosens and colleagues in rat lung.[132] Even though more toxic effects were hypothesized for smaller single NPs than their bigger single or agglomerated counterparts, as previously reported for other nanomaterials,[21,133] no striking differences were observed between 50 nm single NPs and 50 nm agglomerates (of ~200 nm) or between 200 nm single NPs and their agglomerates (~700 nm). However, the authors did mention that specifically for gold NPs, catalytic activity has been previously reported for NPs smaller than 10 nm, thus making 50 nm NPs used in the study unlikely to cause increased reactivity as opposed to submicron particles.[132] Still, the investigation of NPs behavior and agglomeration potential in relevant biological fluids should be pursued to better elucidate their *in vivo* effects.

Systemic Translocation of NPs Across the Air Blood Barrier Following Pulmonary Delivery

Conflicting results have been reported on the *in vivo* biofate of inhaled NPs and specifically their translocation from deposition sites in the pulmonary airways to the systemic circulation as well as the lymph nodes.[41,102,134,135] Furthermore, different mechanisms have been proposed for uptake of NPs into systemic circulation and biological tissues.[39,105] One involves transcytosis across the epithelium of the respiratory tract into the interstitium and access to the blood circulation directly or *via* the lymphatics. Furthermore, lymphatic drainage has also been held responsible for the alveolar clearance of deposited NPs up to a certain particle diameter (500 nm).[39] Translocation of NPs to the circulation may induce systemic side effects beginning from hematological effects up to specific organ toxicities resulting from NPs accumulation in tissues. Particularly, the correlation between cardiovascular diseases and airborne ultrafine particles observed in epidemiological surveys has triggered the discussion on potential risks of nanoparticle exposure to human cardiorespiratory systems.[136] Therefore, for a better understanding of the underlying mechanisms, the study of the biodistribution and clearance behavior *in vivo*, (*i.e.* pulmonary retention, extrapulmonary translocation, redistribution, excretion) and the direct interaction of inhaled nanomaterials with lungs and their secondary target organs has become essential. Indeed, various studies have attempted to elucidate the translocation pathways and the subsequent biodistribution and biofate of inhaled engineered NPs.[137–140]

With regards to biodegradable polymeric NPs, M. Choi *et al.* reported that even though hydrophobically modified glycol chitosan (HGC) NPs exhibited rapid extrapulmonary distribution to several tissues in mice, the distributed levels were low, transient and without accumulation.[129] A recent landmark study by H. S. Choi *et al.* systematically reported the real time *in vivo* biodistribution profiles of various organic NPs (human serum albumin, PEG) as well as hybrid organic/ inorganic NPs (INPs) produced from either quantum dots or silica cores, with different shells and organic coating ligands.[141] The various investigated formulations differed in particle characteristics, most notably composition, size, charge and protein adsorption abilities in serum. The authors reported that there was a size threshold of less than 38 nm for the rapid translocation of NPs from the lungs to lymph nodes within 30 minutes of delivery. Importantly, it was suggested that this translocation was primarily governed by size, irrespective of chemical composition, shape and conformational flexibility. However, below a size of 34 nm, the surface charge was recorded as a critical factor influencing this process with neutral hydrophilic and anionic NPs being permissive of rapid translocation to regional lymph nodes, whereas cationic NPs of the same batch were restrictive with negligible translocation. Additionally, the authors deduced that when the hydrodynamic diameter was below 6 nm with a zwitterionic surface charge, translocation was even more rapid with an ultimate clearance through renal filtration.[141] The authors emphasized the importance of such knowledge on the comprehensive design of drug delivery systems with reasonable biodistribution and safety profiles.

10.3.4 Additional Delivery Routes for NPs

Other delivery routes of NPs have also drawn major interest in recent years as efficient topical tools for the treatment of dermal, ocular and even cochlear disorders.[142] Indeed, colloidal DDS are currently perceived as a suitable strategy to enhance the bioavailability and to prolong the therapeutic action of topically administered drugs.[143] However, despite their designation for topical purposes, these delivery systems may exert potential toxic effects not only locally at their application site but also systemically following possible absorption and translocation.

Different kinds of solid and lipid nano-sized delivery systems have been extensively investigated for cosmetic and therapeutic dermal delivery as well as transdermal drug delivery.[144,145] The safety concerns regarding the dermal and transdermal biological responses to colloidal carriers have been discussed elsewhere.[146,147]

Efficacy and toxicology studies *in vitro* and *in vivo* have also been evaluating the feasibility of ocular delivery of NPs for the treatment of surface inflammatory disorders and intraocular pressure, investigating their general tolerance and influence on specific ocular clinical signs and functions as tear production.[148,149] With systemic toxicity in mind, mucoadhesive polymers such as the biodegradable chitosan, as well as targeting moieties like hyaluronic acid (HA) recognizing CD44 receptors in the cornea and conjunctiva, have been exploited to increase the interaction of NPs with the ocular surface while minimizing systemic absorption following topical application.[148,150,151] Based on *in vivo* tolerance studies, it is suggested that hyaluronic acid-chitosan NPs are safe carriers for ocular surface application.[148]

Intranasal delivery, on the other hand, is often utilized for drug delivery to the brain through the olfactory nerve pathway. NPs can get access to the brain by transcytotic transport through the olfactory epithelium.[152] Despite the tremendous potential concealed in the delivery of NPs to the brain and their possible use for the treatment of brain tumors and neurodegenerative disorders,[153,154] there are only a few reports investigating the toxicity of NPs to the brain. In a recent one, wheat germ agglutinin (WGA) conjugated PEG-PLA NPs were administered intranasally to rats for 7 consecutive days.[155] NPs induced slight toxicity to brain tissue as evidenced by increased glutamate levels in the brain and LDH activity in the olfactory bulb. Still, no significant changes in acetylcholine esterase activity or acetylcholine, GSH, IL-8 or TNF-α levels were observed.[155] Potential effects of NPs on the brain have been reviewed and discussed by Oberdorster *et al.* [156]

To summarize, NPs may be delivered *via* several routes, either topically or systemically. Thus, depending on the port of entry, NPs might elicit different toxic effects. Notably, an allegedly "topical" application does not completely eliminate the possibilities of systemic toxicity, as NPs, due to their small size, are characterized by the propensity to translocate across biological barriers and by para cellular routes to secondary organs.

10.4   Intracellular Mechanisms of NPs Toxicity

Nanoparticle toxicity may present *via* several phenomena, for example as hematotoxic effects (thrombosis and hemolysis), by immunogenic modulation, and even hepatotoxic effects following oxidative metabolic modifications of NPs.[157] So far, four possible pathways responsible for NPs toxicity have been identified by Costigan:[5,158]

1. "Simple" chemical toxicity of one of the constituents following the same mode of action as the bulk source.
2. Toxicity of the degradation products of NPs.
3. Toxicity due to endocytosis of NPs – intracellular targets of NPs (for detailed information please refer to ref. 5 and 40
4. Interactions with membranes resulting in membrane destabilization and lysis.

Molecular Interactions of NPs with the Cellular Machineries

Internalization of NPs to the cells might occur through several mechanisms, as reviewed by Harush-Frenkel *et al.*[40] NPs may be internalized *via* receptor mediated endocytosis (*via* clathrin coated pits), particularly when targeting ligands are attached to their surface, or by other endocytic processes such as caveole or dynamin dependent pathways[40] (Figure 10.4). Nonetheless, NPs may be taken up into cells *via* paracellular transport as recently reported by Goldberg *et al.* who studied the entry of poly(amido amine) dendrimers to caco-2 cells.[159] Next to clathrin and dynamin dependent processes, the tight junctional opening of intestinal epithelia was observed, thus catalyzing their own transport *via* the paracellular route.[159] Similarly, mucoadhesive chitosan can act as an efficient enhancer to increase NPs paracellular transport by opening the tight junctions of epithelial cells, as reported by several authors.[160–162]

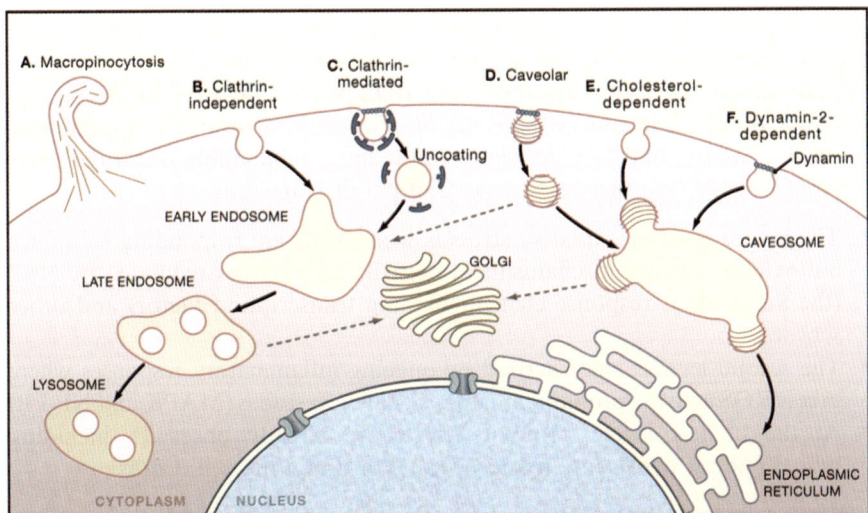

Figure 10.4 Illustration of possible endocytic machineries in mamallian cells, the function of which may differ from cell to cell, leading to distinct regulation and internalization kinetics. (Reprinted from ref. 171. Reproduced with permission from Elsevier Limited.)

When NPs are taken up into cells, they may interact with the cell membrane and intracellular organelles and components, such as the mitochondria and cell nucleus, inducing damage by various mechanisms, including oxidative stress. The plasma membrane functions as a barrier between the cell and the extracellular environment, thus controlling intracellular trafficking and coordinating cellular signaling. First, the interaction of polymeric materials and NPs with the membrane domains might alter its function, for instance, by disturbing the assembly of lipid rafts or caveoles, therefore affecting homeostasis and signaling in the cells. For example, cationic polymers such as PEI were demonstrated to form holes in lipid membranes.[163] This is also true for PAMAM dendrimers.[45]

The flux of the nanocarrier and its cargo between sub-cellular compartments, such as the endoplasmic reticulum, golgi apparatus, endosomes (late and early), lysosomes and the plasma membrane, will dictate its degradation, its recycling and the resulting net therapeutic and toxic effects inside the cell. For example, particles entering the cell through a specific endocytic pathway (fluid phase or clathrin mediated) will be trapped in the endosome, and to some extent will be degraded in the lysosome. Caveolar endocytosis, however, will result in translocation to the endoplasmic reticulum, to the Golgi or simply in transcytosis through the cell.[164] (Figure 10.4).

In vivo, molecular mechanisms of toxicity involve the production of ROS and oxidative stress. This results in the possible oxidation of lipids, proteins and DNA. Sources of ROS production include phagocytic cells' response to foreign materials, possibly with the combination of environmental factors, transition metals, and specific physicochemical properties of NPs.[157] RES organs such as the liver, the lungs and spleen, which are also richly perfused organs, are particularly prone to ROS production as NPs tend to accumulate in these already highly occupied pools of ROS producing cells.[9] In this regard, Nel and colleagues introduced a *hierarchical oxidative stress model* for ambient particulate matter and NPs deposited in the lung, subdividing oxidative stress induced by NPs into three categories and parallel signaling cascades:[41]

- The lowest level of oxidative stress is where cells are responding to induce antioxidant defense mechanisms of detoxifying enzymes regulated by Nrf2 (the antioxidant response element binding transcription factor) and other sensors.
- The second level is characterized by ongoing inflammatory responses where excess ROS activate pro-inflammatory signaling cascades (MAPK and NF-kB)
- At the highest level, cytotoxic effects are already observed, including mitochondrial perturbation and necrotic or apoptotic cell deaths.

With respect to these three steps, Beyerle and colleagues undertook a comprehensive gene expression profiling approach to characterize the underlying mechanisms responsible for the high toxicity of PEI based polymers, using mRNA and protein expression levels, as well as qRT-PCR based gene array analysis (quantitative reverse transcriptase-polymerase chain reaction).[165] The authors reported that upon endocytotic PEI (25 kDa) uptake, endosomal swelling and rupture occurred, resulting in intracellular stress and mitochondrial alteration, eventually leading to apoptotic cell death.[165]

The polymer genomic hypothesis was previously explained and formulated by Kabanov, based on specific experimental observations, as follows:[18,34]

- Transcriptional activation of gene expression by polymers, for example activation of NF-kB and P53 signaling pathways by Pluronic®.
- Alteration of responses to drugs by polymers *via* various mechanisms and up-regulation of apoptosis signaling pathways. For example, a doxorubicin drug-polymer conjugate induced additional caspase-dependent apoptosis pathway compared to the free drug.[18,166]
- MDR inhibition mechanism – polymers such as Pluronic® as well as various kinds of NPs have been shown to directly affect Pgp efflux pumps and to bypass multi drug resistance (MDR) mechanisms, by direct interference with Pgp expression and function, by a decrease in Pgp ATPase activity and by intracellular depletion of ATP.[18,167,168]
- Alteration of genomic profiles affecting or redirecting the development of drug resistance in cancer cells. In this regard, Kabanov mentioned Pluronic® 85 which was reported to alter the gene expression in cancer cells selected with doxorubicin, preventing the amplification of the MDR1 gene, as opposed to cells selected with the drug alone.[169]
- These issues reinforce the need to consider pharmacogenomic effects of both polymer excipients and NPs.[18]

Even though some factors have been elucidated for the design of the desired NPs with tunable circulation, pharmacokinetic and organ biodistribution profiles by the modulation of NPs size and surface properties, the intracellular trafficking and final biofate of most nanocarriers remain largely unknown. Based on the studies reviewed in the present section, it can be concluded that the need to elucidate these issues is becoming essential for the design of safe and efficient nanoparticulate delivery systems. The vivid understanding of the intracellular behavior of NPs will contribute to the deciphering of the molecular basis and mechanisms of their toxicity, and will surely aid in the optimization of NPs physicochemical characteristics to better suit our biological mechanisms of defense.

10.5 Challenges and Hurdles

The translation of NPs from laboratory practice to clinical settings could face various scientific obstacles and regulatory difficulties. Notably, despite the potential therapeutic benefits of NPs, a great deal of concern has been raised in the last few years regarding their safety and possible toxicities. NPs are no longer perceived as merely inert innocuous carriers, and it is widely recognized that these promising delivery systems might possess non desirable, non benign, off target toxic effects. With the acknowledgement of the importance of toxicity challenges that might hurdle nano-sized delivery systems approval by regulatory authorities, scientists are becoming more aware of the urgent need to provide *in vivo* data about safety of these innovative nanotechnologies. Many works have addressed the safety and toxicology concerns of nanotechnology in an attempt to better elucidate the main contributors and mechanisms to possible toxicities. However, the long-term implications of these nanomaterials on human health are still not known, and surely merit further investigation. One of the major challenges of nanotoxicology assessment is the rapidly growing number of possible nanoparticles for nanomedical applications. Naturally, due to the wide versatility of available NPs, hazard and toxicology assessment of every possible nanoparticulate carrier becomes unrealistic and non feasible. Therefore, there is a constantly growing need for the development of a set of standardized tools able to provide a thorough understanding of the pharmacologic and toxicologic mechanistic basis of nanotoxicity and thus predict the toxic potential of diverse nanomaterials.

A rational toxicology assessment approach should be performed on a case by case basis, where the actual testing algorithm should be undertaken step by step, taking into consideration various macroscopic as well as microscopic aspects of toxicity, anywhere from clinical signs up to toxicogenomic effects. Oberdorster *et al.* suggested three key elements in the screening strategy of potential toxicity from exposure to nanomaterials: physicochemical characteristics, *in vitro* assays (cellular and non-cellular) and *in vivo* assays. The authors suggested that there is a strong likelihood that the biological activity of NPs will depend on their physicochemical parameters, *e.g.*, diameter, surface charge, *etc.* Furthermore, they noted that evaluation of the effects of multiple exposures is also important. If we should follow Oberdorster's formula, the following algorithm should be pursued:

1. A prerequisite for a logical assessment of toxicity is the *full physicochemical characterization* of the nanoparticle system. This is crucial for the ability to correlate biological and toxicological responses to specific nanoparticle properties.

2. A paramount question is what is the intended *port of entry* of the nanoformulation to the body? Is it intended for local "topical" delivery or for systemic delivery? This will highly affect the choice of the appropriate methodologies and models for toxicity evaluation, and will focus the search for relevant toxicity markers. The parameters that should be monitored for potential signs of toxicity will thus depend on the delivered dose, delivery route and the exposure time. The *biodistribution profile* of NPs will also dictate the evaluation of organ specific toxicity parameters. For example, if the DDS is intended for intravenous injection, the liver and kidneys can become potential organs for accumulation and elimination of NPs. Thus, liver enzymes and blood urea nitrogen (BUN) or other more sensitive markers could be part of the repertoire of monitored toxicity parameters indicating hepatic and renal functions.

Needless to mention, histologic evaluations of the relevant target tissues of accumulation, metabolism and clearance of NPs are most appropriate. However, one should always keep in mind that tissue remodeling and cells infiltration and their debris clearance rates in murine models might differ from the same processes occurring in human tissues. This might somewhat challenge the reliable use and extrapolation of histopathologic observations to human settings. Systemic delivery will also require the investigation of **hemocompatibility and immunogenicity** profiles of the nanoparticles, as discussed earlier.

3. What are the **relevant doses** that should be administered? A clever choice of dosing should be performed, based on the therapeutic doses that would be used *in vivo* settings to deliver the desired drug dose, thus imitating realistic human clinical exposure scenarios. This is also dictated by the entrapment efficiency and drug content in the nanoformulation.[93] **Dose dependency characterization** is one of the most important characteristics of safety. Each material is potentially hazardous at high doses, and the dose alone might define "toxicity".[25] This is also true for nanoparticles. Thus, it is important to investigate animals' survival as a function of the delivered dose and determine parameters such as LD_{50} and MTD (maximal tolerated dose). General signs of health and well being of the treated animals, such as behavioral changes and body weight changes are also used as signs of tolerability and safety.

4. Once the "correct" or most appropriate *in vivo* dose is found, NPs should be delivered in the **appropriate vehicle** that provides optimal conditions for NPs dispersion and minimal agglomeration, is compatible with physiological fluids (pH, isotonicity, viscosity), while not affecting or masking the surface reactivity of the particle by adsorption of media components, such as proteins, on their surface.[21]

5. Often, the toxicity of the entire formulation is investigated, therefore preventing the discrimination between drug and nanoparticle independent effects. In order to isolate the toxicity of the carrier from that of the drug, it is a mandatory requirement in toxicity assessment to evaluate the behavior and contribution of the blank nanocarrier *per se* without a drug load.

6. As stated by Fischer and Chan,[26] the complete "life cycle" characterization of the carrier should be pursued; that is, its complete physicochemical characterization followed by *in vivo* characterization of biodistribution, accumulation in organs, elimination and toxicities, and the establishment of the correlation between the different components.

7. A critical point that is sometimes disregarded is the need to challenge nanoparticle safety in **disease relevant models**. This is particularly true as epidemiological observations suggest that toxic effects occur predominantly in predisposed subjects with impaired health. Most of the studies investigating *in vivo* toxicity employ healthy animals, while many of the developed nanoparticles are eventually intended for diagnostic or

therapeutic purposes in a disease state. Typical pre-clinical screening in healthy animals or humans may detect risks of particles only at a very late stage. Thus, it should be always kept in mind that the biological responses observed in healthy animals might also be enhanced or modified in a pathological state.

8. *Molecular endpoints* such as whole genome scans by use of micro arrays should be investigated to identify perturbed gene regulations in response to nanoparticle use.

All these issues have also been addressed by De Jong and Borm.[5]

10.6 Conclusions

With nearly three decades of extensive research, drug delivery by nanotechnology has made tremendous advancement towards the rational design of sophisticated and disease targeted drug delivery systems. However, despite the major advancements and gained knowledge as to the factors affecting biodistribution and fate of nanocarriers, the final biofate is still not fully unraveled, with intracellular trafficking and final fate remaining largely non-tackled. Currently, there is limited knowledge of how nanostructures are metabolically processed and the behavior of their breakdown products inside the cells.

Furthermore, the scarcity of relevant and comprehensive *in vivo* toxicity studies as well as the ambiguous approaches and lack of guidelines undertaken by regulatory authorities make it difficult to draw any clear cut conclusions as to safety assessment of NPs and form a basis for rational decision making. Safety regulations of the FDA and the EU will become of actual value only when a real understanding of nanotoxicity will be attained. Until then, the mission and responsibility of toxicity assessment should be pursued by both toxicologists and nanotechnology scientists. Their joint efforts will define the tone and eventually dictate the safety regulations to lay the grounds for future legislation, based on plentiful and solid scientific evidence of carefully designed toxicity studies, using the appropriate assessment tools and clinically relevant study settings. Even though somewhat daunting, the potential toxicity of nanomaterials and nanoparticles should not pose an obstacle or prevent relentless efforts for their development up to their clinical use, as these innovative technologies hold an immense promise to promote patient care and alleviate suffering in complex diseases. Furthermore, in certain cases the hazard risk is possibly outreached by the potential benefit of such treatments particularly when considering the time limited exposure to controlled doses of NPs, restricted to medical settings of episodic rather than chronic treatments.

Indeed, the future prospects of nanomedicine are bright, and the gained nanotoxicology knowledge should pave the way for the development of ground breaking and safe solutions for various pathologies, namely inflammatory, infectious and cancerous diseases.

References

1. M. Wang, and M. Thanou, *Pharmacol. Res.*, 2010, **62**, 90–9.
2. http://clinicaltrials.gov, accessed on April 18, 2011.
3. M. E. Davis, J. E. Zuckerman, C. H. Choi, D. Seligson, A. Tolcher, C. A. Alabi, Y. Yen, J. D. Heidel, and A. Ribas, *Nature*, 2010, **464**, 1067–70.
4. M. Ferrari, *Nat. Rev. Cancer*, 2005, **5**, 161–71.
5. W. H. De Jong, P. J. Borm, *Int. J. Nanomed.*, 2008, **3**, 133–49.
6. J. J. Li, S. Muralikrishnan, C. T. Ng, L. Y. Yung, and B. H. Bay, *Exp. Biol. Med. (Maywood)*, 2010, **235**, 1025–33.
7. X. D. Zhang, H. Y. Wu, D. Wu, Y. Y. Wang, J. H. Chang, Z. B. Zhai, A. M. Meng, P. X. Liu, L. A. Zhang, and F. Y. Fan, *Int. J. Nanomed.*, 2010, **5**, 771–81.
8. J. Lu, M. Liong, Z. Li, J. I. Zink, and F. Tamanoi, *Small*, 2010, **6**, 1794–805.
9. K. L. Aillon, Y. Xie, N. El-Gendy, C. J. Berkland, and M. L. Forrest, *Adv. Drug Delivery Rev.*, 2009, **61**, 457–66.
10. A. Dhawan, and V. Sharma, *Anal. Bioanal. Chem.*, 2010, **398**, 589–605.
11. N. Khlebtsov, and L. Dykman, *Chem. Soc. Rev.*, 2010, **40**, 1647–71.
12. C. Napierska, L. C. Thomassen, D. Lison, J. A. Martens, and P. H. Hoet, *Part. Fibre Toxicol.*, 2010, **7**, 39.
13. W. S. Cho, M. Cho, J. Jeong, M. Choi, H. Y. Cho, B. S. Han, S. H. Kim, H. O. Kim, Y. T. Lim, B. H. Chung, and J. Jeong, *Toxicol. Appl. Pharmacol.*, 2009, **236**, 16–24.
14. M. C. Mancini, B. A. Kairdolf, A. M. Smith, and S. Nie, *J. Am. Chem. Soc.*, 2008, **130**, 10836–7.
15. http://www.fda.gov/ScienceResearch/SpecialTopics/Nanotechnology/NanotechnologyTaskForceReport2007/default.htm, 2007.
16. N. Feliu, and B. Fadeel, *Nanoscale*, 2010, **2**, 2514–20.
17. A. Poma, and M. L. Di Giorgio, *Curr. Genomics*, 2008, **9**, 571–85.
18. A. V. Kabanov, *Adv. Drug Delivery Rev.*, 2006, **58**, 1597–621.
19. G. Oberdorster, *Inhal. Toxicol.*, 1996, **8 Suppl**, 73–89.
20. P. J. Borm, *Inhal. Toxicol.*, 2002, **14**, 311–24.
21. C. L. Geraci, and V. Castranova, *Wiley Interdiscip. Rev. Nanomed. Nanobiotechnol.*, 2010, **2**, 569–77.
22. R. Fernandez-Urrusuno, E. Fattal, D. Porquet, J. Feger, and P. Couvreur, *Toxicol. Appl. Pharmacol.*, 1995, **130**, 272–9.
23. K. M. Al-Hallak, S. Azarmi, A. Anwar-Mohamed, W. H. Roa, and R. Lobenberg, *Eur. J. Pharm. Biopharm.*, 2010, **76**, 112–9.
24. S. Y. Shaw, E. C. Westly, M. J. Pittet, A. Subramanian, S. L. Schreiber, and R. Weissleder, *Proc. Natl. Acad. Sci. U. S. A.*, 2008, **105**, 7387–92.
25. P. Rivera Gil, G. Oberdorster, A. Elder, V. Puntes, and W. J. Parak, *ACS Nano*, 2010, **4**, 5527–31.
26. H. C. Fischer, and W. C. Chan, *Curr. Opin. Biotechnol.*, 2007, **18**, 565–71.
27. M. A. Dobrovolskaia, J. D. Clogston, B. W. Neun, J. B. Hall, A. K. Patri, and S. E. McNeil, *Nano Lett*, 2008, **8**, 2180–7.

28. N. Lewinski, V. Colvin, and R. Drezek, *Small*, 2008, **4**, 26–49.
29. A. Legendre, P. Froment, S. Desmots, A. Lecomte, R. Habert, and E. Lemazurier, *Biomaterials*, 2010, **31**, 4492–505.
30. D. Huh, B. D. Matthews, A. Mammoto, M. Montoya-Zavala, H. Y. Hsin, and D. E. Ingber, *Science*, 2010, **328**, 1662–8.
31. M. Beck-Broichsitter, J. Gauss, C. B. Packhaeuser, K. Lahnstein, T. Schmehl, W. Seeger, T. Kissel, and T. Gessler, *Int. J. Pharm.*, 2009, **367**, 169–78.
32. K. Higashisaka, Y. Yoshioka, K. Yamashita, Y. Morishita, M. Fujimura, H. Nabeshi, K. Nagano, Y. Abe, H. Kamada, S. Tsunoda, T. Yoshikawa, N. Itoh, and Y. Tsutsumi, *Biomaterials*, 2011, **32**, 3–9.
33. S. Akhtar, *Expert Opin. Drug Metab. Toxicol.*, 2010, **6**, 1347–62.
34. A. V. Kabanov, E. V. Batrakova, S. Sriadibhatla, Z. Yang, D. L. Kelly, and V. Y. Alakov, *J. Controlled Release*, 2005, **101**, 259–71.
35. J. Robbens, C. Vanparys, I. Nobels, R. Blust, K. Van Hoecke, C. Janssen, K. De Schamphelaere, K. Roland, G. Blanchard, F. Silvestre, V. Gillardin, P. Kestemont, R. Anthonissen, O. Toussaint, S. Vankoningsloo, C. Saout, E. Alfaro-Moreno, P. Hoet, L. Gonzalez, P. Dubruel, and P. Troisfontaines, *Toxicology*, 2010, **269**, 170–81.
36. J. Rejman, V. Oberle, I. S. Zuhorn, and D. Hoekstra, *Biochem. J.*, 2004, **377**, 159–69.
37. M. P. Desai, V. Labhasetwar, E. Walter, R. J. Levy, and G. L Amidon, *Pharm Res*, 1997, **14**, 1568–73.
38. F. Alexis, E. Pridgen, L. K. Molnar, and O. C. Farokhzad, *Mol. Pharm.*, 2008, **5**, 505–15.
39. P. G. Rogueda, and D. Traini, *Expert Opin. Drug Delivery*, 2007, **4**, 595–606.
40. O. Harush-Frenkel, Y. Altschuler, and S. Benita, *Crit. Rev. Ther. Drug Carrier Syst.*, 2008, **25**, 485–544.
41. A. Nel, T. Xia, L. Madler, and N. Li, *Science*, 2006, **311**, 622–7.
42. O. Harush, Y. Altschuler, and S. Benita, *Nanoparticles for pharmaceutical applications*, ed. A. J. Domb, Y. Tabata, M. N. V. R. Kumar, S. Farber, American Scientific Publishers, Los Angeles CA,, 2007, 85–102.
43. O. Boussif, F. Lezoualc'h, M. A. Zanta, M. D. Mergny, D. Scherman, B. Demeneix, J. P. Behr, *Proc. Natl. Acad. Sci. U. S. A.*, 1995, **92**, 7297–301.
44. K. Rittner, A. Benavente, A. Bompard-Sorlet, F. Heitz, G. Divita, R. Brasseur, and E. Jacobs, *Mol. Ther.*, 2002, **5**, 104–14.
45. S. Hong, A. U. Bielinska, A. Mecke, B. Keszler, J. L. Beals, X. Shi, L. Balogh, B. G. Orr, J. R. Baker, Jr, and M. M. Banaszak Holl, *Bioconjug. Chem.*, 2004, **15**, 774–82.
46. N. A. Stasko, C. B. Johnson, M. H. Schoenfisch, T. A. Johnson, and E. L. Holmuhamedov, *Biomacromolecules*, 2007, **8**, 3853–9.
47. F. Xu, Y. Yuan, X. Shan, C. Liu, X. Tao, Y. Sheng, and H. Zhou, *Int. J. Pharm.*, 2009, **377**, 199–206.

48. J. Geys, A. Nemmar, E. Verbeken, E. Smolders, M. Ratoi, M. F. Hoylaerts, B. Nemery, and P. H. Hoet, *Environ. Health Perspect.*, 2008, **116**, 1607–13.
49. W. Bartsch, G. Sponer, K. Dietmann, and G. Fuchs, *Arzneimittelforschung*, 1976, **26**, 1581–3.
50. P. Montaguti, E. Melloni, and E. Cavalletti, *Arzneimittelforschung*, 1994, **44**, 566–70.
51. R. Gref, Y. Minamitake, M. T. Peracchia, V. Trubetskoy, V. Torchilin, and R. Langer, *Science*, 1994, **263**, 1600–3.
52. M. Sharpe, S. E. Easthope, G. M. Keating, and H. M. Lamb, *Drugs*, 2002, **62**, 2089–126.
53. R. Soloman, and A. A. Gabizon, *Clin. Lymphoma Myeloma*, 2008, **8**, 21–32.
54. V. P. Torchilin, V. G. Omelyanenko, M. I. Papisov, A. A. Bogdanov Jr, V. S. Trubetskoy, J. N. Herron, and C. A. Gentry, *Biochim. Biophys. Acta*, 1994, **1195**, 11–20.
55. W. Wang, W. Xiong, Y. Zhu, H. Xu, and X. Yang, *J. Biomed. Mater. Res. B: Appl. Biomater.*, 2010, **93**, 59–64.
56. W. Tai, R. Mahato, and K. Cheng, *J. Controlled Release*, 2010, **146**, 264–75.
57. O. C. Farokhzad, J. Cheng, B. A. Teply, I. Sherifi, S. Jon, P. W. Kantoff, J. P. Richie, and R. Langer, *Proc. Natl. Acad. Sci. U. S. A.*, 2006, **103**, 6315–20.
58. A. Gschwind, O. M. Fischer, and A. Ullrich, *Nat. Rev. Cancer*, 2004, **4**, 361–70.
59. M. Trzpis, P. M. McLaughlin, L. M. de Leij, and M. C. Harmsen, *Am. J. Pathol.*, 2007, **171**, 386–95.
60. S. Muro, and V. R. Muzykantov, *Curr. Pharm. Des.*, 2005, **11**, 2383–401.
61. A. Albanese, E. A. Sykes, and W. C. Chan, *ACS Nano*, 2010, **4**, 2490–3.
62. S. Liu, L. Wei, L. Hao, N. Fang, M. W. Chang, R. Xu, Y. Yang, and Y. Chen, *ACS Nano*, 2009, **3**, 3891–902.
63. J. A. Champion, and S. Mitragotri, *Pharm. Res.*, 2009, **26**, 244–9.
64. C. Chauvierre, L. Leclerc, D. Labarre, M. Appel, M. C. Marden, P. Couvreur, and C. Vauthier, *Int. J. Pharm.*, 2007, **338**, 327–32.
65. I. Bertholon, C. Vauthier, and D. Labarre, *Pharm. Res.*, 2006, **23**, 1313–23.
66. I. Hamad, O. Al-Hanbali, A. C. Hunter, K. J. Rutt, T. L. Andresen, and S. M. Moghimi, *ACS Nano*, 2010, **4**, 6629–38.
67. S. Stolnik, B. Daudali, A. Arien, J. Whetstone, C. R. Heald, M. C. Garnett, S. S. Davis, and L. Illum, *Biochim. Biophys. Acta*, 2001, **1514**, 261–79.
68. M. Thanou, J. C. Verhoef, and H. E. Junginger, *Adv. Drug Delivery Rev.*, 2001, **52**, 117–26.
69. M. Thanou, J. C. Verhoef, andH. E. Junginger, *Adv. Drug Delivery Rev.*, 2001, **50 Suppl 1**, S91–101.
70. I. Bravo-Osuna, C. Vauthier, A. Farabollini, G. F. Palmieri, and G. Ponchel, *Biomaterials*, 2007, **28**, 2233–43.

71. C. Zhang, G. Qu, Y. Sun, T. Yang, Z. Yao, W. Shen, Z. Shen, Q. Ding, H. Zhou, and Q. Ping, *Eur. J. Pharm. Sci.*, 2008, **33**, 415–23.
72. V. R. Sinha, and A. Trehan, *J. Controlled Release*, 2003, **90**, 261–80.
73. K. Avgoustakis, *Curr. Drug Delivery*, 2004, **1**, 321–33.
74. J. K. Vasir, and V. Labhasetwar, *Adv. Drug Delivery Rev.*, 2007, **59**, 718–28.
75. J. H. Park, M. Ye, and K. Park, *Molecules*, 2005, **10**, 146–61.
76. J. Panyam, M. M. Dali, S. K. Sahoo, W. Ma, S. S. Chakravarthi, G. L. Amidon, R. J. Levy, and V. Labhasetwar, *J. Controlled Release*, 2003, **92**, 173–87.
77. P. Couvreur, B. Kante, M. Roland, P. Guiot, P. Bauduin, and P. Speiser, *J. Pharm. Pharmacol.*, 1979, **31**, 331–2.
78. C. Vauthier, C. Dubernet, E. Fattal, H. Pinto-Alphandary, and P. Couvreur, *Adv. Drug Delivery Rev.*, 2003, **55**, 519–48.
79. J. Kattan, *et al.*, *Invest. New Drugs*, 1992, **10**, 191–9.
80. http://www.drugs.com/clinical_trials/bioalliance-pharma-livatag-doxorubicin-transdrug-follow-up-demonstrates-significant-survival-11412.html, 2011.
81. A. des Rieux, V. Fievez, M. Garinot, Y. J. Schneider, and V. Preat, *J. Controlled Release*, 2006, **116**, 1–27.
82. S. Wadhwa, R. Paliwal, S. R. Paliwal, and S. P. Vyas, *Curr. Pharm. Des.*, 2009, **15**, 2724–50.
83. G. Cevc, and U. Vierl, *J. Controlled Release*, **141**, 277–99.
84. A. Mayer, M. Vadon, B. Rinner, A. Novak, R. Wintersteiger, and E. Frohlich, *Toxicology*, 2009, **258**, 139–47.
85. J. Wilkinson, Medical Device Link, http://www.devicelink.com/mdt/archive/06/06/014.html., 2006.
86. F. Sarti, G. Perera, F. Hintzen, K. Kotti, V. Karageorgiou, O. Kammona, C. Kiparissides, and A. Bernkop-Schnurch, *Biomaterials*, 2011, **32**, 4052–7.
87. T. Nassar, A. Rom, A. Nyska, and S. Benita, *Pharm. Res.*, 2008, **25**, 2019–29.
88. T. Nassar, A. Rom, A. Nyska, and S. Benita, *J. Controlled Release*, 2009, **133**, 77–84.
89. J. R. Kanwar, B. M. Long, and R. K. Kanwar, *Curr. Med. Chem.*, 2011, **18**, 2079–85.
90. J. F. Hillyer, and R. M. Albrecht, *J. Pharm. Sci.*, 2001, **90**, 1927–36.
91. G. M. Hodges, E. A. Carr, R. A. Hazzard, C. O'Reilly, and K. E. Carr, *J. Drug Targeting*, 1995, **3**, 57–60.
92. B. Semete, L. Booysen, Y. Lemmer, L. Kalombo, L. Katata, J. Verschoor, and H. S. Swai, *Nanomedicine*, 2010, **6**, 662–71.
93. B. Semete, L. I. Booysen, L. Kalombo, J. D. Venter, L. Katata, B. Ramalapa, J. A. Verschoor, and H. Swai, *Toxicol. Appl. Pharmacol.*, 2010, **249**, 158–65.
94. D. R. Kalaria, G. Sharma, V. Beniwal, and M. N. Ravi Kumar, *Pharm. Res.*, 2009, **26**, 492–501.
95. K. Sonaje, Y. H. Lin, J. H. Juang, S. P. Wey, C. T. Chen, and H. W. Sung, *Biomaterials*, 2009, **30**, 2329–39.

96. M. C. Chen, H. S. Wong, K. J. Lin, H. L. Chen, S. P. Wey, K. Sonaje, Y. H. Lin, C. Y. Chu, and H. W. Sung, *Biomaterials*, 2009, **30**, 6629–37.

97. U. M. Dhana Lekshmi, G. Poovi, N. Kishore, and P. N. Reddy, *Int. J. Pharm.*, **396**, 194–203.

98. P. Dandekar, R. Dhumal, R. Jain, D. Tiwari, G. Vanage, and V. Patravale, *Food Chem. Toxicol.*, 2010, **48**, 2073–89.

99. P. P. Dandekar, R. Jain, S. Patil, R. Dhumal, D. Tiwari, S. Sharma, G. Vanage, and V. Patravale, *J. Pharm. Sci.*, 2010, **99**, 4992–5010.

100. R. W. Niven, *Crit. Rev. Ther. Drug Carrier Syst.*, 1995, **12**, 151–231.

101. N. Tsapis, D. Bennett, B. Jackson, D. A. Weitz,, and D. A. Edwards, *Proc. Natl. Acad. Sci. U. S. A.*, 2002, **99**, 12001–5.

102. S. Azarmi, W. H. Roa, R. Lobenberg, *Adv. Drug Delivery Rev.*, 2008, **60**, 863–75.

103. E. Rytting, J. Nguyen, X. Wang, and T. Kissel, *Expert Opin. Drug Delivery*, 2008, **5**, 629–39.

104. J. C. Sung, B. L. Pulliam, and D. A. Edwards, *Trends Biotechnol.*, 2007, **25**, 563–70.

105. J. Zhang, L. Wu, H. K. Chan, and W. Watanabe, *Adv. Drug Delivery Rev.*, 2010.

106. H. M. Mansour, Y. S. Rhee, and X. Wu, *Int. J. Nanomed.*, 2009, **4**, 299–319.

107. L. J. Zhang, B. Xing, J. Wu, B. Xu, and X. L. Fang, *Pulmon. Pharmacol. Ther.*, 2008, **21**, 239–46.

108. O. Harush-Frenkel, M. Bivas-Benita, T. Nassar, C. Springer, Y. Sherman, A. Avital, Y. Altschuler, J. Borlak, and S. Benita, *Toxicol. Appl. Pharmacol.*, 2010, **246**, 83–90.

109. J. Fu, J. Fiegel, E. Krauland, and J. Hanes, *Biomaterials*, 2002, **23**, 4425–33.

110. C. L. Tseng, S. Y. Wu, W. H. Wang, C. L. Peng, F. H. Lin, C. C. Lin, T. H. Young, and M. J. Shieh, *Biomaterials*, 2008, **29**, 3014–22.

111. M. Beck-Broichsitter, J. Gauss, T. Gessler, W. Seeger, T. Kissel, and T. Schmehl, *J. Aerosol Med. Pulmon. Drug Delivery*, 2009, **23**, 47–57.

112. W. H. Roa, S. Azarmi, M. H. Al-Hallak, W. H. Finlay, A. M. Magliocco, and R. Lobenberg, *J. Controlled Release*, 2010, **150**, 49–55.

113. O. M. Merkel, A. Beyerle, D. Librizzi, A. Pfestroff, T. M. Behr, B. Sproat, P. J. Barth, and T. Kissel, *Mol. Pharm.*, 2009, **6**, 1246–60.

114. J. Nguyen, R. Reul, T. Betz, E. Dayyoub, T. Schmehl, T. Gessler, U. Bakowsky, W. Seeger, and T. Kissel, *J. Controlled Release*, 2009, **140**, 47–54.

115. J. W. Card, D. C. Zeldin, J. C. Bonner, and E. R. Nestmann, *Am. J. Physiol. Lung Cell. Mol. Physiol.*, 2008, **295**, L400–11.

116. L. A. Dailey, N. Jekel, L. Fink, T. Gessler, T. Schmehl, M. Wittmar, T. Kissel, and W. Seeger, *Toxicol. Appl. Pharmacol.*, 2006, **215**, 100–8.

117. H. M. Courrier, N. Butz, and T. F. Vandamme, *Crit. Rev. Ther. Drug Carrier Syst.*, 2002, **19**, 425–98.

118. M. S. Bakshi, L. Zhao, R. Smith, F. Possmayer, and N. O. Petersen, *Biophys. J.*, 2008, **94**, 855–68.

119. C. Schleh, C. Muhlfeld, K. Pulskamp, A. Schmiedl, M. Nassimi, H. D. Lauenstein, A. Braun, N. Krug, V. J. Erpenbeck, and J. M. Hohlfeld, *Respir. Res.*, 2009, **10**, 90.

120. M. Beck-Broichsitter, C. Ruppert, T. Schmehl, A. Guenther, T. Betz, U. Bakowsky, W. Seeger, T. Kissel, and T. Gessler, *Nanomedicine*, 2010.

121. M. H. Al-Hallak, S. Azarmi, C. Sun, P. Lai, E. J. Prenner, W. H. Roa, and R. Lobenberg, *AAPS J.*, 2010, **12**, 294–9.

122. O. Schmid, W. Moller, M. Semmler-Behnke, G. A. Ferron, E. Karg, J. Lipka, H. Schulz, W. G. Kreyling, and T. Stoeger, *Biomarkers*, 2009, **14 Suppl 1**, 67–73.

123. P. Borm, F. C. Klaessig, T. D. Landry, B. Moudgil, T. Pauluhn, K. Thomas, R. Trottier, and S. Wood, *Toxicol. Sci.*, 2006, **90**, 23–32.

124. R. P. Schins, A. McAlinden, W. MacNee, L. A. Jimenez, J. Ross, K. Guy, S. P. Faux, and K. Donaldson, *Toxicol. Appl. Pharmacol.*, 2000, **167**, 107–17.

125. A. M. Knaapen, P. J. Borm, C. Albrecht, and R. P. Schins, *Int. J. Cancer*, 2004, **109**, 799–809.

126. B. Fahmy, and S. A. Cormier, *Toxicol. In Vitro*, 2009, **23**, 1365–71.

127. G. Oberdorster, *Int. Arch. Occup. Environ. Health*, 2001, **74**, 1–8.

128. A. Natarajan, C. Gruettner, R. Ivkov, G. L. DeNardo, G. Mirick, A. Yuan, A. Foreman, and S. J. DeNardo, *Bioconjug. Chem.*, 2008, **19**, 1211–8.

129. M. Choi, M. Cho, B. S. Han, J. Hong, J. Jeong, S. Park, M. H. Cho, K. Kim, and W. S. Cho, *Toxicol. Lett.*, 2010, **199**, 144–52.

130. G. Oberdorster, A. Maynard, K. Donaldson, V. Castranova, J. Fitzpatrick, K. Ausman, J. Carter, B. Karn, W. Kreyling, D. Lai, S. Olin, N. Monteiro-Riviere, D. Warheit, and H. Yang, *Part Fibre Toxicol.*, 2005, **2**, 8.

131. R. C. Murdock, L. Braydich-Stolle, A. M. Schrand, J. J. Schlager, and S. M. Hussain, *Toxicol. Sci.*, 2008, **101**, 239–53.

132. I. Gosens, J. A. Post, L. J. de la Fonteyne, E. H. Jansen, J. W. Geus, F. R. Cassee, and W. H. de Jong, *Part Fibre Toxicol.*, 2010, **7**, 37.

133. T. M. Sager, C. Kommineni, and V. Castranova, *Part Fibre Toxicol.*, 2008, **5**, 17.

134. P. J. Borm, and W. Kreyling, *J. Nanosci. Nanotechnol.*, 2004, **4**, 521–31.

135. C. Medina, M. J. Santos-Martinez, A. Radomski, O. I. Corrigan, and M. W. Radomski, *Br. J. Pharmacol.*, 2007, **150**, 552–8.

136. G. Oberdorster, E. Oberdorster, and J. Oberdorster, *Environ. Health Perspect.*, 2005, **113**, 823–39.

137. A. Shimada, N. Kawamura, M. Okajima, T. Kaewamatawong, H. Inoue, and T. Morita, *Toxicol. Pathol.*, 2006, **34**, 949–57.

138. X. He, H. Zhang, Y. Ma, W. Bai, Z. Zhang, K. Lu, Y. Ding, Y. Zhao, and Z. Chai, *Nanotechnology*, 2010, **21**, 285103.

139. C. Muhlfeld, P. Gehr, and B. Rothen-Rutishauser, *Swiss Med. Wkly.*, 2008, **138**, 387–91.

140. W. S. Cho, M. Cho, S. R. Kim, M. Choi, J. Y. Lee, B. S. Han, S. N. Park, M. K. Yu, S. Jon, and J. Jeong, *Toxicol. Appl. Pharmacol.*, 2009, **239**, 106–15.

141. H. S. Choi, Y. Ashitate, J. H. Lee, S. H. Kim, A. Matsui, N. Insin, M. G. Bawendi, M. Semmler-Behnke, J. V. Frangioni, and A. Tsuda, *Nat. Biotechnol.*, 2010, **28**, 1300–3.

142. Y. Zhang, W. Zhang, M. Lobler, K. P. Schmitz, P. Saulnier, T. Perrier, I. Pyykko, and J. Zou, *Int. J. Pharm.*, 2011, **404**, 211–9.

143. M. J. Alonso, *Biomed. Pharmacother.*, 2004, **58**, 168–72.

144. M. M. Abdel-Mottaleb, D. Neumann, and A. Lamprecht, *Eur. J. Pharm. Biopharm.*, 2011.

145. A. O. Hassan, and A. H. Elshafeey, *J. Biomed. Nanotechnol.*, 2010, **6**, 621–33.

146. G. J. Nohynek, J. Lademann, C. Ribaud, and M. S. Roberts, *Crit. Rev. Toxicol.*, 2007, **37**, 251–77.

147. A. N. Choksi, T. Poonawalla, and M. G. Wilkerson, *J. Drugs Dermatol.*, 2010, **9**, 475–81.

148. L. Contreras-Ruiz, M. de la Fuente, C. Garcia-Vazquez, V. Saez, B. Seijo, M. J. Alonso, M. Calonge, and Y. Diebold, *Cornea*, 2010, **29**, 550–8.

149. C. Losa, M. J. Alonso, J. L. Vila, F. Orallo, J. Martinez, J. A. Saavedra, and J. C. Pastor, *J. Ocul. Pharmacol.*, 1992, **8**, 191–8.

150. A. Enriquez de Salamanca, Y. Diebold, M. Calonge, C. Garcia-Vazquez, S. Callejo, A. Vila, and M. J. Alonso, *Invest. Ophthalmol. Visual Sci.*, 2006, **47**, 1416–25.

151. T. W. Prow, I. Bhutto, S. Y. Kim, R. Grebe, C. Merges, D. S. McLeod, K. Uno, M. Mennon, L. Rodriguez, K. Leong, and G. A. Lutty, *Nanomedicine*, 2008, **4**, 340–9.

152. S. V. Dhuria, L. R. Hanson, and W. H. Frey 2nd, *J. Pharm. Sci.*, 2010, **99**, 1654–73.

153. J. C. Olivier, *NeuroRx*, 2005, **2**, 108–19.

154. A. Mistry, S. Stolnik, and L. Illum, *Int. J. Pharm.*, 2009, **379**, 146–57.

155. Q. Liu, X. Shao, J. Chen, Y. Shen, C. Feng, X. Gao, Y. Zhao, J. Li, Q. Zhang, and X. Jiang, *Toxicol. Appl. Pharmacol.*, 2011, **251**, 79–84.

156. G. Oberdorster, A. Elder, and A. Rinderknecht, *J. Nanosci. Nanotechnol.*, 2009, **9**, 4996–5007.

157. S. Lanone, and J. Boczkowski, *Curr. Mol. Med.*, 2006, **6**, 651–63.

158. S. Costigan, The website of Medicines and Healthcare Products Regulatory Agency, Department of Health, UK, 2006.

159. D. S. Goldberg, H. Ghandehari, and P. W. Swaan, *Pharm. Res.*, 2010, **27**, 1547–57.

160. K. Bowman, and K. W. Leong, *Int. J. Nanomed.*, 2006, **1**, 117–28.

161. Y. H. Lin, F. L. Mi, C. T. Chen, W. C. Chang, S. F. Peng, H. F. Liang, and H. W. Sung, *Biomacromolecules*, 2007, **8**, 146–52.

162. I. Kadiyala, Y. Loo, K. Roy, J. Rice, and K. W. Leong, *Eur. J. Pharm. Sci.*, 2010, **39**, 103–9.
163. S. Hong, P. R. Leroueil, E. K. Janus, J. L. Peters, M. M. Kober, M. T. Islam, B. G. Orr, J. R. Baker Jr, and M. M. Banaszak Holl, *Bioconjug. Chem.*, 2006, **17**, 728–34.
164. M. N. Moore, *Environ. Int.*, 2006, **32**, 967–76.
165. A. Beyerle, M. Irmler, J. Beckers, T. Kissel, and T. Stoeger, *Mol. Pharm.*, 2010, **7**, 727–37.
166. T. Minko, P. Kopeckova, and J. Kopecek, *J. Controlled Release*, 2001, **71**, 227–37.
167. X. Dong, C. A. Mattingly, M. T. Tseng, M. J. Cho, Y. Liu, V. R. Adams, and R. J. Mumper, *Cancer Res.*, 2009, **69**, 3918–26.
168. J. J. Salomon, and C. Ehrhardt, *Eur. J. Pharm. Biopharm.*, 2011, **77**, 392–7.
169. E. V. Batrakova, D. L. Kelly, S. Li, Y. Li, Z. Yang, L. Xiao, D. Y. Alakhova, S. Sherman, V. Y. Alakhov, and A. V. Kabanov, *Mol. Pharm.*, 2006, **3**, 113–23.
170. N. Debotton, M. Parnes, J. Kadouche, and S. Benita, *J. Controlled Release*, 2008, **127**, 219–30.
171. M. Marsh, and A. Helenius, *Cell*, 2006, **124**, 729–40.

Section 11
A Clinically Relevant Case Study

CHAPTER 11

A Clinically Relevant Case Study: the Development of Livatag® for the Treatment of Advanced Hepatocellular Carcinoma

EMILIENNE SOMA*[a], PIERRE ATTALI[b] AND PHILIPPE MERLE[b]

[a] BioAlliance Pharma, 49 Bd General Martial Valin. 75015 Paris, France;
[b] Croix-Rousse Hospital, Co-Director of the Hepatology and Gastroenterology Unit, 103 Grande Place de la Croix-Rousse. 69004 Lyon, France
*E-mail: emilienne.soma@bioalliancepharma.com

11.1 Introduction and Rationale of Hepatocellular Carcinoma

Hepatocellular carcinoma (HCC) is the fifth most common cancer in men (523 000 cases) and the seventh in women (226 000 cases). Most of the burden is in developing countries, where almost 85% of the cases occur, and particularly in men: the overall sex ratio male/female is around 4. High incidence regions are Eastern and South-Eastern Asia, Middle and Western Africa, but also Melanesia and Micronesia/Polynesia. Low rates are estimated in developed

RSC Drug Discovery Series No. 22
Nanostructured Biomaterials for Overcoming Biological Barriers
Edited by Maria Jose Alonso and Noemi S. Csaba
© The Royal Society of Chemistry 2012
Published by the Royal Society of Chemistry, www.rsc.org

regions, with the exception of Southern Europe where the incidence in men (ASR 10.5 per 100 000) is significantly higher than in other developed regions.[1]

The incidence of primary liver cancer is increasing in several developed countries including the United States, and this increase will likely continue for several decades. The trend has a dominant cohort effect related to exposures to hepatitis B and C viruses. The attributable risk estimates for the combined effects of these infections account for well over 80% of liver cancer cases worldwide.[2] Other risk factors such as alcohol consumption and metabolic syndromes, may explain the residual variations within countries.

There were an estimated 694 000 deaths from liver cancer in 2008 (477 000 in men, 217 000 in women). Because of its high fatality (overall ratio of mortality to incidence of 0.93), liver cancer is the third most common cause of death from cancer worldwide. The geographical distribution of the mortality rates is similar to that observed for incidence.[1]

11.2   Treatments of HCC

Current therapeutic strategies for HCC can be divided into curative treatments such as surgical interventions (tumor resection and liver transplantation) or percutaneous interventions (ethanol injection, radiofrequency thermal ablation) and palliative treatments, such as transarterial interventions (mainly transarterial chemoembolization or TACE), systemic therapies or experimental strategies.[3] In carefully selected patients, resection and transplantation allow a 5 year survival from 60% to 70%. Unfortunately, most patients in Western countries present with an intermediate or advanced HCC at diagnosis, with the consequent inability to use these curative treatments.[4]

Among palliative treatments, the intra-arterial approach with chemoembolization (TACE) has shown to induce objective responses in 16–55% of patients, although many randomized trials did not show any survival benefit. Indeed, TACE is known to be often accompanied by severe side effects like hepatic failure or renal dysfunction.[5,6] A more recent study showed a significant palliative, although modest, effect of TACE in comparison with conservative management on survival, while the overall tumour response rate remained very low.[7] In this study, TACE improved survival of stringently selected patients with unresectable HCC, but only 12% of the patients diagnosed with HCC met the inclusion criteria. This study showed that TACE is beneficial for a relatively small group of patients, likely those with a significant chance of response to ablative therapy. In contrast, a large prospective randomized trial conducted by a French cooperative group failed to show any survival benefit of chemoembolization.[8]

Until recently, for patients with advanced HCCs no therapy was available that prolonged overall survival (OS), indicating the need for new targeted-therapies.[3] In 2007 and for the first time, sorafenib, a multikinase inhibitor, showed an increase, although modest, of the overall survival over placebo in patients with unresectable HCC. Beside this agent, various different molecules are currently tested in advanced stage HCC among which brivanib, another oral multikinase inhibitor, is currently being tested in several phase III studies.[3,9,10]

Even though sorafenib is the standard of care for advanced stage HCC and is registered for the treatment of HCC without restrictions by the European Medicines Agency (EMA) and the Food and Drug Administration (FDA), the narrow inclusion criteria of the clinical trials leave many patients without proven effective treatment with regard to their disease stage. Moreover, since treatment failure will happen eventually in all patients on sorafenib, there still remains a medical need to improve treatment efficacy, drug regimen and overall tolerance, and to overcome resistance.[10]

HCC is known as hypervascular solid cancer characterized by a high degree of drug resistance.[11] The mechanisms of chemoresistance in HCC are multiple. However, the more common mechanism is related to the multidrug resistance (MDR) transporters, P-gp and MRP pumps.[12–14] These pumps allow tumour cells to efflux different types of cytostatic agents into the extracellular environment.[15] Chemoresistance related to the MDR phenotype may be intrinsic or be acquired during cytostatic chemotherapy and "resistance", whether spontaneous or acquired is a serious concern in cancer treatment. HCC is often intrinsic chemoresistant which is the major cause for failure of its therapy.[16–18] As increasing of drug levels to overcome multidrug resistance may induce severe adverse effects, this is a great obstacle to chemotherapy.[19] The poor efficacy of chemotherapeutic agents attributed to the overexpression of the MDR gene underlines the need to develop new treatment strategies for HCC, which could take into account the resistance issues.

In HCC, new therapeutic strategies using cytotoxic agents were developed by hepatic intra-arterial injection in order to reduce the systemic toxicity, to induce important hepatic tumour necrosis and to save the healthy hepatic parenchyma.

11.3 Rationale of Livatag® in Hepatocellular Carcinoma

The proprietary technology Transdrug™ uses polyisohexylcyanoacrylate (PIHCA) polymer to formulate anti-cancer drugs into nanoparticles. Livatag® is a novel drug formulation that associates those PIHCA nanoparticles with doxorubicin molecules loaded onto the nanoparticles by adsorption. Doxorubicin is an antineoplastic antibiotic, which may act by forming a stable complex with DNA and by interfering with the synthesis of nucleic acids. It is an effective antineoplastic against a wide range of tumours, often used in combination with other anticancer drugs.

Pre-clinical pharmacological Pgp- or MRP-related studies have shown that Livatag® was far more cytotoxic than free doxorubicin on resistant cell lines and was able to bypass multidrug resistance, whether this is Pgp or MRP related.[20,21] *In vivo*, Livatag® appears to be more effective than free doxorubicin on both sensitive and resistant tumour models, with significant

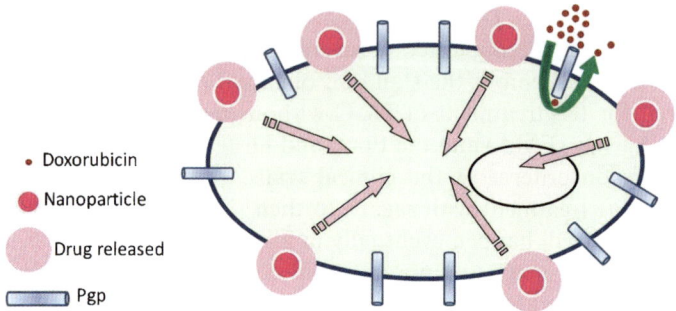

- Doxorubicin
- Nanoparticle
- Drug released
- Pgp

Figure 11.1   Nanoparticle/cell interaction and overcoming of multidrug resistance.

reduction in the number of metastases and increase in tumour cell apoptosis and tumour necrosis.[22–24] Furthermore, after *in vivo* intravenous injection in animals, Livatag® mainly distributed in the liver in opposition to free doxorubicin.[25] The mechanism of action was investigated and it was hypothesized that doxorubicin and polycyanoacrylate, the PIHCA degradation product, formed together an ion pair which hinders doxorubicin from the efflux pumps involved in MDR (Figure 11.1).[26,27]

Finally, the benefit of treatment with Livatag® could also lead to a delayed appearance of acquired resistance, as has been shown after experimental chronic exposure of cancer cells to Livatag®.[28]

With regards to the high degree of resistance of HCC cells and to the hypervascularity of the tumour on the one hand, and to the ability to overcome multidrug resistance, and its preferential distribution in the liver on the other hand, Livatag® may represent a valuable treatment in patients with unresectable HCC. Therefore, it has been decided to conduct human clinical development in the treatment of advanced HCC.

11.4   Livatag® Manufacturing

The Livatag® product developed for use in clinical trials is presented as a sterile lyophilisate for injectable suspension that contains doxorubicin hydrochloride as active ingredient and other excipients including the nanoparticle polymer, PIHCA. Livatag® is manufactured according to current Good Manufacturing Practices for human medicines. The drug-loaded nanoparticles are obtained by aqueous emulsion polymerisation of isohexylcyanoacrylate (IHCA) monomer dropped in the bulk solution containing the active ingredient doxorubicin and the other excipients.[29] No non-aqueous solvent is required during the manufacturing. At the end of polymerisation, a stable suspension of nanoparticles entrapping doxorubicin is obtained. The nanoparticle mean size is about 200 nm. The nanoparticle suspension is then filtered and aseptically filled in glass vials before freeze-drying. The experimental product for clinical study has been manufactured in a pharmaceutical facility authorized to handle cytotoxic drugs. The Livatag® freeze-dried product must be kept protected from light and humidity and stored in a refrigerator at 2–8 °C, for stability purposes.

11.5 Current Clinical Development in Hepatocellular Carcinoma

Before the description of current clinical development in advanced hepatocellular carcinoma, it should be mentioned that Livatag® has primarily been investigated in a phase 1 study in refractory solid tumours[30] and acute myeloid leukemia.

11.5.1 Phase 1–2 Study in Advanced Hepatocellular Carcinoma[31]

This clinical trial conducted with Livatag® was a multicentre, noncontrolled escalating dose study in cirrhotic patients of Child-Pugh A with advanced HCC. For these patients, curative treatments (surgical resection, liver transplantation, or percutaneous ablation) could not be considered. The main objectives of this study performed in France, were to determine the maximum tolerated dose, to define the dose and also to evaluate the efficacy of Livatag® for phase 2. The dose levels ranged from 10 to 40 mg m^{-2}. A single injection of Livatag® was administered through hepatic intra-arterial route. The hepatic intra-arterial route is frequently used for doxorubicin in the treatment of advanced HCC using TACE, in order to provide better tumour responses and improved systemic tolerability.

Twenty-one patients were enrolled and 20 patients were treated and received at least one hepatic intra-arterial injection of Livatag®. Two patients received 2 courses and one patient received 3 courses. The escalating dose was as described in Table 11.1.

Safety

Most adverse events (90%) were grade 1–2, dose-independent, transient, and quickly reversible. No dose-limiting toxicity occurred in the patients treated for

Table 11.1 Livatag® phase 1 study of escalating dose: Treatment doses and number of patients treated at each level.

Livatag® dose level (as equivalent mg doxorubicin per m^2 body area)	Number of patients treated
10	3
20	3
30	6
40	3
35	5

10, 20 and 30 mg m^{-2}. Two out of the three patients treated at 40 mg m^{-2} had a dose-limiting toxicity: reversible grade 4 neutropenia occurring over 7 days. Following these findings, an intermediate dose level of 35 mg m^{-2} was assessed. Two out of the five patients who were treated at 35 mg m^{-2} experienced a severe respiratory adverse reaction. The first one had an acute lung injury 44 hours after injection. The second presented an acute respiratory distress syndrome a few hours after injection of a mixture of Lipiodol® (ethiodized poppyseed oil) and Livatag® followed by hepatic artery embolization with Gelfoam® (absorbable gelatin sponge powder) in deviation to the study protocol. The patient's outcome was favourable, without sequellae in a few days with symptomatic treatment. The relationship with Livatag® was considered as possible and may have been favoured by the combination of Lipiodol® and nanoparticles. As there was no report of respiratory adverse events in the three first patients treated at the dose level of 30 mg m^{-2}, it was decided to treat three more patients at 30 mg m^{-2} to confirm the safety at that dose. No dose-limiting toxicity occurred in these 3 additional patients; no respiratory adverse events were reported. Therefore, the MTD was considered at 35 mg m^{-2} and the recommended dose for further studies was 30 mg m^{-2}, as no patient experienced dose-limiting toxicity at this dose level.

Efficacy

The evaluation of the efficacy was based on the comparison of hepatic computed tomography (CT) scans performed at baseline and 4 weeks after treatment. The results of this phase 1 investigation suggested a benefit of a single hepatic intra-arterial injection of Livatag®, demonstrated by disease control (partial response and stabilised disease based on RECIST criteria) in 60 or 75% patients according to centralised review and investigator's assessment, respectively. One out of the six patients (16.67%) treated at 30 mg m^{-2} dose level showed a partial response (Figure 11.2). Four out of 15 patients with an alpha-fetoprotein value > 400 ng mL^{-1} at the beginning of the study, had a decreased value during the study. According to these results, the dose level of 30 mg m^{-2} was considered for phase 2–3 studies for treatment of patients with cirrhosis Child-Pugh A suffering from advanced HCC.

This phase 1 study showed that 35 mg m^{-2} of Livatag® administered through the hepatic intra-arterial route was the MTD in Child-Pugh A cirrhotic patients with advanced HCC. The efficacy results observed after only one course of treatment in the subgroup of patients treated at 30 mg m^{-2} suggested that Livatag® may be effective in patients suffering from non resectable advanced HCC.

11.5.2 Phase 2–3 Study in Advanced Hepatocellular Carcinoma

The second clinical trial with Livatag® in advanced HCC was a randomized muticenter phase 2–3 study. The main objective was to compare the efficacy and safety of 3 courses of hepatic intra-arterial injection of Livatag® every 4

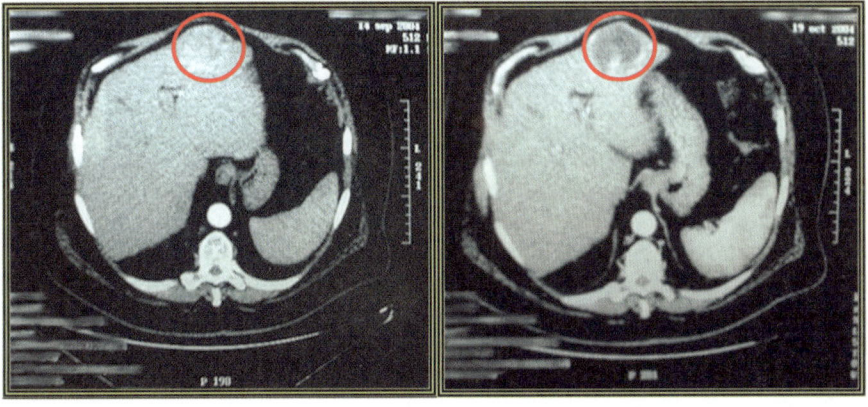

(a)-Baseline

(b)- 4 weeks post treatment
showing tumour necrosis

Figure 11.2   Patient #8: (a) presented with single unresectable tumour in Segment II. Tumour measured 60 × 50 mm (3000 mm^2). (b) After one infusion at 30 mg m^{-2}, tumour necrosis was evident.

weeks with best of care treatment according to each centre's usual practice, including Trans-Arterial ChemoEmbolization (TACE). Twenty-eight patients were enrolled, 17 in the Livatag® group and 11 in the Control group. Thirteen men and 4 women from 51 to 78 years old received a single dose of 30 mg m^{-2} Livatag® by hepatic intra-arterial injection for up to three courses. Ten men and one woman from 52 to 80 years old were randomized in the Control group.

Safety
The study has been prematurely held on following the recommendation from the independent study management boards, as settled in the protocol (the Drug Safety Monitoring Board and the Steering Committee), after the occurrence of more frequent and more severe pulmonary adverse events than expected. Indeed, pulmonary injury was observed previously in the phase 1 clinical trial at the dose of 35 mg m^{-2} which was the MTD, but nothing was observed at 30 mg m^{-2}, the dose choosen for the current clinical trial. The causes of these pulmonary adverse events have been investigated. It is noteworthy that hypersensitivity reactions with impact on the lungs have been reported in the literature, in a relatively large number (up to 7%) of patients with liposomal medicines.[32] With these liposomal drugs, the reactions usually developed at the start of the infusion and included symptoms of cardiopulmonary distress, such as dyspnea, tachypnea, hypo and/or hypertension, chest pain, and back pain. These events are now described as pseudo-allergic reactions mediated through complement activation or CARPA (complement activation-related pseudoallergy).[33] Its mechanism, however, has not been completely clarified to date. In particular, it has been demonstrated in an experimental swine model that

pulmonary effects induced by liposomes could be attenuated or delayed by slowing down the speed of the intravenous administration (slow infusion instead of bolus injection).[32]

Efficacy

The primary efficacy endpoint was to assess the number of patients free of local progression at 3 months after randomization in Livatag®-treated group. Since the study was discontinued prematurely, the overall survival was also followed up for Livatag® and Control group. The results showed that after 18 months of follow-up, a 88.9% survival rate was recorded in patients group who have received three intra-arterial injections of Livatag®, as per protocol. This increased survival rate was higher than the 54.5% survival observed in patients in the Control group (*i.e.* treated with the current standard of care, usually TACE). More recent results showed a median survival of 32 months for Livatag® group as compared with 15 months for patients getting current best of care (TACE). Based on these very encouraging data, new approaches validated in animal models, have been designed for a safer use of Livatag® that could reduce the likelihood of occurrence of pulmonary adverse events, and further improve the benefit/risk ratio.

11.6   Conclusion

Hepatocellular carcinoma is one of the worldwide leading causes of cancer-related death, with a rising incidence. Most patients are untreated with an advanced disease despite the availability or the development of new targeted molecules. There is thus still a medical need for patients that do not benefit from treatments available or under investigation. In this context, Livatag®, a new nanoparticle formulation with preferential liver distribution and ability to overcome multidrug resistance, has demonstrated a strong signal of efficacy in 37 patients with advanced hepatocellular carcinoma. Therefore, a new clinical approach is being designed to overcome safety issues and to confirm the efficacy of Livatag® in advanced HCC. The potential of this new nanoparticle formulation for cancer treatment has been recognized by both of the Health Agencies in Europe (EMA) and in the United States (FDA) with a grant of Orphan Drug Designation.

References

1. Globocan 2008, WHO, International Agency for Research on Cancer (IARC), Cancer Incidence and Mortality Worldwide 2008.
2. F. X. Bosch, J. Ribes, R. Cléries and M. Díaz, *Clin. Liver Dis.,* 2005, **9**(2), 191.
3. H. C. Spangenberg, R. Thimme and H. E. Blum, *Biologics: Targets Therapy,* 2008, **2**(3), 453.

4. L. Faloppi, M. Scartozzi, E. Maccaroni, P. M. Di Pietro, R. Berardi, M. Del Prete and S. Cascinu, *Cancer Treat. Rev.,* 2011, **37**(3), 169.
5. K. Kamada, T. Nakanishi, M. Kitamoto, H. Aikata, Y. Kawakami, K. Ito, T. Asahara and G. Kajiyama, *J. Vasc. Interv. Radiol.,* 2001, **12**(7), 847.
6. H. Yamazaki, H. Oi, M. Matsushita, T. Kim, E. Tanaka, T. Inoue, H. Nakamura, T. Teshima and T. Inoue, *Br. J. Radiol.,* 2001, **74**(884), 695.
7. J.M. Llovet, M.I. Real, X. Montaña, R. Planas, S. Coll, J. Aponte, C. Ayuso, M. Sala, J. Muchart, R. Solà, J. Rodés, J. Bruix and Barcelona Liver Cancer Group, *Lancet,* 2002, **359**(9319), 1734.
8. M. Doffoel, D. Vetter, O. Bouche, F. Bonnetain, A. Abergel, S. Fratte, J. Grange, N. Stremdoerfer, A. Blanchi, L. Bedenne and Fédération Francophone de Cancérologie Digestive (FFCD), *J. Clin. Oncol.*, 2005 ASCO Annual Meeting Proceedings, **23**, No. 16S, Part I of II (June 1 Supplement), 4006.
9. K. Almhanna and P. A. Philip, *Oncol. Targets Ther.,* 2009, **18**(2), 261.
10. M. Peck-Radosavljevic, *Therap. Adv. Gastroenterol.* 2010, **3**(4), 259.
11. T. Wakamatsu, Y. Nakahashi, D. Hachimine, T. Seki and K. Okazaki, *Int. J. Oncol.*, 2007, **31**(6), 1465.
12. A. Aszalos and D. D. Ross, *Anticancer Res.,* 1998, **18**(4C), 2937. Review.
13. D. M. Bradshaw, R. J. Arceci, *J. Clin. Oncol.*, 1998, **16**(11), 3674. Review. Erratum in: *J. Clin. Oncol.,* 1999, **17**(4), 1330.
14. S.V. Ambudkar, S. Dey, C.A. Hrycyna, M. Ramachandra, I. Pastan and M.M. Gottesman, *Annu. Rev. Pharmacol. Toxicol.,* 1999, **39**, 361. Review.
15. Y. Chen and S. M. Simon, *J. Cell Biol.* 2000, **148**(5), 863.
16. P. Merle and C. Trepo, *Cancérol. Auj.,* 2000, **9**(6), 145.
17. I.O. Ng, C.L. Liu, S.T. Fan and M. Ng, *Am. J. Clin. Pathol.,* 2000, **113**(3), 355.
18. R. Pérez-Tomás, *Curr. Med. Chem.,* 2006, **13**(16), 1859. Review.
19. F. Yan, X. M. Wang, Z. C. Liu, C. Pan, S. B. Yuan and Q. M. Ma, *Hepatobiliary Pancreat. Dis. Int.* 2010, 9(3), 287.
20. C. Cuvier, L. Roblot-Treupel, J. M. Millot, G. Lizard, S. Chevillard, M. Manfait, P. Couvreur and M. F. Poupon, *Biochem Pharmacol.* 1992, **44**(3), 509.
21. N. Aouali, H. Morjani, A. Trussardi, E. Soma, B. Giroux and M. Manfait, *Int. J. Oncol.*, 2003, **23**(4), 1195.
22. P. Genne, unpublished data, 2002.
23. L. Barraud , P. Merle, E. Soma, L. Lefrançois, S. Guerret, M. Chevallier, C. Dubernet, P. Couvreur, C. Trépo and L. Vitvitski, *J. Hepatol.*, 2005, **42**(5), 736.
24. N. Chiannilkulchai, Z. Driouich, J.P. Benoit, A.L. Parodi and P. Couvreur, *Sel. Cancer Ther.*, 1989, **5**(1), 1.
25. N. Chiannilkulchai, N. Ammoury, B. Caillou, J.P. Devissaguet and P. Couvreur, *Cancer Chemother. Pharmacol.* 1990, **26**(2), 122.
26. A. C. de Verdière, C. Dubernet, F. Némati, E. Soma, M. Appel, J. Ferté, S. Bernard, F. Puisieux and P. Couvreur, *Br. J. Cancer*, 1997, 76(2), 198.

27. X. Pépin, L. Attali, C. Domrault, S. Gallet, J.M. Metreau, Y. Reault, P. J. Cardot, M. Imalalen, C. Dubernet, E. Soma and P. Couvreur, *J. Chromatogr., B, Biomed. Sci. Appl.*, 1997, **702**(1–2), 181.
28. A. Laurand, A. Laroche-Clary, A. Larrue, S. Huet, E. Soma, J. Bonnet and J. Robert, *Anticancer Res.*, 2004, **24**(6), 3781.
29. P. Couvreur, B. Kanté, M. Roland, P. Guiot, P. Baudhuin and P. Speiser, *J. Pharm. Pharmacol.*, 1979, **31**, 331.
30. J. Kattan, J. P. Droz, P. Couvreur, J.P. Marino, A. Boutan-Laroze, P. Rougier, P. Brault, H. Vranckx, J. M. Grognet, X. Morge and H. Sancho-Garnier, *Invest. New Drugs*, 1992, **10**, 191.
31. P. Merle, S. N. Si Ahmed, F. Habersetzer, A. Abergel, J. Taieb, L. Bonyhay, D. Costantini, J. F. Dufour-Lamartinie and C. Trepo, *7th International Conference of ACOS, Beijing*, September 14–18, 2006.
32. J. Szebeni, L. Baranyi, S. Savay, M. Bodo, D. S. Morse, M. Basta, G. L. Stahl, R. Bünger and C.R. Alving, *Am. J. Physiol. Heart Circ. Physiol.*, 2000, **279**(3), 1319.
33. J. Szebeni, J. L. Fontana, N.M. Wassef, P. D. Mongan, D. S. Morse, D. E. Dobbins, G. L. Stahl, R. Bünger and C. R. Alving, *Circulation,* 1999, **99**(17), 2302.

Section 12
Concluding Remarks

CHAPTER 12
Concluding Remarks

MARÍA JOSÉ ALONSO

Department of Pharmacy and Pharmaceutical Technology, Faculty of
Pharmacy, Campus Sur, University of Santiago de Compostela, 15706
Santiago de Compostela, Spain
E-mail: mariaj.alonso@usc.es

This book aims at providing researchers and students with a realistic view of
the evolution and current status of nanocarriers made of biomaterials, helping
drugs to overcome biological barriers. This focus was chosen because we are
convinced that information on the origin of these ideas, their ups and downs
and the history of their evolution, can help researchers in their attempts to
create revolutionary progress in the field.

The history of nanocarriers for drug delivery started within the early 60s, an
amazing decade in the evolution of medicines from their biopharmaceutical
and pharmaceutical technology stand-point. However, over the last half a
century, we have witnessed an extraordinary and multifocal advancement in
the knowledge of these nanocarriers in conjunction with a unique development
of nanotechnologies. At the beginning, the 60s and early 70s, the original ideas
and discoveries came from a special spectrum of wonderful minds, who are
considered the first generation or the great pioneers of nanomedicines and
include medical doctors (A. Bangham and J. Folkman), chemists (A. Zafaroni,
and H. Ringsdorf) and pharmaceutical technologists (P. Speiser). Fortunately,
their ideas germinated in a very enriched field, represented by the second
generation of wonderful minds, many contributors of this book, who have
paved and extended the path of the development of nanocarriers, most of them
still very active nowadays. Here I refer to the great master Robert Langer, but
also several other chemical engineers, N. Peppas and J. Heller, biologists, such
as G. Gregoriadis, B. Papahohopoulos and V. Torchilin, chemists such as R.

RSC Drug Discovery Series No. 22
Nanostructured Biomaterials for Overcoming Biological Barriers
Edited by Maria Jose Alonso and Noemi S. Csaba
© The Royal Society of Chemistry 2012
Published by the Royal Society of Chemistry, www.rsc.org

Duncan, H. Kopecheck and K. Takaoka, pharmaceutical technologists including P. Couvreur, R. Gurny, J. Kreuter, P. Colombo, J. Robinson, T. Kissel and many others. All, but mainly Robert Langer and Patrick Couvreur, have highly influenced my view and contributions to the field. There are, of course, third (I consider myself in this generation), fourth and probably fifth generations (if we want to depict it this way) of enthusiastic scientists. And it is fair to say that all together we have made it happen: first, by having a significant number of products on the market and under clinical development, second, by promoting the advancement of cooperative knowledge leading to the current paradigmatic multi- and trans-disciplinary field of research of nanotherapeutics, and third, by combining in a spectacular dimension the greatest sciences with these very productive and rewarding technologies.

Bearing this information in mind, it becomes clear that, while the terms "nanomedicines", "nanotherapeutics" or "nanopharmaceuticals" started in the 21st century, they all refer to medicines presented in the form of nanostructures, many of which were developed throughout the second half on the 20th century. Therefore, we are talking about a wide array of nanostructured organizations of drugs and biomaterials, which are expected to grow significantly over the coming decades. In fact, the old paradigm of making a drug molecule able to overcome biological barriers by itself is becoming less and less a reality. It was feasible for simple molecules to modulate their permeability and, to some extent, their biodistribution properties; however, researchers in drug discovery are increasingly aware of the necessity of finding new strategies for improving the absorption as well as the pharmacokinetic and pharmacodynamic behavior. This is particularly true for the macromolecular drugs, which, on the other hand, are taking more and more scope in the pipelines of pharma industries.

There is no room in my mind for being skeptical about the worth of the brain and monetary efforts oriented towards the progress in this field of research. I am convinced that we are only at the beginning of the recollection of the fruits of this activity and that the great nanomedicines revolution is still to come. This will most probably be highly motivated by the growing interest of big pharmas in exploring or at least considering these innovative technologies. Nevertheless, we should also be conscious that the actual evolution of this field might be threatened by some critical changes that are being introduced by health policy makers of wealthy countries. We are all aware of the important economical restrictions, which could have a negative impact on the investment in research in innovative medicines. Very rigorous studies will have to be undertaken in order to persuade politicians of the way to improve the long-term value and cost-reduction of specific nanomedicines. Within this economically uncertain global frame-work, it is also true that there are a number of emerging economies which will surely improve health status and be users of these innovative technologies. Irrespective of the complex economic situation, there is no doubt that the collective and enthusiastic intelligence of young scientist working in the field, better than ever, are paving the way for wonderful discoveries. As a result, I am positive that in the near future we will create a new category of therapies; new therapies that are more effective and more patient-friendly (for cancer, Parkinson's, diabetes, obesity...); new therapies for combating currently untreatable diseases (inflammatory diseases, several types of cancer, multiple sclerosis, Alzheimer's, orphan diseases...); and new therapies for improving global health (infectious diseases).

I consider myself very fortunate in being part of the splendid list of contributors to the field. And I am very grateful to those who decided to join Dr Csaba and myself in this book.

Subject Index